Table 2 **Exponential Functions**

x	e^x	e^{-x}	x	e^x	e^{-x}
0.00	1.0000	1.0000	2.5	12.182	0.0821
0.05	1.0513	0.9512	2.6	13.464	0.0743
0.10	1.1052	0.9048	2.7	14.880	0.0672
0.15	1.1618	0.8607	2.8	16.445	0.0608
0.20	1.2214	0.8187	2.9	18.174	0.0550
0.25	1.2840	0.7788	3.0	20.086	0.0498
0.30	1.3499	0.7408	3.1	22.198	0.0450
0.35	1.4191	0.7047	3.2	24.533	0.0408
0.40	1.4918	0.6703	3.3	27.113	0.0369
0.45	1.5683	0.6376	3.4	29.964	0.0334
0.50	1.6487	0.6065	3.5	33.115	0.0302
0.55	1.7333	0.5769	3.6	36.598	0.0273
0.60	1.8221	0.5488	3.7	40.447	0.0247
0.65	1.9155	0.5220	3.8	44.701	0.0224
0.70	2.0138	0.4966	3.9	49.402	0.0202
0.75	2.1170	0.4724	4.0	54.598	0.0183
0.80	2.2255	0.4493	4.1	60.340	0.0166
0.85	2.3396	0.4274	4.2	66.686	0.0150
0.90	2.4596	0.4066	4.3	73.700	0.0136
0.95	2.5857	0.3867	4.4	81.451	0.0123
1.0	2.7183	0.3679	4.5	90.017	0.0111
1.1	3.0042	0.3329	4.6	99.484	0.0101
1.2	3.3201	0.3012	4.7	109.95	0.0091
1.3	3.6693	0.2725	4.8	121.51	0.0082
1.4	4.0552	0.2466	4.9	134.29	0.0074
1.5	4.4817	0.2231	5	148.41	0.0067
1.6	4.9530	0.2019	6	403.43	0.0025
1.7	5.4739	0.1827	7	1096.6	0.0009
1.8	6.0496	0.1653	8	2981.0	0.0003
1.9	6.6859	0.1496	9	8103.1	0.0001
2.0	7.3891	0.1353	10	22026	0.00005
2.1	8.1662	0.1225			
2.2	9.0250	0.1108			
2.3	9.9742	0.1003			
2.4	11.023	0.0907			

Calculus with Analytic Geometry

A First Course

THIRD EDITION

Calculus with Analytic Geometry

A First Course

THIRD EDITION

MURRAY H. PROTTER
CHARLES B. MORREY, JR.
University of California, Berkeley

ADDISON-WESLEY PUBLISHING COMPANY

Reading, Massachusetts • Menlo Park, California
London • Amsterdam • Don Mills, Ontario • Sydney

This book is in the
ADDISON-WESLEY SERIES IN MATHEMATICS

Second printing, June 1977

This text also appears as part of the larger work *College Calculus with Analytic Geometry, Third Edition* by Murray H. Protter and Charles B. Morrey, Jr., copyright © Addison-Wesley Publishing Company, 1977.

Copyright © 1977, 1970, 1963 by Addison-Wesley Publishing Company, Inc. Philippines copyright 1977, 1970 by Addison-Wesley Publishing Company, Inc.

All rights reserved. No part of this publication may be reproduced, stored in a retrieval system, or transmitted, in any form or by any means, electronic, mechanical, photocopying, recording, or otherwise, without the prior written permission of the publisher. Printed in the United States of America. Published simultaneously in Canada. Library of Congress Catalog Card No. 76-12801.

ISBN 0-201-06037-X
FGHIJKL-HA-89876543

PREFACE

This book is designed to meet the needs of a majority of students taking the first year of calculus at a college or university. The text leans heavily on the intuitive approach, gives many illustrative examples, emphasizes applications wherever suitable, and has a large selection of graded exercises. The changes in this third edition reflect the many suggestions we have received from teachers and students who have used the previous editions.

In Chapter 1 we discuss inequalities, with emphasis on the use of the absolute-value symbol. Set notation, an important concept for students, is introduced in Chapter 2; we use it in the text in only those cases in which ambiguity could result from the employment of classical notation. However, when set notation is cumbersome and does not contribute to the understanding of the subject matter, we continue to use standard notation.

Chapter 4 contains an intuitive development of limits, a geometric interpretation of derivative, and a discussion of continuity and limits of sequences. This material prepares the student for the thorough treatment of the differentiation of algebraic functions and the Chain Rule in Chapter 5, as well as for the applications to problems of maxima and minima, related rates, and approximation in Chapter 6. The section on differential notation in Chapter 6 has been rewritten and simplified.

Chapter 7 begins with a careful treatment of area (Jordan content), an approach which is advantageous in the development of the definition of the integral. We establish various basic properties of the integral, such as the Theorem of the Mean and two forms of the Fundamental Theorem of Calculus. There are also applications to problems of fluid pressure and work, but these may be omitted without loss of continuity in the presentation.

In Chapter 8 we resume the work on analytic geometry which we began in Chapter 3, now using the tools of calculus in conjunction with the geometric theory. In order to achieve flexibility in the use of the text, we have placed the more-detailed topics of analytic geometry in a special appendix. Since many entering students are well prepared in analytic geometry, the instructor may wish to cover the topics in Chapters 3 and 8 very quickly, treating them as a review. On the other hand, he or she could choose to make a thorough presentation of the subject by assigning both Chapters 3 and 8 as well as the appendix on analytic geometry. The total amount of material on conics and related subjects is at least as great in this book as that found in most texts devoted to analytic geometry alone.

We next define the natural logarithm by means of the integral, with the exponential function defined as its inverse. This topic, as well as the differentiation and integration of trigonometric functions and inverse trigonometric functions, is treated in Chapter 9. We have also provided an appendix with a brief review of trigonometry for those students who, in the interim since their study of trigonometry, may have forgotten the definitions of the elementary trigonometric functions, as well as the formulas and identities most frequently used in calculus.

In this edition we provide in Chapter 13 a unified development of vectors in two and three dimensions. In order to do this, we have placed the topics on three-dimensional analytic geometry (except for quadrics) in Chapter 12, so that the student can master the concepts of geometric objects in space before he studies vectors in three dimensions. The study of quadric surfaces, a rather special topic and one not required for vector theory, has been transferred to an appendix.

Chapters 14 and 15 discuss techniques in integration and their applications.

Besides those already mentioned, three other useful appendixes are provided. The one on hyperbolic functions can be used by those students in engineering and technology who are most likely to need such functions for later courses. There is an appendix on the axioms of algebra which should provide interesting and useful additional reading for serious students interested in continuing in mathematics. We have also provided a table of indefinite integrals and a brief description on how it is to be used.

An important feature of this edition is the addition of a large number of challenging exercises which have been inserted at the end of almost every section. These exercises, together with the routine exercises and those retained from the second edition, should give the instructor the flexibility he or she needs to assign large numbers of exercises of moderate difficulty for the average student, as well as a substantial number of challenging exercises for the more capable student.

Because of the widespread use of the metric system in science and technology and the gradual conversion to this system in industry and commerce, we have employed metric units consistently in those problems and applications which involve measurement.

The fifteen chapters cover virtually all the topics traditionally given in the calculus of functions of one variable and analytic geometry.

Each of our volumes, *Calculus with Analytic Geometry, Second Course* (Addison-Wesley, 1971) and *Modern Mathematical Analysis* (Addison-Wesley, 1964), contains more than enough material for a second year's work in calculus and advanced calculus. These volumes cover analytic geometry in three dimensions, calculus of functions of several variables (including Green's and Stokes' theorems), infinite series (including Fourier Series), and linear algebra (including eigenvalues of matrices). *Modern Mathematical Analysis* also has three chapters on ordinary differential equations.

Berkeley, California M. H. P.
January 1977 C. B. M., Jr.

CONTENTS

1. INEQUALITIES   1

1. Inequalities 1
2. Absolute value 7
3. Absolute value and inequalities 11

2. RELATIONS, FUNCTIONS, GRAPHS   16

1. Sets. Set notation 16
2. The number plane. Graphs 18
3. Functions. Functional notation 22
4. Relations. Intercepts. Symmetry 28
5. Domain. Range. Asymptotes 35

3. THE LINE   41

1. Distance formula. Midpoint formula 41
2. Slope of a line. Parallel and perpendicular lines 45
3. The straight line 52

4. INTRODUCTION TO THE CALCULUS. LIMITS   58

1. Limits (intuitive) 58
2. Limits (continued) 65
3. The derivative 69
4. Geometric interpretation of the derivative 72
5. Instantaneous velocity and speed; acceleration 77
6. Definition of limit 83
7. Theorems on limits 88
8. Continuity. One-sided limits 95
9. Limits at infinity; infinite limits 102
10. Limits of sequences 108

5. DIFFERENTIATION OF ALGEBRAIC FUNCTIONS   115

1. Theorems on differentiation 115
2. The Chain Rule. Applications 121

3. The power function 125
4. Implicit differentiation 130

6. APPLICATIONS OF DIFFERENTIATION. THE DIFFERENTIAL 135

1. Tools for applications of the derivative 135
2. Further tools: Rolle's theorem; Theorem of the Mean 141
3. Applications to graphs of functions 146
4. Applications using the second derivative 153
5. The maximum and minimum values of a function on an interval . . . 159
6. Applications of maxima and minima 162
7. The differential. Approximation 170
8. Differential notation 175
9. Related rates 179
10. The definite integral and antiderivatives 185

7. THE DEFINITE INTEGRAL 193

1. Area 193
2. Notations for sums 196
3. The definite integral 198
4. Properties of the definite integral 204
5. Evaluation of definite integrals 208
6. Theorem of the Mean for Integrals 214
7. Indefinite integrals. Change of variable 221
8. Area between curves 225
9. Work 232
10. Fluid pressure 237

8. LINES, CIRCLES, CONICS 243

1. Distance from a point to a line 243
2. Families of lines 246
3. Angle between two lines. Bisectors of angles 249
4. The circle 253
5. The parabola 257
6. The ellipse 265
7. The hyperbola 271
8. Translation of axes 277
9. Rotation of axes. The general equation of the second degree 282

9. THE TRIGONOMETRIC AND EXPONENTIAL FUNCTIONS 291

1. Some special limits 291
2. The differentiation of trigonometric functions 293
3. Integration of trigonometric functions 297
4. Relations and inverse functions 300

5. The inverse trigonometric functions 306
6. Integrations yielding inverse trigonometric functions 314
7. The logarithmic function 316
8. The exponential function 324
9. Differentiation of exponential functions; logarithmic differentiation 329
10. The number e 334
11. Applications 335
12. The hyperbolic functions 340

10. PARAMETRIC EQUATIONS. ARC LENGTH 341

1. Parametric equations 341
2. Derivatives and parametric equations 349
3. Arc length 353
4. Curvature 359

11. POLAR COORDINATES 366

1. Polar coordinates 366
2. Graphs in polar coordinates 368
3. Equations in Cartesian and polar coordinates 373
4. Straight lines, circles, and conics 376
5. Derivatives in polar coordinates 381
6. Area in polar coordinates 385

12. SOLID ANALYTIC GEOMETRY 390

1. The number space R_3. Coordinates. The distance formula 390
2. Direction cosines and numbers 394
3. Equations of a line 400
4. The plane 404
5. Angles. Distance from a point to a plane 408
6. The sphere. Cylinders 413
7. Other coordinate systems 417

13. VECTORS IN THE PLANE AND IN THREE-SPACE 421

1. Directed line segments and vectors in the plane 421
2. Operations with vectors 424
3. Operations with plane vectors, continued. The scalar product 429
4. Vectors in three dimensions 434
5. Linear dependence and independence 440
6. The scalar (inner or dot) product 443
7. The vector or cross product 448
8. Products of three vectors 454
9. Vector functions in the plane and their derivatives 457

10. Vector velocity and acceleration in the plane 461
11. Vector functions in space. Space curves. Tangents and arc length 464
12. Tangential and normal components. The moving trihedral 468

14. FORMULAS AND METHODS OF INTEGRATION 475

1. Integration by substitution 475
2. Integration by substitution. More techniques 480
3. Certain trigonometric integrals 483
4. Trigonometric substitution 487
5. Integrands involving quadratic functions 490
6. Integration by parts 493
7. Integration of rational functions 499
8. Two rationalizing substitutions 506
9. Summary 510

15. APPLICATIONS OF THE INTEGRAL 511

1. Volume of a solid of revolution. Disk method 511
2. Volume of a solid of revolution. Shell method 517
3. Improper integrals 521
4. Arc length 527
5. Area of a surface of revolution 530
6. Center of mass 535
7. Center of mass of a plane region 539
8. Center of mass of a solid of revolution 544
9. Centers of mass of wires and surfaces 549
10. Theorems of Pappus 551
11. Newton's Laws of Motion. Differential equations 553
12. Approximate integration 559

APPENDIX 1 Trigonometry Review App-2

APPENDIX 2 Conic Sections. Quadric Surfaces App-13

APPENDIX 3 Properties of Hyperbolic Functions App-37

APPENDIX 4 Axioms of Algebra. Number Systems App-45

APPENDIX 5 Theorems on Determinants App-56

APPENDIX 6 Proofs of Theorems 6, 10, 16, and 17 of Chapter 13 . . . App-65

APPENDIX 7 Introduction to a Short Table of Integrals App-70

TABLE 4 A Short Table of Integrals App-75

Answers to Odd-Numbered Problems Ans-1

Index Index-1

FRONTPAPERS   Natural Trigonometric Functions
Exponential Functions

REARPAPERS   Natural Logarithms of Numbers
Greek Alphabet

1

INEQUALITIES

1. INEQUALITIES*

Almost all high school students learn plane geometry as a single logical development in which theorems are proved on the basis of a system of axioms or postulates. Unlike plane geometry, however, algebra has traditionally been taught in high school without the aid of a formal logical system. In this method the student simply learns a few rules—or many—for manipulating algebraic quantities; these rules lead to success in solving problems but do not shed any light on the structure of algebra.

The usual rules of algebra are logical consequences of the system of axioms known as the Axioms of Algebra. To prove the rules we use for manipulating algebraic expressions directly from the Axioms, as in Euclidean geometry, would be cumbersome and unwieldy. Therefore we shall assume that the reader is familiar with the usual laws of algebra and begin with a discussion of inequalities. The Axioms of Algebra are given in Appendix 4 at the end of the book, and we recommend their study to students unfamiliar with them.

In elementary algebra and geometry we study equalities almost exclusively. The solution of linear and quadratic algebraic equations, the congruence of geometric figures, and relationships among various trigonometric functions are topics concerned with equality. As we progress in the development of mathematical ideas—especially in that branch of mathematics of which calculus is a part—we shall see that the study of inequalities is both interesting and useful. An inequality is involved when we are more concerned with the approximate size of a quantity than we are with its true value. Since the proofs of some of the most important theorems in calculus depend on certain approximations, it is essential that we develop a facility for working with inequalities.

We shall be concerned with inequalities among real numbers, and we begin by recalling some familiar relationships. Given that a and b are any two real numbers, the symbol

$$a < b$$

---

* This chapter and Chapter 2 consist of review material for many students of calculus. Students who do not have a thorough working knowledge of inequalities should begin here. Readers familiar with inequalities may start with Chapter 2.

means that a **is less than** b.* We may also write the same inequality in the *opposite direction*,

$$b > a,$$

which is read b **is greater than** a.

The rules for handling inequalities can be proved on the basis of the Axioms for Algebra. The rules themselves are only slightly more complicated than the ones we learned in algebra for equalities. However, the differences are so important that we state them as four Theorems about Inequalities and they should be studied carefully.

Theorem 1. *If $a < b$ and $b < c$, then $a < c$. In words: if a is less than b and b is less than c, then a is less than c.*

Theorem 2. *If c is any number and $a < b$, then it is also true that $a + c < b + c$ and $a - c < b - c$. In words: if the same number is added to or subtracted from each side of an inequality, the result is an inequality in the same direction.*

Theorem 3. *If $a < b$ and $c < d$ then $a + c < b + d$. That is, inequalities in the same direction may be added.*

It is important to note that in general inequalities may not be subtracted. For example, $2 < 5$ and $1 < 7$. We can say, by addition, that $3 < 12$, but note that subtraction would state the absurdity that 1 is less than -2.

Theorem 4. *If $a < b$ and c is any positive number, then*

$$ac < bc,$$

while if c is a negative number, then

$$ac > bc.$$

In words: multiplication of both sides of an inequality by the same positive number preserves the direction, while multiplication by a negative number reverses the direction of the inequality.

Since dividing an inequality by a number d is the same as multiplying it by $1/d$, we see that Theorem 4 applies for division as well as for multiplication.

From the geometric point of view we associate a horizontal axis with the totality of real numbers. The origin may be selected at any convenient point, with positive numbers to the right and negative numbers to the left (Fig. 1–1). For every real number there will be a corresponding point on the line and, conversely, every point will represent a real number. Then the inequality $a < b$ may be read: a is to the left of b. This geometric way of looking at inequalities is frequently of

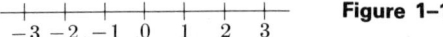

Figure 1–1

---

* Which is true if and only if $b - a$ is positive (see Appendix 4, §2).

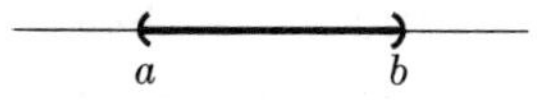

Figure 1-2

Figure 1-3

help in solving problems. It is also helpful to introduce the notion of an *interval of numbers* or *points*. If a and b are numbers (as shown in Fig. 1-2), then the **open interval** from a to b is the collection of all numbers which are both larger than a and smaller than b. That is, an open interval consists of all numbers *between a and b*. A number x is between a and b if *both* inequalities $a < x$ and $x < b$ are true. A compact way of writing this is

$$a < x < b.$$

The **closed interval** from a to b consists of all the points between a and b, *including a and b* (Fig. 1-3). Suppose a number x is either equal to a or larger than a, but we don't know which. We write this conveniently as $x \geq a$, which is read: *x is greater than or equal to a*. Similarly, $x \leq b$ is read: *x is less than or equal to b*, and means that x may be either smaller than b or may be b itself. A compact way of designating a closed interval from a to b is to state that it consists of all points x such that

$$a \leq x \leq b.$$

An interval which contains the endpoint b but not a is said to be **half-open on the left.** That is, it consists of all points x such that

$$a < x \leq b.$$

Similarly, an interval containing a but not b is called **half-open on the right,** and we write

$$a \leq x < b.$$

Parentheses and brackets are used as symbols for intervals in the following way:

(a, b) for the open interval: $a < x < b$,
$[a, b]$ for the closed interval: $a \leq x \leq b$,
$(a, b]$ for the interval half-open on the left: $a < x \leq b$,
$[a, b)$ for the interval half-open on the right: $a \leq x < b$.

We can extend the idea of an interval of points to cover some unusual cases. Suppose we wish to consider *all* numbers larger than 7. This may be thought of as an interval extending to infinity on the right. (See Fig. 1-4.) Of course, *infinity is not a number*, but we use the symbol $(7, \infty)$ to represent all numbers larger than 7. We could also write: all numbers x such that

$$7 < x < \infty.$$

In a similar way, the symbol $(-\infty, 12)$ will stand for all numbers less than 12. The double inequality

$$-\infty < x < 12$$

is an equivalent way of representing all numbers x less than 12.

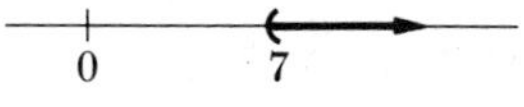

Figure 1-4

4   INEQUALITIES

The first-degree equation $3x + 7 = 19$ has a unique solution, $x = 4$. The quadratic equation $x^2 - x - 2 = 0$ has two solutions, $x = -1$ and $x = 2$. The trigonometric equation $\sin x = \frac{1}{2}$ has an infinite number of solutions: $x = 30°$, $150°$, $390°$, $510°$, ... *The solution of an inequality involving a single unknown, say x, is the collection of all numbers which make the inequality a true statement.* Sometimes this is called the **solution set.** For example, the inequality

$$3x - 7 < 8$$

has as its solution *all* numbers less than 5. To demonstrate this we argue in the following way. If x is a number which satisfies the above inequality we can, by Theorem 2, add 7 to both sides of the inequality and obtain a true statement. That is, we have

$$3x - 7 + 7 < 8 + 7 \quad \text{or} \quad 3x < 15.$$

Now, dividing both sides by 3 (Theorem 4), we obtain

$$x < 5$$

and observe that *if* x is a solution, *then* it is less than 5. Strictly speaking, however, we have not *proved* that every number which is less than 5 is a solution. In an actual proof we would begin by supposing that x is any number less than 5; that is,

$$x < 5.$$

We multiply both sides by 3 (Theorem 4) and then subtract 7 (Theorem 2) to get

$$3x - 7 < 8,$$

the original inequality. Since the condition that x is less than 5 implies the original inequality, we have proved the result. The important thing to notice is that the proof consisted of *reversing* the steps of the original argument which led to the solution $x < 5$ in the first place. So long as each of the steps we take is *reversible*, the above procedure is completely satisfactory so far as obtaining solutions is concerned. The step going from $3x - 7 < 8$ to $3x < 15$ is reversible, since these two inequalities are equivalent. Similarly, the inequalities $3x < 15$ and $x < 5$ are equivalent. Finally, note that the solution set consists of all numbers in the interval $(-\infty, 5)$.

Methods for the solution of various types of simple algebraic inequalities are shown in the following examples, which should be studied carefully.

Example 1. Solve for x:

$$-7 - 3x < 5x + 29.$$

Solution. Subtract $5x$ from both sides, getting

$$-7 - 8x < 29.$$

Multiply both sides by -1, reversing the direction of the inequality, to obtain

$$7 + 8x > -29.$$

Subtracting 7 from both sides yields $8x > -36$, and dividing by 8 gives the

solution
$$x > -\tfrac{9}{2},$$
or, stated in interval form: all x in the interval $(-\tfrac{9}{2}, \infty)$. ◂

To verify the correctness of the result, it is necessary to perform the above steps in reverse order. However, the observation that each individual step is reversible is sufficient to check the validity of the answer.

Example 2. Solve for x ($x \neq 0$):
$$\frac{3}{x} < 5.$$

Solution. We have an immediate inclination to multiply both sides by x. However, since we don't know in advance whether x is positive or negative, we must proceed cautiously. We do this by considering two cases: (1) x is positive, and (2) x is negative.

CASE 1. Suppose $x > 0$. Then multiplying by x preserves the direction of the inequality (Theorem 4), and we get
$$3 < 5x.$$
Dividing by 5, we find that $x > \tfrac{3}{5}$. This means that we must find all numbers which satisfy *both* of the inequalities
$$x > 0 \quad \text{and} \quad x > \tfrac{3}{5}.$$
Clearly, any number greater than $\tfrac{3}{5}$ is also positive, and the solution in Case 1 consists of all x in the interval $(\tfrac{3}{5}, \infty)$.

CASE 2. $x < 0$. Multiplying by x *reverses* the direction of the inequality. We have
$$3 > 5x,$$
and therefore $\tfrac{3}{5} > x$. We seek all numbers x, such that *both* of the inequalities
$$x < 0 \quad \text{and} \quad x < \tfrac{3}{5}$$
hold. The solution in Case 2 is the collection of all x in the interval $(-\infty, 0)$. A way of combining the answers in the two cases is to state that the solution set consists of all numbers x *not* in the closed interval $[0, \tfrac{3}{5}]$. (See Fig. 1–5.) ◂

Figure 1–5

Example 3. Solve for x ($x \neq -2$):
$$\frac{2x - 3}{x + 2} < \frac{1}{3}.$$

Solution. As in Example 2, we must consider two cases, according to whether $x + 2$ is positive or negative.

6 INEQUALITIES

CASE 1. $x + 2 > 0$. We multiply by $3(x + 2)$, which is positive, getting
$$6x - 9 < x + 2.$$
Adding $9 - x$ to both sides, we have
$$5x < 11, \quad \text{from which} \quad x < \tfrac{11}{5}.$$
Since we have already assumed that $x + 2 > 0$, and since we must have $x < \tfrac{11}{5}$, we see that x must be larger than -2 and smaller than $\tfrac{11}{5}$. That is, the solution set consists of all x in the interval $(-2, \tfrac{11}{5})$.

CASE 2. $x + 2 < 0$. Again multiplying by $3(x + 2)$ and *reversing* the inequality, we obtain
$$6x - 9 > x + 2,$$
or $5x > 11$ and $x > \tfrac{11}{5}$. In this case, x must be less than -2 and greater than $\tfrac{11}{5}$, which is impossible. Combining the cases, we get as the solution set all numbers in $(-2, \tfrac{11}{5})$. See Fig. 1–6.

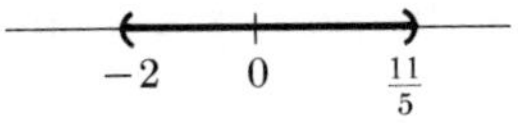

Figure 1–6

PROBLEMS

In Problems 1 through 14, solve for x.

1. $2x - 3 < 7$
2. $2x + 4 < x - 5$
3. $5 - 3x < 14$
4. $2 - 5x < 3 + 4x$
5. $2(3x - 6) < 4 - (2 + 5x)$
6. $x - 4 < \dfrac{2x}{3} + \dfrac{2 - 3x}{5}$
7. $\dfrac{4}{x} < \dfrac{3}{5}$
8. $\dfrac{x - 1}{x} < 4$
9. $\dfrac{x + 3}{x - 2} < 5$
10. $\dfrac{2}{x} - 3 < \dfrac{4}{x} + 1$
11. $\dfrac{x + 3}{x - 4} < -2$
12. $\dfrac{2 - x}{x + 1} < \dfrac{3}{2}$
13. $\dfrac{x}{3 - x} < 2$
*14. $\dfrac{x - 2}{x + 3} < \dfrac{x + 1}{x}$

In Problems 15 through 23, find the values of x, if any, for which both inequalities hold.

15. $2x - 7 < 5 - x$ and $3 - 4x < \tfrac{5}{3}$
16. $3x - 8 < 5(2 - x)$ and $2(3x + 4) - 4x + 7 < 5 + x$
17. $\dfrac{2x + 6}{3} - \dfrac{x}{4} < 5$ and $15 - 3x < 4 + 2x$
18. $3 - 6x < 2(x + 5)$ and $7(2 - x) < 3x + 8$

---

* Starred problems are those that are unusually difficult.

*19. $\dfrac{2}{x-1} < 4$ and $\dfrac{3}{x-2} < 7$

*20. $\dfrac{x-2}{x+1} < 3$ and $\dfrac{3-x}{x-2} < 5$

*21. $\dfrac{3}{x} < 5$ and $x + 2 < 7 - 3x$

*22. $\dfrac{2}{x} < 1$ and $(x - 2)(x + 3) < 0$

*23. $\dfrac{3}{x} < 2$ and $(x - 1)(x + 4) < 0$

24. Show that Theorem 3 for inequalities may be derived from Theorems 1 and 2.

25. Given that a, b, c, and d are all positive numbers, and that $a < b$ and $c < d$; show that $ac < bd$.

*26. a) State the most general circumstances in which the hypotheses $a < b$ and $c < d$ imply that $ac < bd$.
    b) Given that $a < b$ and $c < d$, when is it true that $ac > bd$?

27. If x is a positive number, prove that
$$x + \frac{1}{x} \geq 2.$$

28. a) If x and y are positive numbers, show that
$$\left(\frac{1}{x} + \frac{1}{y}\right)(x + y) \geq 4.$$

    b) If x, y, and z are positive numbers, show that
$$\left(\frac{1}{x} + \frac{1}{y} + \frac{1}{z}\right)(x + y + z) \geq 9.$$

    c) If x, y, z, and w are positive numbers, show that
$$\left(\frac{1}{x} + \frac{1}{y} + \frac{1}{z} + \frac{1}{w}\right)(x + y + z + w) \geq 16.$$

    *d) Generalize the above results to n numbers $x_1, x_2, \ldots, x_n$.

*29. If x and y are any numbers different from zero, show that
$$\frac{x^2}{y^2} + \frac{16y^2}{x^2} + 24 \geq \frac{8x}{y} + \frac{32y}{x}.$$

*30. Let x and y be positive numbers with $x \geq y$. Show that
$$\frac{x}{y} + 3\frac{y}{x} \geq \frac{y^2}{x^2} + 3.$$

Show that the inequality is reversed if $y \geq x$.

2. ABSOLUTE VALUE

If a is any positive number, the **absolute value** of a is defined to be a itself. If a is negative, the absolute value of a is defined to be $-a$. The absolute value of zero is

8 INEQUALITIES

zero. The symbol for the absolute value of a is $|a|$. In other words,

$$|a| = a \quad \text{if} \quad a > 0,$$
$$|a| = -a \quad \text{if} \quad a < 0,$$
$$|0| = 0.$$

For example,

$$|7| = 7, \quad |-13| = 13, \quad |2 - 5| = |-3| = 3.$$

This rather simple idea has important consequences but, before we can consider them, we must discuss methods of solving equations and inequalities involving absolute values.

Example 1. Solve for x: $|x - 7| = 3$.

Solution. This equation, according to the definition of absolute value, expresses the fact that $x - 7$ must be 3 or -3, since in either case the absolute value is 3. If $x - 7 = 3$, we have $x = 10$; and if $x - 7 = -3$, then $x = 4$. We see that there are *two* values of x which solve the equation: $x = 4, 10$. ◄

Example 2. Solve for x: $|2x - 6| = |4 - 5x|$.

Solution. The two possibilities are

$$2x - 6 = 4 - 5x \quad \text{and} \quad 2x - 6 = -(4 - 5x).$$

Solving each of these for x, we obtain the two solutions

$$x = \tfrac{10}{7}, -\tfrac{2}{3}.$$ ◄

Since $|x|$ represents the *distance* of x from the origin, the reader can easily see that the condition that $|x| < 4$ is equivalent to the condition that x is any number in the interval extending from -4 to $+4$ (Fig. 1–7). Or, in terms of the symbol for intervals, x must lie in the interval $(-4, 4)$. Sometimes a statement such as $|x| < 4$ is used to denote the interval $(-4, 4)$. In terms of inequalities without absolute value signs, the statement

$$-4 < x < 4$$

is equivalent to $|x| < 4$. In a similar way, the inequality $|x - 3| < 5$ means that $x - 3$ must lie in the interval $(-5, 5)$. We could also write

$$-5 < x - 3 < 5.$$

This consists of two inequalities, *both* of which x must satisfy. If we add 3 to each member of the double inequality above, then

$$-2 < x < 8.$$

Therefore x must lie in the interval $(-2, 8)$. See Fig. 1–8.

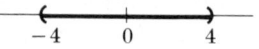

Figure 1–7

Figure 1–8

Example 3. Solve for x: $|3x - 4| \leq 7$.

Solution. We write the inequality in the equivalent form
$$-7 \leq 3x - 4 \leq 7.$$
Now we add 4 to each term:
$$-3 \leq 3x \leq 11,$$
and divide each term by 3:
$$-1 \leq x \leq \tfrac{11}{3}.$$
The solution set consists of all numbers x in the *closed* interval $[-1, \tfrac{11}{3}]$. See Fig. 1-9.

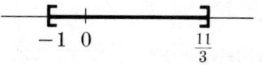

Figure 1-9 ◀

Example 4. Solve for x: $|2 - 5x| < 3$.

Solution. Here we have
$$-3 < 2 - 5x < 3,$$
and subtracting 2 from each member yields
$$-5 < -5x < 1.$$
Dividing by -5 will reverse each of the inequalities, and we get
$$1 > x > -\tfrac{1}{5}.$$
The solution set consists of those x in the open interval $(-\tfrac{1}{5}, 1)$. See Fig. 1-10.

Figure 1-10 ◀

Example 5. Solve for x:
$$\left|\frac{2x - 5}{x - 6}\right| < 3.$$

Solution. Proceeding as before, we see that
$$-3 < \frac{2x - 5}{x - 6} < 3.$$

We would like to multiply by $x - 6$, and in order to do this we must distinguish two cases, depending on whether $x - 6$ is positive or negative.

CASE 1. $x - 6 > 0$. In this case multiplication by $x - 6$ preserves the direction of the inequalities, and we have
$$-3(x - 6) < 2x - 5 < 3(x - 6).$$

10 INEQUALITIES

Now the left inequality states that
$$-3x + 18 < 2x - 5,$$
or that
$$\tfrac{23}{5} < x.$$

The right inequality states that
$$2x - 5 < 3x - 18,$$
or that
$$13 < x.$$

In Case 1 we must have $x - 6 > 0$ and $\tfrac{23}{5} < x$ and $13 < x$. If the third inequality holds, then the other two hold as a consequence. Hence the solution in Case 1 consists of all x in the interval $(13, \infty)$.

CASE 2. $x - 6 < 0$. The inequalities reverse when we multiply by $x - 6$. We then get
$$-3(x - 6) > 2x - 5 > 3(x - 6).$$

The two inequalities now state that $\tfrac{23}{5} > x$ and $13 > x$. The three inequalities
$$x - 6 < 0 \quad \text{and} \quad \tfrac{23}{5} > x \quad \text{and} \quad 13 > x$$
all hold if $x < \tfrac{23}{5}$. In Case 2, the solution consists of all numbers in the interval $(-\infty, \tfrac{23}{5})$. We could also describe the solution set by saying that it consists of all numbers not in the closed interval $[\tfrac{23}{5}, 13]$. See Fig. 1–11. ◀

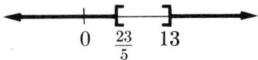

Figure 1–11

PROBLEMS

In Problems 1 through 12, solve for x.

1. $|2x + 1| = 4$
2. $|x - 3| = 1$
3. $|4 - 2x| = 3$
4. $|7 - 5x| = 8$
5. $|x + \tfrac{1}{3}| = 5$
6. $|-2x - 4| = 9$
7. $|3x - 5| = |7x - 2|$
8. $|2x + 1| = |3x|$
9. $|x - 6| = |3 - 2x|$
10. $\left|\dfrac{x + 2}{2x - 5}\right| = 1$
11. $\left|\dfrac{x + 2}{x + 5}\right| = 3$
12. $\dfrac{|3x - 1|}{|4 - 2x|} = 6$

In each of Problems 13 through 23, find the values of x, if any, for which the following inequalities hold. State answers in terms of intervals.

13. $|x + 1| < 4$
14. $|x - 3| < 2$
15. $|2x + 5| < 3$
16. $|3 - 7x| \le 6$
17. $|12 + 5x| \le 1$
18. $\dfrac{|x + 1|}{|3x - 2|} < 5$

19. $\left|\dfrac{3-2x}{2+x}\right| < 4$

20. $|x+3| \le |2x-6|$

21. $|3-2x| \le |x+4|$

22. $\left|\dfrac{x+3}{6-5x}\right| \le 2$

*23. $|x(x+1)| < |x+4|$

24. Find all solutions of the simultaneous equations $|x-y| = 2$, $|2x+y| = 4$.

25. Find all solutions of the simultaneous equations $|x+y| = 1$, $|2x-3y| = 8$.

26. Find all solutions of the simultaneous equations $|x^2 - 2y^2| = 1$, $|2x-y| = 4$.

3. ABSOLUTE VALUE AND INEQUALITIES

If a and b are any two numbers, the reader can easily verify that

$$|ab| = |a| \cdot |b| \quad \text{and} \quad \left|\dfrac{a}{b}\right| = \dfrac{|a|}{|b|}.$$

In words, *the absolute value of a product is the product of the absolute values, and the absolute value of a quotient is the quotient of the absolute values.*

An interesting and useful inequality is stated in the following theorem, which we prove.

Theorem 5. *If a and b are any numbers, then*

$$|a+b| \le |a| + |b|.$$

*Proof.** We consider three cases.

CASE 1. The numbers a and b are both positive. Then $|a| = a$, $|b| = b$, and $|a+b| = a+b$. The conclusion of the theorem is satisfied, since $|a+b| = |a| + |b|$.

CASE 2. The numbers a and b are both negative. Then $|a| = -a$, $|b| = -b$, and $|a+b| = -(a+b)$. As in Case 1, we have $|a+b| = |a| + |b|$.

CASE 3. One number, say a, is positive, and the other is negative. Then $|a| = a$, $|b| = -b$, and we have $|a+b|$ as either $a+b$ or $-(a+b)$, depending on whether $a+b$ is larger or smaller than zero. But we know that

$$a+b < a-b = |a| + |b|,$$

and

$$-(a+b) < a-b = |a| + |b|.$$

Therefore we have shown that in all possible circumstances

$$|a+b| \le |a| + |b|.$$

---

* A shorter although less intuitive proof is given by the following argument. Since $|a| = a$ or $-a$, we may write $-|a| \le a \le |a|$, and similarly, $-|b| \le b \le |b|$. Adding these inequalities, we get $-(|a| + |b|) \le a + b \le |a| + |b|$. The conclusion of Theorem 5 is equivalent to this double inequality.

Corollary. *If a and b are any numbers, then*
$$|a - b| \le |a| + |b|.$$

Proof. We write $a - b$ as $a + (-b)$, and apply Theorem 5 to obtain
$$|a - b| = |a + (-b)| \le |a| + |-b| = |a| + |b|.$$
The final equality holds since, from the definition of absolute value, it is always true that $|-b| = |b|$. ◄

Theorem 5 and the Corollary are invaluable if we wish to estimate how large various expressions can become. We show how this is done by working a few examples.

Example 1. Estimate how large the expression $x^3 - 2$ can become if x is restricted to the interval $[-4, 4]$.

Solution. From the Corollary, we have
$$|x^3 - 2| \le |x^3| + |2|.$$
Since the absolute value of a product is the product of the absolute values, we have $|x^3| = |x \cdot x \cdot x| = |x| \cdot |x| \cdot |x| = |x|^3$. Therefore we get
$$|x^3 - 2| \le |x|^3 + 2.$$
By hypothesis we stated that $|x|$ is always less than 4, and we conclude that
$$|x^3 - 2| \le 4^3 + 2 = 66$$
if x is any number in $[-4, 4]$. ◄

Example 2. Find a positive number M such that
$$|x^3 - 2x^2 + 3x - 4| \le M$$
for all values of x in the interval $[-3, 2]$.

Solution. From Theorem 5 and the Corollary we can write
$$|x^3 - 2x^2 + 3x - 4| \le |x^3| + |2x^2| + |3x| + |4|,$$
and from our knowledge of the absolute value of products, we get
$$|x^3| + |2x^2| + |3x| + |4| = |x|^3 + 2|x|^2 + 3|x| + 4.$$
Since $|x|$ can never be larger than 3, it follows that
$$|x|^3 + 2|x|^2 + 3|x| + 4 \le 27 + 2 \cdot 9 + 3 \cdot 3 + 4 = 58.$$
The positive number M we seek is 58. ◄

Theorem 6. *If a, b, c, and d are all positive numbers and $a \ge b$, $d \ge c$, then*
$$\frac{a}{c} \ge \frac{b}{d}.$$

Proof. If c is less than or equal to d, then the reciprocal of c, the quantity $1/c$, is larger than or equal to the reciprocal of d. This follows from Theorem 4 for inequalities, since dividing both sides of the inequality $c \leq d$ by the positive quantity cd gives

$$\frac{1}{c} \geq \frac{1}{d}.$$

Again from Theorem 4, and from the fact that $a \geq b$, we have

$$a\left(\frac{1}{c}\right) \geq b\left(\frac{1}{c}\right) \geq b\left(\frac{1}{d}\right).$$

The first and last terms yield the result.

Example 3. Find a number M such that

$$\left|\frac{x+2}{x-2}\right| \leq M$$

if x is restricted to the interval $[\frac{1}{2}, \frac{3}{2}]$.

Solution. We know that

$$\left|\frac{x+2}{x-2}\right| = \frac{|x+2|}{|x-2|}.$$

If we can estimate the *smallest* possible value of the denominator and the *largest* possible value of the numerator then, by Theorem 6, we will have estimated the largest possible value of the entire expression. For the numerator we have

$$|x+2| \leq |x| + |2| \leq \tfrac{3}{2} + 2 = \tfrac{7}{2}.$$

For the denominator we note that the smallest value occurs when x is as close as possible to 2. This occurs when $x = \tfrac{3}{2}$, and so

$$|x-2| \geq |\tfrac{3}{2} - 2| = \tfrac{1}{2},$$

if x is in the interval $[\tfrac{1}{2}, \tfrac{3}{2}]$. We conclude finally that

$$\left|\frac{x+2}{x-2}\right| \leq \frac{\tfrac{7}{2}}{\tfrac{1}{2}} = 7. \qquad \blacktriangleleft$$

Example 4. Find an estimate for the largest possible value of

$$\left|\frac{x^2 + 2}{x + 3}\right|$$

if x is restricted to the interval $[-4, 4]$.

Solution. The numerator is simple to estimate, since

$$|x^2 + 2| \leq |x|^2 + 2 \leq 4^2 + 2 = 18.$$

However, we have to find a smallest value for $|x + 3|$ if x is in $[-4, 4]$. We see first of all that the expression is not defined for $x = -3$, since then the denominator would be zero, and division by zero is always excluded. Furthermore, if x is a number near -3, then the denominator is near zero, the numerator has a value near 11, and the quotient will be a "large" number. In this problem there is no largest value of the given expression in $[-4, 4]$. Such questions are discussed further in Chapter 4.

Example 5. Find a number M such that
$$\left|\frac{x+2}{x} - 5\right| \leq M$$
if x is restricted to the interval $(1, 4)$.

Solution

METHOD 1. Since, from elementary algebra,
$$\frac{x+2}{x} - 5 = \frac{-4x+2}{x},$$
we can write
$$\left|\frac{x+2}{x} - 5\right| = \frac{|-4x+2|}{|x|} \leq \frac{|4||x| + |2|}{|x|}.$$
Since $|x| < 4$, we have $4|x| + 2 < 18$, while the denominator, $|x|$, can never be smaller than 1. We obtain
$$\left|\frac{x+2}{x} - 5\right| < 18.$$

METHOD 2. From the Corollary to Theorem 5 it follows that
$$\left|\frac{x+2}{x} - 5\right| \leq \left|\frac{x+2}{x}\right| + |5| = \frac{|x+2|}{|x|} + 5.$$
Now $|x| + 2 < 6$ and $|x| > 1$. Hence we obtain
$$\left|\frac{x+2}{x} - 5\right| < 11.$$

This example shows that some algebraic manipulations lead to better estimates than do others.

PROBLEMS

In Problems 1 through 12, find a positive number M, if there is one, such that the absolute value of the given expression does not exceed M if x is in the interval given.

1. $x^2 - 3x + 4$; x in $[-2, 2]$
2. $x^2 + 4x - 3$; x in $[-2, 4]$
3. $x^3 + 2x^2 - 3x - 6$; x in $[-2, 5]$

4. $x^4 - 2x^3 + x^2 - 3x - 5$; x in $[-3, -1]$

5. $\dfrac{x + 2}{x - 4}$; x in $[5, 8]$

6. $\dfrac{x + 4}{2x + 6}$; x in $(-1, 4)$

7. $\dfrac{x^2 - 6x + 2}{x + 5}$; x in $(-4\tfrac{1}{2}, 4)$

8. $\dfrac{x^3 - 6x + 5}{x^2 + 5}$; x in $(-2, 3)$

9. $\dfrac{x + 7}{x^2 + 4x + 4}$; x in $(-1, 3)$

10. $\dfrac{x^3 - 3x + 5}{(x^2 - 2x - 3)}$; x in $[0, 4]$

11. $\dfrac{2x + 1}{x(x^2 + 2x - 8)}$; x in $(-3, 1)$

12. $\dfrac{2x^3 - 3x + 1}{1 - x}$; x in $(0.7, 0.9)$

13. Prove for any numbers a and b that $|a| - |b| \le |a - b|$.

14. Prove for any numbers a and b that $||a| - |b|| \le |a - b|$.

15. Given that a and b are positive, c and d are negative, and $a > b$, $c > d$, show that

$$\frac{a}{c} < \frac{b}{d}.$$

16. If a_1, a_2, a_3 are any numbers, show that $|a_1 + a_2 + a_3| \le |a_1| + |a_2| + |a_3|$.

17. If a_1, a_2, a_3 are any numbers and $|a_1 + a_2| > |a_3|$, show that

$$\frac{a_3}{a_1 + a_2} < 1.$$

Is the same result true if $|a_1 + a_2| > |a_3|$ is replaced by $a_1 + a_2 > a_3$?

2

RELATIONS, FUNCTIONS, GRAPHS

1. SETS. SET NOTATION

A **set** is a collection of objects. The objects may have any character (numbers, points, lines, etc.) so long as we know which objects are in a given set and which are not. If S is a set and P is an object in it, we write $P \in S$ and say that P **belongs to** S or that P **is an element of** S. If S_1 and S_2 are two sets, their **union,** denoted by $S_1 \cup S_2$, consists of all objects each of which is in at least one of the two sets. The **intersection of** S_1 **and** S_2, denoted by $S_1 \cap S_2$, consists of all objects each of which is in both sets. Schematically, if S_1 is the horizontally shaded set (Fig. 2–1) and S_2 the vertically shaded set, then $S_1 \cup S_2$ consists of the entire shaded area and $S_1 \cap S_2$ consists of the doubly shaded area. Similarly, we may form the union and intersection of any number of sets. When we write $S_1 \cup S_2 \cup \cdots \cup S_7$ for the union of the seven sets $S_1, S_2, \ldots, S_7$, this union consists of all elements each of which is in at least one of the 7 sets. The intersection of these 7 sets is written $S_1 \cap S_2 \cap \cdots \cap S_7$. It may happen that two sets S_1 and S_2 have no elements in common. In such a case we say that their intersection is empty, and we use the term **empty set** for the set which is devoid of members.

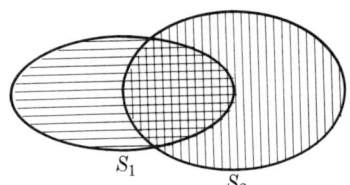

Figure 2–1

Most often we will deal with sets each of which is specified by some property or properties of its elements. For example, we may speak of the set of all even integers or the set of all rational numbers between 0 and 1. We employ the special symbol

$$\{x : x = 2n \quad \text{and} \quad n \text{ is an integer}\}$$

to represent the set of all even integers. In this notation the letter x stands for a generic element of the set, and the properties which determine membership in the set are listed to the right of the colon. The symbol

$$\{x : x \in (0, 1) \quad \text{and} \quad x \text{ is rational}\}$$

represents the rational numbers in the open interval (0, 1). If a set has only a few elements, we may specify it by listing its members between braces. Thus the symbol $\{-2, 0, 1\}$ denotes the set whose elements are the numbers $-2, 0$, and 1. A set may be specified by any number of properties and we may use a variety of notations to determine these properties. If a set of objects has properties A, B, and C, we may denote this set by

$$\{P\colon P \text{ has the properties } A, B, \text{ and } C\}.$$

Other examples are

$$(0, 2) = \{x\colon 0 < x < 2\},$$

$$[2, 14) = \{t\colon 2 \leq t < 14\}.$$

The last set is read as "the set of all numbers t such that t is greater than or equal to two and less than fourteen."

To illustrate the use of the symbols for set union and set intersection, we observe that

$$[1, 3] = [1, 2\tfrac{1}{2}] \cup [2, 3], \qquad (0, 1) = (0, \infty) \cap (-\infty, 1).$$

The words *and* and *or* have precise meanings when used in connection with sets and their properties. The set consisting of elements which have property A *or* property B is the *union* of the set having property A and the set having property B. Symbolically,

$\{x\colon x \text{ has property } A \text{ or property } B\}$

$$= \{x\colon x \text{ has property } A\} \cup \{x\colon x \text{ has property } B\}.$$

The set consisting of elements which have property A *and* property B is the *intersection* of the set having property A with the set having property B. In set notation,

$\{x\colon x \text{ has property } A \text{ and property } B\}$

$$= \{x\colon x \text{ has property } A\} \cap \{x\colon x \text{ has property } B\}.$$

Notation. The expression "if and only if," a technical one used frequently in mathematics, requires some explanation. Suppose A and B stand for propositions which may be true or false. To say that A *is true* **if** B *is true* means that the truth of B implies the truth of A. The statement A *is true* **only if** B *is true* means that the truth of A implies the truth of B. Thus the shorthand statement, "A is true if and only if B is true," is equivalent to the *double implication:* the truth of A implies and is implied by the truth of B. As a further shorthand notation we use the symbol $\Leftrightarrow$ to represent "if and only if," and we write

$$A \Leftrightarrow B$$

for the two implications stated above. The term **necessary and sufficient** is frequently used as a synonym for "if and only if."

2. THE NUMBER PLANE. GRAPHS

In Chapter 1 we saw that the study of inequalities is greatly illuminated when we represent real numbers by means of points on a straight line. In the study of inequalities involving two unknowns, we shall see that a corresponding geometric interpretation is both useful and important.

The determination of solution sets of equations and inequalities in two or more unknowns is an important topic in mathematics, especially from the point of view of applications. The geometric representation of a solution set, called the *graph*, is particularly helpful in the study of analytic geometry.

We now introduce some terminology which will be used throughout the text. It is essential that the student become familiar with these terms as quickly as possible. **The set of all real numbers is denoted by R_1.** Any two real numbers a and b form a **pair.** For example, 2 and 5 are a pair of numbers. When the *order* of the pair is prescribed, we say that we have an **ordered pair.** The ordered pair 2 and 5 is different from the ordered pair 5 and 2. We usually designate an ordered pair in parentheses with a comma separating the first and second elements of the pair: (a, b) is used for the ordered pair consisting of the numbers a (first) and b (second).

Definitions. *The set of all ordered pairs of real numbers is called the **number plane** and is denoted by R_2. Each individual ordered pair is called a **point** in the number plane. The two elements in a number pair are called its **coordinates.***

The number plane can be represented on a geometric or Euclidean plane. It is important that we keep separate the concept of the number plane, R_2, which is an abstract system of ordered pairs, from the concept of the geometric plane, which is the two-dimensional object we studied in Euclidean geometry.

In a Euclidean plane we draw a horizontal and a vertical line, denote them as the x and y axes, respectively, and label their point of intersection O (Fig. 2–2). The point O is called the **origin.** We select a convenient unit of length and,

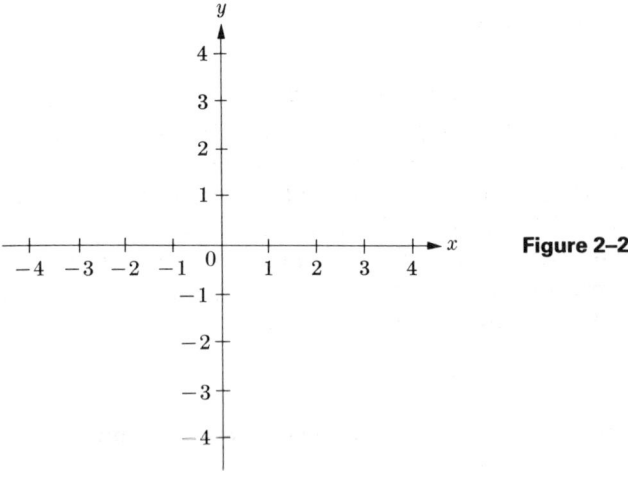

Figure 2–2

starting from the origin as zero, mark off a number scale on the horizontal axis, positive to the right and negative to the left. Similarly, we insert a scale along the vertical axis, with positive numbers extending upward and negative ones downward. It is not necessary that the units along the horizontal axis have the same length as the units along the vertical axis, although we will usually take them as equal.

We now set up a one-to-one correspondence between the points of the number plane, R_2, and the points of the Euclidean (geometric) plane. For each point P in the Euclidean plane we construct perpendiculars from P to the coordinate axes, as shown in Fig. 2–3. The intersection with the x axis is at the point Q, and the intersection with the y axis is at the point R. The distance from the origin to Q (measured positive if Q is to the right of O and negative if Q is to the left of O) is denoted by a. Similarly, the distance OR is denoted by b. Then the point in the number plane corresponding to P is (a, b). It is easy to see that, conversely, to each point in the number plane there corresponds exactly one point in the Euclidean plane. Because of the one-to-one correspondence between the number plane and the plane of Euclidean geometry, we will frequently find it convenient to use geometric terms for the number plane. For example, a "line" in the number plane is used for the set of points corresponding to a line in the geometric plane.

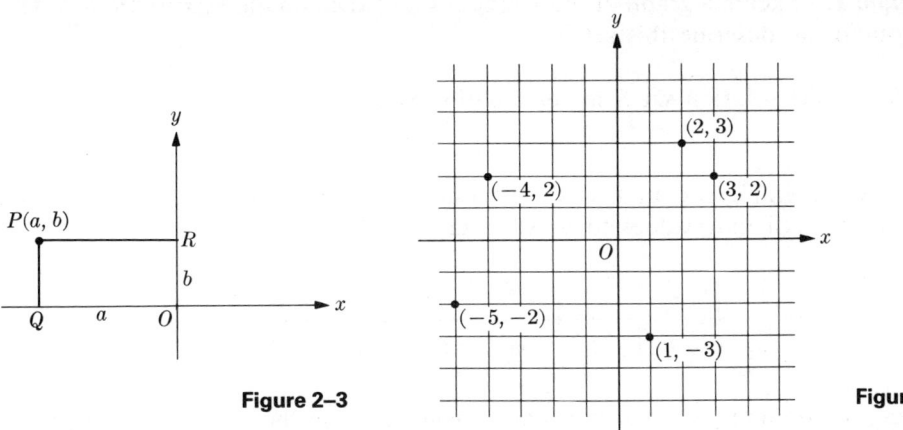

Figure 2–3

Figure 2–4

The description above suggests the method to be used in plotting (that is, representing geometrically) points of the number plane. When plotting points, we enclose the coordinates in parentheses adjacent to the point, as illustrated in Fig. 2–4.

Definitions. *The* **solution set** *of an equation in two unknowns consists of all points in the number plane, R_2, whose coordinates satisfy the equation. A geometric representation of the solution set (that is, the actual drawing) is called the* **graph of the equation.**

The solution set of an equation in two unknowns is, of course, a set in R_2. It is a simple matter to extend the set notation already introduced to include sets in the number plane. If S denotes the set of all (x, y) in R_2 having certain properties,

20 RELATIONS, FUNCTIONS, GRAPHS

which, for example, we call A and B, we describe S as follows:
$$S = \{(x, y):(x, y) \text{ has properties } A \text{ and } B\}.$$
As an illustration, if S is the solution set of the equation
$$2x^2 - 3y^2 = 6,$$
we write
$$S = \{(x, y): 2x^2 - 3y^2 = 6\}.$$

To construct the exact graph of the solution set of some equation is generally impossible, since it would require the plotting of infinitely many points. Usually, in drawing a graph, we select enough points to exhibit the general nature of the graph, plot these points, and then approximate the remaining points by drawing a "smooth curve" through the points already plotted.

It is useful to have a systematic method of choosing points on the graph. We can do this for an equation in two unknowns when we can solve for one of the unknowns in terms of the other. Then, from the formula obtained, we can assign values to the unknown in the formula and obtain values of the unknown for which we solved. The corresponding values are then tabulated as shown in the examples below.

Example 1. Sketch a graph of the solution set of the equation $2x + 3y = 5$. Use set notation to describe this set.

Solution. The solution set S in set notation is
$$S = \{(x, y): 2x + 3y = 5\}.$$
To sketch a graph, we first solve for one of the unknowns, say x. We find $x = \frac{5}{2} - \frac{3}{2}y$. Assigning values to y, we obtain the following table.

x	7	$\frac{11}{2}$	4	$\frac{5}{2}$	1	$-\frac{1}{2}$	-2
y	-3	-2	-1	0	1	2	3

Plotting these points, we see that they appear to lie on the straight line shown in Fig. 2–5. ◀

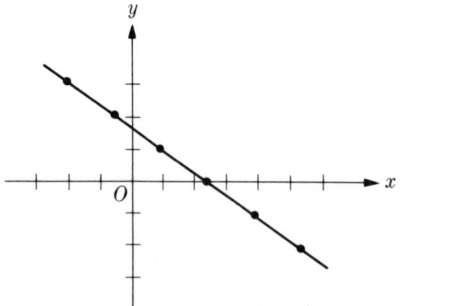

Figure 2–5

Example 2. Sketch a graph of the solution set of the equation

$$\frac{x^2}{25} + \frac{y^2}{16} = 1.$$

Use set notation to describe the solution set S.

Solution. $S = \{(x, y) : (x^2/25) + (y^2/16) = 1\}$. Solving for y in terms of x, we obtain

$$y = \pm \tfrac{4}{5}\sqrt{25 - x^2}.$$

Since x enters into the equation only in the term involving x^2, we may abbreviate the table as indicated below.

x	0	±1	±2	±3	±4	±5
y	±4	$\pm\tfrac{4}{5}\sqrt{24}$	$\pm\tfrac{4}{5}\sqrt{21}$	$\pm\tfrac{16}{5}$	$\pm\tfrac{12}{5}$	0
y (approx.)	±4	±3.92	±3.67	±3.2	±2.4	0

Plotting these points, we see that they appear to lie on the oval curve of Fig. 2-6.

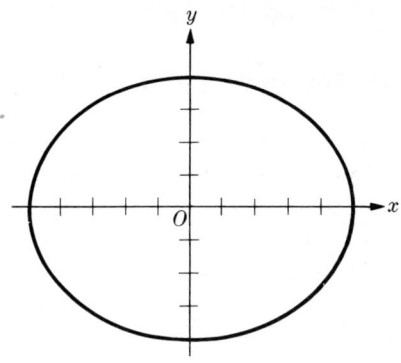

Figure 2-6

Note that each column except the first and last really represents four points. What happens if we choose $x > 5$ or $x < -5$? ◄

PROBLEMS

Sketch a graph of the solution set S of each of the following equations. Describe each set S, using set notation.

1. $x - y = 0$
2. $x + 3y = 2$
3. $y - 2x = 4$
4. $3x - 2y = 6$
5. $x = 4$
6. $y = -1$
7. $x + y = 0$
8. $3x + 2y = 6$
9. $y = \tfrac{1}{2}x^2$
10. $x = \tfrac{1}{2}y^2$
11. $y = x^2 - 4$
12. $x = y^2 - 1$
13. $y = x^2 - 2x - 3$
14. $y = x^2 - 2x + 3$
15. $x^2 + y^2 = 25$

22 RELATIONS, FUNCTIONS, GRAPHS

16. $x^2 + y^2 = 9$
17. $4x^2 + y^2 = 36$
18. $x^2 - y^2 = 9$
19. $2x^2 - y^2 = 17$
20. $x^2 - y^2 + 9 = 0$
21. $y = -x^2 + 2x + 3$
22. $y = \frac{1}{8}x^3$
23. $y = x^3 - 4x$
24. $y = x^3 - 3x + 2$
25. $2x^2 + 3xy + y^2 = 0$ [*Hint:* Write as a quadratic in y.]
26. $x^2 + xy + y^2 = 5$
27. $2x^2 + 3xy + y^2 + 2x + 3y = 10$
28. $x^2 - 5x + 6 = 0$
29. $y = |x|$
30. $y = |x - 1| + 1$
31. $y = 1/(|x| + 1)$

3. FUNCTIONS. FUNCTIONAL NOTATION

In mathematics and many of the physical sciences, simple formulas occur repeatedly. For example, if r is the radius of a circle and A is its area, then

$$A = \pi r^2.$$

If heat is added to an ideal gas in a container of fixed volume, the pressure p and the temperature T satisfy the relation

$$p = a + cT$$

where a and c are fixed numbers with values depending on the properties of the gas, the units used, and so forth.

The relationships expressed by these formulas are simple examples of the concept of function, to be defined precisely later. However, it is not essential that a function be associated with a particular formula. As an example, consider the cost C in cents of mailing a package which weighs x grams. Suppose postal regulations state that the cost is "6¢ per gram or fraction thereof." We can construct the following table.

Weight x in grams	$0 < x \leq 1$	$1 < x \leq 2$	$2 < x \leq 3$	$3 < x \leq 4$	$x \leq 5$
Cost C in cents	6	12	18	24	30

This table could be continued until $x =$ the maximum weight permitted by postal regulations. To each value of x between 0 and the maximum weight there corresponds a precise cost C. We have here an example of a function relating x and C.

It frequently happens that an experimenter finds by measurement that the numerical value y of some quantity depends in a *unique* way on the measured value x of some other quantity. It is usually the case that no known formula expresses the relationship between x and y. All we have is the set of ordered pairs (x, y). In such circumstances, the entire interconnection between x and y is determined by the ordered pairs. We are led to the following definition.

2–3 FUNCTIONS. FUNCTIONAL NOTATION

Definition. *A **function** is a set of ordered pairs* (x, y) *of real numbers in which no two pairs have the same first element. In other words, to each value of x (the first member of the pair) there corresponds exactly one value of y (the second member). The set of all values of x which occur is called the* **domain** *of the function, and the set of all y which occur is called the* **range** *of the function.*

An example of a function is given by the set of all pairs (r, A) obtained from the formula $A = \pi r^2$ when $r > 0$. The domain of this function is the half-infinite interval $(0, \infty)$. The range is $(0, \infty)$.

The set of ordered pairs (T, p) obtained from the formula $p = a + cT$ is also an example of a function. The domain of T and the range of p depend on the particular conditions of the gas, the container, and so forth. Although no simple formula is available, the set of ordered pairs (x, C) as given in Table 1 determines a function, the so-called "postage function."

From the geometric point of view a function is a set of ordered pairs (x, y) in R_2 whose graph may be constructed. The special property implied by the term function assures us that every line parallel to the y axis intersects the graph no more than once (Fig. 2–7). The vertical lines which pass through the graph of a function intersect the x axis. These points of intersection with the x axis form a set called the **projection on the x axis**. This projection is the domain of the function (Fig. 2–7). Horizontal lines through the graph of the function intersect the y axis and these intersections form a set which we recognize as the range of the function. That is, the range is the projection of the graph of the function onto the y axis. Note that horizontal lines may intersect the graph of a function many times.

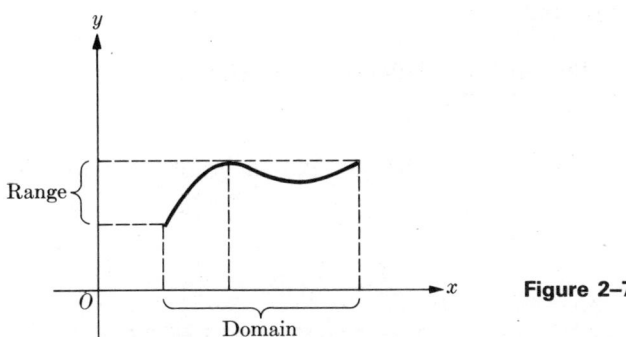

Figure 2–7

It is important to be able to discuss functions and their properties without actually specifying the particular ones we have in mind. For this purpose we introduce a symbol, a letter of the alphabet, to stand for a function. The letters most often used are f, g, F, G, ϕ, Φ. Sometimes, if a problem concerns many different functions, subscripts are employed, so that, for example, f_1, f_2, and f_3 would stand for three different functions.

Suppose that f is a function and x is a number in its domain. To each number x the function f associates a unique number which we denote by y. We also use the symbol $f(x)$ (to be read f of x) as a synonym for the number y. That is, $y = f(x)$. For example, if f is the postage function and the weight of a package is

$3\frac{1}{2}$ grams, then from Table 1, $f(3\frac{1}{2}) = 24$. In set notation, we describe a function f by the statement

$$f = \{(x, y) : x \text{ is in the domain of } f \text{ and } y = f(x)\}.$$

In other words, a function f is just the solution set of the equation $y = f(x)$, and a graph of f is the graph of this equation.

Several symbols for function are now in common use. One such is $f : x \to y$, where x is a generic element of the domain of f and y is the element of the range which is the **image** of x. Another notation for function is $f : D_1 \to D_2$, where D_1 is the set forming the domain of f and D_2 is the set forming the range.

When specifying a function f, we must give its domain and a precise rule for determining the value of $f(x)$ for each x in the domain. For the most part, we shall give functions by means of formulas such as

$$f(x) = x^2 - x + 2 \quad \text{or} \quad f : x \to x^2 - x + 2.$$

Such a formula, by itself, does not give the domain of x. Both in this case and *in general*, we shall take it for granted that if the domain is not specified, then any value of x may be inserted in the formula so long as the result makes sense. The domain shall consist of the set of all such values x.

In prescribing a function by means of a formula, the particular letter used is usually of no importance. The function F determined by the formula

$$F(x) = x^3 - 2x^2 + 5$$

is identical with the function determined by

$$F(t) = t^3 - 2t^2 + 5.$$

The difference is one of notation only.

Example 1. Suppose that f is the function defined by the equation

$$f(x) = x^2 - 2x - 3.$$

Find $f(0)$, $f(-1)$, $f(-2)$, $f(2)$, $f(3)$, $f(t)$, and $f(f(x))$. Plot a graph of f for the portion of the domain in $-2 \le x \le 3$.

Solution. We have

$f(0) = 0^2 - 2 \cdot 0 - 3 = -3,\qquad f(-1) = (-1)^2 - 2(-1) - 3 = 0,$
$f(-2) = (-2)^2 - 2 \cdot (-2) - 3 = 5,\qquad f(2) = 2^2 - 2 \cdot 2 - 3 = -3,$
$f(3) = 3^2 - 2 \cdot 3 - 3 = 0,\qquad f(t) = t^2 - 2t - 3.$

The difficult part is finding $f(f(x))$, and here a clear understanding of the meaning of the symbolism is needed. The formula defining f means that whatever is in the parentheses in $f(\ )$ is substituted in the right side. That is,

$$f(f(x)) = (f(x))^2 - 2 \cdot (f(x)) - 3.$$

However, the right-hand side again has $f(x)$ in it, and we can substitute to get

$$f(f(x)) = (x^2 - 2x - 3)^2 - 2(x^2 - 2x - 3) - 3$$
$$= x^4 - 4x^3 - 4x^2 + 16x + 12.$$

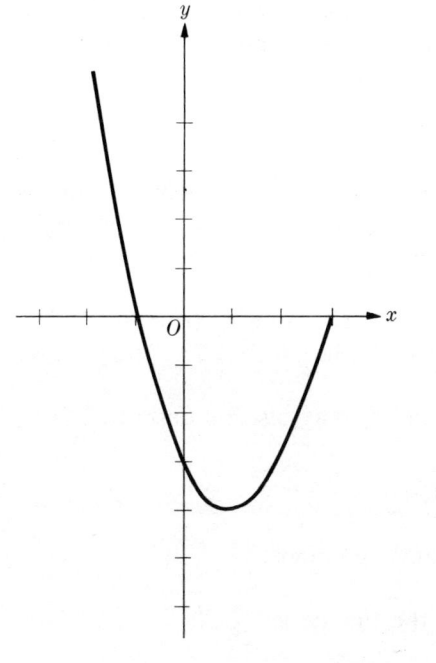

Figure 2–8

To plot the graph we compute

$$f(1) = 1^2 - 2 \cdot 1 - 3 = -4$$

and assemble all the results above to obtain the following table.

x	-2	-1	0	1	2	3
$f(x) = y$	5	0	-3	-4	-3	0

The graph is plotted in Fig. 2–8. ◄

Sometimes successions of parentheses become unwieldy, and we may use brackets or braces with the same meaning: $f[f(x)]$ is the same as $f(f(x))$.

If we write

$$g(x) = x^3 + 2x - 6, \quad -2 \leq x \leq 3,$$

this means that the domain of the function g is the interval $[-2, 3]$. If, in the same formula, the portion $-2 \leq x \leq 3$ were omitted, we would assume that g is defined by that formula for *all* x; the domain would be $(-\infty, \infty)$. This opens up many possibilities. For example, we may define a function F by the following conditions (the function F defined in this way has the interval $[-1, 5]$ for its domain):

$$F(x) = \begin{cases} x^2 + 3x - 2, & -1 \leq x < 2, \\ x^3 - 2x + 4, & 2 \leq x \leq 5. \end{cases}$$

Formulas which define functions may have obvious impossibilities. If these are not explicitly pointed out, the student should be in a position to find them. For example,

$$f_1(x) = \frac{1}{x - 3}$$

is a function defined for all values of x except $x = 3$, since division by zero is always excluded. If

$$f_2(x) = \sqrt{4 - x^2},$$

then it is clear that the domain of f_2 cannot exceed the interval from -2 to $+2$, since imaginary numbers are excluded. Any value of x larger than 2 or less than -2 is impossible.

The formulas for the above functions f_1 and f_2 may also be described by

$$f_1 : x \to \frac{1}{x - 3} \quad \text{and} \quad f_2 : x \to \sqrt{4 - x^2}.$$

These are alternate symbols for exactly the same functions.

Example 2. Discuss the distinction between the functions

$$F(x) = \frac{x^2 - 4}{x - 2} \quad \text{and} \quad G(x) = x + 2.$$

Solution. Since the expression $x^2 - 4$ is factorable into $(x - 2)(x + 2)$, at first glance it appears that the functions are the same. However, the domain of G is all of R_1. That is, x can have any value. For the function F, however, there is difficulty at $x = 2$. If this value is inserted for x, both numerator and denominator are zero. Therefore F and G are identical for all values of x except $x = 2$. For $x = 2$ we have $G(2) = 4$. For $x = 2$ the function F is *not defined*. We could define $F(2)$ to be 4 or any other quantity, at our pleasure. But we would have to specify that fact. If we write

$$F(x) = \frac{x^2 - 4}{x - 2} \quad \text{and} \quad F(2) = 4,$$

then this function is identical with G. This may seem to be a minor point, but we shall see later that it plays an important part in portions of the calculus. ◂

Example 3. Given that $f(x) = x^2$, show that

$$f(x^2 + y^2) = f[f(x)] + f[f(y)] + 2f(x)f(y).$$

Solution

$$\begin{aligned} f(x^2 + y^2) &= (x^2 + y^2)^2 = x^4 + 2x^2y^2 + y^4, \\ f[f(x)] &= f[x^2] = (x^2)^2 = x^4, \\ f[f(y)] &= f[y^2] = (y^2)^2 = y^4, \\ 2f(x)f(y) &= 2x^2y^2. \end{aligned}$$

Adding the last three lines, we obtain
$$f[f(x)] + f[f(y)] + 2f(x)f(y) = x^4 + y^4 + 2x^2y^2,$$
which we have seen is just $f(x^2 + y^2)$. ◄

PROBLEMS

1. Given that $f(x) = x^2 + 3x - 2$, find $f(-3)$, $f(-2)$, $f(-1)$, $f(0)$, $f(1)$, $f(2)$, $f(3)$, and $f(a + 2)$. Plot a graph of the equation $y = f(x)$ for $-3 \le x \le 3$.

2. Given that $f(x) = \frac{1}{4}x^3 - x + 3$, find $f(-3)$, $f(-2)$, $f(-1)$, $f(0)$, $f(1)$, $f(2)$, $f(3)$, and $f(a - 1)$. Plot a graph of the equation $y = f(x)$ for $-3 \le x \le 3$.

3. Given that
$$f(x) = \frac{3x + 2}{2x + 3},$$
find $f(-4)$, $f(-3)$, $f(-2)$, $f(-1)$, $f(0)$, $f(1)$, $f(-1000)$, $f(1000)$, and $f[f(x)]$. Is $-\frac{3}{2}$ in the domain of f? Plot a graph of f for x on $[-4, 1]$ using the values above and additional values near $-\frac{3}{2}$.

4. Given the function
$$f : x \to \frac{2x - 3}{3x - 1}$$
which is an alternate notation for
$$f(x) = \frac{2x - 3}{3x - 1},$$
find $f(x)$ for $x = -1000, -4, -3, -2, -1, 0, 1, 2, 3, 4, 1000$. Also find $f[f(x)]$. Is $\frac{1}{3}$ in the domain of f? Plot a graph of f for x on $[-4, 4]$ using the values above and additional values near $\frac{1}{3}$.

5. Given the function
$$f : x \to \frac{2x}{x^2 + 1},$$
find $f(x)$ for $x = -1000, -3, -2, -1, 0, 1, 2, 3, 1000$. Show that $f(-x) = -f(x)$ for all values of x in R_1. Plot a graph of f for x on $[-3, 3]$.

6. Given that
$$f(x) = \frac{1 + x}{1 - x},$$
show that
$$\frac{f(x) - f(y)}{1 + f(x) \; f(y)} = \frac{x - y}{1 + xy}$$
whenever both sides are defined.

7. Given $f(x) = \sqrt{x + 1}$, find $f(-1)$, $f(0)$, $f(1)$, $f(2)$, $f(3)$. What is the domain of f? Plot its graph.

8. Given $f(x) = |x - 1|$, what is the domain of f? Find $f(x)$ for $x = -2, -1, 0, 1, 2, 3, 4$ and plot a graph of f for x on $[-2, 4]$.

9. Given $f(x) = 3 - |x + 1|$, find $f(x)$ for $x = -4, -3, -2, -1, 0, 1, 2$ and plot a graph of f for x on $[-4, 2]$.

10. Given $f(x) = \sqrt{(x - 1)(x - 3)}$, what is the domain of f? Plot a graph of f for x between -2 and 6, using several values close to 1 and 3.

11. Given $f(x) = \sqrt{2 - 2x - x^2}$, what is the domain of f? [*Hint:* Complete the square under the radical.] Plot a graph of f.

12. Given the function
$$f : x \to \frac{x^2 - 2x - 3}{x - 3},$$
find the domain of f. Give, by formula, a function defined for all x in R_1 which coincides with f wherever f is defined.

13. Same as Problem 12 for the function
$$f(x) = \frac{(x + 2)(x^2 + 6x - 16)(x - 6)}{(x - 2)(x^2 - 4x - 12)}.$$

In Problems 14 through 20, find the value of
$$\frac{f(x + h) - f(x)}{h}.$$
and simplify, assuming that $h \neq 0$.

14. $f(x) = x^2$

15. $f(x) = 2x^2 + x - 3$

16. $f(x) = x^3$

17. $f(x) = \dfrac{1}{x}$

18. $f : x \to \dfrac{1}{x^2}$

19. $f(x) = \sqrt{x}, x > 0$

20. $f : x \to \sqrt{x^2 + 3}$

21. If $f(x) = \sqrt{x}$, show that
$$\frac{f(a) - f(b)}{a - b} = \frac{1}{f(a) + f(b)}.$$

22. A function f is **linear** if it has the form $f : x \to ax + b$ for some numbers a and b. If f and g are linear functions, show that $f + g$ is a linear function. Show that $f[g]$ is a linear function. Under what circumstances is $f \cdot g$ linear?

23. Given $f : x \to 2x - 1$, $g : x \to x^2 + 2$, $h : x \to 3x + 1$, find $f[g[h(x)]]$.

24. The volume of a sphere of radius r is given by the formula
$$f(r) = \frac{4}{3}\pi r^3.$$
Show that if the radius is doubled, the volume is multiplied by a factor of 8.

4. RELATIONS. INTERCEPTS. SYMMETRY

Many of the examples discussed in Section 2 are not functions. In fact, Example 2 on page 21 is not the graph of a function. To see this, observe that in Fig. 2–6 each vertical line for $x \in (-5, 5)$ intersects the graph *twice* and not once, as is required for functions. For convenience in discussing graphs which are not functions, we introduce the following terminology.

Definitions. *Any set in the number plane, R_2, is called a **relation**. More specifically, a set in R_2 is called a **relation from** R_1 **to** R_1. The **domain** of a relation is the set of all x such that (x, y) is in the set forming the relation. The set of all y such that (x, y) is in the relation is the **range** of the relation.*

As in the case for functions, the domain of a relation is its projection on the x axis. The range of a relation is its projection on the y axis. We observe that, unlike the situation in the case of functions, a vertical line may intersect a relation any number of times. Indeed, such a line may intersect it in infinitely many points. After all, according to the definition above, the entire number plane R_2 is a relation.

We shall be interested in relations in which a vertical line intersects the relation in only a finite number of points. Under these circumstances, the relation is composed of the union of a finite number of functions. As Fig. 2–9 shows, the curve going from A to E is the graph of a relation. However, the arc $\widehat{AB}$ is the graph of a function, as are the arcs $\widehat{BC}$, $\widehat{CD}$, and $\widehat{DE}$. The relation is the union of these four functions.

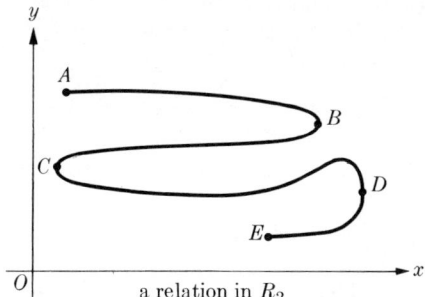

a relation in R_2

Figure 2–9

The solution set of an equation in two unknowns is a relation. A function which forms part of this solution set or relation is said to be **defined implicitly** by this equation.

Example 1. Given the equation $x^2 + xy - 4 = 0$, determine whether or not the solution set S is a function. Plot the graph of S. If S is not a function, show how it can be represented as the union of several functions.

Solution. Solving the above equation for y, we find

$$x^2 + xy - 4 = 0 \quad \Leftrightarrow \quad y = \frac{4-x^2}{x} = \frac{4}{x} - x.$$

The set S is the solution set of the equation $y = (4/x) - x$. Since there is only one value of y for each value of x, S is a function. In fact, S is the function f defined by

$$f(x) = \frac{4-x^2}{x}, \quad \text{or using alternate notation,} \quad f: x \to \frac{4-x^2}{x}.$$

30 RELATIONS, FUNCTIONS, GRAPHS

To plot the graph we make the following table.

x	-4	-3	-2	-1	$-\frac{1}{2}$	$\frac{1}{2}$	1	2	3	4
$f(x) = y$	3	$\frac{5}{3}$	0	-3	$-\frac{15}{2}$	$\frac{15}{2}$	3	0	$-\frac{5}{3}$	-3

We draw a smooth curve through the points plotted from the table. Since f is not defined for $x = 0$, we compute $f(-\frac{1}{2})$ and $f(\frac{1}{2})$ to get an indication of the behavior of f near $x = 0$. (See Fig. 2–10.)

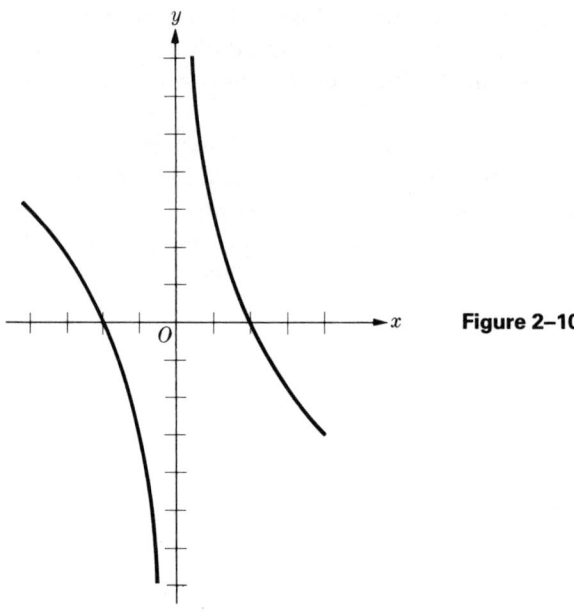

Figure 2–10

Example 2. Let S be the solution set of the equation $x^4 - 4x^2 + y^2 = 0$. Plot the graph of S. If S is not a function, show how it can be represented as the union of several functions.

Solution. Solving for y, we note that

$$x^4 - 4x^2 + y^2 = 0 \quad \Leftrightarrow \quad y = \pm x\sqrt{4 - x^2}.$$

Thus S is a relation which is the union of two functions f_1 and f_2 defined by $f_1(x) = x\sqrt{4 - x^2}$ and $f_2(x) = -x\sqrt{4 - x^2}$. We obtain the following table.

x	± 2	$\pm\frac{3}{2}$	± 1	$\pm\frac{1}{2}$	0
y	0	$\pm\frac{3}{4}\sqrt{7}$	$\pm\sqrt{3}$	$\pm\frac{1}{4}\sqrt{15}$	0
y approx.	0	± 1.98	± 1.73	± 0.97	0

The graph is shown in Fig. 2–11. Note that f_1 and f_2 are defined implicitly by the equation $x^4 - 4x^2 + y^2 = 0$.

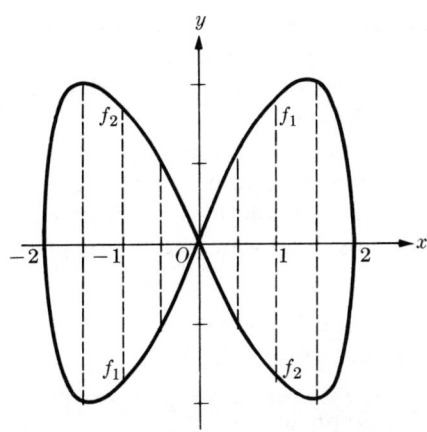

Figure 2–11

Sketching the graph of an equation by plotting a few of its points and drawing a smooth curve through them is an adequate method if the equation is sufficiently simple. However, if the equation is at all complicated, this method not only may be too laborious but, in some cases, may even lead to incorrect graphs. For example, we may try to plot the graph of the equation

$$y = \frac{1}{2x - 1}$$

by letting x take on a sequence of integer values. That is, we construct the following table.

x	-3	-2	-1	0	1	2	3
y	$-\frac{1}{7}$	$-\frac{1}{5}$	$-\frac{1}{3}$	-1	1	$\frac{1}{3}$	$\frac{1}{5}$

Plotting these points and drawing a smooth curve through them leads to the *incorrect* graph shown in Fig. 2–12. However, noting that $x = \frac{1}{2}$ is not in the domain of the function f given by $f: x \to 1/(2x - 1)$ and assigning to x several values near $\frac{1}{2}$ leads to the correct graph shown in Fig. 2–13.

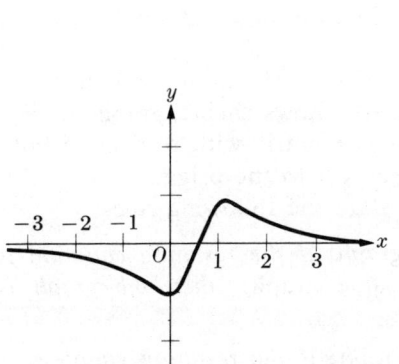

$y = \frac{1}{2x-1}$; incorrect graph

Figure 2–12

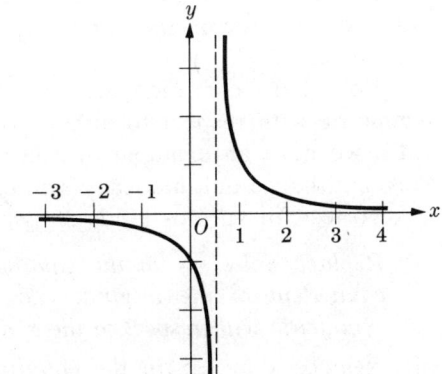

$y = \frac{1}{2x-1}$; correct graph

Figure 2–13

In this section we discuss several facts about graphs which are easily obtained from the equation and which are substantial aids in constructing quick, accurate graphs. Later we shall see how the methods of calculus can be used to get additional information about a graph before it is drawn.

It is useful to know where the graph crosses the x and y axes. A point at which a graph crosses the x axis is called an x **intercept;** a point where it crosses the y axis is called a y **intercept.** It may happen that a graph does not intersect a given axis or that it intersects both axes many times. To locate the intercepts, if any, we have the rule:

> to find the x intercepts, set $y = 0$ and solve for x;
> to find the y intercepts, set $x = 0$ and solve for y.

Example 3. Find the x and y intercepts of the graph of
$$x^2 - 3y^2 + 6x - 3y = 7.$$

Solution. Setting $y = 0$, we get
$$x^2 + 6x - 7 = 0 \quad \text{or} \quad x = -7 \quad \text{or} \quad x = 1.$$
The x intercepts are $(-7, 0)$ and $(1, 0)$.

Setting $x = 0$, we get
$$3y^2 + 3y + 7 = 0 \quad \text{or} \quad y = \frac{-3 \pm \sqrt{-75}}{6}.$$

Since the solutions for y are complex numbers, there are no y intercepts and the graph does not intersect the y axis. ◀

In constructing the graphs discussed in Section 2, we see that some of them have certain kinds of symmetry. For example, Fig. 2–6 exhibits several obvious symmetries. If we had known about these symmetries before drawing the graph, the actual sketching would have been much easier. A graph is said to be **symmetric with respect to the x axis** if and only if whenever a point (a, b) is on the graph the point $(a, -b)$ is also on the graph. A graph is **symmetric with respect to the y axis** if and only if whenever (a, b) is on the graph, the point $(-a, b)$ is also on the graph. We also consider a third type of symmetry. A graph is **symmetric with respect to the origin** if and only if whenever (a, b) is on the graph, $(-a, -b)$ is also on the graph.

Note that a curve may be symmetric with respect to the origin without being symmetric with respect to either axis. Figure 2–10 shows such a graph. In Fig. 2–11, we have an example of a curve which is symmetric with respect to both axes, and hence automatically symmetric with respect to the origin.

To test for various kinds of symmetry, we state the following rules:

i) *Replace y by $-y$ in the equation of the graph; if the resulting equation is equivalent to the original* (i.e., has the same graph), *then the graph is symmetric with respect to the x axis.*

ii) *Replace x by $-x$ in the equation of the graph; if the resulting equation is equivalent to the original, the graph is symmetric with respect to the y axis.*

iii) *Replace x by −x and y by −y in the equation of the graph; if the resulting equation is equivalent to the original, the graph is symmetric with respect to the origin.*

The rules for symmetry reduce the work involved in sketching the graph of an equation. If a curve is symmetric with respect to the x axis, then only the portion above the x axis has to be plotted carefully. The remainder of the graph is obtained as the mirror image in the x axis of the part already sketched. A similar procedure works if the graph is symmetric with respect to the y axis.

Example 4. Test the graph of the equation $9x^2 + 16y^2 = 144$ for symmetry, find its intercepts, and determine whether the graph is or is not a function. If not, find the functions of which it is the union. Sketch the graph.

Solution. (i) Replacing y by $-y$ in the equation yields

$$9x^2 + 16(-y)^2 = 144 \quad \text{or} \quad 9x^2 + 16y^2 = 144.$$

Therefore the graph is symmetric with respect to the x axis. (ii) Replacing x by $-x$ leads similarly to an equivalent equation, so the graph is symmetric with respect to the y axis. (iii) Replacing x by $-x$ and y by $-y$ leads to an equivalent equation, so the graph is symmetric with respect to the origin. Setting $y = 0$, we get

$$9x^2 = 144 \quad \text{or} \quad x = \pm 4,$$

and the x intercepts are $(4, 0)$ and $(-4, 0)$. Setting $x = 0$, we find that the y intercepts are $(0, \pm 3)$. Solving for y, we obtain

$$9x^2 + 16y^2 = 144 \quad \text{or} \quad y = \pm\tfrac{3}{4}\sqrt{16 - x^2},$$

so the graph is the union of the two functions f_1 and f_2 given by

$$f_1 : x \to \tfrac{3}{4}\sqrt{16 - x^2} \quad \text{and} \quad f_2 : x \to -\tfrac{3}{4}\sqrt{16 - x^2}.$$

The graph is sketched in Fig. 2–14. ◂

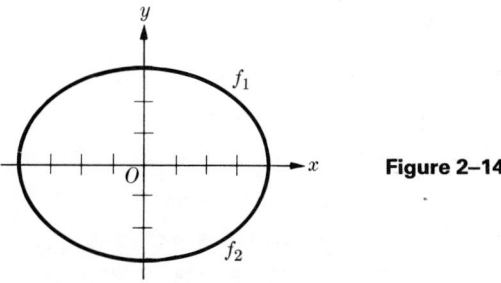

Figure 2–14

Example 5. Test the graph of the equation

$$y = \tfrac{1}{8}(x^3 - 4x)$$

for symmetry and find its intercepts. Sketch the graph.

Solution. Replacing y by $-y$ leads to the equation

$$-y = \tfrac{1}{8}(x^3 - 4x) \quad \text{or} \quad y = -\tfrac{1}{8}(x^3 - 4x),$$

which is not equivalent to the original. Clearly, the graph is not symmetric with respect to the x axis. Similarly, replacing x by $-x$, we see that it is not symmetric with respect to the y axis. However, when we replace x by $-x$ and y by $-y$, we get the equation

$$-y = \tfrac{1}{8}[(-x)^3 - 4(-x)] = -\tfrac{1}{8}(x^3 - 4x),$$

which *is* equivalent to the original equation. Therefore the graph is symmetric with respect to the origin. Setting $y = 0$, we find that the x intercepts are $(-2, 0)$, $(0, 0)$, and $(2, 0)$. Letting $x = 0$, we see that the y intercept is $(0, 0)$. A graph of a portion of the solution set is shown in Fig. 2–15. ◄

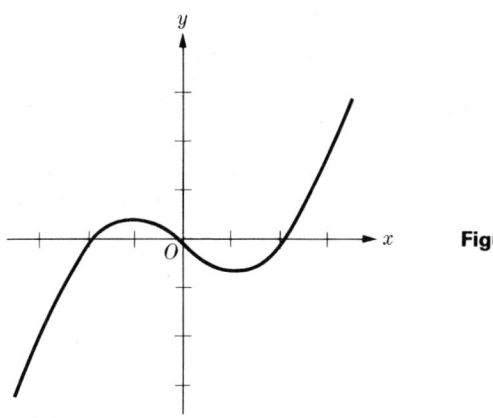

Figure 2–15

PROBLEMS

In Problems 1 through 25, discuss each relation with respect to intercepts and symmetry. When the solution set is not a function, express it as the union of several functions. Sketch the graph.

1. $y = x^2 - 4$
2. $y = 4 - x^2$
3. $y^2 = 3x$
4. $y^2 = -3x$
5. $y^2 = 2x + 2$
6. $y^2 = 4x + 1$
7. $x = y^2 - 1$
8. $4x^2 + y^2 = 64$
9. $2x^2 + 3y^2 = 18$
10. $xy = 4$
11. $y^2 - 2xy = 4$
12. $y^2 = 2 - x$
13. $y^2 = x^2 + 3$
14. $y^2 = x^2 - 4$
15. $x^2 + xy + y^2 = 3$
16. $y = \tfrac{1}{8}x^3$
17. $y = x + \tfrac{1}{4}x^3$
18. $y^2 = 4x^3$
19. $y^2 - 2y + 1 = x$
20. $y^3 = x$
21. $y^4 = x^2$
22. $y^2 + 4y = x - 6$
23. $y^4 - 4y^2 = x^2 + 4x$
24. $y^2 = x^4 - 4x^2$
25. $y^4 - 4y^2 + x^2 = 0$ (See Example 2)
26. Discuss for intercepts and plot the graph of the equation $x = y + \tfrac{1}{4}y^3$. Show that if (x_1, y_1) and (x_2, y_2) are on the graph and $y_1 < y_2$, then $x_1 < x_2$. [*Hint:* Consider the

cases $0 \leq y_1 < y_2$, $y_1 < 0 < y_2$, $y_1 < y_2 \leq 0$.] Would you say that the solution set is a function?

27. Show that if a relation is symmetric with respect to the y axis and the origin, it must be symmetric with respect to the x axis.

28. Let n be any integer. Define
$$f(x) = n \quad \text{for} \quad n \leq x < n + 1, \, n = 0, \pm 1, \pm 2, \ldots .$$
a) Plot the graph of f.
b) Plot the graph of $g(x) = x - f(x)$.
c) Plot the graph of $f(x + \frac{1}{2})$.

29. Let f be a function defined for all x in R_1. Let $g(x) = \frac{1}{2}[f(x) + f(-x)]$, $h(x) = \frac{1}{2}[f(x) - f(-x)]$.
a) Show that g is symmetric with respect to the y axis.
b) Show that h is symmetric with respect to the origin.
c) Show that an arbitrary function on R_1 can always be represented as the sum of two functions, one symmetric with respect to the y axis and the other symmetric with respect to the origin.

30. Let f and g be any functions defined on R_1.
a) If f and g are symmetric with respect to the y axis, show that $f + g$ and fg are also symmetric with respect to the y axis.
b) If f and g are symmetric with respect to the origin, show that $f + g$ is symmetric with respect to the origin and that fg is symmetric with respect to the y axis.

5. DOMAIN. RANGE. ASYMPTOTES

We recall that the *domain* of a relation is the set of all numbers x_0 such that the vertical line $x = x_0$ intersects the graph. The *range* of a relation is the set of all y_0 such that the horizontal line $y = y_0$ intersects the graph. To determine the domain of a relation involving x and y we perform the following two steps: (i) *We solve for y in terms of x.* (We caution that this step is often not possible.) (ii) *In the resulting expression or expressions in x, we determine those values of x for which at least one of the expressions has meaning.* The domain is the totality of such values of x. To find the range, we perform steps (i) and (ii) with the roles of x and y interchanged. Two examples illustrate the method.

Example 1. Find the domain and the range of the relation defined by
$$y^2 = x^2 - 4x + 3.$$

Solution. To find the domain we first solve for y in terms of x. Setting $f(x) = x^2 - 4x + 3$, we see that $y^2 = f(x)$ and
$$y = +\sqrt{f(x)}, \quad y = -\sqrt{f(x)}.$$
Since there is no real number which is the square root of a negative quantity, the domain consists of all x for which $f(x) \geq 0$. To determine when $f(x)$ is nonnegative, we factor the expression $x^2 - 4x + 3$. We obtain
$$f(x) = x^2 - 4x + 3 \equiv (x - 1)(x - 3),$$

and f is positive whenever *both* factors are positive or *both* factors are negative. The factors $x - 1$ and $x - 3$ are both positive for $x > 3$ and both negative for $x < 1$. The domain of the relation is the set

$$(-\infty, 1] \cup [3, +\infty).$$

To find the range we solve for x in terms of y. From the relation

$$x^2 - 4x + 3 - y^2 = 0,$$

we obtain

$$x = 2 \pm \sqrt{1 + y^2}.$$

Since $1 + y^2$ is positive for all y, the range is $(-\infty, +\infty)$. A graph of the relation is sketched in Fig. 2–16.

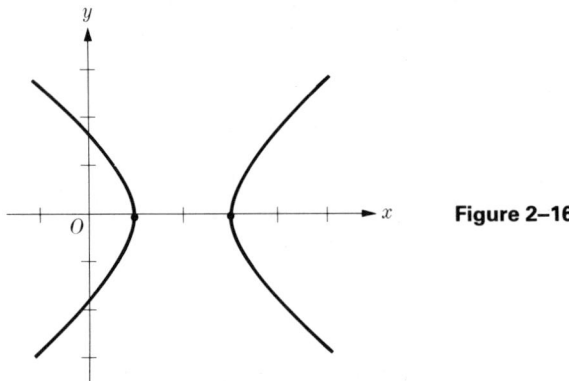

Figure 2–16

◀

Example 2. Given the curve with equation

$$y(x^2 - 1) = 2,$$

find the intercepts, test for symmetry, and sketch the graph.

Solution. Setting $x = 0$, we get $y = -2$, which is the y intercept; $y = 0$ yields the impossible statement $0 = 2$, and the curve has no x intercept.

To test for symmetry with respect to the x axis, we replace y by $-y$, getting

$$-y(x^2 - 1) = 2,$$

which means that the curve is not symmetric with respect to the x axis. Replacing x by $-x$, we find

$$y[(-x)^2 - 1] = 2 \quad \text{or} \quad y(x^2 - 1) = 2,$$

which is the same as the original equation, and the curve is symmetric with respect to the y axis. The curve is not symmetric with respect to the origin, since

$$(-y)[(-x)^2 - 1] = 2$$

is not the same as the original equation.

We make up a table of values for positive x (see Fig. 2–17).

x	0	1	2	3	4	5
y	−2	*	$\frac{2}{3}$	$\frac{1}{4}$	$\frac{2}{15}$	$\frac{1}{12}$

We see that as x increases, the values of y get closer and closer to zero. On the other hand, there is no value of y corresponding to $x = 1$. In order to get a closer look at what happens when x is near 1, we construct an auxiliary table of values for x near 1. We obtain the following set of values.

x	$\frac{1}{2}$	$\frac{3}{4}$	$\frac{7}{8}$	0.9	1.1	1.2	1.3	$1\frac{1}{2}$
y	$-\frac{8}{3}$	$-\frac{32}{7}$	$-\frac{128}{15}$	$-\frac{200}{19}$	$\frac{200}{21}$	$\frac{50}{11}$	$\frac{200}{69}$	$\frac{8}{5}$

As x moves closer to 1 from the left, the corresponding values of y become large negative numbers. As x approaches 1 from the right, the values of y become large positive numbers. The curve is sketched in Fig. 2–17. The portion to the left of the y axis is sketched from our knowledge of the symmetry property; therefore a table of values for negative x is not needed.

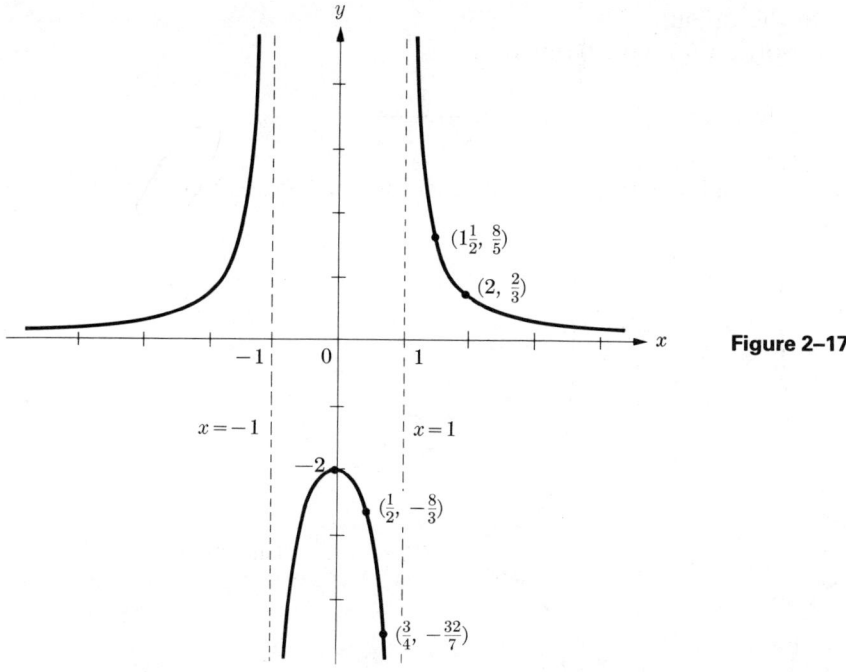

Figure 2–17

◄

In Example 2, the vertical line through the point (1, 0) plays a special role. The curve to the right of the line gets closer and closer to this line as the curve becomes higher and higher. In fact, the distance between the curve and the line tends to zero as the curve continues upward beyond all bound. Such a line is called a **vertical asymptote** to the curve. Similarly, the x axis is called a **horizontal**

asymptote, since the distance between the curve and the x axis tends to zero as x increases beyond all bound. A knowledge of the location of the vertical and horizontal asymptotes is of great help in sketching the curve. We now give a rule (which works in many cases) for finding the asymptotes.

Rule. *To locate the vertical asymptotes, solve the equation for y in terms of x. If the result is a quotient of two expressions involving x, find all those values of x for which the denominator vanishes and the numerator does not. If a is such a value, the vertical line through the point $(a, 0)$ will be a vertical asymptote. To locate the horizontal asymptotes, solve for x in terms of y, and find those values of y for which the denominator vanishes (and the numerator does not). If b is such a value, the horizontal line through the point $(0, b)$ is a horizontal asymptote.*

Two examples illustrate the technique.

Example 3. Find the intercepts, symmetries, domain, range, and asymptotes, and sketch a graph of the equation $(x^2 - 4)y^2 = 1$.

Solution. a) *Intercepts:* The value $y = 0$ yields no x intercept; when $x = 0$, we have $y^2 = -\frac{1}{4}$, and there are no y intercepts.
b) *Symmetry:* Graph is symmetric with respect to both axes (and therefore with respect to the origin).
c) *Domain:* Solving for y in terms of x:

$$y = \pm \frac{1}{\sqrt{x^2 - 4}}.$$

The domain is all x with $|x| > 2$; that is, the set $(-\infty, -2) \cup (2, +\infty)$.

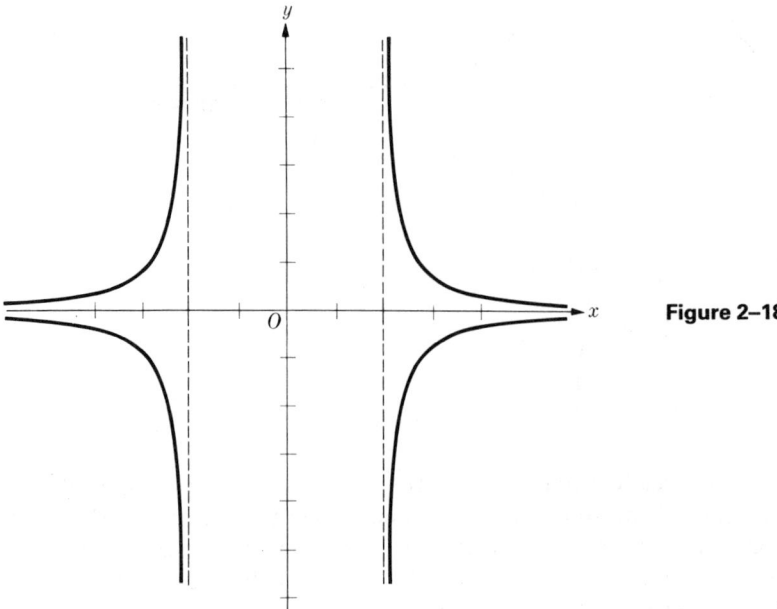

Figure 2-18

d) *Range:* Solving for x in terms of y: $x = \pm\sqrt{1 + 4y^2}/y$. The range is all y except $y = 0$.

e) *Asymptotes:* To find the vertical asymptotes, we use the expression in part (c) for the domain. Setting the denominator equal to zero, we have the vertical lines $x = 2$, $x = -2$ as asymptotes. To find the horizontal asymptotes, we use the expression in part (d) for the range. Setting the denominator equal to zero, we have the horizontal line $y = 0$ as an asymptote. The graph is sketched in Fig. 2–18. ◂

Example 4. Find the intercepts, symmetry, domain, range, and asymptotes, and sketch the graph of the equation

$$x^2 y = x - 3.$$

Solution. a) *Intercepts:* If $y = 0$, then $x = 3$ and the x intercept is 3. Setting $x = 0$ yields no y intercept.

b) *Symmetry:* All symmetry tests fail.

c) *Domain:* Solving for y in terms of x, we find

$$y = \frac{x - 3}{x^2}.$$

The domain is all $x \neq 0$.

d) *Range:* To solve for x, we write the equation $yx^2 - x + 3 = 0$. The solution of this equation in x obtained by the quadratic formula is

$$x = \frac{1 \pm \sqrt{1 - 12y}}{2y} \quad \text{if} \quad y \neq 0.$$

The original equation shows that $x = 3$ when $y = 0$. The range is all $y \leq \frac{1}{12}$; that is, $(-\infty, \frac{1}{12}]$.

e) *Asymptotes:* The line $x = 0$ is a vertical asymptote. The line $y = 0$ is a horizontal asymptote.

To draw the graph we construct the following table. ◂

x	-4	-3	-2	-1	1	2	3	4	5
y	$-\frac{7}{16}$	$-\frac{2}{3}$	$-\frac{5}{4}$	-4	-2	$-\frac{1}{4}$	0	$\frac{1}{16}$	$\frac{2}{25}$

The graph is sketched in Fig. 2–19.

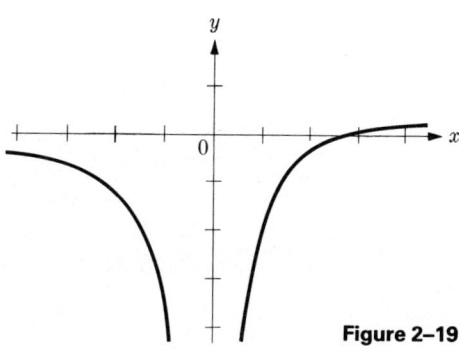

Figure 2–19

PROBLEMS

In Problems 1 through 33, find in each case the intercepts, symmetries, domain, range, and vertical and horizontal asymptotes. Also sketch the graph.

1. $y = 4x^2$
2. $y^2 = 2x$
3. $y^2 = -4x$
4. $y^2 = 2x - 4$
5. $y = 3 - x^2$
6. $xy = 6$
7. $x^2 y = 8$
9. $2x^2 - 3y^2 = 6$
10. $9x^2 - 4y^2 + 36 = 0$
11. $2x^2 + 3y^2 = 6$
12. $3x^2 + 2y^2 = 18$
13. $xy^2 = 8$
14. $y^2 = 1 - 2x^2$
15. $x^2 + xy + y^2 = 12$
16. $y^2 - xy = 2$
17. $y^2(x + 1) = 4$
18. $y(x^2 - 1) = 1$
19. $x^2(y^2 - 4) = 4$
20. $x^2(y - 2) = 2$
21. $(x^2 + 1)y^2 = 4$
22. $y^2(x - 2) + 2 = 0$
23. $y(x - 1)(x - 3) = 4$
24. $y^2(x - 1)(x - 3) = 4$
25. $x(y^2 - 4) = 2y$
26. $x^2(y^2 - 4) = 4y$
27. $y^2 + 2x = x^2 y^2$
28. $y^4 = y^2 - x^2$
29. $y^4 = 4(x^2 - y^2)$
30. $x^2 y^2 = x - 2$
31. $y^2(x^2 - 1) = x + 2$
32. $(x^2 - 4)y^2 = x^2 - 1$
33. $5y(x - 1)(x - 3) = 2(5x + 3)$

*34. a) A curve may cross an asymptote, as demonstrated in Fig. 2–19, where we see that the graph crosses the x axis. Construct an equation such that its graph crosses its asymptote twice.

b) Construct an equation which has the x axis as an asymptote and such that the graph crosses the x axis 3 times. State a method for constructing such equations where the number of crossings is n, where n is any positive integer.

35. Find the range, intercepts, and symmetries, if any, of the solution set of $|y - 1| = |x - 2|$. Draw the graph.

36. Discuss the solution set of

$$y = \frac{|x - 2|}{|x + 3|}$$

for intercepts, symmetry, asymptotes, range, and domain. Draw the graph.

37. Given $|y^2 - 1| = |x + 2|$. Discuss for domain, range, intercepts, and symmetry. Draw the graph.

*38. Discuss the graph of $\sin x + \sec y = 0$.

3

THE LINE

1. DISTANCE FORMULA. MIDPOINT FORMULA

We consider a rectangular coordinate system with the unit of measurement the same along both coordinate axes. When this unit of measurement is equal to the unit of distance in the geometric plane, we call the coordinate system **Cartesian**. The distance between any two points in the plane is the length of the line segment joining them. We shall derive a formula for this distance in terms of the coordinates of the two points.

Let P_1, with coordinates (x_1, y_1), and P_2, with coordinates (x_2, y_2), be two points, and let d denote the length of the segment between P_1 and P_2 (Fig. 3–1). Draw a line parallel to the y axis through P_1, and do the same through P_2. These lines intersect the x axis at the points $A(x_1, 0)$ and $B(x_2, 0)$. [The symbol $A(x_1, 0)$ is to be read "the point A with coordinates $(x_1, 0)$."] Now draw a line through P_1 parallel to the x axis; this line intersects the vertical through P_2 and B at a point $C(x_2, y_1)$.

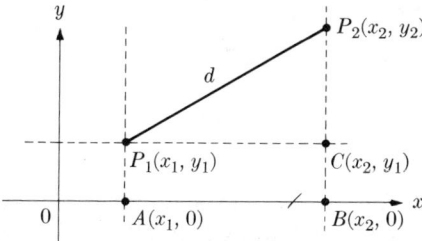

Figure 3–1

The length of the segment between P_1 and C (**denoted** $|P_1C|$) is equal to the length $|AB|$. If x_2 is to the right of x_1 (as shown in the figure), then the length of $|AB|$ is $x_2 - x_1$. If x_2 were to the left of x_1, the distance would be $x_1 - x_2$. In either case, the length of $|AB|$ is easily written by using absolute value notation: $|AB| = |x_2 - x_1|$. Similarly, the length of $|P_2C| = |y_2 - y_1|$. It is a good exercise for the reader to check the correctness of these facts when the points P_1 and P_2 are in various quadrants of the plane.

We note that triangle P_1P_2C is a right triangle, and we recall from plane geometry the Pythagorean theorem: "The sum of the squares of the legs of a right triangle is equal to the square of the hypotenuse." Applying the Pythagorean

theorem to triangle P_1P_2C, we get

$$|P_1P_2|^2 = |P_1C|^2 + |P_2C|^2 \quad \text{or} \quad d^2 = |x_2 - x_1|^2 + |y_2 - y_1|^2.$$

This may also be written

$$d^2 = (x_2 - x_1)^2 + (y_2 - y_1)^2$$

and, extracting the square root, we obtain the **distance formula:**

$$d = \sqrt{(x_2 - x_1)^2 + (y_2 - y_1)^2}.$$

Notation. The distance d between any two distinct points is always *positive* and, in keeping with this, we shall always use the square-root symbol without any plus or minus sign in front of it to mean the positive square root. If we wish to discuss the negative square root of some number, say 3, we will write $-\sqrt{3}$.

Example 1. Find the distance between the points $P_1(3, 1)$ and $P_2(5, -2)$.

Solution. We substitute in the distance formula and get

$$d = \sqrt{(5 - 3)^2 + (-2 - 1)^2} = \sqrt{4 + 9} = \sqrt{13}. \quad \blacktriangleleft$$

Example 2. The point $P_1(5, -2)$ is 4 units away from a second point P_2, whose y coordinate is 1. Locate the point P_2.

Solution. The point P_2 will have coordinates $(x_2, 1)$. From the distance formula we have the equation

$$4 = \sqrt{(x_2 - 5)^2 + (1 - (-2))^2}.$$

To solve for x_2, we square both sides and obtain

$$16 = (x_2 - 5)^2 + 9 \quad \text{or} \quad x_2 - 5 = \pm\sqrt{7}.$$

There are two possibilities for x_2:

$$x_2 = 5 + \sqrt{7}, \quad 5 - \sqrt{7}.$$

In other words, there are two points P_2, one at $(5 + \sqrt{7}, 1)$ and the other at $(5 - \sqrt{7}, 1)$, which have y coordinate 1 and which are 4 units from P_1. $\quad \blacktriangleleft$

Let $A(x_1, 0)$ and $B(x_2, 0)$ be two points on the x axis. We define the **directed distance from** A **to** B to be $x_2 - x_1$. If B is to the right of A, as shown in Fig. 3–2(a), the directed distance is positive. If B is to the left of A, as in Fig. 3–2(b), the direct distance is negative. If two points $C(0, y_1)$ and $D(0, y_2)$ are on the y axis,

Figure 3–2

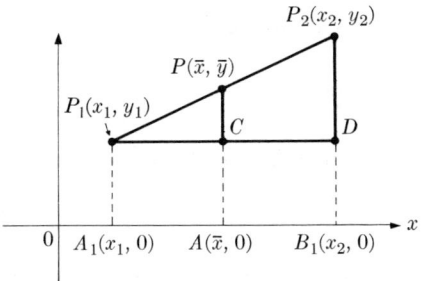

Figure 3–3

the **directed distance from** C **to** D is similarly defined to be $y_2 - y_1$. Directed distances are defined only for pairs of points on a coordinate axis. While ordinary distance between any two points is always positive, note that directed distance may be positive or negative.

Let $P_1(x_1, y_1)$ and $P_2(x_2, y_2)$ be any two points in the plane. We shall show how to find the coordinates of the midpoint of the line segment joining P_1 and P_2. Let P, with coordinates $(\bar{x}, \bar{y})$, be the midpoint. Through P_1, P, and P_2 draw parallels to the y axis, and through P_1 a parallel to the x axis, forming the triangles shown in Fig. 3–3. We recall from plane geometry the statement, "A line parallel to the base of a triangle which bisects one side also bisects the other side." Since PC is parallel to P_2D and P is the midpoint of P_1P_2, it follows that C is the midpoint of P_1D. We see at once that A is the midpoint of A_1B_1. The coordinates of A_1 are $(x_1, 0)$, the coordinates of A are $(\bar{x}, 0)$, and those of B_1 are $(x_2, 0)$. Therefore the directed distance A_1A must be equal to the directed distance AB_1. From the definition of directed distance we have

$$\bar{x} - x_1 = x_2 - \bar{x}$$

and, solving for $\bar{x}$, we obtain

$$\bar{x} = \frac{x_1 + x_2}{2}.$$

If we perform the same argument for the y coordinates, the result by analogy is

$$\bar{y} = \frac{y_1 + y_2}{2}.$$

Example 3. Locate the midpoint of the line segment joining the points $P(3, -2)$ and $Q(-4, 5)$.

Solution. From the above formula,

$$\bar{x} = \frac{3 - 4}{2} = -\frac{1}{2} \quad \text{and} \quad \bar{y} = \frac{-2 + 5}{2} = \frac{3}{2}. \quad \blacktriangleleft$$

Example 4. Find the length of the line segment joining the point $A(7, -2)$ to the midpoint of the line segment between the points $B(4, 1)$ and $C(3, -5)$.

Solution. The midpoint of the line segment between B and C is at

$$\bar{x} = \frac{4+3}{2} = \frac{7}{2}, \quad \bar{y} = \frac{1-5}{2} = -2.$$

From the distance formula applied to A and this midpoint we have

$$d = \sqrt{(7 - \tfrac{7}{2})^2 + (-2 - (-2))^2} = \tfrac{7}{2}.$$ ◀

Example 5. A line segment AB has its midpoint at $C(5, -1)$. Point A has coordinates $(2, 3)$; find the coordinates of B.

Solution. In the midpoint formula, we know that $\bar{x} = 5$, $\bar{y} = -1$, $x_1 = 2$, $y_1 = 3$, and we have to find x_2, y_2. Substituting these values in the midpoint formula, we get

$$5 = \frac{2 + x_2}{2}, \quad -1 = \frac{3 + y_2}{2},$$

and

$$x_2 = 8, \quad y_2 = -5.$$ ◀

PROBLEMS

In Problems 1 through 4, find the lengths of the sides of the triangles with the given points as vertices.

1. $A(4, 1)$, $B(2, -1)$, $C(-1, 5)$
2. $A(-1, 0)$, $B(5, 2)$, $C(3, -2)$
3. $A(3, -4)$, $B(2, 1)$, $C(6, -2)$
4. $A(4, 0)$, $B(0, -2)$, $C(5, 7)$

In Problems 5 and 6, locate the midpoints of the line segments joining the given points.

5. $P_1(2, 5)$, $P_2(7, -4)$
6. $A(3, 7)$, $B(8, -12)$

In Problems 7 and 8, locate the three points which divide the line segment joining P_1 and P_2 into four equal parts.

7. $P_1(7, 1)$, $P_2(-2, 5)$
8. $P_1(4, 0)$, $P_2(-3, -2)$

9. The midpoint of a line segment AB is at the point $P(-4, -3)$. The point A has coordinates $(8, -5)$. Find the coordinates of B.
10. The midpoint of a line segment AB is at the point $P(-7, 2)$. The x coordinate of A is at 5, and the y coordinate of B is at -9. Find the points A and B.
11. Find the lengths of the medians of the triangle with vertices at $A(4, 1)$, $B(-5, 2)$, $C(3, -7)$.
12. Same as Problem 11, for $A(-3, 2)$, $B(4, 3)$, $C(-1, -4)$.
13. Show that the triangle with vertices at $A(1, -2)$, $B(-4, 2)$, $C(1, 6)$ is isosceles.
14. Same as 13, with $A(2, 3)$, $B(6, 2)$, $C(3, -1)$.
15. Show that the triangle with vertices at $A(0, 2)$, $B(3, 0)$, $C(4, 8)$ is a right triangle.
16. Same as 15, with $A(3, 2)$, $B(1, 1)$, $C(-1, 5)$.
17. Show that the quadrilateral with vertices at $(3, 2)$, $(0, 5)$, $(-3, 2)$, $(0, -1)$ is a square.
18. Show that the diagonals of the quadrilateral with the following vertices bisect each other: $(-4, -2)$, $(2, -10)$, $(8, -5)$, $(2, 3)$,

19. Same as 18, for $(5, 3)$, $(2, -3)$, $(-2, 1)$, $(1, 7)$.
20. The four points $A(1, 1)$, $B(3, 2)$, $C(7, 3)$, $D(0, 9)$ form the vertices of a quadrilateral. Show that the midpoints of the sides are the vertices of a parallelogram.
21. a) Show how directed distances may be defined in a natural way for pairs of points on a line parallel to a coordinate axis.
    b) Using the definition in (a), carry through the proof of the midpoint formula, using directed distances along the line P_1D in Fig. 3–3.
22. The formula for the coordinates of the point $Q(x_0, y_0)$ which divides the line segment from $P_1(x_1, y_1)$ to $P_2(x_2, y_2)$ in the ratio p to q is

$$x_0 = \frac{px_2 + qx_1}{p + q}, \quad y_0 = \frac{py_2 + qy_1}{p + q}.$$

Derive this formula.

23. Let $A(x_1, y_1)$, $B(x_2, y_2)$, $C(x_3, y_3)$ be the vertices of any triangle. Show that the length of the line segment joining the midpoints of any two sides of this triangle is $\frac{1}{2}$ the length of the third side.
24. Find the set of all points $P(x, y)$ which are equidistant from $(-1, 2)$ and $(3, 4)$. [*Hint:* Let d_1 be the distance from P to $(-1, 2)$ and d_2 the distance from P to $(3, 4)$. Then the condition $d_1 = d_2$ becomes

$$\sqrt{(x + 1)^2 + (y - 2)^2} = \sqrt{(x - 3)^2 + (y - 4)^2}.$$

Squaring this equation and simplifying yields the equation of the solution set.] Draw the graph.

25. Find the set of all points $P(x, y)$ which are at a distance of 2 from $(1, 1)$. Draw the graph.
26. Find the set of all points $P(x, y)$ which are at a distance of 5 from $(3, 4)$. Draw the graph.
27. Find the set of all points $P(x, y)$ such that the distance of P from $(4, -1)$ equals the distance of P from the y axis. Draw the graph.
28. Find the set of all points $P(x, y)$ such that the distance of P from $(1, 2)$ equals the distance of P from the x axis. Draw the graph.
29. Find the set of all points $P(x, y)$ such that the distance of P from $(1, -1)$ is twice the distance of P from $(4, -1)$. Draw the graph.
30. Find the set of all points $P(x, y)$ such that the sum of the squares of the distances of P from the coordinate axes is 16. Draw the graph.

2. SLOPE OF A LINE. PARALLEL AND PERPENDICULAR LINES

A line L, not parallel to the x axis, intersects it. Such a line and the x axis form two angles which are supplementary. To be definite, we denote by α the angle formed by starting on the positive side of the x axis and going counterclockwise until we reach the line L. The angle α will have a value between 0 and 180°. Two examples are shown in Fig. 3–4. The angle α is called the **inclination** of the line L. All lines parallel to the x axis are said to make a zero angle with the x axis and therefore have inclination zero. From plane geometry we recall the statement: "If, when two lines are cut by a transversal, corresponding angles are equal, the lines

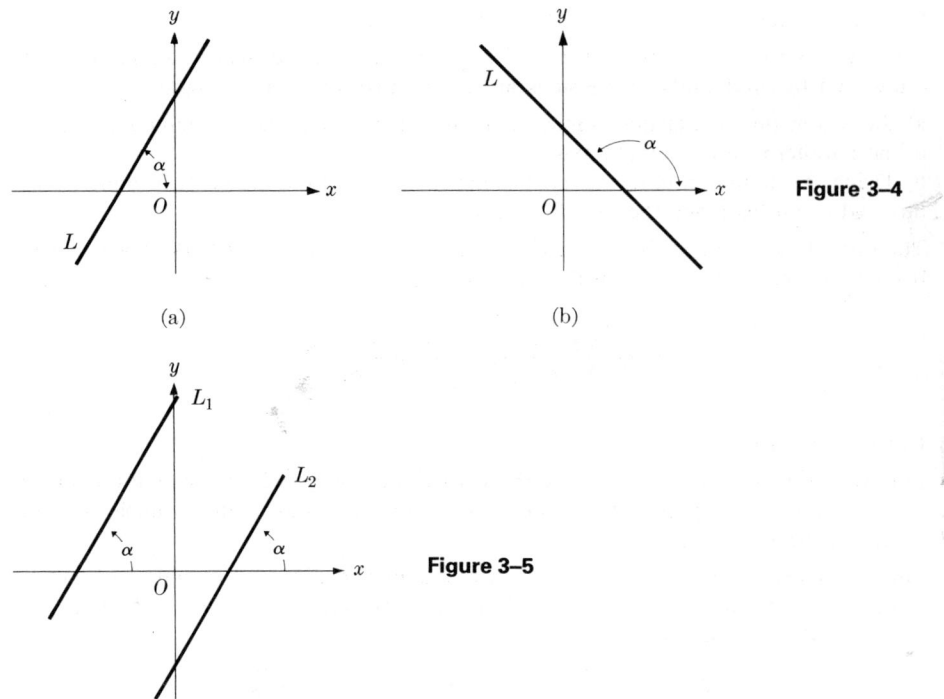

Figure 3-4

Figure 3-5

are parallel, and conversely." In Fig. 3–5, lines L_1 and L_2 each have inclination α. Applying the theorem of plane geometry, with the x axis as the transversal, we conclude that L_1 and L_2 are parallel. More generally, *all lines with the same inclination are parallel* and, conversely, *all parallel lines have the same inclination.*

The notion of inclination is a simple one which is easy to understand. However, for purposes of analytic geometry and calculus it is cumbersome and difficult to use. For this reason we introduce the notion of the **slope of a line** L. The slope is usually denoted by m and is defined in terms of the inclination by

$$m = \tan \alpha.$$

Before discussing slope, let us recall some of the properties of the tangent function. The tangent function is positive for angles in the first quadrant. It starts at zero for 0° and increases steadily, reaching 1 when $\alpha = 45°$. It continues to increase until, as the angle approaches 90°, the function increases without bound. The tangent of 90° is not defined. Loosely speaking, we sometimes say that the tangent of 90° is infinite. In the second quadrant, the tangent function is negative, and its values are obtained with the help of the relation

$$\tan (180° - \alpha) = -\tan \alpha.$$

That is, the tangent of an obtuse angle is the negative of that of the corresponding acute supplementary angle.

From these facts about the tangent function, we see that any line parallel to the x axis has zero slope, while a line with an inclination between 0° and 90° has positive slope. A line parallel to the y axis has no slope, strictly speaking,

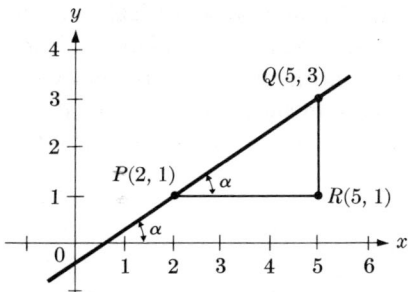

Figure 3–6

although we sometimes say such a line has infinite slope. If the inclination is an obtuse angle, the slope is negative. The line in Fig. 3–4(a) has positive slope, and the one in Fig. 3–4(b) has negative slope.

Consider the line passing through the points $P(2, 1)$ and $Q(5, 3)$, as shown in Fig. 3–6. To find the slope of this line, draw a parallel to the x axis through P and a parallel to the y axis through Q, forming the right triangle PQR. The angle α at P is equal to the inclination, by corresponding angles of parallel lines. The definition of tangent function in a right triangle is "opposite over adjacent." The slope is therefore given by

$$m = \tan \alpha = \frac{|RQ|}{|PR|}.$$

Since $|RQ| = 3 - 1 = 2$ and $|PR| = 5 - 2 = 3$, the slope m is $\frac{2}{3}$.

Suppose that $P(x_1, y_1)$ and $Q(x_2, y_2)$ are any two points in the plane, and we wish to find the slope m of the line passing through these two points. The procedure is exactly the same as the one just described. In Fig. 3–7(a) we see that

$$m = \tan \alpha = \frac{y_2 - y_1}{x_2 - x_1},$$

and in Fig. 3–7(b)

$$m = \tan \alpha = -\tan(180 - \alpha) = -\frac{y_2 - y_1}{x_1 - x_2} = \frac{y_2 - y_1}{x_2 - x_1}.$$

A difficulty may arise if the points P and Q are on a vertical line, since then

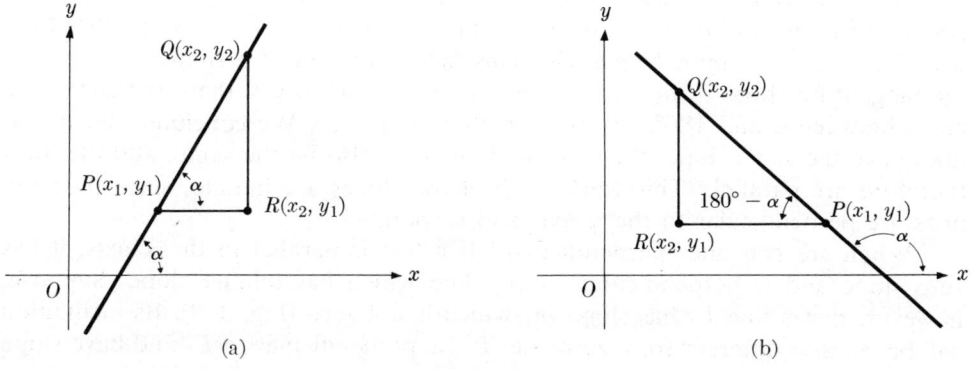

Figure 3–7

$x_1 = x_2$ and the denominator is zero. However, we know that a vertical line has no slope, and we state that the formula holds for all cases except when $x_1 = x_2$. Note that there is no difficulty if P and Q lie on a horizontal line. In that case, $y_1 = y_2$ and the slope is zero, as it should be. We conclude: the *slope of a line through the points* $P(x_1, y_1)$ *and* $Q(x_2, y_2)$ *with* $x_1 \neq x_2$ *is given by the formula*

$$m = \frac{y_2 - y_1}{x_2 - x_1}.$$

Example 1. Find the slope of the line through the points $(4, -2)$ and $(7, 3)$.

Solution. First we note that in the formula for slope it doesn't matter which point we label (x_1, y_1) [the other being labeled (x_2, y_2)]. We let $(4, -2)$ be (x_1, y_1) and $(7, 3)$ be (x_2, y_2). This gives

$$m = \frac{3 - (-2)}{7 - 4} = \frac{5}{3}.$$ ◀

Example 2. Through the point $P(4, 1)$ construct a line with slope equal to $\frac{3}{2}$.

Solution. Starting at P, draw a parallel to the x axis extending to the point R one unit to the right (Fig. 3–8). Now draw a parallel to the y axis, stopping $\frac{3}{2}$ units above R. The coordinates of this point Q are $(5, \frac{5}{2})$. The line through P and Q has slope $\frac{3}{2}$. ◀

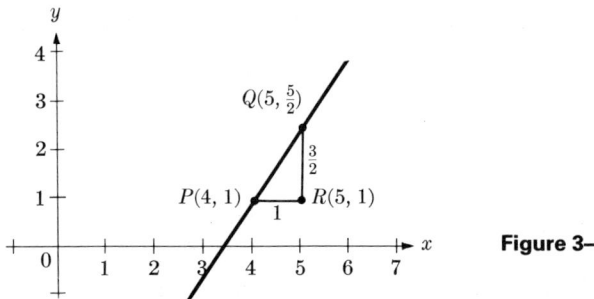

Figure 3–8

We have seen that parallel lines always have the same inclination. Therefore, *two parallel lines will always have the same slope.* Conversely, we show that *lines with the same slope must be parallel.* This fact is a result of a simple property of the tangent function: if this function has a given value, say b, there is exactly one angle between 0 and 180°, say β, such that $\tan \beta = b$. We conclude that if two lines have the same slope, their inclinations must also be the same, and the lines therefore are parallel. (This works even if the slopes are infinite, since then the lines are perpendicular to the x axis and so parallel.)

When are two lines perpendicular? If a line is parallel to the x axis, it has zero slope and is perpendicular to any line which has infinite slope. Suppose, however, that a line L_1 has slope m_1 which is not zero (Fig. 3–9). Its inclination will be α_1, also different from zero. Let L_2 be perpendicular to L_1 and have slope m_2 and inclination α_2, as shown; it is assumed here that $m_1 > 0$. We recall from

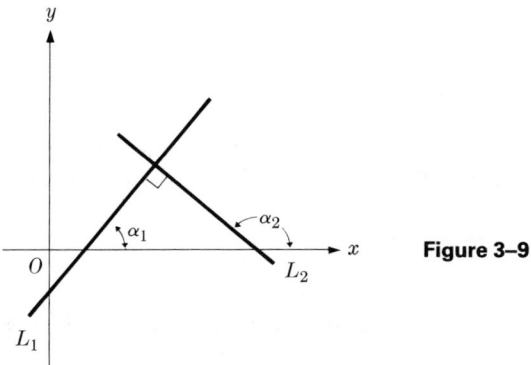

Figure 3-9

plane geometry: "An exterior angle of a triangle is equal to the sum of the remote interior angles." This means that (in Fig. 3–9)

$$\alpha_2 = 90° + \alpha_1 \quad \text{and} \quad \tan \alpha_2 = \tan(90° + \alpha_1).$$

We recall from trigonometry the formula

$$\tan(90° + A) = -\cot A,$$

from which we obtain

$$\tan \alpha_2 = -\cot \alpha_1 = -\frac{1}{\tan \alpha_1}.$$

If m_1 were negative, we would obtain the same result by interchanging the lines L_1 and L_2. The last formula, in terms of slopes, states that

$$m_2 = -\frac{1}{m_1}$$

or, in words: *Two lines are perpendicular if their slopes are the negative reciprocals of each other, and conversely.*

Example 3. Show that the line through $P_1(3, -4)$ and $Q_1(-2, 6)$ is parallel to the line through $P_2(-3, 6)$ and $Q_2(9, -18)$.

Solution. The slope of the line through P_1 and Q_1, according to the formula, is

$$m_1 = \frac{6 - (-4)}{-2 - 3} = \frac{10}{-5} = -2.$$

Similarly, the line through P_2 and Q_2 has slope

$$m_2 = \frac{-18 - 6}{9 + 3} = -2.$$

Since the slopes are the same, the lines must be parallel. ◂

Example 4. Determine whether or not the three points $P(-1, -5)$, $Q(1, 3)$, and $R(7, 12)$ lie on the same straight line.

THE LINE

Solution. The line through P and Q has slope

$$m_1 = \frac{3-(-5)}{1-(-1)} = 4.$$

If R were on this line, the line joining R and Q would have to be the very same line; therefore it would have the same slope. The slope of the line through R and Q is

$$m_2 = \frac{12-3}{7-1} = \frac{3}{2},$$

and this is different from m_1. Therefore, P, Q, and R do not lie on the same line. ◄

Example 5. Is the line through the points $P_1(5, -1)$ and $Q_1(-3, 2)$ perpendicular to the line through the points $P_2(-3, 1)$ and $Q_2(0, 9)$?

Solution. The line through P_1 and Q_1 has slope

$$m_1 = \frac{2-(-1)}{-3-5} = -\frac{3}{8}.$$

The line through P_2 and Q_2 has slope

$$m_2 = \frac{9-1}{0-(-3)} = \frac{8}{3}.$$

The slopes are the negative reciprocals of each other, and the lines are perpendicular. ◄

Example 6. Given the isosceles triangle with vertices at the points $P(-1, 4)$, $Q(0, 1)$, and $R(2, 5)$, show that the median drawn from P is perpendicular to the base QR (Fig. 3-10).

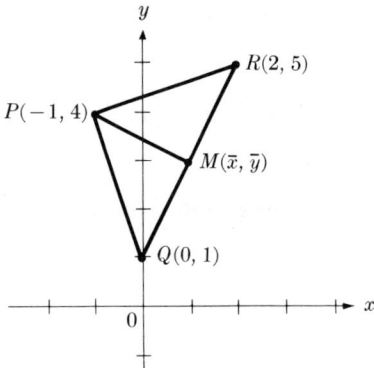

Figure 3-10

Solution. Let M be the point where the median from P intersects the base QR. From the definition of median, M must be the midpoint of the segment QR. The coordinates of M, from the midpoint formula, are

$$\bar{x} = \frac{0+2}{2} = 1, \qquad \bar{y} = \frac{5+1}{2} = 3.$$

Now we check the slopes of PM and QR. The slope of PM is

$$m_1 = \frac{3-4}{1-(-1)} = -\frac{1}{2}.$$

The slope of QR is

$$m_2 = \frac{5-1}{2-0} = 2.$$

Since

$$m_1 = -\frac{1}{m_2},$$

the median is perpendicular to the base. ◀

PROBLEMS

In Problems 1 through 6, check to see whether the line through the pair of points P_1, Q_1 is parallel or perpendicular to the line through the pair of points P_2, Q_2.

1. $P_1(-5, 2)$, $Q_1(3, -1)$ and $P_2(4, 2)$, $Q_2(12, -1)$
2. $P_1(3, 1)$, $Q_1(-2, 7)$ and $P_2(5, -3)$, $Q_2(-1, -8)$
3. $P_1(5, 3)$, $Q_1(8, 3)$ and $P_2(7, 4)$, $Q_2(7, -4)$
4. $P_1(-4, 5)$, $Q_1(14, 2)$ and $P_2(6, 0)$, $Q_2(9, -4)$
5. $P_1(12, 8)$, $Q_1(4, 8)$ and $P_2(3, 1)$, $Q_2(-6, 1)$
6. $P_1(7, -1)$, $Q_1(10, 2)$ and $P_2(0, -4)$, $Q_2(1, -5)$

In Problems 7 through 10, determine whether or not the three points all lie on the same straight line.

7. $P(3, 4)$, $Q(8, 5)$, $R(13, 6)$
8. $P(2, -1)$, $Q(5, 3)$, $R(-7, 4)$
9. $P(7, 1)$, $Q(-7, 2)$, $R(4, -5)$
10. $P(-3, 0)$, $Q(4, 1)$, $R(11, 2)$
11. Construct a line passing through the point $(5, -2)$ and having slope $\frac{3}{4}$.
12. Construct a line passing through the point $(-3, 1)$ and having slope $-\frac{2}{3}$.

In Problems 13 through 23, the points P, Q, R, S are the vertices of a quadrilateral. In each case, determine whether the figure is a trapezoid, parallelogram, rhombus, rectangle, square, or none of these.

13. $P(1, 3)$, $Q(2, 5)$, $R(6, 17)$, $S(5, 15)$
14. $P(1, 3)$, $Q(3, 7)$, $R(3, 11)$, $S(1, 7)$
15. $P(2, 3)$, $Q(4, 1)$, $R(0, -8)$, $S(-5, -3)$
16. $P(5, 6)$, $Q(-8, -7)$, $R(-8, -10)$, $S(5, 3)$
17. $P(-1, 2)$, $Q(2, -2)$, $R(\frac{38}{13}, -\frac{8}{13})$, $S(-\frac{17}{13}, \frac{8}{13})$
18. $P(\frac{4}{3}, \frac{20}{3})$, $Q(4, 2)$, $R(1, -4)$, $S(5, 12)$
19. $P(3, -1)$, $Q(2, -6)$, $R(1, 7)$, $S(0, -8)$
20. $P(5, 3)$, $Q(8, 3)$, $R(4, 4)$, $S(7, 4)$
21. $P(3, 0)$, $Q(3, 5)$, $R(0, 9)$, $S(0, 4)$

22. $P(2, 1)$, $Q(7, 13)$, $R(-5, 18)$, $S(-10, 6)$
23. $P(6, 2)$, $Q(3, -1)$, $R(7, 5)$, $S(8, -4)$
24. The points $A(3, -2)$, $B(4, 1)$, and $C(-3, 5)$ are the vertices of a triangle. Show that the line through the midpoints of the sides AB and AC is parallel to the base BC of the triangle.
25. The points $A(0, 0)$, $B(a, 0)$, and $C(\frac{1}{2}a, b)$ are the vertices of a triangle. Show that the triangle is isosceles. Prove that the median from C is perpendicular to the base AB. How general is this proof?
26. The points $A(0, 0)$, $B(a, 0)$, $C(a + b, c)$, $D(b, c)$ form the vertices of a parallelogram. Prove that the diagonals bisect each other.
27. The points $A(0, 0)$, $B(a, 0)$, $C(b, c)$, $D(e, f)$ are the vertices of a quadrilateral. Show that the line segments joining the midpoints of opposite sides bisect each other. How general is this proof?
28. Show that in any triangle the length of one side is no larger than the sum of the lengths of the other two sides.
29. Let $P(-\frac{1}{2}a, 0)$ and $Q(\frac{1}{2}a, 0)$ be two adjacent vertices of a regular hexagon above the side PQ. Find the coordinates of the remaining vertices and the length of the diagonal joining two opposite vertices.
30. Let $P(-\frac{1}{2}a, 0)$ and $Q(\frac{1}{2}a, 0)$ be two adjacent vertices of a regular octagon situated above the side PQ. Find the coordinates of the remaining six vertices and the length of the diagonal joining two opposite vertices.

3. THE STRAIGHT LINE

It is easy to verify that the equation $y = 4$ represents all points on a line parallel to the x axis and four units above it. Similarly, the equation $x = -2$ represents all points on a line parallel to the y axis and two units to the left. In general, any line parallel to the y axis has an equation of the form

$$x = a,$$

where a is the number denoting how far the line is to the right or left of the y axis. The equation

$$y = b$$

describes a line parallel to the x axis and b units from it. In this way we obtain the equations of all lines with zero or infinite slope. A line which is not parallel to either axis has a slope m which is different from zero. Suppose that the line passes through a point denoted $P(x_1, y_1)$. To be specific, we consider the slope m to be $-\frac{2}{3}$ and the point P to have coordinates $(4, -3)$. If a point $Q(x, y)$ is on this line, then the slope as calculated from P to Q must be $-\frac{2}{3}$, as shown in Fig. 3-11. That is,

$$\frac{y + 3}{x - 4} = -\frac{2}{3} \quad \text{or} \quad y + 3 = -\frac{2}{3}(x - 4).$$

This is the equation of the line passing through the point $(4, -3)$ with slope $-\frac{2}{3}$. In the general case of a line with slope m passing through $P(x_1, y_1)$, the statement that $Q(x, y)$ is on the line is the same as the statement that the slope m as

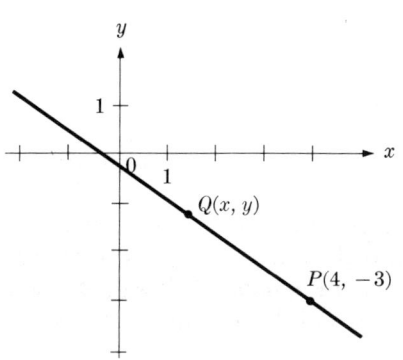

Figure 3–11

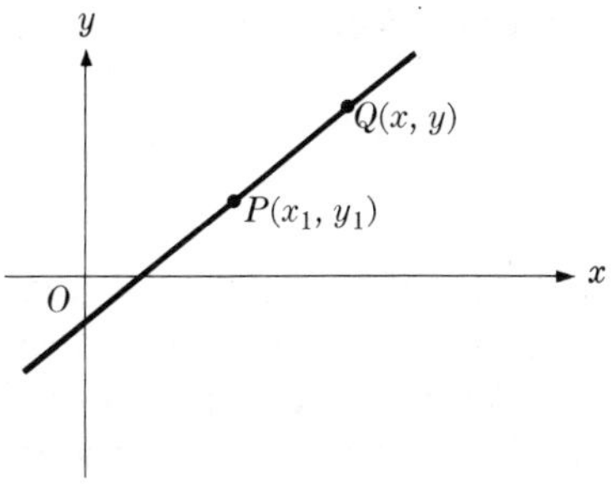

Figure 3–12

computed from P to Q (Fig. 3–12) is

$$\frac{y - y_1}{x - x_1} = m \quad \text{or} \quad y - y_1 = m(x - x_1).$$

This last equation is called **the point-slope form for the equation of a line.** That is, if we are given the coordinates of a point and the numerical value of the slope, substitution in the above formula yields the equation of the line going through the point and having the given slope.

Example 1. Find the equation of the line passing through the point $(-2, 5)$ and having slope $\frac{4}{3}$.

Solution. Substitution in the above formula gives

$$y - 5 = \tfrac{4}{3}[x - (-2)] \quad \text{or} \quad 3y = 4x + 23. \quad \blacktriangleleft$$

We know that two points determine a line. The problem of finding the equation of the line passing through the points $(3, -5)$ and $(-7, 2)$ can be solved in two steps. First we employ the formula for the slope of a line, as given in Section 2, to obtain the slope of the line through the given points. We get

$$m = \frac{2 - (-5)}{-7 - 3} = -\frac{7}{10}.$$

Then, knowing the slope, we use *either* point, together with the slope, in the point-slope formula. This gives [using the point $(3, -5)$]

$$y - (-5) = -\tfrac{7}{10}(x - 3) \quad \text{or} \quad 10y = -7x - 29.$$

We verify quite easily that the same equation is obtained if the point $(-7, 2)$ is used instead of $(3, -5)$.

The above process may be transformed into a general formula by applying it to two points, $P_1(x_1, y_1)$ and $P_2(x_2, y_2)$. The slope of the line through these points is

$$m = \frac{y_2 - y_1}{x_2 - x_1}.$$

Substituting this value for the slope into the point-slope form, we get the

two-point form for the equation of a line:

$$y - y_1 = \frac{y_2 - y_1}{x_2 - x_1}(x - x_1).$$

Note that this is really not a new formula, but merely the point-slope form with an expression for the slope substituted into it.

Another variation of the point-slope form is obtained by introducing a number called the y intercept. Every line not parallel to the y axis must intersect it; if we denote by $(0, b)$ the point of intersection, **the number b is called the y intercept.*** Suppose a line has slope m and y intercept b. We substitute in the point-slope form to get

$$y - b = m(x - 0) \quad \text{or} \quad y = mx + b.$$

This is called the **slope-intercept form** for the equation of a straight line.

Example 2. A line has slope 3 and y intercept -4. Find its equation.

Solution. Substitution in the slope-intercept formula gives

$$y = 3x - 4. \qquad \blacktriangleleft$$

The important thing to notice is that the point-slope form is the basic one for the equation of a straight line. The other formulas were derived as simple variations or particular cases.

Examples 1 and 2 led to equations of lines which could be put in the form

$$Ax + By + C = 0,$$

where A, B, and C are any numbers. This equation is the most general equation of the first degree in x and y. We shall establish the theorem:

Theorem 1. *Every equation of the form*

$$Ax + By + C = 0,$$

so long as A and B are not both zero, is the equation of a straight line.

Proof. We consider two cases, according as $B = 0$ or $B \neq 0$. If $B = 0$, then we must have $A \neq 0$, and the above equation becomes

$$x = -\frac{C}{A},$$

which we know is the equation of a straight line parallel to the y axis and $-C/A$ units from it.

If $B \neq 0$, we divide by B and solve for y, getting

$$y = -\frac{A}{B}x - \frac{C}{B}.$$

---

* The point $(0, b)$ is also called **the y intercept**.

From the slope-intercept form for the equation of a line, we recognize this as the equation of a line with slope $-A/B$ and y intercept $-C/B$.

In the statement of the theorem it is necessary to make the requirement that A and B are not both zero. If both of them vanish and C is zero, the linear equation reduces to the triviality $0 = 0$, which is satisfied by every point $P(x, y)$ in the plane. If $C \neq 0$, then no point $P(x, y)$ satisfies the equation $0 = C$.

Example 3. Given the linear equation
$$3x + 2y + 6 = 0,$$
find the slope and y intercept.

Solution. Solving for y, we have
$$y = -\tfrac{3}{2}x - 3.$$
From this we simply read off that $m = -\tfrac{3}{2}$ and $b = -3$. ◀

An equation of the first degree in x and y is called a **linear equation.** When we solve for y in terms of x, as in the point-slope form, y becomes a function of x. Such a function is called a **linear function.** Any line not parallel to the y axis may be thought of as a function. Similarly, the equation of any line not parallel to the x axis may be solved for x in terms of y.

Example 4. Find the equation of the line which is the perpendicular bisector of the line segment joining the points $P(-3, 2)$ and $Q(5, 6)$. (See Fig. 3–13.)

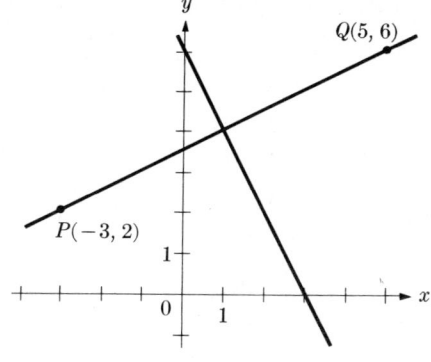

Figure 3–13

Solution. We give two methods.

METHOD 1. We first find the slope m of the line through P and Q. It is
$$m = \frac{6 - 2}{5 + 3} = \frac{1}{2}.$$
The slope of the perpendicular bisector must be -2, the negative reciprocal. Next we get the coordinates of the midpoint of the line segment PQ. They are
$$\bar{x} = \frac{5 - 3}{2} = 1, \qquad \bar{y} = \frac{6 + 2}{2} = 4.$$

The equation of the line through (1, 4) with slope −2 is
$$y - 4 = -2(x - 1) \quad \Leftrightarrow \quad 2x + y - 6 = 0,$$
which is the desired equation.

METHOD 2. We start by noting that any point on the perpendicular bisector must be equidistant from P and Q. Let $R(x, y)$ be any such point, and let d_1 denote the distance from R to P, and d_2 the distance from R to Q. From the formula for the distance between two points and the condition $d_1 = d_2$, we have
$$\sqrt{(x + 3)^2 + (y - 2)^2} = \sqrt{(x - 5)^2 + (y - 6)^2}.$$
Squaring both sides and multiplying out, we obtain
$$x^2 + 6x + 9 + y^2 - 4y + 4 = x^2 - 10x + 25 + y^2 - 12y + 36.$$
The terms of the second degree cancel and the remaining terms combine to give
$$6x - 4y + 13 = -10x - 12y + 61 \quad \Leftrightarrow \quad 16x + 8y - 48 = 0.$$
Dividing by 8, we obtain the same answer as in the first method:
$$2x + y - 6 = 0. \quad \blacktriangleleft$$

PROBLEMS

In Problems 1 through 17, find the equation of the line with the given requirements.
1. Slope 2 and passing through (−1, 4)
2. Slope 4 and passing through (3, −1)
3. Slope $-\frac{2}{3}$ and passing through (−2, 5)
4. Passing through the points (5, 2) and (−1, −6)
5. Passing through the points (2, −3) and (0, −4)
6. Passing through the points (1, 4) and (−2, −7)
7. Slope 0 and passing through (−2, −7)
8. Passing through the points (3, 8) and (3, −7)
9. Passing through the points (4, −2) and (−7, −2)
10. Slope $-\frac{4}{5}$ and y intercept 3
11. Slope 0 and y intercept −5
12. Slope $\frac{4}{5}$ and y intercept 8
13. Slope 2 and passing through the midpoint of the line segment connecting (3, −2) and (4, 7)
14. Slope $-\frac{3}{2}$ and passing through the midpoint of the line segment connecting (2, −7) and (5, −1)
15. Slope −3 and y intercept 0
16. Parallel to the y axis and passing through the point (4, −3)
17. Parallel to the x axis and passing through the point (6, −5)
18. Find the slope and y intercept of the line $2x + 7y + 4 = 0$.

19. Find the slope and y intercept of the line $2x - 3y - 7 = 0$.
20. Find the equation of the line through the point $(1, -4)$ and parallel to the line $x + 5y - 3 = 0$.
21. Find the equation of the line through the point $(-2, -3)$ and parallel to the line $3x - 7y + 4 = 0$.
22. Find the equation of the line through the point $(3, -2)$ and perpendicular to the line $2x + 3y + 4 = 0$
23. Find the equation of the line through the point $(1, 5)$ and perpendicular to the line $5x - 4y + 1 = 0$.
24. Find the equation of the line through the point $(-1, -3)$ and parallel to the line through the points $(3, 2)$ and $(-5, 7)$.
25. Find the equation of the line passing through $(4, -2)$ and parallel to the line through the points $(2, -1)$ and $(5, 7)$.
26. Find the equation of the line through the midpoint of the line segment joining $(2, 1)$ and $(6, -4)$ and through the point which is $\frac{1}{4}$ of the way from $(3, 2)$ to $(7, -6)$.
27. Find the equation of the line passing through $(-5, 3)$ and perpendicular to the line through the points $(7, 0)$ and $(-8, 1)$.
28. Find the equation of the line passing through $(2, -1)$ and perpendicular to the line through the points $(3, 1)$ and $(-2, 5)$.
29. Find the equation of the perpendicular bisector of the line segment joining $(6, 2)$ and $(-1, 3)$.
30. Find the equation of the perpendicular bisector of the line segment joining $(3, -1)$ and $(5, 2)$.
31. The points $P(0, 0)$, $Q(a, 0)$, $R(a, b)$, $S(0, b)$ are the vertices of a rectangle. Show that if the diagonals meet at right angles, then $a = \pm b$ and the rectangle must be a square.
32. Let P, Q, R, S be the vertices of a parallelogram. Show, by analytic geometry, that if the diagonals are equal the parallelogram must be a rectangle.
33. P, Q, R, S are the vertices of a parallelogram. Show, by analytic geometry, that if the diagonals are perpendicular the figure is a rhombus.
34. The points $A(1, 0)$, $B(7, 0)$, $C(3, 4)$ are the vertices of a triangle. Find the equations of the three medians. Show that the three medians intersect in a point.
35. The points $A(0, 0)$, $B(a, 0)$, $C(b, c)$ are vertices of a triangle. Show that the three medians meet in a point.
36. The points $A(1, 0)$, $B(5, 1)$, $C(3, 8)$ are the vertices of a triangle. Find the equation of the perpendicular from each vertex to the opposite side. Show that these three lines intersect in a point.
37. The points $A(0, 0)$, $B(a, 0)$, $C(b, c)$ are the vertices of a triangle. Find the equation of the perpendicular from each vertex to the opposite side. Show that these three lines intersect in a point.
38. Let $a > 0$, $c > 0$ be given numbers. Show that the points $A(0, 0)$, $B(a, 0)$, $C(a + b, c)$, $D(b, c)$ are the vertices of a parallelogram, and show that its area is ac.
39. Let $c > 0$ and $d > b$ be given numbers. Show that the points $A(0, 0)$, $B(a, 0)$, $C(d, c)$, $D(b, c)$ are the vertices of a trapezoid and that its area is $\frac{1}{2}c(a + d - b)$.
40. Let $a > 0$, $c > 0$ be given numbers. The points $A(0, 0)$, $B(a, 0)$, $C(b, c)$ are the vertices of a triangle. Suppose that D and E are points on sides AC and BC, respectively, with line segment DE parallel to AB. Show that $|CD|/|DE| = |CA|/|AB|$.

INTRODUCTION TO THE CALCULUS. LIMITS

1. LIMITS (INTUITIVE)

The single most important idea in calculus is that of limit. The limit concept is at the foundation of almost all of mathematical analysis, and an understanding of it is absolutely essential. The reader should not be surprised to find the discussion rather complex and perhaps difficult. This should not prove discouraging, however, for once a precise understanding of the limit concept is achieved, the reward is a good grasp of the basic processes of calculus.

We shall begin with an intuitive discussion of limit and postpone a formal description to Section 6. It may seem strange that it is possible to work with and apply an idea without defining it precisely, but historically the limit concept evolved in just this way. In the early development of calculus precise statements in the modern sense were seldom made. Yet progress was achieved because there was a certain degree of understanding. When imprecise ideas led to trouble, a later, more careful approach overcame the difficulty, and a more solid basis for calculus was attained. The same process continues today in mathematical research.

Let us begin by considering a function f and its graph. We concentrate on a particular value of x, say $x = a$. For example, let the function be

$$y = f(x) = -2x^2 + 8x - 4$$

and take the particular value $x = 3$. A graph of this function is shown in Fig. 4–1. We are interested in this function not only when $x = 3$ but also when x takes on values in various intervals containing 3.

Suppose that we select the interval from $x = 1$ to $x = 4$. A graph of the function in this region shows that the highest value is attained at $x = 2$ when y is 4, and the lowest value occurs at $x = 4$ when y is -4 (Fig. 4–2). In other words, the graph of the function lies in the rectangle bounded by the lines $x = 1$, $x = 4$ and $y = 4$, $y = -4$. The next step is to select a smaller interval about $x = 3$. Suppose we take the interval from $x = 2\frac{1}{2}$ to $x = 3\frac{1}{2}$ and draw the graph in this region (Fig. 4–3). The graph of the function is now in the rectangle bounded by the lines $x = 2\frac{1}{2}$, $x = 3\frac{1}{2}$, $y = -\frac{1}{2}$, $y = 3\frac{1}{2}$, as shown. Proceeding further, we take a still smaller interval about $x = 3$, say from $x = 2.9$ to $x = 3.1$. We draw the graph (enlarged) in this region, as shown in Fig. 4–4. The function values are now

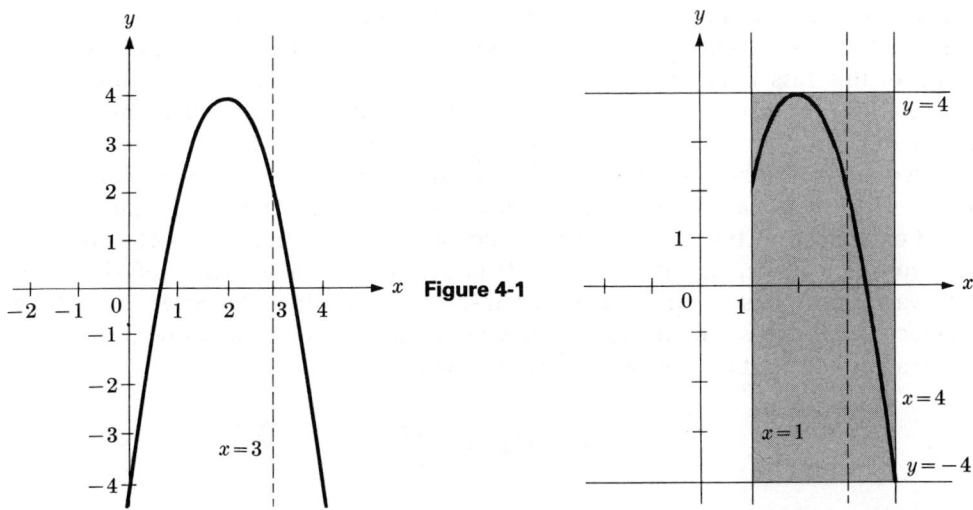

Figure 4-1

Figure 4-2

situated in the rectangle bounded by the lines $x = 2.9$, $x = 3.1$, $y = 1.58$, $y = 2.38$.

The main point we wish to emphasize concerns the height of these rectangles. As the widths of the rectangles become narrower, the heights also shrink in size. If we went further and took an x-interval from 2.99 to 3.01, the corresponding rectangle containing the graph of the function would be bounded by the lines $x = 2.99$, $x = 3.01$, $y = 1.9598$, $y = 2.0398$. A width of 0.02 unit containing the value 3 leads to a rectangle of height only 0.08 unit. In addition to the heights

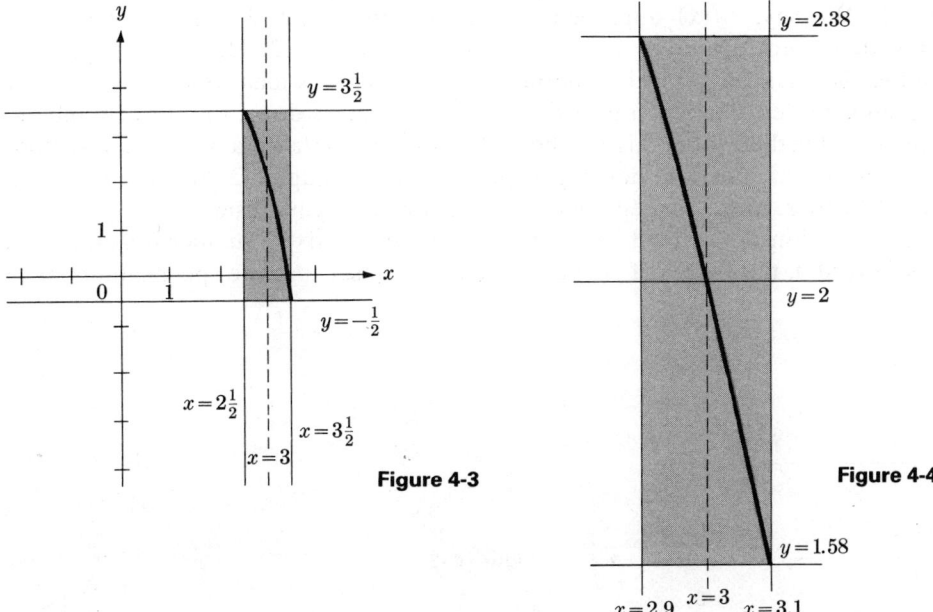

Figure 4-3

Figure 4-4

becoming narrower and narrower as the widths squeeze in to the value $x = 3$, we observe that the heights cluster about the value $y = 2$. The reader may wonder what all this fuss is about, since the formula for the equation yields, by direct substitution, the result that $y = 2$ when $x = 3$. Note, however, that throughout the discussion we never made use of this fact (Fig. 4–5). Indeed, we carefully avoided any consideration at all of what happens when x is 3. We are concerned solely with the behavior of y when x is in some *interval* about the value 3.

For almost all functions we have studied up to this time, the behavior of a function at a point, say at $x = 3$, and its behavior in a sequence of shrinking intervals about this point were never distinguished. Now, however, a striking change occurs as we start to study functions whose behavior cannot be discovered by straight substitution. For example, the function

$$y = f(x) = \frac{\sin x}{x}$$

is defined for all values of x except $x = 0$. A straight substitution at $x = 0$ would tell us that

$$y = f(0) = \tfrac{0}{0},$$

which is completely meaningless. However, we shall see (Chapter 9, Section 1) that by studying a sequence of intervals about $x = 0$ which get smaller and smaller, we find that the corresponding rectangles containing the function get thinner and thinner, and the heights cluster about a particular value of y. Nothing is ever stated about the value of y when x *is* zero. We just study the value of y when x gets closer and closer to zero.

Returning to the first example of the function $f(x) = -2x^2 + 8x - 4$, we see that as x comes closer and closer to the value 3, $f(x)$ gets closer and closer to the value 2. We say, "$f(x)$ approaches 2 as x approaches 3." This sentence is abbreviated even further by the statement $\lim_{x \to 3} f(x) = 2$. Here once again we have the shorthand so typical of mathematics. A complex idea which takes several paragraphs to describe, even in a simple case, is boiled down to a brief symbolic expression. Furthermore, this symbolic expression contains a shorthand symbol for the notion of function, developed previously in Chapter 2, Section 3. As we proceed further, more concepts will be built on this symbolism.

If a function f is defined for values of x about the fixed number a, and if, as x tends toward a, the values of $f(x)$ get closer and closer to some specific number L,

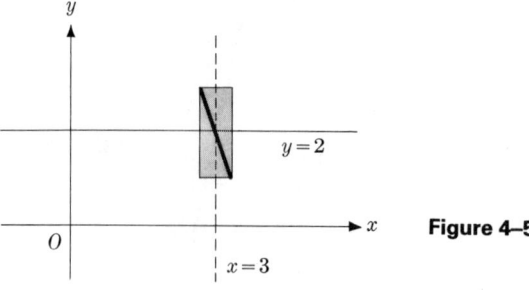

Figure 4–5

we write

$$\lim_{x \to a} f(x) = L,$$

and we read it, "The limit of $f(x)$ as x approaches a is L." Geometrically, this means that the series of rectangles which surround a and which have narrower and narrower widths become smaller and smaller in height and cluster about the point (a, L).

Of course all the above statements containing terms such as "closer," "nearer," "narrower," etc., are quite imprecise and are intended to give only an intuitive idea of what occurs. As mentioned, a precise definition of the expression $\lim_{x \to a} f(x) = L$ is given in Section 6.

Now let us take an example in which the result is quite different from that of the first example. We define the function

$$y = f(x) = \frac{(x-2)}{(2|x-2|)},$$

which is well determined for all values of x except $x = 2$. At $x = 2$ the function is not defined, since straight substitution yields $y = \frac{0}{0}$, a meaningless expression. The graph of the function, shown in Fig. 4–6, is quite simple. If x is larger than 2, then $|x - 2| = x - 2$ and the function has the value $+\frac{1}{2}$. If x is less than 2, then $|x - 2| = -(x - 2)$, and the function is equal to $-\frac{1}{2}$. We wish to study the behavior of the function as x tends to 2. We select an interval containing $x = 2$, say from $x = 1.4$ to $x = 2.3$. We see that the function is contained in the rectangle bounded by the lines $x = 1.4$, $x = 2.3$, $y = -\frac{1}{2}$, $y = +\frac{1}{2}$ (Fig. 4–7). In fact, no matter how narrow the interval about $x = 2$ becomes, the *height* of the rectangle will always be 1 unit. *There is no limit as x approaches* 2. We say that

$$\lim_{x \to 2} \frac{x-2}{2|x-2|} \quad \text{does not exist.}$$

We now examine a number of examples of functions, with a view toward discovering what happens in the neighborhood of a particular value when the function is not defined at that value by straight substitution.

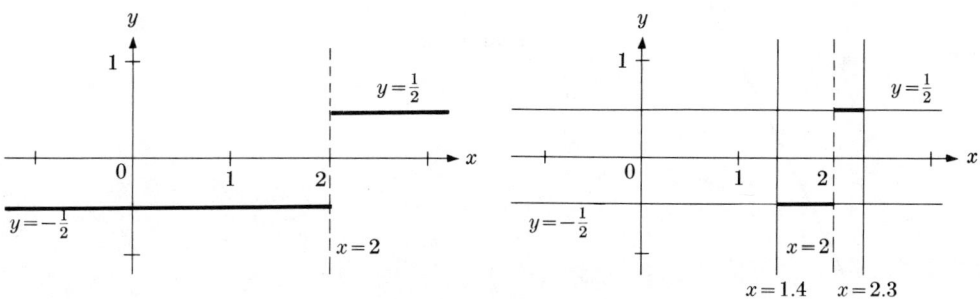

Figure 4–6   Figure 4–7

Example 1. The function
$$f(x) = \frac{2x^2 - x - 3}{x + 1}$$
is defined for all values of x except $x = -1$, since at $x = -1$ both numerator and denominator vanish. Does
$$\lim_{x \to -1} f(x)$$
exist and, if so, what is its value?

Solution. To get an idea of what is happening, we construct a table of values. Then we draw the graph, which appears to be a straight line with a "hole" at the point $(-1, -5)$ (Fig. 4–8). From the geometric discussion of rectangles given at the beginning of this section we see that
$$\lim_{x \to -1} f(x) = -5.$$
However, we would like a more systematic method of obtaining limits, without relying on pictorial representation and intuition. By means of factoring, we can write $f(x)$ in the form
$$f(x) = \frac{(2x - 3)(x + 1)}{x + 1}.$$
Now if $x \neq -1$, we are allowed to divide both the numerator and the denominator by $(x + 1)$. Then
$$f(x) = 2x - 3, \quad \text{if} \quad x \neq -1.$$
This function tends to -5 as x tends to -1, since simple substitution now works.

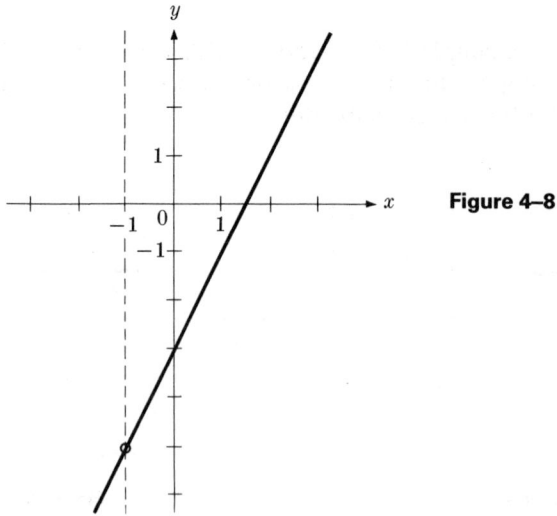

Figure 4–8

We conclude that

$$\lim_{x \to -1} f(x) = -5.$$

Note that we never substituted the value $x = -1$ in the original expression. ◄

Example 2. Find the limit of the function

$$f(x) = \frac{x - 4}{3(\sqrt{x} - 2)}, \quad x \neq 4,$$

as x tends to 4. (This function is defined for nonnegative values of x only.)

Solution. Note that straight substitution of $x = 4$ fails, since $f(4) = 0/0$, which is meaningless. We could proceed graphically, as in Example 1, but instead we introduce a useful algebraic trick. We rationalize the denominator by multiplying both numerator and denominator by $\sqrt{x} + 2$. So long as $x \neq 4$, we can write

$$\frac{(x-4)}{3(\sqrt{x}-2)} \frac{(\sqrt{x}+2)}{(\sqrt{x}+2)} = \frac{(x-4)(\sqrt{x}+2)}{3(x-4)},$$

and furthermore the common factor can be canceled if $x \neq 4$. We get

$$f(x) = \frac{\sqrt{x} + 2}{3}, \quad \text{if} \quad x \neq 4.$$

The limit of this expression can be found by straight substitution of $x = 4$. We find

$$\lim_{x \to 4} f(x) = \frac{\sqrt{4} + 2}{3} = \frac{4}{3}.$$ ◄

Example 3. Find the limit of the function

$$f : x \to \frac{\sqrt{2 + x} - 1}{x + 1}, \quad x \neq -1,$$

as x tends to -1. (This function is defined only for values of x larger than or equal to -2.)

Solution. Since straight substitution fails, we employ the device of "rationalizing the numerator." Multiplication of numerator and denominator by $\sqrt{2 + x} + 1$ yields

$$f(x) = \frac{(\sqrt{2+x} - 1)(\sqrt{2+x} + 1)}{(x+1)(\sqrt{2+x} + 1)} = \frac{x + 1}{(x+1)(\sqrt{2+x} + 1)}, \quad x \neq -1.$$

We now divide numerator and denominator by $x + 1$ and find that

$$f(x) = \frac{1}{\sqrt{2 + x} + 1} \quad \text{for} \quad x \neq -1, \ x \geq -2.$$

As x tends to -1, $f(x)$ tends to

$$\frac{1}{\sqrt{2-1}+1} = \frac{1}{2}.$$

PROBLEMS

In Problems 1 through 12, in each case find the limit (if any) approached by $f(x)$ as x tends to a. Note all values where $f(x)$ is not defined.

1. $f(x) = \dfrac{x^2 - 4}{x^2 - 5x + 6}, \ a = 2$

2. $f(x) = \dfrac{x^2 - 9}{x^2 - 7x + 12}, \ a = 3$

3. $f(x) = \dfrac{x^3 + 2x - 1}{x + 2}, \ a = 3$

4. $f(x) = \dfrac{x^2 - x - 2}{x^2 - 4x + 4}, \ a = 2$

5. $f(x) = \dfrac{x^3 + 1}{x + 1}, \ a = -1$

6. $f(x) = \dfrac{x^4 - 1}{x^2 - 1}, \ a = 1$

7. $f: x \to \dfrac{\sqrt{x + 2} - 2}{x - 2}, \ a = 2, \ x \geq -2$

8. $f: x \to \dfrac{x^4 - 2x - 3}{x + 1}, \ a = -1$

9. $f: x \to \dfrac{\sqrt{2x + 3} - x}{x - 3}, \ a = 3, \ x \geq -\dfrac{3}{2}$

10. $f: x \to \dfrac{\sqrt{x^2 + 3} - 2}{x + 1}, \ a = -1$

11. $f: x \to \dfrac{x - 9}{\sqrt{x} - 3}, \ a = 9, \ x \geq 0$

12. $f: x \to \dfrac{x + 3}{\sqrt{x^2 + 7} - 4}, \ a = -3$

In Problems 13 through 26, in each case find the limit.

13. $\lim\limits_{x \to 3} (x^2 + 4x - 6)$

14. $\lim\limits_{x \to 2} (2x^4 - 3x^2 + 7x + 5)$

15. $\lim\limits_{x \to 4} \dfrac{x^2 - 16}{x - 4}$

16. $\lim\limits_{x \to -2} \dfrac{\sqrt{4 - x^2}}{x + 2}$

17. $\lim\limits_{x \to 0} \dfrac{\sqrt{5 + x} - \sqrt{5}}{2x}$

18. $\lim\limits_{h \to 0} \dfrac{\sqrt{9 + 2h} - 3}{h}$

19. $\lim\limits_{h \to 0} \dfrac{h^2}{\sqrt{4 + 3h^2} - 2}$

20. $\lim\limits_{h \to 0} \dfrac{\sqrt{3x + h} - \sqrt{3x}}{h}$

21. $\lim\limits_{h \to 1} \dfrac{\sqrt{b + 2(h - 1)} - \sqrt{b}}{h - 1}$

22. $\lim\limits_{x \to 2} \dfrac{\sqrt[3]{x} - \sqrt[3]{2}}{x - 2}$

23. $\lim\limits_{x \to -4} \dfrac{x^6 - 4096}{x + 4}$

24. $\lim\limits_{x \to -2} \dfrac{x^3 + 4x^2 + x - 6}{x^2 + 6x + 8}$

25. $\lim\limits_{x \to 8} \dfrac{\sqrt[4]{x} - \sqrt[4]{8}}{\sqrt{x} - \sqrt{8}}$

26. $\lim\limits_{x \to 0} \dfrac{\sqrt{3 + x} - \sqrt{3 - x}}{\sqrt{4 + x^2} - \sqrt{4 - 2x}}$

27. Let $f: x \to (1 - |x|)/(1 + x)$ be given with domain all $x \neq -1$. Plot the graph of f. Find
$$\lim_{x \to -1} \frac{1 - |x|}{1 + x}.$$

28. Given
$$f(x) = \begin{cases} 1 & \text{for } -\infty < x < 3 \\ 2 & \text{for } 3 \leq x < \infty \end{cases}.$$
Plot the graph. Show that $\lim_{x \to 3} f(x)$ does not exist.

29. Given $f(x) = (\sin x)/x$ for all $x \neq 0$. Sketch the graph of f for x between -1 and 1 by plotting the values of f for $x = -1, -0.9, -0.8, \ldots, 0.9, 1$. Can any conclusion be suggested about the value of
$$\lim_{x \to 0} \frac{\sin x}{x}?$$

2. LIMITS (continued)

In evaluating limits, functional notation plays a convenient and important part. Consider the function
$$f(x) = x^2 + 3,$$
and suppose that we wish to evaluate the limit
$$\lim_{x \to 0} \frac{f(5 + x) - f(5)}{x}, \qquad x \neq 0.$$
A straight substitution of $x = 0$ leads to the meaningless expression
$$\frac{f(5) - f(5)}{0} = \frac{0}{0},$$
and we must penetrate a little deeper. We have $f(5) = 28$ and $f(5 + x) = (5 + x)^2 + 3 = x^2 + 10x + 28$. Therefore
$$\frac{f(5 + x) - f(5)}{x} = \frac{x^2 + 10x + 28 - 28}{x} = x + 10.$$
The cancellation of x from numerator and denominator is most fortunate, since now
$$\lim_{x \to 0} (x + 10)$$
may be found by direct substitution; the answer is 10.

Example 1. Find the limit

$$\lim_{h \to 0} \frac{f(4 + h) - f(4)}{h}, \qquad h \neq 0,$$

where

$$f(x) = \frac{1}{(x + 1)^2}, \qquad x \neq -1.$$

Solution. Direct substitution of $h = 0$ gives

$$\frac{f(4) - f(4)}{0} = \frac{0}{0},$$

which is meaningless. However,

$$f(4) = \frac{1}{25}, \qquad f(4 + h) = \frac{1}{(4 + h + 1)^2} = \frac{1}{(5 + h)^2}.$$

Therefore

$$\frac{f(4 + h) - f(4)}{h} = \frac{\frac{1}{(5 + h)^2} - \frac{1}{25}}{h}.$$

We might try direct substitution at this point, but again we would fail, since the result is again $\frac{0}{0}$. We proceed by finding the lowest common denominator, getting

$$\frac{f(4 + h) - f(4)}{h} = \frac{25 - (5 + h)^2}{25h(5 + h)^2} = \frac{-(10h + h^2)}{25h(5 + h)^2}.$$

So long as $h \neq 0$, this factor may be canceled in both numerator and denominator. Then

$$\lim_{h \to 0} \frac{-(10 + h)}{25(5 + h)^2}$$

may be obtained by direct substitution. The answer is $-2/125$. ◀

Example 2. Find the limit

$$\lim_{x \to 3} f(x),$$

where

$$f(x) = \frac{1}{(x - 3)^2}, \qquad x \neq 3.$$

Solution. Sketching the graph of this function about $x = 3$, we see that it increases without bound as x tends to 3 (Fig. 4–9). According to the notion of limit as given in Section 1, we take an interval of x values about 3 and then see in what rectangle the function values are contained. From the figure it is clear that there is no such rectangle, regardless of how small an interval is chosen about

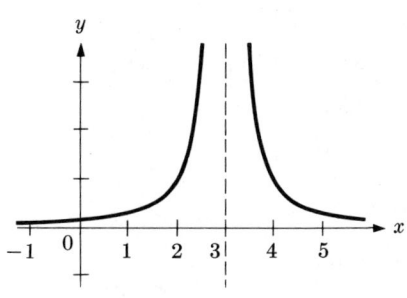

Figure 4–9   Figure 4–10

$x = 3$. In such a case we say that
$$\lim_{x \to 3} f(x)$$
does not exist.   ◀

Example 3. Find the limit
$$\lim_{t \to 0} \frac{f(2 + t) - f(2)}{t^2}$$
where $f(x) = x^2$.

Solution. Direct substitution of $t = 0$ fails, and we proceed as before to find that
$$f(2) = 4 \quad \text{and} \quad f(2 + t) = (2 + t)^2 = 4 + 4t + t^2.$$
Therefore, if $t \neq 0$,
$$\frac{f(2 + t) - f(2)}{t^2} = \frac{4 + 4t + t^2 - 4}{t^2} = \frac{4 + t}{t}.$$
However, this function increases without bound as t tends to zero through positive values, and it decreases without bound as t tends to zero through negative values (Fig. 4–10). The limit does not exist.   ◀

A convenient symbol used throughout calculus to designate a real number is Δx, which is read "delta x." This expression is to be treated as a single quantity such as t in the above Example 3 or h in Example 1 (and *not* as the product of Δ and x). Later we shall see that such symbols actually help in understanding rather complicated limiting processes.

Example 4. Find the limit
$$\lim_{\Delta x \to 0} \frac{f(-1 + \Delta x) - f(-1)}{\sqrt{\Delta x}}, \quad \Delta x > 0,$$
where $f(x) = x^3$.

Solution. We have $f(-1) = -1$, $f(-1 + \Delta x) = (-1 + \Delta x)^3$ and

$$\frac{f(-1 + \Delta x) - f(-1)}{\sqrt{\Delta x}} = \frac{-1 + 3(\Delta x) - 3(\Delta x)^2 + (\Delta x)^3 - (-1)}{\sqrt{\Delta x}}$$

$$= \frac{(\Delta x)[3 - 3(\Delta x) + (\Delta x)^2]}{\sqrt{\Delta x}}$$

$$= \sqrt{\Delta x}\,[3 - 3(\Delta x) + (\Delta x)^2].$$

As Δx tends to zero, this expression tends to $\sqrt{0}\,[3 - 3(0) + (0)^2]$, and the answer is zero. ◀

PROBLEMS

Evaluate the following limits.

1. $\lim\limits_{x \to 0} \dfrac{f(4 + x) - f(4)}{x}$, $\quad f(x) = 3x^2 - 1$

2. $\lim\limits_{h \to 0} \dfrac{f(-3 + h) - f(-3)}{h}$, $\quad f(x) = x^2 + 2$

3. $\lim\limits_{h \to 0} \dfrac{f(1 + h) - f(1)}{h}$, $\quad f(x) = 2x^3 - 2$

4. $\lim\limits_{k \to 0} \dfrac{f(3 + k) - f(3)}{k}$, $\quad f(x) = \dfrac{2}{x}, \quad x \neq 0$

5. $\lim\limits_{h \to 0} \dfrac{f(a + h) - f(a)}{h}$, $\quad f(x) = 2x^2 - 3$

6. $\lim\limits_{h \to 0} \dfrac{f(c + h) - f(c)}{h}$, $\quad f(x) = -\dfrac{2}{x^2}, \quad x \neq 0$

7. $\lim\limits_{h \to 0} \dfrac{g(2 + h) - g(2)}{h}$, $\quad g : x \to 12 - x^2$

8. $\lim\limits_{h \to 0} \dfrac{g(-4 + h) - g(-4)}{h}$, $\quad g : x \to \dfrac{1}{x^2 + 1}$

9. $\lim\limits_{h \to 0} \dfrac{F(-2 + h) - F(-2)}{h}$, $\quad F : x \to \dfrac{1}{x - 1}, \quad x \neq 1$

10. $\lim\limits_{h \to 0} \dfrac{F(a + h) - F(a)}{h}$, $\quad F(x) = \dfrac{1}{\sqrt{x}}, \quad x > 0, \quad a > 0$

11. $\lim\limits_{h \to 0} \dfrac{F(x + h) - F(x)}{h}$, $\quad F(x) = -x^2 + 5$

12. $\lim\limits_{k \to 0} \dfrac{G(b + k) - G(b)}{k}$, $\quad G(x) = x^2 - 2x + 3$

13. $\lim_{t \to 0} \dfrac{H(3+t) - H(3)}{t}$, $\quad H(x) = x^3 + 2x + 5$

14. $\lim_{h \to 0} \dfrac{F(4+h) - F(4)}{h}$, $\quad F(x) = \dfrac{1}{x^2 + 4}$

15. $\lim_{h \to 0} \dfrac{F(2+h) - F(2)}{\sqrt{h}}$, $\quad F(x) = x^4 - 2x + 5, \quad (h > 0)$

16. $\lim_{h \to 0} \dfrac{L(3+h) - L(3)}{h^{3/2}}$, $\quad L(x) = \dfrac{1}{x^2 + 2} - 3, \quad (h > 0)$

17. $\lim_{h \to 0} \dfrac{f(x+h) - f(x)}{h}$, $\quad f(x) = x^{3/2}, \quad x > 0$

18. $\lim_{h \to 0} \dfrac{F(t+h) - F(t)}{h}$, $\quad F(x) = x^2 - \dfrac{1}{x^2}, \quad x \neq 0$

19. $\lim_{h \to 0} \dfrac{g(c+h) - g(c)}{h}$, $\quad g: x \to x^2 + 3x - c$

20. $\lim_{h \to 0} \dfrac{g(x + 3h) - g(x)}{h}$, $\quad g(x) = 3x^2 + 2x - 1$

21. $\lim_{h \to 0} \dfrac{f(x+h) - f(x)}{h}$, $\quad f(x) = \dfrac{1}{x^3 + 1}, \quad x \neq -1$

22. $\lim_{h \to 0} \dfrac{f(x+h) - f(x)}{h}$, $\quad f(x) = \dfrac{1}{x^{3/2}}, \quad x > 0$

23. $\lim_{\Delta x \to 0} \dfrac{f(2 + \Delta x) - f(2)}{\Delta x}$, $\quad f(x) = x + \sqrt{x}, \quad x > 0$

24. $\lim_{\Delta x \to 0} \dfrac{f(x + \Delta x) - f(x)}{\Delta x}$, $\quad f: x \to \sqrt[3]{x}$

25. $\lim_{\Delta x \to 0} \dfrac{G(x + \Delta x) - G(x)}{\Delta x}$, $\quad G: x \to \dfrac{x-1}{x+1}, \quad x \neq -1$

3. THE DERIVATIVE

If f is a function, **the derivative of the function f, denoted by f'** (read "f prime"), **is defined by the formula**

$$f'(x) = \lim_{h \to 0} \dfrac{f(x+h) - f(x)}{h}.$$

In this definition, x remains fixed, while h tends to zero. If the limit does not exist for a particular value of x, the function has no derivative for that value.

Looking back at the previous section, we see that many of the problems consisted of finding the derivative at a particular value of x. We now give a systematic procedure for obtaining derivatives, a method which we call the **three-step rule.** The technique is illustrated by some examples.

Example 1. Given $f(x) = x^2$. Find the derivative, $f'(x)$, by the three-step rule.

Solution

Step 1: Write the formula for the expression $f(x + h) - f(x)$. We have $f(x) = x^2$ and $f(x + h) = (x + h)^2 = x^2 + 2xh + h^2$. Therefore

$$f(x + h) - f(x) = x^2 + 2xh + h^2 - x^2$$
$$= 2xh + h^2.$$

Step 2: Divide the expression for $f(x + h) - f(x)$ by h. Since h is a factor of this difference, we get

$$\frac{f(x + h) - f(x)}{h} = \frac{2xh + h^2}{h} = 2x + h.$$

Step 3: Take the limit as h tends to zero:

$$\lim_{h \to 0} (2x + h) = 2x.$$

Answer. $f'(x) = 2x$. ◀

In the three-step rule, the first two steps are purely mechanical and are carried out in a routine manner. It is the third and last step that frequently requires ingenuity and algebraic manipulation.

Example 2. Given $f(x) = 1/x$, $x \neq 0$. Find $f'(x)$ by the three-step rule.

Solution. We have the following.

Step 1: $f(x + h) - f(x) = \dfrac{1}{x + h} - \dfrac{1}{x}$.

Step 2: $\dfrac{f(x + h) - f(x)}{h} = \dfrac{\dfrac{1}{x + h} - \dfrac{1}{x}}{h}$.

It is at this point that the difficulty arises, since letting h tend to zero yields $\tfrac{0}{0}$. Before applying the third step, we manipulate the expression by finding the lowest common denominator. We obtain

$$\frac{f(x + h) - f(x)}{h} = \frac{x - (x + h)}{hx(x + h)} = \frac{-1}{x(x + h)}.$$

Step 3: $\lim\limits_{h \to 0} \dfrac{-1}{x(x + h)} = -\dfrac{1}{x^2}$.

Answer. $f'(x) = -\dfrac{1}{x^2}$. ◀

Example 3. Given $f: x \to \sqrt{x}$ for $x > 0$. Find $f'(x)$ by the three-step rule.

Solution. In the three-step rule, it is not essential that the letter h be used in computing the limit. Any symbol, such as t, k, Δx, and so forth, may be used. We employ Δx in solving this problem.

Step 1: $f(x + \Delta x) - f(x) = \sqrt{x + \Delta x} - \sqrt{x}$.

Step 2: $\dfrac{f(x + \Delta x) - f(x)}{\Delta x} = \dfrac{\sqrt{x + \Delta x} - \sqrt{x}}{\Delta x}$.

Now the trick of *rationalizing the numerator* is used. The result is

$$\frac{\sqrt{x + \Delta x} - \sqrt{x}}{\Delta x} \cdot \frac{\sqrt{x + \Delta x} + \sqrt{x}}{\sqrt{x + \Delta x} + \sqrt{x}} = \frac{(x + \Delta x) - x}{\Delta x(\sqrt{x + \Delta x} + \sqrt{x})}$$

$$= \frac{1}{\sqrt{x + \Delta x} + \sqrt{x}}.$$

Now taking the limit, we get

Step 3: $\lim\limits_{\Delta x \to 0} \dfrac{1}{\sqrt{x + \Delta x} + \sqrt{x}} = \dfrac{1}{\sqrt{x} + \sqrt{x}} = \dfrac{1}{2\sqrt{x}}$.

Answer: $f'(x) = \dfrac{1}{2\sqrt{x}}$. ◀

PROBLEMS

Find the derivatives of the following by the three-step rule.

1. $f(x) = 3x^2$
2. $f(x) = -2x^2 + 1$
3. $f(x) = 2x^2 - 3x + 4$
4. $f(x) = 3x^3$
5. $f(x) = -x^4$
6. $f(x) = x^3 - 2x$
7. $f(x) = \dfrac{1}{x - 1}$
8. $f(x) = \dfrac{1}{x^2}$
9. $f: x \to \dfrac{1}{x^3}$
10. $f: x \to \dfrac{2x + 3}{3x - 2}$
11. $f: x \to \dfrac{x}{1 - x}$
12. $f: x \to \dfrac{1}{x^2 + 1}$
13. $f(x) = \dfrac{x^2 - 1}{x^2 + 1}$
14. $f(x) = \dfrac{x^2 - 2x}{2}$
15. $f(x) = \dfrac{1}{\sqrt{x}}$
16. $f(x) = \dfrac{1}{\sqrt{x + 3}}$

17. $f: x \to x\sqrt{x}$

18. $f: x \to x\sqrt{x+1}$

19. $f(x) = \dfrac{x}{\sqrt{x-1}}$

20. $f(x) = \dfrac{\sqrt{x-1}}{x}$

21. $f(x) = \sqrt{4-x^2}$

22. $f(x) = \dfrac{1}{\sqrt{4-x^2}}$

23. $f(x) = \dfrac{x^2 - 2x + 2}{x^2 + x + 2}$

24. $f(x) = 2x - \dfrac{1}{3x} + \dfrac{2}{x^2}$

25. $f(x) = \sqrt{x^2 + x + 1}$

26. $f: x \to x^5 + 2x - 1$

4. GEOMETRIC INTERPRETATION OF THE DERIVATIVE

It is easy to define a *tangent line* to a circle at a point P on the circle. A radius from the center O to P is drawn and then, at P, a line L perpendicular to this radius is constructed (Fig. 4–11). **The line L is called the tangent to the circle at the point P.** We next draw the curve given by the equation $y = x^2 + 1$, as shown in Fig. 4–12. How do we define a tangent line to this curve at a point P?

If the point P is $(0, 1)$, the lowest point on the curve, it seems natural to take the line $y = 1$, as shown, as the tangent line. However, at a point such as Q, it is not evident what line is tangent to the curve. Intuitively, the definition should state that a tangent line should "touch" the curve at one point only. But this "definition" is not good enough, since there may be several such lines and, in fact, a vertical line through Q has only one point in common with the curve. While the definition of tangent line to a circle is quite elementary, the definition for other kinds of curves requires the sophisticated notion of limit.

Let a curve be given by a function $y = f(x)$, a section of which is shown in Fig. 4–13. Draw a line L_h through P_0 and P, two points on the curve. Any line through two points on a curve is called a **secant line**. The line L_h is such a secant line. Suppose that the coordinates of P_0 are (x_0, y_0). This means that $y_0 = f(x_0)$.

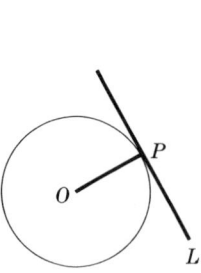

Figure 4–11

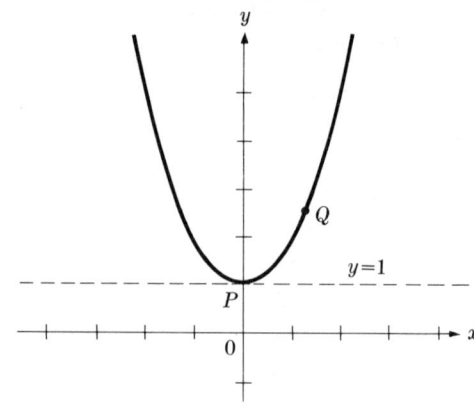

Figure 4–12

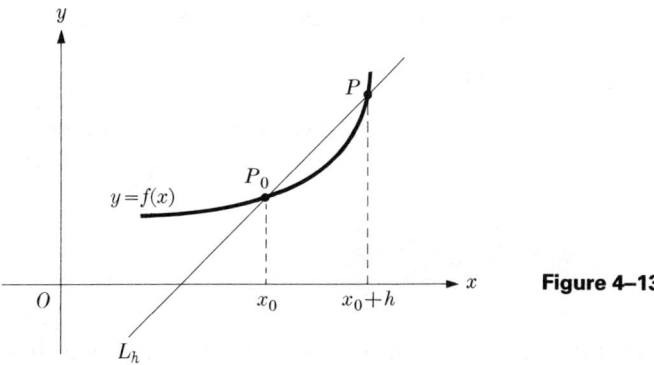

Figure 4-13

The coordinates of P are taken to be $(x_0 + h, f(x_0 + h))$, where h is some positive or negative quantity. Figure 4-13 is drawn with h positive. If P were to the left of P_0, then h would be negative. The slope of the line L_h is obtained from the usual formula for slope when two points are known:

$$\text{slope of } L_h = \frac{f(x_0 + h) - f(x_0)}{(x_0 + h) - x_0} = \frac{f(x_0 + h) - f(x_0)}{h}.$$

As h tends to zero, we see geometrically that the point P tends to the point P_0. The secant line L_h rotates about P_0. Intuitively it seems that L_h approaches a limiting line as $h \to 0$, and this limiting line must be the tangent line to the curve at P_0. Figure 4-14 shows several positions for L_h and the limiting position as $h \to 0$. On the other hand, if the function f possesses a derivative, then

$$\lim_{h \to 0} \frac{f(x_0 + h) - f(x_0)}{h} = f'(x_0).$$

In other words, *the slopes of the lines L_h tend to a limiting slope which is the derivative of the function f at P_0.* Suppose that we now construct, at P_0, the line passing through P_0 and having slope $f'(x_0)$. This can be done by the point-slope formula for the straight line.

Definition. *The* **tangent** *to the curve with equation* $y = f(x)$ *at* $P_0(x_0, f(x_0))$ *is the line through* $P_0(x_0, f(x_0))$ *with slope* $f'(x_0)$.

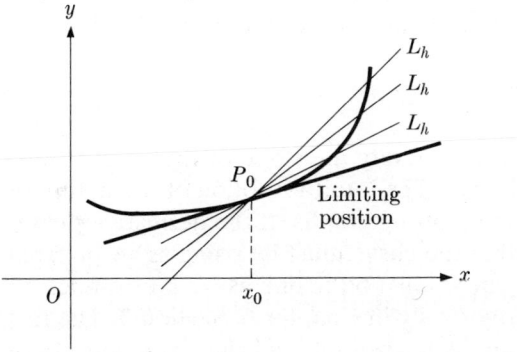

Figure 4-14

We recall that the point-slope form for the equation of a straight line is

$$y - y_0 = m(x - x_0).$$

If in this formula we take $y_0 = f(x_0)$ and $m = f'(x_0)$, we obtain the equation of the line tangent to the curve at the point P_0. That is, the equation of the tangent line is

$$y - y_0 = f'(x_0)(x - x_0).$$

Example 1. Find the equation of the line which is tangent to the curve that has equation $y = x^2 + 1$ at the point on the curve where $x = 2$.

Solution. In this problem $f(x) = x^2 + 1$, $x_0 = 2$. Therefore $y_0 = f(x_0) = f(2) = 5$. To find $f'(x_0)$, we apply the three-step rule to $f(x) = x^2 + 1$.

Step 1: $f(x + h) - f(x) = x^2 + 2xh + h^2 + 1 - x^2 - 1.$

Step 2: $\dfrac{f(x + h) - f(x)}{h} = \dfrac{2xh + h^2}{h} = 2x + h.$

Step 3: Taking the limit, we get $f'(x) = 2x$.

Then $f'(x_0) = f'(2) = 4$ and $m = 4$. The desired equation is

$$y - 5 = 4(x - 2). \qquad \blacktriangleleft$$

Definition. *The* **normal** *to the curve with equation* $y = f(x)$ *at* $P_0(x_0, f(x_0))$ *is the line through P_0 which is perpendicular to the tangent line at P_0.*

Example 2. In the problem of Example 1, find the equation of the line normal to the curve at the point where $x = 3$.

Solution. Since $x_0 = 3$ and $f(x) = x^2 + 1$, we have $y_0 = f(x_0) = 10$. The tangent line at $x_0 = 3$ has slope $f'(x_0) = f'(3)$. We have already found that $f'(x) = 2x$ for any value of x. Therefore the tangent line at $x = 3$ has slope 6. The normal line must have a slope which is the negative reciprocal of the slope of the tangent line, namely $-\frac{1}{6}$. The equation of the normal line at $(3, 10)$ is

$$y - 10 = -\frac{1}{6}(x - 3). \qquad \blacktriangleleft$$

We have just seen that the derivative is merely the slope of the tangent line. In fact, the tangent line is defined in this way. The intuitive notion of the nature of a line tangent to a curve is of great help in plotting graphs. If the derivative $f'(x)$ is positive, the slope is positive, showing that the curve must be rising as we go from left to right. Similarly, if $f'(x) < 0$, the curve must be falling as we go from left to right. These statements will be proved rigorously in Chapter 6, Section 3. The fact that $f'(x) = 0$ means that the tangent line is horizontal, and the curve *may* have either a low point or a high point at such a location. (However, other possibilities exist, as we shall see later in Chapter 6, Section 3.)

Example 3. Find the intervals in which the function $y = f(x) = x^3 - 3x + 2$ is increasing and those in which it is decreasing. Sketch the graph.

Solution. We first find the derivative by the three-step rule.

Step 1: $f(x + h) - f(x) = (x + h)^3 - 3(x + h) + 2 - (x^3 - 3x + 2)$
$= x^3 + 3x^2h + 3xh^2 + h^3 - 3x - 3h + 2$
$\quad - x^3 + 3x - 2$
$= 3x^2h + 3xh^2 + h^3 - 3h.$

Step 2: $\dfrac{f(x + h) - f(x)}{h} = 3x^2 - 3 + 3xh + h^2.$

Step 3: Taking the limit as $h \to 0$, we get $f'(x) = 3x^2 - 3$. We first find the places at which $f'(x) = 0$. That is, we solve the equation $3x^2 - 3 = 0$ to get $x = 1, -1$. An examination of the function $f'(x) = 3x^2 - 3$ shows that $f'(x) > 0$ for $x < -1$; also we see that $f'(x) < 0$ for $-1 < x < 1$ and $f'(x) > 0$ for $x > 1$. We conclude that

$f(x)$ is increasing for $x < -1$,
$f(x)$ is decreasing for $-1 < x < 1$,
$f(x)$ is increasing for $x > 1$.

A table of values gives the following points.

x	-2	-1	0	1	2
$y = f(x)$	0	4	2	0	4

The graph is sketched in Fig. 4–15. ◀

The fact that the curve reached a low point at exactly $(1, 0)$ and a high point at exactly $(-1, 4)$ was discovered from studying the derivative. The location of

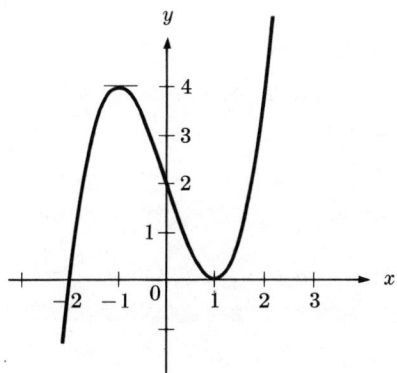

Figure 4–15

high and low points could not easily be obtained merely from plotting points and sketching the graph. A low point such as $(1, 0)$ is a minimum point only if we restrict our attention to a small interval about this point. Such a point is called a **relative minimum.** It is clearly not the minimum point on the entire curve, since the curve falls below the x axis as we go to the left beyond $x = -2$. Similarly, the point $(-1, 4)$ is called a **relative maximum.** At a relative maximum or minimum it is always true that if f has a derivative, then $f'(x) = 0$. (The converse is not necessarily true.)

Notation. We saw earlier that the symbol Δx can be used instead of h in calculating the derivative of a function f. Writing $y = f(x)$, we define the symbol Δy by the formula

$$\Delta y = f(x + \Delta x) - f(x).$$

Then Step 2 in the three-step rules states that

$$\frac{\Delta y}{\Delta x} = \frac{f(x + \Delta x) - f(x)}{\Delta x}.$$

The derivative consists of taking the limit of the above expression as $\Delta x \to 0$. The symbol we have been using is $f'(x)$. However, it is helpful to use a symbol related to the quantity $\frac{\Delta y}{\Delta x}$. It turns out to be most useful to write

$$f'(x) = \frac{dy}{dx}$$

where, at this time, dy and dx will not be defined separately. We simply use the ratio $\frac{dy}{dx}$ as an alternate expression for $f'(x)$. That is, we write

$$\lim_{\Delta x \to 0} \frac{\Delta y}{\Delta x} = \frac{dy}{dx} = f'(x).$$

PROBLEMS

In Problems 1 through 12, find the equation of the tangent line and the normal line at the point corresponding to the given value of x_0.

1. $y = x^2 + x - 1$, $x_0 = 1$
2. $y = x^2 + 3x$, $x_0 = -1$
3. $y = x^3 - 2x^2 + 3$, $x_0 = 2$
4. $y = x^3 - x + 2$, $x_0 = -2$
5. $y = -x^3 + 3x + 1$, $x_0 = 2$
6. $y = \frac{x + 1}{x - 2}$, $x_0 = 3$
7. $y = \sqrt{2x}$, $x_0 = 2$
8. $y = \sqrt{3 - x}$, $x_0 = -1$
9. $y = \frac{1}{\sqrt{x - 2}}$, $x_0 = 6$
10. $y = 2x + 3\sqrt{x}$, $x_0 = 4$
11. $y = 2x + \frac{2}{x}$, $x_0 = 1$
12. $y = x^4 - 2x + 5$, $x_0 = -1$

In Problems 13 through 20, find the intervals in which f is increasing and those in which it is decreasing. Plot the graph of $y = f(x)$. Note relative maxima and minima.

13. $f(x) = x^2 - 4x + 5$
14. $f(x) = 1 + 2x - x^2$
15. $f(x) = \frac{1}{3}x^3 - \frac{1}{2}x^2 - 2x$
16. $f(x) = \frac{1}{3}x^3 - x^2 - 3x + 2$
17. $f(x) = 1 + 3x^2 - x^3$
18. $f(x) = x^3 + 3x - 2$
19. $f(x) = 3x^4 - 8x^3 - 6x^2 + 24x + 2$
20. $f(x) = 3x^5 - 25x^3 + 60x + 10$

In Problems 21 through 23, a function $y = f(x)$ is given. Compute $\Delta y = f(x + \Delta x) - f(x)$ and $\Delta y / \Delta x$ for the given values of x and Δx.

21. $y = f(x) = x^2 - 3x + 4$, $x = 1$, $\Delta x = 0.1$
22. $y = f(x) = \dfrac{x - 3}{x + 3}$, $x = 2$, $\Delta x = -0.2$
23. $y = f(x) = \dfrac{x^3 - 9}{x^2 + 1}$, $x = 0$, $\Delta x = 0.01$

In Problems 24 through 26, find dy/dx and evaluate the derivative at the given value of x.

24. $y = x^4 + 2x - 1$, $x = -2$
25. $y = \dfrac{x^2 - 1}{x^2 + 1}$, $x = 1$
26. $y = \sqrt[3]{x}$, $x = 8$

5. INSTANTANEOUS VELOCITY AND SPEED: ACCELERATION

If the needle of the speedometer in a car is pointing to 60, we say that the car is traveling at 60 kilometers per hour*. An analysis of this statement requires a definition of the phrase "60 km per hour." As a start, we might say that if the car were to continue in exactly the same manner, then in one hour it would have traveled exactly 60 kilometers. But such a statement avoids the problem completely, since the needle pointing to 60 does so without any knowledge of the way the car behaved some minutes ago, or how it intends to behave in the future.

Before any further discussion of this question we introduce two simplifying assumptions.

1) Motion will always be assumed to take place along a *straight line*, although the object in motion may go in either direction. One direction will arbitrarily be selected as positive, and the other direction will then be negative.

2) The object in motion is idealized to be a point or particle. Such a simplification is necessary, since a mathematical treatment of the exact motion of every portion of a complicated object such as an automobile would defy analysis.

Suppose that the path of a particle is along a horizontal line L, with distance to the right designated as positive and that to the left as negative (Fig. 4–16). We let t denote time in some convenient unit such as seconds or minutes, and we measure the distance s of the object from the point O in units such as centimeters, meters, or kilometers. If the particle in motion is at the point s_1 at time t_1

---

* The metric system will be used throughout.

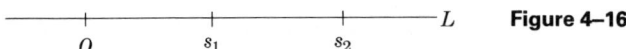

Figure 4–16

and at the point s_2 at time t_2, then it took $t_2 - t_1$ units of time to travel the distance $s_2 - s_1$. We define the **average velocity** of the particle over this time interval to be

$$\frac{s_2 - s_1}{t_2 - t_1}.$$

If s is measured in meters and t in seconds, the units of this average velocity would be denoted as **meters per second** (m/sec). If a car travels 40 kilometers in one hour, the average velocity of the car is 40 kilometers per hour (km/hr). But we still know very little about how fast the car is traveling at various times during the hour. The needle of a speedometer pointing to 60 is an instantaneous event, and we need a notion of velocity at a given instant of time along the path. The development of this notion might proceed in the following way.

We mark off two points along a straight stretch of road, accurately measure the distance between them, and provide a timing device to indicate the exact moment the car passes each location. The average velocity is then computed. We believe that if the distance between the two points is small, the car's motion cannot change too radically as it travels from one point to the other; we therefore obtain a good indication of the velocity of the car at any time during this interval. For example, if the points are 100 meters apart and the average velocity is 60 km/hr, the variation in the motion along such a short stretch of road is probably not too great. If we desire greater accuracy, we can shorten the interval to ten meters, improve the measurement of the distance, and increase the precision of the timing device. Now we again argue that, in such a short distance, the motion of the car varies so slightly that the average velocity closely approximates the velocity indicated on the speedometer. On the other hand, some variation may still be possible, and so we cut the interval again, this time to one meter, say. The continuation of this process shows that instantaneous velocity is the result of computing the average velocity over a sequence of smaller and smaller intervals.

To make the above discussion precise, we first note that the distance s traveled along the straight line is a function of the time t; that is, $s = f(t)$. Then the **average velocity** between the times t_1 and t_2 is simply expressed as

$$\frac{f(t_2) - f(t_1)}{t_2 - t_1} = \frac{\text{distance}}{\text{time}}.$$

The **instantaneous velocity** at time t_1 is defined to be

$$\lim_{t_2 \to t_1} \frac{f(t_2) - f(t_1)}{t_2 - t_1}.$$

That is, the instantaneous velocity is the limit of the average velocities as the time over which these averages are taken tends to zero. Since t_2 must be different from t_1, we can write $t_2 = t_1 + \Delta t$, where Δt may be *positive or negative*. The above

4-5 INSTANTANEOUS VELOCITY AND SPEED: ACCELERATION

expression for instantaneous velocity then becomes

$$\lim_{t_1+\Delta t \to t_1} \frac{f(t_1 + \Delta t) - f(t_1)}{t_1 + \Delta t - t_1} \quad \text{or} \quad \lim_{\Delta t \to 0} \frac{f(t_1 + \Delta t) - f(t_1)}{\Delta t}.$$

We note that this is precisely the derivative of $f(t)$ evaluated at t_1. In other words, **the instantaneous velocity of a particle moving in a straight line according to the law $s = f(t)$, describing the motion, is** $f'(t)$. We also use the symbol $\frac{ds}{dt}$ to describe this derivative. The quantity ds/dt may be positive or negative, according to whether the particle is moving along the line in the positive or negative direction. The **speed** of the particle is defined to be $|ds/dt| = |f'(t)|$. The speed is merely the magnitude of the velocity and is always positive or zero. It may seem strange to introduce a special term specifically for the absolute value of the velocity. However, the notion of speed is especially useful when one is studying motion along curved paths. The speed tells us how fast the particle is moving but gives no information about its direction.

Example 1. A projectile shot directly upward with a speed of 100 meters/sec moves according to the law

$$y = 100t - 5t^2,$$

where y is the height in meters above the starting point, and t is the time in seconds after it is thrown. Find the velocity of the projectile after 2 seconds. Is it still rising or is it falling? For how many seconds does it continue to rise? How high does the projectile go?

Solution It is assumed that the motion of the projectile (particle) is in a straight vertical line. If we let $f(t) = 100t - 5t^2$, the velocity is given by ds/dt. Applying the three-step rule, we obtain

$$\frac{ds}{dt} = 100 - 10t.$$

Let v denote the velocity at any time t; then

$$v = \frac{ds}{dt} = 100 - 10t.$$

If $t = 2$, then $v = 100 - 10(2) = 80$ m/sec. Since v is positive, the projectile is still rising. The velocity is zero when $100 - 10t = 0$, or $t = 10$. When t is larger than 10, the velocity is negative and the motion is downward, i.e., the projectile is falling. When $t = 10$, we see that $f(10) = 100(10) - 5(10)^2 = 500$ meters, and this is the highest point the projectile reaches. See Fig. 4–17. ◄

In motion along a straight line, whenever a particle changes direction the velocity goes from positive to negative or from negative to positive. *Therefore at the instant the particle reverses direction the velocity must be zero.*

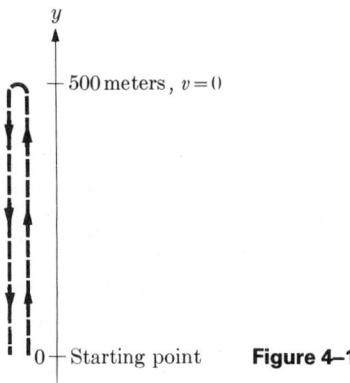

500 meters, $v = 0$

0 Starting point   **Figure 4–17**

Example 2. A particle moves along a horizontal line (positive to the right) according to the law

$$s = t^3 - 3t^2 - 9t + 5.$$

During which intervals of time is the particle moving to the right, and during which is it moving to the left?

Solution. It is moving to the right whenever the velocity is positive and to the left when the velocity is negative. The velocity v is just the derivative of $f(t) = t^3 - 3t^2 - 9t + 5$. Applying the three-step rule, we have

$$v = \frac{ds}{dt} = 3t^2 - 6t - 9 = 3(t + 1)(t - 3).$$

If $t = -1, 3$, the velocity is zero. By examining the signs of the factors $t + 1$ and $t - 3$, we find that:

If $t < -1$, v is positive and the motion is to the right.
If $-1 < t < 3$, v is negative and the motion is to the left.
If $t > 3$, v is positive and the motion is to the right.

Motion to the left and motion to the right are separated by points of zero velocity, i.e., at $t = -1, 3$ (Fig. 4–18). Although the motion is along the line L, its schematic behavior is shown by the path above this line. Zero velocity occurs at points P, Q. The point P corresponds to $t = -1$, $s = 10$, $v = 0$; Q corresponds to $t = 3$, $s = -22$, $v = 0$.

Acceleration is a measure of the change in velocity. If a particle moves along a straight line with constant velocity, the acceleration is zero. In an automobile race, the cars pass the starting position traveling at a uniform velocity—say 30 km/hr. Within ten seconds one of the cars is traveling at 120 km/hr. The **average**

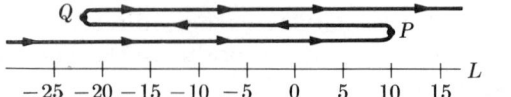

Figure 4–18

acceleration of this car is

$$\frac{120 - 30}{10} = 9 \text{ (km/hr)/sec}.$$

The units appear rather strange, since velocity is in kilometers per hour and the time in the denominator is in seconds. We can convert 120 km/hr to $33\frac{1}{3}$ m/sec if we multiply 120 by 1000/3600. Similarly, 30 km/hr = $8\frac{1}{3}$ m/sec. With this change the average acceleration is expressed as

$$\frac{33\frac{1}{3} - 8\frac{1}{3}}{10} = 2.5 \text{ (m/sec)/sec}.$$

If the velocity v of the particle is varying according to the law $v = F(t)$, where t is the time, then the **instantaneous acceleration** a, or simply the **acceleration**, is defined as the limit of the average acceleration:

$$a = \lim_{h \to 0} \frac{F(t + h) - F(t)}{h} = F'(t).$$

This is in complete analogy with the definition of instantaneous velocity as the limit of the average velocity.

If a particle is moving (along a straight line) so that the distance s is given by $s = f(t)$, then the velocity v is found by taking the derivative $f'(t)$. The derivative f' is the law which describes the velocity. The derivative of this derivative is the acceleration. We call it the **second derivative** and use the notation $f''(t)$. That is, if $s = f(t)$, then

$$v = f'(t) \quad \text{and} \quad a = \frac{dv}{dt} = f''(t).$$

We could continue this process and take third derivatives, fourth derivatives, etc., and indeed we sometimes do. The notation for third derivative is $f'''(t)$ or $f^{(3)}(t)$. [Beyond the third derivative, primes become unwieldy, and a numeral in parentheses is used. For example, the seventh derivative is $f^{(7)}(t)$.] It should be mentioned that, from the point of view of physics, the first and second derivatives are the most important—especially in the study of the motion of objects. Derivatives higher than the second are of occasional interest only.

Example 3. A particle moves along a straight line according to the law

$$s = 132 + 108t - 16t^2 + 3t^3,$$

s being the distance in meters and t the time in seconds. Find the velocity and acceleration at any time t. What is the velocity when $t = 2$? What is the acceleration when $t = 1$ and when $t = 3$?

Solution. The velocity is found by taking the derivative, which we do according to the three-step rule. Then

$$v = 108 - 32t + 9t^2.$$

The acceleration is obtained by taking the derivative of this function, which we also do by the three-step rule. We have

$$a = -32 + 18t.$$

When $t = 2$, we substitute in the equation for v to get $v = 108 - 64 + 36 = 80$ m/sec. When $t = 1$, we see that $a = -32 + 18 = -14$ (m/sec)/sec, and when $t = 3$, $a = -32 + 54 = 22$ (m/sec)/sec.

Since the acceleration is negative when $t = 1$, we conclude that the particle is slowing down, while a is positive when $t = 3$ and therefore the particle is speeding up.

PROBLEMS

In Problems 1 through 6, a particle is moving along a horizontal line (positive to the right), according to the stated law. Determine whether the particle is moving to the right or to the left at the given time.

1. $s = t^2 - 2t + 7$, $\quad t = 3$
2. $s = t^2 + t - 5$, $\quad t = 1$
3. $s = t^3 - 2t^2 + 7t - 6$, $\quad t = -1$
4. $s = t^3 - 5t^2 + 2t + 4$, $\quad t = 3$
5. $s = t^3 - 4t^2 - 2t - 1$, $\quad t = -2$
6. $s = t^4 + 2t - 1$, $\quad t = 1$

In Problems 7 through 12, calculate the velocity and acceleration of the given laws of motion. Determine the times, if any, when the velocity is zero.

7. $s = t^2 - 4t + 2$
8. $s = 6 - 2t - 16t^2$
9. $s = \frac{1}{3}t^3 - 2t^2 + 3t - 5$
10. $s = t^3 + 2t^2 - t - 1$
11. $s = bt^2 + ct + d$ (b, c, d constants)
12. $s = bt^3 + ct^2 + dt + e$ (b, c, d, e constants)

In Problems 13 through 20, motion is along a horizontal line, positive to the right. When is the particle moving to the right and when is it moving to the left? Discuss the acceleration.

13. $s = t^2 + 3t - 1$
14. $s = t^2 - 2t + 3$
15. $s = 8 - 4t + t^2$
16. $s = t^3 + 3t^2 - 9t + 4$
17. $s = 2t^3 - 3t^2 - 12t + 8$
18. $s = t^3 - 3t^2 + 3t$
19. $s = \dfrac{t}{1 + t^2}$
20. $s = \dfrac{1 + t}{4 + t^2}$

In Problems 21 and 22, find expressions for the velocity and acceleration.

21. $s = bt^2 + ct + d$ (b, c, d constants)
22. $s = bt^3 + ct^2 + dt + e$ (b, c, d, e constants)

23. A stone is thrown upward from the top of a building 50 meters high with a velocity of 15 meters/sec. The height s at any time t is given by the formula $s = 50 + 15t - 5t^2$. Find the maximum height the stone will reach and how long it takes to reach this height. When will the stone hit the ground?

24. A projectile hurled straight up from the earth travels according to the law $s = x_0 + v_0 t - 5t^2$, where x_0 is the height above the ground at the start of the motion, v_0 is the initial velocity, t is the time traveled, and s is the height at time t. Distance is measured in meters, time in seconds.

A man on a tower throws a ball upwards. He observed that it passes him on the way down exactly 3 seconds later and that it hits the ground in one more second. How high is the tower? What was the initial velocity?

25. A projectile hurled upward from the surface of the moon follows the law $s = v_0 t - \frac{1}{2}gt^2$ where v_0 is the initial velocity and g is the gravitational constant for the moon. A stone thrown up with an initial velocity of 20 meters/sec is observed hitting the surface 25 seconds later. What is the value of the gravitational constant?

6. DEFINITION OF LIMIT*

In Section 1 we introduced the notion of limit in an informal way. We spoke of intervals as being "small," numbers as being "close," quantitites "approaching" zero, and so forth. However, these nonmathematical words vary widely in meaning from person to person and cannot be the basis for a mathematical structure. Therefore we give the following statement as a precise definition of limit:

Definition. *Given a function f and numbers a and L, we say that f(x)* **tends to** *L* **as a limit as** *x* **tends to** *a if for each positive number ϵ there is a positive number δ such that $f(x)$ is defined and*

$$|f(x) - L| < \epsilon \quad \text{whenever} \quad 0 < |x - a| < \delta.$$

In abbreviated notation, we write

$$f(x) \to L \quad \text{as} \quad x \to a$$

for this definition of limit.

We now elaborate on the meaning of the above definition. First of all, the definition implies that there can be at most one limit L. (This fact is proved at the beginning of the next section.) Next we recall that

$$|x - a| < \delta$$

is the same as the two inequalities

$$a - \delta < x < a + \delta.$$

This double inequality states that x must lie in an interval of length 2δ having a as its center (Fig. 4–19). The part of the inequality which states that $0 < |x - a|$ merely means that x is not allowed to be equal to a itself. This is done for convenience. The inequality

$$|f(x) - L| < \epsilon$$

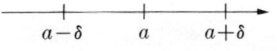

Figure 4–19

---

* Except for the actual definition of limit, this section may be omitted without loss of continuity. The problems at the end of this section are above average in difficulty.

is equivalent to
$$L - \epsilon < f(x) < L + \epsilon,$$
which asserts that the function f lies above the line $y = L - \epsilon$ and below the line $y = L + \epsilon$ (Fig. 4–20).

The definition itself may be interpreted as a test. If I am given any positive number whatsoever (call it ϵ), the test consists of finding a number δ such that $f(x)$ lies between the values $L - \epsilon$ and $L + \epsilon$ if x is in the interval $(a - \delta, a + \delta)$ and $x \neq a$. If such a δ can be found for *every* positive number ϵ, then we say that $f(x)$ has the limit L as x approaches a. Note that the value of delta will be different for different epsilons. Also the test must be performed for *every* positive epsilon, which means in general that it is an extremely difficult thing to check.

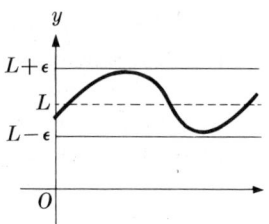

Figure 4–20

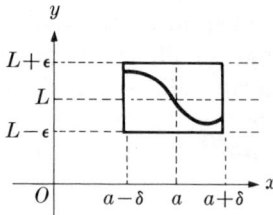

Figure 4–21

The geometric explanation, one similar to that given in Section 1, states that if an ϵ is given, a δ can be found such that the graph of the function f lies in the rectangle bounded by the lines $x = a - \delta$, $x = a + \delta$, $y = L - \epsilon$, $y = L + \epsilon$ (Fig. 4–21). Nothing at all is said about the value of f when x is a.

It is good to get some practice in finding the δ which corresponds to a given ϵ; the δ can actually be found in very simple cases. To consider an easy case, we let $f(x) = 3x - 2$ and take $a = 5$. We know intuitively from our earlier work that
$$\lim_{x \to 5} f(x) = 13.$$
We wish to show that, given an ϵ, we can find a δ such that
$$|3x - 2 - 13| < \epsilon \quad \text{whenever} \quad 0 < |x - 5| < \delta.$$
But $|3x - 15| = |3(x - 5)|$. If someone gives us an ϵ, we simply take $\delta = \epsilon/3$. Then, if $|x - 5| < \delta = \epsilon/3$, we find (by multiplying through by 3) that $3|x - 5| < \epsilon$, which is the same as $|3x - 15| < \epsilon$, as was required.

Example 1. Draw a graph of the function
$$f(x) = \frac{1}{x + 1},$$
$$x \neq -1.$$
Find a δ such that $|f(x) - \tfrac{1}{2}| < 0.01$ if $0 < |x - 1| < \delta$.

Solution. Part of the graph is sketched in Fig. 4–22. In the definition of limit we have $L = \tfrac{1}{2}$ and $a = 1$. We must find an interval of the x axis about $x = 1$ such

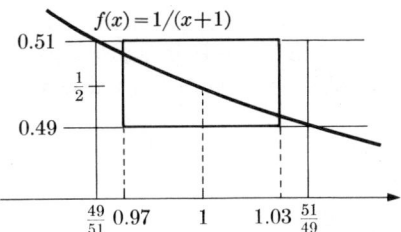

Figure 4–22

that the graph lies in the proper rectangle. The function decreases steadily as we go to the right and therefore, when we erect vertical lines where the lines $y = 0.51$, $y = 0.49$ intersect the curve, the largest possible interval on the x axis is obtained. When we solve

$$\frac{1}{x+1} = 0.51$$

for x, we get $x = \frac{49}{51}$, and similarly, solving $1/(x+1) = 0.49$ gives $x = \frac{51}{49}$. In Fig. 4–22 these values are shown with units greatly exaggerated. Fortunately, once we find a δ, then any *smaller* δ will also be valid; for if the function lies in a rectangle, it certainly lies in a similar rectangle which is of the same height but narrower. So we may take $\delta = 0.03$, since $\frac{49}{51} < 0.97$ and $\frac{51}{49} > 1.03$. ◀

Example 2. Draw a graph of

$$f(x) = \frac{2(x-4)}{\sqrt{x}-2}, \qquad x \geq 0, \quad x \neq 4,$$

and find a δ such that

$$|f(x) - 8| < 0.01 \qquad \text{whenever} \qquad 0 < |x - 4| < \delta.$$

Solution. The function f is not defined at $x = 4$, but for $x \neq 4$ we can multiply numerator and denominator by $\sqrt{x} + 2$ to obtain

$$f(x) = \frac{2(x-4)(\sqrt{x}+2)}{(x-4)} = 2\sqrt{x} + 4 \qquad \text{for} \qquad x \neq 4.$$

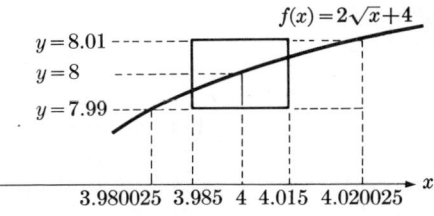

Figure 4–23

The graph of this function is shown in Fig. 4–23. We construct the lines $y = 7.99$ and $y = 8.01$, shown with greatly enlarged units in the figure. The intersections of these lines with the function are found by solving the equations

$$2\sqrt{x} + 4 = 7.99 \qquad \text{and} \qquad 2\sqrt{x} + 4 = 8.01.$$

We get $x = 3.980025$ and $x = 4.020025$. Since the function steadily increases to the right, an adequate selection for δ is 0.015. In other words, it is true that

$$|f(x) - 8| < 0.01 \qquad \text{whenever} \qquad 0 < |x - 4| < 0.015. \qquad ◀$$

86 INTRODUCTION TO THE CALCULUS. LIMITS

Remark. To establish the existence of a limit we must find a δ for each positive ϵ. It is not necessary that we find the largest possible δ and, if work can be avoided by selecting a δ smaller than the largest possible one, we usually do so.

Example 3. Show directly from the definition that

$$\lim_{x \to 1} \frac{x}{x+1} = \frac{1}{2},$$

and draw the graph of the function $f: x \to x/(x+1)$.

Solution. We have $L = \frac{1}{2}$ and $a = 1$. We must show that for every $\epsilon > 0$ we can find a $\delta > 0$ such that

$$\left| \frac{x}{x+1} - \frac{1}{2} \right| < \epsilon \quad \text{whenever} \quad 0 < |x - 1| < \delta.$$

In order to get an idea of the appearance of the function we sketch the graph (Fig. 4–24). From the graph we see that the function is a steadily increasing one. We verify this fact by writing the identity

$$\frac{x}{x+1} = 1 - \frac{1}{x+1},$$

and noting that as x gets larger, $1/(x+1)$ gets smaller and, as a result, $1 - [1/(x+1)]$ increases.

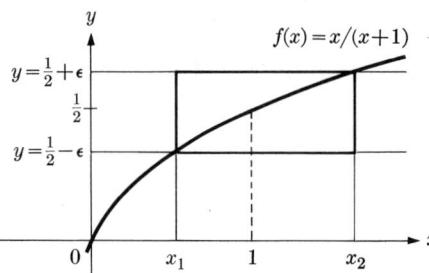

Figure 4–24

Let us first suppose that $\epsilon < \frac{1}{2}$. Then $L + \epsilon < 1$ and $L - \epsilon > 0$, since $L = \frac{1}{2}$. Next, to see where the lines $y = \frac{1}{2} - \epsilon$ and $y = \frac{1}{2} + \epsilon$ intersect the curve, we solve the equations

$$\frac{x}{x+1} = \frac{1}{2} - \epsilon \quad \text{and} \quad \frac{x}{x+1} = \frac{1}{2} + \epsilon.$$

The first equation gives

$$x = (\tfrac{1}{2} - \epsilon)(x+1) \quad \Leftrightarrow \quad (\tfrac{1}{2} + \epsilon)x = \tfrac{1}{2} - \epsilon \quad \Leftrightarrow \quad x = \frac{\tfrac{1}{2} - \epsilon}{\tfrac{1}{2} + \epsilon} \equiv x_1.$$

Similarly, the second equation yields

$$x = \frac{\tfrac{1}{2} + \epsilon}{\tfrac{1}{2} - \epsilon} \equiv x_2.$$

We select as δ the smaller of the distances between 1 and x_1 and between 1 and x_2. The student can check the fact that $1 - x_1$ is smaller than $x_2 - 1$. Therefore

$$\delta = 1 - x_1 = 1 - \frac{\frac{1}{2} - \epsilon}{\frac{1}{2} + \epsilon} = \frac{2\epsilon}{\frac{1}{2} + \epsilon} = \frac{4\epsilon}{1 + 2\epsilon}.$$ ◀

Remark. We restricted ϵ to a value smaller than $\frac{1}{2}$ and then found a δ for every positive $\epsilon < \frac{1}{2}$. While it is true that the basic definition states that a δ has to be found for *every* ϵ, in actuality this is not so. Once we have found a δ for a specific ϵ, we can use the same δ for *all* larger ϵ. Geometrically, this means that once the function is known to lie in a rectangle, clearly it lies in every rectangle which has the same sides but is taller.

PROBLEMS

In Problems 1 through 17, the numbers a, L, and ϵ are given. Determine a number δ so that $|f(x) - L| < \epsilon$ for all x such that $0 < |x - a| < \delta$. Draw a graph.

1. $f(x) = 2x + 3$, $a = 1$, $L = 5$, $\epsilon = 0.001$
2. $f(x) = 1 - 2x$, $a = -1$, $L = 3$, $\epsilon = 0.01$
3. $f(x) = (x^2 - 9)/(x + 3)$, $a = -3$, $L = -6$, $\epsilon = 0.005$
4. $f(x) = \sqrt{x}$, $a = 1$, $L = 1$, $\epsilon = 0.01$
5. $f(x) = \sqrt[3]{x}$, $a = 1$, $L = 1$, $\epsilon = 0.01$
6. $f(x) = \sqrt[3]{x}$, $a = 0$, $L = 0$, $\epsilon = 0.1$
7. $f(x) = \sqrt{2x}$, $a = 2$, $L = 2$, $\epsilon = 0.02$
8. $f(x) = 1/x$, $a = 2$, $L = \frac{1}{2}$, $\epsilon = 0.002$
9. $f(x) = 2/\sqrt{x}$, $a = 4$, $L = 1$, $\epsilon = 0.1$
10. $f(x) = 3/(x + 2)$, $a = 1$, $L = 1$, $\epsilon = 0.001$
11. $f(x) = 1/x$, $a = -1$, $L = -1$, $\epsilon = 0.01$
12. $f(x) = (x - 1)/(x + 1)$, $a = 0$, $L = -1$, $\epsilon = 0.01$
13. $f(x) = (\sqrt{x} - 1)/(x - 1)$, $a = 1$, $L = \frac{1}{2}$, $\epsilon = 0.01$
14. $f(x) = (\sqrt{2x} - 2)/(x - 2)$, $a = 2$, $L = \frac{1}{2}$, $\epsilon = 0.01$
15. $f(x) = x^2$, $a = 1$, $L = 1$, $\epsilon = 0.01$
16. $f(x) = x^3 - 6$, $a = 1$, $L = -5$, $\epsilon = 0.1$
*17. $f(x) = x^3 + 3x$, $a = -1$, $L = -4$, $\epsilon = 0.5$

In Problems 18 through 22, show that

$$\lim_{x \to a} f(x) = L$$

directly by finding the δ corresponding to every positive ϵ. (Use the method of Example 3.)

18. $f(x) = (x^2 - 4)/(x - 2)$, $a = 2$, $L = 4$
19. $f(x) = \sqrt{x}$, $a = 2$, $L = \sqrt{2}$
20. $f(x) = \sqrt[3]{x}$, $a = 3$, $L = \sqrt[3]{3}$
21. $f(x) = 1/(x + 1)$, $a = 2$, $L = \frac{1}{3}$
22. $f(x) = 1/(x + 2)$, $a = -3$, $L = -1$

7. THEOREMS ON LIMITS

In Sections 1 through 5 we discussed limits without any attempt at rigorous or exact mathematical statements. In the process of doing this, we performed all sorts of algebraic manipulations. The skeptical student realizes that each of these needs justification, even though on the surface many appear obvious. The first step in such a justification requires the precise definition of limit which we have just given in Section 6. The next step should be the statements and the proofs of the theorems which allow us to manipulate limits. However, with one exception we shall restrict ourselves to statements of the theorems only, since many of the proofs are beyond the scope of a course in elementary calculus. The statements themselves will help the student understand the kind of work that has to be done.

As an example of the kind of statement that should be proved even though it is completely obvious in character, we give the following theorem, which says that a function cannot approach two different limits at the same time.

Theorem 1 (Uniqueness of Limits). *Suppose that $f(x) \to L_1$ as $x \to a$, and $f(x) \to L_2$ as $x \to a$. Then $L_1 = L_2$.*

This theorem is so simple that we can easily give a proof.

*Proof.** We shall suppose $L_1 \neq L_2$ and show it to be impossible. If $L_1 \neq L_2$, we let $\epsilon = \frac{1}{2}|L_1 - L_2|$ and ϵ will be *positive*. Since

$$\lim_{x \to a} f(x) = L_1,$$

we know from the definition of limit that there is a δ such that

$$|f(x) - L_1| < \epsilon \quad \text{if} \quad 0 < |x - a| < \delta.$$

But if, in addition,

$$\lim_{x \to a} f(x) = L_2,$$

we also know from the definition that there is a δ' (perhaps different from the δ) such that

$$|f(x) - L_2| < \epsilon \quad \text{when} \quad 0 < |x - a| < \delta'.$$

Then either $\delta \leq \delta'$ or $\delta' \leq \delta$. For convenience, suppose that $\delta \leq \delta'$. We now use the trick of writing a simple expression in a complicated way:

$$L_1 - L_2 = L_1 - f(x) + f(x) - L_2.$$

Therefore

$$|L_1 - L_2| = |(L_1 - f(x)) + (f(x) - L_2)| \leq |L_1 - f(x)| + |f(x) - L_2|,$$

the inequality arising from the fact—which we recall—that

$$|a + b| \leq |a| + |b|.$$

---

*This proof may be skipped without inconvenience. It depends on the definition of limit given in Section 6 of this chapter.

Let us now divide by 2 to obtain
$$\tfrac{1}{2}|L_1 - L_2| \leq \tfrac{1}{2}|L_1 - f(x)| + \tfrac{1}{2}|f(x) - L_2|.$$
But $|f(x) - L_2| < \epsilon$, and $|f(x) - L_1| < \epsilon$, and so
$$\tfrac{1}{2}|L_1 - L_2| < \tfrac{1}{2}\epsilon + \tfrac{1}{2}\epsilon = \epsilon.$$
However, we defined $\epsilon = \tfrac{1}{2}|L_1 - L_2|$, and we now have the absurd assertion that $\epsilon < \epsilon$. Therefore the assumption that $L_1 \neq L_2$ must be false. ◀

We have just seen that if f is a function and a is a number, *there is at most one number L such that $f(x) \to L$ as $x \to a$.* When this number exists we denote it by the symbol
$$\lim_{x \to a} f(x)$$
and we write
$$\lim_{x \to a} f(x) = L.$$
If no such number L exists, the symbol is not defined.

Theorem 2 (Limit of a Constant). *If c is a constant and $f(x) = c$ for all values of x, then for any number a*
$$\lim_{x \to a} f(x) = c.$$

Theorem 2 is established by applying the definition of limit to the particular function $f(x) = c$. Geometrically, the function $f(x) = c$ represents a line parallel to the x axis and c units from it.

Theorem 3 (Obvious Limit). *If a is a real number and $f(x) = x$ for all x, then*
$$\lim_{x \to a} f(x) = a.$$

This self-evident theorem is logically necessary, since we defined only the expression "$f(x) \to L$ as $x \to a$." We can prove it by applying the definition of limit and taking $\delta = \epsilon$. Theorem 3 says that for the special function $f: x \to x$, the limit is always obtained by *direct substitution* of the value a for x.

Theorem 4 (Limit of Equal Functions). *Suppose that there is a number $h > 0$ such that $f(x) = g(x)$ for all x for which $0 < |x - a| < h$. Suppose also that*
$$\lim_{x \to a} g(x) = L.$$
Then
$$\lim_{x \to a} f(x) = L.$$

This theorem is useful whenever the limit of $f(x)$ cannot be found by direct substitution but where a "simplified function" g can be obtained with the property that $g(x) = f(x)$ for $0 < |x - a| < h$ and such that the limit of g can be found.

Examples 2 and 3 of Section 1 make use of Theorem 4. In Example 2, for instance, the functions are

$$f(x) = \frac{x-4}{3(\sqrt{x}-2)} \quad \text{and} \quad g(x) = \frac{\sqrt{x}+2}{3}.$$

By rationalizing the denominator we see that these functions are identical except when $x = 4$. Therefore

$$\lim_{x \to 4} f(x) = \lim_{x \to 4} g(x) = \tfrac{4}{3}.$$

Theorem 5 (Limit of a Sum). *If f and g are two functions, with*

$$\lim_{x \to a} f(x) = L_1 \quad \text{and} \quad \lim_{x \to a} g(x) = L_2,$$

then

$$\lim_{x \to a} (f(x) + g(x)) = L_1 + L_2.$$

The hypothesis states that $|f(x) - L_1|$ can be made "small" if x is "close to" a; the same is true about $|g(x) - L_2|$. The conclusion asserts that $|f(x) + g(x) - L_1 - L_2|$ can be made "small" if x is "close to" a.

Once Theorem 5 is established, we can use it over and over to add the limits of any number of functions. For example, if

$$\lim_{x \to a} f(x) = L_1, \quad \lim_{x \to a} g(x) = L_2, \quad \text{and} \quad \lim_{x \to a} h(x) = L_3,$$

then

$$\lim_{x \to a} (f(x) + g(x) + h(x)) = L_1 + L_2 + L_3.$$

To demonstrate this, we first apply Theorem 5 to f and g, designating $f(x) + g(x) = F(x)$. Then we apply Theorem 5 again to $F(x)$ and $h(x)$. This technique of combining and using the same theorem over and over occurs frequently in the study of limits. The limit of the sum of any (finite) number of functions is the sum of the limits of each of the functions.

Theorem 6 (Limit of a Product). *If f and g are two functions, with*

$$\lim_{x \to a} f(x) = L_1 \quad \text{and} \quad \lim_{x \to a} g(x) = L_2,$$

then

$$\lim_{x \to a} [f(x) \cdot g(x)] = L_1 \cdot L_2.$$

In analogy with the discussion following Theorem 5, we note that the limit of the product of any number of functions is the product of the limits.

Example 1. Given

$$\lim_{x \to a} f(x) = L_1, \quad \lim_{x \to a} g(x) = L_2, \quad \text{and} \quad \lim_{x \to a} h(x) = L_3,$$

find the value of
$$\lim_{x \to a} [f(x) \cdot g(x) + h(x)].$$

Solution. We define $F(x) = f(x) \cdot g(x)$. From Theorem 6 we know that
$$\lim_{x \to a} f(x) \cdot g(x) = L_1 \cdot L_2.$$
That is,
$$\lim_{x \to a} F(x) = L_1 \cdot L_2.$$
Applying Theorem 5 to $F(x) + h(x)$, we now find that
$$\lim_{x \to a} (F(x) + h(x)) = L_1 \cdot L_2 + L_3. \qquad \triangleleft$$

Theorem 7 (Limit of a Quotient). *If f and g are two functions, with*
$$\lim_{x \to a} f(x) = L_1, \qquad \lim_{x \to a} g(x) = L_2, \qquad \text{and} \qquad L_2 \neq 0,$$
then
$$\lim_{x \to a} \frac{f(x)}{g(x)} = \frac{L_1}{L_2}.$$

It is necessary to assume that $L_2 \neq 0$ if the expression L_1/L_2 is to have a meaning.

Example 2. Given $F(x) = x^2$, show that
$$\lim_{x \to a} F(x) = a^2.$$

Solution. It is possible to obtain this result by appealing to the definition of limit. However, we can also prove the statement simply by correct application of the theorems on limits. Using Theorem 3, we have
$$\lim_{x \to a} x = a.$$
Let $f(x) = x$ and $g(x) = x$. Then
$$\lim_{x \to a} f(x) = a \qquad \text{and} \qquad \lim_{x \to a} g(x) = a.$$
We can apply Theorem 6, with $L_1 = a$, $L_2 = a$, to get
$$\lim_{x \to a} f(x) \cdot g(x) = a \cdot a = a^2,$$
which says that
$$\lim_{x \to a} F(x) = a^2. \qquad \triangleleft$$

Example 3. Given $F(x) = x^2/(3x - 2)$, show that
$$\lim_{x \to a} \frac{x^2}{3x - 2} = \frac{a^2}{3a - 2} \qquad \text{if} \qquad a \neq \tfrac{2}{3}.$$

Solution. From Example 2 we know that
$$\lim_{x \to a} x^2 = a^2,$$
and from Theorems 2 and 3, we know that
$$\lim_{x \to a} 3 = 3, \quad \lim_{x \to a} (-2) = -2, \quad \lim_{x \to a} x = a.$$
From Theorem 6 with $f(x) = 3$ and $g(x) = x$, we see that
$$\lim_{x \to a} 3x = 3a.$$
An application of Theorems 5 and 2 yields
$$\lim_{x \to a} (3x - 2) = \lim_{x \to a} [3x + (-2)] = 3a + (-2) = 3a - 2.$$

Finally, using Theorem 7 with $f(x) = x^2$, $g(x) = 3x - 2$, $L_1 = a^2$, and $L_2 = 3a - 2$, we conclude that
$$\lim_{x \to a} \frac{x^2}{3x - 2} = \frac{a^2}{3a - 2}. \quad \blacktriangleleft$$

In working the problems at the end of this section the student should give the reason for each step. An abbreviation for the *name*, rather than the number of the theorem being used, helps in the development of a clear understanding of the processes. We illustrate this procedure in the next two examples.

Example 4. Find the value of
$$\lim_{x \to 2} \frac{x^3 + 3}{2x^2 + 5}$$
and justify each step.

Solution

Step 1: $\lim_{x \to 2} 2 = 2, \quad \lim_{x \to 2} 3 = 3, \quad \lim_{x \to 2} 5 = 5$ (lim. const.)

Step 2: $\lim_{x \to 2} x = 2$ (obv. lim.)

Step 3: $\lim_{x \to 2} 2x^2 = \lim_{x \to 2} (2 \cdot x \cdot x) = 2 \cdot 2 \cdot 2 = 8,$

$\lim_{x \to 2} x^3 = \lim_{x \to 2} (x \cdot x \cdot x) = 8$ (lim. prod.)

Step 4: $\lim_{x \to 2} (x^3 + 3) = 8 + 3 = 11,$

$\lim_{x \to 2} (2x^2 + 5) = 8 + 5 = 13$ (lim. sum)

Step 5: $\lim_{x \to 2} \frac{x^3 + 3}{2x^2 + 5} = \frac{11}{13}$ (lim. quot.)

$\blacktriangleleft$

Theorem 8 (Limit of a Composite Function). *Suppose f and g are functions, a and b are numbers, f(b) is defined, and*

$$\lim_{x \to b} f(x) = f(b) \quad \text{and} \quad \lim_{x \to a} g(x) = b.$$

Then

$$\lim_{x \to a} f[g(x)] = f(b).$$

Theorem 9. *If n is a positive integer and a > 0, then*

$$\lim_{x \to a} \sqrt[n]{x} = \sqrt[n]{a}.$$

The next result can be obtained by combining Theorems 8 and 9.

Theorem 10. *If n is a positive integer, L > 0, and*

$$\lim_{x \to a} f(x) = L, \quad \text{then} \quad \lim_{x \to a} \sqrt[n]{f(x)} = \sqrt[n]{L}.$$

Example 5. Evaluate

$$\lim_{x \to 2} \sqrt{\frac{x^4 - 16}{x^3 - 8}}.$$

Solution. Straight substitution shows that the function is undefined when $x = 2$. If we denote the quantity under the radical by $f(x)$, we find that for $x \neq 2$:

$$f(x) = \frac{(x - 2)(x^3 + 2x^2 + 4x + 8)}{(x - 2)(x^2 + 2x + 4)}.$$

Setting

$$g(x) = \frac{x^3 + 2x^2 + 4x + 8}{x^2 + 2x + 4},$$

we observe that $g(x)$ is defined for all values of x. Note that $f(x) = g(x)$ except when $x = 2$.

Step 1: $\lim_{x \to 2} 2 = 2,$ $\quad \lim_{x \to 2} 4 = 4,$ $\quad \lim_{x \to 2} 8 = 8$ (lim. const.)

Step 2: $\lim_{x \to 2} x = 2$ (obv. lim.)

Step 3: $\lim_{x \to 2} x^3 = 8,$ $\quad \lim_{x \to 2} 2x^2 = 8,$ $\quad \lim_{x \to 2} 4x = 8,$

$\lim_{x \to 2} x^2 = 4,$ $\quad \lim_{x \to 2} 2x = 4$ (lim. prod.)

Step 4: $\lim_{x \to 2} (x^3 + 2x^2 + 4x + 8) = 32,$

$\lim_{x \to 2} (x^2 + 2x + 4) = 12$ (lim. sum)

Step 5: $\lim_{x \to 2} \dfrac{x^3 + 2x^2 + 4x + 8}{x^2 + 2x + 4} = \dfrac{32}{12}$ (lim. quot.)

Step 6: $\lim_{x \to 2} g(x) = \lim_{x \to 2} f(x) = \dfrac{32}{12} = \dfrac{8}{3}$ (lim. = fcts.)

Step 7: $\lim_{x \to 2} \sqrt{\dfrac{x^4 - 16}{x^3 - 8}} = \sqrt{\dfrac{8}{3}}$ (lim. $\sqrt[n]{f(x)}$)

◀

PROBLEMS

In Problems 1 through 20, evaluate the limits by following the methods given in Examples 4 and 5. Give the reason for each step as in those examples.

1. $\lim_{x \to 2} (x^2 + 4x - 3)$

2. $\lim_{x \to 3} (x^3 + 2x^2 - 7x + 4)$

3. $\lim_{x \to -1} \dfrac{2x + 1}{x^2 + 3x + 4}$

4. $\lim_{x \to 1} \dfrac{x^2 + 6x - 5}{x^3 + 2x + 7}$

5. $\lim_{x \to 2} \dfrac{x^3 - 8}{x - 2}$

6. $\lim_{x \to -2} \dfrac{x^3 + 8}{x + 2}$

7. $\lim_{x \to -2} \dfrac{x^2 + x - 2}{x^2 - 4}$ $\dfrac{(x-1)(x+2)}{(x-2)(x+2)}$

8. $\lim_{x \to -2} \dfrac{x^2 - 4}{x^3 + 8}$

9. $\lim_{t \to 2} \sqrt{\dfrac{2t + 5}{3t - 2}}$

10. $\lim_{r \to 1} \sqrt{\dfrac{2r^2 + 3r - 1}{r^2 + 1}}$

11. $\lim_{y \to 2} \sqrt{\dfrac{y^2 - 4}{y^2 - 3y + 2}}$

12. $\lim_{x \to 3} \sqrt[3]{\dfrac{x^3 - 27}{x^2 - 2x - 3}}$

13. $\lim_{h \to 2} \sqrt{\dfrac{h^3 - 8}{h^2 - 4}}$

14. $\lim_{h \to 0} \dfrac{\sqrt{1 + h} - 1}{h}$

15. $\lim_{h \to 0} \dfrac{\sqrt{x + h} - \sqrt{x}}{h}$, $x > 0$

16. $\lim_{h \to 0} \dfrac{1}{h}\left(\dfrac{1}{x + h} - \dfrac{1}{x}\right)$

17. $\lim_{h \to 0} \dfrac{1}{h}\left(\dfrac{1}{\sqrt{1 + h}} - 1\right)$

18. $\lim_{h \to 0} \dfrac{(1 + h)^{3/2} - 1}{h}$

19. $\lim_{x \to 2} \dfrac{x^5 - 32}{x - 2}$

20. $\lim_{x \to 3} \dfrac{x^6 - 729}{x + 3}$

*21. Prove Theorem 2 by employing the definition of limit.

*22. Prove Theorem 3 by employing the definition of limit.

23. Given

$\lim_{x \to a} f(x) = L_1, \quad \lim_{x \to a} g(x) = L_2, \quad \lim_{x \to a} h(x) = L_3, \quad \lim_{x \to a} p(x) = L_4,$

state the theorems which justify the following statements:

a) $\lim_{x \to a} \dfrac{f(x) + g(x)}{h(x)} = \dfrac{L_1 + L_2}{L_3}$, if $L_3 \neq 0$.

b) $\lim_{x \to a} \dfrac{f(x)g(x) - h(x)}{g(x) + p(x)} = \dfrac{L_1 L_2 - L_3}{L_2 + L_4}$, if $L_2 + L_4 \neq 0$.

c) $\lim_{x \to a} \left(\dfrac{f(x) - g(x) + h(x)p(x)}{g(x) + h(x)} \right)^{2/3} = \left(\dfrac{L_1 - L_2 + L_3 L_4}{L_2 + L_3} \right)^{2/3}$, if $L_2 + L_3 \neq 0$.

24. Use the definition of limit to show that if a is any number then
$$\lim_{x \to a} x^2 = a^2.$$

25. Let $P(x) = a_n x^n + a_{n-1} x^{n-1} + \cdots + a_1 x + a_0$ be any polynomial. Show that
$$\lim_{x \to b} P(x) = P(b).$$

26. a) Suppose that $\lim_{x \to a} f(x) = L$. Show that $\lim_{x \to a} |f(x)| = |L|$.
    b) Suppose that $\lim_{x \to a} |f(x)| = |L|$. Show that it is not necessarily true that $\lim_{x \to a} f(x) = L$. Give an example.

8. CONTINUITY. ONE-SIDED LIMITS

In Section 6 we analyzed the meaning of
$$\lim_{x \to a} f(x) = L.$$

In doing so we made a point of ignoring the actual value of the function f when $x = a$. In fact, for many expressions the function was not even defined at $x = a$. Suppose we have a function f which is defined at $x = a$, and suppose the limit L, which f approaches when x tends to a, is the value of f when x is a—just the quantity we call $f(a)$. When this happens, we say that the function is **continuous** at $x = a$.

Definition. "The function f is continuous at the number a" means that

i) $f(a)$ is defined, and

ii) $\lim_{x \to a} f(x) = f(a)$.

When a function is not continuous, we say that it is **discontinuous at** a. Most, but not all, of the functions we have been studying are continuous everywhere. For example, the functions

$$f(x) = 3x - 2, \quad g(x) = 5 - 2x + 4x^2, \quad \text{and} \quad h : x \to \dfrac{3x + 1}{2 + x^2}$$

are defined for all values of x and are continuous for all values of x. On the other hand, we shall show that the function

$$F(x) = \begin{cases} 2x + 1, & -\infty < x \leq 1 \\ \tfrac{1}{2} x^2 - 3, & 1 < x < \infty \end{cases}$$

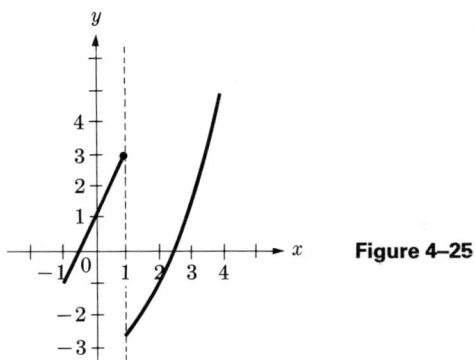

Figure 4–25

is discontinuous at $x = 1$. The graph of part of F is shown in Fig. 4–25. We define the functions

$$G(x) = 2x + 1, \qquad H(x) = \tfrac{1}{2}x^2 - 3,$$

in each case the domain being all of R_1; that is, G and H are given by the above formulas for all values of x. Now we apply Theorem 4 on the limit of equal functions. We observe that F and G are identical for all values of x in the interval $-\infty < x < 1$; therefore the limit as x tends to 1 from the left (that is, as x tends to 1 through values less than 1) of the functions F and G must be the same. Since G is continuous everywhere, we see that $F(1) = 2 \cdot 1 + 1 = 3$ is equal to the limit of $G(x)$ as x tends to 1 from the left. We employ a similar argument for F and H; these functions are identical for $1 < x < \infty$. Also, H is continuous for all values of x. Therefore the limit as x tends to 1 from the right (that is, as x tends to 1 through values larger than 1) of the functions F and H must be the same. We have $H(1) = -\tfrac{5}{2}$ and so $F(x)$ tends to $-\tfrac{5}{2}$ as x tends to 1 through values larger than 1. These statements are intuitively clear in the graph shown in Fig. 4–25. The definition of continuity for F is not satisfied at $x = 1$; this assertion is justified in Theorem 11 on page 97.

In the definition of limit, when we state that

$$f(x) \to L \qquad \text{as} \qquad x \to a,$$

we must consider values of x larger than a and values smaller than a as x tends to a. However, in the above discussion on the discontinuity of F, we saw that it is important to consider the limit of F as the number 1 is approached through values entirely on one side of 1. This leads to the following definition of "one-sided limit." This definition parallels word-for-word the definition of limit given in Section 6, except that x tends to a through values on one side of a.

Definition. *Given a function f and numbers a and L, we say that $f(x)$* **tends to L as a limit as x tends to a from the right** *if for each positive number ϵ there is a positive number δ such that*

$$|f(x) - L| < \epsilon \qquad \text{whenever} \qquad 0 < |x - a| < \delta,$$

and furthermore, x is larger than a. In abbreviated notation, we write

$$f(x) \to L \quad \text{as} \quad x \to a^+.$$

The plus sign after the letter a indicates that x tends to a through values larger than a. We also say that L is the **one-sided limit from the right of** f.

The definition of **one-sided limit from the left** is entirely analogous to that above, with the change that x is smaller than a instead of larger than a. We use the notation

$$f(x) \to L \quad \text{as} \quad x \to a^-,$$

the minus sign after the letter a indicating that x tends to a through values smaller than a. Additional notations for one-sided limits are

$$\lim_{x \to a^+} f(x) = L, \quad \lim_{x \to a^-} f(x) = L.$$

All the theorems on limits stated in Section 7 carry over to one-sided limits with the exception of Theorem 8 on composite functions. In that theorem we must assume that the limit of f exists, although one-sided limits of g are allowed. In addition, we have the following extension of Theorem 9.

Theorem 9'. *If n is a positive integer, then*

$$\lim_{x \to 0^+} \sqrt[n]{x} = 0.$$

We note that the restriction to one-sided limits is necessary in Theorem 9' since $\sqrt[n]{x}$ is not defined for negative values of x when n is an even integer. An extended theorem similar to Theorem 10 holds for

$$\lim_{x \to a^+} \sqrt[n]{f(x)} \quad \text{if} \quad f(x) \to 0^+ \quad \text{as} \quad x \to a^+.$$

The next theorem is an immediate consequence of the definitions of one-sided limits.

Theorem 11. *If f is a function and a and L are numbers, then*

$$\lim_{x \to a} f(x) = L$$

if and only if

$$\lim_{x \to a^+} f(x) = L \quad \text{and} \quad \lim_{x \to a^-} f(x) = L.$$

Theorem 11 asserts that the limit of f exists if and only if *both* one-sided limits exist *and they have the same value*. For this reason, we sometimes say that the ordinary limit is the **two-sided limit**.

Using Theorem 11, we can justify the conclusion that the function F given on page 95 is not continuous at $x = 1$. Both one-sided limits of F exist, but they have different values. Therefore the two-sided limit (ordinary limit) does not exist and the function cannot be continuous.

Example 1. Find the value of the one-sided limit

$$\lim_{x \to 3^-} \frac{9 - x^2}{\sqrt{3 + 2x - x^2}}$$

and give a reason for each step.

Solution. We factor the expression, getting

$$\frac{9 - x^2}{\sqrt{3 + 2x - x^2}} = \frac{(3 + x)(3 - x)}{\sqrt{(1 + x)(3 - x)}}$$

$$= \frac{(3 + x)\sqrt{3 - x}}{\sqrt{1 + x}} \quad \text{for} \quad -1 < x < 3.$$

From the theorems on limit of a constant and obvious limit, applied to one-sided limits, we have

$$\lim_{x \to 3^-} 1 = 1, \qquad \lim_{x \to 3^-} 3 = 3, \qquad \lim_{x \to 3^-} x = 3.$$

Therefore, from the limit of a sum (one-sided case), we find

$$\lim_{x \to 3^-} (3 + x) = 6, \qquad \lim_{x \to 3^-} (1 + x) = 4, \qquad \lim_{x \to 3^-} (3 - x) = 0.$$

Also,

$$\lim_{x \to 3^-} \sqrt{3 - x} = 0, \qquad \lim_{x \to 3^-} \sqrt{1 + x} = 2, \qquad \text{(lim. } \sqrt[n]{f(x)}\text{)}$$

$$\lim_{x \to 3^-} \frac{(3 + x)\sqrt{3 - x}}{\sqrt{1 + x}} = \frac{6 \cdot 0}{2} = 0, \qquad \text{(lim. prod.; lim. quot.)}$$

$$\lim_{x \to 3^-} \frac{9 - x^2}{\sqrt{3 + 2x - x^2}} = 0. \qquad \text{(lim. equal func.)}$$

◀

Example 2. For what values is the function

$$f : x \to \frac{x^2 - 9}{x + 3}$$

continuous?

Solution. This function is defined for all numbers except $x = -3$. Straight substitution at -3 yields $\frac{0}{0}$ and f is undefined. Therefore f is not continuous at -3. For all other numbers in the domain of f, the function has the same value as the function $g(x) = x - 3$, and so is continuous. In fact,

$$\lim_{x \to -3} f(x) = -6. \qquad ◀$$

Remark. The important distinction between the function f in the above example and the function $g : x \to x - 3$ is found in the *domain* of these functions. The domain of g is all of R_1, while the domain of f is all of R_1 with the exception of -3. The functions coincide wherever they are both defined, but the functions are not identical. These fine points occur frequently in mathematics and they cannot be overlooked.

Example 3. Find the values of x where the function

$$h(x) = \begin{cases} 2x^2, & -1 \leq x < 1 \\ 3 - x, & 1 \leq x < 2 \end{cases}$$

is continuous.

Solution. The domain of h is the half-open interval $[-1, 2)$. Clearly h is continuous except possibly at -1 and at 1. Taking one-sided limits, we find that

$$\lim_{x \to 1^-} 2x^2 = 2 \quad \text{and} \quad \lim_{x \to 1^+} (3 - x) = 2.$$

Since both one-sided limits are equal and since $h(1) = 2$, we conclude that h is continuous at $x = 1$. (See Fig. 4-26.)

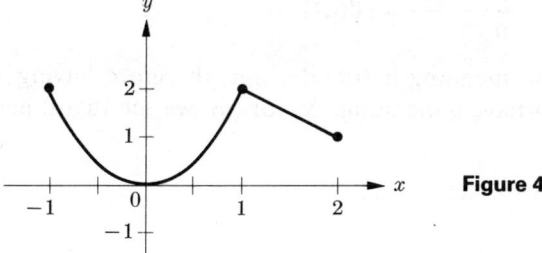

Figure 4-26

At $x = -1$, the function h has the value 2. Also, the limit as x tends to -1 from the right exists and the limiting value is 2. Under these circumstances, we say that h is **continuous on the right at $x = -1$**. ◀

Remark. If a function f is defined on a closed interval $[a, b]$, then continuity of f may be defined at every interior point. The function f is **continuous on the right at the endpoint a** if and only if

$$\lim_{x \to a^+} f(x) = f(a).$$

Similarly, f is **continuous on the left at b** if and only if

$$\lim_{x \to b^-} f(x) = f(b).$$

A function f is **continuous on the closed interval** $[a, b]$ if and only if f is continuous at every interior point, continuous on the right at a, and continuous on the left at b.

To say that a function has a derivative at the value a means that a tangent to the curve can be drawn at the point $(a, f(a))$. In fact, the value of the derivative,

defined as
$$\lim_{h \to 0} \frac{f(a+h) - f(a)}{h},$$
is simply the slope of this tangent line. As we have seen in Example 3, a continuous function may have corners, and it is hard to imagine how a tangent to a curve can be constructed at a corner; in general, it cannot. Not every continuous function possesses a tangent at each point. If a function has a derivative, however, it is continuous, as we shall prove.

Theorem 12. *If the function f possesses a derivative at the value a, it is continuous at a.*

Proof. We recall that two things must be shown:

i) $f(a)$ is defined   and   ii) $\lim_{x \to a} f(x) = f(a)$.

If a function has a derivative at a, then
$$\lim_{h \to 0} \frac{f(a+h) - f(a)}{h} = f'(a)$$
exists. This statement would have no meaning if $f(a)$ did not; therefore having a derivative at a means that $f(a)$ must have a meaning. As for (ii), we see that it has the same meaning as
$$\lim_{h \to 0} f(a+h) = f(a).$$
We can write
$$f(a+h) - f(a) = \left(\frac{f(a+h) - f(a)}{h} \right) \cdot h.$$

On the right, the term in parentheses approaches $f'(a)$, a finite number, as h tends to zero. The second part, namely h, tends to zero. Therefore, as h tends to zero, the right side—according to the theorem on the limit of a product—tends to zero. We conclude that
$$f(a+h) - f(a) \to 0 \quad \text{as} \quad h \to 0,$$
and (ii) is established; $f(x)$ is continuous at a.

PROBLEMS

In each of Problems 1 through 8, evaluate the one-sided limit, giving a reason for each step.

1. $\lim_{x \to 1^+} \dfrac{x - 1}{\sqrt{x^2 - 1}}$

2. $\lim_{x \to 1^+} (x + \sqrt{x^2 - 1})$

3. $\lim_{x \to 2^-} \dfrac{4 - x^2}{\sqrt{2 + x - x^2}}$

4. $\lim_{x \to 2^+} \dfrac{x^2 - 3x + 2}{\sqrt{x^2 - 4}}$

5. $\lim_{x \to 0^+} x\sqrt{1 + \dfrac{1}{x^2}}$

6. $\lim_{x \to 0^-} x\sqrt{1 + \dfrac{1}{x^2}}$

7. $\lim_{x \to 4^-} (x - \sqrt{16 - x^2})$

8. $\lim_{x \to 2^-} \dfrac{\sqrt{4 - x^2}}{\sqrt{6 - 5x + x^2}}$

9. Evaluate
$$\lim_{x \to 1^-} |x - 1| \quad \text{and} \quad \lim_{x \to 1^+} |x - 1|.$$

Does (the two-sided limit) $\lim_{x \to 1} |x - 1|$ exist? Draw the graph.

10. Evaluate
$$\lim_{x \to 0^+} \dfrac{x}{|x|} \quad \text{and} \quad \lim_{x \to 0^-} \dfrac{x}{|x|}.$$

Does
$$\lim_{x \to 0} \dfrac{x}{|x|}$$

exist? Give a reason for your answer.

In each of Problems 11 through 25, a function is defined in a certain domain. Determine those values of x at which the function is continuous. Sketch the graph.

11. $f(x) = \dfrac{1}{x^2 + 3}$ for $-4 < x < 7$

12. $f(x) = \dfrac{x + 3}{x^2 - 2x - 15}$ for $-7 < x < 4,\ x \neq -3$
    $f(-3) = 2$

13. $f(x) = \dfrac{x^2 - 4}{x^2 - 8x + 12},\quad 0 < x < 5,\ x \neq 2$
    $f(2) = 1$

14. $f(x) = \dfrac{x + 2}{x^2 - 2x - 8}$ for $-1 < x < 3$

15. $f(x) = \dfrac{x^3 - 1}{x^2 + x - 2},\quad 0 < x < 2,\ x \neq 1$
    $f(1) = 1$

16. $f: x \to \begin{cases} x - 4, & -1 < x \leq 2 \\ x^2 - 6, & 2 < x < 5 \end{cases}$

17. $f: x \to \begin{cases} x^2 - 6, & -\infty < x < -1 \\ -5, & -1 \leq x \leq 10 \\ x - 15, & 10 < x < \infty \end{cases}$

18. $f(x) = \begin{cases} \dfrac{x^2 - 1}{x^4 - 1}, & -1 < x < 2,\ x \neq 1 \\ x^2 + 3x - 2, & 2 \leq x < 5 \end{cases}$
    $f(1) = \tfrac{1}{2}$

19. $f(x) = |x + 3|$ for all values of x

20. $f: x \to |2x - 5|$ for all values of x

21. $f(x) = \dfrac{x-2}{|x-2|}$ for all x except $x = 2$

$f(2) = 0$

22. $f(x) = \dfrac{3 + |x-2|}{1 + x^2}$ for all values of x

23. $f(x) = \dfrac{x^2 - x - 6}{x - 3}$ for all values of x except $x = 3$

$f(3) = 5$

*24. $f(x) = \sin \dfrac{\pi}{x}$ for $-1 < x < 1$, $x \neq 0$

$f(0) = 0$

*25. $f(x) = x \sin \dfrac{\pi}{x}$ for $-1 < x < 1$, $x \neq 0$

$f(0) = 0$

9. LIMITS AT INFINITY; INFINITE LIMITS

The graph of the function $f(x) = x^2/(1 + x^2)$ is shown in Fig. 4–27.

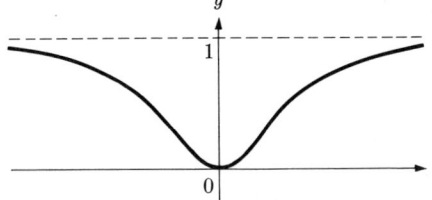

Figure 4–27

The values of this function are always less than 1; in particular, we note the following:

$$f(1) = \dfrac{1}{2}, \quad f(10) = \dfrac{100}{101}, \quad f(100) = \dfrac{10{,}000}{10{,}001}, \quad f(1000) = \dfrac{10^6}{10^6 + 1}, \ldots$$

As x gets larger and larger, the graph of f gets closer and closer to the line $y = 1$, as shown in Fig. 4–27. We use the term "x tends to infinity" when referring to values which increase or decrease without bound. If x is becoming larger through positive values only, we write $x \to +\infty$; if through negative values only, we write $x \to -\infty$. The symbol $x \to \infty$ means that $|x|$ increases without bound.† In the above example, we would write

$$f(x) \to 1 \quad \text{as} \quad x \to \infty.$$

More generally, we formulate the following definition.

---

† Some texts use the symbols ∞, $-\infty$, $+\infty$ corresponding to our use in this section of $+\infty$, $-\infty$, ∞, respectively.

Definition. We say that $f(x) \to c$ as $x \to \infty$, if for each $\epsilon > 0$ there is a number $A > 0$, such that $|f(x) - c| < \epsilon$ for all x for which $|x| > A$.

If we want to define
$$f(x) \to c \quad \text{as} \quad x \to +\infty$$
(i.e., $f(x)$ tends to c as x tends to infinity through positive values only), the only change in the definition would be to write $x > A$ instead of $|x| > A$. Further, the definition of
$$f(x) \to c \quad \text{as} \quad x \to -\infty$$
would require $x < -A$ instead of $|x| > A$.

The theorems on uniqueness of limit and on limit of a constant, equal functions, sum, product, and quotient remain unchanged. In place of the theorem on limits of functions of the form $f(x) = x$, we have the following theorem on obvious limits.

Theorem 13. $\lim_{x \to \infty} \frac{1}{x} = 0$, $\quad \lim_{x \to +\infty} \frac{1}{x} = 0$, $\quad \lim_{x \to -\infty} \frac{1}{x} = 0$.

Example 1. Evaluate
$$\lim_{x \to +\infty} \frac{3x - 2}{5x + 4},$$
giving the reason for each step.

Solution. We first employ a standard trick in working with infinite limits by dividing each term in the numerator and denominator by x:
$$\frac{3x - 2}{5x + 4} = \frac{3 - (2/x)}{5 + (4/x)}, \quad x \neq 0, \; -\frac{4}{5}.$$
For the numerator, we find that
$$\lim_{x \to +\infty} \left(3 - \frac{2}{x}\right) = \lim_{x \to +\infty} 3 + \lim_{x \to +\infty} \left(-\frac{2}{x}\right),$$
since the limit of a sum is the sum of the limits. We now reason that
$$\lim_{x \to +\infty} 3 = 3 \quad \text{(using limit of a constant)}$$
and
$$\lim_{x \to +\infty} \left(-\frac{2}{x}\right) = -2 \lim_{x \to +\infty} \frac{1}{x} = 0 \quad \text{(using limit of a product and Theorem 13)}.$$
Therefore
$$\lim_{x \to +\infty} \left(3 - \frac{2}{x}\right) = 3.$$
In exactly the same way,
$$\lim_{x \to +\infty} \left(5 + \frac{4}{x}\right) = 5.$$

Now we can say that

$$\lim_{x \to +\infty} \frac{3 - (2/x)}{5 + (4/x)} = \frac{\lim_{x \to +\infty} (3 - (2/x))}{\lim_{x \to +\infty} (5 + (4/x))},$$

since the limit of a quotient is the quotient of the limits. We conclude that (limit of equal functions)

$$\lim_{x \to +\infty} \frac{3x - 2}{5x + 4} = \frac{3}{5}. \quad \triangleleft$$

Example 2. Evaluate

$$\lim_{x \to +\infty} \frac{\sqrt{x^2 - 1}}{2x + 1},$$

giving a reason for each step.

Solution. We write

$$\frac{\sqrt{x^2 - 1}}{2x + 1} = \frac{\sqrt{1 - (1/x^2)}}{2 + (1/x)}.$$

Since

$$\frac{1}{x^2} = \frac{1}{x} \cdot \frac{1}{x},$$

we have

$$\lim_{x \to +\infty} \frac{1}{x^2} = \lim_{x \to +\infty} \frac{1}{x} \cdot \lim_{x \to +\infty} \frac{1}{x} = 0,$$

using the limit of a product theorem and Theorem 13. Then

$$\lim_{x \to +\infty} \left(1 - \frac{1}{x^2}\right) = \lim_{x \to +\infty} 1 - \lim_{x \to +\infty} \frac{1}{x^2} = 1 \quad \text{(using limit of a sum).}$$

Further, by using Theorem 10, we find that

$$\lim_{x \to +\infty} \sqrt{1 - \frac{1}{x^2}} = 1.$$

We now see that

$$\lim_{x \to +\infty} \left(2 + \frac{1}{x}\right) = 2,$$

as in Example 1, and when we apply the quotient rule and the theorem on the limit of equal functions, we have

$$\lim_{x \to +\infty} \frac{\sqrt{x^2 - 1}}{2x + 1} = \frac{1}{2}. \quad \triangleleft$$

The graph of the function $f(x) = x/(x - 1)$ is shown in Fig. 4–28. We note the following values of this function:

$f(2) = 2, \quad f(\tfrac{3}{2}) = 3, \quad f(1.1) = 11, \quad f(1.01) = 101, \quad f(1.001) = 1,001, \ldots;$
$f(0) = 0, \quad f(\tfrac{1}{2}) = -1, \quad f(0.9) = -9, \quad f(0.99) = -99, \quad f(0.999) = -999, \ldots.$

4–9 LIMITS AT INFINITY; INFINITE LIMITS

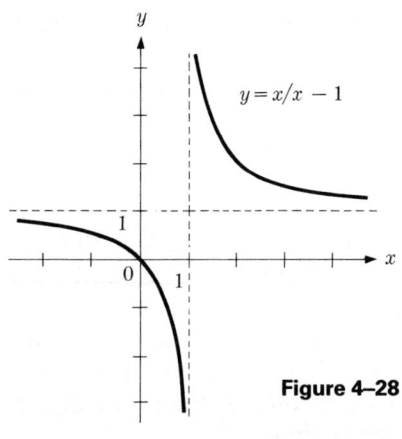

Figure 4–28

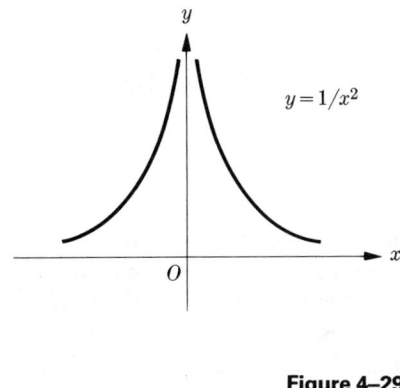

Figure 4–29

This type of behavior implies that the function grows without bound as x gets closer and closer to 1. We use the expression

$$f(x) \to \infty \quad \text{as} \quad x \to 1,$$

which we define in the following way.

Definition. *We say that $f(x)$ **becomes infinite as** $x \to a$, and write $f(x) \to \infty$ as $x \to a$, if for each number $A > 0$ there is a $\delta > 0$ such that $|f(x)| > A$, for all x for which $0 < |x - a| < \delta$.*

This definition does not state whether the function goes off to infinity "upward" or "downward." In other words, we may have $f(x) \to +\infty$ or $f(x) \to -\infty$ as special cases of $f(x) \to \infty$. For example, if $f(x) \to +\infty$ as $x \to a$, we say that "*f* **becomes positively infinite as** x **tends to** a." The only change required in the above definition is the replacement of $|f(x)| > A$ by $f(x) > A$. Figure 4–29 shows the function $g(x) = 1/x^2$, which becomes positively infinite as $x \to 0$.

Similarly, a function may become **negatively infinite as** x **tends to** a. In this case, we replace $|f(x)| > A$ by $f(x) < -A$ in the above definition.

We also consider functions which become infinite as a value a is approached from one side. In Fig. 4–28 it is clear that $f(x) \to +\infty$ as $x \to 1^+$ and $f(x) \to -\infty$ as $x \to 1^-$. We now give a precise definition of one-sided limit from the right.

Definition. *We say that $f(x)$ **becomes positively infinite as** x **tends to** a **from the right** if for each number A there is a $\delta > 0$ such that $f(x) > A$ for all x for which $0 < |x - a| < \delta$ and x is larger than a. We use the notations*

$$f(x) \to +\infty \quad \text{as} \quad x \to a^+ \quad \text{and} \quad \lim_{x \to a^+} f(x) = +\infty$$

for this one-sided limit.

Analogous definitions and notations are used for functions which become negatively infinite and for one-sided limits from the left.

Figure 4–29 shows a function which becomes positively infinite as x tends to zero both from the left and from the right. There is an analog to the "two-sided limit" theorem for functions which become infinite. We can conclude that $g(x) = 1/x^2$ becomes positively infinite as $x \to 0$.

Some functions have infinite limits from one side only. For example, the function

$$h(x) = \frac{1}{\sqrt{x-2}}$$

becomes positively infinite as $x \to 2^+$, but h is not defined for x less than 2. Therefore the limit from the left cannot exist and consequently there can be no (two-sided) limit as x tends to 2.

The rules for operating with limits are somewhat tricky when one of the quantities becomes infinite. We must remember that ∞ *is not a number* and cannot be treated as such. There are some simple facts about infinite limits which the reader will easily recognize. For example, in a case such as

$$f(x) \to +\infty \quad \text{as} \quad x \to a,$$
$$g(x) \to c \quad \text{as} \quad x \to a,$$

where c is any number, then

$$f(x) + g(x) \to +\infty \quad \text{as} \quad x \to a.$$

If $c \neq 0$, we have

$$f(x) \cdot g(x) \to +\infty \quad \text{when} \quad c > 0$$

and

$$f(x) \cdot g(x) \to -\infty \quad \text{when} \quad c < 0.$$

On the other hand, if $c = 0$, further investigation is necessary. To see this, we let $f(x) = 1/(x-3)^2$ and $g(x) = (x-3)^2$. Then $f(x) \to +\infty$ as $x \to 3$ and $g(x) \to 0$ as $x \to 3$. However,

$$f(x) \cdot g(x) \equiv 1,$$

and so, $f \cdot g \to 1$ as $x \to 3$. If we replace $f(x)$ by the function $A/(x-3)^2$, where A is any number, the product $f \cdot g$ tends to A. The reader can easily construct examples in which $f \cdot g \to +\infty$ and $f \cdot g \to 0$ as $x \to 3$. (See Problems 25 and 26 at the end of this section.)

In cases in which $f(x) \to +\infty$ and $g(x) \to -\infty$ as $x \to a$, nothing can be said about the behavior of

$$f(x) + g(x)$$

without a closer examination of the particular functions.

PROBLEMS

In each of Problems 1 through 15, evaluate the limit or show that the function tends to ∞, $+\infty$, or $-\infty$. Give a reason for each step.

1. $\lim\limits_{x \to \infty} \dfrac{2x+4}{3x+1}$

2. $\lim\limits_{x \to \infty} \dfrac{x^2 + 2x + 5}{2x^2 - 6x + 1}$

3. $\lim\limits_{x \to \infty} \dfrac{2x^2 + 7x + 5}{x^3 + 2x + 1}$

4. $\lim\limits_{x \to +\infty} \dfrac{\sqrt{2x^2 + 1}}{x + 3}$

5. $\lim_{x \to -\infty} \dfrac{\sqrt{x^2 + 4}}{x + 2}$

6. $\lim_{x \to +\infty} (\sqrt{x^2 + a^2} - x)$

7. $\lim_{x \to 1} \dfrac{x^2}{1 - x^2}$

8. $\lim_{x \to 2} \dfrac{x}{4 - x^2}$

9. $\lim_{x \to 0} \dfrac{\sqrt{1 + x}}{x}$

10. $\lim_{x \to \infty} \dfrac{x^2 + 1}{x}$

11. $\lim_{x \to 2^+} \dfrac{\sqrt{x^2 - 4}}{x - 2}$

12. $\lim_{x \to +\infty} (\sqrt{x^2 + 2x} - x)$

13. $\lim_{x \to -\infty} (\sqrt{x^2 + 2x} - x)$

14. $\lim_{x \to 2^-} \dfrac{\sqrt{4 - x^2}}{\sqrt{6 - 5x + x^2}}$

15. $\lim_{x \to 0} \dfrac{x + 1}{|x|}$

In each of Problems 16 through 19, determine those values of x at which the function is continuous. Sketch the graph.

16. $\begin{cases} f(x) = \dfrac{1}{x + 3} & \text{for } -7 < x < -1, \quad x \neq -3 \\ f(-3) = 2 \end{cases}$

17. $\begin{cases} f(x) = \dfrac{x + 3}{x^2 - 9}, & -6 < x < 6, \quad x \neq -3, 3 \\ f(-3) = 2, \quad f(3) = 1 \end{cases}$

18. $f(x) = \begin{cases} \dfrac{2x}{x^2 - 4} & \text{for } 0 < x < 2 \\ 3x - 5 & \text{for } 2 \leq x \leq 5 \\ x^2 + 6 & \text{for } 5 < x < 7 \end{cases}$

19. $\begin{cases} f : x \to \dfrac{x + 1}{|x - 2|} & \text{for all } x \text{ except } x = 2 \\ f(2) = 3 \end{cases}$

In each of Problems 20 through 23, sketch the graph of the given function. Then decide what limit, if any, is approached.

*20. $\lim_{x \to 0^+} f(x)$ where $f(x) = x \sin \dfrac{\pi}{x}$

*21. $\lim_{x \to +\infty} f(x)$ where $f(x) = \dfrac{1}{x} \sin \dfrac{\pi}{x}$

*22. $\lim_{x \to 0^+} f(x)$ where $f(x) = \dfrac{1}{x} \sin \dfrac{\pi}{x}$

*23. $\lim_{x \to 0} f(x)$ where $f(x) = \sin \dfrac{\pi}{x}$

*24. Given the function f defined by the conditions

$$f(x) = \begin{cases} 2x & \text{if } x \text{ is a rational number,} \\ 0 & \text{if } x \text{ is an irrational number.} \end{cases}$$

Find $\lim_{x \to 0^+} f(x)$.

25. Find functions f and g such that $f(x) \to +\infty$ and $g(x) \to 0$ as $x \to a$ and $f(x) \cdot g(x) \to +\infty$ as $x \to a$. Also find functions f and g such that $f(x) \to +\infty$, $g(x) \to 0$ as $x \to a$ and $f(x) \cdot g(x) \to 0$ as $x \to a$. Is it possible that $f(x) \to +\infty$, $g(x) \to 0$, and $f(x) \cdot g(x) \to A$ as $x \to a$, where A is negative? Justify the answer.

26. Find functions f and g such that $f(x) \to +\infty$ and $g(x) \to -\infty$ as $x \to a$ and $f(x) + g(x) \to +\infty$ as $x \to a$. Find functions f and g such that $f(x) \to +\infty$, $g(x) \to -\infty$ as $x \to a$ and $f(x) + g(x) \to -\infty$. If A is any given number, find functions f and g such that $f(x) \to +\infty$, $g(x) \to -\infty$ as $x \to a$, and $f(x) + g(x) \to A$ as $x \to a$.

*27. Let $P(x) = a_n x^n + a_{n-1} x^{n-1} + \cdots + a_1 x + a_0$ and $Q(x) = b_m x^m + b_{m-1} x^{m-1} + \cdots + b_1 x + b_0$ be two polynomials with $a_n \neq 0$ and $b_m \neq 0$. Find the value of

$$\lim_{x \to +\infty} \frac{P(x)}{Q(x)}$$

by considering three cases: (i) $m > n$, (ii) $m = n$, (iii) $m < n$.

*28. Let P and Q be defined as in Problem 27. Find the value of

$$\lim_{x \to -\infty} \frac{P(x)}{Q(x)}.$$

*29. Suppose that $P(x)$ and $Q(x)$ in Problem 27 are written in the form $P(x) = a_n(x - c_1)(x - c_2)(x - c_3) \cdots (x - c_n)$, $Q(x) = b_m(x - d_1)(x - d_2) \cdots (x - d_m)$. Describe the various possible results in evaluating the

$$\text{limit of } \frac{P(x)}{Q(x)} \quad \text{as} \quad x \to d_i$$

where i is a number between 1 and m.

10. LIMITS OF SEQUENCES

The numbers

$$3, \ 8, \ 17, \ -12, \ 15$$

form a sequence of numbers. Since this set contains both a first and a last element, the sequence is termed **finite**. The numbers

$$a_1, \ a_2, \ a_3, \ldots, \ a_{25}, \ a_{26}, \ a_{27}$$

form a sequence with 27 elements. The subscripts used here to identify the location of each element are more than a convenience; they provide a way of associating a number with each of 27 positive integers. The process of determining one number when another is given reminds us of the idea of function. A **sequence** is a function the **domain** of which is a portion of, or all of, the positive integers. The **range** may be in any part of the real number system. The subscripts form the domain and the members of the sequence make up the range.

A sequence such as

$$2, \ 4, \ 6, \ 8, \ldots,$$

(here consisting of the even positive integers), where the dots indicate that it is nonterminating, is called an **infinite** sequence. In general, a finite sequence shows the last term, as in

$$a_1, \ a_2, \ldots, \ a_{56},$$

indicating that there are 56 terms. An infinite sequence is written

$$a_1, a_2, \ldots, a_n, \ldots,$$

where the final dots exhibit the never-ending character of the sequence. The domain of this sequence is the set of all positive integers.

A simple example of an infinite sequence is

$$1, \frac{1}{2}, \frac{1}{3}, \frac{1}{4}, \frac{1}{5}, \ldots, \frac{1}{n}, \ldots.$$

In this sequence

$$a_1 = 1, \quad a_2 = \frac{1}{2}, \quad a_3 = \frac{1}{3}, \ldots, \quad a_n = \frac{1}{n}, \ldots.$$

When we draw a horizontal axis we see graphically that the successive terms in the sequence come closer and closer to zero, and yet no term in the sequence actually is zero (Fig. 4–30). Intuitively, it appears that the further along one gets in this sequence, the more closely the terms approach zero.

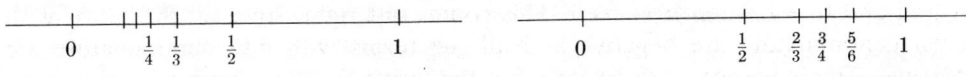

Figure 4–30   Figure 4–31

A second example is the sequence

$$\frac{1}{2}, \frac{2}{3}, \frac{3}{4}, \frac{4}{5}, \frac{5}{6}, \ldots, \frac{n}{n+1}, \ldots,$$

where

$$a_1 = \frac{1}{1+1}, \quad a_2 = \frac{2}{2+1}, \quad a_3 = \frac{3}{3+1}, \ldots, \quad a_n = \frac{n}{n+1}, \ldots.$$

Graphically it is readily seen that these terms approach 1 as n gets larger, although no individual element in the sequence actually has the value 1 (Fig. 4–31). We write $a_n \to 1$ as $n \to 1$ as $n \to \infty$ in this case* and say that the limit of the sequence is 1.

Definition. *Given the infinite sequence $a_1, a_2, \ldots, a_n, \ldots$*, **we say that** $a_n \to c$ **as** $n \to \infty$ *if for each $\epsilon > 0$ there is a positive integer N such that $|a_n - c| < \epsilon$ for all $n > N$.*

We can visualize this definition by first marking off the quantity c on a number scale (taking $c = 0$ in Fig. 4–30, and $c = 1$ in Fig. 4–31), as in Fig. 4–32. The definition asserts that given any positive number, ϵ, then, after a certain stage in the sequence is reached, all the terms lie in the interval $(c - \epsilon, c + \epsilon)$. That is,

$$c - \epsilon < a_n < c + \epsilon$$

for all n larger than some particular integer N. The first few (or few million) terms

---

* For sequences of positive integers tending to infinity, we write $n \to \infty$ to conform with widespread usage. Actually, we should write $n \to +\infty$, but there is no danger of confusion as to meaning.

Figure 4–32 **Figure 4–33**

may be scattered anywhere. But if c is to be the limit, then eventually all the terms must be in this interval; the quantity ϵ may have any value. If ϵ is quite "small," then N, the place in the sequence where the terms must begin to be in the interval about c, may be required to be very "large."

Another interesting sequence is given by

$$1, \ -\tfrac{1}{2}, \ +\tfrac{1}{3}, \ -\tfrac{1}{4}, \ +\tfrac{1}{5}, \ldots,$$

where

$$a_1 = 1, \quad a_2 = -\frac{1}{2}, \quad a_3 = \frac{1}{3}, \quad a_4 = -\frac{1}{4}, \ldots, \quad a_n = \frac{(-1)^{n+1}}{n}, \ldots.$$

The "general term" is worth examining, since $(-1)^{n+1}$ is just equal to $+1$ when n is odd and to -1 when n is even. This comes out right, since all the terms with even denominators are negative and all the terms with odd denominators are positive. The sequence tends to zero, but the terms oscillate about the value zero, as shown in Fig. 4–33.

Suppose that we have a sequence of numbers in which each term in the sequence is larger than the preceding term. We usually think that there are only two possibilities: either (1) the terms increase without bound, i.e., they go off to $+\infty$ as does, for example, the sequence of odd integers

$$1, \ 3, \ 5, \ 7, \ 9, \ldots,$$

or (2) they cluster about a point which is a limit of the sequence as does, for example, the sequence

$$\tfrac{1}{2}, \ \tfrac{3}{4}, \ \tfrac{7}{8}, \ \tfrac{15}{16}, \ \tfrac{31}{32}, \ldots,$$

in which the successive terms are

$$a_1 = \frac{1}{2}, \quad a_2 = \frac{2^2 - 1}{2^2}, \quad a_3 = \frac{2^3 - 1}{2^3}, \quad a_4 = \frac{2^4 - 1}{2^4}, \quad a_n = \frac{2^n - 1}{2^n}, \ldots,$$

and the limit is 1. These facts cannot be proved on the basis of the axioms of the number system given in Appendix 4. An additional axiom, the **Axiom of Continuity,** is required.

Axiom C (Axiom of Continuity). *Suppose that an infinite sequence $a_1, a_2, \ldots, a_n, \ldots$ has the properties* (1) $a_{n+1} \geq a_n$ *for all n, and* (2) *there is a number M such that $a_n \leq M$ for all n. Then there is a number $b \leq M$ such that*

$$\lim_{n \to \infty} a_n = b \quad \text{and} \quad a_n \leq b$$

for all n.

Figure 4–34 shows the situation. The numbers a_n move steadily to the right, and yet they can never get beyond M. It is reasonable to have an axiom which

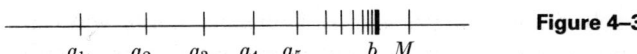

Figure 4–34

assumes that there must be some number b (perhaps M itself) toward which the a_n cluster. Axiom C is usually stated in the form: **Every bounded, nondecreasing sequence of numbers tends to a limit.**

The evaluation of limits of sequences is quite similar to that of limits of functions. For example, we know that

$$\frac{1}{n} \to 0 \quad \text{as} \quad n \to \infty$$

in much the same way that

$$\frac{1}{x} \to 0 \quad \text{as} \quad x \to \infty.$$

Example. Evaluate the limit

$$\lim_{n \to \infty} \frac{3n^2 - 2n + 1}{4n^2 + 1}.$$

Solution. When we divide both numerator and denominator by n^2 (a customary device in evaluating the limit of sequences), we have

$$\frac{3n^2 - 2n + 1}{4n^2 + 1} = \frac{3 - (2/n) + (1/n^2)}{4 + (1/n^2)}.$$

The theorems on sums, products, quotients, etc., for limits apply equally well to sequences, and so we get

$$\lim_{n \to \infty} \frac{3 - (2/n) + (1/n^2)}{4 + (1/n^2)} = \frac{\lim_{n \to \infty} [3 - (2/n) + (1/n^2)]}{\lim_{n \to \infty} [4 + (1/n^2)]}$$

$$= \frac{\lim_{n \to \infty} 3 - \lim_{n \to \infty} (2/n) + \lim_{n \to \infty} (1/n^2)}{\lim_{n \to \infty} 4 + \lim_{n \to \infty} (1/n^2)}.$$

We now see that

$$\lim_{n \to \infty} 3 = 3; \quad \lim_{n \to \infty} \frac{2}{n} = 2 \lim_{n \to \infty} \frac{1}{n} = 0;$$

$$\lim_{n \to \infty} \frac{1}{n^2} = \lim_{n \to \infty} \frac{1}{n} \cdot \lim_{n \to \infty} \frac{1}{n} = 0 \cdot 0 = 0,$$

and so on. This gives us

$$\lim_{n \to \infty} \frac{3n^2 - 2n + 1}{4n^2 + 1} = \frac{3}{4}. \quad \blacktriangleleft$$

What does it mean to say that a sequence does not approach a limit as n tends to infinity? The definition of the limit of a sequence contains a test for deciding when a limit is approached and when it is not. There are many ways in which a sequence may fail to approach a limit, some of which we illustrate with examples.

Perhaps the simplest type of sequence which does not tend to a limit is the **arithmetic progression.** Such a sequence has the property that the difference between successive terms always has the same value. That is, there is a number d, called the **common difference,** such that $a_{n+1} - a_n = d$ for all n. The sequence

$$1,\ 4,\ 7,\ 10, \ldots,\ 3n - 2, \ldots,$$

in which $d = 3$, is an example of an arithmetic progression. When d is positive, the nth term of such a sequence tends to $+\infty$, and when d is negative, it tends to $-\infty$.

A sequence such as

$$\tfrac{1}{2},\ \tfrac{3}{4},\ \tfrac{1}{4},\ \tfrac{7}{8},\ \tfrac{1}{8},\ \tfrac{15}{16},\ \tfrac{1}{16},\ \tfrac{31}{32},\ \tfrac{1}{32}, \ldots,$$

in which

$$a_1 = \tfrac{1}{2},\ a_2 = \tfrac{3}{4},\ a_3 = \tfrac{1}{4},\ a_4 = \tfrac{7}{8},\ a_5 = \tfrac{1}{8}, \ldots,$$

has as its formula for the general term

$$a_{2n-1} = \frac{1}{2^n} \qquad \text{(for odd-numbered terms),}$$

and

$$a_{2n} = \frac{2^{n+1} - 1}{2^{n+1}} \qquad \text{(for even-numbered terms).}$$

This sequence does not tend to a limit, since there are *two* numbers toward which the terms cluster (Fig. 4–35). In order for a sequence to approach a limit, there must be *exactly one number* about which the terms cluster (uniqueness of limits).

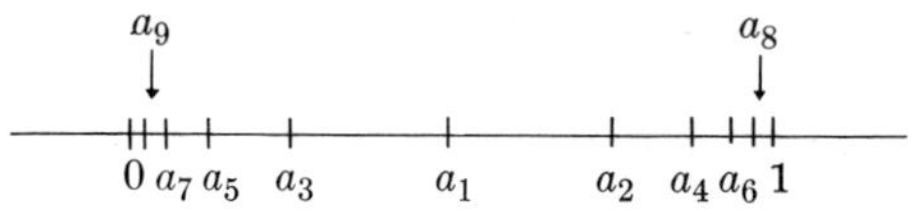

Figure 4–35

A **geometric progression** is a sequence in which there is some number r, called the **common ratio,** with the property that

$$\frac{a_{n+1}}{a_n} = r \qquad \text{for all } n.$$

Examples of geometric progressions are

$$\tfrac{3}{2},\ \tfrac{3}{4},\ \tfrac{3}{8},\ \tfrac{3}{16}, \ldots,\ \tfrac{3}{2^n}, \ldots \qquad \left(r = \tfrac{1}{2}\right);$$

and

$$6,\ 18,\ 54,\ 162,\ 486, \ldots,\ 2 \cdot 3^n, \ldots \qquad (r = 3).$$

The following facts about geometric progressions can be proved.

Theorem 14. *If $-1 < r < 1$ in a geometric progression, the limit of the sequence is zero. If $r > 1$, the sequence does not tend to a limit. If $r = 1$, the terms are all* identical and the limit is this common term. If $r = -1$, the sequence is of the form

$$a, \quad -a, \quad a, \quad -a, \quad a, \quad -a, \ldots,$$

which has no limit except in the trivial case in which $a = 0$.

PROBLEMS

In each of Problems 1 through 16, evaluate the limit of the sequence (or show that $a_n \to \infty$, $+\infty$, or $-\infty$).

1. $\lim\limits_{n \to \infty} \dfrac{3}{\sqrt{n}}$

2. $\lim\limits_{n \to \infty} \dfrac{2n - 3}{5n + 2}$

3. $\lim\limits_{n \to \infty} \dfrac{3n - 1}{4 - 2n}$

4. $\lim\limits_{n \to \infty} \dfrac{5n - 6}{n^2 + 3n}$

5. $\lim\limits_{n \to \infty} \dfrac{2n^2 + 3n - 1}{n + 4}$

6. $\lim\limits_{n \to \infty} \dfrac{2n^2 + 3n - 5}{3n^2}$

7. $\lim\limits_{n \to \infty} \dfrac{\sqrt[3]{n + 1}}{n}$

8. $\lim\limits_{n \to \infty} \dfrac{1}{\sqrt[3]{n^2 + 2}}$

9. $\lim\limits_{n \to \infty} \dfrac{2n^2 + n + 1}{n^2 + 1}$

10. $\lim\limits_{n \to \infty} \dfrac{2n^2 + 3n + 4}{n^3}$

11. $\lim\limits_{n \to \infty} \dfrac{n + 1}{n^2 + 3}$

12. $\lim\limits_{n \to \infty} (\sqrt{2n + 1} - \sqrt{2n - 1})$

13. $\lim\limits_{n \to \infty} (\sqrt{n^2 + 1} - n)$

14. $\lim\limits_{n \to \infty} (\sqrt{n^2 + n + 1} - n)$

15. $\lim\limits_{n \to \infty} \dfrac{\sqrt{n} + \sqrt[3]{n + 1}}{3\sqrt[2]{2n} - \sqrt[4]{n + 5}}$

16. $\lim\limits_{n \to \infty} \dfrac{\sqrt[4]{n^3 + 2n} + \sqrt[4]{n}}{\sqrt[3]{n^2 + 1} + \sqrt[5]{n}}$

17. Prove that an arithmetic progression with $d \neq 0$ cannot tend to a limit.
18. Prove that a geometric progression with $0 < r < 1$ must tend to the limit zero.
19. Show that
$$0 < \frac{n}{3^n} < \left(\frac{2}{3}\right)^n$$
for every $n \geq 1$ and, therefore, that
$$\lim_{n \to \infty} \frac{n}{3^n} = 0.$$

20. Show that
$$\lim_{n \to \infty} \frac{n}{2^n} = 0.$$

21. Does the sequence
$$\frac{1}{2}, \frac{3}{2}, \frac{2}{3}, \frac{4}{3}, \ldots, \frac{n-1}{n}, \frac{n+1}{n}, \ldots$$
tend to a limit?

22. Does the sequence
$$2, \frac{3}{2}, 2, \frac{7}{4}, 2, \frac{15}{8}, 2, \frac{31}{16}, \ldots, 2, \frac{2^{n+1} - 1}{2^n}, \ldots$$
tend to a limit?

*23. Consider all positive solutions of the equation $\sin x = \frac{1}{2}$ arranged in order of size. Denote these numbers $r_1, r_2, r_3, \ldots$. Show that

$$\lim_{n \to \infty} \left(\frac{1}{r_n}\right) = 0.$$

*24. Let $P(x) = a_r x^r + a_{r-1} x^{r-1} + \cdots + a_1 x + a_0$ and $Q(x) = b_s x^s + b_{s-1} x^{s-1} + \cdots + b_1 x + b_0$ be two polynomials with $a_r \neq 0$, $b_s \neq 0$. Find the value of

$$\lim_{n \to \infty} \frac{P(n)}{Q(n)}$$

according as $r > s$, $r = s$, and $r < s$.

25. Show that every nonincreasing sequence of numbers which is bounded from below must tend to a limit.

*26. Consider the geometric progression

$$a, ar, ar^2, \ldots, ar^n, \ldots,$$

and define $s_n = a + ar + \cdots + ar^n$. Show that

$$s_n = a \frac{1 - r^{n+1}}{1 - r}.$$

If $|r| < 1$, show that

$$\lim_{n \to \infty} s_n = \frac{a}{1 - r}.$$

5

DIFFERENTIATION OF ALGEBRAIC FUNCTIONS

1. THEOREMS ON DIFFERENTIATION

The process of finding the derivative is called **differentiation.** The reader will recall that use of the three-step rule for differentiation, as shown in Chapter 4, Section 3, frequently involves a lengthy and complicated process. There are certain simple expressions and combinations of expressions which occur repeatedly; it pays to be able to differentiate these at sight. In this section we establish some theorems which help us in the process of differentiation so that we do not always have to use the three-step rule.

Theorem 1. *If $f(x) = c$, a constant, for all x, then $f'(x) = 0$ for all x.* (**The derivative of a constant is zero.**)

Proof. We prove this by the three-step rule. Since f has the value c for all values of x, we have

Step 1: $f(x + h) - f(x) = c - c = 0.$

Step 2: $\dfrac{f(x + h) - f(x)}{h} = \dfrac{0}{h} = 0.$

Step 3: $\lim\limits_{h \to 0} 0 = 0.$

That is, $f'(x) = 0$.

Expressions such as

$$3x + 7, \qquad x^2 - 6x + 5, \qquad x^3 + 2x^2 - 7$$

are examples of polynomials in x. In general, an expression of the form

$$a_n x^n + a_{n-1} x^{n-1} + a_{n-2} x^{n-2} + \cdots + a_1 x + a_0,$$

where n is a positive integer and the coefficients $a_n, a_{n-1}, a_{n-2}, \ldots, a_1, a_0$ are numbers, is called a polynomial in x. The **degree** of a polynomial is the highest exponent which appears. In the above examples the polynomials are of degree 1, 2, and 3, respectively.

Combinations of the form

$$\frac{2x-6}{x^2+1}, \quad \frac{3x^3-2x+5}{7x-6}, \quad \frac{x^2+5x-2}{x^5-3x^2+1}$$

are called *rational functions*. More generally, any function which can be written as the ratio of two polynomials is, by definition, a **rational function.** That is, it must be expressible as

$$\frac{P(x)}{Q(x)}$$

where $P(x)$ and $Q(x)$ are polynomials.

Theorem 2. *If n is a positive integer and $f(x) = x^n$, then $f'(x) = nx^{n-1}$.*

Proof. We use the three-step rule. Recalling that Δf is a shorthand expression for $f(x+h) - f(x)$, we have

Step 1: $\Delta f = (x+h)^n - x^n$.

At this point we make use of the binomial theorem and write

$$(x+h)^n = x^n + nx^{n-1}h + \frac{n(n-1)}{2!}x^{n-2}h^2$$
$$+ \frac{n(n-1)(n-2)}{3!}x^{n-3}h^3 + \cdots + h^n,$$

and so we find

$$\Delta f = nx^{n-1}h + \frac{n(n-1)}{2!}x^{n-2}h^2 + \frac{n(n-1)(n-2)}{3!}x^{n-3}h^3 + \cdots + h^n.$$

Note that every term on the right has h as a factor.

Step 2: $\dfrac{\Delta f}{h} = nx^{n-1} + \dfrac{n(n-1)}{2!}x^{n-2}h$
$$+ \frac{n(n-1)(n-2)}{3!}x^{n-3}h^2 + \cdots + h^{n-1}.$$

Now we observe that every term on the right *except* the first has h in it. Since the limit of a sum is the sum of the limits, we take the limit of each of the terms on the right as $h \to 0$ and add the result. However, each term on the right, except the first, tends to zero. The term nx^{n-1} is completely unaffected as $h \to 0$, since it has no h in it at all.

Step 3: $f'(x) = \lim\limits_{h \to 0} \dfrac{\Delta f}{h} = nx^{n-1}$.

Theorem 2 asserts that whenever we see an expression such as x^8 or x^{17}, the derivative may be found by inspection. The derivative of x^8 is $8x^7$, and the derivative of x^{17} is $17x^{16}$.

Theorem 3. *If $f(x)$ has a derivative $f'(x)$, then the derivative of $g(x) = cf(x)$ is $cf'(x)$, where c is any constant.* **(The derivative of a constant times a function is the constant times the derivative of the function.)**

Proof. We use the three-step rule.

Step 1: $\Delta g = g(x + h) - g(x) = cf(x + h) - cf(x) = c\,\Delta f.$

Step 2: $\dfrac{\Delta g}{h} = c\dfrac{\Delta f}{h}.$

Step 3: $g'(x) = \lim\limits_{h \to 0} c\dfrac{\Delta f}{h} = c \lim\limits_{h \to 0} \dfrac{\Delta f}{h} = cf'(x).$

The justification for Step 3 comes from the theorem that the limit of a product is the product of the limits.

Theorems 2 and 3 can be combined to find derivatives of expressions such as $4x^7$, $-2x^5$, etc. We know from Theorem 2 that x^7 has the derivative $7x^6$ and therefore, from Theorem 3, $4x^7$ has the derivative $4(7x^6) = 28x^6$. Similarly, the derivative of $-2x^5$ is $-10x^4$.

Theorem 4. *If $f(x)$ and $g(x)$ have derivatives and $F(x) = f(x) + g(x)$, then $F'(x) = f'(x) + g'(x)$.* **(The derivative of the sum is the sum of the derivatives.)**

Proof. We use the three-step rule.

Step 1: $\Delta F = F(x + h) - F(x) = f(x + h) + g(x + h) - f(x) - g(x)$
$\qquad\quad = \Delta f + \Delta g.$

Step 2: $\dfrac{\Delta F}{h} = \dfrac{\Delta f}{h} + \dfrac{\Delta g}{h}.$

Since the limit of a sum is the sum of the limits, we see that the right side tends to $f'(x) + g'(x)$. Therefore,

Step 3: $F'(x) = f'(x) + g'(x).$

While Theorem 4 has proved for the sum of two functions, the same proof works for the sum of any (finite) number of functions. Theorems 2, 3, and 4 combine to enable us to differentiate any polynomial.

Example 1. Find the derivative of

$$f(x) = 3x^5 + 2x^4 - 7x^2 + 2x + 5.$$

Solution. The quantity x^5 has the derivative $5x^4$, as we know from Theorem 2. Using this fact and Theorem 3, we find the derivative of $3x^5$ to be $15x^4$. Similarly, the derivative of

$\qquad 2x^4$ is $8x^3$, $\qquad -7x^2$ is $-14x$, $\qquad 2x$ is 2, $\qquad 5$ is 0.

Theorem 4 tells us that the derivative of the sum of these terms is the sum of the derivatives. We obtain

$$f'(x) = 15x^4 + 8x^3 - 14x + 2. \qquad \triangleleft$$

Theorem 5. If $u(x)$ and $v(x)$ are any two functions which have a derivative and if $f(x) = u(x) \cdot v(x)$, then

$$f'(x) = u(x) \cdot v'(x) + v(x) \cdot u'(x).$$

(The derivative of the product of two functions is the first times the derivative of the second plus the second times the derivative of the first.)

Proof. We use the three-step rule, as usual.

Step 1: $\Delta f = f(x + h) - f(x) = u(x + h) \cdot v(x + h) - u(x) \cdot v(x).$

At this point we use a trick of a type we have used before. We write the quantity $f(x + h)$ in a more complicated way, as follows:

$$f(x + h) = [u(x) + u(x + h) - u(x)] \cdot [v(x) + v(x + h) - v(x)].$$

With the symbols Δu and Δv this becomes

$$f(x + h) = (u + \Delta u)(v + \Delta v).$$

Therefore

$$\Delta f = f(x + h) - f(x) = (u + \Delta u)(v + \Delta v) - uv.$$

Multiplying out the right side, we obtain

$$\Delta f = u\,\Delta v + v\,\Delta u + \Delta u \cdot \Delta v.$$

Step 2: $\dfrac{\Delta f}{h} = u\dfrac{\Delta v}{h} + v\dfrac{\Delta u}{h} + \Delta u \cdot \dfrac{\Delta v}{h}.$

We now examine each term on the right as h tends to zero. The first term tends to $u(x)v'(x)$, since

$$\frac{\Delta v}{h} \to v'(x) \quad \text{as} \quad h \to 0.$$

The second term tends to $vu'(x)$. The last term has two parts, $\Delta v/h$ and Δu. The quantity $\Delta v/h$ tends to $v'(x)$. However, the function $u(x)$ is continuous (according to the theorem which says that if a function has a derivative it is continuous), and so $u(x + h) - u(x) = \Delta u \to 0$ as $h \to 0$. This means that the last term on the right tends to zero as $h \to 0$. We conclude

Step 3: $f'(x) = \lim\limits_{h \to 0} \dfrac{\Delta f}{h} = u(x)v'(x) + v(x)u'(x).$

Example 2. Find the derivative of

$$f(x) = (x^2 - 3x + 2)(x^3 + 2x^2 - 6x).$$

What is the value of $f'(3)$? the value of $f'(2)$?

Solution. One way to do this would be to multiply the two expressions in parentheses, thus obtaining a polynomial of the 5th degree, and then to find the derivative as in Example 1. However, Theorem 5 gives us an idea for a simpler way. We write

$$u(x) = x^2 - 3x + 2 \quad \text{and} \quad v(x) = x^3 + 2x^2 - 6x.$$

This means that $f(x) = u(x)v(x)$. We can now readily find $u'(x)$ and $v'(x)$, using the method in Example 1: $u'(x) = 2x - 3$ and $v'(x) = 3x^2 + 4x - 6$. Then

$$\begin{aligned}
f'(x) &= u(x)v'(x) + v(x)u'(x) \\
&= (x^2 - 3x + 2)(3x^2 + 4x - 6) + (x^3 + 2x^2 - 6x)(2x - 3), \\
f'(3) &= (9 - 9 + 2)(27 + 12 - 6) + (27 + 18 - 18)(6 - 3) \\
&= 66 + 81 = 147, \\
f'(2) &= (4 - 6 + 2)(12 + 8 - 6) + (8 + 8 - 12)(4 - 3) \\
&= 0 + 4 = 4.
\end{aligned}$$
◄

Theorem 6. *If $u(x)$ and $v(x)$ are any two functions which have a derivative and if*

$$f(x) = \frac{u(x)}{v(x)}$$

with $v(x) \neq 0$, then

$$\boxed{f'(x) = \frac{v(x)u'(x) - u(x)v'(x)}{[v(x)]^2}.}$$

(The derivative of a quotient of two functions is the denominator times the derivative of the numerator minus the numerator times the derivative of the denominator, all divided by the square of the denominator.)

Proof. We proceed by the three-step rule.

Step 1: $\Delta f = f(x + h) - f(x) = \dfrac{u(x + h)}{v(x + h)} - \dfrac{u(x)}{v(x)}.$

We may write $u(x + h) = u(x) + u(x + h) - u(x) = u + \Delta u.$ Similarly, $v(x + h) = v + \Delta v.$ Therefore

$$\Delta f = \frac{u + \Delta u}{v + \Delta v} - \frac{u}{v} = \frac{v(u + \Delta u) - u(v + \Delta v)}{v(v + \Delta v)} = \frac{v\Delta u - u\Delta v}{v(v + \Delta v)}.$$

Step 2: $\dfrac{\Delta f}{h} = \dfrac{v\dfrac{\Delta u}{h} - u\dfrac{\Delta v}{h}}{v(v + \Delta v)}.$

Step 3: $\displaystyle\lim_{h \to 0} \dfrac{\Delta f}{h} = f'(x) = \dfrac{vu'(x) - uv'(x)}{v^2}.$

Example 3. Find the derivative of

$$f(x) = \frac{2x^2 - 3x}{x^2 + 3}.$$

Solution. We set $u = 2x^2 - 3x$ and $v = x^2 + 3$. Then $u'(x) = 4x - 3$ and $v'(x) = 2x$. Now, by Theorem 6,

$$f'(x) = \frac{(x^2 + 3)(4x - 3) - (2x^2 - 3x) \cdot 2x}{(x^2 + 3)^2} = \frac{3x^2 + 12x - 9}{(x^2 + 3)^2}.$$

The last step was merely an algebraic simplification and, strictly speaking, not a part of the differentiation process. ◂

Theorem 7. *If n is a positive integer and $f(x) = x^{-n}$, with $x \neq 0$, then*

$$f'(x) = -nx^{-n-1}.$$

Proof. A negative exponent simply means that we may write $f(x)$ in the form

$$f(x) = \frac{1}{x^n}.$$

We can apply Theorem 6: Let $u(x) = 1$ and $v(x) = x^n$. Then

$$f(x) = \frac{u(x)}{v(x)}.$$

From Theorem 1, $u'(x) = 0$ and, from Theorem 2, $v'(x) = nx^{n-1}$. We substitute in the formula of Theorem 6 to get

$$f'(x) = \frac{x^n \cdot 0 - 1 \cdot nx^{n-1}}{(x^n)^2},$$

which gives us

$$f'(x) = \frac{-nx^{n-1}}{x^{2n}} = -nx^{-n-1}.$$

Combining Theorems 2 and 7, we see that **if k is any positive or negative integer and**

if $f(x) = x^k$, then $f'(x) = kx^{k-1}$.

PROBLEMS

In Problems 1 through 30, differentiate the functions by the methods given in this section.

1. $x^2 + 3x - 4$
2. $2x^3 + 5x^2 - 3x + 7$
3. $x^{14} + 10x^2 + 7x - 4$
4. $x^8 + 2x^6 - 4x^3 + 6x + 9$
5. $x^{-2} - 4x^{-5} + 3x^{-8}$
6. $2x^{-1} - 3x^{-5} + 2x^2 - 7$
7. $x^3 + 2x^2 - 3x + 5 - 2x^{-1} + 4x^{-2}$
8. $x^5 + 12x - 1 + 3x^{-4} - 5x^{-7}$
9. $x^2 + 2x - \dfrac{1}{x^2}$
10. $x^3 - 3x - \dfrac{2}{x^4}$
11. $\dfrac{3}{4x^7}$
12. $2x^4 - 3x + \dfrac{5}{8x^3}$
13. $(x^2 + 2x)(3x + 1)$
14. $(2x^3 + 6x)(7x - 5)$

15. $(x^3 + 6x^2 - 2x + 1)(x^2 + 3x - 5)$

16. $(x^4 + 2x - 3)(x^6 - 7x^5 + 8x^3 + 9x^2 + 1)$

17. $\dfrac{2x^2 - 3x + 4}{x}$
18. $\dfrac{x^3 - 3x + 5}{x^2}$
19. $\dfrac{2x^3 - 3x^2 + 4x - 2}{x^3}$

20. $\dfrac{x^3 - 4x^2 + 3x - 2}{x^4}$
21. $\dfrac{x - 1}{x + 1}$
22. $\dfrac{2x + 3}{3x - 2}$

23. $\dfrac{3x - 2}{2x + 3}$
24. $\dfrac{5x - 2}{4x + 3}$
25. $\dfrac{x}{x^2 + 1}$

26. $\dfrac{x^2 + 1}{x^2 - 1}$
27. $\dfrac{x + 1}{x^2 + 2x + 2}$
28. $\dfrac{2x^2 - 3x + 4}{x^2 - 2x + 3}$

29. $\dfrac{x + 1}{2x + 3}(2x - 5)$
30. $\dfrac{x^2 - x + 6}{x^2 + 1}(x^2 + x + 1)$

*31. Using Theorem 5, find a formula for the derivative of $f(x) = [u(x)]^2$. Apply this to the functions
   a) $(x^2 + 2x - 1)^2$,
   b) $(x^3 + 7x^2 - 8x - 6)^2$,
   c) $(x^7 - 2x + 3x^{-2})^2$.

32. Given that $f(x) = u(x) \cdot v(x) \cdot w(x)$, find a formula for $f'(x)$. [*Hint:* Apply Theorem 5 twice.] Using this formula, find the derivative of the following functions:
   a) $(x^2 + 2x - 6)(3x - 2)(x^2 + 5)$,
   b) $(2x^2 + x^{-2})(x^2 - 3)(4x + 1)$,
   c) $(2x + 3)^2(x^2 + 1)$,
   d) $(x^2 + x + 1)^3$.

33. Given that
$$f(x) = \dfrac{u(x)}{v(x)} w(x),$$
find a formula for $f'(x)$. [*Hint:* Apply Theorems 5 and 6.] Using this formula, find the derivative of the functions
   a) $\dfrac{x + 2}{3x + 1}(x - 6)$,
   b) $\dfrac{x^2 - 1}{2x + 6}(x^2 + 5)$.

34. Let $P(x) = a_n x^n + a_{n-1} x^{n-1} + \cdots + a_1 x + a_0$ and $Q(x) = b_m x^m + b_{m-1} x^{m-1} + \cdots + b_1 x + b_0$ be two polynomials. Let $y = P(x)/Q(x)$. Find a formula for dy/dx.

*35. Suppose $f(x) = u(x) \cdot v(x)$ where u and v are given functions. We denote the kth derivative of u by $u^{(k)}(x)$, and similarly for v and f. Show that
$$f^{(k)}(x) = u^{(k)}(x)v + \dfrac{k}{1!}u^{(k-1)}(x)v'(x) + \dfrac{k(k-1)}{2!}u^{(k-2)}(x)v^{(2)}(x) + \cdots + u(x)v^{(k)}(x).$$

36. Suppose that $f(x) = u(x)/v(x)$ and $v(x) \neq 0$. Find a formula for $f''(x)$.

2. THE CHAIN RULE. APPLICATIONS

The chain rule is one of the most important and useful tools in differentiation. The proof of the theorem establishing this rule is somewhat difficult to follow but, if the reader studies it carefully, he will achieve a deeper understanding of differentiation which will make the effort well worth while. The examples should be gone over thoroughly, as a means of studying the applications of this rule.

Theorem 8 (Fundamental Lemma of Differentiation). *Suppose that F has a derivative at a value u so that $F'(u)$ exists. We define the function*

$$G(h) = \begin{cases} \dfrac{F(u+h) - F(u)}{h} - F'(u), & \text{if } h \neq 0, \\ 0, & \text{if } h = 0. \end{cases}$$

Then (a) *G is continuous at $h = 0$, and* (b) *the formula*

$$F(u+h) - F(u) = [F'(u) + G(h)]h \tag{1}$$

holds.

Proof. From the definition of derivative, we know that

$$\lim_{h \to 0} \frac{F(u+h) - F(u)}{h} = F'(u)$$

or

$$\lim_{h \to 0} \left[\frac{F(u+h) - F(u)}{h} - F'(u) \right] = 0.$$

Hence $G(h) \to 0$ as $h \to 0$. Therefore G is continuous at 0, and (a) holds. To establish (b), we observe that for $h \neq 0$, the formula (1) is a restatement of the definition of G. For $h = 0$, both sides of (1) are zero.

Theorem 9 (Chain Rule). *Suppose that f, g, and u are functions with $f(x) = g[u(x)]$, and suppose that g and u are differentiable. Then f is differentiable and the following formula holds:*

$$f'(x) = g'[u(x)]u'(x).$$

Proof. We use the three-step rule to find the derivative of f.

Step 1: $\Delta f = f(x+h) - f(x) = g[u(x+h)] - g[u(x)]$
$= g[u(x) + u(x+h) - u(x)] - g[u(x)]$
$= g(u + \Delta u) - g(u).$

We now apply Theorem 8 to the right-hand side, where g is used instead of F and Δu instead of h. Then Step 1 can be written

$$\Delta f = [g'(u) + G(\Delta u)]\Delta u.$$

Step 2: $\dfrac{\Delta f}{h} = [g'(u) + G(\Delta u)]\dfrac{\Delta u}{h}.$

Step 3: Since $\Delta u/h \to u'(x)$ as $h \to 0$, since $\Delta u \to 0$ as $h \to 0$, and since $G(\Delta u) \to 0$ as $\Delta u \to 0$, we conclude that

$$\lim_{h \to 0} \frac{\Delta f}{h} = f'(x) = g'(u)u'(x).$$

This is exactly the formula given in the statement of the theorem.

Before illustrating the chain rule by examples, we establish one of the most important special cases.

Corollary. If $f(x) = [u(x)]^n$ and n is an integer, then
$$f'(x) = n[u(x)]^{n-1}u'(x).$$

Proof. In the chain rule we take $g(u) = u^n$. Then $f = g[u(x)]$ means that $f(x) = [u(x)]^n$. We obtain
$$f'(x) = g'(u)u'(x) = nu^{n-1}u'(x). \quad \blacktriangleleft$$

We can also write schematically
$$f(x) = (\text{expression in } x)^n,$$
$$f'(x) = n(\text{expression in } x)^{n-1}(\text{derivative of expression in } x).$$

Remark. If we write $y = f(x)$ then, in Theorem 9, y is a function of u in the form $y = g(u)$. Also, u is a function of x which is expressed as $u = u(x)$. The symbol $\dfrac{dy}{dx}$ may be used for $f'(x)$, the symbol $\dfrac{dy}{du}$ may be used for $g'(u)$, and the symbol $\dfrac{du}{dx}$ may be used for $u'(x)$. The chain rule then takes the particularly simple form

$$\boxed{\dfrac{dy}{dx} = \dfrac{dy}{du} \cdot \dfrac{du}{dx}.}$$

We recall that the individual symbols dx, du, and dy (called **differentials**) have not as yet been defined. In Chapter 6 we shall give a precise meaning to each of these quantities, and show that the above formula is obtained by dividing and multiplying by du.

Example 1. Find $f'(x)$, given that $f(x) = (x^2 + 3x - 2)^4$.

Solution. From the Corollary, with $n = 4$ and the expression in x being $x^2 + 3x - 2$, we find
$$f'(x) = 4(x^2 + 3x - 2)^3 \cdot (2x + 3),$$
since the derivative of $x^2 + 3x - 2$ is $2x + 3$. $\quad \blacktriangleleft$

Example 2. Find dy/dx, given that
$$y = \frac{1}{x^3 + 3x^2 - 6x + 4}.$$

Solution. We write $y = (x^3 + 3x^2 - 6x + 4)^{-1}$ and apply the Corollary with $n = -1$:
$$\frac{dy}{dx} = -1(x^3 + 3x^2 - 6x + 4)^{-2} \cdot (3x^2 + 6x - 6)$$
$$= \frac{-3(x^2 + 2x - 2)}{(x^3 + 3x^2 - 6x + 4)^2}.$$

We note that this example could also have been worked by using the formula for the derivative of a quotient. ◀

Example 3. Find dy/dx, given that

$$y = \left(\frac{3x-2}{2x+1}\right)^7.$$

Solution. This requires a combination of formulas. First, from the Corollary, we see that

$$\frac{dy}{dx} = 7\left(\frac{3x-2}{2x+1}\right)^6 \cdot \left(\text{derivative of } \frac{3x-2}{2x+1}\right).$$

To find the derivative of $(3x-2)/(2x+1)$ we apply the quotient formula to get

$$\frac{dy}{dx} = 7\left(\frac{3x-2}{2x+1}\right)^6 \frac{(2x+1)(3) - (3x-2)(2)}{(2x+1)^2}$$

$$= \frac{49(3x-2)^6}{(2x+1)^8}.$$
◀

Example 4. Find $f'(x)$, given that

$$f: x \to (x^2 + 2x - 3)^{16}(2x+5)^{13}.$$

Solution. We first observe that if we let $u(x) = (x^2 + 2x - 3)^{16}$ and $v(x) = (2x+5)^{13}$, we may use Theorem 5, on the product of two functions:

$$f'(x) = u(x)v'(x) + v(x)u'(x).$$

To find $u'(x)$ we employ the chain rule:

$$u'(x) = 16(x^2 + 2x - 3)^{15}(2x+2);$$

and similarly,

$$v'(x) = 13(2x+5)^{12}(2).$$

Substituting in the formula for $f'(x)$, we get

$$f'(x) = (x^2 + 2x - 3)^{16} 26(2x+5)^{12}$$
$$+ (2x+5)^{13} 32(x^2 + 2x - 3)^{15}(x+1)$$
$$= (x^2 + 2x - 3)^{15}(2x+5)^{12}[26(x^2 + 2x - 3) + 32(2x+5)(x+1)].$$ ◀

PROBLEMS

In each of Problems 1 through 22 find the derivative.

1. $(2x+3)^9$
2. $(x^2 + 4x - 3)^6$
3. $(2 - 5x)^4$
4. $(x^3 + 4x^2 - 3x + 7)^5$
5. $(2x+1)^{-4}$
6. $(x^2 - 3x + 2)^{-5}$
7. $(x^3 + 2x - 3 + x^{-2})^4$
8. $(x^4 + 5x - 6x^{-1})^3$
9. $(x^2 + 2 - x^{-2})^{-1}$
10. $(3x^3 + 2x^2 - 6x^{-4})^{-4}$

11. $(x^2 + 1)^2(x^3 - 2x)^2$

12. $(x^3 + 2x - 6)^3(x^2 - 4x + 5)^7$

13. $(x^2 - x^{-1} + 1)(x^3 + 2x - 6)^7$

14. $\dfrac{(3x - 2)^2}{(2x - 6)^2}$

15. $\dfrac{(x^2 + 1)^3}{(x^2 + 2)^2}$

16. $\dfrac{(2x - 6)^4}{(x + 1)^7}$

17. $\dfrac{(x^2 + 2)^2}{(x^2 + x^{-1})^3}$

18. $\dfrac{(2x - 6)^{-1}}{(x^2 + 3)^{-2}}$

19. $\dfrac{(x^{-1} + x^2)^{-1}}{(x^3 - 2x^{-2})^{-2}}$

20. $\dfrac{(x^2 + 2x - 1)(x^3 + 3x - 4)^2}{(2x + 6)^2}$

21. $(x + 5)^2(3x - 6)^3(7x^2 + 1)^4$

22. $(2x^2 + 6x - 1)^3(x^2 + 5)^5(x^2 - 7)^6$

23. Suppose that v is a function of x so that $v = v(x)$, that u is a function of v, and that g is a function of u. Writing $f(x) = g\{u[v(x)]\}$, show that

$$f'(x) = g'u'v'$$

or, in differential notation with $y = f(x)$, establish the formula

$$\frac{dy}{dx} = \frac{dy}{du} \cdot \frac{du}{dv} \cdot \frac{dv}{dx}.$$

24. Let $P(u) = a_n u^n + a_{n-1} u^{n-1} + \cdots + a_1 u + a_0$ be a polynomial in u. Suppose that u is a function of x. Writing $f(x) = P[u(x)]$, find $f'(x)$ in terms of u and u'.

25. Let u and v be functions of x. Suppose that

$$f(x) = g(u + v), \quad F(x) = G(u \cdot v), \quad \text{and} \quad H(x) = h\left(\frac{u}{v}\right)$$

where g, G, and h are differentiable functions. Find formulas for $f'(x)$, $F'(x)$, and $H'(x)$.

3. THE POWER FUNCTION

The function

$$f(x) = x^n,$$

where n is some number, is called the **power function.** That is, x is raised to an exponent called the **power.** If n is an *even positive integer*, the graph of the function appears as in Fig. 5–1. If n is an odd positive integer the curve is below the x axis for negative values of x, as shown in Fig. 5–2.

For negative exponents, the y axis is a vertical asymptote and the x axis is a horizontal asymptote. The general behavior of these functions is shown in Fig. 5–3.

If n is of the form $1/q$, where q is a positive integer, the graph of the function has a completely different character. For example, the function

$$y = x^{1/2} = \sqrt{x}$$

is not even defined for negative values of x. In fact, if q is any even number, the function

$$y = x^{1/q} = \sqrt[q]{x}$$

126 DIFFERENTIATION OF ALGEBRAIC FUNCTIONS

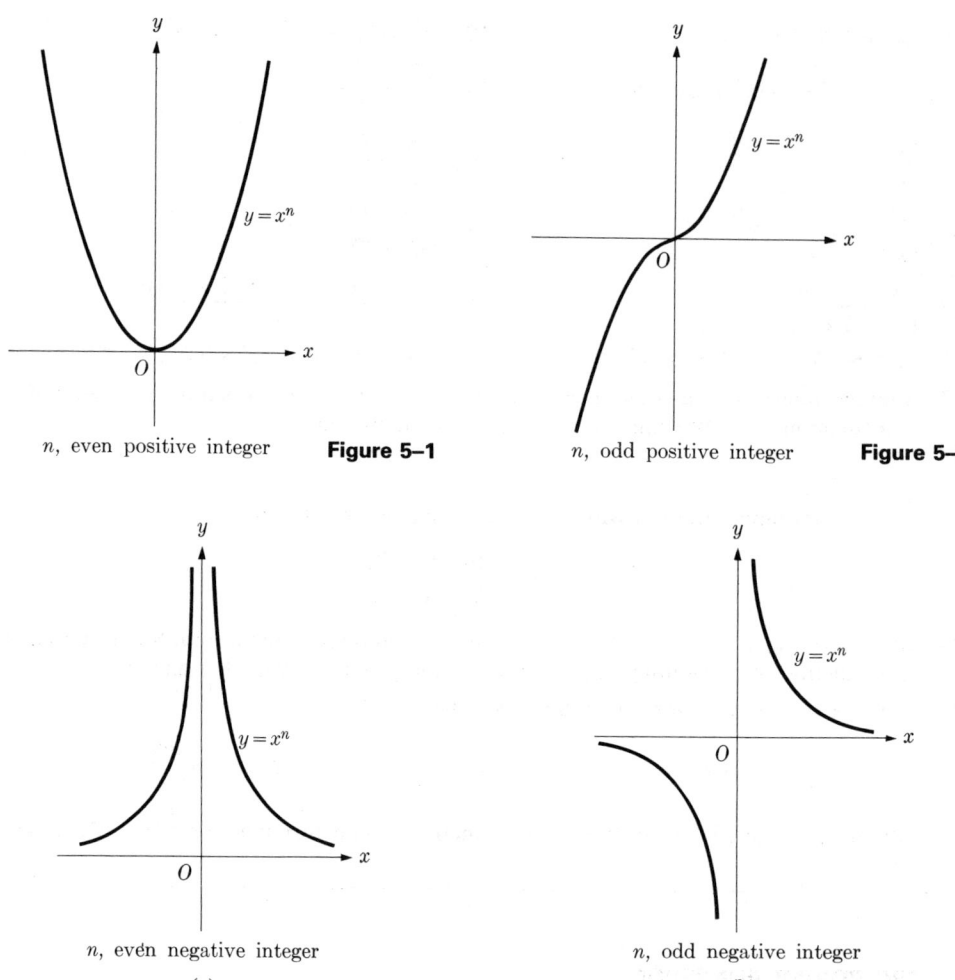

n, even positive integer **Figure 5–1**

n, odd positive integer **Figure 5–2**

n, even negative integer
(a)

Figure 5–3

n, odd negative integer
(b)

is defined only for x positive or zero. On the other hand, if q is odd, we know that because the odd root of a negative number can be found (for example, $\sqrt[5]{-32} = -2$), the function is defined for all values of x. The graphs of $\sqrt[q]{x}$ are shown in Fig. 5–4.

We follow the same rule for the derivative of $x^{1/q}$ as for the derivative of the power function when the exponent is an integer. The following theorem establishes this fact.

Theorem 10. *If $f(x) = x^{1/q}$ and q is an integer, then*

$$f'(x) = \frac{1}{q} x^{(1/q)-1}.$$

This theorem is a special case of Theorem 19 in Chapter 9, Section 9. The result established there shows that if s is *any real number* then the derivative of

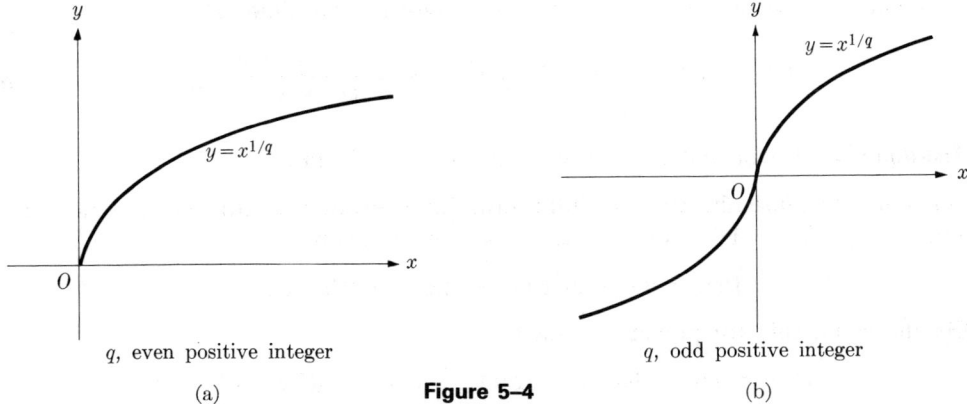

Figure 5–4
(a) q, even positive integer
(b) q, odd positive integer

$f(x) = x^s$ is $f'(x) = sx^{s-1}$. In the meantime we shall use the result of Theorem 10 above and the corollaries below. For a direct proof of Theorem 10 see Problem 51 at the end of this section and the hint given there.

A **rational number** is one which can be written as one integer over another. "The number r is rational" means that $r = p/q$, where p and q are integers. Theorem 10 can be combined with the chain rule to give the general rule for differentiating the power function when the exponent is any rational number.

Corollary 1. *If $f(x) = x^r$ and r is any rational number, then $f'(x) = rx^{r-1}$.*

Proof. Since r is rational, we can write

$$f(x) = x^{p/q} = (x^{1/q})^p,$$

where p and q are integers. By the special case of the chain rule, we have

$$f'(x) = p(x^{1/q})^{p-1} \cdot (\text{derivative of } x^{1/q});$$

and so, by Theorem 10,

$$f'(x) = p(x^{1/q})^{p-1} \frac{1}{q} x^{(1/q)-1} = \frac{p}{q} x^{(p/q)-(1/q)+(1/q)-1} = rx^{r-1}.$$

Corollary 2. *If $f(x) = [u(x)]^r$ and r is rational, then $f'(x) = r[u(x)]^{r-1} u'(x).$*

Proof. This is an immediate consequence of Corollary 1 and the chain rule.

Example 1. Given that $f(x) = 2\sqrt[5]{x^3}$, find $f'(x)$.

Solution. $f(x) = 2x^{3/5}$ and, by the rule for rational exponents,

$$f'(x) = 2(\tfrac{3}{5})x^{(3/5)-1} = \tfrac{6}{5}x^{-2/5}. \quad \blacktriangleleft$$

Example 2. Given that $f(t) = \sqrt[3]{t^3 + 3t + 1}$, find $f'(t)$.

Solution. We write $f(t) = (t^3 + 3t + 1)^{1/3}$ and use Corollary 2:

$$f'(t) = \tfrac{1}{3}(t^3 + 3t + 1)^{-2/3}(3t^2 + 3) = \frac{t^2 + 1}{(t^3 + 3t + 1)^{2/3}}.$$ ◂

Example 3. Given that $f(x) = (x + 1)^3(2x - 1)^{4/3}$, find $f'(x)$.

Solution. We use the rule for differentiating a product, setting $u(x) = (x + 1)^3$ and $v(x) = (2x - 1)^{4/3}$, so that $f(x) = u(x)v(x)$. Then

$$f'(x) = (x + 1)^3 v'(x) + (2x - 1)^{4/3} u'(x).$$

By the chain rule for powers we have

$$u'(x) = 3(x + 1)^2 \cdot 1 \quad \text{and} \quad v'(x) = \tfrac{4}{3}(2x - 1)^{1/3} \cdot 2.$$

Therefore

$$\begin{aligned} f'(x) &= \tfrac{8}{3}(x + 1)^3(2x - 1)^{1/3} + 3(2x - 1)^{4/3}(x + 1)^2 \\ &= (x + 1)^2(2x - 1)^{1/3}[\tfrac{8}{3}(x + 1) + 3(2x - 1)] \\ &= \tfrac{1}{3}(x + 1)^2(2x - 1)^{1/3}(26x - 1). \end{aligned}$$

Example 4. Given the function

$$f: s \to \frac{s}{\sqrt{s^2 - 1}},$$

find $f'(s)$.

Solution. We write

$$f(s) = \frac{s}{(s^2 - 1)^{1/2}}$$

and use Theorem 6 for the derivative of a quotient:

$$f'(s) = \frac{(s^2 - 1)^{1/2} \cdot 1 - s\tfrac{1}{2}(s^2 - 1)^{-1/2}(2s)}{s^2 - 1}.$$

This expression may be simplified in several ways. One easy method is to multiply through both the numerator and the denominator by $(s^2 - 1)^{1/2}$. This gives

$$f'(s) = \frac{(s^2 - 1)^1 - s^2(s^2 - 1)^0}{(s^2 - 1)^{3/2}}.$$

Since $(s^2 - 1)^0 = 1$, we get

$$f'(s) = \frac{-1}{(s^2 - 1)^{3/2}}.$$ ◂

PROBLEMS

Find the derivative in each of Problems 1 through 28.

1. $x^{5/3} + 2x^{4/3} - 3x^{-1/3}$
2. $x^{3/2} + 2x^{3/5} - 4x^{4/7}$
3. $x^{-2/3} + x^{-4/3} - 2x^{4/7}$
4. $2x^{2/3} + 5x^{-1/4} - \tfrac{2}{5}x^{5/2}$
5. $\tfrac{3}{4}x^{4/3} + 2x^{1/2} - 2x^{-1}$
6. $\tfrac{3}{8}x^{8/3} + \tfrac{1}{2}x^{-1} - 4x^{1/4}$

7. $\dfrac{x^{3/2} - 2x^{1/2} + 4x^{-1/2}}{5}$

8. $\dfrac{x^{5/2} - 3x^{3/2} + 2x^{1/2}}{x^2}$

9. $2x\sqrt{x} + 3\sqrt[3]{x^2} - 5x\sqrt[5]{x^2}$

10. $\tfrac{1}{4}x\sqrt[3]{x} - \tfrac{1}{7}x^2\sqrt[3]{x} + \tfrac{1}{10}x^3\sqrt[3]{x}$

11. $\dfrac{x^2 - 3x + 2}{2\sqrt{x}}$

12. $\dfrac{x^3 - 3x^2 - 5x + 2}{5x\sqrt[3]{x^2}}$

13. $(2x + 3)^{10/3}$

14. $(3x - 2)^{4/3}$

15. $(x^2 + 2x + 3)^{3/2}$

16. $(x^3 + 3x^2 + 6x + 5)^{2/3}$

17. $(2x^3 - 3x^2 + 3x - 1)^{-1/3}$

18. $(x^4 + 2x^2 + 1)^{-7/4}$

19. $\sqrt[3]{(x^3 + 3x + 2)^2}$

20. $\dfrac{1}{\sqrt[4]{x^2 + 2x + 3}}$

21. $\dfrac{(x^2 + x - 3)\sqrt{x^2 + x - 3}}{5}$

22. $\dfrac{1}{2(x^2 + 2x + 3)\sqrt[3]{(x^2 + 2x + 3)^2}}$

23. $(2x + 3)^4(3x - 2)^{7/3}$

24. $(3 - 2x)^{5/2}(2 - 3x)^{1/3}$

25. $(2x - 1)^{5/2}(7x - 3)^{3/7}$

26. $(2 - 3x)^{4/3}(5x + 2)^3$

27. $\sqrt[3]{(x + 1)^2}\sqrt{x - 1}$

28. $\sqrt{(1 - 2x)^3}\sqrt[3]{x^2 + 1}$

29. Find $g'(t)$, given $g(t) = \dfrac{\sqrt[3]{2 + 6t}}{t}$

30. Find $\dfrac{dy}{ds}$, given $y = \dfrac{\sqrt{s^2 + 1}}{s}$

31. Find $f'(y)$, given $f(y) = \dfrac{2y + 3}{\sqrt{y^2 + 3y + 4}}$

32. Find $\dfrac{dy}{dr}$, given $y = \dfrac{\sqrt{1 + 4r^2}}{r^2}$

33. Find $f'(\tau)$, given $f : \tau \to \sqrt{\dfrac{\tau - 1}{\tau + 1}}$

34. Find $K'(x)$, given $K : x \to \sqrt{\dfrac{3x - 2}{2x + 3}}$

35. Find $\phi'(t)$, given $\phi : t \to \sqrt[3]{\dfrac{3t + 4}{3t - 2}}$

36. Find $\psi'(y)$, given $\psi : y \to \sqrt[3]{\dfrac{y^2 - 1}{y^2 + 1}}$

In Problems 37 through 40, find the value of $f'(x)$ for the given value of x.

37. $f(x) = (x^2 + 2x + 3)^{2/3}$, $x = -1$

38. $f(x) = \dfrac{2x}{\sqrt{4x^2 + 1}}$, $x = 2$

39. $f(x) = (x^2 + 1)^3(2x + 4)^{1/3}$, $x = 2$

40. $f(x) = (x^2 + 6)^2\sqrt[3]{5x + 12}$, $x = 3$

In Problems 41 through 46, find the equations of the tangent line and normal line to the curve $y = f(x)$, at the point having the given value of x.

41. $f(x) = (2x - 3)^{5/2}$, $x = 2$
42. $f(x) = \dfrac{1}{\sqrt[3]{2 - 3x}}$, $x = 1$
43. $f(x) = x\sqrt{25 - x^2}$, $x = 4$
44. $f(x) = x\sqrt[3]{2 - x}$, $x = 1$
45. $f(x) = \dfrac{\sqrt{5 + 2x}}{x}$, $x = 2$
46. $f(x) = \dfrac{\sqrt{3 - x}}{x}$, $x = -1$

47. Find a formula for the derivative of
$$f(x) = [u(x)]^m[v(x)]^n$$
and apply it to the functions
a) $f(x) = (2x + 1)^6(x^3 - 6)^{4/3}$,
b) $f(x) = (x^2 + 2x - 1)^{3/2}(x^4 - 3)^3$.

48. Find a formula for the derivative of
$$f(x) = [u(x)]^r[v(x)]^s[w(x)]^t,$$
where r, s, and t are rational numbers. Apply it to the functions
a) $f(x) = (x^2 + 1)^{3/2}(2x - 5)^{1/3}(x^2 + 4)^2$,
b) $f(x) = (x - 1)^3(3 - 2x)^2(x^2 - 5)^{4/3}$.

49. Use the three-step rule to find the derivative of $f(x) = x^{1/2}$.
*50. Use the three-step rule to find the derivative of $f(x) = x^{1/3}$. [*Hint:* Use the formula
$$a^3 - b^3 = (a - b)(a^2 + ab + b^2).]$$

*51. Use the three-step rule to find the derivation of $f(x) = x^{1/q}$ where q is a positive integer. [*Hint:* Use the formula
$$a^m - b^m = (a - b)(a^{m-1} + a^{m-2}b + a^{m-3}b^2 + \cdots + ab^{m-2} + b^{m-1})$$
and set $a = (x + h)^{1/q}$, $b = x^{1/q}$, $m = q$. Then "rationalize the numerator" in the formula obtained in Step 2 of the three-step rule by multiplying numerator and denominator by $a^{q-1} + a^{q-2}b + \cdots + ab^{q-2} + b^{q-1}$.]

4. IMPLICIT DIFFERENTIATION

Before discussing implicit differentiation, we shall introduce additional symbolism which will be particularly helpful. Suppose that f is a function of x, such as
$$f(x) = x^4 - 2x^2 + 3.$$
If y is used as an element in the range of the function, we can write
$$y = f(x) = x^4 - 2x^2 + 3.$$
Sometimes, instead of $f(x)$, it is useful to write $y(x)$, which means the same thing. The derivative of f is denoted by f', and, similarly, we may denote this derivative as y' or as $y'(x)$. The symbol $\dfrac{dy}{dx}$ is interchangeable with y' and $y'(x)$. Thus we

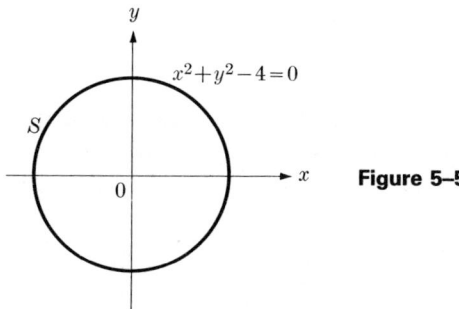

Figure 5–5

could write
$$y = y(x) = x^4 - 2x^2 + 3, \quad \text{and} \quad y' = y'(x) = 4x^3 - 4x.$$

In Chapter 2, Section 4, we discussed relations in the plane. The graph of an equation such as
$$x^2 + y^2 - 4 = 0$$
is an example of such a relation. The solution set S defined by
$$S = \{(x, y) : x^2 + y^2 - 4 = 0\}$$
is a circle, as shown in Fig. 5–5. Although this relation is not a function, we may describe S by means of two functions. Defining
$$S_1 = \{(x, y) : = y = \sqrt{4 - x^2}\}$$
and
$$S_2 = \{(x, y) : y = -\sqrt{4 - x^2}\},$$
we see that $S = S_1 \cup S_2$. The functions describing S_1 and S_2 are
$$y_1 = \sqrt{4 - x^2} \quad \text{and} \quad y_2 = -\sqrt{4 - x^2}.$$
See Figs. 5–6 and 5–7.

Examples of relations which may be defined by decomposition into one or more functions are
$$x^2 - 2xy + 3y^2 - 7 = 0, \quad x^3 y^2 - 4 = 0, \quad y^3 - x - 5 = 0.$$

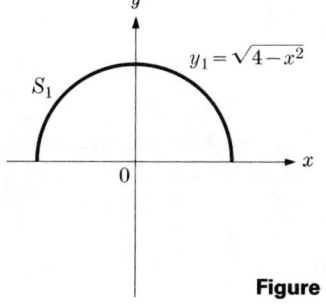

Figure 5–6

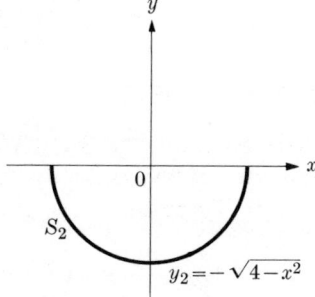

Figure 5–7

Note that all these expressions are of the general form
$$f(x, y) = 0.$$
When this situation occurs we say that the functions, if any, are **defined implicitly** by the equation. We say "if any" because the equation may not define any function at all. For example, there are no values of x and y which satisfy an equation such as
$$x^2 + y^2 + 9 = 0.$$
We could easily compose other examples.

Returning to the equation
$$x^2 + y^2 - 4 = 0,$$
we see that this consists of two functions, each function having for its domain the interval $-2 \leq x \leq 2$. In this interval the functions possess derivatives which can readily be found by differentiating the expressions
$$\sqrt{4 - x^2} \quad \text{and} \quad -\sqrt{4 - x^2}.$$
However, it is possible to proceed directly, without solving for y in terms of x. In the equation
$$x^2 + y^2 - 4 = 0,$$
we first differentiate x^2; its derivative is simply $2x$. The derivative of y^2 is more complicated, since $y = y(x)$ and we must use the chain rule. The derivative of y^2 is $2yy'$. The derivative of -4 is zero. We have now found the derivative of the left side, and we know that the right side, being zero, has derivative zero. We conclude that
$$2x + 2yy' = 0.$$
This may be solved for y' to yield (when $y \neq 0$),
$$y' = -\frac{x}{y}.$$
We now ask ourselves which function of x has been differentiated. The answer is *both*. If y_1 is substituted for y in the right side of the expression for y', we have the derivative of y_1; if y_2 is substituted, we have the derivative of y_2. Further examples illustrate the method.

Example 1. Assuming that y is a differentiable function defined implicitly by the equation
$$y^3 + 3xy + x^3 - 5 = 0,$$
find dy/dx in terms of x and y.

Solution. The derivative of y^3 is $3y^2(dy/dx)$. The term $3xy$ must be treated as a product. The derivative of $3xy$ is $3x(dy/dx) + 3y$. The derivative of x^3 is $3x^2$. The derivative of -5 is 0. Therefore
$$3y^2 \frac{dy}{dx} + 3x \frac{dy}{dx} + 3y + 3x^2 = 0.$$

We can now solve for $\dfrac{dy}{dx}$:

$$\frac{dy}{dx} = -\frac{y + x^2}{y^2 + x}.$$ ◀

Remark. If it turns out that the implicit relation is determined by several functions, then the answer gives the appropriate derivative according to which $y(x)$ is put into the right-hand side of the equation.

Example 2. Find the derivative of

$$4x^2 + 9y^2 = 36$$

implicitly, and check the result by solving for y and differentiating explicitly.

Solution. The implicit method gives

$$8x + 18yy' = 0 \quad \text{and} \quad y' = -\frac{4x}{9y}, \quad y \neq 0.$$

On the other hand, using the explicit method, we find that

$$y_1 = +\tfrac{1}{3}\sqrt{36 - 4x^2}, \quad y_2 = -\tfrac{1}{3}\sqrt{36 - 4x^2},$$

$$y_1'(x) = \tfrac{1}{3} \cdot \tfrac{1}{2}(36 - 4x^2)^{-1/2}(-8x) = -\frac{4x}{3\sqrt{36 - 4x^2}},$$

and $y_2'(x)$ is the same expression preceded by a plus sign.

From the implicit method,

$$y_1' = -\frac{4x}{9y_1} = -\frac{4x}{9 \cdot \tfrac{1}{3}\sqrt{36 - 4x^2}} = -\frac{4x}{3\sqrt{36 - 4x^2}},$$

which is identical with the explicit result; similar results are obtained for $y_2(x)$. ◀

Example 3. Assuming that y is a differentiable function, find the derivative when $y^5 + 3x^2y^3 - 7x^6 - 8 = 0$.

Solution. In this problem we have no choice as to whether we shall use the explicit or the implicit method, since it is impossible to solve for y in terms of x or for x in terms of y. We obtain

$$5y^4 \frac{dy}{dx} + 3x^2 \cdot 3y^2 \frac{dy}{dx} + 3y^3 \cdot 2x - 42x^5 = 0,$$

and it follows that

$$\frac{dy}{dx} = \frac{42x^5 - 6xy^3}{5y^4 + 9x^2y^2}.$$ ◀

PROBLEMS

In each of Problems 1 through 12, find y' in terms of x and y by implicit differentiation, assuming that y is a differentiable function.

1. $x^2 - 2y^2 + 5 = 0$ ~~(crossed out)~~
2. $4x^2 + 8y^3 - 5 = 0$
3. $x^4 + 2y^4 - 4 = 0$ ~~(crossed out)~~
4. $2x^3 + 3y^3 + 6 = 0$
5. $x^2 - 3xy + y^2 = 6$
6. $x^2 + 2xy - 2y^2 + 3x - y - 9 = 0$
7. $x^3 + 6xy + 5y^3 = 3$ ~~(crossed out)~~
8. $x^3 + 2x^2y - xy^2 + 2y^3 = 4$
9. $2x^3 - 3x^2y + 2xy^2 - y^3 = 2$
10. $x^4 + 2x^2y^2 + xy^3 + 2y^4 = 6$
11. $x^6 + 2x^3y - xy^7 = 10$ ~~(crossed out)~~
12. $x^5 - 2x^3y^2 + 3xy^4 - y^5 = 5$

In Problems 13 through 23, find dy/dx in terms of x and y by implicit differentiation; then solve for y in terms of x, and show that each solution and its derivative satisfy the equation obtained.

13. $x^2 - 2xy = 3$
14. $y^2 = 4x$
15. $xy - 4 = 0$
16. $x^{1/2} + y^{1/2} = 1$
17. $x^{2/3} + y^{2/3} = a^{2/3}$, $a = $ const
18. $y^2 - 2x = 4$
19. $y^2 + 2x = 8$
20. $2x^2 + 3y^2 = 5$
21. $2x^2 - y^2 = 1$
22. $x^2 - xy + y^2 = 1$
23. $2x^2 - 3xy - 4y^2 = 5$

In Problems 24 through 29, find y' by implicit differentiation; then solve for y in terms of x and select the particular function passing through the given point. Find the general form for y' for this function and evaluate y' at the given point.

24. $x^2 + y^2 = 25$ $(4, 3)$
25. $y^2 = 2x$ $(2, -2)$
26. $xy = 3$ $(1, 3)$
27. $x^2 + y^2 = 10$ $(3, -1)$
28. $3x^2 - 2xy - y^2 = 3$ $(1, 0)$
29. $2x^2 - 3xy + 2y^2 = 2$ $(-1, -\frac{3}{2})$

30. Find the derivative, dy/dx, of the relation $|x - 3y| - 4 = 0$. Sketch the graph.
31. Find the derivative, y', of the relation $|y^2 - 3xy| - 1 = 0$. Sketch the graph.
32. Given the relation $|x^2 - y^2| - 10(x^2 + y^2) = 0$, find dy/dx. For what values of x is there a corresponding value of dy/dx?
33. Let r be a rational number of the form p/q where p and q are integers. Let $y = x^r$. By raising both sides of this equation to the qth power, use implicit differentiation to obtain a proof of Theorem 10 of Section 3. What facts about implicit functions are used in this proof?

APPLICATIONS OF DIFFERENTIATION. THE DIFFERENTIAL

1. TOOLS FOR APPLICATIONS OF THE DERIVATIVE

As we learned earlier, a **closed interval** along the x axis is an interval which includes its endpoints, an **open interval** is one which excludes the endpoints, and a **half-open** interval contains one endpoint but not the other.

The basic theorem of this section is the **Extreme Value Theorem.**

Theorem 1 (Extreme Value Theorem). *If f is a continuous function defined on the closed interval $[a, b]$, there is (at least) one point in $[a, b]$ (call it x_1) where f has a largest value, and there is (at least) one point (call it x_2) where f has a smallest value.*

This theorem is fairly clear intuitively if we think of a continuous function as one with no breaks or gaps. As we move along the curve from the point corresponding to $x = a$ to the point corresponding to $x = b$, there must be a place where the curve has a high point (called the **maximum value**) and there must also be a place where it has a low point (called the **minimum value**). In spite of the simplicity of the situation from the intuitive point of view, the proof of Theorem 1 is hard. Consequently we shall restrict ourselves to a discussion of the meaning of this theorem.

A good way to examine a theorem critically is to see what happens if some of the hypotheses are altered. In Theorem 1 there are two principal hypotheses: (1) the interval $[a, b]$ is closed, and (2) the function f is continuous. We shall show by example that if we tamper with either hypothesis, the conclusion of the theorem may be false.

Suppose the assumption that the interval is closed is replaced by the assumption that the interval is open. The function $f(x) = 1/x$ is continuous on the open interval $0 < x < 1$ (Fig. 6–1), and it has no maximum value in this open interval. A more subtle example is given by the function $f(x) = x^2$ defined in the open interval $0 < x < 2$ (Fig. 6–2). This function has no maximum or minimum value on the *open* interval but does have a maximum of 4 and a minimum of 0 on the *closed* interval. This situation comes about because $f(x) = x^2$ is continuous on the closed interval $0 \le x \le 2$. The first example of the function $1/x$ is continuous on $0 < x < 1$ but not on the closed interval $0 \le x \le 1$. Even making the interval half-open is not good enough. The first example of the function $1/x$ is continuous

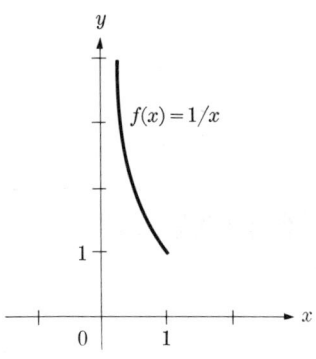

Figure 6–1

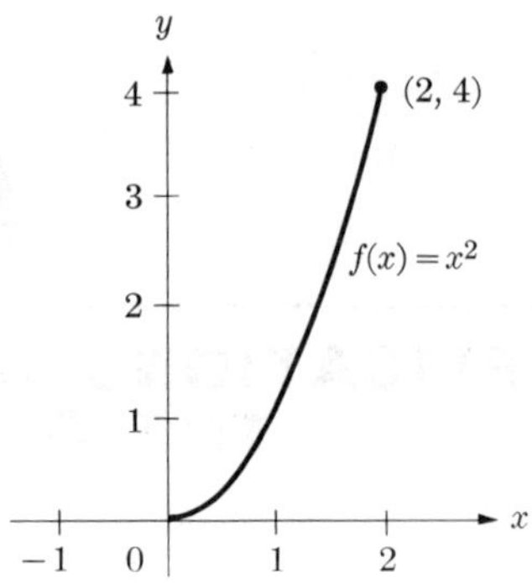

Figure 6–2

on the half-open interval $0 < x \leq 1$ and still has no maximum value there. The second example, $f(x) = x^2$, is continuous on the half-open interval $0 < x \leq 2$ and does not have a minimum in this half-open interval.

The second hypothesis, that of continuity, is also essential. The function

$$f(x) = \frac{1}{x-1}, \qquad 0 \leq x \leq 2, \quad x \neq 1$$

$$f(1) = 2$$

is continuous except at $x = 1$ (Fig. 6–3). As the graph clearly shows, the function has no maximum and no minimum on the closed interval $[0, 2]$. A second, more sophisticated, example is the function

$$f(x) = \begin{cases} x^2 + 1, & 0 \leq x < 1 \\ x - 1, & 1 \leq x \leq 2 \end{cases}$$

whose graph is shown in Fig. 6–4. This function is continuous except at $x = 1$, where its minimum occurs with the value zero. There is no point on the closed

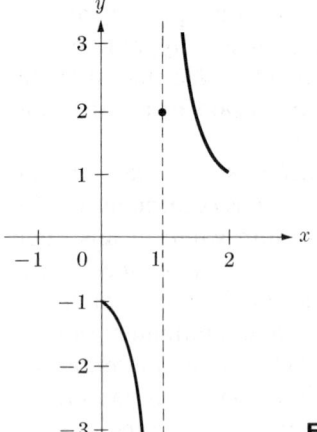

Figure 6–3

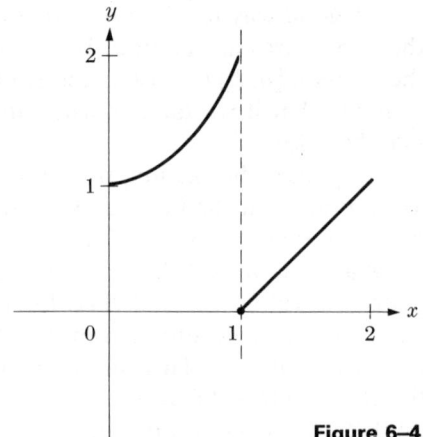

Figure 6–4

interval [0, 2] where it has a maximum value. Note that the function gets close to the value 2 as $x \to 1$ from the left. But there is no point where it actually *has* the value 2.

If f is continuous on $[a, b]$, with its maximum value M at x_1 and its minimum value m at x_2, then Theorem 1 implies that

$$m \leq f(x) \leq M$$

for all x in $[a, b]$. We could equally well write $f(x_1)$ instead of M and $f(x_2)$ instead of m. In words, Theorem 1 states that *a function which is continuous on a closed interval takes on its maximum and minimum values there.*

We shall now prove a theorem concerning the value of the derivative at a maximum or minimum point.

Theorem 2. *Suppose that f is continuous on an interval and takes on its maximum (or minimum) at some point x_0 which is in the interior of the interval. If $f'(x_0)$ exists, then*

$$f'(x_0) = 0.$$

Proof. We prove the theorem for the case where $f(x_0)$ is a maximum. The proof for a minimum is similar. If $f(x_0)$ is the maximum value, then

$$f(x_0 + h) \leq f(x_0)$$

for every possible h, both positive and negative. The only restriction is that $x_0 + h$ must be in the interval in order for $f(x_0 + h)$ to have a meaning. We can also write (see Fig. 6–5)

$$f(x_0 + h) - f(x_0) \leq 0. \tag{1}$$

If h is positive we may divide by h to get

$$\frac{f(x_0 + h) - f(x_0)}{h} \leq 0, \qquad h > 0.$$

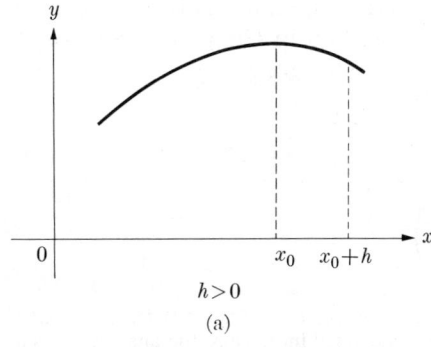
(a) $h > 0$

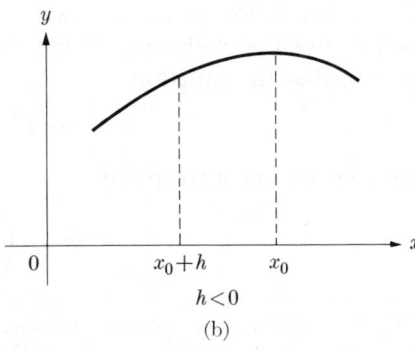
(b) $h < 0$

Figure 6–5

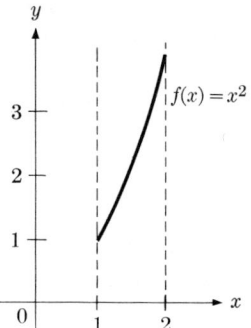

Figure 6-6

Taking the one-sided limit as $h \to 0^+$, we conclude* that

$$\lim_{h \to 0^+} \frac{f(x_0 + h) - f(x_0)}{h} \leq 0. \qquad (2)$$

If h is negative, *the inequality* (1) *reverses* when we divide by h, so that

$$\frac{f(x_0 + h) - f(x_0)}{h} \geq 0, \qquad h < 0.$$

Taking the one-sided limit as $h \to 0^-$, we conclude* that

$$\lim_{h \to 0^-} \frac{f(x_0 + h) - f(x_0)}{h} \geq 0. \qquad (3)$$

Since the ordinary limit (two-sided limit) exists as h tends to zero, the one-sided limits exist and are equal (see Theorem 11 of Chapter 4). Examining (2) and (3), we see that they can only be equal if $f'(x_0) = 0$, which is what we wished to prove.

Discussion. The important hypotheses in Theorem 2 are (1) that x_0 is an interior point, and (2) that f has a derivative at x_0. If the first hypothesis is neglected the theorem is false, as is shown by the example $f(x) = x^2$ on the interval $1 \leq x \leq 2$ (Fig. 6-6). The maximum occurs at $x = 2$ (not an interior point) and the minimum occurs at $x = 1$ (not an interior point). The derivative of $f(x) = x^2$ is $f'(x) = 2x$, and this is different from zero throughout the interval $[1, 2]$. The fact that x_0 is **interior** to the interval is important; whether the interval is open, closed, or half-open is irrelevant. Using the methods described in Chapter 4, Section 4, we see that the function

$$f : x \to x^3 - x, \qquad -1 < x < 1$$

has a maximum at the point

$$\left(-\frac{\sqrt{3}}{3}, \frac{2}{9}\sqrt{3} \right)$$

---

* We are employing the following (intuitively clear) theorem on limits: *if $F(x) \leq 0$ for all x and if $F(x) \to L$ as $x \to a$, then $L \leq 0$.* The same result holds for one-sided limits; also the analogous result holds for functions $F(x) \geq 0$, in which case $L \geq 0$.

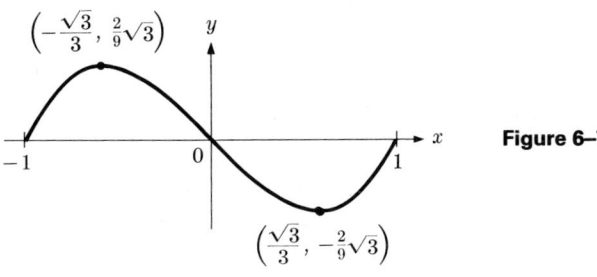

Figure 6–7

and a minimum at

$$\left(\frac{\sqrt{3}}{3}, -\frac{2}{9}\sqrt{3}\right)$$

(see Fig. 6–7). To find these points we first get the derivative of f:

$$f'(x) = 3x^2 - 1.$$

Then, setting this equal to zero and solving for x, we obtain $3x^2 - 1 = 0$, $x = \frac{1}{3}\sqrt{3}, -\frac{1}{3}\sqrt{3}$. Substitution of the x values into the original expression yields the desired points. We shall prove the conclusions above in Section 3.

The second hypothesis in Theorem 2, namely, that f has a derivative at x_0, is also essential, as is shown by the following example. Let

$$f(x) = \begin{cases} -\frac{1}{3}x^2 + 2x - \frac{5}{3}, & 1 \le x \le 2 \\ -x^2 + 2x + 1, & 2 < x \le 3. \end{cases}$$

This function is continuous on the interval $1 \le x \le 3$, since both expressions approach the same value at $x = 2$. As Fig. 6–8 shows, however, there is a "corner" at $x = 2$, and the derivative does not exist at this point. Nevertheless, the point $P(2, 1)$ is exactly where the maximum occurs.

The derivative may not exist by virtue of being infinite at a maximum or minimum point. The relation $x^{2/3} + y^{2/3} = 1$ for $-1 \le x \le 1$, has the graph shown in Fig. 6–9. The portion of the graph above the x axis represents a

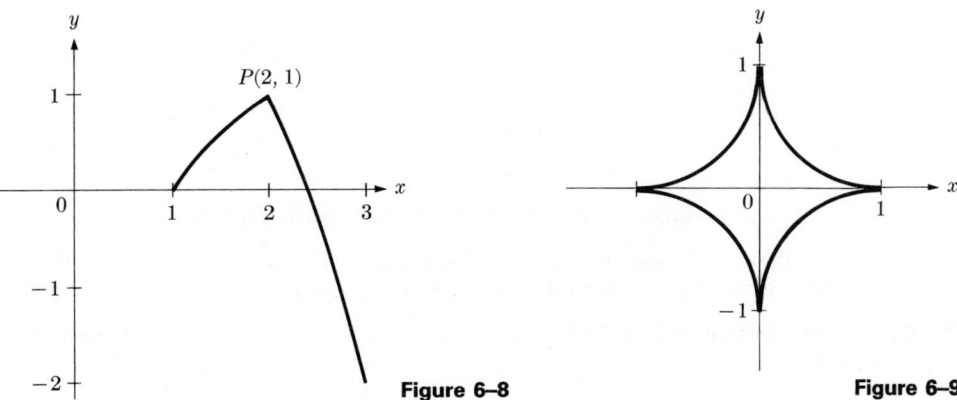

Figure 6–8

Figure 6–9

function. The maximum occurs at $x = 0$. We can compute the derivative implicitly and obtain

$$\frac{2}{3}x^{-1/3} + \frac{2}{3}y^{-1/3}\frac{dy}{dx} = 0 \quad \text{or} \quad \frac{dy}{dx} = -\frac{y^{1/3}}{x^{1/3}},$$

and we see that this tends to infinity as x tends to zero. Therefore there is no derivative at the maximum point.

PROBLEMS

1. Show that the continuous function

$$f(x) = \begin{cases} x, & 0 \le x < 1 \\ 2 - x, & 1 \le x \le 2 \\ x - 2, & 2 < x \le 3 \end{cases}$$

   has its maximum value at two different points. Also show that the minimum value is taken on at two points. Sketch.

2. Construct by formula (as in Problem 1) a continuous function on the interval $[-2, 5]$ which assumes its maximum value at two different points and its minimum value at three different points.

3. Construct a continuous function (by formula) defined on the interval $[1, 6]$ which assumes its maximum value at every point of the subinterval $[2, 4.5]$.

4. Construct a continuous function (by formula) defined on the interval $[0, 5]$ which assumes its minimum at $x = 1$ and at every point of the interval $[2, 3]$ but at no other points.

5. Construct a continuous function (by formula) which is defined on $[-1, 4]$, which has its maximum at $x = 0$, its minimum at $x = 2$ and is such that f' does not exist at $x = 0$ and at $x = 2$.

6. The function

$$f(x) = \frac{x^2 - 1}{x^2 + 1}$$

   is continuous for all values of x from $-\infty$ to $+\infty$. Sketch the graph. Show that f has the range $[-1, 1)$, and that the function never assumes its maximum value. Why does Theorem 1 not apply?

7. After studying the function in Problem 6, construct a continuous function defined for $-\infty < x < \infty$ with range $(-2, 3]$ and which never assumes its minimum value.

8. Given the continuous function

$$f(x) = \begin{cases} x^2/(x^2 + 1), & 0 \le x < +\infty \\ -x^2/(x^2 + 1), & -\infty < x < 0. \end{cases}$$

   Show that f never assumes its maximum or minimum value. Sketch.

In Problems 9 through 14, use the method described in Chapter 4, Section 4. These methods will be justified rigorously in Section 3 of this chapter.

9. Given the function $f(x) = 3x^2 + 6x + 1$ on the interval $-2 \le x \le 0$, find the minimum point.

10. Given the function $f(x) = 3x^2 + 6x + 1$ on the interval $0 \le x \le 3$, find the minimum point.

11. Find the maximum value of
$$f(x) = \begin{cases} x + 1, & 0 \le x \le 1 \\ -x^2 + 3, & 1 < x \le 3 \end{cases}$$
on the interval $0 \le x \le 3$. Where is the minimum value?

12. Find the maximum point of $f(x) = \frac{1}{3}x^3 - x + 2$ on the interval $[-2, 2]$. What is the maximum point on the interval $[-3, 3]$?

13. Given the function
$$f(x) = \begin{cases} -x^2 + 4x - 2, & 1 \le x \le 2 \\ -\frac{1}{2}x^2 + 2x, & 2 < x \le 3, \end{cases}$$
find the maximum point. Does the function have a derivative at this point? If so, find the value of the derivative.

14. Examine the relation
$$x^{1/3} + y^{1/3} = 2, \quad -8 \le x \le 8$$
with regard to maximum and minimum points.

*15. The function $y = \arctan x$ has domain $(-\infty, +\infty)$ and range $(-\pi/2, \pi/2)$. Sketch the graph. Does the function have maximum and minimum values?

*16. Prove the following result: If $F(x) \ge 0$ for all x on an interval $[a, b]$ containing the point c and if $F(x) \to L$ as $x \to c$, then $L \ge 0$. [*Hint*: Assume that L is negative and use the definition of limit to show there must be a value of x near c where $F(x) < 0$, thereby reaching a contradiction. (See the footnote on page 138).]

17. Write out a proof of Theorem 2 (page 137) for the case where f takes on its minimum value.

18. Suppose that a function f defined on the closed interval $[a, b]$ assumes its maximum and minimum values on this interval. Is f continuous on $[a, b]$? Justify your answer.

19. A function f defined on $[a, b]$ has a **left-hand derivative** at b if
$$\lim_{h \to 0^-} \frac{f(b + h) - f(b)}{h}$$
exists. The **right-hand derivative** at a is defined similarly. Show that the Extreme Value Theorem holds for functions which are differentiable on (a, b) and have one-sided derivatives at a and b.

2. FURTHER TOOLS: ROLLE'S THEOREM; THEOREM OF THE MEAN

Theorems 1 and 2 of Section 1 are the basis for the following result, known as Rolle's Theorem.

Theorem 3 (**Rolle's Theorem**). *Suppose that f is continuous for $a \le x \le b$ and that $f'(x)$ exists for each x between a and b. If*
$$f(a) = f(b) = 0,$$

then there must be (at least) one point, call it x_0, between a and b such that
$$f'(x_0) = 0.$$

Proof. There are three possibilities:

CASE 1. (The trivial case.) $f(x) = 0$ for all x between a and b; then $f'(x) = 0$ for all x, and x_0 can be chosen to be any value between a and b.

CASE 2. $f(x)$ is positive somewhere between a and b. Then the maximum of f is positive, and we choose x_0 (Theorem 1) to be a place where this maximum occurs. (See Fig. 6–10a.) According to Theorem 2 of the previous section, $f'(x_0) = 0$, since x_0 must be interior to the interval.

CASE 3. $f(x)$ is negative somewhere between a and b. Then the minimum of f is negative, and we choose x_0 to be a place where this minimum occurs. (See Fig. 6–10b.) According to Theorem 2 of the previous section, $f'(x_0) = 0$.

Since every function which is zero at a and b must fall into one of the three cases, the theorem is proved.

Figure 6–10(c) shows that a function may fall into both Case 2 and Case 3, and Fig. 6–10(d) illustrates the possibility of several choices for x_0 even though it satisfies only Case 2.

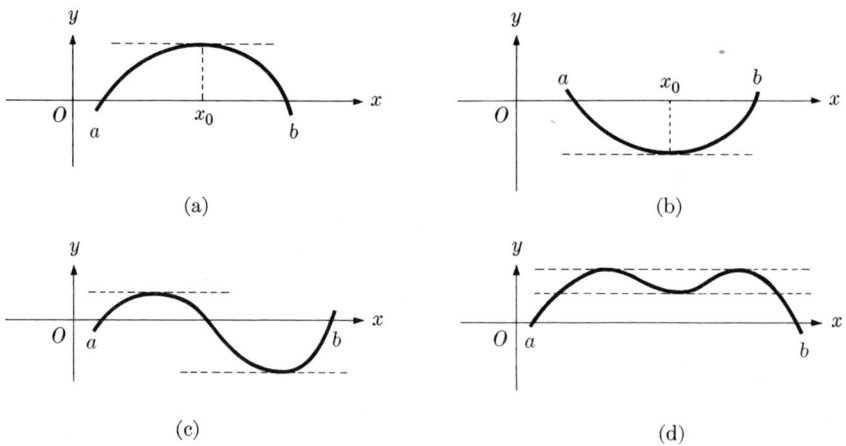

Figure 6–10

We can state Rolle's Theorem in a simple way: If a differentiable curve crosses the x axis twice there must be a point between successive crossings at which the tangent to the curve is parallel to the x axis.

We have been building up to the proof of the theorem known as the **Theorem of the Mean,** one which is used over and over again throughout the branch of mathematics known as analysis. Every mathematician who works in analysis encounters it often, uses it, and feels at home with it. We shall see that, important though it is, it is merely a slight variation of Rolle's Theorem.

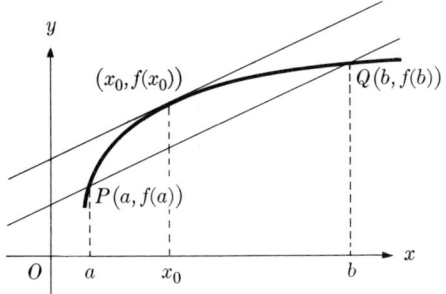

Figure 6-11

Theorem 4 (Theorem of the Mean). *Suppose that f is continuous for $a \leq x \leq b$ and that $f'(x)$ exists for each x between a and b. Then there is an x_0 between a and b (that is, $a < x_0 < b$) such that*

$$f'(x_0) = \frac{f(b) - f(a)}{b - a}.$$

Before proving this theorem we shall discuss its meaning from a geometric standpoint. Figure 6-11 shows a typical function f between the points a and b. The point P has coordinates $(a, f(a))$ and Q has coordinates $(b, f(b))$. We construct the straight line through PQ and calculate its slope. We know that the slope m is the difference of the y values over the difference of the x values, that is,

$$m = \frac{f(b) - f(a)}{b - a}.$$

This is exactly the same expression that occurs in the statement of the Theorem of the Mean. The theorem says there is a point $(x_0, f(x_0))$ on the curve where the slope has such a value; that is, the tangent line at $(x_0, f(x_0))$ is parallel to the line through PQ. Glancing at the figure, we see that there must be such a point. In fact, by means of the following device, we can see that there must be one: We look at the figure and tilt it so that the line through PQ appears horizontally as the x axis. Then the Theorem of the Mean resembles Rolle's Theorem.

Proof of Theorem 4. The equation of the line through PQ (according to the two-point formula for the equation of a straight line) is

$$y - f(a) = \frac{f(b) - f(a)}{b - a}(x - a).$$

We construct the function

$$F(x) = f(x) - \frac{f(b) - f(a)}{b - a}(x - a) - f(a).$$

By straight substitution with $x = a$ and then $x = b$, we find

$$F(a) = f(a) - \frac{f(b) - f(a)}{b - a}(a - a) - f(a) = 0,$$

$$F(b) = f(b) - \frac{f(b) - f(a)}{b - a}(b - a) - f(a) = 0.$$

Therefore $F(x)$ satisfies all the hypotheses of Rolle's Theorem. There must be a value x_0 such that $F'(x_0) = 0$. But (by differentiation) we see that
$$F'(x) = f'(x) - \frac{f(b) - f(a)}{b - a}.$$
This implies that
$$f'(x_0) = \frac{f(b) - f(a)}{b - a},$$
which is what we wished to prove.

Example 1. Given that
$$f(x) = \frac{x + 2}{x + 1} \quad \text{and} \quad a = 1, b = 2,$$
find all values x_0 in the interval $1 < x < 2$ such that
$$f'(x_0) = \frac{f(b) - f(a)}{b - a}.$$

Solution
$$f'(x) = \frac{(x + 1) \cdot 1 - (x + 2) \cdot 1}{(x + 1)^2} = \frac{-1}{(x + 1)^2},$$
$$f(1) = \frac{3}{2}, \quad f(2) = \frac{4}{3}, \quad \frac{f(b) - f(a)}{b - a} = -\frac{1}{6}.$$

We solve the equation $f'(x_0) = -\frac{1}{6}$, which yields
$$\frac{-1}{(x_0 + 1)^2} = -\frac{1}{6} \quad \text{or} \quad x_0^2 + 2x_0 - 5 = 0,$$
and
$$x_0 = \frac{-2 \pm \sqrt{24}}{2} = -1 \pm \sqrt{6}.$$

The value $x_0 = -1 + \sqrt{6}$ is in the interval $(1, 2)$, while the value $x_0 = -1 - \sqrt{6}$ is rejected since it is outside this interval. ◂

Example 2. Given that $f(x) = x^3 - 2x^2 + 3x - 2$ and $a = 0, b = 2$, find all possible values for x_0 in the interval $0 < x < 2$ such that
$$f'(x_0) = \frac{f(b) - f(a)}{b - a}.$$

Solution
$$f'(x) = 3x^2 - 4x + 3, \quad f(0) = -2, \quad f(2) = 4,$$
so that
$$\frac{f(b) - f(a)}{b - a} = \frac{4 + 2}{2} = 3.$$

We solve the equation
$$3x_0^2 - 4x_0 + 3 = 3,$$

which has two solutions: $x_0 = 0$ and $x_0 = \frac{4}{3}$. But $x_0 = 0$ is not *between* 0 and 2, and therefore the only answer is $x_0 = \frac{4}{3}$. ◀

Example 3. Given the function
$$f(x) = \frac{x^2 - 4x + 3}{x - 2}$$
and
$$a = 1, b = 3,$$
discuss the validity of the Theorem of the Mean.

Solution. If we proceed formally, we see that $f(1) = 0$ and $f(3) = 0$, so we must find an x_0 in the interval $(1, 3)$ such that $f'(x_0) = 0$. Computing the derivative by the quotient rule, we write
$$f'(x) = \frac{(x - 2)(2x - 4) - (x^2 - 4x + 3)}{(x - 2)^2} = \frac{x^2 - 4x + 5}{(x - 2)^2}.$$

Setting this equal to zero, we obtain
$$x_0^2 - 4x_0 + 5 = 0$$
and
$$x_0 = \frac{4 \pm \sqrt{16 - 20}}{2} = \frac{4 \pm \sqrt{-4}}{2},$$

which is impossible! We look once again at the function f and see that it becomes infinite at $x = 2$. Therefore the hypotheses of the Theorem of the Mean are not satisfied, and the theorem is not applicable. There is no value x_0. ◀

PROBLEMS

In each of Problems 1 through 12, find all numbers x_0 between a and b which satisfy the equation
$$f'(x_0) = \frac{f(b) - f(a)}{b - a}.$$

1. $f(x) = x^2 - 3x - 2$, $a = 0, b = 2$
2. $f(x) = x^3 - 5x^2 + 4x - 2$, $a = 1, b = 3$
3. $f(x) = x^3 - 7x^2 + 5x$, $a = 1, b = 5$
4. $f(x) = x^3 + 2x^2 - x$, $a = -3, b = 2$
5. $f(x) = \dfrac{x - 2}{x + 2}$, $a = 0, b = 1$
6. $f(x) = \dfrac{2x + 3}{3x - 2}$, $a = 1, b = 5$
7. $f(x) = \sqrt{25 - x^2}$, $a = -3, b = 4$
8. $f(x) = \sqrt{x^2 + 81}$, $a = 12, b = 40$
9. $f(x) = x^4 - 2x^3 + x^2 - 2x$, $a = -1, b = 2$

10. $f: x \to x^4 + x^3 - 3x^2 + 2x$, $a = -2, b = -1$

11. $f: x \to \dfrac{x^2 - 3x - 4}{x + 5}$, $a = -1, b = 4$

12. $f: x \to \dfrac{x^2 + 6x + 5}{x - 6}$, $a = 1, b = 5$

13. Given that $f(x) = (2x + 3)/(3x + 2)$, $a = -1$, $b = 0$, show that there is no number x_0 between a and b which satisfies the Theorem of the Mean. Sketch the graph.

14. Given that $f(x) = 2x^{2/3}$, $a = -1$, $b = 1$, show that there is no number x_0 between a and b which satisfies the Theorem of the Mean. Sketch the graph.

15. Given that
$$f(x) = 1/(x - 1)^2, \quad a = -1, b = 2.$$
Is there a number x_0 between a and b which satisfies the equation
$$(b - a)f'(x_0) = f(b) - f(a)?$$
Can the Theorem of the Mean be used?

16. Given that f is a quadratic function of x, and a and b are any numbers, show that $x_0 = \tfrac{1}{2}(a + b)$ is the value that satisfies the Theorem of the Mean. [*Hint:* Assume that $f(x) = cx^2 + dx + e$, where c, d, and e are any constants.]

*17. Given that f is a general equation of the third degree. That is, suppose $f(x) = \alpha x^3 + \beta x^2 + \gamma x + \delta$, with $\alpha \neq 0$. Show that there are at most two values of x_0 which satisfy the Theorem of the Mean for any given numbers a and b. Prove that these are given by the formula
$$x_0 = -\frac{\beta}{3\alpha} \pm \frac{1}{3\alpha}\sqrt{\beta^2 + 3\alpha^2(a^2 + ab + b^2) + 3\alpha\beta(a + b)}.$$

18. Given the function f defined for $0 \leq x \leq 2$ by the formula
$$f(x) = \begin{cases} x & \text{for } 0 \leq x < 1 \\ \tfrac{1}{2}x^2 + \tfrac{1}{2} & \text{for } 1 \leq x \leq 2. \end{cases}$$
Does the Theorem of the Mean hold for $0 \leq x \leq 2$? Justify your answer.

19. Same as Problem 18 for the function
$$f(x) = \begin{cases} 2x^3 - 1 & \text{for } 0 \leq x < 1 \\ 6x - 5 & \text{for } 1 \leq x \leq 2. \end{cases}$$

20. Same as Problem 18 for the function
$$f(x) = \begin{cases} x^2 + 2x - 1 & \text{for } 0 \leq x < 1 \\ x^4 + 2 & \text{for } 1 \leq x \leq 2. \end{cases}$$

3. APPLICATIONS TO GRAPHS OF FUNCTIONS

Knowing how the derivative of a function behaves helps us in obtaining an accurate idea of the graph of the function. In this and the following section we establish some rules (in the form of theorems) which are most useful for graphing functions; we also give some examples to illustrate the technique.

Definition. *A function f is said to be* **increasing on the interval** *I if $f(x_2) > f(x_1)$ whenever $x_2 > x_1$, so long as both x_1 and x_2 are in I. It is* **decreasing** *if*

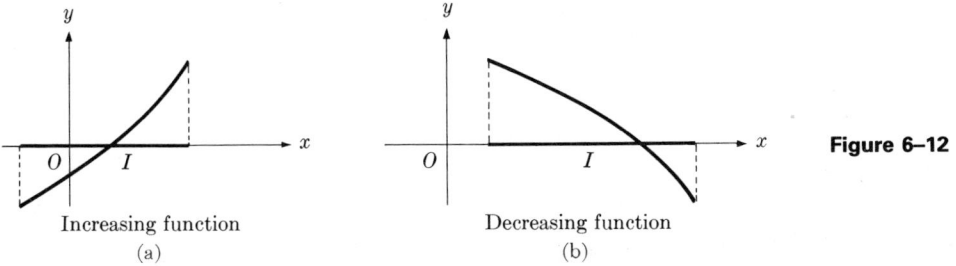

Increasing function
(a)

Decreasing function
(b)

Figure 6–12

$f(x_2) < f(x_1)$ *whenever* $x_2 > x_1$ (Fig. 6–12). *The interval I may contain one endpoint, both endpoints, or neither endpoint.*

Definition. *A function f is said to have a* **relative maximum** *at x_0 if there is some interval with x_0 as an interior point, such that $f(x_0)$ is the true maximum of f in this interval. Similarly, it has a* **relative minimum** *at x_1 if there is some interval with x_1 as interior point, such that $f(x_1)$ is the true minimum of f in this interval.*

The latter definition needs some explanation, as provided in Fig. 6–13, which shows a typical situation. The function f has its maximum on $[a, b]$ at x_0, a point which is a relative maximum. However, f has a relative maximum at x_2, although $f(x_2)$ is not the maximum of the function on $[a, b]$. The function f has a relative minimum at x_1. The true minimum on $[a, b]$ occurs at b. However, f does not have a relative minimum at b, since the definition for relative minimum is not satisfied there.

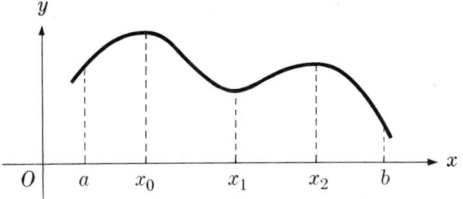

Figure 6–13

Theorem 5. *If f is continuous on an interval I and if $f'(x) > 0$ for each x in the interior of I, then f is increasing on I.*

Proof. We apply the Theorem of the Mean to two points x_1, x_2 in I. We find that

$$f(x_2) - f(x_1) = f'(x_0)(x_2 - x_1),$$

where x_0 is between x_1 and x_2 and hence interior to I. For $x_2 > x_1$ and f' positive, we obtain $f'(x_0)(x_2 - x_1) > 0$, and so $f(x_2) - f(x_1) > 0$. This means, by definition, that f is increasing.

Corollary. *If f is continuous on I and $f'(x) < 0$ for each x interior to I, then f is decreasing on I.*

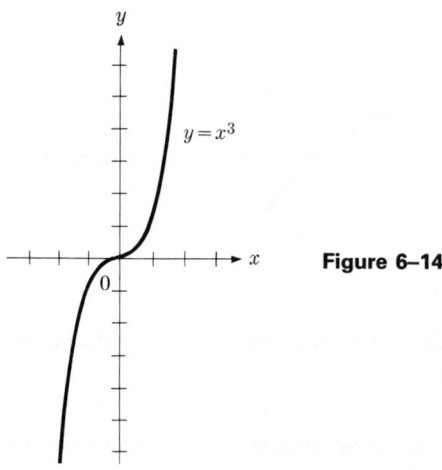

Figure 6-14

Theorem 5 is useful for graphing; we can find out whether the graph of a function is rising or falling by observing whether the derivative is positive or negative. This theorem (together with Theorem 2 of Section 1, which says that at a relative maximum or minimum the derivative of a function is zero) helps us find the peaks and troughs.

There is a complication in using Theorem 2 because, although the derivative of a function is zero at a maximum or minimum point (assuming the derivative exists there), the converse may not hold true. For example, the function $f(x) = x^3$ (Fig. 6-14) has derivative $f'(x) = 3x^2$, which is zero at $x = 0$. But $(0, 0)$ is neither a maximum nor a minimum point of the function. Thus the knowledge that the derivative of a function vanishes at a certain point is not enough to guarantee that the function has a maximum or minimum there.

Definitions. *A* **critical value** *of a function f is a value of x where $f'(x) = 0$. A* **critical point** *of a function f is the point $(x, f(x))$ on the graph corresponding to the critical value x.*

A critical point occurs at any relative maximum or minimum point of a function which has a derivative at that point. However, the function $f(x) = x^3$ has a critical point at $(0, 0)$, but this point is neither a maximum nor a minimum of the function.

Suppose a function f and its derivative f' are continuous. If the equation $f'(x) = 0$ has only a finite number of solutions, we may use Theorem 5 and its corollary as an aid in graphing. To do so, we let $x_1, x_2, \ldots, x_n$ be *all* the values of x for which $f'(x) = 0$, arranged in order of increasing size. In each of the intervals $(-\infty, x_1), (x_1, x_2), (x_2, x_3), \ldots, (x_{n-1}, x_n), (x_n, +\infty)$, the quantity $f'(x)$ must remain either *positive throughout* or *negative throughout*. This fact is a result of Theorem 6, stated below without proof. As a result, to determine the sign of $f'(x)$ in any one of the intervals, we need only find its sign at a single interior point in that interval.

Theorem 6. *Suppose f is continuous on an interval I and f is not zero at any point of I. Then f is either positive on all of I or negative on all of I.*

This theorem is usually proved in a course in advanced calculus. It is a consequence of the Intermediate Value Theorem, stated in Chapter 7 (Theorem 10, page 214).

We give an example showing how the determination of the zeros of f' can be used as an aid in graphing.

Example 1. Study the derivative of the function $f(x) = \frac{1}{3}x^3 - x^2 - 3x + 3$ and use the resulting knowledge to sketch the graph of the function.

Solution. The derivative is $f'(x) = x^2 - 2x - 3 = (x + 1)(x - 3)$. The derivative is zero for $x_1 = -1$ and $x_2 = 3$. We consider the intervals $(-\infty, -1)$, $(-1, 3)$, and $(3, +\infty)$. We pick a single point in each interval, selecting the point -3 in $(-\infty, -1)$, the point 0 in $(-1, 3)$, and the point 5 in $(3, +\infty)$. We determine the sign of $f'(x)$ at each of these points and the value of f at each of them, as shown in the table below. With these aids we are ready to graph the function (Fig. 6–15).

x	-3	-1	0	3	5
$f'(x)$	$+$	0	$-$	0	$+$
$f(x)$	-6	$4\frac{2}{3}$	3	-6	$4\frac{2}{3}$

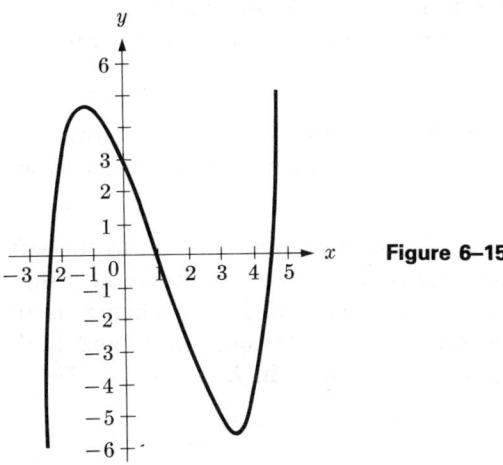

Figure 6–15

The critical value $x = -1$ corresponds to a relative maximum, and the critical value $x = 3$ corresponds to a relative minimum. ◀

The theorems and definitions presented in this section give us a method for finding relative maxima and minima. We state the method in the form of a test.

Test 1 (Theorem 7). i) *If f is increasing ($f' > 0$) on some interval to the left of x_0 with x_0 as endpoint of this interval, and if f is decreasing ($f' < 0$) on some interval to the*

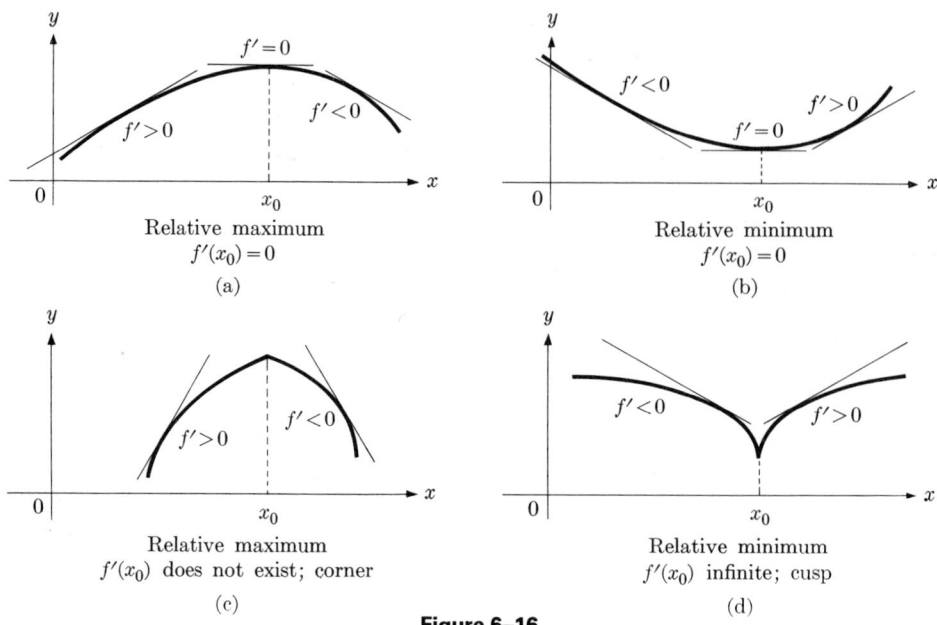

Figure 6-16

right of x_0 (with x_0 as endpoint), then f has a relative maximum at x_0 if it is continuous there.

ii) *If f is decreasing ($f' < 0$) in some interval to the left of x_0 with x_0 as endpoint of this interval, and if f is increasing ($f' > 0$) in some interval to the right of x_0 (with x_0 as endpoint), then f has a relative minimum at x_0 if it is continuous there.*

Remarks. If f has a derivative at x_0, then the derivative will be zero. However, we include the possibility that there may be a cusp or a corner. Figure 6–16 shows various possibilities.

Proof of Theorem 7. To prove part (i), we denote by I' the interval to the left (including x_0) and by I'' the interval to the right of x_0; then we combine them into one interval and call this interval I (Fig. 6–17). That is, $I' \cup I'' = I$. Then x_0 is an interior point of I. Since f is increasing on I' and decreasing on I'' and both contain the point x_0, we must have $f(x_0) \geq f(x)$ for all x in I.

The proof of (ii) is the same.

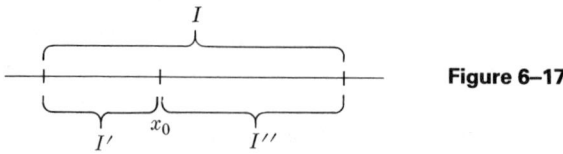

Figure 6-17

Example 2. Discuss the function f defined by
$$f(x) = x^{5/3} + 5x^{2/3}$$

for relative maxima and minima, and determine the intervals in which f is increasing and those in which f is decreasing. Sketch the graph.

Solution. The derivative is
$$f'(x) = \tfrac{5}{3}x^{2/3} + \tfrac{10}{3}x^{-1/3} = \tfrac{5}{3}x^{-1/3}(x + 2).$$
The critical value is $x = -2$. The derivative is not defined at $x = 0$. We construct the following table (* means undefined):

x	-5	-2	-1	0	1
$f'(x)$	$+$	0	$-$	$*$	$+$
$f(x)$	0	$3(2)^{2/3}$	4	0	6

We conclude that
f increases for $x \leq -2$;
f decreases for $-2 \leq x \leq 0$;
f increases for $x \geq 0$.

We now apply Test I and conclude that there is a relative maximum at $x = -2$ and a relative minimum at $x = 0$. The graph is sketched in Fig. 6–18. ◀

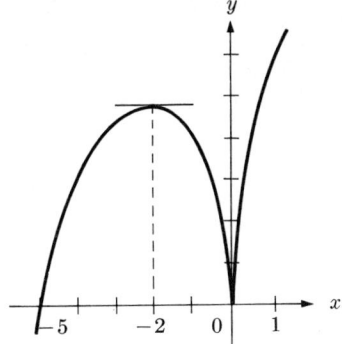

Figure 6–18

Example 3. Discuss the function f defined by
$$f(x) = \frac{x^2 + x + 7}{\sqrt{2x + 1}}, \qquad x > -\frac{1}{2},$$
for relative maxima and minima, and determine where f is increasing and where it is decreasing. Sketch the graph.

Solution. The derivative is
$$f'(x) = \frac{(2x + 1)^{1/2}(2x + 1) - (x^2 + x + 7)\tfrac{1}{2}(2x + 1)^{-1/2}(2)}{(2x + 1)}.$$
We simplify by multiplying numerator and denominator through by $(2x + 1)^{1/2}$

and obtain

$$f'(x) = \frac{(2x+1)^2 - (x^2+x+7)}{(2x+1)^{3/2}} = \frac{3(x+2)(x-1)}{(2x+1)^{3/2}}.$$

We set this equal to zero and find that $x = 1$ is a critical value. The apparent solution $x = -2$ is excluded, since the function is defined only for $x > -\frac{1}{2}$. We easily discover that

$$f(x) \text{ decreases for } -\tfrac{1}{2} < x \leq 1; \quad f(x) \text{ increases for } x \geq 1.$$

Test I now applies, and we find that the value $x = 1$ yields a relative minimum. The graph is sketched in Fig. 6–19. ◀

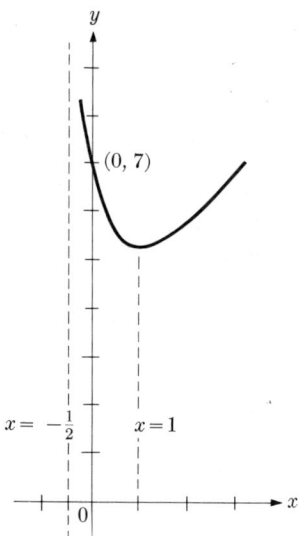

Figure 6–19

PROBLEMS

In Problems 1 through 32, discuss each function for relative maxima and minima, and determine those intervals in which the function is increasing and those in which it is decreasing. Sketch the graphs.

1. $x^2 + 3x + 4$
2. $x^2 - 4x + 5$
3. $-2x^2 + x - 6$
4. $-3x^2 - 3x + 2$
5. $2x^3 + 3x^2 - 12x$
6. $x^3 - 6x^2 + 9x + 5$
7. $x^3 + 2x^2 - 3x - 2$
8. $x^3 - 3x^2 + 6x - 3$
9. $x^3 + 6x^2 + 12x - 5$
10. $x^3 + 3x^2 + 6x - 3$
11. $-x^3 + 2x^2 - x + 1$
12. $-x^3 - 2x^2 + 3x - 6$
13. $x^4 - \tfrac{4}{3}x^3 - 4x^2 + \tfrac{2}{3}$
14. $x^4 + \tfrac{4}{3}x^3 - 12x^2$
15. $x^4 + 2x^3$
16. $x^4 + 4x^3 + 6x^2$
17. $x + \dfrac{1}{x}$
18. $\dfrac{2x}{x^2 + 1}$

19. $\dfrac{x-1}{x+1}$

20. $\dfrac{3x-2}{2x+3}$

21. $\dfrac{x^2}{\sqrt{x+1}}, \quad x > -1$

22. $3x^{1/2} - x^{3/2}$

23. $x^{2/3}(x+3)^{1/3}$

24. $x/(x+1)$

25. $x\sqrt{3-x}, \quad x \le 3$

26. $x^2\sqrt{5-x}, \quad x \le 5$

27. $x\sqrt{2-x^2}, \quad |x| \le \sqrt{2}$

28. $\dfrac{x^2 - 2x + 1}{x + 1}$

29. $\dfrac{x^2 - 3x - 4}{x - 2}$

30. $|x^2 - 2x| + 1$

31. $x^2 + |2x + 2|$

32. $\dfrac{x-2}{|x^2 - 2x| + 1}$

*33. Given the function $f(x) = ax^3 + bx^2 + cx + d$ which is the most general equation of the third degree. Figure 6–15 exhibits the shape of this function when $a = \frac{1}{3}$, $b = -1, c = -3, d = 3$. Figure 6–14 exhibits the shape when $a = 1, b = 0, c = 0, d = 0$. Show that there are basically four possible shapes for a third-degree equation and sketch the remaining two shapes. (Assume $a \ne 0$.)

*34. The general equation of the fourth degree has the form $f(x) = ax^4 + bx^3 + cx^2 + dx + e$. Assuming that $a \ne 0$, decide how many different general shapes the graph of such an equation can exhibit. Sketch one of each type.

4. APPLICATIONS USING THE SECOND DERIVATIVE

The more information we have about a function the more accurately we can construct its graph. In Section 3 we saw that a knowledge of the first derivative helps us decide when the graph is increasing and when it is decreasing. Furthermore, we learned that Test I enables us to locate the relative maxima and minima. The second derivative yields additional facts which are helpful in determining the nature of the graph. The first of these facts is a test for relative maxima and minima which employs the second derivative.

Definition. *If, at each point of an interval, the graph of a function f always remains above the line tangent to the curve at this point, we say that the curve is* **concave upward** *on the interval* (see Fig. 6–20). *If the curve always remains below its tangent line, we say it is* **concave downward** (see Fig. 6–21).

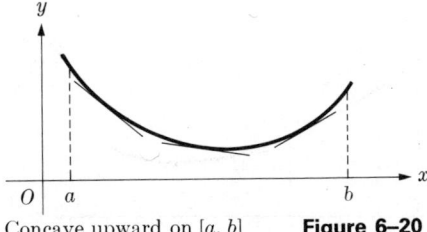

Concave upward on [a, b]   **Figure 6–20**

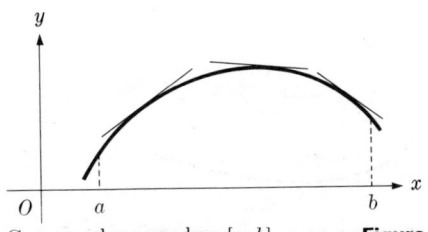

Concave downward on [a, b]   **Figure 6–21**

Before stating and proving the basic theorem of this section we recall the equation of the line tangent to a curve. At any value x_0 the function f has value

$f(x_0)$ and the slope of the curve at this point is $f'(x_0)$. The equation of a line through the point $(x_0, f(x_0))$ with slope $f'(x_0)$ is, according to the point-slope formula,

$$y - f(x_0) = f'(x_0)(x - x_0) \quad \text{or} \quad y = f(x_0) + f'(x_0)(x - x_0).$$

This is the tangent line.

Theorem 8. *Assume that f has a second derivative on an interval I.*

a) *If $f''(x) > 0$ for x interior to I, then the curve is concave upward on I.*

b) *If $f''(x) < 0$ for x interior to I, then the curve is concave downward on I.*

Proof. To prove (a), we must show that the curve lies above the tangent line at any point. Let x_0 be any (fixed) point in I (see Fig. 6-22). The tangent line at x_0 has equation

$$y = f(x_0) + f'(x_0)(x - x_0),$$

and we therefore have to establish (for all x in I) the inequality

$$f(x) \geq f(x_0) + f'(x_0)(x - x_0).$$

If $x = x_0$, this becomes $f(x_0) \geq f(x_0)$, which is true. If x is some other value, say x_1, then we can apply the Theorem of the Mean to the function f between the two points x_0 and x_1. This theorem says that there is a value $\bar{x}$ such that

$$f'(\bar{x}) = \frac{f(x_1) - f(x_0)}{x_1 - x_0},$$

where $\bar{x}$ is between x_0 and x_1. We have two possibilities: (1) $x_1 > x_0$, and (2) $x_1 < x_0$. In Case (1) we write the Theorem of the Mean in the form

$$f(x_1) = f(x_0) + f'(\bar{x})(x_1 - x_0). \tag{1}$$

Since $f'' > 0$ by hypothesis, we know from Theorem 5 that f' is increasing on I. If $x_1 > x_0$, then

$$f'(\bar{x}) > f'(x_0).$$

We multiply this last inequality through by the positive quantity $(x_1 - x_0)$ to get

$$f'(\bar{x})(x_1 - x_0) > f'(x_0)(x_1 - x_0),$$

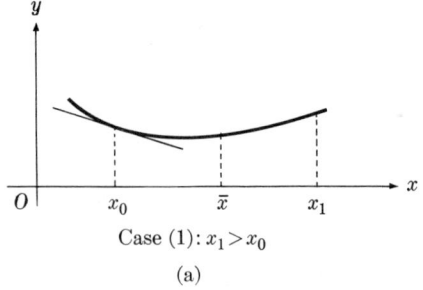

Case (1): $x_1 > x_0$

(a)

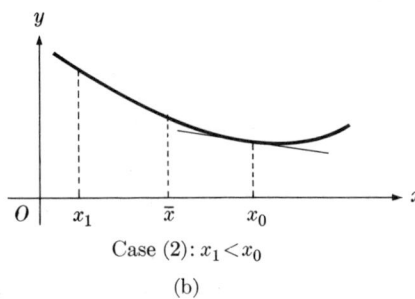

Case (2): $x_1 < x_0$

(b)

Figure 6-22

and therefore, by substitution in Eq. (1), we obtain
$$f(x_1) > f(x_0) + f'(x_0)(x_1 - x_0),$$
which is what we wished to show.

In Case (2), since $x_0 > \bar{x}$, we must have $f'(x_0) > f'(\bar{x})$. But now we multiply through by the negative number $(x_1 - x_0)$, which reverses the inequality, giving
$$f'(\bar{x})(x_1 - x_0) > f'(x_0)(x_1 - x_0)$$
as before, and therefore
$$f(x_1) > f(x_0) + f'(x_0)(x_1 - x_0).$$

The proof of part (b) of Theorem 8 is similar. ◀

Theorem 8 shows that knowledge of the second derivative gives us an even clearer picture of the appearance of the curve. In addition, it gives us the following useful test, known as the **Second Derivative Test,** for relative maxima and minima.

Test II (Theorem 9). *Assume that f has a second derivative, that f″ is continuous, and that x_0 is a critical value ($f'(x_0) = 0$). Then*

a) *If $f''(x_0) > 0$, f has a relative minimum at x_0.*

b) *If $f''(x_0) < 0$, f has a relative maximum at x_0.*

c) *If $f''(x_0) = 0$, the test fails.*

Proof. To prove part (a), we see from Theorem 8 that the curve is concave upward and must lie above the tangent line at x_0. But this line is horizontal, since $f'(x_0) = 0$. Therefore $f(x_0)$ must be a minimum value. The proof of (b) is the same. Part (c) is added for the sake of completeness.

Example 1. Discuss the function
$$f(x) = x^3 - \tfrac{21}{4}x^2 + 9x - 4$$
for relative maxima and minima. Sketch the graph.

Solution. The derivative is
$$f'(x) = 3x^2 - \tfrac{21}{2}x + 9 = 3(x^2 - \tfrac{7}{2}x + 3),$$
and the critical values are solutions of
$$x^2 - \tfrac{7}{2}x + 3 = 0 \quad \text{or} \quad x = \tfrac{3}{2}, 2.$$
We apply the Second Derivative Test.
$$f''(x) = 6x - \tfrac{21}{2}; \quad f''(\tfrac{3}{2}) = -\tfrac{3}{2} < 0, \quad f''(2) = \tfrac{3}{2} > 0.$$
Therefore $x = \tfrac{3}{2}$ corresponds to a maximum, and $x = 2$ corresponds to a minimum. Also, the second derivative is positive if $x \geq \tfrac{7}{4}$ and negative if $x \leq \tfrac{7}{4}$. That is, the curve is concave upward for $x \geq \tfrac{7}{4}$ and concave downward for $x \leq \tfrac{7}{4}$. We construct the following table.

x	0	$\frac{3}{2}$	$\frac{7}{4}$	2	3
f	-4	$\frac{17}{16}$	$1\frac{1}{32}$	1	$\frac{11}{4}$
f'	$+$	0	$-$	0	$+$
f''	$-$	$-$	0	$+$	$+$

This problem indicates the difficulty that may arise when a relative maximum and a relative minimum are close together. Without knowledge of the derivatives given in the table, we would normally plot the points at $x = 0, 1, 2, 3$, etc. This would miss the maximum at $x = \frac{3}{2}$ and give a misleading picture of the curve. We know that the curve is concave downward to the left of $x = \frac{7}{4}$ and concave upward to the right of $\frac{7}{4}$. These facts prevail indefinitely far to the left and right. The graph is sketched in Fig. 6–23. In this example the point $x = \frac{7}{4}$ is important, as it separates a concave downward region from a concave upward region. ◀

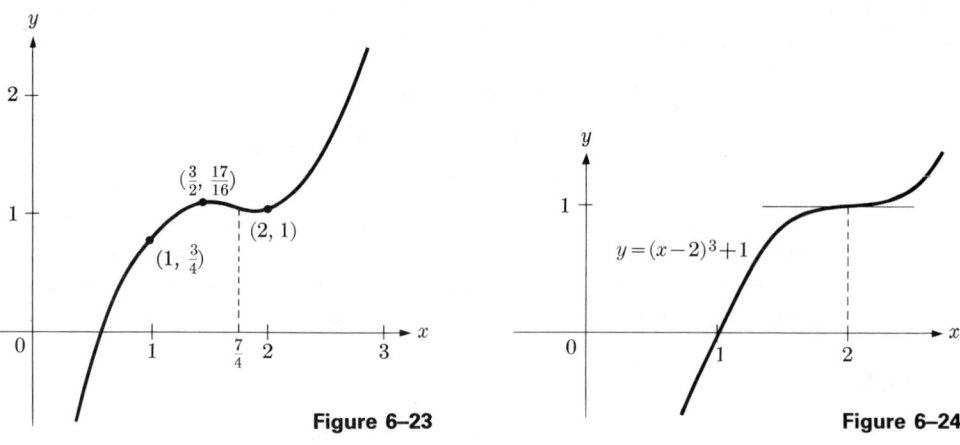

Figure 6–23

Figure 6–24

Definition. *A point on a curve is a* **point of inflection** *if* $f''(x_0) = 0$ *at this point and if the graph is concave upward on one side and concave downward on the other.*

We make two remarks which the reader should keep in mind: The first derivative may or may not vanish at a point of inflection. In Example 1 the value $x = \frac{7}{4}$ corresponds to a point of inflection, and $f'(\frac{7}{4}) = -\frac{3}{16}$. On the other hand, the function $f(x) = (x - 2)^3 + 1$ has a point of inflection (see Fig. 6–24) at $x = 2$, since $f''(x) = 6(x - 2)$ and $f''(2) = 0$. However, we notice that $f'(x) = 3(x - 2)^2$ and $f'(2) = 0$, also.

The second remark: It is not enough to know that $f''(x_0) = 0$ to guarantee that x_0 corresponds to a point of inflection. We must also know that $f''(x) > 0$ on one side and that $f''(x) < 0$ on the other. An example which shows this difficulty is the function $f(x) = x^4$ (see Fig. 6–25). In this case, $f''(x) = 12x^2$ and $f''(0) = 0$.

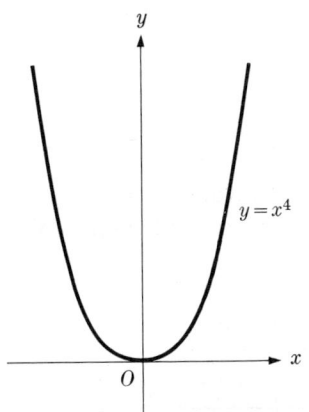

Figure 6–25

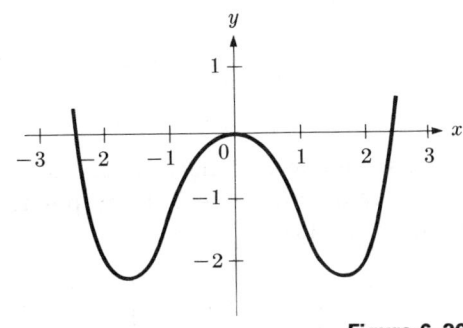

Figure 6–26

However, $f''(x)$ is always positive and, in fact, $x = 0$ is the value for a minimum. (Part (c) of Theorem 9 for the Second Derivative Test is the rule which holds here.) Test I, however, works to give a minimum at $x = 0$.

Example 2. Discuss the function

$$f(x) = \tfrac{1}{4}x^4 - \tfrac{3}{2}x^2$$

for relative maxima, relative minima, and points of inflection. Sketch the graph.

Solution. The derivatives are

$$f'(x) = x^3 - 3x \quad \text{and} \quad f''(x) = 3x^2 - 3.$$

The critical values are solutions of $x^3 - 3x = 0$, and so we get $x = 0, \sqrt{3}, -\sqrt{3}$. The Second Derivative Test tells us that

$x = 0$ is at a relative maximum;

$x = \sqrt{3}, -\sqrt{3}$ are at relative minima.

The possible points of inflection are obtained from solutions of $3x^2 - 3 = 0$; that is, $x = +1, -1$. Since $f''(x)$ is negative for $-1 < x < 1$ and positive for $|x| > 1$, both $x = 1$ and $x = -1$ yield points of inflection. We construct the table:

x	-2	$-\sqrt{3}$	-1	0	1	$\sqrt{3}$	2
f	-2	$-\tfrac{9}{4}$	$-\tfrac{5}{4}$	0	$-\tfrac{5}{4}$	$-\tfrac{9}{4}$	-2
f'	$-$	0	$+$	0	$-$	0	$+$
f''	$+$	$+$	0	$-$	0	$+$	$+$

The graph is symmetric with respect to the y axis (Fig. 6–26). ◀

Example 3. Discuss the function

$$f(x) = x^{4/3} + 4x^{1/3}$$

for relative maxima, relative minima, and points of inflection. Sketch the graph.

Solution. We have

$$f'(x) = \tfrac{4}{3}x^{1/3} + \tfrac{4}{3}x^{-2/3} = \tfrac{4}{3}x^{-2/3}(x+1);$$

$$f''(x) = \tfrac{4}{9}x^{-2/3} - \tfrac{8}{9}x^{-5/3} = \tfrac{4}{9}x^{-5/3}(x-2).$$

The first and second derivatives do not exist at $x = 0$. The value $x = -1$ is a critical value and $x = 2$ corresponds to a possible point of inflection. We construct the table:

x	-4	-1	$-\tfrac{1}{2}$	0	1	2	3
f	0	-3	$-\tfrac{7}{4}\sqrt[3]{4}$	0	5	$6\sqrt[3]{2}$	$7\sqrt[3]{3}$
f'	$-$	0	$+$	$*$	$+$	$+$	$+$
f''	$+$	$+$	$+$	$*$	$-$	0	$+$

From the table we see that $x = -1$ yields a relative minimum and that $x = 2$ yields a point of inflection. Furthermore, $x = 0$ yields a point where the curve changes from concave upward to concave downward. The graph is shown in Fig. 6–27. ◀

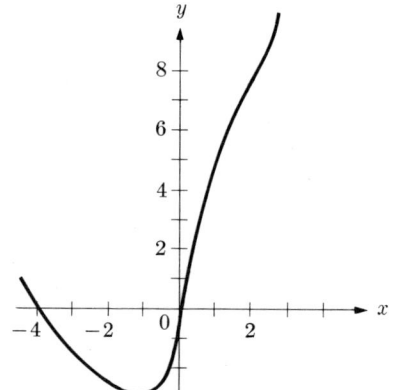

Figure 6–27

PROBLEMS

In Problems 1 through 28, discuss each of the functions for relative maxima and minima, concavity, and points of inflection. Sketch the graphs.

1. $f(x) = x^2 - 3x + 4$
2. $f(x) = -2x^2 + 6x + 5$
3. $f(x) = x^3 - 27x + 4$
4. $f(x) = x^3 + 3x^2 + 6x - 4$
5. $f(x) = x^4 + 4x^3$
6. $f(x) = x^4 - 4x^3 - 2x^2 + 12x - 8$
7. $f(x) = x^2 + \dfrac{1}{x^2}$
8. $f(x) = 2x^2 - \dfrac{1}{x^2}$

9. $f(x) = \dfrac{2x}{x^2 + 1}$

10. $f(x) = x\sqrt{x + 3}$

11. $f(x) = x^3 - \tfrac{3}{2}x^2 - 6x + 2$

12. $f(x) = x^3 + x^2 - x - 1$

13. $f(x) = x^3 - 4x^2 + 4x - 1$

14. $f(x) = x^3 + 3x^2 - 3x - 5$

15. $f(x) = x^3 - x^2 + x - 1$

16. $f(x) = x^4 + \tfrac{4}{3}x^3 - 4x^2 - \tfrac{4}{3}$

17. $f(x) = (x + 2)(x - 2)^3$

18. $f(x) = x^4 - 3x^3 + 3x^2$

19. $f(x) = x^4 + 5x^3 + 6x^2$

20. $f(x) = x - 3 + \dfrac{2}{x + 1}$

21. $f(x) = \dfrac{4x}{x^2 + 4}$

22. $f(x) = 5x^{2/3} - x^{5/3}$

23. $f(x) = x\sqrt{8 - x^2},\ |x| \leq \sqrt{8}$

24. $f(x) = x^{2/3}(x + 2)^{-1}$

25. $f(x) = x^2\sqrt{5 + x},\ x \geq -5$

26. $f(x) = x^2\sqrt{3 - x^2},\ |x| \leq \sqrt{3}$

27. $f(x) = x^{1/3}(x + 2)^{-2/3}$

28. $f(x) = |x^2 - 6x| + 2$

In Problems 29 and 30, find the relative maxima and minima.

29. $f(x) = x^3 + |2x^2 - 2x| + 1$

30. $f(x) = \dfrac{x - 1}{|x^2 - 4x| + 2}$

*31. Given the function $f : x \to x^n$, where n is a positive integer. Show that f has a minimum at $x = 0$ if n is an even integer and that f has a point of inflection at $x = 0$ if n is an odd integer.

*32. Using the results of Problem 31 as a guide, devise a "Higher Derivative Test" for relative maxima and minima when both $f'(x_0) = 0$ and $f''(x_0) = 0$. Describe the situation if the first k derivatives of f vanish at x_0 but the derivative of order $k + 1$ is different from zero.

33. If f is a given function, devise a definition of a point of inflection at a point at which the second derivative does not exist. Construct an example of a function which has such a point of inflection.

34. Show that if the function $f(x) = a_0x^3 + a_1x^2 + a_2x + a_3$ has exactly one critical point, then it is a point of inflection.

35. Show that the function

$$f(x) = \dfrac{ax + b}{cx + d}$$

has no relative maxima or minima, where a, b, c, d are any numbers with $ad - bc \neq 0$.

5. THE MAXIMUM AND MINIMUM VALUES OF A FUNCTION ON AN INTERVAL

In the preceding sections we studied ways of finding relative maxima and minima of a function. One of these relative maxima may be the true maximum of the function, or it may not be. As Fig. 6–28 shows, the function exhibited has a relative maximum at P and a relative minimum at Q. As we go off to the right, however, the function gets larger than the value at P, and as we go off to the left, the function gets smaller than the value at Q. Suppose that we look at the above function only on some interval $[a, b]$, as shown. Then the maximum occurs at b and is $f(b)$, while the minimum occurs at a and is $f(a)$. If $[a, b']$ were the interval

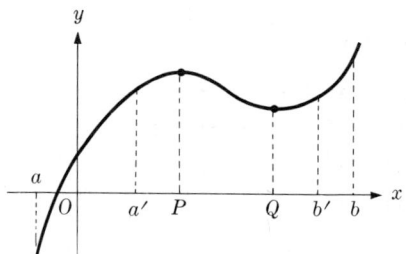

Figure 6-28

of interest, as shown, then the maximum of f on $[a, b']$ is at P, while the minimum is at a. Further, on the interval $[a', b']$ the maximum is at P, the minimum at Q.

To obtain the maximum and minimum of a continuous function on an interval $[a, b]$, we employ the following procedure (which is based on Theorem 5 and Test I):

Rule. a) *Find the relative maximum and minimum values of f.*
b) *Find the value of the function at each of the endpoints.*

The largest of the values of (a) and (b) is the maximum. The smallest value is the minimum.

Example. Given the function

$$f(x) = \tfrac{1}{3}x^3 + \tfrac{1}{2}x^2 - 2x,$$

find the maximum and minimum on the interval $[-3, 4]$.

Solution. We follow the Rule and find relative maxima and minima:

$$f'(x) = x^2 + x - 2,$$

and the critical values are $x = -2, 1$. The Second Derivative Test gives us

$$f''(x) = 2x + 1; \qquad f''(-2) = -3,$$

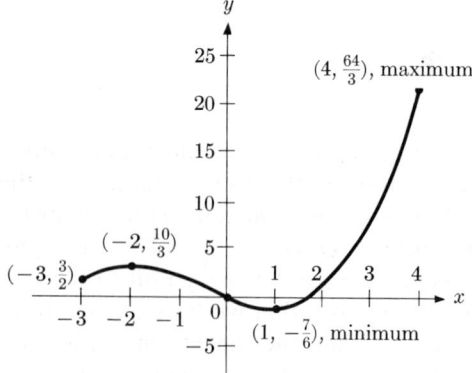

Figure 6-29

making -2 a relative maximum, and

$$f''(1) = 3,$$

which makes 1 a relative minimum. That is, the point $(1, -\frac{7}{6})$ is a relative minimum, and $(-2, \frac{10}{3})$ is a relative maximum. Now we go on to part (b) of the Rule: $f(-3) = \frac{3}{2}$ and $f(4) = \frac{64}{3}$. Since $\frac{64}{3}$ is larger than $\frac{10}{3}$, the maximum of the function occurs at $x = 4$, the right endpoint. Since $-\frac{7}{6}$ is smaller than $\frac{3}{2}$, the minimum of the function is at $x = 1$, the relative minimum. Figure 6–29 shows the various points. ◂

PROBLEMS

In Problems 1 through 10, find the maximum and minimum of the given function on the interval indicated.

1. $f(x) = x^2 + 2x - 4, \quad -4 \le x \le 3$
2. $f(x) = -2x^2 + x - 3, \quad -1 \le x \le 6$
3. $f(x) = 2x^3 - x^2 + 2, \quad -2 \le x \le 1$
4. $f(x) = x^3 + 2x^2 + 18x, \quad -1 \le x \le 2$
5. $f(x) = x^3 + 3x^2 - 9x + 4, \quad -5 \le x \le 6$
6. $f(x) = x^4 - 2x^2 + 1, \quad -2 \le x \le 1$
7. $f(x) = x^4 - 2x^2 + 1, \quad 0 \le x \le 5$
8. $f(x) = x^4 - 2x^2 + 1, \quad -5 \le x \le 5$
9. $f(x) = \dfrac{x}{x+1}, \quad -\frac{1}{2} \le x \le 1$
10. $f: x \to \dfrac{x+3}{x-2}, \quad -3 \le x \le 1$
11. Prove that the Rule of this section fails if the function considered is not continuous.

In each of Problems 12 through 18, find the maxima and minima of the functions on the interval indicated, or show that there are none.

12. $f(x) = x^2, \quad -2 \le x < 1$
13. $f(x) = x^3 - 2x^2 + x - 1, \quad -3 < x < 3$
14. $f(x) = \dfrac{x-1}{x+1}, \quad -2 \le x \le 2$
15. $f(x) = \dfrac{x^2}{x-2}, \quad 1 \le x \le 3$
16. $f(x) = \dfrac{x^2+2}{x+2}, \quad -2 < x \le 0$
17. $f(x) = \dfrac{1}{x+1}, \quad 2 \le x < \infty$
18. $f(x) = \dfrac{x-1}{x+1}, \quad -\infty < x \le -2$

19. Given $f: x \to x^3 + ax^2 + bx + c$ defined on an interval $[d_1, d_2]$. Show that if $a^2 < 3b$ then the maximum and minimum of f must occur at the endpoints.

20. Let r be a rational number. Let $f: x \to (1 + x)^r - (1 + rx)$ be defined for $-1 \leq x \leq a$. Locate the maximum and minimum values of f. Show that if $r \geq 1$, then the minimum must occur at $x = 0$.

6. APPLICATIONS OF MAXIMA AND MINIMA

Until now we have encountered functions in the form of polynomials, rational functions, and so on. Once we know these functions, we set out to find the various relative maxima and minima, maxima at endpoints, and such other properties as concavity and points of inflection.

Now we shall complicate matters by treating problems in which we must discover the function itself before we can discuss its properties. Having found the function, we then use the methods described in the previous sections of this chapter to determine the function's characteristics. In order to be equal to the task, the student should memorize the formulas for areas and volumes of simple geometric shapes, as listed below. In many of the problems these formulas turn out to be likely candidates for functions.

i) Circle of radius r. Circumference $= 2\pi r$. Area $= \pi r^2$.
ii) Circular sector. Area $= \frac{1}{2}r^2\alpha$, α being the central angle, measured in radians.
iii) Trapezoid of height h and bases b and B. Area $= \frac{1}{2}h(b + B)$.
iv) Right circular cylinder of height h, radius of base r. Volume $= \pi r^2 h$. Lateral surface area $= 2\pi rh$.
v) Right circular cone of height h, radius of base r. Volume $= \frac{1}{3}\pi r^2 h$. Lateral surface area $= \pi rL$, where $L = \sqrt{r^2 + h^2}$.
vi) Sphere of radius r. Volume $= \frac{4}{3}\pi r^3$. Surface area $= 4\pi r^2$.

If a manufacturer wants to make a tin can at the lowest possible cost, he faces the problem of finding a *minimum*. Making the strongest possible bridge of a certain size, type and span presents a problem of finding a *maximum*. Whenever we use words such as largest, most, least, smallest, best, etc., we can easily translate them into mathematical language in terms of maxima and minima. If we have a specific formula for the function in question (a situation which is very often impossible in actual practice), then we may be able to use the methods of the calculus to find the required maximum or minimum.

We shall start by giving several examples of the operating technique and then outline the procedure by giving a set of rules consisting of five steps.

Example 1. A man has a stone wall alongside a field. He has 1200 meters of fencing material and he wishes to make a rectangular pen, using the wall as one side. What should the dimensions of the pen be in order to enclose the largest possible area?

Solution. Here we see that the problem is to find the largest area; clearly this is a maximum problem. If we can find the area as a function of something or other

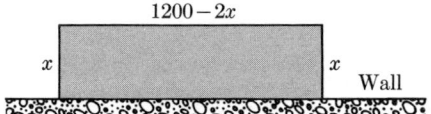

Figure 6-30

and then differentiate it, we may be able to find the maximum of this function and so get the answer. Let us then draw a figure (Fig. 6–30), and call the area of the pen by some letter, say A. The length and width of the pen are both unknown. However, if x is the width, then the length must be $1200 - 2x$, since there are 1200 m of fencing to be used. Now we can get an expression for the area, namely, length times width or

$$A = A(x) = x(1200 - 2x) = 1200x - 2x^2.$$

We note that x must lie between 0 and 600.

The derivative gives us

$$A'(x) = 1200 - 4x,$$

and this vanishes when $x = 300$.

The Second Derivative Test tells us that

$$A'' = -4,$$

and so a relative maximum occurs at $x = 300$. Actually, it is *the* maximum, since $A(x)$ is a quadratic function which is concave downward everywhere. In fact, $A = 0$ when $x = 0$ and $x = 600$, so the endpoints give a minimum. We conclude that the width of the pen must be 300 meters and the length $1200 - 600 = 600$ meters. ◄

Example 2. A rectangular box with an open top is to be made in the following way. A piece of tin 10 cm by 16 cm has a small square cut from each corner (shaded portion in Fig. 6–31) and then the edges (dashed lines) are folded vertically. What should be the size of the squares cut out if the box is to have as large a volume as possible?

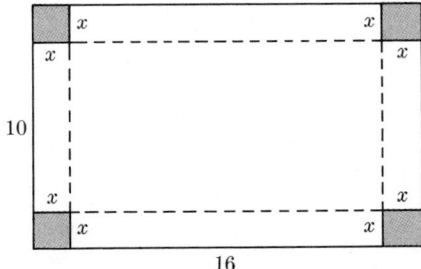

Figure 6-31

Solution. We see that if we cut out exceptionally small squares the box will have practically no height at all and so will have a small volume. Similarly, if the squares are too large the base of the box will be tiny and again the volume will be small. We seek a formula for the volume. We let x be the edge of one of the squares cut out. Then:

$$16 - 2x \text{ will be the length of the base of the box,}$$
$$10 - 2x \text{ will be the width of the base of the box,}$$
$$x \text{ will be the height of the box.}$$

The volume, of course, is length times width times height.
Therefore, calling the volume V, we have

$$V = V(x) = (16 - 2x)(10 - 2x)x \quad \text{and} \quad V(x) = 4(40x - 13x^2 + x^3),$$

which gives volume as a function of x. Now we must note that there are restrictions on x. First of all, x must be positive in order to make any sense at all. Secondly, x must be less than 5. For, once it gets to 5, the width of the base is zero and we have no box at all. We write

$$0 < x < 5.$$

With this in mind we calculate the derivative:

$$V'(x) = 4(40 - 26x + 3x^2),$$

and by solving

$$40 - 26x + 3x^2 = 0,$$

we obtain

$$x = 2, 6\tfrac{2}{3}.$$

But $6\tfrac{2}{3}$ is outside the interval of interest and we reject it. We note that

$$V''(x) = (-26 + 6x) \quad \text{and} \quad V''(2) < 0,$$

which gives a relative maximum at $x = 2$. The endpoints of the interval $[0, 5]$ both give zero volume. Therefore $x = 2$ cm yields the true maximum, and this is the answer.* The volume of the box is 144 cm^3. ◀

Example 3. The sum of one number and three times a second number is 60. Among the possible numbers which satisfy this condition, find the pair whose product is as large as possible.

Solution. We start by letting x be one of the numbers and y the other. We want to find the maximum of the product, which we shall call P. So we write

$$P = xy.$$

We have expressed P in terms not of one quantity but of two. We can correct this difficulty by recalling that the first sentence of the problem asserts that

$$x + 3y = 60.$$

We can eliminate either x or y; if we eliminate y, we obtain

$$P = x\tfrac{1}{3}(60 - x) = 20x - \tfrac{1}{3}x^2.$$

Since P is a function of x, we can differentiate. We write

$$P' = P'(x) = 20 - \tfrac{2}{3}x.$$

---

* Based on the Rule in Section 5.

Setting this equal to zero, we find that $x = 30$ and $y = 10$. It is easy to verify that this value of x gives the maximum value to P. The result is the same if we eliminate x and express P as a function of y. ◄

Using these three examples as a guide, we now list the steps the student should take in attacking this type of problem in maxima and minima.

Step 1: Draw a figure when appropriate.

Step 2: Assign a letter to each of the quantities mentioned in the problem.

Step 3: Select the quantity which is to be made a maximum or minimum and express it as a function of the other quantities.

Step 4: Use the information in the problem to eliminate all quantities but one so as to have a function of one variable. Determine the possible domain of this function.

Step 5: Use the methods of Sections 4 and 5 to get the maximum or minimum.

With this five-step procedure in mind, we now study two more examples.

Example 4. Find the dimensions of the right circular cylinder of maximum volume which can be inscribed in a sphere of radius 12.

Solution. When a cylinder is inscribed in a sphere we mean that the upper and lower bases of the cylinder have their bounding circles on the surface of the sphere. The axis of the cylinder is along a diameter of the sphere. We first draw a cross section of the inscribed cylinder, as in Fig. 6–32 (Step 1). The cylinder might be short and fat or tall and thin. We label the radius of the base r and the height of the cylinder h; the volume is denoted by V (Step 2). Then (Step 3),

$$V = \pi r^2 h.$$

We have now expressed the volume V in terms of two quantities, so we must eliminate one (Step 4). Referring to the figure, we use the Pythagorean Theorem to get

$$r^2 + \frac{h^2}{4} = 144.$$

This means that

$$h = 2\sqrt{144 - r^2},$$

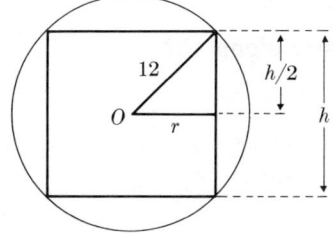

Figure 6–32

and
$$V = 2\pi r^2 \sqrt{144 - r^2}, \quad 0 < r < 12.$$
By differentiating (Step 5), we obtain
$$V'(r) = 4\pi r\sqrt{144 - r^2} + 2\pi r^2 \cdot \tfrac{1}{2}(144 - r^2)^{-1/2}(-2r)$$
$$= 4\pi r\sqrt{144 - r^2} - \frac{2\pi r^3}{\sqrt{144 - r^2}}$$
$$= \frac{2\pi r(288 - 2r^2 - r^2)}{\sqrt{144 - r^2}}.$$

Setting this equal to zero, we find that $r = 0$ or $r = \pm 4\sqrt{6}$. We reject $r = 0$ and the negative value of r. Since the Second Derivative Test is rather messy, we reason as follows: V is continuous for $0 \le r \le 12$; when $r = 0$ and $r = 12$, we get $V = 0$. The volume is positive for r in $(0, 12)$. V takes on its maximum on $[0, 12]$ at some r_0 which must be interior to the interval. Thus $V'(r_0) = 0$. But $V'(r) = 0$ for only one value of r in $(0, 12)$, so that value must be $4\sqrt{6}$. Therefore the critical value $r_0 = 4\sqrt{6}$ must give the maximum. From the relation between r and h we find that $h = 8\sqrt{3}$. ◂

Example 5. A lighthouse is at point A, 4 km offshore from the nearest point O of a straight beach; a store is at point B, 4 km down the beach from O. If the lighthouse keeper can row 4 km/hr and walk 5 km/hr, how should he proceed in order to get from the lighthouse to the store in the least possible time?

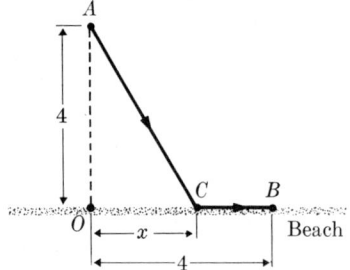

Figure 6–33

Solution. Clearly he will row to some point on the beach between O (the nearest point) and the store at B and then walk the rest of the way. We draw a figure such as Fig. 6–33 (Step 1), where C is the spot on the beach where he lands and x is the distance from O to C. We denote by T the time of the trip from A to B (Step 2); T is the quantity we wish to make a minimum. To get a formula for T, we use the fact that *rate times time equals distance.* We then have
$$T = \text{time of trip along } AC + \text{time of trip along } CB$$
$$= \frac{\text{distance } AC}{4} + \frac{\text{distance } CB}{5}.$$
To express this in terms of one quantity is not hard, since
$$CB = 4 - x,$$

and, using right triangle AOC,
$$AC = \sqrt{16 + x^2}.$$
Therefore (Step 3),
$$T(x) = \tfrac{1}{4}\sqrt{16 + x^2} + \tfrac{1}{5}(4 - x).$$
The only part of Step 4 we need is the determination of the domain of the function T. Since C is between O and B we have the restriction
$$0 \le x \le 4.$$
Taking the derivative (Step 5), we obtain
$$T'(x) = \tfrac{1}{8}(16 + x^2)^{-1/2}(2x) - \tfrac{1}{5}$$
and, setting this equal to zero, we find that
$$\frac{x}{4\sqrt{16 + x^2}} = \frac{1}{5}$$
or
$$x = \pm \frac{16}{3}.$$
But these values of x are outside the interval $[0, 4]$. The conclusion is that there are no relative maxima or minima in the interval $0 \le x \le 4$. Therefore the minimum must occur at one of the endpoints. We have
$$T(0) = \tfrac{9}{5}$$
and
$$T(4) = \sqrt{2}.$$
Since $\sqrt{2} < \tfrac{9}{5}$, the fastest method for the lighthouse keeper is to row directly to the store B and do no walking. ◀

PROBLEMS

1. A rectangle has a perimeter of 120 meters. What length and width yield the maximum area? What is the result when the perimeter is L units?
2. Find the dimensions of the rectangle of maximum area that can be inscribed in a circle of radius 6. What is the result for a circle of radius R?
3. Find the dimensions of the right circular cylinder of maximum volume which can be inscribed in a sphere of radius R.
4. A horizontal gutter is to be made from a long piece of sheet iron 8 cm wide by turning up equal widths along the edges into vertical position. How many centimeters should be turned up at each side to yield the maximum carrying capacity?
5. The difference between two numbers is 20. Select the numbers so that the product is as small as possible.
6. A box with a square base is to have an open top. The area of the material in the box is to be 100 cm². What should the dimensions be in order to make the volume as large as possible? What is the result for an area of S square centimeters?

7. A window is in the shape of a rectangle surmounted by a semicircle. Find the dimensions when the perimeter is 12 meters and the area is as large as possible.

8. Find the radius and central angle (in radians) of the circular sector of maximum area having a perimeter of 16 cm.

9. The top and bottom margins of a page are each $1\frac{1}{2}$ cm and the side margins are each 1 cm. If the area of the printed material per page is fixed at 30 cm^2, what are the dimensions of the page of least area?

10. A closed right circular cylinder (i.e., top and bottom included) has a surface area of 100 cm^2. What should the radius and altitude be in order to provide the largest possible volume? Find the result if the surface area is S cm^2.

11. A right circular cone has a volume of 120 cm^3. What shape should it be in order to have the smallest lateral surface area? Find the result if the volume is V cm^3.

12. At midnight, ship B was 90 km due south of ship A. Ship A sailed east at 15 km/hr and ship B sailed north at 20 km/hr. At what time were they closest together?

13. Suppose that in Example 5 on page 166, the lighthouse is 5 km from shore, the store is 6 km down the beach from O, the lighthouse keeper can row 2 km/hr, and he can walk 4 km/hr. Where should he land in order to get from the lighthouse to the store in the shortest time?

14. Find the dimensions of the rectangle having the largest area which can be inscribed in the ellipse

$$\frac{x^2}{a^2} + \frac{y^2}{b^2} = 1.$$

15. Find the coordinates of the point or points on the curve $y = 2x^2$ which are closest to the point $(9, 0)$.

16. Find the coordinates of the point or points on the curve $x^2 - y^2 = 16$ which are nearest to the point $(0, 6)$.

17. (a) A right triangle has hypotenuse of length 13 and one leg of length 5. Find the dimensions of the rectangle of largest area which has one side along the hypotenuse and the ends of the opposite side on the legs of this triangle. (b) What is the result for a hypotenuse of length H with an altitude to it of length h?

18. A trough is to be made from a long strip of sheet metal 12 cm wide by turning up strips 4 cm wide on each side so that they make the same angle with the bottom of the trough (trapezoidal cross section). Find the width across the top such that the trough will have maximum carrying capacity.

19. The sum of three positive numbers is 30. The first plus twice the second plus three times the third add up to 60. Select the numbers so that the product of all three is as large as possible.

20. The stiffness of a given length of beam is proportional to the product of the width and the cube of the depth. Find the shape of the stiffest beam which can be cut from a cylindrical log (of the given length) with cross-sectional diameter of 4 cm.

21. (a) A manufacturer makes aluminum cups of a given volume (16 cm^3) in the form of right circular cylinders open at the top. Find the dimensions which use the least material. (b) What is the result for a given volume V?

22. In Problem 21, suppose that the material for the bottom is $1\frac{1}{2}$ times as expensive as the material for the sides. Find the dimensions which give the lowest cost.

23. Find the shortest segment with ends on the positive x and y axes which passes through the point $(1, 8)$.

24. The product of two numbers is 16. Determine them so that the square of one plus the cube of the other is as small as possible.
25. Find the dimensions of the cylinder of greatest lateral area which can be inscribed in a sphere of given radius R.
26. A piece of wire of length L is cut into two parts, one of which is bent into the shape of a square and the other into the shape of a circle. (a) How should the wire be cut so that the sum of the enclosed areas is a minimum? (b) How should it be cut to get the maximum enclosed areas?
27. Solve Problem 26 if one part is bent into the shape of an equilateral triangle and the other into a circle.
28. Solve Problem 26 if one part is bent into the shape of an equilateral triangle and the other into a square.
29. Find the dimensions of the right circular cone of maximum volume which can be inscribed in a sphere of given radius R.
30. Find the dimensions of the right circular cone of minimum volume which can be circumscribed about a sphere of radius R.
31. A right circular cone is to have a volume V. Find the dimensions so that the lateral surface area is as small as possible.
32. A silo is to be built in the form of a right circular cylinder surmounted by a hemisphere. If the cost of the material per square meter is the same for floor, walls, and top, find the most economical proportions for a given capacity V.
33. Work Problem 32, given that the floor costs twice as much per square meter as the sides and the hemispherical top costs three times as much per square meter as the sides.
34. A tank is to have a given volume V and is to be made in the form of a right circular cylinder with hemispheres attached at each end. The material for the ends costs twice as much per square meter as that for the sides. Find the most economical proportions.
35. Find the length of the longest rod which can be carried horizontally around a corner from a corridor 8 m wide into one 4 m wide. [*Hint:* Observe that this length is the minimum value of certain lengths.]
36. Suppose the velocity of light is V_1 in air and V_2 in water. A ray of light traveling from a point P_1 above the surface of the liquid to a point P_2 below the surface will travel by the path which requires the least time. Show that the ray will cross the surface at the point Q in the vertical plane through P_1 and P_2 so placed that

$$\frac{\sin \theta_1}{V_1} = \frac{\sin \theta_2}{V_2},$$

where θ_1 and θ_2 are the angles shown in Fig. 6–34.

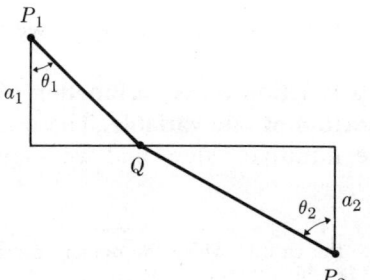

Figure 6–34

37. A manufacturer makes widgets to sell so that an order of 1000 costs \$3 per hundred. For each additional 100 in the order, the price is reduced by 6¢ per hundred. What size order will yield the maximum dollar value?

38. Wheat is sold at \$4 per bushel for the first 1000 bushels. On larger orders, the price is reduced by 2¢ per bushel on the portion over 1000 except that there is an additional shipping charge of 50 cents for each 100 bushels (or fraction thereof) which exceeds 1000. Find the number of bushels in each shipment which maximizes income.

7. THE DIFFERENTIAL. APPROXIMATION

Until now we have been dealing with functions whose domain is a set of real numbers (i.e., a set in R_1) and whose range is also a set of real numbers (a set in R_1). Symbolically, if D is the set of numbers comprising the domain of a function f and S is the set of numbers comprising the range, we may write $f: D \to S$. In the case of functions given by formulas, we write $f(x) = x^2 - 2x + 7$ or $f: x \to x^2 - 2x + 7$ to indicate the particular function.

We now enlarge the notion of function by considering domains which are ordered pairs of numbers instead of numbers in R_1. That is, we consider the case in which an element in the domain is a point in R_2. The range is the same as before: a collection of real numbers (a set in R_1). These functions are defined as **functions on** R_2, and we give the following precise statement.

Definition. *Consider a collection of ordered pairs (A, w), in which the elements A are themselves ordered pairs of real numbers and the elements w are real numbers. If no two members of the collection have the same item A as a first element—i.e., if it can never happen that there are two members (A_1, w_1) and (A_1, w_2) with $w_1 \neq w_2$—then we call this collection a **function on** R_2. The totality of possible ordered pairs A is called the **domain** of the function. The totality of possible values for w is called the **range** of the function.*

We now revise our terminology and call a function as defined on page 23 a **function on** R_1. We use the same letters f, F, ϕ, g, H, etc. to represent functions on R_2 that we used for functions on R_1. However, the formulas used for describing particular functions on R_2 appear quite different from those we have used until now. If the pair (x, y) is used to represent a typical element in the domain of a function F on R_2, a particular function may be written

$$F(x, y) = x^2 - 2y^2 + 3x + 4y$$

or

$$F:(x, y) \to x^2 - 2y^2 + 3x + 4y.$$

If the letter z is used as a typical element in the range of a function F, we write $F:(x, y) \to z$ or $z = F(x, y)$.

Employing classical terminology, we call a function on R_2 a **function of two variables,** while a function on R_1 is called a **function of one variable.** These terms, although not quite precise, have considerable intuitive value, and we shall use them from time to time.*

---

* It is not difficult to extend the definition of function to the case in which the domain consists of ordered triples, ordered quadruples, etc., of numbers and in which the range also consists of ordered pairs, triples, etc., of numbers.

The letters used for elements of the domain in R_2 are not restricted to x and y. Simple examples of functions of two variables (functions on R_2) are

$$z = u^2 + 2v^2 + 3v^4, \qquad w = \frac{x^2 + y^2 - 2}{1 + x^2}, \qquad t = \frac{x^2 + u^2 + 1}{4 - 2u^2 - x^2}.$$

The domain in each of these formulas is not described explicitly. In each such case we shall always assume that the domain consists of all possible values which may be substituted in the right-hand side of the formula.

We are now ready to define a function which we call *the differential*. Suppose that f is a function on R_1 (of one variable).

Definition. *We designate by df* **the differential of** *f which is a function on R_2 (of two variables) given by the formula*

$$df(x, h) = f'(x) \cdot h.$$

In this definition, the symbol df is considered to be a single entity and does *not* mean d times f. Since df is the symbol for a function on R_2, we recognize $df(x, h)$ as the symbol for the value of the function when x and h are substituted into the right side of the above formula.

Example 1. Given that $f(x) = x^2(1 - x^2)^{1/2}$, find the formula for df. Also, find the value of $df(\frac{1}{2}, 3)$.

Solution. By differentiation, we have

$$f'(x) = 2x(1 - x^2)^{1/2} - x^3(1 - x^2)^{-1/2}.$$

Therefore

$$\begin{aligned}df(x, h) &= [2x(1 - x^2)^{1/2} - x^3(1 - x^2)^{-1/2}] \cdot h \\ &= (1 - x^2)^{-1/2}[2x(1 - x^2) - x^3]h \\ &= \frac{x(2 - 3x^2)}{\sqrt{1 - x^2}} h.\end{aligned}$$

Substituting $x = \frac{1}{2}$ and $h = 3$ in this expression, we get

$$df(\tfrac{1}{2}, 3) = \tfrac{5}{4}\sqrt{3}. \qquad \blacktriangleleft$$

Earlier (in Chapter 5, Section 1) we introduced the symbol Δf by defining

$$\Delta f = f(x + h) - f(x).$$

We did not say so at the time, but the reader can now see that Δf is a function on R_2. That is, we may write $\Delta f = \Delta f(x, h)$.

It will help our understanding to compare the functions df and Δf. We can do this geometrically, as shown in Fig. 6–35, where P represents a point on the graph of $y = f(x)$. If h has any value,* then the point $(x + h, f(x + h))$ is on the graph

---

*Figure 6–35 is drawn with h positive. If h is negative, the point $x + h$ lies to the left of x. The discussion is unchanged.

172 APPLICATIONS OF DIFFERENTIATION. THE DIFFERENTIAL

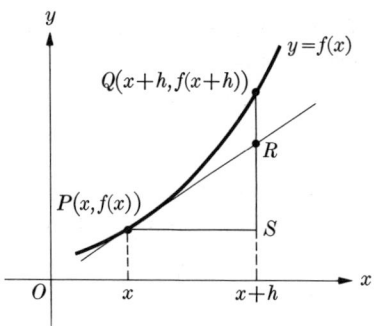

Figure 6–35

and is denoted by Q. We see that

$$\Delta f(x, h) = \text{height } \overline{SQ}.$$

The tangent at P intersects the line through Q and S at R, and we know that $\overline{SR}/\overline{PS}$ is the slope of this line, or simply $f'(x)$. Since $\overline{PS} = h$, we have

$$df(x, h) = \text{height } \overline{SR}.$$

This tells us that the height $\overline{RQ}$ is just the difference between Δf and df. That is,

$$\Delta f - df = \text{height } \overline{RQ}.$$

Also we see from the graph that as h tends to zero (thus implying that the point Q slides along the curve to P), the difference between Δf and df tends to zero. In fact, we have

$$\lim_{h \to 0} \frac{\Delta f - df}{h} = 0.$$

The proof of this statement is easy. We know from the definition of derivative that

$$\frac{\Delta f}{h} \to f'(x) \quad \text{as} \quad h \to 0.$$

On the other hand, df/h **is** the derivative $f'(x)$. Therefore

$$\lim_{h \to 0} \frac{\Delta f - df}{h} = \lim_{h \to 0} \frac{\Delta f}{h} - \lim_{h \to 0} \frac{df}{h} = f'(x) - f'(x) = 0.$$

If h is small, Δf and df must be close together. The function Δf measures the *change* in f as we go from one point to another. The differential df can be used as a good approximation for Δf if the points are near each other. We write

$$df \approx \Delta f$$

for this approximation.

Example 2. Given that $f(x) = 1/x$, compute df and Δf when $x = 1$ and h is 0.1, 0.01, 0.001. Do this to four significant figures.

Solution. $f(1) = 1$; $f'(x) = -1/x^2$ and $f'(1) = -1$. For $h = 0.1$,

$$f(1.1) = 0.909091 \quad \text{and} \quad \Delta f = 0.909091 - 1 = -0.09091.$$

On the other hand, $df = f'(1) \cdot h = -1 \cdot (0.1) = -0.10000$. In this way we obtain the table:

h	df	Δf
0.1	-0.10000	-0.09091
0.01	-0.010000	-0.009901
0.001	-0.0010000	-0.0009990

◀

The point of this example—as the student can readily see if he has worked along with it—is that the quantities df require practically no computing, while finding Δf is quite a bit of work. Since the error made in using df instead of Δf is small, the saving in computation time may make it worthwhile.

Example 3. Use the differential to compute $\sqrt[5]{33}$ approximately.

Solution. To work this problem we need an idea; it is simply that we know $\sqrt[5]{32}$ *exactly* and 33 is "close to" 32. The function we are talking about is

$$f(x) = \sqrt[5]{x} = x^{1/5}.$$

If we think of $x = 32$ and $h = 1$, then $f(33)$ is what we want to find. We know that

$$\Delta f = f(33) - f(32),$$

while

$$df = f'(32) \cdot (1).$$

The derivative is

$$f'(x) = \tfrac{1}{5} x^{-4/5},$$

and

$$f'(32) = \tfrac{1}{5} \cdot \tfrac{1}{16} = \tfrac{1}{80}, \qquad df = \tfrac{1}{80}.$$

Using the fact that df is almost the same as Δf, we write

$$f(33) = f(32) + \Delta f \approx f(32) + df.$$

(The symbol $\approx$ means "nearly equal.") Therefore

$$\sqrt[5]{33} = f(33) \approx 2 + \tfrac{1}{80} = 2.0125,$$

approximately. The actual error in this computation is less than $\tfrac{1}{6400}$. ◀

A measurement of a length of pipe is made, and it is found that there is an error of $\tfrac{1}{2}$ cm. Is this error large or small? If the length of the pipe is about 8 cm we would say that the error is pretty large. If the total length measured were $\tfrac{1}{2}$ km, we would consider the error extremely small. The error is the difference between the planned length and the actual length.

If a quantity is being measured and the true value is a but there is an error of an amount h, we define the **proportional error** as

$$\frac{h}{a}$$

and the **percentage error** as

$$\frac{h}{a} \cdot 100\%.$$

In the above example of the $\frac{1}{2}$-cm error, when $a = 8$ cm and $h = \frac{1}{2}$, the proportional error is $\frac{1}{16}$, and the percentage error is $\frac{1}{16} \cdot 100 = 6\frac{1}{4}\%$. If $a = \frac{1}{2}$ km and $h = \frac{1}{2}$ cm., then the proportional error is $1/100,000$ and the percentage error is 0.0001%—one ten-thousandth of one percent.

Suppose that f is a function and we wish to find $f(a)$, the value a to be determined by measurement. This measurement is not precise (measurements never are) and there is an error h in measuring the value a. The error in the function f is $\Delta f = f(a + h) - f(a)$ and the *proportional error* in the function is simply

$$\frac{\Delta f}{f} = \frac{f(a + h) - f(a)}{f(a)}.$$

However, in such computations, we conveniently use the **approximation to the proportional error:**

$$\frac{df}{f} = \frac{f'(a) \cdot h}{f(a)}.$$

The **approximate percentage error** is then found by multiplying by 100.

Example 4. The radius of a sphere is found by measurement to be 3 cm, but there is a possible error of ±0.03 cm in the measurement. Find, approximately, the error and percentage error in the value of the surface area of the sphere that might occur because of the error in the radius.

Solution. The surface area S is given by the formula $S = 4\pi r^2$, and for $r = 3$ the area is 36π cm². The error is approximated by $dS = S'(r)h = 8\pi rh$. The quantity $h = \pm 0.03$, and so the approximate error is

$$dS = 8\pi(3)(\pm 0.03) = \pm 0.72\pi \text{ cm}^2.$$

The approximate proportional error is

$$\frac{dS}{S} = \frac{\pm 0.72\pi}{36\pi} = \pm 0.02,$$

and the approximate percentage error is 2%. ◂

PROBLEMS

In Problems 1 through 6, find $df(x, h)$.

1. $f(x) = x^4 + 3x^2 - 2x$

2. $f(x) = \dfrac{x^2 + 2}{x^2 + 4}$

3. $f(x) = x^2\sqrt{x^4 + 1}$

4. $f(x) = \dfrac{(x + 1)^2}{x^2 + 4x + 1}$

5. $f : x \to (x^2 + 2)^{1/2}(2x + 1)^{1/3}$

6. $f : x \to (x^2 + 1)^{1/4}(x + 2)^{1/5}$

In Problems 7 through 14, find df and Δf and evaluate them for the quantities given.

7. $f(x) = x^2 + x - 1$, $\quad x = 1$, $\quad h = 0.01$
8. $f(x) = x^2 - 2x - 3$, $\quad x = -1$, $\quad h = -0.02$
9. $f(x) = x^3 + 3x^2 - 6x - 3$, $\quad x = 2$, $\quad h = 0.01$
10. $f(x) = x^3 - 2x^2 + 3x + 4$, $\quad x = -1$, $\quad h = 0.02$
11. $f(x) = 1/x$, $\quad x = 2$, $\quad h = 0.05$
12. $f(x) = x^{1/2}$, $\quad x = 1$, $\quad h = -0.1$
13. $f(x) = x^{-1/2}$, $\quad x = 1$, $\quad h = 0.1$
14. $f(x) = \dfrac{x}{1 + x}$, $\quad x = 0$, $\quad h = 0.1$

In Problems 15 through 22, calculate (approximately) the given quantity by means of the differential.

15. $\sqrt{65}$
16. $\sqrt[3]{124}$
17. $(0.98)^{-1}$
18. $\sqrt[4]{80}$
19. $(31)^{1/5}$
20. $(17)^{-1/4}$
21. $\sqrt{0.0024}$
22. $\sqrt{82} + \sqrt[4]{82}$

23. The diameter of a sphere is to be measured and its volume computed. The diameter is 9 cm, with a possible error of ±0.05 cm. Find (approximately) the maximum possible percentage error in the volume.

24. The radius of a sphere is to be measured and its volume computed. If the diameter can be accurately measured to within 0.1%, find (approximately) the maximum percentage error in the determination of the volume.

25. A coat of paint of thickness t cm is applied evenly to the faces of a cube of edge a cm. Use differentials to find approximately the number of cubic centimeters of paint used. Compare this with the exact amount used by computing volumes before and after painting.

26. Work Problem 25 with the cube replaced by a sphere of radius R.

27. State precisely the definition of a function when the domain consists of elements of R_3 and the range is a set in R_1. Do the same when the domain is a set in R_3 and the range is a set in R_2.

8. DIFFERENTIAL NOTATION

Let f be a function of one variable. Writing $y = f(x)$, we recall that in Chapter 4, Section 4 the symbol $\dfrac{dy}{dx}$ was introduced as an alternate notation for the derivative $f'(x)$. At this point we identify dx with the number h used in the definition of the differential, and we identify dy with the differential df. That is, setting

$$dx = h \quad \text{and} \quad dy = df,$$

we obtain the equivalent formula for the differential

$$dy = f'(x)\,dx.$$

We call dx the **differential of** x and dy **the differential of** y. Then dx is merely an independent variable and dy is defined by the above formula. As expected, the ratio dy/dx is the derivative whenever $dx \neq 0$, and the use of this notation for the derivative is justified in terms of differentials.

Suppose that $y = f(x)$ is a function and that $x = g(t)$ is a second function with the range of g in the domain of f. Then we may consider y as a function of t by writing

$$y = f[g(t)].$$

To get the derivative of $f[g(t)]$ we apply the chain rule to obtain

$$f'[g(t)]g'(t).$$

We also have the following formulas for differentials:

$$dy = f'(x)\,dx \quad \text{and} \quad dx = g'(t)\,dt.$$

By direct substitution, we find

$$dy = f'(x)g'(t)\,dt.$$

The chain rule may now be expressed in terms of differentials. If $dx \neq 0$ and $dt \neq 0$, then

$$\frac{dy}{dt} = \frac{dy}{dx} \cdot \frac{dx}{dt}.$$

Differentials may be multiplied, divided (whenever different from zero), added, and subtracted. One differential divided by another may be thought of as a derivative. We can now write all the elementary rules for derivatives as differentials. Since they become derivative formulas merely by division by dx, there is really not much new in them.

If c is a constant, $dc = 0$ and $d(cu) = c\,du$. Also,

$$d(u + v) = du + dv,$$
$$d(u \cdot v) = u\,dv + v\,du,$$
$$d\left(\frac{u}{v}\right) = \frac{v\,du - u\,dv}{v^2},$$
$$d(u^n) = nu^{n-1}\,du.$$

Example 1. Given that $y = (x^2 + 2x + 1)^3$ and $x = 3t^2 + 2t - 1$, find dy/dt.

Solution. Here we have

$$\frac{dy}{dt} = \frac{dy}{dx} \cdot \frac{dx}{dt}$$

and

$$\frac{dy}{dx} = 3(x^2 + 2x + 1)^2(2x + 2), \qquad \frac{dx}{dt} = 6t + 2.$$

Therefore
$$\frac{dy}{dt} = 6(x+1)(x^2+2x+1)^2 \cdot 2(3t+1).$$ ◀

Example 2. Given that $y = \sqrt{x^2+1}/(2x-3)$, find dy.

Solution. One way would be to find the derivative and multiply by dx. But we may also use the formula for the differential of a quotient, and we do it here for practice:
$$dy = \frac{(2x-3)d(\sqrt{x^2+1}) - \sqrt{x^2+1} \cdot d(2x-3)}{(2x-3)^2}.$$

We see now that $d(\sqrt{x^2+1}) = \frac{1}{2}(x^2+1)^{-1/2} 2x\,dx$ and $d(2x-3) = 2\,dx$. We obtain
$$dy = \frac{x(2x-3)(x^2+1)^{-1/2}\,dx - 2\sqrt{x^2+1}\,dx}{(2x-3)^2}.$$

We multiply both the numerator and the denominator by $\sqrt{x^2+1}$ to get
$$dy = \frac{-3x-2}{\sqrt{x^2+1}(2x-3)^2}\,dx.$$ ◀

The method of differentials is particularly helpful in implicit differentiation. Two functions, each defined on R_1, may satisfy a relation. We may take differentials if one of the functions is expressible in terms of the other. However, the actual process of solving for one quantity in terms of the other need not be carried out. An example will illustrate the idea.

Example 3. Suppose that u and v are functions defined on R_1 and that they satisfy the relation
$$u^2 + 2uv^2 + v^3 - 6 = 0.$$
Find dv/du.

Solution. Take differentials:
$$2u\,du + 2u \cdot 2v\,dv + 2v^2\,du + 3v^2\,dv = 0.$$

Divide through by du and solve:
$$\frac{dv}{du} = -\frac{2(u+v^2)}{4uv+3v^2},$$

if $du \neq 0$ and if $4uv + 3v^2 \neq 0$. ◀

If $y = f(x)$, there are now two symbols for the derivative: $f'(x)$ and dy/dx. These are the most prevalent symbols, commonly used in texts and papers on various related subjects. Another symbol, not quite so common but nevertheless used often, is $D_x f$. The only notation we learned for the **second derivative** is $f''(x)$.

The expression d^2y/dx^2 (read: d second y by dx second) is a classical one for the second derivative. The numerator, d^2y, and the denominator, dx^2, have absolutely no meaning by themselves. (In elementary calculus there is no such thing as the differential of a differential.) We just use d^2y/dx^2 as an equivalent for $f''(x)$. Similarly, third, fourth, and fifth derivatives are written

$$\frac{d^3y}{dx^3}, \quad \frac{d^4y}{dx^4}, \quad \frac{d^5y}{dx^5},$$

and so on. In each case, the expression is to be thought of not as a fraction dividing two quantities but as an inseparable symbol representing the appropriate derivative.

PROBLEMS

In Problems 1 through 14, find the differential dy.

1. $y = (x^3 - 3x^2 + 2)^5$
2. $y = (x^2 + 1)^{1/3}$
3. $y = (x^3 + 4)^{-5}$
4. $y = \dfrac{x+1}{x^2+1}$
5. $y = x^2\sqrt{2x+3}$
6. $y = (x-1)\sqrt{x+1}$
7. $y = (x+1)^2(2x-1)^3$
8. $y = (x+2)^{2/3}(x-1)^{1/3}$
9. $y = \dfrac{2x}{x^2+1}$
10. $y = \dfrac{x}{\sqrt{x^2+1}}$
11. $y = \dfrac{x^{2/3}}{(x+1)^{2/3}}$
12. $y = \dfrac{\sqrt{x^2+1}}{x}$
13. $y = \sqrt{\dfrac{x+1}{x-1}}$
14. $y = \dfrac{x}{\sqrt{x^2+x+1}}$

In Problems 15 through 19, find the derivative dy/dx by the method of differentials.

15. $2x^2 + xy - y^2 + 2x - 3y + 5 = 0$
16. $x^3 + x^2y - 2y^3 = 0$
17. $\sqrt{x} + \sqrt{y} = 2$
18. $x^{2/3} + y^{2/3} = a^{2/3}$
19. $2x^3 - xy^2 - y^3 + 2x - y = 0$
20. Find dy/dt, given that $y = (x^2 + 2x + 5)^{3/4}$ and $x = \sqrt{t^2 - 2t + 1}$.
21. Find dz/dr, given that $z = (2u+1)/(u^2+u)$ and $u = (r^2+5)^4$.
22. Find ds/dt, given that $s = x^2 + 3x - 6$, $x = r^3 - 8r + 5$, $r = \sqrt{t^3+5}$.
23. Find dz/ds, given that

$$z = \frac{y+1}{y-1}, \quad y = \frac{x^3 - 8x + 1}{\sqrt{x+1}}, \quad x = s^3 - 8s + 5.$$

24. Find dy/dt, given that $x^3 + 2xy - y^3 + 8 = 0$ and $x = t^3 - 2t + 1$.
25. Find dy/dt, given that

$$x^4 + 2x^2y - y^4 - 3y = 0, \quad x = \sqrt{2t+3}.$$

26. Find dz/dt, given that

$$z^3 + 2y^3 - 3z^2 = 1, \quad y = x^2 + 2x - 6, \quad x = (t+1)^4.$$

27. Find dy/dx, given that
$$x^3 + t^2 - 2\sqrt{t} - 4 = 0, \qquad y^4 - t^3 + 2y - 7 = 0.$$

*28. Find d^2y/dx^2, given that
$$x^3 - 4y^4 + 7x - 6y - 3 = 0.$$

29. a) *Given* $y = f(x)$, $x = g(t)$, $t = h(s)$. State appropriate conditions which make the formula
$$\frac{dy}{ds} = \frac{dy}{dx} \cdot \frac{dx}{dt} \cdot \frac{dt}{ds}$$
valid.

b) Extend the above formula to the case of n functions $f_1, f_2, \ldots, f_n$.

*30. Given the relations
$$x^2 + 2xt + y^3 - t^2 + 3xy - 2t^3 = 1,$$
$$2x^3 - y^3 + 2xy^2 + 3yt^3 + x = 3.$$

Use differentials to find a formula (in terms of x, y, t) for the derivative dx/dt. [*Hint:* Observe that *theoretically* y may be eliminated between the two equations and x expressed in terms of t.]

9. RELATED RATES

In Chapter 4, Section 5, we discussed motion along a straight line and developed the ideas of velocity and speed. We learned that if a particle moves in a straight line so that the distance traveled, s, depends on the time t, according to some law $s = f(t)$, the velocity is obtained by finding the derivative $f'(t)$. We can also write

$$f'(t) = \frac{ds}{dt}.$$

The velocity may be thought of as the *rate of change* of distance with respect to time. If there are several particles, each moving in a straight line according to some law, then we can talk about the rate of change of each of the particles. Suppose the motion of these particles is related in some way. (One may go up as the other goes down, as in a lever, for example.) Then we say that we have a problem in **related rates**.

We are not limited to particles moving in a straight line. If a tank is being filled with water, the level of the surface is rising with time. We talk about the rate of change of the depth of the water. If the depth is denoted by h, then dh/dt is the rate of change of the depth. Similarly, the volume V is increasing; dV/dt measures the rate of this increase. Any quantity which grows or diminishes with time is a possible candidate for a problem in related rates.

In the problems we shall consider, it is important to remember that **every quantity is a function of time.** Therefore we can take derivatives of each quantity with respect to t, the time. This derivative is called the **rate of change**. We first give two examples and then map out general rules of procedure.

Example 1. One airplane flew over an airport at the rate of 300 km/hr. Ten minutes later another airplane flew over the airport at 240 km/hr. If the first airplane was flying west and the second flying south (both at the same altitude), determine the rate at which they were separating 20 minutes after the second plane flew over the airport. (We assume the airplanes are travelling at constant speed.)

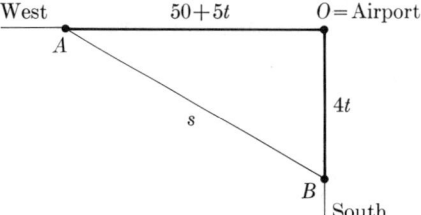

Figure 6–36

Solution. Draw a diagram, as shown in Fig. 6–36. Let s be the distance between the airplanes at time t, let x be the distance the westbound airplane has traveled at time t, and let y be the distance the southbound airplane has traveled at time t. We know that $dx/dt = 5$ km/min, $dy/dt = 4$ km/min. We wish to find ds/dt when $t = 20$ min, measured from the time the second airplane passes over O. We have the relation

$$s^2 = x^2 + y^2.$$

Take differentials:

$$2s\, ds = 2x\, dx + 2y\, dy.$$

Divide by dt:

$$s \frac{ds}{dt} = x \frac{dx}{dt} + y \frac{dy}{dt}.$$

When $t = 20$, then $y = 80$, $x = 150$, and $s = 170$, and substituting all these values we get

$$170 \cdot \frac{ds}{dt} = 150 \cdot 5 + 80 \cdot 4 \quad \Leftrightarrow \quad \frac{ds}{dt} = \frac{107}{17} \text{ km/min.} \quad \blacktriangleleft$$

We begin the solution of such a problem by using a letter to denote each of the quantities which change. In the above example we use x, y, and s; each of these is a function of t. Some relation is found among the letters ($s^2 = x^2 + y^2$, as we saw), and differentials by implicit methods are used to get derivatives with respect to t. Then the numerical value for each quantity is substituted to obtain the answer. Let us work another example.

Example 2. Water is flowing at the rate of 5 cubic meters/min into a tank (Fig. 6–37) in the form of a cone of altitude 20 meters and base radius 10 meters and with its vertex in the downward direction. How fast is the water level rising when the water is 8 meters deep?

Solution. Let h be the depth, r the radius of the surface, and V the volume of the water at an arbitrary time t. We wish to find dh/dt. We know $dV/dt = 5$. The volume of water is given by

$$V = \tfrac{1}{3}\pi r^2 h,$$

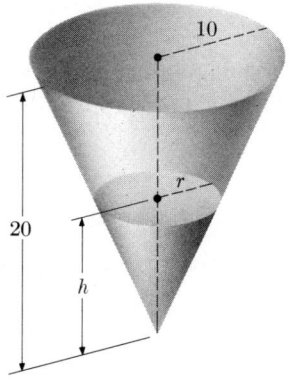

Figure 6–37

where all quantities depend on t. By similar triangles,

$$\frac{r}{h} = \frac{10}{20} \quad \text{or} \quad r = \tfrac{1}{2}h,$$

and so

$$V = \tfrac{1}{3}\pi \frac{h^3}{4} = \tfrac{1}{12}\pi h^3.$$

We take differentials:

$$dV = \tfrac{1}{4}\pi h^2\, dh,$$

and, dividing by dt:

$$\frac{dV}{dt} = \tfrac{1}{4}\pi h^2 \frac{dh}{dt}.$$

We want to find dh/dt when $h = 8$, and we obtain

$$\frac{dh}{dt} = \frac{5}{16\pi} \text{ m/min.} \qquad \blacktriangleleft$$

The major trap to avoid in this problem is the premature use of the fact that $h = 8$ at the instant we want to find dh/dt. The height h *changes* with time and must be denoted by a letter. If $h = 8$ is put in the diagram, it usually leads to disaster. **All quantities which change with time must be denoted by letters.**

The rules of procedure illustrated in Examples 1 and 2 are now outlined in the form of four steps.

Step 1: Draw a diagram. Label any numerical quantities which remain fixed throughout the problem (such as the dimensions of the cone in Example 2).

Step 2: Denote all quantities which change with time by letters. A relation (or relations) is found among the quantities which vary; these relations must hold for all time.

Step 3: Take differentials of the relation (or relations) found in Step 2. Divide by dt to obtain a relation among the derivatives.

Step 4: Insert the special numerical values of all quantities to get the desired result.

182 APPLICATIONS OF DIFFERENTIATION. THE DIFFERENTIAL

We apply these rules in the following example.

Example 3. An airplane at an altitude of 3000 meters, flying horizontally at 300 km/hr, passes directly over an observer. Find the rate at which it is approaching the observer when it is 5000 meters away.

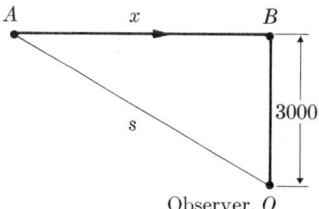

Figure 6–38

Solution. We draw Fig. 6–38. The airplane is at point A flying toward point B. The distance OB is labeled as 3000 (Step 1). Note that this distance does not change in the course of the problem. Let x be the distance of the airplane A from the point B directly over the observer, O, and s be the distance from the airplane to the observer. We want to find ds/dt when $s = 5000$. We are given $dx/dt = -300$ km/hr. The negative sign is used because x is decreasing. A change of units gives us $dx/dt = -83\frac{1}{3}$ m/sec.

We have (Step 2)
$$s^2 = x^2 + (3000)^2,$$
and, applying differentials (Step 3),
$$2s\,ds = 2x\,dx.$$
Dividing by dt, we get
$$s\frac{ds}{dt} = x\frac{dx}{dt}.$$

We now insert the values (Step 4), noting that $x = 4000$ when $s = 5000$:

$$5000\frac{ds}{dt} = 4000(-83\tfrac{1}{3}) \quad \Leftrightarrow \quad \frac{ds}{dt} = -66\tfrac{2}{3} \text{ m/sec} = -240 \text{ km/hr}. \blacktriangleleft$$

PROBLEMS

1. Water is flowing into a vertical cylindrical tank of radius 2 meters at the rate of 8 cubic meters per minute. How fast is the water level rising?

2. A launch whose deck is 7 meters below the level of a wharf is being pulled toward the wharf by a rope attached to a ring on the deck. If a winch pulls in the rope at the rate of 15 m/min, how fast is the launch moving through the water when there are 25 m of rope out?

3. Two automobiles start from a point A at the same time. One travels west at 60 km/hr and the other travels north at 35 km/hr. How fast is the distance between them increasing 3 hr later?

4. At noon of a certain day, ship A is 60 km due north of ship B. If A sails east at 12 km/hr and B sails north at 9 km/hr, determine how rapidly the distance between them is changing 2 hr later. Is it increasing or decreasing?

5. At a given instant the legs of a right triangle are 16 cm and 12 cm, respectively. The first leg decreases at $\frac{1}{2}$ cm/min and the second increases at 2 cm/min. At what rate is the area increasing after 2 min?

6. At a certain instant a small balloon is released (from ground level) at a point 75 m away from an observer (on ground level). If the balloon goes straight up at a rate of 4 m/sec, how rapidly will it be receding from the observer 30 sec later?

7. An arc light is 5 meters above a sidewalk. A man 2 meters tall walks away from the point under the light at the rate of 2 m/sec. How fast is his shadow lengthening when he is 7 m away from the point under the light?

8. A balloon is being inflated at the rate of 15 m^3/min. At what rate is the diameter increasing after 5 min? Assume that the diameter is zero at time zero.

9. A baseball diamond is 90 ft on a side. (It is really a square.) A man runs from first base to second base at 25 ft/sec. At what rate is his distance from third base decreasing when he is 30 ft from first base? At what rate is his distance from home plate increasing at the same instant?

10. A man starts walking eastward at 2 m/sec from a point A. Ten minutes later a second man starts walking west at the rate of 2 m/sec from a point B, 3000 meters north of A. How fast are they separating 10 min after the second man starts?

11. A point moves along the curve $y = \sqrt{x^2 + 1}$ in such a way that $dx/dt = 4$. Find dy/dt when $x = 3$.

12. A point moves along the upper half of the curve $y^2 = 2x + 1$ in such a way that $dx/dt = \sqrt{2x + 1}$. Find dy/dt when $x = 4$.

13. The variables x, y, and z are all functions of t and satisfy the relation $x^3 - 2xy + y^2 + 2xz - 2xz^2 + 3 = 0$. Find dz/dt when $x = 1$, $y = 2$, if $dx/dt = 3$ and $dy/dt = 4$ for all times t.

14. A ladder 5 meters long leans against a vertical wall of a house. If the bottom of the ladder is pulled horizontally away from the house at 4 m/sec, how fast is the top of the ladder sliding down when the bottom is 3 meters from the wall?

15. A light is on the ground 40 m from a building. A man 2 m tall walks from the light toward the building at 2 m/sec. How rapidly is his shadow on the building growing shorter when he is 20 m from the building?

16. A trough is 10 m long and its ends are isosceles triangles with altitude 2 m and base 2 m, their vertices being at the bottom. If water is let into the trough at the rate of 3 m^3/min, how fast is the water level rising when it is 1 m deep?

17. A trough 10 m long has as its ends isosceles trapezoids, altitude 2 m, lower base 2 m, upper base 3 m. If water is let in at the rate of 3 m^3/min, how fast is the water level rising when the water is 1 m deep?

18. A swimming pool is 9 m wide, 15 m long, 1 m deep at the shallow end, and 3 m deep at the deep end, the bottom being an inclined plane. If water is pumped into the pool at the rate of 3 m^3/min, how fast is the water level rising when it is 2 m deep at the deep end?

19. Sand is issuing from a spout at the rate of 3 m^3/min and falling on a conical pile whose diameter at the base is always three times the altitude. At what rate is the altitude increasing when the altitude is 4 m?

20. Water is leaking out of a conical tank (vertex down) at the rate of 0.5 m³/min. The tank is 30 m across at the top and 10 m deep. If the water level is rising at the rate of $1\frac{1}{2}$ m/min, at what rate is water being poured into the tank from the top?

21. In Example 2, find the rate at which the uncovered surface of the conical tank is decreasing at the instant in question.

22. A boat is anchored in such a way that its deck is 25 m above the level of the anchor. If the boat drifts directly away from the point above the anchor at the rate of 5 m/min, how fast does the anchor rope slip over the edge of the deck when there are 65 m of rope out? (Assume that the rope forms a straight line from deck to anchor.)

23. A man lifts a bucket of cement to a scaffold 30 m above his head by means of a rope which passes over a pulley on the scaffold. The rope is 60 m long. If he keeps his end of the rope horizontal and walks away from beneath the pulley at 4 m/sec, how fast is the bucket rising when he is $22\frac{1}{2}$ m away?

24. Water is flowing into a tank in the form of a hemisphere of radius 10 m with flat side up (Fig. 6–39) at the rate of 4 m³/min. At any instant let h denote the depth of the water, r the radius of the surface, and V the volume of the water. Assuming $dV = \pi r^2\, dh$, find how fast the water level is rising when $h = 5$ m.

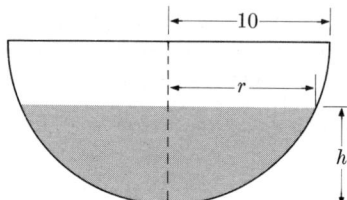

Figure 6–39

25. A bridge is 10 m above a canal. A motorboat going 3 m/sec passes under the center of the bridge at the same instant that a man walking 2 m/sec reaches that point. How rapidly are they separating 3 sec later?

26. If in Problem 25 the man reaches the center of the bridge 5 sec before the boat passes under it, find the rate at which the distance between them is changing 4 sec after the man crosses.

27. When a gas expands or contracts adiabatically, it obeys the law $pv^\gamma = K$ where p is the pressure, v the volume, and γ and K are constants. At a certain instant a container of air is under pressure of 40 gram/cm², the volume is 32 cm³, and the volume is increasing at the rate of 5 cm³ per second. Assuming that $\gamma = 1.4$, find how rapidly the pressure is decreasing at that instant.

28. A light is at the top of a pole which is h meters high. A ball is dropped from a height $\frac{1}{2}h$ at a point which is at a horizontal distance d meters from the pole. Assume that the ball falls according to the law $s = gt^2$, where t is the time in seconds, s is the distance in meters, and g is a constant. Find how fast the tip of the shadow of the ball is moving along the ground t_0 seconds after it is dropped.

29. Ohm's law for a certain electrical circuit states that $E = IR$ where E is the voltage in volts, I the current in amperes, and R the resistance in ohms. If the circuit heats up and the voltage is kept constant, the resistance increases at the rate of 0.5 ohms per second. Find the rate at which the current decreases when $I = 2$ amps and E is kept constant at 10 volts.

10. THE DEFINITE INTEGRAL AND ANTIDERIVATIVES

What is meant by the area enclosed in an irregularly shaped region, such as the one shown in Fig. 6–40? Area is a measure of the size of a region and, in developing this concept, we begin by determining the sizes of regions having simple shapes. Then we proceed step by step to define the areas of more complicated regions.

Almost all notions of area start with the idea that a rectangle of length l and width w has, *by definition*, area

$$A = lw.$$

From this formula we obtain immediately the formula for the area of a right triangle as half the area of a rectangle; then the usual formulas for the area of any triangle are developed as in elementary geometry. Any polygon may be decomposed into triangles, and therefore the area of a polygon is simply the sum of the areas of the triangles which comprise it.

Finding the area of a circle without the use of calculus is hard but not impossible. The method consists first of finding the area of inscribed and circumscribed regular polygons. Then, as the number of sides of the polygons increases without bound, the polygons approximate the circle more and more closely. Finally, the area of the circle is defined as the limit of the areas of these polygons.

The area of an irregular region (as in Fig. 6–40) may be explained intuitively in the following way. Suppose the region were made of some uniform material of uniform thickness which could be weighed with great accuracy. We could then make a rectangular region from the same kind of material of the same thickness and having the same weight; we would like to be able to say that the two regions then have the same area. However, such a process could never be used as the basis of a mathematical definition. Not only is it impossible to reproduce the irregular region exactly, but also nonuniformities of all sorts appear in every material. Moreover, while physical processes have intuitive value, they can never yield mathematical definitions.

The concept of limit has one of its most fruitful applications in the development of the notion of area. The limiting process for area is different from that for derivative and, while precise statements will be postponed to Chapter 7, we shall give some introductory material here.

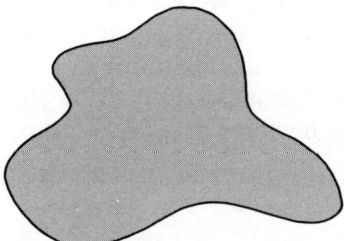

Figure 6–40

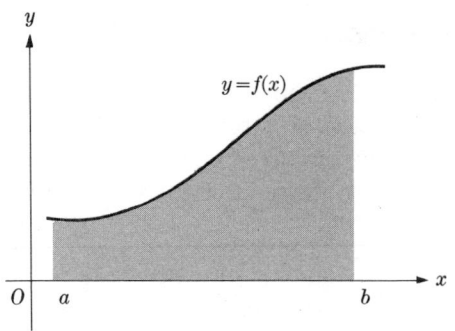

Figure 6–41

Figure 6–42

Suppose that $y = f(x)$ is a function which happens to lie above the x axis, as shown in Fig. 6–41. At two points on the x axis, a and b, vertical lines are drawn. We shall concern ourselves with the problem of calculating the area of the region bounded by the x axis, the vertical lines through a and b, and the graph of the function f. The fact that three sides of the region are straight lines simplifies the problem. At this stage we assume that the only kind of region for which we can calculate the area is a rectangle. The problem seems fairly hopeless until we realize that we can get an *approximate* idea of how large the area is in the following way.

First, we take the interval $[a, b]$ on the x axis and divide it into a number of subintervals (Fig. 6–42). We shall be astute and not say exactly how many at this time; we do so by letting n be the number of subintervals and by not saying exactly how large n shall be. We label the points of subdivision x_1, x_2, $x_3, \ldots, x_{n-1}$ and we stipulate that $a = x_0$ and $b = x_n$. The subintervals do not have to be of equal size. The first interval is $[x_0, x_1] = [a, x_1]$, and consists of all points x such that $x_0 \leq x \leq x_1$. The length of this interval is just $x_1 - x_0$. The length of the second interval is $x_2 - x_1$, and so on. In fact, if i is any integer between 1 and n, then $x_i - x_{i-1}$ is the length of the ith subinterval. We introduce a special symbol for this:

$$\Delta_i x = \text{length of } i\text{th interval} = x_i - x_{i-1}.$$

Note that $\Delta_i x$ (read "delta sub i of x") does not mean Δ times x or any such thing. It is merely a complicated (but useful) symbol used to represent a simple quantity.

In each of the intervals into which we divided the interval $[a, b]$ we now select a point. This may be done in any way we please. We designate these points $\xi_1, \xi_2, \ldots, \xi_n$, as shown in Fig. 6–43. At each of the points ξ_i we erect a vertical segment to the curve $y = f(x)$. In Fig. 6–44 this is shown for the case $n = 6$. The height of the dashed line through ξ_1 is $f(\xi_1)$, the height of the line through ξ_2 is $f(\xi_2)$, and so on. Now we calculate the areas of the rectangles formed by the subintervals and the heights $f(\xi_i)$. The first rectangle has area $f(\xi_1)(x_1 - x_0)$ or $f(\xi_1) \Delta_1 x$. The second has area

$$f(\xi_2)(x_2 - x_1) = f(\xi_2) \Delta_2 x,$$

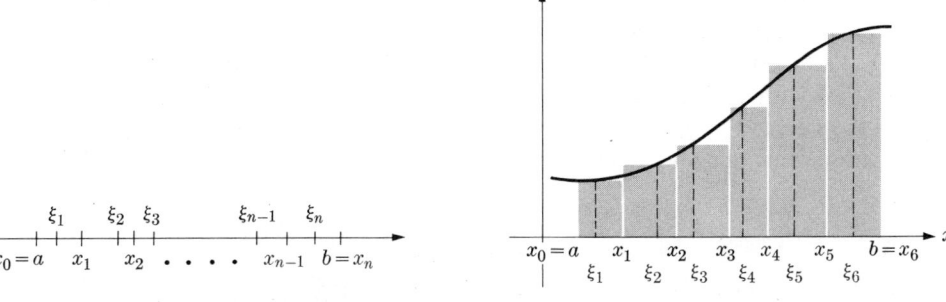

Figure 6–43 **Figure 6–44**

and so on. If we add the area of all these rectangles, we get

$$f(\xi_1)\Delta_1 x + f(\xi_2)\Delta_2 x + f(\xi_3)\Delta_3 x + \cdots + f(\xi_n)\Delta_n x,$$

which is the desired *approximation* to the area under the curve.

Figure 6–44 shows such an approximation for six intervals of subdivision. It is natural to expect that if we take 60 intervals of subdivision, each of them much narrower than the ones shown, the approximation to the area would be much better. And if we take 600 subintervals, it would be better still. Here the notion of limit appears. Suppose that these approximations approach a limiting value (call it L) as the number of subintervals tends to infinity, and suppose that the following statements are true:

1) The value of L does not depend on how the subdivisions are selected, so long as the maximum length of any subinterval tends to zero as the number of such subintervals tends to infinity.

2) The value of L does not depend on where in each subinterval the points ξ_i are chosen.

In these circumstances, we call this limit L the **definite integral of f from a to b** and denote it by

$$\int_a^b f(x)\,dx$$

(read "the integral from a to b of f of x, dx"). We must say a few words about this notation. The dx is just a juxtaposition of letters which are considered inseparable. From the way we defined "definite integral," the value of L depends only on f and a and b. Therefore the choice of the letter x has no particular meaning in the above notation; we could have used any symbol. For example, if we had the *same* function f but had used the letter t, the expression

$$\int_a^b f(t)\,dt$$

would have exactly the same meaning and the value of L would be unchanged. The dt, in this case, merely identifies the variable (letter) being used in making evaluations of the function.

188 APPLICATIONS OF DIFFERENTIATION. THE DIFFERENTIAL

We shall show in Chapter 7 that the number L may be used to define the area under the curve and, indeed, that this coincides with the usual definition of area for elementary regions.

The concepts of definite integral and area under a curve have an intimate connection with the idea of derivative. Consider the problem of finding the area under the curve $y = f(x)$ between the points a and X, where X is between a and b, as shown in Fig. 6–45. We would write this area as the definite integral

$$\int_a^X f(x)\,dx.$$

As X changes, this area assumes different values. If X is a, the value is zero. Let's call the area A and, since A is a function of X, we write

$$A = F(X).$$

We now select an $h > 0$ and compute $F(X + h)$. That is, we calculate the area under the curve between a and $X + h$. This is just the integral

$$\int_a^{X+h} f(x)\,dx.$$

The difference $F(X + h) - F(X)$ is the area of the shaded region shown in Fig. 6–45. If h is "small," the shaded region is almost a rectangle and, if we let Y be some average value of the function $f(x)$ between X and $X + h$, we could say that

$$F(X + h) - F(X) = Y \cdot h \quad \text{or} \quad Y = \frac{F(X + h) - F(X)}{h}.$$

If $h \to 0$, we know from our earlier experience that the limit approached is the derivative $F'(X)$. On the other hand, "geometrically" we see that as $h \to 0$, the average height Y must approach the height of the function f at the point X, namely $f(X)$. We conclude by this "argument" that

$$F'(X) = f(X),$$

which states that the function f is the derivative of the area function F. We also say that F is an **antiderivative** of f. In other words, *the process of finding area and the process of differentiation are the inverses of each other.*

Suppose we know that the derivative of a function f is

$$f'(x) = 4x^3 + 2x - 5.$$

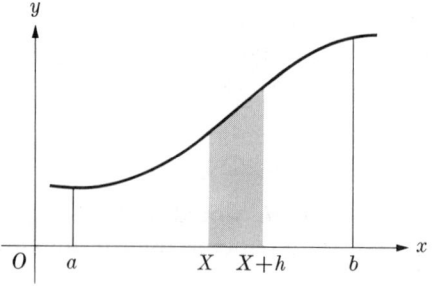

Figure 6–45

What is the function? It's easy to guess an answer. One possibility is
$$f(x) = x^4 + x^2 - 5x + 7.$$
Another is $f(x) = x^4 + x^2 - 5x + 4$; still another is $f(x) = x^4 + x^2 - 5x - 3$. We see that there is no single solution to the problem of finding the function if the derivative is given. That is, *the problem of finding the antiderivative does not have a unique solution.* It is easy to see that any function f of the form
$$f(x) = x^4 + x^2 - 5x + c$$
is a solution to the above problem, where c is a number which may have any value whatsoever.* If, in addition to being given the derivative f', we are given one more fact about the function f, we can usually determine the quantity c. In the above illustration, if we also know that $f(2) = 4$, we can write
$$f(2) = 4 = 16 + 4 - 10 + c,$$
and conclude that $c = -6$. The function f is then
$$f(x) = x^4 + x^2 - 5x - 6.$$

Example 1. Find the antiderivative of the general polynomial
$$P(x) = a_0 x^n + a_1 x^{n-1} + \cdots + a_{n-1} x + a_n.$$

Solution. Since $a_0, a_1, \ldots, a_{n-1}, a_n$ are constants, we find the antiderivative of each term and add the results. The answer is
$$\frac{a_0 x^{n+1}}{n+1} + \frac{a_1 x^n}{n} + \frac{a_2 x^{n-1}}{n-1} + \cdots + \frac{a_{n-1} x^2}{2} + a_n x + c,$$
where c is a number which may have any value. By noting that the derivative of the answer is just $P(x)$, we verify that the result is correct. ◀

Remark. In Chapter 5 we developed systematic methods for finding the derivatives of algebraic functions. So far, we can only find antiderivatives by guessing at the result and then verifying the correctness of the guess by differentiating the answer. Systematic techniques for finding antiderivatives are given in Chapters 7 and 14.

Let us work a second example by the guessing method.

Example 2. Suppose that f is a function whose derivative is
$$f'(x) = \tfrac{1}{2} x^3 (x + 1)^{-1/2} + 3x^2 (x + 1)^{1/2}.$$
Given that $f(3) = 5$, find the function f.

Solution. We observe that if $u(x) = x^3$, then $u'(x) = 3x^2$, and if $v(x) = (x + 1)^{1/2}$, then $v'(x) = \tfrac{1}{2}(x + 1)^{-1/2}$. Guessing that the expression for $f'(x)$ is in the form $uv' + u'v$, we observe that $f(x) = u(x) \cdot v(x)$. Therefore
$$f(x) = x^3 (1 + x)^{1/2} + c.$$

---
* We shall see in Chapter 7 that all solutions are of this form.

Setting $f(3) = 5$, we find
$$5 = 27 \cdot \sqrt{4} + c \quad \Leftrightarrow \quad c = -49.$$
We conclude that
$$f(x) = x^3(1 + x)^{1/2} - 49. \quad \blacktriangleleft$$

The next example employs the definitions of velocity and acceleration as given in Chapter 4, Section 5.

Example 3. A ball is thrown upward, and we know that the acceleration is -10(m/sec)/sec. (a) What is the velocity at any instant of time if the ball was launched with a velocity of 60 m/sec? (b) Find the height of the ball above the launching point at any time t.

Solution. Since the acceleration a is given, finding the velocity v is a problem in antiderivatives. We know that
$$a = -10, \quad v = -10t + c,$$
where t is the time (starting at $t = 0$ when the ball is thrown). We have $v = 60$ when $t = 0$, which means that $60 = -10 \cdot 0 + c$, or $c = 60$. We obtain the formula
$$v = -10t + 60.$$
To get the distance traveled, we again have a problem in antiderivatives. We find
$$s = -5t^2 + 60t + k.$$
To evaluate k, we note that at time $t = 0$ the distance traveled is also zero. This makes $k = 0$, and we have the formula
$$s = -5t^2 + 60t. \quad \blacktriangleleft$$

Example 4. Approximate, by rectangles, the area under the curve $f(x) = x^2$ if $a = 0$, $b = 1$, the number of rectangles is 5, the subintervals are all equal in size, and the ξ_i are all taken at the midpoints of the subintervals. Draw a graph. Compute the exact area by the antiderivative method.

Solution. Each $\Delta_i x = 0.2$, and $\xi_1 = 0.1$, $\xi_2 = 0.3$, $\xi_3 = 0.5$, $\xi_4 = 0.7$, $\xi_5 = 0.9$. We have (Fig. 6–46)
$$f(\xi_1) \Delta_1 x = 0.002$$
$$f(\xi_2) \Delta_2 x = 0.018$$
$$f(\xi_3) \Delta_3 x = 0.050$$
$$f(\xi_4) \Delta_4 x = 0.098$$
$$f(\xi_5) \Delta_5 x = 0.162$$
$$\overline{\text{Sum} = 0.330}$$

Since $F'(x) = f(x) = x^2$, we see from our knowledge of antiderivatives that
$$F(x) = \tfrac{1}{3}x^3 + c,$$
which gives the area under the curve at any point x. We must, however, use the fact that we are starting at the point a. This is equivalent to the statement that

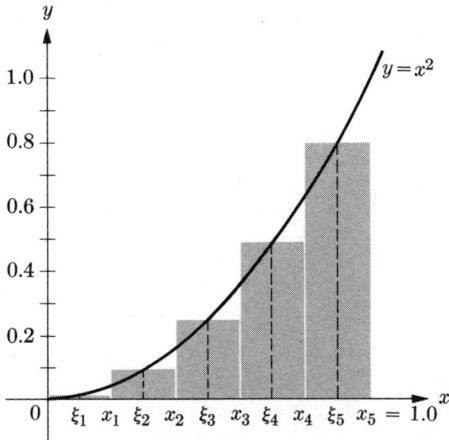

Figure 6–46

$F(a) = 0$. In our case $a = 0$, and so $F(0) = 0$. Substituting in the above expression for $F(x)$, we get $c = 0$. The area we seek extends from $a = 0$ to $b = 1$; substituting $x = 1$ in the formula for $F(x) = \frac{1}{3}x^3$, we obtain

$$F(1) = \tfrac{1}{3} = \text{area under curve.} \quad \blacktriangleleft$$

PROBLEMS

In each of Problems 1 through 6, find the antiderivative. Use the additional condition to determine the solution uniquely.

1. $f'(x) = x^4 - 2x^3 + 7x - 6$; $f(1) = 2$
2. $f'(x) = -2x^5 + 7x^3 - 5x^2 + 2x$; $f(2) = 1$
3. $f'(x) = (x + 2)^7$; $f(-2) = 1$
4. $f'(x) = 3x^2(x + 1)^2 + 2x(x + 1)^3$; $f(3) = -1$
5. $f'(x) = 2(2x + 1)^3(x + 2) + 6(2x + 1)^2(x + 2)^2$; $f(-\tfrac{1}{2}) = 0$
6. $f'(x) = (x + 3)^{-2}[2(x + 3) - (2x + 1)]$; $f(1) = 1$

In Problems 7 through 10, find the distance s traveled along a straight line by a particle if the velocity $v = v(t)$ follows the law given. The additional fact required for the result is also given.

7. $v = t^2 + t - 2$ and $s = 2$ when $t = -1$
8. $v = -t^2 + 2t + 1$ and $s = 0$ when $t = 2$
9. $v = 6 - 2t - 3t^2$ and $s = 0$ when $t = 0$
10. $v = 5 - 7t^2$ and $s = 5$ when $t = 0$
11. A ball is thrown upward from the ground with a launching velocity of 80 m/sec. Its acceleration is constant and equal to -10(m/sec)/sec. How long does the ball continue to rise? How long does the ball stay in the air?
12. A man driving an automobile in a straight line at a speed of 80 m/sec applies the brakes at a certain instant (which we take to be $t = 0$). If the brakes furnish a constant acceleration of -20(m/sec)/sec (actually a deceleration), how far will he go before he stops?

192 APPLICATIONS OF DIFFERENTIATION. THE DIFFERENTIAL

In Problems 13 through 16, approximate by rectangles the area under the curve $y = f(x)$ from the data given. Then compute each term to three decimal places as in Example 4. Sketch a graph.

13. $f(x) = 2x + 3$, $n = 5$, $a = -1$, $b = 0$.

i	0	1	2	3	4	5
x_i	-1	-0.8	-0.6	-0.4	-0.2	0
ξ_i		-1.0	-0.8	-0.5	-0.2	0

(In this problem, compute the exact area.)

14. $f(x) = x^2$, $n = 5$, $a = -1$, $b = 0$

i	0	1	2	3	4	5
x_i	-1	-0.8	-0.6	-0.4	-0.2	0
ξ_i		-1.00	-0.75	-0.50	-0.25	0

15. $f(x) = x^2$, $n = 7$, $a = -2.0$, $b = 0$

i	0	1	2	3	4	5	6	7
x_i	-2	-1.8	-1.6	-1.4	-1.2	-0.8	-0.4	0
ξ_i		-2.0	-1.6	-1.5	-1.3	-1.0	-0.7	-0.1

16. $f(x) = 2x - x^2$, $n = 5$, $a = 0$, $b = 2$

i	0	1	2	3	4	5
x_i	0	0.4	0.8	1.2	1.6	2.0
ξ_i		0	0.5	1.0	1.5	2.0

In Problems 17 through 23, find the exact area under the curve $y = f(x)$ from a to b by finding $F(x)$, the area function, and then $F(b)$.

17. $f(x) = x^2$, $a = -1$, $b = 0$
18. $f(x) = x^2$, $a = -2$, $b = 0$
19. $f(x) = 2x - x^2$, $a = 0$, $b = 2$
20. $f(x) = x^2 - 2x + 2$, $a = 1$, $b = 3$
21. $f(x) = 3x - x^2$, $a = 1$, $b = 2$
22. $f(x) = x^2 + x + 1$, $a = -2$, $b = 1$
23. $f(x) = 6 + x - x^2$, $a = -2$, $b = -1$

In Problems 24 through 28, approximate the area by rectangles with the number n given. Take the ξ_i at the midpoints.

24. $f(x) = x + 1$, $a = 1$, $b = 3$, $n = 1$
25. $F(x) = x^2$, $a = -2$, $b = -1$, $n = 5$
26. $f(x) = \dfrac{1}{1+x}$, $a = 0$, $b = 1$, $n = 5$
27. $f(x) = \dfrac{1}{1+x^2}$, $a = 0$, $b = 1$, $n = 5$
28. Compute the exact area in Problems 24 and 25 by using the antiderivative method.

7
THE DEFINITE INTEGRAL

1. AREA

In the last Section of Chapter 6, we discussed the notion of the area of a region in the plane in an informal way. Now we shall go into more detail and treat the topic in a precise manner. The area of a region is a measure of its size. A rectangle of length l and width w has area $A = l \cdot w$. The area of a triangle of base b and altitude h is just $\frac{1}{2}bh$. Similarly, we may obtain the area of a polygon simply by decomposing it into triangles and adding the areas of the component parts. Moreover, we can readily see that the area of a polygon is independent of the way in which the polygon is cut up into triangles.

Suppose we have an irregularly shaped region in the plane, such as the one shown in Fig. 7–1. How do we go about defining its area? The process is rather complicated and depends intrinsically on the idea of limit, which we have previously learned. In this section the procedure for defining area will be given, although the proofs of some of the statements will be omitted.

We start with the notion of a **square grid.** The entire plane is divided into squares by constructing equally spaced lines parallel to the coordinate axes. Figure 7–2 shows a square grid with squares $\frac{1}{2}$ unit on each side, and Fig. 7–3 shows a square grid with squares $\frac{1}{4}$ unit on each side. We always select the coordinate axes themselves as lines of the grid, and then the size of the squares determines the location of all other grid lines. We may have a grid with squares of any size, but for convenience we consider only grids of size $\frac{1}{2}$, $\frac{1}{4}$, $\frac{1}{8}$, $\frac{1}{16}$, ..., $1/2^n$, ..., where n is any positive integer. We obtain a sequence of grids with the property that the lines of any grid in the sequence are also lines in all the grids farther along in the sequence.

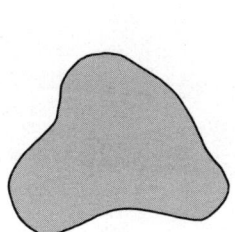

Figure 7–1

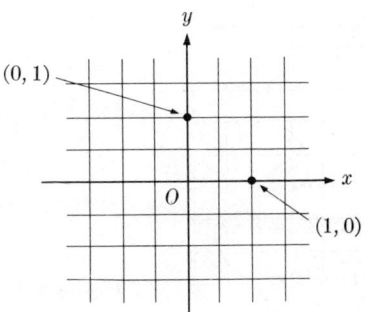

Figure 7–2

194 THE DEFINITE INTEGRAL

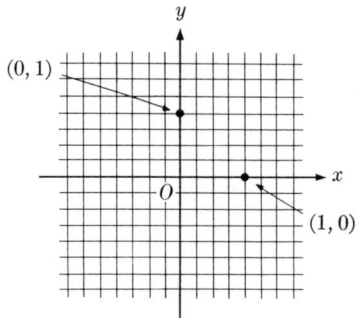

Figure 7–3

We want to *define* the area of an irregularly shaped region R, such as the one shown in Fig. 7–4. We begin by constructing a square grid (one of our sequence of grids) in the plane. All the squares of this grid can be divided into three types: (1) squares which are completely inside the region R; (2) squares which have some points in the region and some points not in the region; (3) squares which are completely outside the region.

If the grid we select is the 5th in the sequence ($n = 5$), then the length of the side of each square is $1/2^5 = 1/32$ and the area of each square is $\frac{1}{32} \cdot \frac{1}{32} = \frac{1}{1024}$ square units. We define for the square grid with $n = 5$;

A_5^L = sum of the areas of all squares of type (1)

A_5^U = sum of the areas of all squares of type (1) plus type (2).

(The L stands for *lower* and the U stands for *upper*.) According to the way we have defined these quantities, if the region we started with is to have something we call "area," then we would expect that A_5^L would be less than this area, while A_5^U would be greater. We have the inequality

$$A_5^L \leq A_5^U,$$

which comes from the very definition. The quantity A_5^U contains everything in A_5^L and more.

We have described a typical step in a process which consists of a sequence of steps. The next step consists of constructing a grid with $n = 6$. Each square is of side $\frac{1}{64}$, the area being $\frac{1}{64} \cdot \frac{1}{64} = \frac{1}{4096}$ square units. Each square of the $n = 5$ grid is divided into exactly 4 subsquares to form the $n = 6$ grid. We now define the quantities:

A_6^L = sum of areas of all squares of type (1) of the $n = 6$ grid;

A_6^U = sum of areas of all squares of type (1) plus type (2) of the $n = 6$ grid.

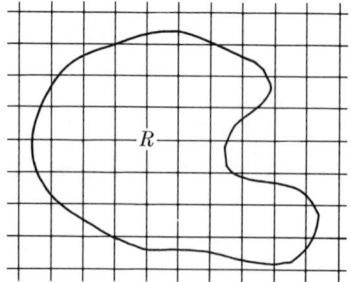

Figure 7–4

As before, the definition tells us that

$$A_6^L \leq A_6^U.$$

But now we can see that there is a relation between A_5^L and A_6^L. Any square in A_5^L is the same as 4 squares of A_6^L, with the same total area. However, A_6^L may have some squares of type (1) near the boundary which are part of type (2) regions for the $n = 5$ grid. Figure 7–5 shows that the lower left corner is of type (1) for $n = 6$, while the large square is of type (2) for $n = 5$. It can be proved (although our discussion has not done so) that for *any* region, we have

$$A_5^L \leq A_6^L.$$

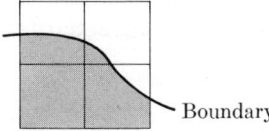
Boundary
Figure 7–5

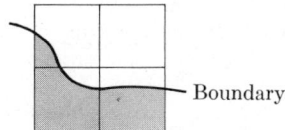
Boundary
Figure 7–6

Similarly, A_5^U may have a type (2) region (see Fig. 7–6) which, when considered in the $n = 6$ grid, gives three type (2) regions and one type (3) region. This means that for such a square the contribution to A_6^U (three smaller squares) is less than the corresponding contribution to A_5^U (the larger square). *It can be shown in a general way that*

$$A_6^U \leq A_5^U.$$

The inequalities may also be combined to give

$$A_5^L \leq A_6^L \leq A_6^U \leq A_5^U.$$

When the above process is performed for $n = 1, 2, 3, \ldots$ we get a sequence of numbers

$$A_1^L \leq A_2^L \leq A_3^L \leq A_4^L \leq \cdots \leq A_n^L \leq \cdots.$$

and a corresponding sequence

$$A_1^U \geq A_2^U \geq A_3^U \geq A_4^U \geq \cdots \geq A_n^U \geq \cdots.$$

By Axiom C (of Chapter 4, Section 10) we know that the A_n^L tend to a limit, since the sequence is bounded (by A_1^U—in fact by all A_n^U) and is steadily increasing (or at least nondecreasing). The limit of A_n^L is denoted by A^- and is called the **inner area** of the region. Since the region is designated by R, we also write $A^-(R)$. In the same way it can be shown that the A_n^U tend to a limit which we denote by A^+, and this number is called the **outer area** of the region. We also write $A^+(R)$.

It is always true that

$$A^-(R) \leq A^+(R),$$

but it is not always true that these numbers are equal. However, the sets R for which they are unequal are weird, and for the kinds of regions we shall be concerned with it can be shown that

$$A^-(R) = A^+(R).$$

This common value is then **defined to be the area of the region,** *and we shall denote it* $A(R)$.

Note that the above discussion did not *prove* the validity of the inequalities for A_n^L and A_n^U. The arguments may or may not have appeared convincing, but arguing "pictorially" is not a proof. The material given here is intended only to establish the reasonableness of the results.

The fundamental unit for determining areas is the square. A completely analogous development could be made with rectangles instead of squares. A sequence of *rectangular grids* is then used, the lengths and widths of the rectangles tending to zero as the nth term in the sequence tends to infinity. We would naturally expect, and it can be proved, that both the rectangular and the square grids yield the same number for the area of R. If a region R is formed as the sum of two nonoverlapping regions R_1 and R_2, we can easily prove the expected fact that

$$A(R) = A(R_1) + A(R_2).$$

Finally, it can be shown that the value of $A(R)$ is not influenced by the location of the coordinate axes.

2. NOTATIONS FOR SUMS

The sum of seven terms,

$$a_1 + a_2 + a_3 + a_4 + a_5 + a_6 + a_7,$$

can be written in abbreviated form using a symbol which we shall now introduce. The Greek letter sigma and certain subscripts are combined in the following way:

$$\sum_{i=1}^{7} a_i \quad \text{means} \quad a_1 + a_2 + a_3 + a_4 + a_5 + a_6 + a_7.$$

We read it: "the sum from 1 to 7 of a sub i." The lower number (1 in this case) indicates where the sum starts, and the upper number (7 in this case) indicates where it ends. The expression

$$\sum_{i=3}^{8} b_i \quad \text{means} \quad b_3 + b_4 + b_5 + b_6 + b_7 + b_8.$$

The symbol i is a "dummy" symbol (called the **index of summation**), since

$$\sum_{k=3}^{8} b_k \quad \text{means} \quad b_3 + b_4 + b_5 + b_6 + b_7 + b_8,$$

which is exactly the same thing.

Carrying the process one step further, we have

$$\sum_{i=1}^{5} (2i + 1) = (2 \cdot 1 + 1) + (2 \cdot 2 + 1) + (2 \cdot 3 + 1) + (2 \cdot 4 + 1)$$
$$+ (2 \cdot 5 + 1) = 3 + 5 + 7 + 9 + 11,$$

and

$$\sum_{k=2}^{5} [(k + 1)^3 - k^3] = [(2 + 1)^3 - 2^3] + [(3 + 1)^3 - 3^3]$$
$$+ [(4 + 1)^3 - 4^3] + [(5 + 1)^3 - 5^3] = -2^3 + 6^3.$$

The **summation notation,** as it is called, is of great help in manipulating sums. As an example, consider

$$\sum_{i=1}^{6} 2i = 2 \cdot 1 + 2 \cdot 2 + 2 \cdot 3 + 2 \cdot 4 + 2 \cdot 5 + 2 \cdot 6$$
$$= 2(1 + 2 + 3 + 4 + 5 + 6)$$
$$= 2 \sum_{i=1}^{6} i.$$

More generally, we have the rule that for any sum

$$\sum_{k=1}^{n} ca_k = c \sum_{k=1}^{n} a_k, \quad c = \text{const},$$

where the symbol means that the sum starts with a_1 and ends with a_n. The number at the bottom is called the **lower limit** of the sum and the number at the top the **upper limit.** We also have

$$\sum_{k=1}^{n} (a_k + b_k) = \sum_{k=1}^{n} a_k + \sum_{k=1}^{n} b_k.$$

PROBLEMS

In Problems 1 through 8, find in each case the value of a given sum.

1. $\sum_{i=1}^{12} i$

2. $\sum_{i=3}^{7} (2i - 3)$

3. $\sum_{i=-4}^{3} (i + 2)$

4. $\sum_{k=3}^{7} b_k$, given that $b_j = 2^j$ for every j.

5. $\sum_{j=1}^{5} \frac{1}{j}$

6. $\sum_{k=-4}^{-1} \frac{1}{1 + k^2}$

7. $\sum_{j=1}^{10} (2^{j+1} - 2^j)$

8. $\sum_{k=-1}^{2} \cos k\pi$

9. Show that $\sum_{k=1}^{8} (3k + 2) = \sum_{k=0}^{7} (3k + 5)$.

10. Find the value of $\sum_{k=1}^{6} (a_{k+1} - a_k)$, given that $a_k = 10^k$.

11. Find the value of $\sum_{k=2}^{7} (a_k - a_{k+1})$, given that $a_k = \frac{1}{k}$.

3. THE DEFINITE INTEGRAL

In Chapter 6, Section 10, we introduced the definite integral and discussed its relationship to the notion of antiderivative. We are now in a position to make the treatment more precise and to develop a rigorous definition of the definite integral.

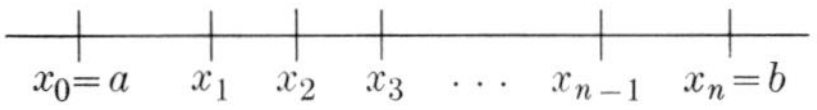

Figure 7-7

Suppose that f is a function defined on an interval $[a, b]$. We make a *subdivision* of this interval by introducing $n - 1$ intermediate points; call them $x_1, x_2, x_3, \ldots, x_{n-1}$ (Fig. 7-7). These points are not necessarily equally spaced. Letting $x_0 = a$ and $x_n = b$, we see that there are exactly n subintervals. Since it is convenient to have a symbol for such a subdivision, we shall use the Greek letter Δ for this purpose. That is, Δ stands for the process of introducing $n - 1$ points between a and b. The subdivision can be made in an infinite variety of ways for each value of n and can also be made for every positive integer n. However, we will simply use the symbol Δ to indicate such a subdivision without attempting to indicate the way in which it is made. An important symbol which we introduced in Chapter 6 is

$$\Delta_i x = x_i - x_{i-1} = \text{length of } i\text{th subinterval in the subdivision } \Delta.$$

There are exactly n subintervals, and among them there is always one which is largest. (Of course there may be several of equal size.) We introduce

$$\|\Delta\| = \text{length of largest subinterval in subdivision } \Delta.$$

This quantity $\|\Delta\|$ is called the **norm** of the subdivision. When we want to know how "fine" a subdivision is, the norm $\|\Delta\|$, the maximum distance between points, is a measure of this "fineness." For example, if $a = 2$, $b = 5$, and $\|\Delta\| = 0.05$, we know that n has to be at least 60. This reasoning holds because it takes 60 intervals of length 0.05 to add up to $b - a = 3$. Of course, $\|\Delta\|$ could be 0.05 and $n = 100$, with many points quite close together and only one or two intervals actually of length 0.05.

In each of the subintervals of a subdivision Δ we select a point. *Such a selection may be made in any way whatsoever.* Let ξ_1 be the point selected in $[x_0, x_1]$; let ξ_2 be the point in $[x_1, x_2]$, and, in general, let ξ_k be the point in $[x_{k-1}, x_k]$.

We now form the sum

$$f(\xi_1)(x_1 - x_0) + f(\xi_2)(x_2 - x_1) + \cdots$$
$$+ f(\xi_k)(x_k - x_{k-1}) + \cdots + f(\xi_n)(x_n - x_{n-1}).$$

We can abbreviate this to

$$f(\xi_1)\Delta_1 x + f(\xi_2)\Delta_2 x + \cdots + f(\xi_k)\Delta_k x + \cdots + f(\xi_n)\Delta_n x,$$

and even further, by using the summation notation, to

$$\sum_{k=1}^{n} f(\xi_k)\Delta_k x.$$

7-3 THE DEFINITE INTEGRAL

Definition. *A **function f is said to be integrable on the interval** $[a, b]$ if there is a number A with the following property: for each $\epsilon > 0$ there is a $\delta > 0$ such that*

$$\left| \sum_{k=1}^{n} f(\xi_k)\, \Delta_k x - A \right| < \epsilon,$$

for every subdivision Δ with $\|\Delta\| < \delta$ and for any choices of the ξ_k in $[x_{k-1}, x_k]$.

A simpler but less precise way of writing the above definition nevertheless gives the idea of what is happening:

$$\lim_{\|\Delta\| \to 0} \sum_{k=1}^{n} f(\xi_k)\, \Delta_k x = A.$$

The lack of precision comes about because, for any given value of the norm, $\|\Delta\|$, the $\Delta_k x$ can vary greatly and the ξ_k can wander throughout the subinterval. This limiting process is quite different from the one we studied earlier.

It is natural to ask if there are several possible values for the number A in any particular case. *It can be shown that if there is a number A which satisfies the definition, then it is unique.* There cannot be two different values. The proof of this fact follows the same outline as the proof of uniqueness of limits (see Chapter 4, Section 7).

Definitions. *The number A in the above definition is called **the definite integral of f from** a **to** b and is denoted by*

$$\int_a^b f(x)\, dx.$$

*The function f is the **integrand**, and the numbers a and b are the **lower** and **upper limits of integration**, respectively. The letter x is the **variable of integration**. This is a "dummy" variable and may be replaced by any other letter (although to prevent confusion we would avoid using, a b, d, or f for the variable of integration).*

When does a function f satisfy the conditions of the definition? To find out in any particular case, we would have to investigate all possible subdivisions with all possible choices for the ξ_i and be sure that the same number A is approached in every instance. The task is a hopeless one. It is imperative that we have some theorems which tell us when a function is integrable. We have seen that not all functions possess derivatives. If a function has a "corner" or a discontinuity at a point, it has no tangent and therefore is not differentiable at that point. Similarly, we wish to know which of the various types of functions are integrable. Is a function with a corner integrable? Does a function with a "jump" discontinuity satisfy the definition of integrability?

In order to gain further insight into the process of integration we use a method of selecting special sums of the type which appear in the definition of integral. Suppose that f is a continuous function on a closed interval $[a, b]$. The first theorem in Chapter 6 states that there is a place in $[a, b]$ where f has a largest value and another place where f has a smallest value (Extreme Value Theorem). Starting with this fact, we make a subdivision Δ of the interval $[a, b]$. That is,

$$\Delta \text{ is } \{a = x_0 < x_1 < x_2 < \cdots < x_{n-1} < x_n = b\}.$$

Since f is continuous on $[x_0, x_1]$, there is a point, call it ξ_1', where f has a minimum value and a point, call it ξ_1'', where f has a maximum value. On $[x_1, x_2]$ f is continuous. Let ξ_2' and ξ_2'' be the places where f has minimum and maximum values, respectively. Continue this process. The values ξ_k', ξ_k'' are respectively the places where f has minimum and maximum values in the interval $[x_{k-1}, x_k]$. In this way corresponding to the subdivision Δ we get *two* sums:

$$\sum_{k=1}^{n} f(\xi_k') \Delta_k x \quad \text{and} \quad \sum_{k=1}^{n} f(\xi_k'') \Delta_k x.$$

We abbreviate this even further by writing

$$\underline{S}(\Delta) = \sum_{k=1}^{n} f(\xi_k') \Delta_k x, \qquad \bar{S}(\Delta) = \sum_{k=1}^{n} f(\xi_k'') \Delta_k x.$$

From our method of selection we know that $\underline{S}(\Delta) \leq \bar{S}(\Delta)$, always. But we know even more: since in each subinterval the function f is always between its minimum and maximum, then for *any* choice ξ_k in a particular subdivision Δ we always have

$$\underline{S}(\Delta) \leq \sum_{k=1}^{n} f(\xi_k) \Delta_k x \leq \bar{S}(\Delta). \tag{1}$$

If we can show that $\underline{S}(\Delta)$ and $\bar{S}(\Delta)$ tend to the same limit, A, for all possible Δ, with $\|\Delta\| \to 0$, then the same will be true for every possible choice of the ξ_i and therefore for all subdivisions. The function f will satisfy the definition of integrability.

Example. Given that $f(x) = x^2$ and given the subdivision

$$\Delta: \{a = x_0 = -1, \, x_1 = -0.6, \, x_2 = -0.2, \, x_3 = 0.3, \, x_4 = 0.7, \, x_5 = b = 1\}.$$

Find $\underline{S}(\Delta)$ and $\bar{S}(\Delta)$.

Solution. We construct the graph shown in Fig. 7–8, and the corresponding table:

	i	1	2	3	4	5
$\Delta_i x = x_i - x_{i-1}$	$\Delta_i x$	0.4	0.4	0.5	0.4	0.3
Place where minimum occurs	ξ_i'	−0.6	−0.2	0	0.3	0.7
Minimum value	$f(\xi_i')$	0.36	0.04	0	0.09	0.49
Place where maximum occurs	ξ_i''	−1	−0.6	0.3	0.7	1
Maximum value	$f(\xi_i'')$	1	0.36	0.09	0.49	1

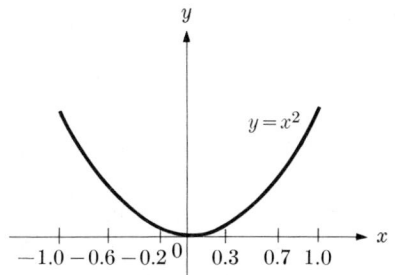

Figure 7-8

Then we can obtain

$$\underline{S}(\Delta) = \sum_{i=1}^{5} f(\xi_i') \Delta_i x$$
$$= 0.36(0.4) + 0.04(0.4) + 0(0.5) + 0.09(0.4) + 0.49(0.3)$$
$$= 0.343;$$

$$\bar{S}(\Delta) = \sum_{i=1}^{5} f(\xi_i'') \Delta_i x$$
$$= 1(0.4) + 0.36(0.4) + 0.09(0.5) + (0.49)(0.4) + 1(0.3)$$
$$= 1.085. \qquad \blacktriangleleft$$

We now sketch a proof of a theorem which indicates when certain functions are integrable.

Theorem 1. *If $f(x)$ is defined and increasing (or at least nondecreasing) on the closed interval $a \leq x \leq b$, then it is integrable there.*

Proof. Let Δ be any subdivision. On each subinterval $[x_{i-1}, x_i]$ the minimum must occur at the left endpoint and the maximum must occur at the right endpoint. That is, $\xi_i' = x_{i-1}$ and $\xi_i'' = x_i$. We construct the sums

$$\underline{S}(\Delta) = \sum_{i=1}^{n} f(x_{i-1}) \Delta_i x, \qquad \bar{S}(\Delta) = \sum_{i=1}^{n} f(x_i) \Delta_i x.$$

We know that $\underline{S}(\Delta) \leq \bar{S}(\Delta)$. We now subtract:

$$\bar{S}(\Delta) - \underline{S}(\Delta) = \sum_{i=1}^{n} [f(x_i) - f(x_{i-1})] \Delta_i x.$$

Each $\Delta_i x$ is positive and the quantities $f(x_i) - f(x_{i-1})$ are all positive (or at least nonnegative). If we replace each $\Delta_i x$ by its largest possible value, the right side becomes larger, and we get the inequality

$$\bar{S}(\Delta) - \underline{S}(\Delta) \leq \sum_{i=1}^{n} [f(x_i) - f(x_{i-1})] \|\Delta\|.$$

The quantity $\|\Delta\|$, by definition, stands for the largest $\Delta_i x$. We now note that the terms on the right all "telescope," and we have:

$$\bar{S}(\Delta) - \underline{S}(\Delta) \leq [f(x_n) - f(x_0)] \|\Delta\| = [f(b) - f(a)] \|\Delta\|.$$

Then, given any $\epsilon > 0$, if we select a subdivision with

$$\|\Delta\| < \frac{\epsilon}{f(b) - f(a)}, \quad \text{we have} \quad \bar{S}(\Delta) - \underline{S}(\Delta) < \epsilon.$$

From the definition of area as given in Section 1 and inequality (1) on page 200, it is possible to show that there is a limit A to which both $\bar{S}(\Delta)$ and $\underline{S}(\Delta)$ tend. Furthermore, every intermediate sum also tends to A. The function f is integrable. ◀

Theorem 1 illustrates how careful we must be in stating the hypotheses of a theorem. To make the sloppy statement that an increasing function in an interval is integrable is to make a false statement. For example, consider the function

$$f(x) = \frac{1}{2 - x}$$

in the interval $0 \leq x < 2$. (See Fig. 7–9.) This function is increasing everywhere from 0 to 2, and it tends to infinity as $x \to 2$. This function is not integrable on $[0, 2]$. On the other hand, Theorem 1 tells us that the function is integrable on any interval $[0, c]$ if $c < 2$.

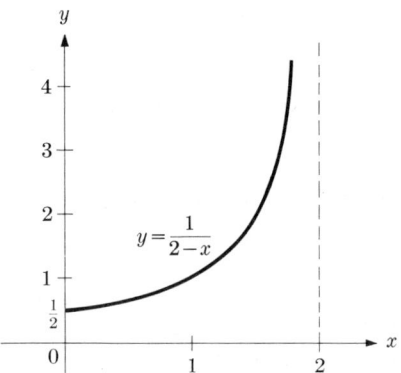

Figure 7–9

In Theorem 1, the hypothesis that f is increasing is a fairly restrictive one. Actually functions are integrable under much more general hypotheses. The following theorem, which we state without proof, is an example of the type of result which establishes the integrability of a large class of functions—namely, those which are continuous on a closed interval.

Theorem 2. *If f is continuous on $[a, b]$, then it is integrable on $[a, b]$.*

The proof of this theorem depends on the notion of *uniform continuity*, a concept usually taken up in more advanced courses in analysis. However, the idea behind the proof can be discussed here. As in the discussion of the preceding theorem, we consider

$$\bar{S}(\Delta) - \underline{S}(\Delta) = \sum_{i=1}^{n} [f(\xi_i'') - f(\xi_i')] \Delta_i x.$$

We apply the "triangle inequality" for absolute values ($|a + b| \leq |a| + |b|$) and, since $|\Delta_i x| = \Delta_i x$, we can write

$$\bar{S}(\Delta) - \underline{S}(\Delta) \le \sum_{i=1}^{n} |f(\xi_i'') - f(\xi_i')| \cdot \Delta_i x.$$

Uniform continuity on an interval $[a, b]$ means that, given any $\epsilon > 0$, there is a $\delta > 0$ such that

$$|f(x') - f(x'')| < \epsilon$$

for *any* two points in $[a, b]$ such that $|x' - x''| < \delta$. This concept is used to establish the inequality

$$\bar{S}(\Delta) - \underline{S}(\Delta) \le \sum_{i=1}^{n} \epsilon \Delta_i x = \epsilon \sum_{i=1}^{n} \Delta_i x = \epsilon(b - a),$$

from which it follows that f is integrable.

PROBLEMS

In Problems 1 through 10, a function f and a subdivision are given. Sketch the graph of the function and find the value of $\underline{S}(\Delta)$ and $\bar{S}(\Delta)$. Use knowledge of $f'(x)$ when necessary.

1. $f(x) = x^2 + 1$; Δ: $a = x_0 = 1$, $x_1 = 1.2$, $x_2 = 1.5$, $x_3 = 1.6$, $x_4 = 2$, $x_5 = 2.5$, $x_6 = b = 3$.
2. $f(x) = 1/(1 + x)$; Δ: $a = x_0 = 0$, $x_1 = 0.2$, $x_2 = 0.4$, $x_3 = 0.6$, $x_4 = 0.8$, $x_5 = b = 1$.
3. $f(x) = x^2 - x + 1$; Δ: $a = x_0 = 0$, $x_1 = 0.1$, $x_2 = 0.3$, $x_3 = 0.5$, $x_4 = 0.7$, $x_5 = 0.8$, $x_6 = 0.9$, $x_7 = b = 1$.
4. $f(x) = 1/(1 + x^2)$; Δ: $a = x_0 = -1$, $x_1 = -0.9$, $x_2 = -0.5$, $x_3 = -0.2$, $x_4 = 0$, $x_5 = 0.2$, $x_6 = 0.3$, $x_7 = 0.7$, $x_8 = b = 1$.
5. $f(x) = x^3 + x$; Δ: $a = x_0 = 0$, $x_1 = 0.1$, $x_2 = 0.2$, $x_3 = 0.3$, $x_4 = 0.5$, $x_5 = b = 0.6$.
6. $f(x) = x^2 + 2x - 1$; Δ: $a = x_0 = -2$, $x_1 = -1.8$, $x_2 = -1.6$, $x_3 = -1.4$, $x_4 = -1.2$, $x_5 = b = -1$.
7. $f(x) = (x^2 - 2)/(x + 1)$; Δ: $a = x_0 = 0$, $x_1 = 0.5$, $x_2 = 1$, $x_3 = 1.5$, $x_4 = 2$, $x_5 = 2.5$, $x_6 = 3$, $x_7 = b = 3.5$.
8. $f(x) = x/(x^2 + 1)$; Δ: $a = x_0 = 0$, $x_1 = 1$, $x_2 = 2$, $x_3 = 3$, $x_4 = 4$, $x_5 = 4.5$, $x_6 = 5$, $x_7 = b = 6$.
9. $f(x) = x^3 - 3x + 1$; Δ: $a = x_0 = -2$, $x_1 = -1.5$, $x_2 = -1$, $x_3 = -0.5$, $x_4 = 0$, $x_5 = 0.5$, $x_6 = 1$, $x_7 = 1.5$, $x_8 = b = 2$.
10. $f(x) = x^3/(x^3 + 1)$; Δ: $a = x_0 = 0$, $x_1 = 1$, $x_2 = 2$, $x_3 = 3$, $x_4 = b = 4$.
11. Prove that if a function f is decreasing (or nonincreasing) on a closed interval it is integrable.
12. Suppose that a function f is differentiable on a closed interval $[a, b]$. If f has only a finite number of maximum and minimum points on $[a, b]$, use the results of Theorem 1 and Problem 11 to show that f is integrable there.
*13. Show that every polynomial function is integrable on any interval $[a, b]$ of finite length. [*Hint:* Use the result in Problem 12.]
14. Why is the "postage" function (see Chapter 2, Section 3) integrable even though it is not continuous? Find the value of the integral from $x = 0$ to $x = 4$.

15. Suppose that f and f' are continuous on $[a, b]$ and that $|f'(x)| \leq M$ on $[a, b]$. Use the Theorem of the Mean to show that for any subdivision Δ,
$$\bar{S}(\Delta) - \underline{S}(\Delta) \leq M(b - a) \cdot \|\Delta\|.$$
16. In what intervals is the function $f(x) = -1/x$ integrable?
17. Let Δ be a subdivision of the interval $[0, 1]$ into n equal parts. Compute $\underline{S}(\Delta)$ for $f(x) = 1/x^2$. What happens to $\underline{S}(\Delta)$ as $n \to \infty$? What can be said about $\bar{S}(\Delta)$ and therefore about the integrability of f?

4. PROPERTIES OF THE DEFINITE INTEGRAL

Later we shall develop techniques for evaluating integrals without recourse to the definition. This is analogous to what we did for derivatives: first we showed how to find the derivative directly from the definition, and then we developed methods for differentiation which bypassed the definition entirely. The properties we shall now discuss will be useful in learning methods for integrating various kinds of functions.

The simplest properties are given in the two following theorems.

Theorem 3. *If c is any number and f is integrable on $[a, b]$, then the function $cf(x)$ is integrable on $[a, b]$ and*
$$\int_a^b cf(x)\, dx = c \int_a^b f(x)\, dx.$$

Theorem 4. *If $f(x)$ and $g(x)$ are integrable on $[a, b]$, then $f(x) + g(x)$ is integrable on $[a, b]$ and*
$$\int_a^b [f(x) + g(x)]\, dx = \int_a^b f(x)\, dx + \int_a^b g(x)\, dx.$$

We shall prove Theorem 4.

Proof. If M and N are any numbers, we know that $|M + N| \leq |M| + |N|$. Let Δ be any subdivision and take
$$M = \sum_{i=1}^n f(\xi_i)\, \Delta_i x - \int_a^b f(x)\, dx, \qquad N = \sum_{i=1}^n g(\xi_i)\, \Delta_i x - \int_a^b g(x)\, dx.$$

From the definition of integrability, we have $|M| < \epsilon$ and $|N| < \epsilon$ if $\|\Delta\|$ is sufficiently small. But
$$M + N = \sum_{i=1}^n [f(\xi_i) + g(\xi_i)]\, \Delta_i x - \left[\int_a^b f(x)\, dx + \int_a^b g(x)\, dx\right].$$

We conclude that $|M + N| < 2\epsilon$. This means that $M + N$ tends to zero as $\|\Delta\|$ tends to zero and, therefore, the function $f + g$ is integrable and its integral is equal to the sum of the integrals of f and g.

The proof of Theorem 3 is similar.

Theorem 5. *If $f(x)$ is integrable on an interval $[a, b]$, then f is bounded there.*

This theorem is proved in more advanced courses in mathematics. The theorem means that if a function is integrable there must be two numbers m and M (m may be negative and M may be very large, positive) such that for all values x in $[a, b]$, $f(x)$ lies between m and M. That is,

$$m \leq f(x) \leq M.$$

Theorem 5 is of considerable importance in developing the theory of integration.

Theorem 6. *If $f(x)$ is integrable on an interval $[a, b]$ and m and M are numbers such that*

$$m \leq f(x) \leq M \quad \text{for} \quad a \leq x \leq b,$$

then it follows that

$$m(b - a) \leq \int_a^b f(x)\, dx \leq M(b - a).$$

Proof. Let Δ be a subdivision of $[a, b]$. We know that

$$\sum_{i=1}^n f(\xi_i)\, \Delta_i x$$

tends to the integral if $x_{i-1} \leq \xi_i \leq x_i$ and if $\|\Delta\| \to 0$ as $n \to \infty$. By hypothesis, $m \leq f(\xi_i) \leq M$ for each i. Multiplying through by $\Delta_i x$ yields

$$m\, \Delta_i x \leq f(\xi_i)\, \Delta_i x \leq M\, \Delta_i x.$$

Summing from 1 to n, we now write

$$\sum_{i=1}^n m\, \Delta_i x \leq \sum_{i=1}^n f(\xi_i)\, \Delta_i x \leq \sum_{i=1}^n M\, \Delta_i x.$$

On the left and the right we can factor out the m and M, respectively, to find

$$m \sum_{i=1}^n \Delta_i x \leq \sum_{i=1}^n f(\xi_i)\, \Delta_i x \leq M \sum_{i=1}^n \Delta_i x.$$

But

$$\sum_{i=1}^n \Delta_i x$$

"telescopes" to give $b - a$. Therefore

$$m(b - a) \leq \sum_{i=1}^n f(\xi_i)\, \Delta_i x \leq M(b - a).$$

The term in the middle tends to the integral and, since the inequalities hold in the limit, the result follows.

Theorem 7. *If f and g are integrable on the interval $[a, b]$ and $f(x) \leq g(x)$ for each x in $[a, b]$, then*

$$\int_a^b f(x)\, dx \leq \int_a^b g(x)\, dx.$$

Proof. The proof furnishes an excellent illustration of the way in which mathematics is composed of building blocks. We show how earlier theorems form the blocks which produce the result. We define $F(x) = g(x) - f(x)$. Then by hypothesis $F(x) \geq 0$ on $[a, b]$. By Theorem 3, if $f(x)$ is integrable, so is $-f(x)$, $(c = -1)$. According to Theorem 4, if $g(x)$ and $-f(x)$ are integrable, so is $g(x) + (-f(x)) = F(x)$. We recall that Theorem 6 tells us that

$$m(b - a) \leq \int_a^b F(x)\, dx$$

and, since $F(x) \geq 0$ on $[a, b]$, we can select $m = 0$. Therefore

$$\int_a^b [g(x) - f(x)]\, dx \geq 0,$$

or

$$\int_a^b g(x)\, dx - \int_a^b f(x)\, dx \geq 0 \quad \text{(by Theorem 4)},$$

or

$$\int_a^b g(x)\, dx \geq \int_a^b f(x)\, dx,$$

which is what we wanted to prove.

Theorem 8. *If $f(x)$ is continuous on the interval $[a, b]$ and c is any number in $[a, b]$, then*

$$\int_a^b f(x)\, dx = \int_a^c f(x)\, dx + \int_c^b f(x)\, dx.$$

Sketch of proof. Consider a subdivision Δ of the interval $[a, b]$. This subdivision may or may not contain c as one of its points x_i. If it does not, we insert c into it and obtain a new subdivision Δ', with one more point and one more interval. The new subdivision has a norm $\|\Delta'\|$ which is smaller than or the same as the norm $\|\Delta\|$; at least it is not larger. Then the part of the subdivision of Δ' from a to c gives a sum of the form

$$\sum_{i=1}^{n'} f(\xi_i)\, \Delta_i x,$$

where n' is the number of subintervals in $[a, c]$; similarly, if n'' is the number of subintervals from c to b there will be a sum of the form

$$\sum_{i=1}^{n''} f(\xi_i)\, \Delta_i x$$

for the interval $[c, b]$. The first sum tends to $\int_a^c f(x)\,dx$, the second sum tends to $\int_c^b f(x)\,dx$, and the two added together tend to $\int_a^b f(x)\,dx$. This is exactly the statement of the result of the theorem.

Two useful definitions,

$$\int_a^a f(x)\,dx = 0 \quad \text{and} \quad \int_a^b f(x)\,dx = -\int_b^a f(x)\,dx,$$

imply that the relation

$$\int_a^b f(x)\,dx = \int_a^c f(x)\,dx + \int_c^b f(x)\,dx$$

holds for any value of c whether or not it is between a and b. If $c = b$ the result is clear, while if $c > b$ or $c < a$, the appropriate reversal of sign reduces the result to the one stated in Theorem 8.

PROBLEMS

In Problems 1 through 15, in each case study the function in the given interval and apply Theorem 6 to find largest and smallest values the integral can have.

1. $\int_1^4 2x\,dx$

2. $\int_0^3 x^2\,dx$

3. $\int_{-1}^2 x^3\,dx$

4. $\int_0^5 \dfrac{2}{1+2x}\,dx$

5. $\int_3^4 \dfrac{x}{1+x}\,dx$

6. $\int_{-2}^2 \dfrac{1}{1+x^2}\,dx$

7. $\int_0^4 \dfrac{x^2}{1+x^2}\,dx$

8. $\int_{-3}^{-1} \dfrac{1}{1-x}\,dx$

9. $\int_2^5 \sqrt{1+x}\,dx$

10. $\int_0^3 \sqrt{x^2+1}\,dx$

11. $\int_0^4 \dfrac{2\sqrt{x}}{1+x}\,dx$

12. $\int_1^2 \dfrac{1-x^2}{1+x^2}\,dx$

13. $\int_{-2}^3 \dfrac{2+x}{\sqrt{1+x^2}}\,dx$

14. $\int_0^{\pi/2} \sin x\,dx$

15. $\int_0^{\pi/4} \tan x\,dx$

16. Prove Theorem 3.

17. Assuming that $\int_a^b x^n\,dx$ is integrable for every nonnegative integer n, use Theorems 3 and 4 to write out a proof that every polynomial is integrable on every finite interval $[a, b]$.

18. Given the function

$$f: x \to \begin{cases} x, & -1 \le x \le 0, \\ x^2 - x + 1, & 0 < x \le 1, \end{cases}$$

show, by combining Theorems 3 and 4, that f is integrable on $[-1, 1]$.

19. Write out the details of the proof that

$$\int_a^b f(x)\,dx = \int_a^c f(x)\,dx + \int_c^b f(x)\,dx$$

when c is not between a and b.

20. Let $c_1, c_2, \ldots, c_k$ be any real numbers. Show that

$$\int_{c_1}^{c_k} f(x)\,dx = \sum_{i=1}^{k-1} \int_{c_i}^{c_{i+1}} f(x)\,dx.$$

21. Let $a_0 < a_1 < \cdots < a_k$ be any $k+1$ real numbers. Let $P_i(x)$, $i = 1, 2, \ldots, k$ be any polynomial functions. Define $F(x) = P_i(x)$ for $a_{i-1} \leq x < a_i$. Show that F is integrable on the interval $[a_0, a_k]$. (See Problems 17 and 18.)

5. EVALUATION OF DEFINITE INTEGRALS

It is apparent from the work in Chapter 6, Section 10, and the methods developed in Section 3 of this chapter, that the definite integral is intimately connected with area. In fact, *if $f(x) \geq 0$ on an interval $[a, b]$ and if it is integrable there, then*

$$\int_a^b f(x)\,dx$$

is the area bounded by the lines $x = a$, $x = b$, the x axis, and the curve of $y = f(x)$ (shaded region in Fig. 7–10). This is certainly not startling in the light of all our discussions of area as an interpretation of integral. If $f(x)$ is entirely *below* the x axis, then the sums

$$\sum_{i=1}^n f(\xi_i)\,\Delta_i x$$

are always *negative*. If f is integrable on $[a, b]$, this integral will have a *negative value*. The value of

$$\int_a^b f(x)\,dx$$

will be *the negative of the area* bounded by the lines $x = a$, $x = b$, the x axis, and the curve of $y = f(x)$ (shaded region in Fig. 7–11). If a function f is partly positive and partly negative on $[a, b]$ so that there is a point c where it crosses the x axis (see Fig. 7–12), the value of

$$\int_a^b f(x)\,dx$$

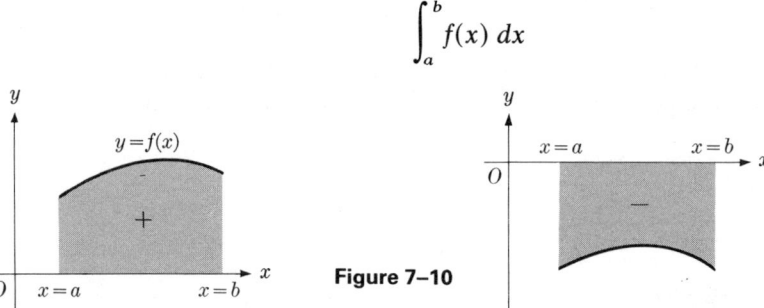

Figure 7–10   Figure 7–11

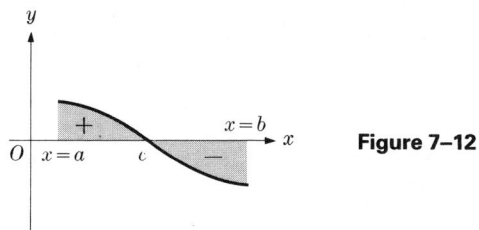

Figure 7–12

is the difference between the area above the x axis and the area below it. To obtain the true area of a region such as the shaded one shown in Fig. 7–12, we must first find the point c and compute

$$\int_a^c f(x)\,dx \quad \text{and} \quad \int_c^b f(x)\,dx$$

separately. The value of the first integral is positive and that of the second negative. We add the absolute value of the second integral to the value of the first, and the resulting quantity is the total area.

The next theorem we shall establish, known as the **Fundamental Theorem of the Calculus,** is useful for actually finding the value of a definite integral. In Chapter 6, Section 10, we saw in an intuitive way the relationship between derivative and integral. The integral was interpreted as an area (we considered positive functions only) and the derivative of the integral turned out to be the integrand. Now we establish the relation between derivative and integral purely on the basis of the material in this chapter.

Theorem 9 (Fundamental Theorem of the Calculus, first form). *Suppose that F is continuous and that F' is integrable on the interval $[a, b]$. Then*

$$\int_a^b F'(x)\,dx = F(b) - F(a).$$

Proof. We make a subdivision

$$\Delta: \{a = x_0 < x_1 < x_2 < \cdots < x_{n-1} < x_n = b\}$$

of the interval $[a, b]$. We assemble the following three facts:

1) $$F(b) - F(a) = \sum_{i=1}^n [F(x_i) - F(x_{i-1})].$$

   This comes about from "telescoping."

2) From the Theorem of the Mean (Chapter 6, Theorem 4) we know that for any two points $\bar{x}, \bar{\bar{x}}$, we have $F(\bar{\bar{x}}) - F(\bar{x}) = F'(\xi)(\bar{\bar{x}} - \bar{x})$, where ξ is some number between $\bar{x}$ and $\bar{\bar{x}}$.

3) From the definition of integral, if ϵ is any positive number, then there is a number $\delta > 0$ such that if the norm of the subdivision, $\|\Delta\|$, is less than δ,

then
$$\left|\sum_{i=1}^{n} F'(\xi_i)\,\Delta_i x - \int_a^b F'(x)\,dx\right| < \epsilon.$$

We apply the Theorem of the Mean (fact (2)) to $F(x_i) - F(x_{i-1})$, calling the intermediate point ξ_i, to obtain
$$F(x_i) - F(x_{i-1}) = F'(\xi_i)(x_i - x_{i-1}).$$

We use fact (1) and the customary abbreviation $\Delta_i x = x_i - x_{i-1}$, to write
$$F(b) - F(a) = \sum_{i=1}^{n} F'(\xi_i)\,\Delta_i x.$$

Fact (3) allows us to substitute $F(b) - F(a)$ for the sum and so obtain
$$\left|F(b) - F(a) - \int_a^b F'(x)\,dx\right| < \epsilon.$$

We now make use of a typical mathematical argument which seems to give us something for nothing. Since the above inequality must hold for *every possible* $\epsilon > 0$, the part within the absolute-value sign must be *equal* to zero. (If it were not, we would simply choose an ϵ smaller than that value to get a contradiction.) Thus the proof is complete. ◂

The Fundamental Theorem states that if we are given any integrable function f we can compute the definite integral over $[a, b]$ by finding *any* antiderivative of f and then evaluating this antiderivative at b, evaluating it at a, and then subtracting. A function f may have more than one antiderivative; in fact it has infinitely many of them. However, as the following Corollary shows, any two antiderivatives differ by a constant.

Corollary. *If $F_1(x)$ and $F_2(x)$ are antiderivatives of the same function f on the interval $[a, b]$, then*
$$F_1(x) = F_2(x) + \text{const} \quad \text{on } [a, b].$$

Proof. Since $F_1(x)$ and $F_2(x)$ are antiderivatives of the same function, the function $F(x) = F_1(x) - F_2(x)$ has a derivative which is identically zero. It follows from the Theorem of the Mean that $F(x)$ must be constant. ◂

To get antiderivatives of functions we simply work backward from our knowledge of derivatives. If
$$f(x) = x^n, \quad n \text{ rational}, \neq -1,$$
then
$$\text{antiderivative of } f = \frac{x^{n+1}}{n+1} + \text{const},$$
since the derivative of
$$\frac{x^{n+1}}{n+1}$$

is exactly x^n. We illustrate the procedure with some examples.

Example 1. Evaluate $\int_1^2 x^2 \, dx$.

Solution. The antiderivative of x^2 is

$$\frac{x^3}{3} + C, \quad \text{where } C \text{ is a constant.}$$

We evaluate this at $x = 2$ and at $x = 1$ and then subtract:

$$\frac{x^3}{3} + C \quad \text{at} \quad x = 2 \quad \text{is } \tfrac{8}{3} + C;$$

$$\frac{x^3}{3} + C \quad \text{at} \quad x = 1 \quad \text{is } \tfrac{1}{3} + C.$$

Difference $= \tfrac{8}{3} - \tfrac{1}{3} = \tfrac{7}{3}$. Then $\int_1^2 x^2 \, dx = \tfrac{7}{3}$, which is the desired evaluation. ◄

Observe that the constant in the antiderivative canceled. *This is always the case when we evaluate definite integrals; therefore, from now on, the constant of integration will be omitted in evaluating definite integrals.*

Example 2. Evaluate $\int_0^4 (x^3 - 2) \, dx$.

Solution. An antiderivative of $x^3 - 2$ is $\tfrac{1}{4}x^4 - 2x$.

$$\tfrac{1}{4}x^4 - 2x \quad \text{at} \quad x = 4 \quad \text{is} \quad 64 - 8 = 56;$$
$$\tfrac{1}{4}x^4 - 2x \quad \text{at} \quad x = 0 \quad \text{is} \quad 0.$$

Therefore

$$\int_0^4 (x^3 - 2) \, dx = 56. \quad \blacktriangleleft$$

There is a convenient and simple notation which reduces the work of evaluating integrals. We write

$$F(x)]_a^b \quad \text{or} \quad [F(x)]_a^b \quad \text{for} \quad F(b) - F(a).$$

The way in which the notation works is exhibited in the following example.

Example 3. Evaluate $\int_{-1}^2 (2x^3 - 3x^2 + x - 1) \, dx$.

Solution

$$\int_{-1}^2 (2x^3 - 3x^2 + x - 1) \, dx = \left[\frac{x^4}{2} - x^3 + \frac{x^2}{2} - x\right]_{-1}^2$$
$$= (8 - 8 + 2 - 2) - (\tfrac{1}{2} + 1 + \tfrac{1}{2} + 1)$$
$$= -3. \quad \blacktriangleleft$$

If the integrand is a complicated function, it may be difficult or even impossible to get an expression for the antiderivative. In such cases we may be interested in obtaining an *approximate value* for the definite integral. The very definition of a definite integral yields useful methods for finding an approximate value. One such process, which we shall describe, is called the **midpoint rule.** We first subdivide the interval of integration into n *equal* parts so that each subinterval has length $(b - a)/n$. Then we obtain the value $f(\xi_i)$, where ξ_i is the *midpoint* of the ith subinterval. The quantity

$$\sum_{i=1}^{n} f(\xi_i) \frac{b - a}{n}$$

is an approximation of $\int_a^b f(x)\, dx$.

Example 4. Using the midpoint rule, compute

$$\int_1^2 \frac{1}{x}\, dx$$

approximately, with $n = 5$

Solution. Each $\Delta_i x = 0.2$, since there are five subintervals from 1 to 2. Then $x_0 = 1.0$, $x_1 = 1.2$, $x_2 = 1.4$, $x_3 = 1.6$, $x_4 = 1.8$, $x_5 = 2.0$. Using midpoints, we find that $\xi_1 = 1.1$, $\xi_2 = 1.3$, $\xi_3 = 1.5$, $\xi_4 = 1.7$, $\xi_5 = 1.9$. We obtain

$$f(\xi_1)\, \Delta_1 x = \frac{1}{1.1}(0.2) = 0.18182,$$

$$f(\xi_2)\, \Delta_2 x = \frac{1}{1.3}(0.2) = 0.15385,$$

$$f(\xi_3)\, \Delta_3 x = \frac{1}{1.5}(0.2) = 0.13333,$$

$$f(\xi_4)\, \Delta_4 x = \frac{1}{1.7}(0.2) = 0.11765,$$

$$f(\xi_5)\, \Delta_5 x = \frac{1}{1.9}(0.2) = 0.10526,$$

$$\sum_{i=1}^{5} f(\xi_i)\, \Delta_i x = 0.69191.$$

The true value of the integral, correct to three decimal places, is 0.693^+. ◀

PROBLEMS

In Problems 1 through 14, in each case evaluate the given definite integral.

1. $\int_1^2 (x^2 + x - 5)\, dx$

2. $\int_1^3 (x^3 + 4x^2 - 6)\, dx$

3. $\int_{-1}^{0} (3x^4 + 2x^2 - 5)\, dx$

4. $\int_{-2}^{2} (x^5 + 4x^3 - 3x)\, dx$

5. $\int_{1}^{3} (x^{3/2} + 2x^{1/2}) \, dx$ 6. $\int_{0}^{2} (2x^3 - 4x^{2/3} + 1) \, dx$

7. $\int_{1}^{9} \frac{x^2 + x + 1}{\sqrt{x}} \, dx$ 8. $\int_{1}^{2} (2t + 1)^2 \, dt$

9. $\int_{0}^{2} (x + 1)(2x + 6) \, dx$ 10. $\int_{1}^{2} (2x + 1)^2 (x - 6)^2 \, dx$

11. $\int_{-1}^{1} \frac{(x^2 + 6x - 2)^2}{9} \, dx$ 12. $\int_{1}^{3} \frac{t^2 + 2t - 1}{\sqrt[3]{t}} \, dt$

13. $\int_{0}^{x} t^2 \, dt$ 14. $\int_{1}^{x} \frac{1}{\sqrt{t}} \, dt, \quad x > 0$

In Problems 15 through 24, compute the integrals approximately, using the midpoint rule. Compute each term $f(\xi_i) \Delta_i x$ to four decimals and round off the result to three decimal places. *Also* compute the exact value in each problem by the rule for integrating.

15. $\int_{-2}^{-1} x^2 \, dx, \quad n = 5$ 16. $\int_{0}^{1} x^3 \, dx, \quad n = 5$

17. $\int_{1}^{2} \frac{1}{x^2} \, dx, \quad n = 5$ 18. $\int_{0.5}^{1} \frac{1}{x^3} \, dx, \quad n = 5$

19. $\int_{1}^{3} (2x + 1) \, dx, \quad n = 1$ 20. $\int_{-1}^{0} (2 - 3y) \, dy, \quad n = 1$

21. $\int_{a}^{b} (At + B) \, dt, \quad n = 1$ 22. $\int_{1}^{4} \sqrt{u} \, du, \quad n = 6$

23. $\int_{1}^{2} (x + 1)(2x + 1) \, dx, \quad n = 4$ 24. $\int_{1}^{3} \frac{x^2 + 2}{x^2} \, dx, \quad n = 2$

In Problems 25 through 30, use the midpoint rule to compute the integrals approximately as in Problems 15–24. The answers given are the exact values rounded off to the number of decimals indicated. They are *not* necessarily the answers which the student should obtain by following the procedure given.

25. $\int_{0}^{1} \frac{dx}{x + 1}, \quad n = 5;$ answer, 0.693^+

26. $\int_{0}^{1} \frac{dx}{x^2 + 1}, \quad n = 5$ answer, $\frac{\pi}{4} = 0.7854^-$

27. $\int_{0}^{1} \sqrt{1 - x^2} \, dx, \quad n = 5;$ answer, $\frac{\pi}{4} = 0.7854^-$

28. $\int_{0}^{0.5} \frac{dx}{\sqrt{1 - x^2}}, \quad n = 5;$ answer, $\frac{\pi}{6} = 0.5236^-$

29. $\int_{0}^{\pi} \sin x \, dx, \quad n = 2;$ answer, 2

30. $\int_{-\pi/2}^{\pi/2} \cos x \, dx, \quad n = 3;$ answer, 2

*31. Show that if f is integrable then

$$\lim_{n \to \infty} \frac{1}{n} \sum_{k=1}^{n} f\left(\frac{k}{n}\right) = \int_{0}^{1} f(x) \, dx.$$

214 THE DEFINITE INTEGRAL

*32. Let m be a nonnegative integer and set $f(x) = x^m$. Use the formula in Problem 31 to prove that

$$\lim_{n \to \infty} \frac{1^m + 2^m + \cdots + n^m}{n^{m+1}} = \frac{1}{1 + m}.$$

33. Use the formula in Problem 31 to prove that

$$\lim_{n \to \infty} \frac{(n + 1) + (n + 2) + \cdots + (n + n)}{n^2} = \frac{3}{2}.$$

6. THEOREM OF THE MEAN FOR INTEGRALS

In Chapter 6 we took up the Extreme Value Theorem (Theorem 1), which says that *a function which is continuous on a closed interval takes on its maximum and minimum values there*. A companion to this theorem is the **Intermediate Value Theorem** given below.

Theorem 10 (Intermediate Value Theorem). *Suppose that f is continuous on an interval $[a, b]$ and that $f(a) = A$, $f(b) = B$. If C is any number between A and B, there is a number c between a and b such that $f(c) = C$.*

This theorem is usually proved in advanced analysis courses. However, we shall discuss the plausibility of the result. Figure 7–13 shows a typical function defined on $[a, b]$ with $f(a) = A$, $f(b) = B$. We give a geometrical interpretation of Theorem 10: If C is any point on the y axis between A and B and we draw the line $y = C$, then this line must intersect the curve representing $y = f(x)$. Furthermore, the x value of the point of intersection is in (a, b). Figure 7–13 shows the intersection which occurs when $x = c$. That is, we have $f(c) = C$. Intuitively, the theorem asserts that if the function is continuous and if it extends from a point below the line $y = C$ to a point above the line $y = C$, it must cross this line. Note that for some values of C (say C', as shown in Fig. 7–13) there may be several possible values of c. The theorem says that there is always *at least one*.

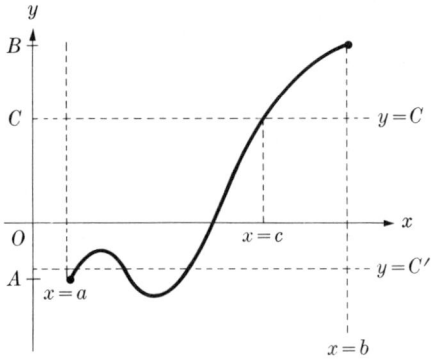

Figure 7–13

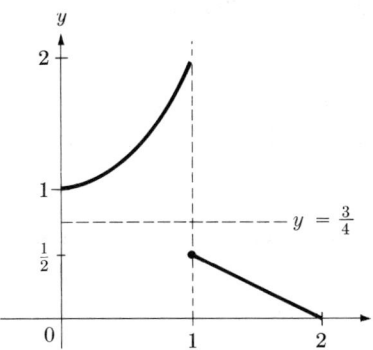

Figure 7–14

If the function is not continuous the theorem is false, as the following example shows. Define

$$f(x) = \begin{cases} x^2 + 1, & 0 \le x < 1, \\ \frac{1}{2}(2 - x), & 1 \le x \le 2. \end{cases}$$

The graph of this function is shown in Fig. 7–14. If we select $a = 0$, $b = 2$, then $f(0) = 1$, $f(2) = 0$, and $A = 1$, $B = 0$. We see that if C is chosen so that $\frac{1}{2} < C < 1$, there is no value of c such that $f(c) = C$. The function never assumes any value between $\frac{1}{2}$ and 1. The discontinuity at $x = 1$ causes the trouble.

Another example in which Theorem 10 fails is given by the function (Fig. 7–15)

$$f: x \to \frac{1}{x - 3}, \qquad x \ne 3, f(3) = L,$$

where L is any number. As the graphs shows, f is not continuous at $x = 3$. Any interval $[a, b]$ which contains 3 will have values C for which we cannot find numbers c such that $f(c) = C$. If $a = 1$, $b = 4$, then $f(1) = -\frac{1}{2}$, $f(4) = 1$, and a value of $C = \frac{1}{2}$ which is between $A = -\frac{1}{2}$, $B = 1$ is taken on only when

$$\frac{1}{x - 3} = \frac{1}{2} \qquad \text{or} \qquad x = 5.$$

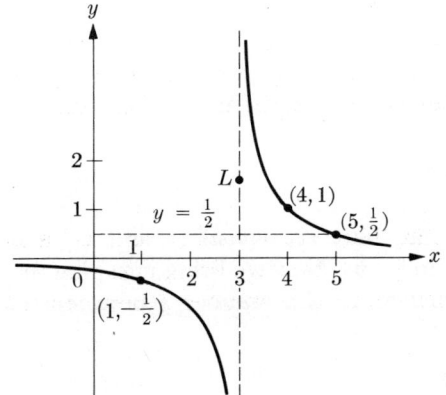

Figure 7–15

Then $f(5) = \frac{1}{2}$, but 5 is not in the inverval $[1, 4]$, and Theorem 10 is not applicable.

Example 1. Given the function $f(x) = (x - 1)/(x^2 + 1)$, $a = 0$, $b = 2$. Select a value of C between the A and B of the Intermediate Value Theorem and verify the validity of the result.

Solution. $f(0) = -1$, $f(2) = \frac{1}{5}$. We select C between -1 and $\frac{1}{5}$, say $-\frac{1}{2}$. Then we solve for x:

$$\frac{x - 1}{x^2 + 1} = -\frac{1}{2}.$$

This yields

$$x = -1 \pm \sqrt{2}.$$

The value $-1 + \sqrt{2}$ is in the interval $[0, 2]$ and is the value for which

$$f(-1 + \sqrt{2}) = -\frac{1}{2}.$$

The number $-1 - \sqrt{2}$ is rejected, since it falls outside $[0, 2]$. ◀

The Intermediate Value Theorem is used in establishing the **Theorem of the Mean for Integrals.**

Theorem 11 (Theorem of the Mean for Integrals). *If f is continuous on a closed interval with endpoints a and b, there is a number ξ between a and b (i.e., in $[a, b]$) such that*

$$\int_a^b f(x)\,dx = f(\xi)(b - a).$$

We shall first prove this theorem and then discuss it geometrically.

Proof. We know by the Extreme Value Theorem (Chapter 6, Theorem 1) that there are two numbers m and M such that $m \leq f(x) \leq M$ for all x in $[a, b]$. Furthermore, in Theorem 6 of this chapter we showed that

$$m(b - a) \leq \int_a^b f(x)\,dx \leq M(b - a).$$

Since the integral is between two numbers, there must be some number which gives the exact value. That is, there is a number D between m and M such that

$$\int_a^b f(x)\,dx = D(b - a).$$

By the Extreme Value Theorem, we know that there are values x_0 and x_1 such that $f(x_0) = m$ and $f(x_1) = M$ with x_0 and x_1 in $[a, b]$. Now we bring into play the Intermediate Value Theorem, which states that there is a number ξ between x_0 and x_1 such that

$$f(\xi) = D.$$

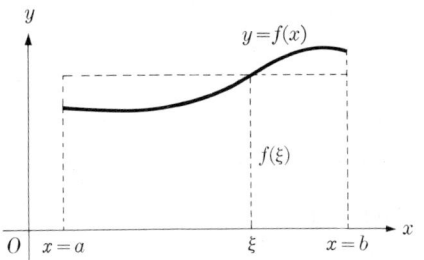

Figure 7–16

In other words,*

$$\int_a^b f(x)\,dx = f(\xi)(b - a).$$

In Fig. 7–16 we exhibit a geometric interpretation for a nonnegative function f. Then $\int_a^b f(x)\,dx$ is the area bounded by the lines $x = a$, $x = b$, the x axis, and the curve of $y = f(x)$.

In effect, Theorem 11 states that there is a point ξ (between a and b) such that the area determined by the integral is equal to that of the rectangle of height $f(\xi)$ and width $b - a$, namely, $f(\xi)(b - a)$. If $f(x)$ is considered to be made of rubber and if the low parts are pushed up and the high parts are pushed down, always keeping the area under the curve the same, the result would be a rectangle of height $f(\xi)$ and width $b - a$.

Finding the number ξ is usually very difficult. However, Theorem 11 proves its worth (particularly in the development of mathematical theory) by demonstrating that such a value ξ does exist. It may seem strange that a use can be found for the knowledge of the existence of something without knowledge of its value or even of how to obtain its value. But that is exactly the state of affairs, as is shown in the proof of Theorem 12 on page 218.

In some cases the value of ξ can be determined, and the following example shows one way to do it.

Example 2. Find a value of ξ such that

$$\int_1^3 f(x)\,dx = f(\xi)(3 - 1),$$

where f is given by $f: x \to x^2 - 2x + 1$.

Solution. We find the value of the integral

$$\int_1^3 (x^2 - 2x + 1)\,dx = \tfrac{1}{3}x^3 - x^2 + x \Big]_1^3$$

$$= (9 - 9 + 3) - (\tfrac{1}{3} - 1 + 1) = \tfrac{8}{3}.$$

---

* Of course, if $D = m$ we may take $\xi = x_0$ and if $D = M$ we may take $\xi = x_1$.

Therefore $f(\xi) \cdot (2) = \tfrac{8}{3}$ and $f(\xi) = \tfrac{4}{3}$. We solve for ξ:
$$\xi^2 - 2\xi + 1 = \tfrac{4}{3},$$
and the result is
$$\xi = 1 \pm \tfrac{2}{3}\sqrt{3}.$$
Of these numbers, the value $1 + \tfrac{2}{3}\sqrt{3}$ is in the interval $[1, 3]$. We conclude that
$$\int_1^3 f(x)\,dx = f(1 + \tfrac{2}{3}\sqrt{3})(3 - 1).$$
◀

Theorem 12 (Fundamental Theorem of Calculus, second form). *Suppose that f is continuous on an interval $[a, b]$ and c is some number in this interval. Define the function F by*
$$F(x) = \int_c^x f(t)\,dt$$
for each x in the interval (a, b). Then
$$\boxed{F'(x) = f(x).}$$

Proof. We know by Theorem 8 (see also Problem 19 of Section 4) that
$$\int_c^{x+h} f(t)\,dt = \int_c^x f(t)\,dt + \int_x^{x+h} f(t)\,dt,$$
so long as h is a number sufficiently small that $x + h$ does not fall outside the interval. The way in which we defined F tells us that this relationship is nothing but
$$F(x + h) = F(x) + \int_x^{x+h} f(t)\,dt.$$
We apply the Theorem of the Mean for Integrals (Theorem 11) to the integral $\int_x^{x+h} f(t)\,dt$, getting
$$\int_x^{x+h} f(t)\,dt = f(\xi) \cdot h,$$
where ξ is some value between x and $x + h$. We now have
$$\frac{F(x + h) - F(x)}{h} = f(\xi).$$
The hypothesis that f is continuous states (by the definition of continuity) that given $\epsilon > 0$ there is a $\delta > 0$ such that
$$|f(\xi) - f(x)| < \epsilon$$
whenever $|\xi - x| < \delta$. Note that as $h \to 0$, then $\xi \to x$, since ξ is always between x and $x + h$. Therefore, if h is sufficiently close to zero,* we may write
$$\left|\frac{F(x + h) - F(x)}{h} - f(x)\right| < \epsilon.$$

---

* In fact, if $-\delta < h < \delta$ and $h \neq 0$.

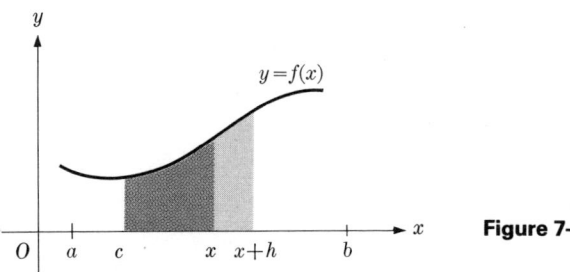

Figure 7-17

Since, by definition,

$$\frac{F(x+h) - F(x)}{h} \to F'(x) \quad \text{as} \quad h \to 0,$$

we conclude that

$$F'(x) = f(x).$$

In the proof of Theorem 12, we employed the Theorem of the Mean for Integrals (Theorem 11) without ever finding the value of ξ which occurs. This is a common practice in the proof of theorems in mathematical analysis.

A geometric interpretation of the above result for nonnegative functions f was given in Chapter 6, Section 10. We shall repeat it here. From Fig. 7-17, where $F(x)$ is the shaded area on the left, $F(x + h)$ is the sum of the shaded areas, and $F(x + h) - F(x)$ is the shaded area on the right, we find that

$$\frac{F(x+h) - F(x)}{h}$$

is the *average* height of $f(x)$ in the right-hand area. As $h \to 0$ this average height tends to the height at x, namely $f(x)$. On the other hand, as $h \to 0$, the expression above tends to $F'(x)$. ◂

The real meaning of the Fundamental Theorem of Calculus is that **differentiation and integration are inverse processes.** This is meant in the same sense that addition and subtraction are inverse processes, as are multiplication and division (excluding division by zero). What about squaring and taking the square root? These are inverse to each other if an additional condition is tacked on: only positive numbers are allowed. The same situation prevails for derivatives and integrals; they are inverse if some additional condition is added (such as continuity of f in Theorem 12). In more advanced courses theorems similar to Theorem 9 and Theorem 12 are established, which show under much less restrictive hypotheses that differentiation and integration are inverse processes.

PROBLEMS

In each of Problems 1 through 8, a function f and numbers a, b, and C are given. Verify the Intermediate Value Theorem, or show that this theorem does not apply.

1. $f(x) = x^2 + x + 3, \quad a = 1, b = 3, C = 9$

2. $f(x) = 2 + 4x - x^2$, $\quad a = 0, b = 2, C = 4$
3. $f(x) = (x + 2)/(x^2 + 2)$, $\quad a = 0, b = 3, C = 2$
4. $f(x) = (x^2 + 1)/(x - 2)$, $\quad a = 3, b = 5, C = 12$
5. $f: x \to (x - 1)/(x + 1)$, $\quad a = -2, b = 1, C = 1$
6. $f: x \to 3x^2 + x - 6$, $\quad a = 0, b = 2, C = -4$
7. $f(x) = \begin{cases} 2x, & -1 \le x < 1 \\ 3 - x, & 1 \le x < 5 \end{cases}$, $\quad a = 0, b = 2, C = \frac{1}{2}$
8. $f(x) = \begin{cases} 1 - x^2, & -10 \le x < 0 \\ x - 3, & 0 \le x < 5 \end{cases}$, $\quad a = -1, b = 2, C = -\frac{1}{2}$

In Problems 9 through 15, verify the Theorem of the Mean for Integrals by finding appropriate values of ξ. Sketch graphs.

9. $\int_2^5 f(x)\, dx$, with $f(x) = 3x - 2$

10. $\int_{-1}^2 f(x)\, dx$, with $f(x) = 2 + 4x$

11. $\int_2^7 f(x)\, dx$, with $f(x) = x^2 - 2x + 1$

12. $\int_0^1 f(x)\, dx$, with $f(x) = (x + 1)^2$

13. $\int_{-1}^1 f(x)\, dx$, with $f(x) = 3 - 2x + 4x^2$

14. $\int_0^2 f(x)\, dx$, with $f(x) = 4x^3 - 3x^2 - 4x$

15. $\int_{-2}^2 f(x)\, dx$, with $f(x) = x^5 - 4x^3 + 3x$

16. Given that $f(x) = (x^2 + 2)^3$, verify that $f'(x) = 6x(x^2 + 2)^2$. Use the Fundamental Theorem of Calculus to evaluate $\int_0^2 6x(x^2 + 2)^2 \, dx$.

17. Given that $f(x) = (x^2 + 5)^{1/2}$, verify that $f'(x) = x(x^2 + 5)^{-1/2}$. Use the Fundamental Theorem of Calculus to evaluate

$$\int_1^3 \frac{x\, dx}{\sqrt{x^2 + 5}}.$$

18. Given that $f(x) = 6(x + 2)^{3/2}(x^2 + 1)^{4/3}$, verify that

$$f'(x) = (x + 2)^{1/2}(x^2 + 1)^{1/3}(25x^2 + 32x + 9).$$

Evaluate

$$\int_0^2 (x + 2)^{1/2}(x^2 + 1)^{1/3}(25x^2 + 32x + 9)\, dx.$$

19. Given that $f(x) = (x + 3)/(x - 2)$, verify that $f'(x) = -5/(x - 2)^2$. Evaluate

$$\int_{-1}^1 \frac{dx}{(x - 2)^2}.$$

20. If a function is odd, that is, if $f(-x) = -f(x)$ for all x, what can be said about $\int_{-b}^b f(x)\, dx$? [*Hint:* Draw a sketch.]

21. If a function is even, that is, if $f(-x) = f(x)$ for all x, what can be said about the relation between $\int_{-b}^b f(x)\, dx$ and $\int_0^b f(x)\, dx$?

22. Let f be a given function. Define F and G by the formulas

$$F(x) = \int_c^x f(t)\, dt, \qquad G(x) = \int_d^x f(t)\, dt,$$

where c and d are given numbers. Show that F and G differ by a constant and find the value of this constant.

7. INDEFINITE INTEGRALS. CHANGE OF VARIABLE

To evaluate the definite integral of a function, we must first obtain an antiderivative of the function and then substitute the appropriate limits of integration into this antiderivative. Every function has an infinite number of antiderivatives, since a constant of arbitrary magnitude may be added to one antiderivative to yield another. A common notation used for all the antiderivatives of a function f is

$$\int f(x)\, dx,$$

in which there are no upper and lower limits. This is called the **indefinite integral** of f. For example, the formula for the antiderivative of x^n, n rational, $\neq -1$, is written

$$\int x^n\, dx = \frac{x^{n+1}}{n+1} + C, \qquad n \neq -1,$$

where the constant C is the **constant of integration.**

The following seemingly trivial variation of the above formula has many useful consequences. We write

$$\int u^n\, du = \frac{u^{n+1}}{n+1} + C, \qquad n \neq -1.$$

True, this *is* the same formula. However, according to the chain rule, we know that

$$d\left(\frac{u^{n+1}}{n+1}\right) = u^n\, du$$

holds if u is *any function* of x. Schematically we write

$$\int (\text{expression in } x)^n\, d(\text{same expression in } x) = \frac{(\text{expression in } x)^{n+1}}{n+1} + C.$$

We shall illustrate the significance of this formula with two examples.

Example 1. Find $\int (3x + 2)^7\, dx$.

Solution. The integrand could be multiplied out and then each term could be evaluated separately. Instead, we observe that $(3x + 2)^7$ is an expression in x raised to the 7th power. So we write

$$u = 3x + 2, \qquad du = 3\, dx \quad \text{or} \quad dx = \tfrac{1}{3}\, du.$$

Substituting, we get

$$\int (3x + 2)^7 \, dx = \int u^7 \cdot \tfrac{1}{3} \, du.$$

A constant (*and only a constant*) may be moved in and out of integrals at will. We obtain

$$\int u^7 \tfrac{1}{3} \, du = \tfrac{1}{3} \int u^7 \, du = \tfrac{1}{3}\left(\frac{u^8}{8} + C\right).$$

Since C represents *any* constant, so does $\tfrac{1}{3}C$, and we write, after substituting the value of u in terms of x,

$$\int (3x + 2)^7 \, dx = \tfrac{1}{24}(3x + 2)^8 + C. \qquad \blacktriangleleft$$

Example 2. Find $\int (x^2 + 1)^{5/2} x \, dx$.

Solution. Let $u = x^2 + 1$. Then $du = 2x \, dx$ and, after substitution, we obtain

$$\int (x^2 + 1)^{5/2} x \, dx = \int u^{5/2} \tfrac{1}{2} \, du = \tfrac{1}{2} \int u^{5/2} \, du$$

$$= \tfrac{1}{2}\left(\frac{2u^{7/2}}{7} + C\right) = \tfrac{1}{7}(x^2 + 1)^{7/2} + C,$$

which is our answer. $\qquad \blacktriangleleft$

Remark. The integrand in Example 2 was rigged so that the result is readily obtainable by the method we are learning. For example, the integral $\int (x^2 + 1)^{5/2} \, dx$, which appears "simpler" than the one in Example 2, cannot be evaluated by this method. For, if we let $u = x^2 + 1$, then $du = 2x \, dx$ and, substituting, we get

$$\int (x^2 + 1)^{5/2} \, dx = \int u^{5/2} \frac{1}{2x} \, du.$$

There is no way to get rid of the x in the denominator. Later we shall study other methods which will enable us to perform this type of integration.

The technique illustrated in Examples 1 and 2, known as **change of variable**, may be used for evaluating definite integrals as well. However, *great care* must be exercised in evaluating the limits. There are two possible methods of procedure, and we illustrate each with an example.

Example 3. Evaluate $\int_1^2 (x + 1)(x^2 + 2x + 2)^{1/3} \, dx$.

Solution. Let $u = x^2 + 2x + 2$. Then

$$du = (2x + 2) \, dx = 2(x + 1) \, dx.$$

Substituting, we find for the *indefinite integral*

$$\int (x + 1)(x^2 + 2x + 2)^{1/3} \, dx = \int u^{1/3} \frac{1}{2} \, du = \frac{1}{2} \frac{3u^{4/3}}{4} + C.$$

Method I involves substituting back for u in terms of x, thereby obtaining for the *definite integral*

$$\int_1^2 (x+1)(x^2+2x+2)^{1/3}\,dx = \tfrac{3}{8}(x^2+2x+2)^{4/3}\Big]_1^2$$
$$= \tfrac{3}{8}(10^{4/3} - 5^{4/3}).\quad\blacktriangleleft$$

The next example will be done by the second method.

Example 4. Evaluate

$$\int_{-2}^0 x\sqrt{2x^2+1}\,dx.$$

Solution. Let $u = 2x^2 + 1$, $du = 4x\,dx$.

Method II involves changing the limits of the definite integral to values of u instead of values of x. Since $u = 2x^2 + 1$, we easily see that when $x = -2$, then $u = 9$, and when $x = 0$, $u = 1$. Therefore we can write

$$\int_{-2}^0 x\sqrt{2x^2+1}\,dx = \int_9^1 u^{1/2}\cdot\frac{1}{4}\,du = \frac{1}{4}\cdot\frac{2u^{3/2}}{3}\bigg]_9^1$$
$$= \tfrac{1}{6}(1^{3/2} - 9^{3/2}) = -\tfrac{13}{3}.\quad\blacktriangleleft$$

Most of the time Method II is simpler and shorter. Occasionally there are problems in which it pays to go back to the original variables before evaluating the limits of integration. The justification of Method II depends on the formula

$$\int_a^b f[u(x)]u'(x)\,dx = \int_{u(a)}^{u(b)} f(u)\,du, \tag{1}$$

which the next theorem establishes.

Theorem 13. *Suppose that u and u' are continuous functions defined on $[a, b]$ and that f is continuous on an interval $[c, d]$. Suppose, also, that the range* of u is in (c, d). Then the above substitution formula (1) holds.*

Proof. We define functions F and G by the formulas

$$F(u) = \int_{x_0}^u f(t)\,dt, \tag{2}$$

where x_0 is any fixed number in $[c, d]$, and

$$G(x) = F[u(x)].$$

Then, from the Fundamental Theorem of Calculus (Theorem 12), we see that

$$F'(u) = f(u) \quad \text{for each } u \text{ in } (c, d).$$

---
* We omit the technical discussion which occurs if for some x, $u(x) = c$ or $u(x) = d$.

Also, the chain rule applied to G gives
$$G'(x) = F'[u(x)]u'(x) = f[u(x)]u'(x).$$
We integrate this last equation between the limits a and b, getting
$$\int_a^b G'(x)\,dx = \int_a^b f[u(x)]u'(x)\,dx.$$
Now using the Fundamental Theorem of Calculus, first form (Theorem 9), we find
$$\int_a^b f[u(x)]u'(x)\,dx = G(b) - G(a)$$
$$= F[u(b)] - F[u(a)].$$
However,
$$F[u(b)] - F[u(a)] = \int_{x_0}^{u(b)} f(t)\,dt - \int_{x_0}^{u(a)} f(t)\,dt = \int_{u(a)}^{u(b)} f(t)\,dt.$$
Therefore
$$\int_{u(a)}^{u(b)} f(t)\,dt = \int_a^b f[u(x)]u'(x)\,dx,$$
and formula (1) is established.

PROBLEMS

In each of Problems 1 through 20, find the indefinite integral and check the result by differentiation.

1. $\int (3x - 2)^5\,dx$
2. $\int (2x + 5)^4\,dx$
3. $\int (2 - x)^{-4}\,dx$
4. $\int (4 - 6x)^{-7}\,dx$
5. $\int (2y + 1)^{3/2}\,dy$
6. $\int (2x^2 + 3)^4 x\,dx$
7. $\int (2x^2 + 3)^{7/3} x\,dx$
8. $\int (4 - x^2)^3 x\,dx$
9. $\int (3 - 2x^2)^{-2/3} x\,dx$
10. $\int \dfrac{x\,dx}{\sqrt{x^2 - 1}}$
11. $\int \dfrac{x\,dx}{(3x^2 + 2)^2}$
12. $\int x\sqrt{2x^2 - 1}\,dx$
13. $\int x^2 \sqrt{x^3 + 1}\,dx$
14. $\int u^3 \sqrt{u^4 + 1}\,du$
15. $\int \dfrac{(x^2 + 2x)\,dx}{\sqrt[3]{x^3 + 3x^2 + 1}}$
16. $\int \left(1 + \dfrac{1}{t}\right)^2 \dfrac{dt}{t^2}$

17. $\int (x^2 + 1)^3 x^3 \, dx$

18. $\int (2x^2 - 3)^{4/3} x^3 \, dx$

19. $\int (x^3 + 1)^{7/5} x^5 \, dx$

20. $\int (x^2 - 2x + 1)^{4/3} \, dx$

Evaluate the following definite integrals by Method I.

21. $\int_0^3 \sqrt[3]{(3t - 1)^2} \, dt$

22. $\int_{-2}^1 \sqrt{2 - x} \, dx$

23. $\int_{-3}^{-1} \dfrac{dx}{(x - 1)^2}$

24. $\int_0^1 x\sqrt{x^2 + 1} \, dx$

In Problems 25 through 28, evaluate each integral by Method II.

25. $\int_0^2 \dfrac{(x^2 + 1) \, dx}{\sqrt{x^3 + 3x + 1}}$

26. $\int_1^2 (2x + 1)\sqrt{x^2 + x + 1} \, dx$

27. $\int_1^2 \dfrac{(x^{1/3} + 2)^4 \, dx}{\sqrt[3]{x^2}}$

28. $\int_1^3 \left(x + \dfrac{1}{x}\right)^{3/2} \left(\dfrac{x^2 - 1}{x^2}\right) dx$

29. Verify that a straightforward evaluation by Method II of the integral

$$\int_{-1}^3 \left(x - \dfrac{3}{x}\right)^5 \left(1 + \dfrac{3}{x^2}\right) dx$$

yields the value zero. What is wrong with the result?

8. AREA BETWEEN CURVES

Suppose that f and g are two continuous functions defined on the interval $[a, b]$ and, furthermore, suppose that

$$f(x) \geq g(x), \quad a \leq x \leq b.$$

Figure 7–18 shows a typical situation. Let R denote the region bounded by the lines $x = a$, $x = b$, and the two curves. The area A of the region R is given by

$$A = \int_a^b f(x) \, dx - \int_a^b g(x) \, dx = \int_a^b [f(x) - g(x)] \, dx.$$

The interesting point about this formula is that *it holds whether the curves are*

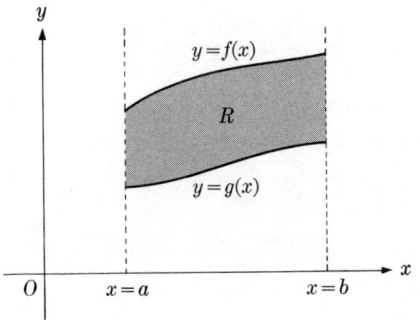

Figure 7–18

226 THE DEFINITE INTEGRAL

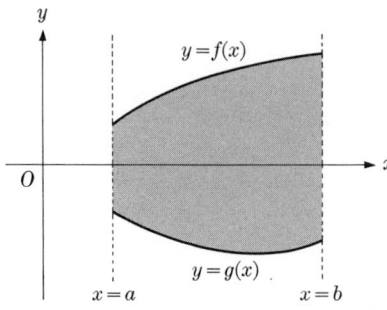

Figure 7–19

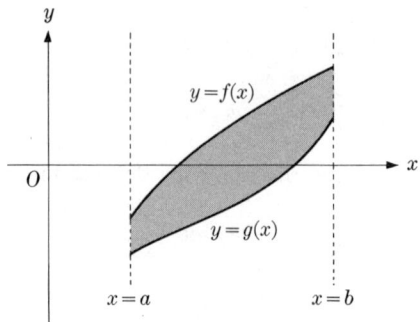

Figure 7–20

above or below the x axis. Figures 7–19 and 7–20 show other possible situations. A moment's thought about Fig. 7–19 discloses the fact that $\int_a^b f(x)\,dx$ is the area below $f(x)$ and above the x axis. Also, $-\int_a^b g(x)\,dx$ is the area between the x axis and $g(x)$. Adding these, we get for the total area A between the curves

$$A = \int_a^b [f(x) - g(x)]\,dx.$$

Figure 7–20 appears slightly more complicated, but the student can easily satisfy himself that the same formula holds.

When we apply the formula for the area between two curves, the main problem lies in verifying the fact that $f(x)$ is larger than $g(x)$ *throughout the interval.* Once this verification is made, we perform the integration in the usual way.

Example 1. Find the area of the region R bounded by the lines $x = 1$, $x = 2$, $y = 3x$, and the curve $y = x^2$.

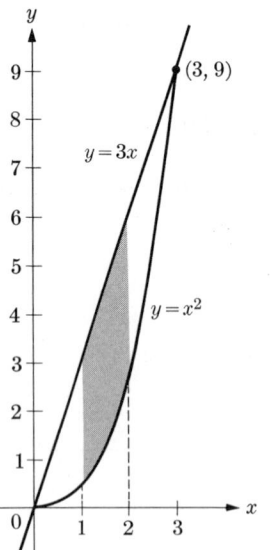

Figure 7–21

Solution. We draw a sketch, as shown in Fig. 7–21, which makes it obvious that $y = 3x$ is above $y = x^2$ on the interval $[1, 2]$. This can be shown analytically first by finding where the curves intersect and then by noting that between intersection points one curve must *always remain above* the other; $y = 3x$ and $y = x^2$ meet when $3x = x^2$ or $x = 0, 3$. Between $x = 0$ and $x = 3$, say at $x = 1$, we know that $3x > x^2$, since $3 > 1$. This means that $y = 3x$ is above $y = x^2$ throughout the interval $[0, 3]$.

The area is given by

$$A = \int_1^2 (3x - x^2)\, dx = \tfrac{3}{2}x^2 - \tfrac{1}{3}x^3 \Big]_1^2 = (6 - \tfrac{8}{3}) - (\tfrac{3}{2} - \tfrac{1}{3}) = \tfrac{13}{6}.$$ ◂

A similar situation, but one with a slight complication, is exhibited next.

Example 2. Find the area bounded by the curves $y = x$ and $y = x^3$.

Solution. To determine the nature of the region bounded by the curves, we begin by finding the points of intersection of the two curves. Solving the equation $x = x^3$, we get $x = 0, 1, -1$; therefore the curves intersect at $(0, 0)$, $(1, 1)$, and $(-1, -1)$. After plotting a few points, we see that the graph appears as in Fig. 7–22. There are two regions, one in the first quadrant (denoted S_1) and one in the third (denoted S_2). We label their areas A_1 and A_2, respectively. For S_1 the line $y = x$ is above the curve $y = x^3$, since for $x = \tfrac{1}{2}$ (a typical value between 0 and 1), $\tfrac{1}{2} > (\tfrac{1}{2})^3$. We obtain

$$A_1 = \int_0^1 (x - x^3)\, dx = \tfrac{1}{2}x^2 - \tfrac{1}{4}x^4 \Big]_0^1 = (\tfrac{1}{2} - \tfrac{1}{4}) - 0 = \tfrac{1}{4}.$$

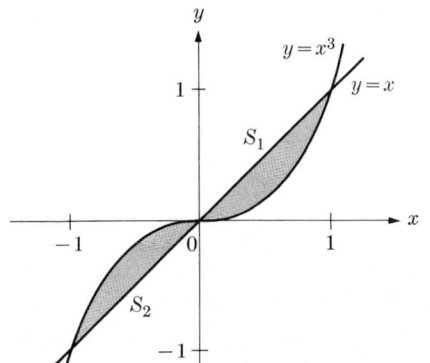

Figure 7–22

To find A_2 we note that $y = x^3$ is above $y = x$ for S_2 (since $(-\tfrac{1}{2})^3 > -\tfrac{1}{2}$), and so

$$A_2 = \int_{-1}^0 (x^3 - x)\, dx = \tfrac{1}{4}x^4 - \tfrac{1}{2}x^2 \Big]_{-1}^0 = 0 - (\tfrac{1}{4} - \tfrac{1}{2}) = \tfrac{1}{4}.$$

The total area between the curves is $A_1 + A_2 = \tfrac{1}{2}$. ◂

Remark. We could have observed that since the function $h(x) = x - x^3$ is an **odd function**, that is, $h(-x) = -h(x)$, the areas A_1 and A_2 must be equal. Such knowledge should be exploited whenever possible.

The next example will be worked by two methods and, in the process, a number of ideas will be developed which will show how the scope of the method for finding the area between curves may be expanded considerably.

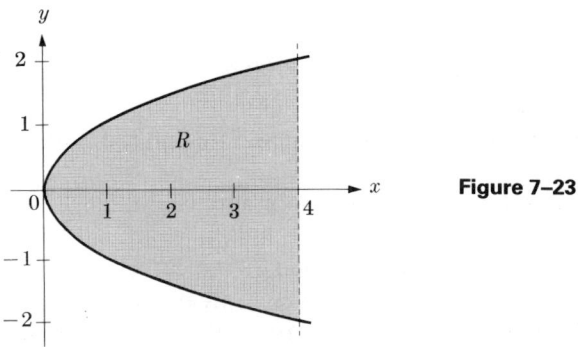

Figure 7–23

Example 3. Find the area bounded by the curve $y^2 = x$ and the line $x = 4$.

Solution. A sketch of the region R is shown in Fig. 7–23. The first difficulty is that the relation $y^2 = x$ does not express y as a function of x. The upper boundary of R is given by the function $y = \sqrt{x}$ and the lower boundary by the function $y = -\sqrt{x}$. Applying the method used in the previous examples, we obtain

$$A = \int_0^4 [\sqrt{x} - (-\sqrt{x})]\, dx$$
$$= \int_0^4 (\sqrt{x} + \sqrt{x})\, dx$$
$$= 2\int_0^4 x^{1/2}\, dx = \tfrac{4}{3} x^{3/2}\big]_0^4 = \tfrac{32}{3}.$$

In the second method of solving this problem, we change our point of view. We observe that the equation $x = y^2$ expresses x as a function of y. Therefore we may integrate along the y axis. The region bounded by the curve $x = y^2$, the lines $y = \pm 2$, and the y axis is shown as the shaded region in Fig. 7–24. Its area is

$$\int_{-2}^{2} y^2\, dy = \tfrac{1}{3} y^3 \big]_{-2}^{2} = \tfrac{8}{3} - (-\tfrac{8}{3}) = \tfrac{16}{3}.$$

The area of the rectangle bounded by the four lines $y = -2$, $y = 2$, $x = 0$, and $x = 4$ is 16. The area of the region R is obtained by subtraction:

$$A = 16 - \tfrac{16}{3} = \tfrac{32}{3}. \quad \blacktriangleleft$$

We could have performed the integration of this second method in one step by finding the area between the curves $x = 4$ and $x = y^2$. Integration along the y

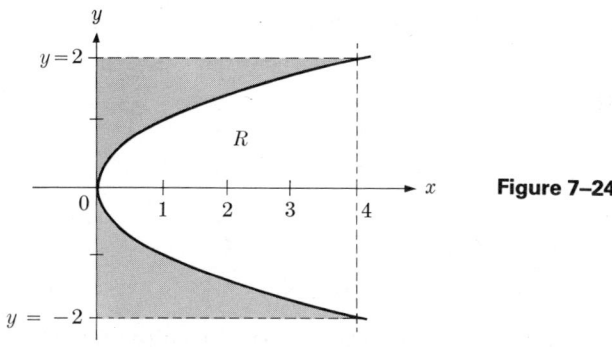

Figure 7-24

axis then yields

$$\int_{-2}^{2} (4 - y^2)\, dy = 4y - \tfrac{1}{3}y^3 \big]_{-2}^{2}$$
$$= (8 - \tfrac{8}{3}) - (-8 + \tfrac{8}{3}) = \tfrac{32}{3}.$$

The technique of looking at the same problem from several points of view is a favorite trick of mathematicians. One way of looking at a problem often yields an insight which other ways miss. If in Example 3 we note that the region is symmetric with respect to the x axis, the problem becomes simpler. In that case we integrate

$$\int_0^4 \sqrt{x}\, dx,$$

double the result, and get the answer.

Example 4. Find the area bounded by $y^2 = x - 1$ and $y = x - 3$.

Solution. The curves are sketched in Fig. 7-25. They intersect where

$$y^2 + 1 = y + 3,$$

or $y = -1, 2$. The corresponding values of x are 2, 5.

There are two ways of proceeding. In the first we write the equations

$$x = y^2 + 1 \quad \text{and} \quad x = y + 3$$

and observe that in each case x is a function of y. Then the area between the

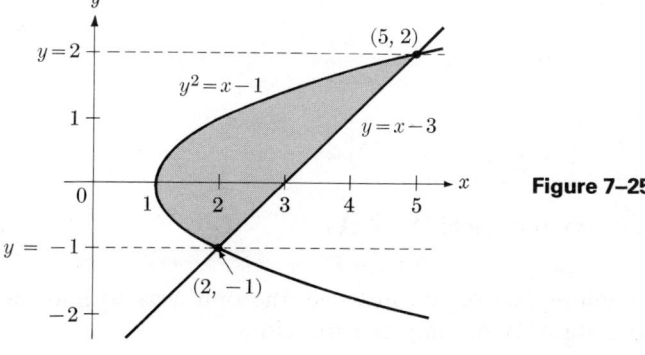

Figure 7-25

curves is obtained by integrating along the y axis from $y = -1$ to $y = 2$. We get

$$A = \int_{-1}^{2} [(y + 3) - (y^2 + 1)] \, dy$$
$$= \int_{-1}^{2} (-y^2 + y + 2) \, dy$$
$$= -\tfrac{1}{3} y^3 + \tfrac{1}{2} y^2 + 2y \big]_{-1}^{2}$$
$$= (-\tfrac{8}{3} + 2 + 4) - (\tfrac{1}{3} + \tfrac{1}{2} - 2) = \tfrac{9}{2}.$$

The second method employs integration along the x axis. From the figure, we see that the upper portion of the region is bounded by

$$y = \sqrt{x - 1}.$$

The lower boundary is made up of two parts:

$$y = -\sqrt{x - 1} \quad \text{if } x \text{ is between 1 and 2;}$$
$$y = x - 3 \quad \text{if } x \text{ is between 2 and 5.}$$

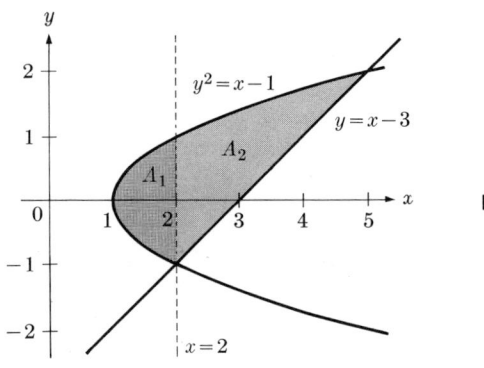

Figure 7–26

Figure 7–26 shows how we solve the problem in two stages. The area A is divided into two parts A_1 and A_2, as shown.

$$A_1 = \int_{1}^{2} [\sqrt{x - 1} - (-\sqrt{x - 1})] \, dx$$
$$= 2 \int_{1}^{2} \sqrt{x - 1} \, dx;$$
$$A_2 = \int_{2}^{5} [\sqrt{x - 1} - (x - 3)] \, dx.$$

These integrations can be performed to yield $A_1 + A_2 = \tfrac{9}{2}$. ◀

It is easy to imagine a complicated region in which the only way to find the area is to decompose it into a number of simpler subregions.

PROBLEMS

In Problems 1 through 25, in each case sketch the bounded region formed by the given curves and find its area.

1. $x = 2, x = 4, y = x^2, y = -x$
2. $x = 2, x = -2, y = x^2, y = 16 - x^2$
3. $x = 0, x = 3, y = x^2, y = 12x$
4. $y = 0, y = \frac{3}{4}, x = y^2, x = y$
5. $x = -2, x = 0, y = x^3, y = -x$
6. $x = 0, x = 3, y = \sqrt{x+1}, y = \frac{1}{2}x$
7. $x = 1, x = 2, x^2y = 2, x + y = 4$
8. $x = 0, x = \frac{1}{2}, y = x^2, y = x^4$
9. $y = x^2, y = x^4$
10. $x = 0, x = 3, y = 0, y = 1/\sqrt{x+1}$
11. $x = 1, x = 2, y^2 = x$
12. $y = 0, y = 2, x = 0, x = y^2$
13. $y = -1, y = 1, x - y + 1 = 0, x = 1 - y^2$
14. $y = 1, x = y, xy^2 = 4$
15. $x = y^2, x = 18 - y^2$
16. $y = x^2, y^2 = x$
17. $y = \sqrt{x}, y = x^3$
18. $y^2 = 4x + 1, x + y = 1$
19. $y = x^2 + 4x + 2, 2x - y + 5 = 0$
20. $y = 1, y = 2, x = 0, x = y/\sqrt{9 - y^2}$
21. $y^2 = 2 - x, y = 2x + 2$
22. $y = x^2, y = 3x, y = 1, y = 2$
23. $y = 2x, 4y = x, y = 2/x^2, x > 0$
24. $x = 1, x = 2, x^2y = 1, y = 0$
25. $x = 1, x = 2, y = \frac{1}{2}x, x^2y = 1$. (Find area of all bounded regions.)

In Problems 26 through 28, find the areas of the triangles having the given vertices, using the method of this section.

26. $(0, 0), (4, 1), (2, 4)$
27. $(-2, -1), (2, 2), (3, -2)$
28. $(0, 0), (a, 0), (b, c)$
29. Find the area enclosed by the curves
$$y = x^2 - x, \qquad y = x - x^2.$$
30. A trapezoid connects the four points $(0, 0), (B, 0), (a, h), (a + b, h)$. Use the methods of this section to obtain the formula for the area $A = \frac{1}{2}h(B + b)$.
31. Find the area of the region bounded by $x = 1, y = 0, y = 1/x^2$, and the line $x = a$, where a is some number greater than 1. The result will depend on the number a. What happens to the value of the area as $a \to +\infty$?
32. Same as Problem 31, with the equation $y = 1/x^2$ replaced by $y = 1/x^{1/2}$.
33. Same as Problem 31, with the equation $y = 1/x^2$ replaced by $y = 1/x^p$, where p is a fixed positive number larger than 1. What happens for p between 0 and 1 ($p \neq 0, 1$)?
34. Let $f(x)$ be integrable on $[a, b]$. Define
$$g(x) = \tfrac{1}{2}[|f(x)| + f(x)], \qquad h(x) = \tfrac{1}{2}[|f(x)| - f(x)].$$
Show that
$$\int_a^b f(x)\,dx = \int_a^b g(x)\,dx - \int_a^b h(x)\,dx,$$
$$\int_a^b |f(x)|\,dx = \int_a^b g(x)\,dx + \int_a^b h(x)\,dx.$$

35. Let $f(x)$ be continuous on $[a, b]$ with $f(a) = f(b) = 0$. Define
$$\text{sign}\,[f(x)] = \begin{cases} 1 \text{ if } f(x) > 0 \\ -1 \text{ if } f(x) < 0 \\ 0 \text{ if } f(x) = 0. \end{cases}$$

Suppose that f is zero only at the points
$$a = x_0 < x_1 < x_2 < \cdots < x_n = b.$$

Find an expression in terms of the lengths of the subintervals for the value of
$$\int_a^b \text{sign}\,[f(x)]\,dx.$$

9. WORK*

We shall consider only *the motion of objects along a straight line in one direction*. It is convenient to suppose that the motion is along the x axis or the y axis in the positive direction. The restriction to motions along a line is essential, since the study of motion along curved paths requires types of derivatives and integrals of a nature more complicated than those we have taken up so far.

The term **work** is a technical one which requires for its definition the notion of **force**. Using mass, distance, and time as *undefined terms*, we know from Newton's law that

$$\text{force} = \text{mass} \times \text{acceleration}.$$

Force is measured in units of dynes, newtons, etc. To take the simplest possible case, suppose that an object (assumed to occupy a single point) is moving along the x axis (to the right) and suppose that this object is subject to a *constant force* (to the right) of B newtons. Let d be the distance it moves, measured in meters. We define the **work** done, W, as

$$W = B \cdot d.$$

The units of W are measured in joules or newton-meters. If the force were in dynes and the distance in centimeters, the work would be measured in ergs. The definition above is fine so long as the force exerted is constant, but what is to be the definition when the force is variable (the most common situation)? As a start toward the definition we state the following principle:

Principle 1. *If a body is moved from position A_0 to position A_1, then from A_1 to A_2, then from A_2 to A_3, etc., and finally from A_{n-1} to A_n, the total work done in moving the body from A_0 to A_n is the sum of the amounts of work done in moving the body from one position to the next.*

If the force exerted in moving a body from one position to another is constant, the work is known; otherwise it is not defined as yet. If the positions are

---

* Readers not interested in the physical applications may skip Sections 9 and 10 without loss of continuity in the development of either the theory or the techniques of calculus.

sufficiently close together and if the force acting is continuous, then the force is *almost constant.* Suppose that the body is moving along the x axis from a to b ($b > a$) and suppose that there is a law $F(x)$ which gives the amount of force (exerted to the right) at point x. According to Principle 1, if we subdivide the interval $[a, b]$ into n parts: $\{a = x_0 < x_1 < \cdots < x_n = b\}$, and if we *assume that the force is constant* in each subinterval, we can compute the total work in moving the body from point a to point b. Suppose that on the subinterval $[x_{i-1}, x_i]$ the force has a constant value equal to $F(\xi_i)$, where ξ_i is some number in the interval $[x_{i-1}, x_i]$. Then, writing $\Delta_i x = x_i - x_{i-1}$, the *total work done* is

$$\sum_{i=1}^{n} F(\xi_i) \Delta_i x.$$

This expression is suggestive of the way we defined integral and leads to the following definition of work.

Definition. **The total work** W done in moving an object along the x axis from a to b, if the force exerted at point x obeys the law $F = F(x)$, is

$$W = \int_a^b F(x)\, dx.$$

Example 1. Suppose that an object is moved along the x axis from $x = 2$ to $x = 5$ (units in centimeters), and suppose that the force exerted obeys the law $F(x) = x^2 + x$ dynes. Find the total work done.

Solution

$$W = \int_2^5 (x^2 + x)\, dx = \tfrac{1}{3}x^3 + \tfrac{1}{2}x^2 \Big]_2^5 = 49\tfrac{1}{2} \text{ ergs}. \qquad \blacktriangleleft$$

Example 2. A spring has a natural length of 12 cm. When it is stretched x cm **Hooke's law** states that it pulls back with a force equal to kx, where k is a constant. The value of k depends on the material, thickness of the wire, temperature, etc. If 10 dynes of force are required to hold it stretched $\tfrac{1}{2}$ cm, how much work is done in stretching it from its natural length to a length of 16 cm?

Solution. We locate the spring along the x axis and place the origin at the position where we start stretching (Fig. 7–27). The fact that the force is 10 dynes when $x = \tfrac{1}{2}$ is used to find k. We have $10 = k \cdot \tfrac{1}{2}$ or $k = 20$. From the force law,

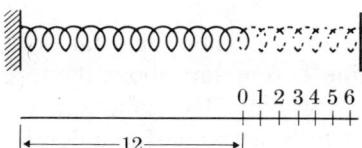

Figure 7–27

234   THE DEFINITE INTEGRAL

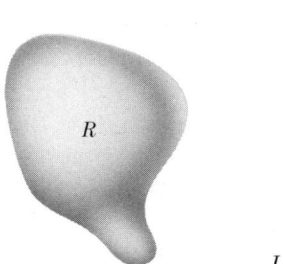

Figure 7-28

we may write $F = 20x$. From the definition of work,

$$W = \int_0^4 20x\, dx = 10x^2\Big]_0^4 = 160 \text{ ergs.} \qquad \blacktriangleleft$$

Another type of work problem requires a somewhat different principle. Figure 7-28 represents a substance which occupies a large region R. Suppose the material in this region (sand, water, etc.) is to be moved to some location L. The following principle tells how to find the work done in moving this mass. It is supposed that each particle in R moves until it reaches the location L.

Principle 2.   *If a mass M is moved to a position L, the total work done may be obtained by (i) subdividing the mass into smaller masses*

$$M_1 + M_2 + \cdots + M_n = M,$$

by (ii) finding the work done in moving each of the smaller masses to the position L, and by (iii) adding the results.

Principle 2 is used as a guide for setting up an integral for the total work done in the same way that Principle 1 served as a guide for defining work. We illustrate the method in the following example.

Example 3.   A tank having the shape of a right circular cylinder of altitude 8 meters and radius of base 5 meters is full of water. Find the amount of work done in pumping all the water in the tank up to a level 6 meters above the top of the tank.

Solution.   Figure 7-29 shows the tank. Since the motion is in a vertical direction we set up the x axis so that it is vertically downward, as shown. According to Principle 2, we divide the tank into n portions, in the form of slabs or disks. That is, we divide the x axis into n parts,

$$0 = x_0 < x_1 < x_2 < \cdots < x_n = 8,$$

and consider the ith disk to be of thickness $\Delta_i x = x_i - x_{i-1}$. Then the total work done is the work required to move each disk up to line L, 6 meters above the top. The only force exerted on the water is that due to gravity. By definition, *the weight of any object is the measure of the force of gravity*. Since water has a density

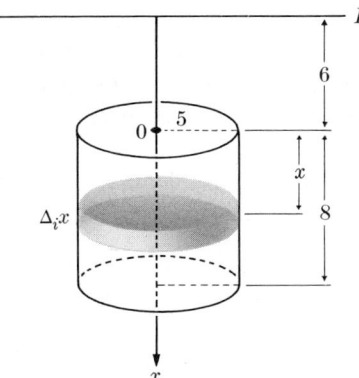

Figure 7–29

of 1,000 kg/m³, raising one cubic meter of water requires a force of 1,000 newtons.

This volume of the disk of width $\Delta_i x$ is $\pi \cdot 5^2 \cdot \Delta_i x = 25\pi \Delta_i x$. The force required to raise this volume is its weight $(1{,}000)(25\pi \Delta_i x)$. Since the force is *constant*, the work done is merely the product of the force and the distance that it moves. Here is the real crux of the problem: each disk moves a distance which is different from that of any other disk. In fact, even within one disk different particles are raised different distances.

A disk with midpoint at location x meters below the top of the tank moves *approximately* $(6 + x)$ meters. Since x is between x_{i-1} and x_i, the work done in moving one disk is approximately $1{,}000\,(25\pi \Delta_i x)(6 + x)$. We now recall that Principle 2 says that we may add the work done in moving in each of these disks to find the total work, and so (approximately)

$$W = \sum_{i=1}^{n} 1000(25\pi \Delta_i x)(6 + x), \qquad x \text{ between } x_{i-1} \text{ and } x_i.$$

When we proceed to the limit, this formula suggests the integral

$$W = \int_0^8 1000(25\pi)(6 + x)\,dx = 25{,}000\pi[6x + \tfrac{1}{2}x^2]_0^8 = 2 \times 10^6 \pi \text{ joules.} \blacktriangleleft$$

Such an integral is used to define the work done in moving a large mass. Its evaluation yields the total work performed.

The ideas about work follow the outline:

1) A physical law is given in the simplest case. (Work is defined when the force is constant and the object occupies a single point.)
2) A principle is stated which says that a complicated situation may be decomposed into a large number of simple ones; each of these may be solved by the simple physical law (1), and then the results may be added.
3) The sum obtained in (2) is of the form used in defining an integral.
4) We proceed to the limit, obtain the integral, and use this expression as the *definition* of the physical law in the complicated case.

Now it is plain that no calculus students can be expected to be able to master this procedure immediately when he is confronted with an unfamiliar physical law.

236 THE DEFINITE INTEGRAL

However, practice in carrying out the process in familiar situations makes possible an understanding of the physics, the definition, and the integral concept. Invention is a modification of experience. Once the student has a backlog of experience in solving problems involving familiar situations, he is capable of attacking an unfamiliar physical process right from the start.

With this in mind, the student will appreciate the benefits of mastering the exercises in this and the following section. Additional material of the same kind is given in subsequent chapters.

PROBLEMS

In Problems 1 through 6, a particle is moving along the x axis from a to b according to the force law, F, which is given. Find the work done.

1. $F(x) = x^3 - 2x^2 + 6x + 4$, $\quad a = 0, b = 3$
2. $F(x) = x^4 + 2x + 7$, $\quad a = -1, b = 2$
3. $F(x) = x\sqrt{1 + x^2}$, $\quad a = 1, b = 3$
4. $F(x) = (x^2 + 2x + 1)^3(x + 1)$, $\quad a = 0, b = 1$
5. $F: x \to (x + 3)^{-4}$, $\quad a = -1, b = 1$
6. $F: x \to x^3 + x$, $\quad a = -2, b = 2$. Interpret the result physically in this problem.

In Problems 7 through 11, assume that each spring obeys Hooke's law ($F = kx$) as stated in Example 2. Find the work done in each case.

7. Natural length, 8 cm; 20 dynes force stretches spring $\frac{1}{2}$ cm. Find the work done in stretching the spring from 8 to 11 cm.
8. Natural length 6 cm; 500 dynes stretches spring $\frac{1}{4}$ cm. Find the work done in stretching it 1 cm.
9. Natural length 10 cm; 30 dynes stretches it to $11\frac{1}{2}$ cm. Find the work done in stretching it (a) from 10 to 12 cm.; (b) from 12 to 14 cm.
10. Natural length 6 cm.; 12,000 dynes compresses it $\frac{1}{2}$ cm. Find the work done in compressing it from 6 cm to 5 cm. (Hooke's law works for compression as well as for extension.)
11. Natural length 6 cm; 1200 dynes compresses it $\frac{1}{2}$ cm. Find the work done in compressing it from 6 cm to $4\frac{1}{2}$ cm. What is the work required to bring the spring to 9 cm from its compressed state of $4\frac{1}{2}$ cm?
12. A cable 100 meters long and weighting 5 kg/m is hanging from a winch. Find the work done in winding it up.
13. A boat is anchored so that the anchor is 100 meters directly below the post about which the anchor chain is wound. The anchor weighs 3000 kg and the chain weighs 20 kg/m. How much work is done in bringing up the anchor?
14. A tank full of water is in the form of a right circular cylinder of altitude 5 m and radius of base 3 m. How much work is done in pumping the water up to a level 10 m above the top of the tank?
15. A swimming pool full of water is in the form of a rectangular parallelepiped 5 m deep, 15 m wide, and 25 m long. Find the work required to pump the water up to a level 1 m above the surface of the pool.
16. The base of one cylindrical tank of radius 6 m and altitude 10 m is 3 m above the top of another tank of the same size and shape. The lower tank is full of oil (900 kg/m^3)

and the upper tank is empty. How much work is done in pumping all the oil from the lower tank into the upper through a pipe which enters the upper tank through its base?

17. A trough full of water is 10 m long, and its cross section is in the shape of an isosceles triangle 2 m wide across the top and 2 m high. How much work is done in pumping all the water out of the trough over one of its ends?

18. A trough full of water is 12 m long and its cross section is in the shape of an isosceles trapezoid 1 m high, 2 m wide at the bottom, and 3 m wide at the top. Find the work done in pumping the water to a point 2 m above the top of the trough.

19. A trough full of water is 6 m long and its cross section is in the shape of a semicircle with the diameter at the top 2 m wide. How much work is required to pump the water out over one of its ends?

20. A steam shovel is excavating sand. Each load weighs 500 kg when excavated. The shovel lifts each load to a height of 15 m in $\frac{1}{2}$ min, then dumps it. A leak in the shovel lets sand drop out while the shovel is being raised. The rate at which sand leaks out is 160 kg/min; find the amount of work done by the shovel in raising one load.

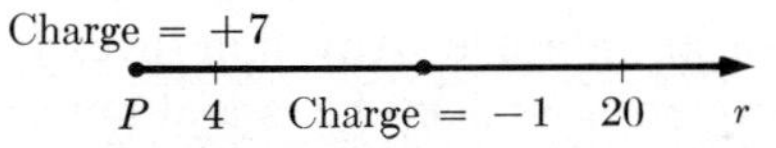

Figure 7–30

21. Two unlike charges of amount e_1 and e_2 electrostatic units attract each other with a force $e_1 e_2 / r^2$, where r is the distance between them (assuming appropriate units). A positive charge e_1 of 7 units is held fixed at a certain point P. (See Fig. 7–30.) How much work is done in moving a negative charge e_2 of 1 unit from a point 4 units away from the positive charge to a point 20 units away from that charge along a straight line pointing directly away from the positive charge?

22. The base of a tank of radius 5 m and altitude 8 m is directly above a tank of the same radius but of altitude 6 m. The lower tank is full of water. How much work is done in pumping the water from the lower tank to the upper tank if the pipe enters the upper tank halfway up its side?

23. A metal bar of length L and cross section A obeys Hooke's laws when stretched. If it is stretched x units, the force F required is given by

$$F = \frac{EA}{L} x,$$

where E is a constant known as *Young's modulus* or the *modulus of elasticity*. The quantity E depends only on the type of material used, not on the shape or size of the bar. Given that a steel bar 12 cm long and of uniform cross section 4 cm² is stretched $\frac{1}{2}$ cm, find the work done. (Give answer in terms of E.)

10. FLUID PRESSURE*

The definition of pressure, like the definition of work, depends on the notion of force. In the simplest case, suppose that a fluid (liquid or gas) is in a container and, at some point in the container, a small flat plate is inserted. The mere presence of the fluid exerts a force on this plate both from the weight of the fluid and from the action of the molecules. The **pressure** is defined as the *force per unit area* of the fluid on the plate.

---
* See the footnote at the beginning of Section 9.

When a liquid is in an open container so that there is a *free surface* at the top, the pressure at any depth of the liquid is given by a particularly simple law. Suppose that the liquid weighs w kg/m³ (i.e., its density is w). **The pressure p at a depth h meters below the surface is**

$$p = w \cdot h.$$

This is a remarkable formula in many ways; it says that the size of the vessel is irrelevant. At a depth of 3 m in a swimming pool the pressure is the same as it is at a depth of 3 m in an ocean (assuming that the pool is filled with salt water). The pressure is determined by depth alone, all other dimensions of the vessel being of no consequence.

The following principle is also needed in the study of liquid pressure:

Principle 1. *At any point in a liquid the pressure is the same in all directions.*

This means that a plate below the surface has the same pressure on it whether it is located vertically, horizontally, or at an angle. Any skin diver knows that the pressure on his eardrums depends only on how deep he is and not on the angle at which his head happens to be tilted.

The problem we shall concern ourselves with is the determination of the *total force* on a plate situated in a fluid. In the simplest case, if the plate has area A m² and if it is at a depth where the pressure is p nt/m², then the **total force** F is given by

$$F = p \cdot A.$$

If the density of the liquid is w and the plate is h ft below the surface, we may also write

$$F = w \cdot h \cdot A,$$

since $p = w \cdot h$.

If a large plate is located in a fluid either vertically or at an angle with the horizontal, then different portions of it are at different depths. The elementary rule given above for finding the force will not work, since we do not know what depth h to use. To handle problems of this sort we need the following principle:

Principle 2. *If a flat plate is divided into several parts, then the total force on the entire plate is the sum of the forces on each of the parts.*

Note the similarity between this principle and those in the previous section on work. We shall now show how Principles 1 and 2 lead to an integral which will define the total force on a plate in a liquid. A specific example will be worked first.

Example 1. The face of a dam adjacent to the water is vertical and rectangular in shape with width 30 m and height 10 m. Find the total force exerted by the

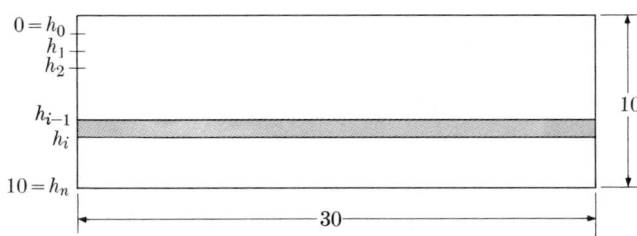

Figure 7–31

liquid on the face of the dam when the water surface is level with the top of the dam.

Solution. The depth of water varies from 0 to 10 m and the pressure changes with depth. We divide the dam into horizontal strips $0 = h_0 < h_1 < h_2 < \cdots < h_n = 10$. (See Fig. 7–31.) If the number of subdivisions is large, then the pressure on a strip such as the ith one will be *almost* constant, since all of its points are at approximately the same depth. The width of the strip is

$$\Delta_i h = h_i - h_{i-1}.$$

If $\bar{h}_i$ is an average value between h_i and h_{i-1}, then the force on the ith strip is the area ($= 30\,\Delta_i h$) times the pressure ($= w\bar{h}_i$). We have

$$\text{Force on the } i\text{th strip} = 30 w \bar{h}_i \, \Delta_i h.$$

According to Principle 2, we obtain the total force by adding the forces on the individual strips:

$$\text{Total force} = \sum_{i=1}^{n} 30 w \bar{h}_i \, \Delta_i h.$$

This expression is suggestive of the integral

$$\int_0^{10} 30 w h \, dh,$$

which is the *definition of the* **total force** on the dam. Therefore, by integration,

$$\text{Total force} = 30 w \tfrac{1}{2} h^2 \big]_0^{10} = 1500 w. \quad \blacktriangleleft$$

The technique for working Example 1 follows the outline given at the end of the previous section:

1. In the simple case of constant pressure, force = pressure × area. (We know additionally that, for a liquid, pressure = weight × depth.)
2. A principle is given which says that the total force is the sum of the forces on the various parts.
3. The sum obtained is of the type used in defining integrals.
4. We proceed to the limit to obtain an integral defining the total force.

240 THE DEFINITE INTEGRAL

The preceding suggests the formula

$$F = \int_a^b wp(h)\, dh$$

where w is the density of the fluid, $p(h)$ is the pressure as a function of the height h, and a, b are the lower and upper limits, respectively, of the region in which the pressure is exerted.

Another example illustrates the procedure in a slightly more complicated case.

Example 2. The face of a dam adjacent to the water has the shape of an isosceles trapezoid of altitude 20 m, upper base 50 m, and lower base 40 m. Find the total force exerted by the water on the dam when the water is 15 m deep.

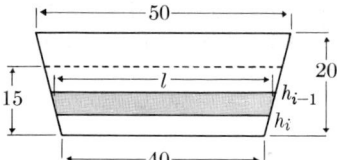

Figure 7–32

Solution. Figure 7–32 shows the dam and a typical strip along which the pressure is almost constant. The width of the strip is $\Delta_i h = h_i - h_{i-1}$. The force on this strip is

$$\text{Force on } i\text{th strip} = w \cdot \bar{h}_i \cdot \text{area},$$

where $\bar{h}_i$ is between h_{i-1} and h_i. The main problem is to find the area of the strip. For practical purposes we suppose it to be a rectangle. Then the area is (length) $\times (\Delta_i h)$. But what is the length l? Here the shape of the dam plays a role. In order to find l *as a function of the depth* h, we use the method of similar triangles (Fig. 7–33). From triangles ABC and DEC we have

$$\frac{l - 40}{15 - h} = \frac{10}{20} \quad \text{or} \quad l = \frac{95}{2} - \frac{1}{2}h,$$

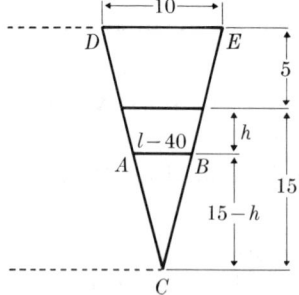

Figure 7–33

which is the length of the ith strip. Its area is

$$(\tfrac{95}{2} - \tfrac{1}{2}h)\,\Delta_i h,$$

and the total force on the ith strip is (approximately)

$$w\bar{h}_i(\tfrac{95}{2} - \tfrac{1}{2}\bar{h}_i)\,\Delta_i h.$$

The total force on the dam, according to Principle 2, is (approximately)

$$\tfrac{1}{2}\sum_{i=1}^{n} w\bar{h}_i(95 - \bar{h}_i)\,\Delta_i h.$$

We proceed to the limit and obtain

$$\text{Total force} = \int_0^{15} \tfrac{1}{2}wh(95 - h)\,dh = w[\tfrac{95}{4}h^2 - \tfrac{1}{6}h^3]_0^{15} = \frac{19{,}125}{4}\,w. \quad \blacktriangleleft$$

The essential step in Example 2 is finding the area of the ith strip in terms of the depth h, which is necessary in order that the sum we obtain be of the type used in approximating an integral.

PROBLEMS

In Problems 1 through 8, the face of a dam adjacent to the water is vertical and has the shape indicated. Find the total force on the dam due to fluid pressure.

1. A rectangle 200 m wide, 15 m high; water 10 m deep.
2. A rectangle 150 m wide, 12 m high; water 8 m deep.
3. An isosceles triangle 30 m wide at the top, 20 m high in the center; dam full of water.
4. An isosceles triangle 30 m wide at the top, 15 m high in the center; water 10 m deep in center.
5. An isosceles triangle 250 m across the top, 100 m high in the center; water 60 m deep in center.
6. A right triangle, one leg horizontal (at top) and 10 m wide, one leg vertical and 20 m high; dam full of water.
7. An isosceles trapezoid 90 m across the top, 60 m wide at the bottom, 20 m high; full of water.
8. An isosceles trapezoid 60 m across the top, 30 m wide at the bottom, 10 m high; water 8 m deep.
9. A square of side 5 m is submerged in water so that its diagonal is perpendicular to the surface. Find the total force on the square (due to the liquid on one side) if one vertex is at the surface.
10. The vertical ends of a trough are in the shapes of semicircles of diameter 2 m with diameter horizontal. Find the total force on one end when the trough is full of water.
11. An oil tank is in the shape of a right circular cylinder of diameter 4 m with axis horizontal. Find the total force on one end when the tank is half full of oil weighing 900 kg/m^3.

In Problems 12 through 16, it is assumed that in a dam the face of a gate of the shape indicated is vertical. If the gate is adjacent to the water, find the total force on the gate in each case.

12. A rectangle 4 m wide, 3 m high, with upper edge 20 m below the surface of the water.

13. An isosceles triangle 4 m wide at the top, 3 m high, with upper edge 15 m below the water surface.
14. An isosceles triangle 2 m wide at the bottom, 2 m high, with top vertex 10 m below the surface of the water.
15. An isosceles trapezoid 3 m wide at the top, 4 m wide at the bottom, and 3 m high, with upper base 20 m below the water surface.
16. An isosceles right triangle, length of leg 5 m, with one leg vertical and the horizontal leg at the bottom. The horizontal leg is 30 m below the water surface.

In Problems 17 through 19 it is assumed that the face of the dam adjacent to the water is a plane figure of the given shape and is inclined at an angle of 30° from the vertical. Find the total force on that face in each case.

17. A rectangle 50 m wide with slant height 30 m; dam full of water.
18. An isosceles trapezoid 60 m wide at the top, 40 m wide at the bottom, slant height 20 m; dam full of water.
19. An isosceles trapezoid 200 m wide at the top, 120 m wide at the bottom, slant height 50 m; water $20\sqrt{3}$ m deep.
20. A swimming pool is 25 m wide, 40 m long, 2 m deep at one end, and 8 m deep at the other, the bottom being an inclined plane. Find the total force on the bottom.
21. Consider a gas confined to a right circular cylinder closed at one end by a movable piston. Assume that the pressure of the gas is p nt/m^2 when the volume is v m^3. The pressure depends on the volume and we write $p = p(v)$. Show that the *work done* by the gas in expanding from a volume v_1 to a volume v_2 may be defined by

$$\int_{v_1}^{v_2} p(v)\, dv \text{ joules.}$$

[*Hint:* Let the area of the cross section be A m^2, and let x be the variable distance from the piston to the fixed end of the cylinder.]

22. Suppose that v and p are related by the equation $p \cdot v^{1.4} = $ const. Find the work done by such a gas as it expands from a volume of 5 m^3 to 200 m^3 if the pressure is 2500 nt/m^2 when $v = 5$ m^3.
23. Suppose that v and p are related by the equation $p \cdot v^{1.4} = $ const. If $p = 60$ nt/m^2 when $v = 10$ m^3, (a) find v when $p = 15$ nt/m^2, and (b) find the work done by the gas as it expands until the pressure reaches this value (15 nt/m^2).

8

LINES, CIRCLES, CONICS

1. DISTANCE FROM A POINT TO A LINE

In this chapter we continue the study of analytic geometry which we started in Chapter 3. The tools of the calculus which we have developed in the interim turn out to be powerful aids in obtaining properties of curves and regions in the plane.

A first-degree equation of the form

$$Ax + By + C = 0$$

represents a straight line. Suppose that $P(x_0, y_0)$ is a point not on the line. **The distance from a point to a line** is defined as the length of the perpendicular segment dropped from this point to the line. The number representing this distance is always positive. The following theorem gives a simple formula for this distance.

Theorem 1. *The distance d from the point $P(x_0, y_0)$ to the line L whose equation is $Ax + By + C = 0$ is given by the formula*

$$d = \frac{|Ax_0 + By_0 + C|}{\sqrt{A^2 + B^2}}.$$

Proof. We shall use a proof which is easy to devise but which requires some complicated algebraic manipulation. The steps of the proof are: (1) Find the equation of the line through $P(x_0, y_0)$ which is perpendicular to L (Fig. 8–1). (2) Find the point of intersection of this line with L. Call the point $Q(\bar{x}, \bar{y})$. (3) Find the distance $|PQ|$. (4) Treat vertical and horizontal lines separately. (The proof will help us review some of the facts and formulas concerning straight lines.)

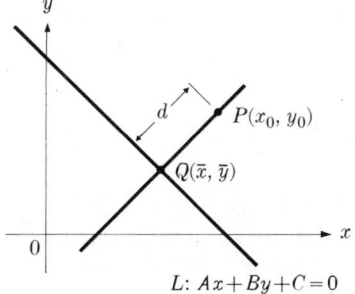

Figure 8–1

The slope of the line $Ax + By + C = 0$ is $-A/B$. If the line is neither vertical nor horizontal, the slope of the line through P and perpendicular to L is $+B/A$, the negative reciprocal. The equation of this line, according to the point-slope formula, is

$$y - y_0 = (B/A)(x - x_0) \quad \Leftrightarrow^* \quad Bx - Ay + Ay_0 - Bx_0 = 0.$$

To find $Q(\bar{x}, \bar{y})$ we solve simultaneously the equations

$$Ax + By + C = 0, \quad Bx - Ay + Ay_0 - Bx_0 = 0.$$

We multiply the first equation by A and the second by B and add, to get

$$\bar{x} = \frac{B^2 x_0 - ABy_0 - AC}{A^2 + B^2}.$$

Now we multiply the first equation by B and the second by A and subtract, to find

$$\bar{y} = \frac{-ABx_0 + A^2 y_0 - BC}{A^2 + B^2}.$$

The distance d, according to the formula for the distance between two points, is

$$d = \sqrt{(x_0 - \bar{x})^2 + (y_0 - \bar{y})^2}.$$

Substitution of the expressions for $\bar{x}$ and $\bar{y}$ in this formula yields

$$d^2 = \left[x_0 - \frac{B^2 x_0 - ABy_0 - AC}{A^2 + B^2}\right]^2 + \left[y_0 - \frac{-ABx_0 + A^2 y_0 - BC}{A^2 + B^2}\right]^2.$$

We find the least common denominator and simplify, to obtain

$$d^2 = \frac{A^2(Ax_0 + By_0 + C)^2}{(A^2 + B^2)^2} + \frac{B^2(Ax_0 + By_0 + C)^2}{(A^2 + B^2)^2}.$$

This combines even further to yield

$$d^2 = \frac{(Ax_0 + By_0 + C)^2}{A^2 + B^2},$$

and finally, taking the positive square root, we write the result:

$$d = \frac{|Ax_0 + By_0 + C|}{\sqrt{A^2 + B^2}}.$$

If the line L is horizontal, its equation is of the form $y = -C/B$ and the distance from $P(x_0, y_0)$ to L is the difference of the y values (draw a sketch, if necessary). Therefore,

$$d = \left|y_0 - \left(-\frac{C}{B}\right)\right| = \left|\frac{By_0 + C}{B}\right|.$$

---

*The symbol $\Leftrightarrow$ stands for "if and only if." For ease in reading the student may translate the symbol as "is equivalent to." Also, in a definition such as we have here, the symbol is used to connect the definition with the term being defined. In these definitions we may read "$\Leftrightarrow$" as "means that."

Similarly, if the line is vertical,
$$d = \left|\frac{Ax_0 + C}{A}\right|.$$

Note that the distance formula in the statement of the theorem works in *all* cases so long as A and B are not both zero. Furthermore if, by accident, the point $P(x_0, y_0)$ happened to be on the line L the distance d would be zero, since (x_0, y_0) would satisfy the equation $Ax_0 + By_0 + C = 0$.

Example 1. Find the distance d from the point $(2, -1)$ to the line
$$3x + 4y - 5 = 0.$$

Solution. We see that $A = 3$, $B = 4$, $C = -5$, $x_0 = 2$, $y_0 = -1$. Substituting in the formula, we get
$$d = \frac{|3 \cdot 2 + 4(-1) - 5|}{\sqrt{3^2 + 4^2}} = \frac{|-3|}{5} = \frac{3}{5}.$$ ◀

Definition. The **distance between parallel lines** *is the shortest distance from any point on one of the lines to the other line.*

Example 2. Find the distance between the parallel lines
$$L_1: 2x - 3y + 7 = 0 \quad \text{and} \quad L_2: 2x - 3y - 6 = 0.$$

Solution. First find any point on one of the lines, say L_1. To do this, let x be any value and solve for y in the equation for L_1. If $x = 1$ in L_1, then $y = 3$ and the point $(1, 3)$ is on L_1. The distance from $(1, 3)$ to L_2 is
$$d = \frac{|2(1) - 3(3) - 6|}{\sqrt{2^2 + (-3)^2}} = \frac{|-13|}{\sqrt{13}} = \frac{13}{\sqrt{13}} = \sqrt{13}.$$

The distance between the parallel lines is $\sqrt{13}$. The answer could be checked by taking any point on L_2 and finding its distance to L_1. ◀

PROBLEMS

In Problems 1 through 8, in each case find the distance from the given point to the given line.

1. $(3, 2)$, $2x + 4y - 4 = 0$
2. $(1, 0)$, $x - 5y + 1 = 0$
3. $(2, -1)$, $2x - 2y + 9 = 0$
4. $(-3, 1)$, $5x + y + 2 = 0$
5. $(-1, -4)$, $-2x + 3y - 6 = 0$
6. $(2, -4)$, $x - 3 = 0$
7. $(-1, 0)$, $y + 2 = 0$
8. $(2, 1)$, $x - y = 0$

In Problems 9 through 12, find the distance between the given parallel lines. Check your results.

9. $3x + 4y - 7 = 0$ and $3x + 4y + 3 = 0$
10. $x + 2y + 4 = 0$ and $2x + 4y - 5 = 0$

11. $5x - 6y = 0$ and $5x - 6y + 1 = 0$

12. $Ax + By + 10 = 0$ and $Ax + By + 15 = 0$

In Problems 13 through 16, the vertices of triangles are given. Find the lengths of the altitudes from the first vertex to the side joining the other two. Then find the areas of the triangles.

13. $(3, 2), (-1, 1), (-2, 3)$
14. $(4, -2), (-1, -3), (3, 2)$
15. $(-2, -1), (-1, 4), (3, 2)$
16. $(-3, 4), (1, -2), (3, 4)$

17. Find the two points on the x axis which are at a distance 2 from $3x - 4y - 6 = 0$.

18. Find the two points on the y axis which are at a distance 3 from the line $12x + 5y + 9 = 0$.

19. Find the two points on the line $2x + 3y + 4 = 0$ which are at a distance 2 from the line $3x + 4y - 6 = 0$.

20. Find the equations of the two lines parallel to $3x + 4y - 5 = 0$ and at a distance 2 from it.

21. Find the equations of the two lines parallel to $x - 2y + 1 = 0$ and at a distance 3 from it.

22. Find the distance from the midpoint of the segment joining $(3, 2)$ and $(-4, 6)$ to the line $2x - 3y + 5 = 0$.

23. Find the point on the curve $y = x^2$ which is closest to the line $2x - y - 4 = 0$.

*24. Find the point on the curve $(x - 1)^2 + y^2 - 4 = 0$ which is closest to the line $4x - 3y + 12 = 0$.

25. By using the formula for the distance from a point to a line, find the equations of the two lines which bisect the angles made by

$$L_1 : x + 2y - 3 = 0 \quad \text{and} \quad L_2 : 2x - y + 3 = 0.$$

26. The lines $L_1 : Ax + By + C_1 = 0$ and $L_2 : Ax + By + C_2 = 0$ are parallel. Show that the distance between them is given by the formula

$$d = \frac{|C_1 - C_2|}{\sqrt{A^2 + B^2}}.$$

2. FAMILIES OF LINES

The equation

$$y = 2x + k$$

represents a line of slope 2 for each value of k. The totality of lines obtained by letting k have any value—positive, negative, or zero—is called the **family of lines** of slope 2. In this instance, the family consists of a collection of parallel lines. More generally, a family of lines is a collection of lines which have a particular geometric property. The equation

$$y = kx + 3$$

represents for each value of k a line which has y intercept 3. The totality of such lines obtained by letting k take on all possible values is another example of a family of lines. Note that there is one line which passes through the point $(0, 3)$

which is not in this family: the y axis itself. Its equation is $x = 0$ and it is not included in the above family. It occasionally happens that families of lines have exceptions, and these must be taken into account. The following examples illustrate other families.

Example 1. Write the equation for the family of lines parallel to

$$3x + 2y + 4 = 0.$$

Solution. The slope of every member of the family must be $-\frac{3}{2}$. Therefore $3x + 2y + k = 0$, where k can have any value, represents the family. ◀

Example 2. Write the equation for the family of all lines passing through $(3, -2)$.

Solution. The point-slope form for the equation of a line through $(3, -2)$ is

$$y + 2 = m(x - 3),$$

and this represents every line through $(3, -2)$, with one exception. The line $x = 3$ must be added to the above family to get *all* lines through $(3, -2)$. ◀

The two lines

$$L_1: 3x - 2y + 7 = 0 \quad \text{and} \quad L_2: 2x + y - 6 = 0$$

intersect at a point (call it P) which may be found by solving the equations simultaneously. The first-degree equation

$$(3x - 2y + 7) + k(2x + y - 6) = 0$$

represents a family of lines passing through the intersection of L_1 and L_2. To see this we first note that no matter what k is, the above line must pass through P because $3x - 2y + 7 = 0$ at P and $2x + y - 6 = 0$ at P. As k varies, the equation changes, so the above system represents a family of lines through P. The family consists of *all* lines through P, with one exception. No value of k gives L_2 itself. Recombining the terms, we conclude that the equation

$$(3 + 2k)x + (k - 2)y + 7 - 6k = 0$$

represents the family of all lines through P except for L_2.

Example 3. Find the equation of the line passing through $(2, -3)$ and the intersection of the lines

$$L_1: 3x - y + 8 = 0 \quad \text{and} \quad L_2: 2x + 5y + 7 = 0.$$

Solution. We could find the intersection point of L_1 and L_2 and then use the two-point form of the equation of a line to get the answer. Instead we shall use the idea of a family of lines. The equation

$$3x - y + 8 + k(2x + 5y + 7) = 0$$

represents the family through the intersection of L_1 and L_2. The particular member passing through $(2, -3)$ must satisfy

$$3(2) - (-3) + 8 + k[2(2) + 5(-3) + 7] = 0.$$

This equation yields $k = \frac{17}{4}$ and, substituting this value of k in the equation of the family, we obtain

$$3x - y + 8 + \tfrac{17}{4}(2x + 5y + 7) = 0 \quad \Leftrightarrow \quad 46x + 81y + 151 = 0. \quad \blacktriangleleft$$

PROBLEMS

In Problems 1 through 8, in each case write the equation of the family of lines satisfying the given condition.

1. Parallel to the line $2x + 3y + 4 = 0$
2. Perpendicular to the line $x - 2y + 3 = 0$
3. Having y intercept 5
4. Having x intercept -3
5. Passing through the point $(2, -3)$
6. Passing through the point $(-4, 5)$
7. Having the sum of the x and y intercepts equal to 10
8. The area of the right triangle formed by the line and the axes is 12
9. Find the equation of the line passing through $(5, -1)$ and through the intersection of $2x + y - 3 = 0$ and $x + 4y - 7 = 0$.
10. Find the equation of the line passing through $(2, -4)$ and through the intersection of $3x - 4y + 1 = 0$ and $2x - 6y + 3 = 0$.
11. Find the equation of the line passing through the intersection of $2x - y + 3 = 0$ and $x + 3y - 2 = 0$ and having slope 4.
12. Find the equation of the line passing through the intersection of $3x + 4y - 2 = 0$ and $2x + y - 6 = 0$ and having slope -2.
13. Find the equation of the line passing through the intersection of $2x + y - 1 = 0$ and $x - y + 3 = 0$ and having y intercept -3.
14. Find the equations of the lines which bisect both of the angles formed by the lines $L_1: x + y - 5 = 0$ and $L_2: 2x - y + 3 = 0$. [*Hint:* Write the condition that the distances from $P(x, y)$ to L_1 and L_2 are equal.]
15. Find the equations of the lines which pass through the intersection of the lines $L_1: 2x + 3y - 1 = 0$ and $L_2: x - y + 12 = 0$ and which are three units from the origin.
16. Find the equation of the family of lines parallel to the line tangent to the curve $y = x^2 - 7x + 5$ at the point $(2, -5)$.
17. Find the equation of the family of lines parallel to the line which is perpendicular to the curve $y = x^3 - 6x^2 + 2x + 3$ at the point $(-1, -6)$.
*18. Find the equation of the family of lines through the point $(3, -2)$. Determine the member or members of the family which are tangent to the curve $y = x^2 + 4x + 7$.
19. Find the equation of the line which belongs to both the families $2x + 3y - k = 0$ and $3x + my - 6 = 0$.
20. Find the member (or members) of the family $y = 2x + k$ which intersect the curve $x^2 + y^2 = \frac{4}{5}$ in exactly one point.

3. ANGLE BETWEEN TWO LINES. BISECTORS OF ANGLES

We recall that the inclination α of a line L is the angle that the upward-pointing ray of L makes with the positive x axis. If the line is parallel to the x axis, its inclination is zero. The slope m of the line L is $m = \tan \alpha$ ($\alpha \neq 90°$). Two nonparallel lines L_1 and L_2 make four angles at their intersection: two equal obtuse angles and two equal acute angles (unless the lines are perpendicular). The obtuse angle is the supplement of the acute angle.

The angle swept out when the line L_1 is rotated counterclockwise to L_2 about the point of intersection is called the **angle from L_1 to L_2**. Let ϕ be the angle (measured in radians) from L_1 to L_2 (see Fig. 8–2). Then, clearly, the angle from L_2 to L_1 is $\pi - \phi$.

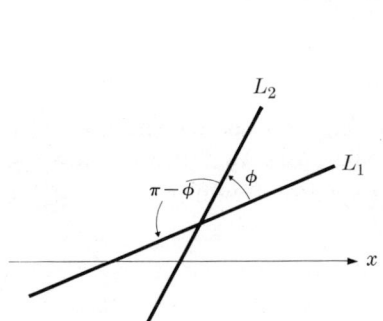

Figure 8–2

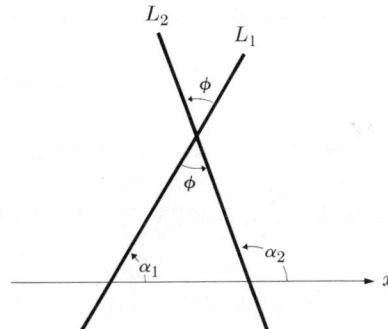

Figure 8–3

Let L_1 and L_2 be two intersecting lines (neither of which is vertical) with inclinations α_1 and α_2 and slopes $m_1 = \tan \alpha_1$ and $m_2 = \tan \alpha_2$, respectively. Let ϕ be the angle from L_1 to L_2 and suppose $\alpha_1 < \alpha_2$. Then, since the exterior angle of a triangle is the sum of the remote interior angles (see Fig. 8–3), we have

$$\alpha_2 = \alpha_1 + \phi, \qquad \phi = \alpha_2 - \alpha_1, \qquad \tan \phi = \tan(\alpha_2 - \alpha_1).$$

Using the formula for the tangent of the difference of two angles,* we get

$$\tan \phi = \frac{\tan \alpha_2 - \tan \alpha_1}{1 + \tan \alpha_1 \tan \alpha_2}.$$

In terms of slopes, we obtain

$$\tan \phi = \frac{m_2 - m_1}{1 + m_1 m_2}, \qquad (1)$$

which is the formula for **the tangent of the angle from L_1 to L_2**. The derivation above assumed that $\alpha_2 > \alpha_1$. In case $\alpha_1 > \alpha_2$, we see that

$$\pi - \phi = \alpha_1 - \alpha_2, \qquad \tan(\pi - \phi) = \frac{m_1 - m_2}{1 + m_1 m_2}.$$

---

* A review of all the basic formulas in trigonometry as well as the graphs of all the elementary trigonometric functions may be found in Appendix 1 at the end of the book.

Since $\tan(\pi - \phi) = -\tan\phi$, the formula (1) holds in this case also.

Example 1. Find the tangent of the angle from the line
$$L_1 = \{(x, y): 2x + 3y = 5\}$$
to the line $L_2 = \{(x, y): 4x - 3y = 2\}$.

Solution. Here $m_1 = -\frac{2}{3}$ and $m_2 = \frac{4}{3}$. The formula yields
$$\tan\phi = \frac{\frac{4}{3} - (-\frac{2}{3})}{1 + (-\frac{2}{3})(\frac{4}{3})} = 18.$$
◄

Formula (1) fails if one of the lines is vertical. We divide both the numerator and the denominator by m_2 in the formula for $\tan\phi$, getting
$$\tan\phi = \frac{1 - (m_1/m_2)}{(1/m_2) + m_1}.$$

Consider now that L_2 becomes vertical; that is, $m_2 \to \infty$ as L_2 approaches a vertical line. Then $m_1/m_2 \to 0$ and $1/m_2 \to 0$. Therefore the formula for the tangent of the angle ϕ from L_1 to L_2 when L_2 is vertical is simply
$$\tan\phi = \frac{1}{m_1}.$$

Example 2. Given the curve with equation $y = 2x^2 - 3x - 4$. Find the equations of the lines tangent to the curve at the points where $x = 1$ and $x = 2$. Find the tangent of the angle between these two tangent lines.

Solution. The slope at any point is found by obtaining the derivative:
$$\frac{dy}{dx} = 4x - 3.$$

At $x = 1$ the slope $m_1 = 1$; at $x = 2$ the slope $m_2 = 5$. If $x = 1$, then $y = -5$ and if $x = 2$, then $y = -2$. The first line has equation $y + 5 = 1(x - 1)$ and the second has equation $y + 2 = 5(x - 2)$. Using the formula for the tangent of the angle between two lines, we get
$$\tan\phi = \frac{5 - 1}{1 + 5 \cdot 1} = \frac{2}{3}.$$

Figure 8–4 shows the result. ◄

From plane geometry we recall that the angle bisectors of two given intersecting lines represent the graph of all points which are equidistant from the given lines. We can find the equations of these angle bisectors by using the formula for the distance from a point to a line. The following example illustrates the technique.

Example 3. Find the equations of the bisectors of the angles formed by the lines $3x + 4y - 7 = 0$ and $4x + 3y + 2 = 0$.

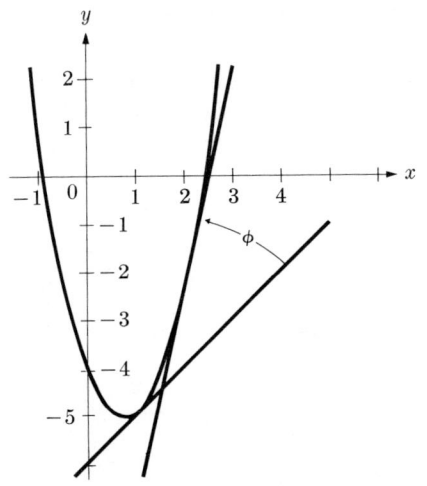
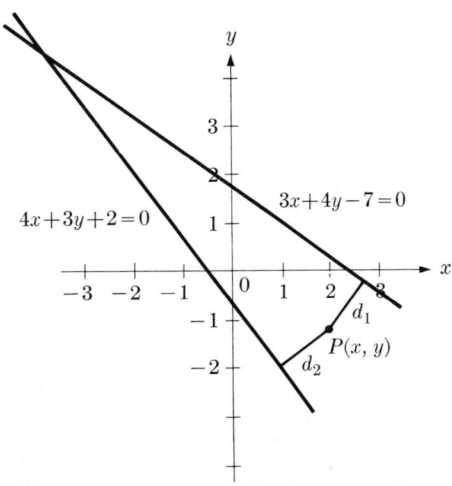

Figure 8–4   Figure 8–5

Solution. We draw a figure, as shown (Fig. 8–5), and let $P(x, y)$ be a point on the graph (angle bisector, in this case). Let d_1 be the distance from P to the line $3x + 4y - 7 = 0$ and d_2 the distance from P to the line $4x + 3y + 2 = 0$. The required condition is $d_1 = d_2$. Then, from the formula for the distance from a point to a line, we have

$$d_1 = \frac{|3x + 4y - 7|}{\sqrt{3^2 + 4^2}}$$

and

$$d_2 = \frac{|4x + 3y + 2|}{\sqrt{4^2 + 3^2}}.$$

The condition $d_1 = d_2$ becomes

$$|3x + 4y - 7| = |4x + 3y + 2|.$$

At this time we make use of our knowledge of absolute values. If $|a| = |b|$, then either $a = b$ or $a = -b$. This tells us that either

$$3x + 4y - 7 = 4x + 3y + 2$$

or

$$3x + 4y - 7 = -(4x + 3y + 2).$$

Simplifying, we obtain

$$x - y + 9 = 0$$

and

$$7x + 7y - 5 = 0$$

as the equations of the two angle bisectors. Note that the angle bisectors have slopes 1 and -1 and are therefore perpendicular, as we know they should be.

To distinguish the bisectors from each other, we have merely to find the tangent of the angle from one of the given lines to the other, and then the tangent of the angle from one of the given lines to the angle bisectors. A comparison of the sizes indicates where the lines fall. ◂

PROBLEMS

In Problems 1 through 7, find in each case $\tan \phi$, where ϕ is the angle from L_1 to L_2.

1. $L_1: 2x + 3y - 4 = 0;\quad L_2: x - 4y + 1 = 0$
2. $L_1: x + 3y - 2 = 0;\quad L_2: 3x - y + 1 = 0$
3. $L_1: 2x - 5y + 7 = 0;\quad L_2: 2x - 4y - 3 = 0$
4. $L_1: 5x + 7y - 6 = 0;\quad L_2: 8x + 5y - 2 = 0$
5. $L_1: 2x + 3y = 5;\quad L_2: 3x + 2y = 1$
6. $L_1: 5x + y = 0;\quad L_2: 3x + 2y = 5$
7. $L_1: 3x - 2y = 2;\quad L_2: y = 2$

In Problems 8 through 11, find the tangents of the angles of the triangles ABC with the vertices given.

8. $A(3, 0), B(1, 2), C(5, 6)$
9. $A(-1, 2), B(4, 1), C(0, 5)$
10. $A(-1, -3), B(2, -1), C(-3, 4)$
11. $A(-1, 4), B(5, 1), C(5, 6)$
12. Given a triangle with vertices at $A(0, 0), B(5, 0), C(3, 4)$; show that the three angle bisectors meet in a point.
13. Show that the bisectors of the angles of any triangle meet in a point.

In Problems 14 through 18, find the equations of the lines through the given point P_0 making the given angle θ with the given line L.

14. $P_0 = (2, -3), \theta = \pi/4,\quad L = \{(x, y): 3x - 2y = 5\}$
15. $P_0 = (-1, 2), \tan \theta = \frac{1}{2},\quad L = \{(x, y): 2x + y - 3 = 0\}$
16. $P_0 = (4, 1), \tan \theta = 3,\quad L = \{(x, y): 3x + 4y - 5 = 0\}$
17. $P_0 = (-1, -2), \theta = \pi/6,\quad L = \{(x, y): x + y = 1\}$
18. $P_0 = (3, 1), \theta = \pi/2,\quad L = \{(x, y): 2x - y = 3\}$
19. Tangent lines are drawn to the curve of $y = x^2 - 2x + 4$ at the points $(1, 3)$ and $(-2, 12)$. Find the tangent of the angle at which the first tangent intersects the second. Is it necessary to find the equations of the lines?
20. Tangent lines are drawn to the curve given by $y = x^3 - 2x^2 + 3x - 4$ at $(2, 2)$ and $(-1, -10)$. Find the tangent of the angle at which these lines intersect.
21. Two tangent lines are drawn to the curve (a circle) K given by

$$K = \{(x, y): x^2 + y^2 - 2x + 5y + 7 = 0\}$$

from the point $(-1, -2)$. Find the tangent of the angle at which these lines intersect.

In Problems 22 through 26, find the equations of the two bisectors of the angles formed by the given lines. Draw figures.

22. $x - 2y = 0, \quad 11x + 2y = 0$
23. $2x + 3y = 0, \quad 18x + y = 0$
24. $2x + y + 5 = 0, \quad x + 2y - 3 = 0$
25. $y = 0, \quad 4x - 3y = 0$
26. $x = 4, \quad 2x + 3y - 1 = 0$

In Problems 27 through 29, find the equation of the bisector of $\angle AVB$. Draw a figure.

27. $A(6, -2), V(2, 1), B(5, 5)$
28. $A(3, -3), V(1, -1), B(8, 0)$
29. $A(2, 4), V(-1, 3), B(7, 3)$
*30. Given the curve K with equation $y = x^2 + 2x + 4$. Find the two lines through the point $P(3, -2)$ which are tangent to K.
31. Find the equation of the family (or families) of lines which make an angle of $\pi/4$ with the line $2x + y - 5 = 0$.
32. The points $A(1, 0), B(5, 0), C(7, 4), D(3, 8)$ are the vertices of a quadrilateral. Find the tangent of the angle the diagonals make with each other.

4. THE CIRCLE*

Straight lines correspond to equations of the first degree in x and y. A systematic approach to the study of equations in x and y would proceed to equations of the second degree, third degree, and so on. Curves of the second degree are sufficiently simple that we can study them thoroughly in this course. Curves corresponding to equations of the third degree and higher will not be treated systematically, since the complications rapidly become too great. However, the study of equations of any degree—the theory of algebraic curves—is an interesting subject on its own and may be studied in advanced courses.

The set of all points (x, y) which are at a given distance r from a fixed point (h, k) represents a **circle**. We also use the statement, *the graph of all such points is a circle*. From the formula for the distance between two points we know that

$$r = \sqrt{(x - h)^2 + (y - k)^2},$$

or, upon squaring,

$$(x - h)^2 + (y - k)^2 = r^2.$$

This is the equation of a circle with center at (h, k) and radius r.

---

* Additional material on the circle, as well as another problem set, may be found in Appendix 2, Section 1, at the end of the book.

Example 1. Find the equation of the circle with center at $(-3, 4)$ and radius 6.

Solution. We have $h = -3$, $k = 4$, $r = 6$. Substituting in the equation of the circle, we obtain
$$(x + 3)^2 + (y - 4)^2 = 36.$$
We can multiply out to get
$$x^2 + y^2 + 6x - 8y - 11 = 0$$
as another form for the answer. ◀

We now start the other way around. Suppose that we are given an equation such as
$$x^2 + y^2 - 4x + 7y - 8 = 0;$$
we ask whether or not this represents a circle and, if so, what the center and radius are. The process for determining these facts consists of "completing the square." We write
$$(x^2 - 4x \quad) + (y^2 + 7y \quad) = 8,$$
in which we leave the appropriate spaces, as shown. We add the square of half the coefficient of x and the square of half the coefficient of y to both sides and so obtain
$$(x^2 - 4x + 4) + (y^2 + 7y + \tfrac{49}{4}) = 8 + 4 + \tfrac{49}{4}$$
or
$$(x - 2)^2 + (y + \tfrac{7}{2})^2 = \tfrac{97}{4}.$$
This represents a circle with center at $(2, -\tfrac{7}{2})$ and radius $\tfrac{1}{2}\sqrt{97}$.

A line tangent to a circle is defined in the usual way by the methods of calculus. We also have the elementary definition that a line perpendicular to a radius and passing through the point where the radius meets the circle is tangent to the circle at this point. We consider an example which illustrates the fact that the two definitions agree. The point $(1, 2)$ is on the circle
$$x^2 + y^2 + 2x + 3y - 13 = 0.$$
We pose the problem of finding the equation of the line tangent to the circle at the point $(1, 2)$. (See Fig. 8–6.) By implicit differentiation we can find the slope of the tangent line at any point. We have
$$2x + 2y\frac{dy}{dx} + 2 + 3\frac{dy}{dx} = 0$$
and
$$\frac{dy}{dx} = -\frac{2x + 2}{2y + 3}.$$

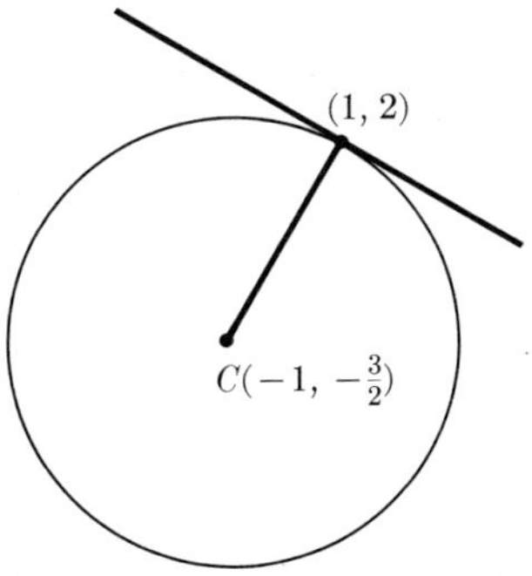

[Figure 8-6]

At the point $(1, 2)$ we obtain

$$\frac{dy}{dx} = -\frac{2+2}{4+3} = -\frac{4}{7},$$

and the equation of the line, according to the point-slope formula, is

$$y - 2 = -\tfrac{4}{7}(x - 1) \quad \Leftrightarrow \quad 4x + 7y - 18 = 0.$$

Any line perpendicular to this tangent line has slope $\tfrac{7}{4}$. The particular perpendicular passing through $(1, 2)$ has the equation

$$y - 2 = \tfrac{7}{4}(x - 1) \quad \Leftrightarrow \quad 7x - 4y + 1 = 0. \tag{1}$$

The center C of the circle is at $(-1, -\tfrac{3}{2})$. Since

$$7(-1) - 4(-\tfrac{3}{2}) + 1 = 0,$$

the line (1) passes through the center C. Therefore we see that the tangent line is perpendicular to the radius at the point of contact. Thus the two definitions for a tangent to a circle agree.

Let $P(x_1, y_1)$ be a point outside the circle K which has its center at $C(h, k)$ and has radius r. In set notation, we may write

$$K = \{(x, y) : (x - h)^2 + (y - k)^2 - r^2 = 0\}.$$

A tangent from P to the circle has contact at a point Q (Fig. 8-7). Since triangle PQC is a right triangle, we see that $|PQ| = \sqrt{|PC|^2 - r^2}$ and, using the distance formula for the length $|PC|$, we have

$$|PQ| = \sqrt{(x_1 - h)^2 + (y_1 - k)^2 - r^2}.$$

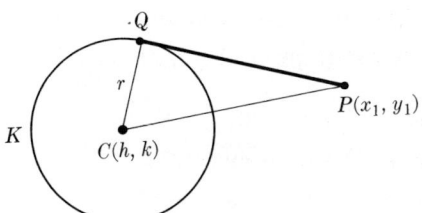

Figure 8-7

From the procedure for completing the square, it follows that if the circle having equation

$$x^2 + y^2 + Dx + Ey + F = 0 \tag{2}$$

has center at (h, k) and radius $r > 0$, then

$$(x - h)^2 + (y - k)^2 - r^2 \equiv x^2 + y^2 + Dx + Ey + F.$$

Hence if the circle in Fig. 8–7 has the equation (2), then

$$|PQ| = \sqrt{x_1^2 + y_1^2 + Dx_1 + Ey_1 + F}.$$

Example 2. A tangent is drawn from $P(8, 4)$ to the circle K given by

$$K = \{(x, y) : x^2 + y^2 + 2x + y - 3 = 0\}.$$

Find the distance from P to the point of tangency Q.

Solution. Using the second formula above for $|PQ|$, we obtain

$$|PQ| = \sqrt{8^2 + 4^2 + 2 \cdot 8 + 4 - 3} = \sqrt{97}. \qquad \blacktriangleleft$$

PROBLEMS

In Problems 1 through 14, in each case find the equation of the circle determined by the given conditions, where C denotes the center and r the radius.

1. $C = (-3, 4), r = 5$
2. $C = (0, 0), r = 4$
3. $C = (2, 3), r = 2$
4. $C = (2, -1), r = 3$
5. A diameter is the segment from $(4, 2)$ to $(8, 6)$
6. A diameter is the segment from $(-2, 3)$ to $(4, -1)$
7. $C = (3, 4)$, tangent to the x axis
8. $C = (-2, 3)$, tangent to the y axis
9. $C = (2, 1)$, passing through $(3, 4)$
10. $C = (-1, -2)$, passing through $(-2, 2)$
11. $C = (2, 3)$, tangent to $3x + 4y + 2 = 0$
12. $C = (3, -2)$, tangent to $5x - 12y = 0$
13. Tangent to both axes, radius 6, in second quadrant
14. Tangent to the x axis, to the line $y = 2x$ and radius 5 (four solutions)

In Problems 15 through 22, in each case determine the graph of the given equation by completing the square.

15. $x^2 + y^2 + 6x - 8y = 0$
16. $x^2 + y^2 + 2x - 4y - 11 = 0$
17. $x^2 + y^2 - 4x + 2y + 5 = 0$
18. $x^2 + y^2 + 6x - 4y + 15 = 0$
19. $x^2 + y^2 + 3x - 5y - \frac{1}{2} = 0$
20. $x^2 + y^2 + 4x - 3y + 4 = 0$
21. $2x^2 + 2y^2 + 3x + 5y + 2 = 0$
22. $2x^2 + 2y^2 - 5x + 7y + 10 = 0$

In Problems 23 through 26, in each case find the equation of the line tangent to the given circle at the point P on the circle.

23. $x^2 + y^2 + 2x - 4y = 0, P(1, 3)$
24. $x^2 + y^2 - 3x + 2y - 7 = 0, P(-1, 1)$
25. $x^2 + y^2 + 5x - 6y - 21 = 0, P(2, -1)$
26. $x^2 + y^2 + 2x - 19 = 0, P(-3, 4)$

In Problems 27 through 30, in each case find the equation of the line normal to the given circle at the given point P on the circle. Do each problem in two ways.

27. $x^2 + y^2 + 4y - 1 = 0, P(2, -3)$
28. $x^2 + y^2 - 3x - 8y + 18 = 0, P(1, 4)$
29. $x^2 + y^2 - 4x - 7y - 6 = 0, P(6, 1)$
30. $x^2 + y^2 + 8x + 2y + 16 = 0, P(-4, -2)$

In Problems 31 through 34, in each case lines are drawn tangent to the given circle through the given point P, not on the circle. Find the equations of these lines.

31. $x^2 + y^2 + 4x + 6y - 21 = 0, P(-4, 5)$
32. $x^2 + y^2 - 2x + 5y + 7 = 0, P(-1, -2)$
33. $x^2 + y^2 - x - 4y - 7 = 0, P(3, 0)$
34. $x^2 + y^2 + 2x + 6y + 2 = 0, P(-3, 3)$
35. The slope of the tangent to

$$x^2 + y^2 + 3x + y + 25 = 0 \quad \text{is} \quad \frac{dy}{dx} = -\frac{2x + 3}{2y + 1}, y \neq -\tfrac{1}{2}.$$

What is wrong with this statement? or is it correct?

5. THE PARABOLA*

In Section 4, dealing with the circle, we began a systematic study of equations of the second degree in x and y. We now continue the analysis of second-degree curves.

Definitions. A **parabola** *is the graph of all points whose distances from a fixed point equal their distances from a fixed line. The fixed point is called the* **focus** *and the fixed line the* **directrix.**

The definition of the parabola is purely geometric in the sense that it says nothing about coordinates—that is, the values of x and y. Recall that the definition of the circle as the graph of points at a given distance from a fixed point is also a geometric one. It will turn out that parabolas are always given by second-degree equations, as is true in the case of the circle.

---

* Additional material on the parabola, as well as another problem set, may be found in Appendix 2, Section 2, at the end of the book.

258   LINES, CIRCLES, CONICS

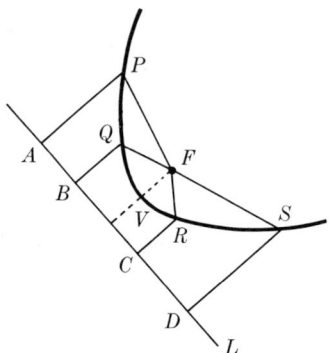

Figure 8–8

Figure 8–8 shows a typical situation. The line L is the directrix, the point F is the focus, and P, Q, R, S are points which satisfy the conditions of the graph. That is,

$$|AP| = |PF|, \quad |BQ| = |QF|, \quad |CR| = |RF|, \quad \text{and} \quad |DS| = |SF|.$$

The perpendicular to L through the focus intersects the parabola at a point V. This point is called the **vertex** of the parabola.

To find an equation for a parabola, we set up a coordinate system and place the directrix and the focus at convenient places. Suppose that the distance from the focus F to the directrix L is p units. We place the focus at $(p/2, 0)$ and let the line L be $x = -p/2$, as shown in Fig. 8–9. If $P(x, y)$ is a typical point on the graph, the conditions are such that the distance $|PF| = \sqrt{[x - (p/2)]^2 + (y - 0)^2}$ is equal to the distance from P to the line L. From an examination of the figure (or from the formula for the distance from a point to a line), we easily deduce that the distance from P to the line L is $|x - (-p/2)|$. Therefore we have

$$\sqrt{\left(x - \frac{p}{2}\right)^2 + y^2} = \left|x + \frac{p}{2}\right|,$$

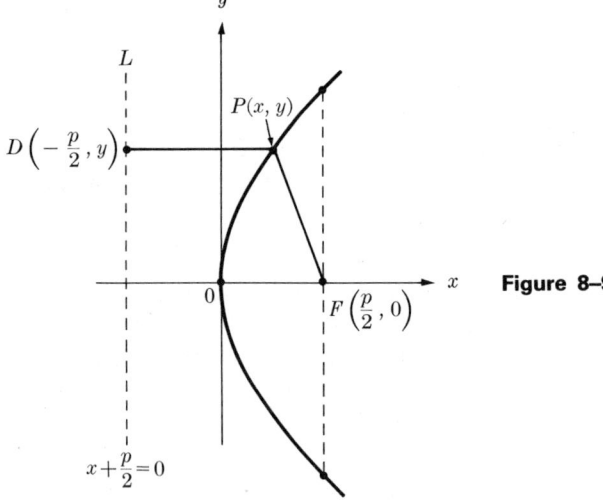

Figure 8–9

and, since both sides are positive, we may square to obtain

$$\left(x - \frac{p}{2}\right)^2 + y^2 = \left(x + \frac{p}{2}\right)^2$$

and

$$x^2 - px + \frac{p^2}{4} + y^2 = x^2 + px + \frac{p^2}{4}.$$

This equation yields

$$y^2 = 2px, \qquad p > 0.$$

Each point on the graph satisfies this equation. Conversely, by reversing the steps, it can be shown that every point which satisfies this equation is on the graph. We call this *the equation of the parabola with focus at $(p/2, 0)$ and with the line $x = -p/2$ as directrix*. The vertex of the parabola is at the origin. It is apparent from the equation that the x axis is an axis of symmetry, for if y is replaced by $-y$, the equation is unchanged. The line of symmetry of a parabola is called the **axis of the parabola.**

If the focus and directrix are placed in other positions, the equation will of course be different. There are four *standard positions* for the focus-directrix combination, of which the first has just been described. With p always a positive number, the second standard position is defined so that its focus is at $(-p/2, 0)$ and the line $x = p/2$ is its directrix. Using the same method as before, we solve the problem which defines the parabola. We obtain the equation

$$y^2 = -2px, \qquad p > 0,$$

and a parabola in the position shown in Fig. 8–10. The third position has its focus at $(0, p/2)$ and the line $y = -p/2$ as the directrix. The resulting equation is

$$x^2 = 2py, \qquad p > 0,$$

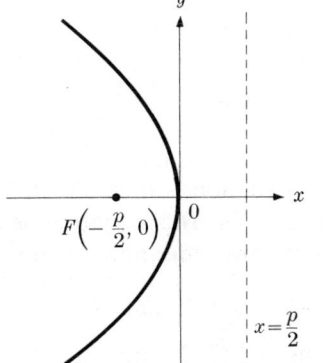

Figure 8–10

260 LINES, CIRCLES, CONICS

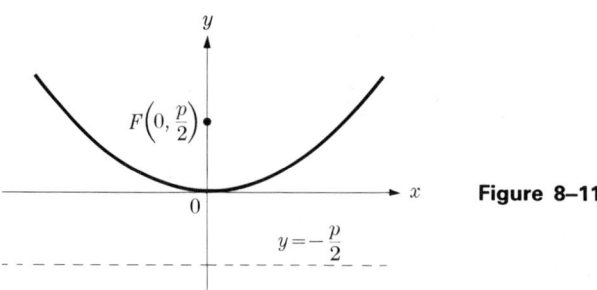

Figure 8–11

and the parabola is in the position shown in Fig. 8–11. In standard position (4) the focus is at $(0, -p/2)$ and the directrix is the line $y = p/2$. The equation then becomes

$$x^2 = -2py, \quad p > 0,$$

and the parabola is in the position shown in Fig. 8–12. In all four positions the vertex is at the origin.

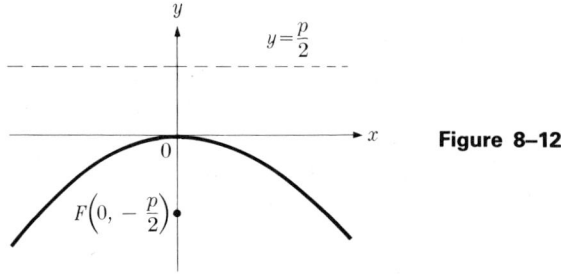

Figure 8–12

Example 1. A parabola is given by

$$S = \{(x, y): y^2 = -12x\}.$$

Find the focus, directrix, and axis. Sketch the graph.

Solution. Since $2p = 12$, we have $p = 6$ and, since the equation is in the second of the standard forms, the focus is at $(-3, 0)$. The directrix is the line $x = 3$, and the x axis is the axis of the parabola. The curve is sketched in Fig. 8–13. ◄

When the parabola is not in one of the standard positions, the equation is more complicated. If the directrix is horizontal or vertical, the equation is only slightly different; however, if the directrix is parallel to neither axis, the equation is altered considerably. The following example shows how the equation of the parabola may be obtained directly from the definition.

Example 2. Find the equation of the parabola with focus at $F(3, 2)$ and with the line $x = -4$ as directrix. Locate the vertex and the axis of symmetry.

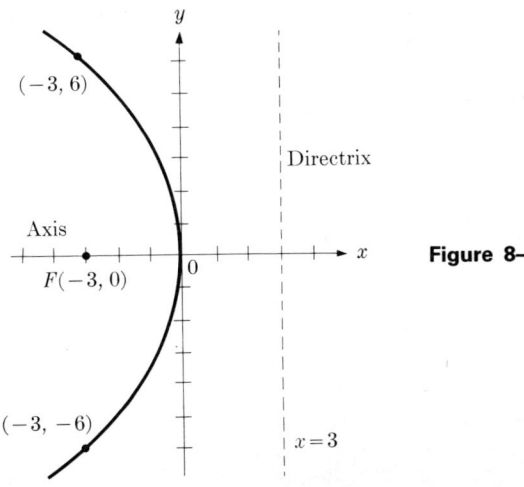

Figure 8–13

Solution. From the definition of the parabola, $P(x, y)$ is a point on the curve if and only if $|PF| = \sqrt{(x - 3)^2 + (y - 2)^2}$ is equal to the distance from P to the directrix, which is $|x + 4|$. We have

$$\sqrt{(x - 3)^2 + (y - 2)^2} = |x + 4|.$$

Since both sides are positive, we square both sides and find

$$x^2 - 6x + 9 + y^2 - 4y + 4 = x^2 + 8x + 16.$$

This yields

$$(y - 2)^2 = 14x + 7 \quad \Leftrightarrow \quad (y - 2)^2 = 14(x + \tfrac{1}{2}).$$

A sketch of the equation (Fig. 8–14) shows that the axis is the line $y = 2$ and the vertex is at the point $(-\tfrac{1}{2}, 2)$. ◀

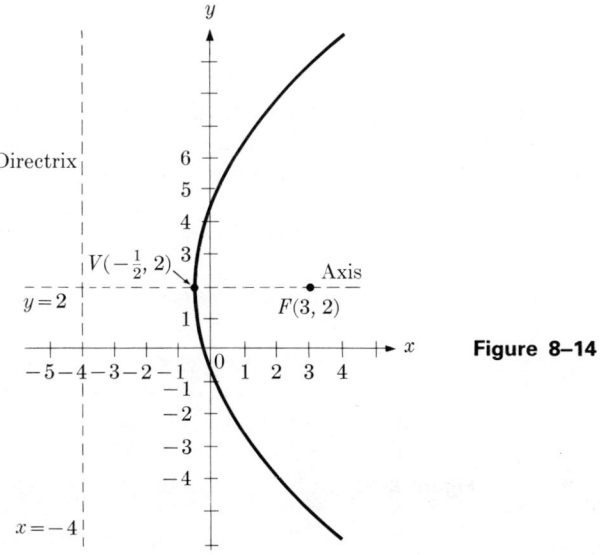

Figure 8–14

The four forms for the parabola when the axis is vertical or horizontal and the vertex is at the point (a, b) instead of the origin are:

$$(y - b)^2 = 2p(x - a), \quad p > 0, \tag{1}$$

with focus at $(a + p/2, b)$; directrix, $x = a - p/2$;

$$(y - b)^2 = -2p(x - a), \quad p > 0, \tag{2}$$

with focus at $(a - p/2, b)$; directrix, $x = a + p/2$;

$$(x - a)^2 = 2p(y - b), \quad p > 0, \tag{3}$$

with focus at $(a, b + p/2)$; directrix, $y = b - p/2$;

$$(x - a)^2 = -2p(y - b), \quad p > 0, \tag{4}$$

with focus at $(a, b - p/2)$; directrix, $y = b + p/2$.

The four formulas above may be derived directly from the definition of parabola (as in Example 2). See Problem 23 at the end of this section.

Example 3. Given the parabola $x^2 = -4y - 3x + 2$, find the vertex, focus, directrix, and axis. Sketch the curve.

Solution. To complete the square in x, we write

$$x^2 + 3x \quad = -4y + 2,$$

and then obtain

$$(x + \tfrac{3}{2})^2 = -4y + 2 + \tfrac{9}{4} = -4y + \tfrac{17}{4}.$$

This yields

$$(x + \tfrac{3}{2})^2 = -4(y - \tfrac{17}{16}),$$

which is in the fourth form above. We read off that the vertex is at $(-\tfrac{3}{2}, \tfrac{17}{16})$. Since $2p = 4$ we know that $p = 2$ and the focus is at $(-\tfrac{3}{2}, \tfrac{1}{16})$; the directrix is the line $y = \tfrac{33}{16}$, and the axis is the line $x = -\tfrac{3}{2}$. The curve is sketched in Fig. 8-15. ◀

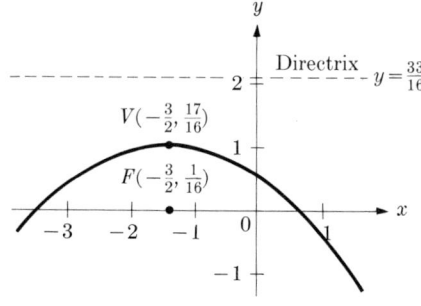

Figure 8-15

In the next two examples, we discuss problems which combine a number of ideas we have already studied.

Example 4. Given the family of lines

$$l(k) = \{(x, y): 3x + 2y - k = 0\},$$

find the particular member of this family which is tangent to the parabola

$$S = \{(x, y): y = 3x^2 - 2x + 1\}.$$

Solution. When $l(k) \cap S$ consists of one point, the line is tangent to the parabola or is parallel to the axis of the parabola. If the line is tangent to the parabola, the slope of the line and the parabola must be the same at the point of contact. The family $l(k)$ consists of all lines of slope $-\frac{3}{2}$. The slope of the parabola at any point is

$$\frac{dy}{dx} = 6x - 2.$$

The slope of the parabola and the slope of the line are the same when

$$6x - 2 = -\tfrac{3}{2} \quad \Leftrightarrow \quad x = \tfrac{1}{12}.$$

Substituting $x = \tfrac{1}{12}$ in the equation for S, we get

$$y = 3(\tfrac{1}{144}) - \tfrac{2}{12} + 1 = \tfrac{123}{144}.$$

The point on the parabola at which the slope is $-\tfrac{3}{2}$ is $(\tfrac{1}{12}, \tfrac{123}{144})$. This point must also be on the line. Therefore

$$3(\tfrac{1}{12}) + 2(\tfrac{123}{144}) - k = 0$$

or $k = \tfrac{47}{24}$. The desired member of $l(k)$ is

$$l(\tfrac{47}{24}) = \{(x, y): 3x + 2y - \tfrac{47}{24} = 0\}. \quad \blacktriangleleft$$

Example 5. Find the equation of the line tangent to the parabola

$$S = \{(x, y): y^2 + 3y - 2x + 4 = 0\},$$

at the point on the parabola at which $y = 2$.

Solution. When $y = 2$, by substitution in the equation for S, we obtain $x = 7$. The slope at any point on S is given by the relation

$$2y\frac{dy}{dx} + 3\frac{dy}{dx} - 2 = 0,$$

and at $(7, 2)$ we find

$$\frac{dy}{dx} = \frac{2}{2y + 3} = \frac{2}{7}.$$

The equation of the tangent line is $y - 2 = \tfrac{2}{7}(x - 7)$. $\quad \blacktriangleleft$

PROBLEMS

In Problems 1 through 4, find the coordinates of the foci and the equations of the directrices. Sketch the curves.

1. $y^2 = 8x$
2. $x^2 = 6y$
3. $y^2 = -12x$
4. $x^2 = -16y$

In Problems 5 through 10, find the equation of the parabola with vertex at the origin which satisfies the given additional condition.

5. Focus at $(-4, 0)$
6. Focus at $(0, 2)$
7. Directrix: $y = 2$
8. Directrix: $x = \frac{7}{3}$
9. Passing through $(2, 3)$ and axis along the x axis
10. Passing through $(2, 3)$ and axis along the y axis

In Problems 11 through 16, find the focus, vertex, directrix, and axis of the parabola. Sketch the curve.

11. $y = x^2 - 2x + 3$
12. $x = y^2 + 2y - 4$
13. $y = -4x^2 + 3x$
14. $x = -y^2 + 2y - 7$
15. $x^2 + 2y - 3x + 5 = 0$
16. $y^2 + 2x - 4y + 7 = 0$

In Problems 17 through 22, in each case find the equation of the parabola from the definition.

17. Directrix $x = 0$, focus at $(6, 0)$
18. Directrix $y = 0$, focus at $(0, 5)$
19. Vertex $(0, 4)$, focus at $(0, 2)$
20. Vertex $(-2, 0)$, directrix $x = 1$
21. Focus $(-1, -2)$, directrix $x - 2y + 3 = 0$
22. Focus $(2, 3)$, directrix $x + y + 1 = 0$

23. (a) Use the definition of a parabola to establish formula (1) on page 262. (b) Same as part (a) for formula (2) on page 262. (c) Same as part (a) for formula (3) on page 262. (d) Same as part (a) for formula (4) on page 262.

24. Find the member of the family

$$l(k) = \{(x, y): 4x - y - k = 0\}$$

which is tangent to the parabola $y = 2x^2 - x + 1$.

25. Find a member of the family

$$l(k) = \{(x, y): kx - y - 5 = 0\}$$

which is tangent to the parabola $y = 3x^2 + 2x + 4$.

26. Find the member of the family of parabolas $y = kx^2 + 4x - 3$ which is tangent to the line $2x - y + 1 = 0$.

In Problems 27 through 31, in each case find the equation of the line tangent to the parabola at the given point P on the parabola.

27. $y^2 = 9x$, $P(1, 3)$
28. $x^2 = 12y + 7$, $P(1, -\frac{1}{2})$
29. $y^2 - 3x + 2y + 4 = 0$, $P(\frac{4}{3}, 0)$
30. $(y - 2)^2 = -4(x + 3)$, $P(-4, 4)$
31. $x^2 - 2x + y = 0$, $P(-3, -15)$

32. The two towers of a suspension bridge are 300 meters apart and extend 80 meters above the road surface. If the cable (in the shape of a parabola) is tangent to the road at the center of the bridge, find the height of the cable above the road at 50 meters

and also at 100 meters from the center of the bridge. (Assume that the road is horizontal.)

33. At a point P on a parabola a tangent line is drawn. This line intersects the axis of the parabola at a point A. If F denotes the focus, show that $\triangle APF$ is isosceles.

34. Let $P(x_1, y_1)$ be a point (not the vertex) on the parabola $y^2 = 2px$. Find the equation of the line normal to the parabola at this point. Show that this normal line intersects the x axis at the point $Q(x_1 + p, 0)$.

6. THE ELLIPSE*

Continuing our discussion of curves of the second degree, we now take up the ellipse.

Definitions. *An* **ellipse** *is the graph of all points the sum of whose distances from two fixed points is constant. The two fixed points are called the* **foci.**

For the definition to make sense, the constant (the sum of the distances from the foci) must be larger than the distance between the foci. Note that the definition is purely geometric and says nothing about coordinates or equations. It will turn out that the equation of an ellipse is of the second degree.

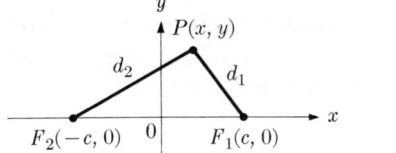

Figure 8–16

Suppose that the distance between the foci is $2c$ ($c > 0$). We place the foci—label them F_1 and F_2—at convenient points, say F_1 at $(c, 0)$ and F_2 at $(-c, 0)$. Let $P(x, y)$ be a point on the ellipse, d_1 the distance from P to F_1, and d_2 the distance from P to F_2 (Fig. 8–16). The conditions state that $d_1 + d_2$ is always constant. As we said, this constant (which we shall denote by $2a$) must be larger than $2c$. Thus we have

$$d_1 + d_2 = 2a,$$

or, from the distance formula,

$$\sqrt{(x-c)^2 + (y-0)^2} + \sqrt{(x+c)^2 + (y-0)^2} = 2a.$$

We simplify by transferring one radical to the right and squaring both sides:

$$(x-c)^2 + y^2 = 4a^2 - 4a\sqrt{(x+c)^2 + y^2} + (x+c)^2 + y^2.$$

---

* Additional material on the ellipse, as well as another problem set, may be found in Appendix 2, Section 3, at the end of the book.

By multiplying out and combining, we obtain

$$\sqrt{(x+c)^2 + y^2} = a + \frac{c}{a}x.$$

We square again to find

$$(x+c)^2 + y^2 = a^2 + 2cx + \frac{c^2}{a^2}x^2,$$

and this becomes

$$\frac{x^2}{a^2} + \frac{y^2}{a^2 - c^2} = 1.$$

Since $a > c$, we can introduce a new quantity,

$$b = \sqrt{a^2 - c^2},$$

and we can write

$$\boxed{\frac{x^2}{a^2} + \frac{y^2}{b^2} = 1.}$$

This is **the equation of an ellipse.**

We have shown that every point on the graph satisfies this equation. Conversely (by reversing the steps*), it can be shown that each point which satisfies this equation is on the ellipse. A sketch of such an equation is shown in Fig. 8–17.

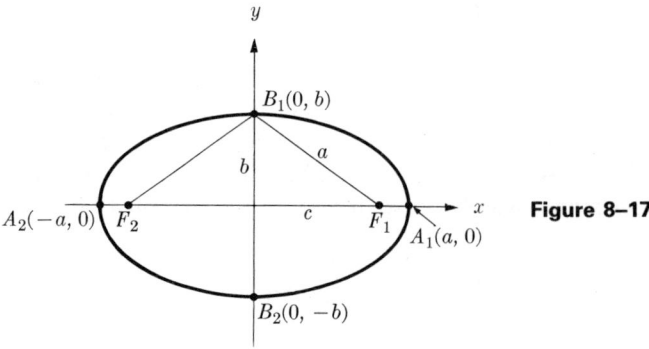

Figure 8–17

---

* In the process of reversing the steps, we must verify (because of the squaring operations) that

$$a + \frac{c}{a}x \geq 0 \quad \text{and} \quad 2a - \sqrt{(x+c)^2 + y^2} \geq 0.$$

However,

$$2a - \sqrt{(x+c)^2 + y^2} = 2a - \left(a + \frac{c}{a}x\right) = a - \frac{c}{a}x.$$

Since $-a \leq x \leq a$ for every point satisfying the equation, and since $c < a$, the verification follows.

From the equation, we see immediately that the curve is symmetric with respect to both the x and y axes.

Definitions. *The line passing through the two foci F_1 and F_2 is called the **major axis**. The perpendicular bisector of the line segment $F_1 F_2$ is called the **minor axis**. The intersection of the major and minor axes is called the **center**.*

The equation of the ellipse as we have obtained it indicates that the ellipse intersects the x axis at $(a, 0)$, $(-a, 0)$. These points are called the **vertices** of the ellipse. The distance between them, $2a$, is called the *length of the major axis*. (Sometimes this segment is called simply the major axis.) The ellipse intersects the y axis at $(0, b)$ and $(0, -b)$. The distance between them, $2b$, is called the *length of the minor axis*. (Sometimes this segment is called the minor axis.)

The **eccentricity** e of an ellipse is defined as

$$e = c/a.$$

Note that since $c < a$, the eccentricity is always between 0 and 1. The eccentricity measures the flatness of an ellipse. If a is kept fixed and c is very "small," then e is close to zero. But this means that the foci are close together, and the ellipse is almost a circle, as in Fig. 8–18. On the other hand, if c is close to a, then e is near 1 and the ellipse is quite flat, as shown in Fig. 8–19. The limiting position of an ellipse as $e \to 0$ is a circle of radius a. The limiting position as $e \to 1$ is a line segment of length $2a$.

In developing the equation of an ellipse we placed the foci at $(c, 0)$ and $(-c, 0)$. If instead we put them on the y axis at $(0, c)$ and $(0, -c)$ and carry through the same argument, then the equation of the ellipse we obtain is

$$\frac{x^2}{b^2} + \frac{y^2}{a^2} = 1.$$

The y axis is now the major axis, the x axis is the minor axis, and the vertices are at $(0, a)$ and $(0, -a)$. The quantity $b = \sqrt{a^2 - c^2}$ and the eccentricity $e = c/a$ are defined as before. Figure 8–20 shows an ellipse with foci on the y axis.

Example 1. Given the ellipse with equation $9x^2 + 25y^2 = 225$, find the major and minor axes, the eccentricity, the coordinates of the foci and the vertices. Sketch the ellipse.

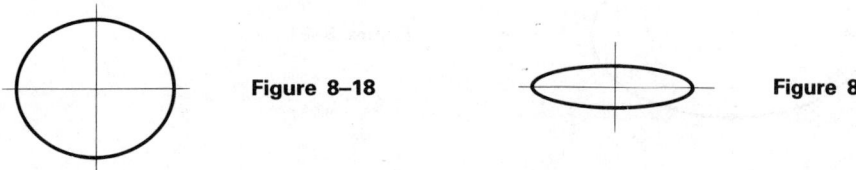

Figure 8–18

Figure 8–19

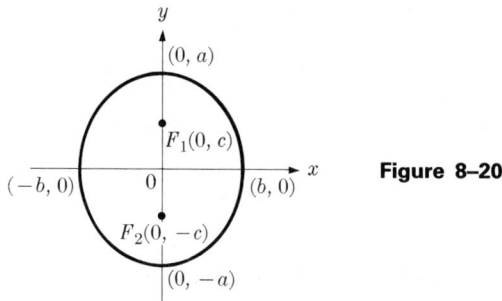

Figure 8–20

Solution. We put the equation in "standard form" by dividing by 225. We obtain

$$\frac{x^2}{25} + \frac{y^2}{9} = 1,$$

which tells us that $a = 5, b = 3$. Since $a^2 = b^2 + c^2$, we find that $c = 4$. The eccentricity $e = c/a = \frac{4}{5}$. The major axis is along the x axis, the minor axis is along the y axis, the vertices are at $(\pm 5, 0)$, and the foci are at $(\pm 4, 0)$. The ellipse is sketched in Fig. 8–21. ◂

Example 2. Given the ellipse with equation $16x^2 + 9y^2 = 144$, find the major and minor axes, the eccentricity, the coordinates of the foci, and the vertices. Sketch the ellipse.

Solution. We divide by 144 to get

$$\frac{x^2}{9} + \frac{y^2}{16} = 1.$$

We note that *the number under the y term is larger.* Therefore $a = 4, b = 3$, $c^2 = a^2 - b^2$, and $c = \sqrt{7}$. The eccentricity $e = \sqrt{7}/4$, the major axis is along the y axis, the minor axis is along the x axis, the foci are at $(0, \pm\sqrt{7})$, and the vertices at $(0, \pm 4)$. The ellipse is sketched in Fig. 8–22. ◂

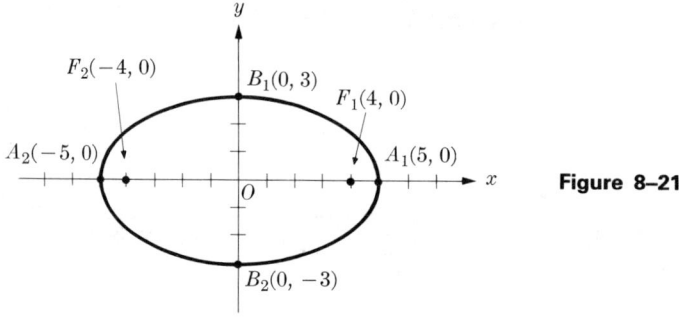

Figure 8–21

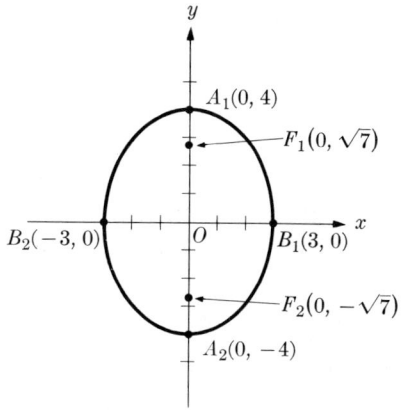

Figure 8-22

The foregoing examples illustrate the fact that when we have an equation of the form

$$\frac{x^2}{(\ )^2} + \frac{y^2}{(\ )^2} = 1,$$

the equation represents an ellipse (or a circle if the denominators are equal in size), and the *larger denominator* determines whether the foci, vertices, and major axes are along the x axis or the y axis.

Given that $P(x_1, y_1)$ is a point on an ellipse, it is a simple matter to find the equation of the line tangent to the ellipse at P. In fact, we can develop a simple formula for the equation of the line when the ellipse is given by

$$\frac{x^2}{a^2} + \frac{y^2}{b^2} = 1.$$

The slope of the ellipse is obtained by implicit differentiation:

$$\frac{2x}{a^2} + \frac{2y}{b^2}\frac{dy}{dx} = 0$$

and

$$\frac{dy}{dx} = -\frac{b^2}{a^2}\frac{x}{y}.$$

The equation of the line tangent at $P(x_1, y_1)$ is

$$y - y_1 = -\frac{b^2 x_1}{a^2 y_1}(x - x_1) \quad \Leftrightarrow \quad a^2 y_1 y - a^2 y_1^2 = -b^2 x_1 x + b^2 x_1^2.$$

But $b^2 x_1^2 + a^2 y_1^2 = a^2 b^2$, since (x_1, y_1) is on the ellipse. Therefore

$$\frac{x_1 x}{a^2} + \frac{y_1 y}{b^2} = 1 \tag{1}$$

is the equation of the tangent line.

Example 3. Find the equation of the line tangent to the ellipse

$$\frac{x^2}{9} + \frac{y^2}{16} = 1$$

at the point $(2, -\frac{4}{3}\sqrt{5})$.

Solution. The point $(2, -\frac{4}{3}\sqrt{5})$ is on the ellipse and, even though the major axis is along the y axis, it is easily seen that the formula analogous to (1) for the tangent line holds. We obtain

$$\frac{2x}{9} + \frac{-\frac{4}{3}\sqrt{5}\,y}{16} = 1 \quad \Leftrightarrow \quad 8x - 3\sqrt{5}\,y - 36 = 0. \quad \blacktriangleleft$$

PROBLEMS

In Problems 1 through 10, find the lengths of the major and minor axes, the coordinates of the foci and vertices, and the eccentricity. Sketch the curve.

1. $16x^2 + 25y^2 = 400$
2. $25x^2 + 169y^2 = 4225$
3. $9x^2 + 16y^2 = 144$
4. $25x^2 + 16y^2 = 400$
5. $3x^2 + 2y^2 = 6$
6. $3x^2 + 4y^2 = 12$
7. $5x^2 + 2y^2 = 10$
8. $5x^2 + 3y^2 = 15$
9. $2x^2 + 3y^2 = 11$
10. $5x^2 + 4y^2 = 17$

In Problems 11 through 21, find the equation of the ellipse satisfying the given conditions.

11. Vertices at $(\pm 5, 0)$, foci at $(\pm 4, 0)$
12. Vertices at $(0, \pm 5)$, foci at $(0, \pm 3)$
13. Vertices at $(0, \pm 10)$, eccentricity 4/5
14. Vertices at $(\pm 6, 0)$, eccentricity 2/3
15. Foci at $(0, \pm 4)$, eccentricity 4/5
16. Vertices at $(0, \pm 5)$, passing through $(3, 3)$
17. Axes along the coordinate axes, passing through $(4, 3)$ and $(-1, 4)$
18. Axes along the coordinate axes, passing through $(-5, 2)$ and $(3, 7)$
19. Foci at $(\pm 3, 0)$, passing through $(4, 1)$
20. Eccentricity 3/4, foci along x axis, center at origin, and passing through $(6, 4)$
21. Eccentricity 3/4, foci on the y axis, center at origin, and passing through $(6, 4)$
22. Find the equation of the graph of all points the sum of whose distances from $(6, 0)$ and $(-6, 0)$ is 20.
23. Find the equation of the graph of all points the sum of whose distances from $(3, 0)$ and $(9, 0)$ is 12.
24. Find the equation of the graph of all points whose distances from $(3, 0)$ are 1/3 of their distances from the line $x = 27$.
25. Find the equation of the graph of all points whose distances from $(0, 4)$ are 2/3 of their distances from the line $y = 9$.
26. Find the equation of the graph of all points whose distances from $(4, 0)$ are 1/2 of their distances from the line $x = 6$. Locate the major and minor axes.

27. Find the equation of the graph of all points whose distances from (0, 5) are 3/4 of their distances from the line $y = 8$. Locate the major and minor axes.
28. Find the equation of the line tangent to the ellipse.
$$\frac{x^2}{9} + \frac{y^2}{4} = 1$$
at the point $P(1, -\frac{4}{3}\sqrt{2})$.
29. Find the equation of the line tangent to the ellipse $4x^2 + 12y^2 = 1$ at the point $P(1/4, 1/4)$.
30. Find the equation of the line normal to the ellipse $4x^2 + 9y^2 = 36$ at the point $P(2, \frac{2}{3}\sqrt{5})$.
31. Find the equation of the line normal to the ellipse $16x^2 + 25y^2 = 400$ at the point $P(-3, 16/5)$.
32. Find the equation of the line normal to the ellipse $(x^2/a^2) + (y^2/b^2) = 1$ which passes through the point $P(x_1, y_1)$ on the ellipse.
33. The tangent to an ellipse at a point P_0 meets the tangent at a vertex in a point Q. Prove that the line joining the other vertex to P_0 is parallel to the line joining the center to Q.
34. A square is inscribed in the ellipse whose equation is $(x^2/16) + (y^2/9) = 1$. Find the coordinates of the vertices of the square; find the perimeter and area of the square.
35. The orbit in which the earth travels about the sun is approximately an ellipse with the sun at one focus. The major axis of the elliptical orbit is 300,000 kilometers and the eccentricity is 0.017 approximately. Find the maximum and minimum distances from the earth to the sun.

7. THE HYPERBOLA*

A study of the hyperbola will complete our discussion of second-degree curves in the plane.

Definitions. *A* **hyperbola** *is the graph of all points the difference of whose distances from two fixed points is a positive constant. The two fixed points are called the* **foci.**

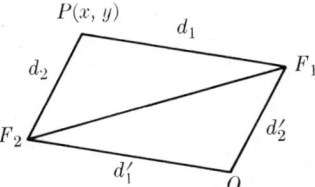

Figure 8–23

A typical situation is shown in Fig. 8–23, where the two fixed points (foci) are labeled F_1 and F_2. For a point P to be on the graph, the distance $|PF_1|$ minus the

---
* Additional material on the hyperbola as well as another problem set may be found in Appendix 2, Section 4, at the end of the book.

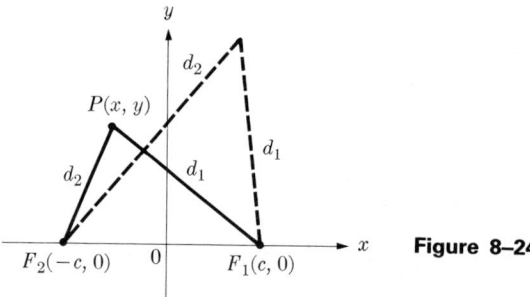

Figure 8-24

distance $|PF_2|$ must be equal to a positive constant. Also, a point, say Q, will be on the graph if the distance $|QF_2|$ minus $|QF_1|$ is this same constant. We remark that the definition of a hyperbola, like that of a parabola and an ellipse, is purely geometric and makes no mention of coordinate systems. It is true that hyperbolas are second-degree curves.

Suppose that the distance between the foci is $2c$ ($c > 0$). We place the foci at convenient points, F_1 at $(c, 0)$ and F_2 at $(-c, 0)$. Let $P(x, y)$ be a point on the hyperbola; then the conditions of the graph assert that $d_1 - d_2$ or $d_2 - d_1$ is always a positive constant (Fig. 8–24), which we label $2a$ ($a > 0$). We write the conditions of the graph as

$$d_1 - d_2 = 2a \quad \text{or} \quad d_2 - d_1 = 2a,$$

which may be combined as

$$d_1 - d_2 = \pm 2a.$$

Using the distance formula, we obtain

$$\sqrt{(x - c)^2 + y^2} - \sqrt{(x + c)^2 + y^2} = \pm 2a.$$

We transfer one radical to the right side and square both sides to get

$$(x - c)^2 + y^2 = (x + c)^2 + y^2 \pm 4a\sqrt{(x + c)^2 + y^2} + 4a^2.$$

Some terms may be canceled to yield

$$\pm 4a\sqrt{(x + c)^2 + y^2} = 4a^2 + 4cx.$$

We divide by $4a$ and once again square both sides, to find

$$x^2 + 2cx + c^2 + y^2 = a^2 + 2cx + \frac{c^2}{a^2}x^2 \quad \Leftrightarrow \quad \left(\frac{c^2}{a^2} - 1\right)x^2 - y^2 = c^2 - a^2.$$

We now divide through by $c^2 - a^2$, with the result

$$\frac{x^2}{a^2} - \frac{y^2}{c^2 - a^2} = 1.$$

We shall show that the quantity c must be larger than a. We recall that the sum of the lengths of any two sides of a triangle must be greater than the length of the third side. When we refer to Fig. 8–24, we observe that

$$2c + d_2 > d_1 \quad \text{and} \quad 2c + d_1 > d_2.$$

Therefore

$$2c > d_1 - d_2 \quad \text{and} \quad 2c > d_2 - d_1;$$

combined, these inequalities give $2c > |d_1 - d_2|$. But $|d_1 - d_2| = 2a$, and so $c > a$.

We define the positive number b by the relation

$$b = \sqrt{c^2 - a^2},$$

and we write *the equation of the hyperbola:*

$$\boxed{\frac{x^2}{a^2} - \frac{y^2}{b^2} = 1.}$$

Conversely, by showing that all the steps above are reversible, we could have demonstrated that every point which satisfies the above equation is on the graph. (We shall not enter into the discussion required for the proof of this fact.)

The equation of the hyperbola in the above form shows at once that it is symmetric with respect to both the x axis and the y axis.

Definitions. *The line passing through the foci F_1 and F_2 of a hyperbola is called the* **transverse axis.** *The perpendicular bisector of the segment F_1F_2 is called the* **conjugate axis.** *The intersection of these axes is called the* **center.**

[See Fig. 8–25, where the foci are at $(c, 0)$ and $(-c, 0)$, the transverse axis is the x axis, the conjugate axis is the y axis, and the center is at the origin.]

The points of intersection of a hyperbola with the transverse axis are called its **vertices.** In Fig. 8–25, corresponding to the equation

$$\frac{x^2}{a^2} - \frac{y^2}{b^2} = 1,$$

these vertices occur at $(a, 0)$ and $(-a, 0)$. The length $2a$ is called the *length of the transverse axis.* Even though the points $(0, b)$ and $(0, -b)$ are not on the locus, the length $2b$ is a useful quantity. It is called the *length of the conjugate axis.*

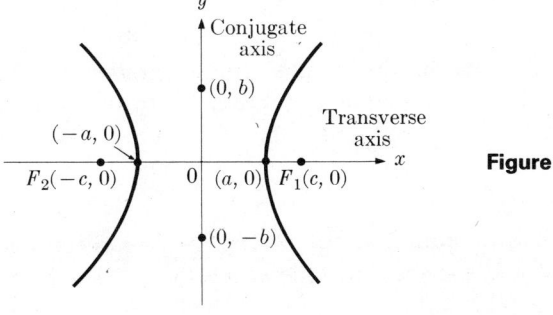

Figure 8–25

The **eccentricity** of a hyperbola is defined as

$$e = c/a.$$

Note that since $c > a$, the eccentricity of a hyperbola is always larger than 1.

If the foci of the hyperbola are placed along the y axis at the points $(0, c)$ and $(0, -c)$, the equation takes the form

$$\frac{y^2}{a^2} - \frac{x^2}{b^2} = 1,$$

where, as before,

$$b = \sqrt{c^2 - a^2}.$$

The curve has the general appearance of the curve shown in Fig. 8–26.

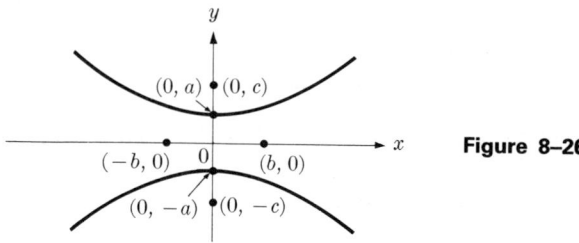

Figure 8–26

In contrast with the equation of an ellipse, the equation for a hyperbola indicates that the relative sizes of a and b play no role in determining where the foci and axes are. An equation of the form

$$\frac{x^2}{(\ )^2} - \frac{y^2}{(\ )^2} = 1$$

always has its transverse axis (and foci) on the x axis, while an equation of the form

$$\frac{y^2}{(\ )^2} - \frac{x^2}{(\ )^2} = 1$$

always has its transverse axis (and foci) on the y axis. In the first case, the quantity under the x^2 term is a^2 and that under the y^2 term is b^2. In the second case the quantity under the y^2 term is a^2 and that under the x^2 term is b^2. In *both cases* we have $c^2 = a^2 + b^2$.

Example 1. Given the hyperbola with equation $9x^2 - 16y^2 = 144$. Find the axes, the coordinates of the vertices and the foci, and the eccentricity. Sketch the curve.

Solution. Dividing by 144, we find that

$$\frac{x^2}{16} - \frac{y^2}{9} = 1,$$

and, therefore,

$$a = 4, \quad b = 3, \quad c^2 = a^2 + b^2 = 25, \quad c = 5.$$

The transverse axis is along the x axis; the conjugate axis is along the y axis. The vertices are at $(\pm 4, 0)$. The foci are at $(\pm 5, 0)$. The eccentricity $e = c/a = 5/4$. The curve is sketched in Fig. 8–27, where we have constructed a rectangle by drawing parallels to the axes through the points $(a, 0)$, $(-a, 0)$, $(0, b)$, and $(0, -b)$. This is called the **central rectangle.** Note that the diagonal from the origin to one corner of this rectangle has length c. ◀

Example 2. Find the equation of the hyperbola with vertices at $(0, \pm 6)$ and eccentricity $\frac{5}{3}$. Locate the foci.

Solution. Since the vertices are along the y axis, the equation is of the form

$$\frac{y^2}{a^2} - \frac{x^2}{b^2} = 1,$$

with $a = 6$, $e = c/a = \frac{5}{3}$, and $c = (5 \cdot 6)/3 = 10$. Therefore

$$b = \sqrt{100 - 36} = 8,$$

and the equation is

$$\frac{y^2}{36} - \frac{x^2}{64} = 1.$$

The foci are at the points $(0, \pm 10)$. ◀

If $P(x_1, y_1)$ is a point on the hyperbola

$$\frac{x^2}{a^2} - \frac{y^2}{b^2} = 1,$$

we can get a formula for the line tangent to the hyperbola at P in a way that precisely parallels the work we did to obtain a formula for the line tangent to the ellipse. The slope of the hyperbola at any point, by implicit differentiation, is

$$\frac{2x}{a^2} - \frac{2y}{b^2}\frac{dy}{dx} = 0 \quad \Leftrightarrow \quad \frac{dy}{dx} = \frac{b^2 x}{a^2 y}.$$

The equation of the line tangent at $P(x_1, y_1)$ is

$$y - y_1 = \frac{b^2 x_1}{a^2 y_1}(x - x_1) \quad \Leftrightarrow \quad a^2 y_1 y - a^2 y_1^2 = b^2 x_1 x - b^2 x_1^2.$$

But since P is on the hyperbola, we know that $b^2 x_1^2 - a^2 y_1^2 = a^2 b^2$, and so

$$\frac{x_1 x}{a^2} - \frac{y_1 y}{b^2} = 1$$

is the equation of the tangent line.

Example 3. Find the equations of the tangent line and the normal line to the hyperbola

$$\frac{x^2}{9} - \frac{y^2}{16} = 1$$

at the point $(5, -\frac{16}{3})$.

Solution. The point $(5, -\frac{16}{3})$ is on the hyperbola, and so the equation of the tangent line is

$$\frac{5x}{9} + \frac{\frac{16}{3}y}{16} = 1 \quad \Leftrightarrow \quad 5x + 3y - 9 = 0.$$

The normal line will have slope $\frac{3}{5}$, and the equation is

$$y + \tfrac{16}{3} = \tfrac{3}{5}(x - 5). \quad \blacktriangleleft$$

PROBLEMS

In Problems 1 through 10, find in each case the lengths of the transverse and conjugate axes, the coordinates of the foci and the vertices, and the eccentricity. Sketch the curve.

1. $16x^2 - 9y^2 = 144$
2. $16x^2 - 9y^2 + 576 = 0$
3. $25x^2 - 144y^2 + 3600 = 0$
4. $25x^2 - 144y^2 - 900 = 0$
5. $2x^2 - 3y^2 - 6 = 0$
6. $2x^2 - 3y^2 + 6 = 0$
7. $5x^2 - 2y^2 = 10$
8. $5x^2 - 4y^2 + 13 = 0$
9. $3x^2 - 2y^2 = 1$
10. $2x^2 - 3y^2 = -1$

In Problems 11 through 20, find for each case the equation of the hyperbola (or hyperbolas) satisfying the given conditions.

11. Vertices at $(\pm 5, 0)$, foci at $(\pm 7, 0)$
12. Vertices at $(\pm 4, 0)$, foci at $(\pm 6, 0)$
13. Vertices at $(0, \pm 7)$, eccentricity $\frac{4}{3}$
14. Foci at $(\pm 6, 0)$, eccentricity $\frac{3}{2}$
15. Eccentricity $\sqrt{5}$, foci on x axis, center at origin, passing through the point $(3, 2)$

16. Vertices at $(0, \pm 4)$, passing through $(-2, 5)$
17. Foci at $(0, \pm\sqrt{10})$, passing through $(2, 3)$
18. Axes along the coordinate axes, passing through $(4, 2)$ and $(-6, 7)$
19. Axes along the coordinate axes, passing through $(-3, 4)$ and $(5, 6)$
20. Foci at $(\pm 8, 0)$, length of conjugate axis 6
21. Find the equation of the graph of all points such that the difference of their distances from $(4, 0)$ and $(-4, 0)$ is always equal to 2.
22. Find the equation of the graph of all points such that the difference of their distances from $(0, 7)$ and $(0, -7)$ is always equal to 3.
23. Find the equation of the graph of all points such that the difference of their distances from $(10, 0)$ and $(2, 0)$ is always 1.
24. Find the equation of the graph of all points such that the difference of their distances from $(2, 0)$ and $(2, 12)$ is always equal to 3.
25. Find the equations of the tangent and normal lines to the hyperbola

    $$(x^2/16) - (y^2/4) = 1 \quad \text{at the point} \quad (-5, \tfrac{3}{2}).$$

26. Find the point of intersection of the lines tangent to the hyperbola

    $$(y^2/36) - (x^2/9) = 1 \quad \text{at the points} \quad (1, -2\sqrt{10}), (-4, 10).$$

27. Given the hyperbola $(x^2/6) - (y^2/2) = 1$. Find the equations of the lines tangent to this hyperbola passing through the point $(1, -1)$.
28. Find the equation of the graph of all points such that their distances from $(5, 0)$ are always $\tfrac{5}{3}$ times their distances from the line $x = 9/5$.
29. A *focal radius* is the line segment from a point on a hyperbola to one of the foci. The foci of a hyperbola are at $(4, 0)$ and $(-4, 0)$. The difference in the lengths of the focal radii from any point is always ± 6. Find the equation of the hyperbola.
30. A hyperbola has its foci at $(c, 0)$ and $(-c, 0)$. The point $P(2, 4)$ is on the hyperbola and the focal radii (see Problem 29) from P are perpendicular. Find the equation of the hyperbola.

8. TRANSLATION OF AXES

Coordinate systems were not employed in the definitions and descriptions of the principal properties of second-degree curves. However, when we want to find equations for such curves and wish to use the methods of analytic geometry, coordinate systems become essential.

When we use the method of coordinate geometry we place the axes at a position "convenient" with respect to the curve under consideration. In the examples of ellipses and hyperbolas which we studied, the foci were located on one of the axes and were situated symmetrically with respect to the origin. But now suppose that we have a problem in which the curve (hyperbola, parabola, ellipse, etc.) is *not* situated so conveniently with respect to the axes. We would then like to change the coordinate system in order to have the curve at a convenient and familiar location. The process of making this change is called a **transformation of coordinates**.

The first type of transformation we consider is one of the simplest, and is called

278 LINES, CIRCLES, CONICS

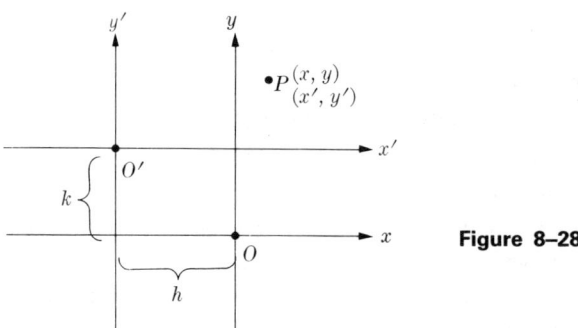

Figure 8–28

the **translation of axes.** Figure 8–28 shows the usual Cartesian xy coordinate system. We now introduce an additional coordinate system with axis x' parallel to x and k units away, and axis y' parallel to y and h units away. This means that the origin O' of the new coordinate system has coordinates (h, k) in the original system. The positive x' and y' directions are taken to be the same as the positive x and y directions. When P is any point in the plane, what are its coordinates in each of the systems? Suppose that P has coordinates (x, y) in the original system and (x', y') in the new system and *suppose that the new origin has coordinates (h, k) in the original system*. An inspection of Fig. 8–29 suggests that if $x > h > 0$ and $y > k > 0$, then

$$x = x' + h \quad \text{and} \quad y = y' + k \tag{1}$$

or equivalently

$$x' = x - h \quad \text{and} \quad y' = y - k. \tag{2}$$

The proof of these formulas follows at once by subtraction of the appropriate directed distances shown in Fig. 8–29. See Problem 29 at the end of this section.

Two coordinate systems xy and $x'y'$ which satisfy equations (1) or, equivalently, equations (2) are said to be related by a **translation of axes.**

The most general equation of the second degree has the form

$$Ax^2 + Bxy + Cy^2 + Dx + Ey + F = 0,$$

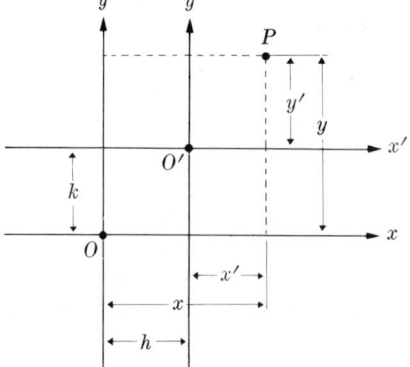

Figure 8–29

where $A, B, C, D, E,$ and F represent numbers. We assume that $A, B,$ and C are not all zero, since then the equation would be of the first degree. We recognize that the equations we have been discussing are all special cases of the above equation. For example, the circle $x^2 + y^2 - 16 = 0$ has $A = 1, B = 0, C = 1, D = E = 0, F = -16$. The ellipse $(x^2/9) + (y^2/4) = 1$ has $A = \frac{1}{9}, B = 0, C = \frac{1}{4}, D = E = 0, F = -1$. Similarly, parabolas and hyperbolas are of the above form.

An interesting case is $A = 1, B = 0, C = -1, D = E = F = 0$. We get $x^2 - y^2 = 0$ or $y = \pm x$, and the graph is two intersecting lines. Second-degree curves whose graphs reduce to lines or points are frequently called **degenerate**.

We now illustrate how to use translation of axes to reduce an equation of the form

$$Ax^2 + Cy^2 + Dx + Ey + F = 0$$

to an equation of the same form but with new letters (x', y'), and with D and E both equal to zero (with certain exceptions). Although presented in a new way, the equation will be easily recognized as one of the second-degree curves we have studied. *The principal tool in this process is "completing the square."*

Example 1. Given the equation

$$9x^2 + 25y^2 + 18x - 100y - 116 = 0,$$

by using a translation of axes determine whether the graph of the equation is a parabola, ellipse, or hyperbola. Determine foci (or focus), vertices (or vertex), and eccentricity. Sketch the curve.

Solution. To complete the square in x and y, we write the equation in the form

$$9(x^2 + 2x \quad) + 25(y^2 - 4y \quad) = 116.$$

We add 1 in the parentheses for x, which means adding 9 to the left side, and we add 4 in the parentheses for y, which means adding 100 to the left side. We obtain

$$9(x^2 + 2x + 1) + 25(y^2 - 4y + 4) = 116 + 9 + 100$$
$$\Leftrightarrow 9(x + 1)^2 + 25(y - 2)^2 = 225.$$

Now we have the clue for translation of axes. We define

$$x' = x + 1 \quad \text{and} \quad y' = y - 2.$$

That is, the translation is made with $h = -1, k = 2$. The equation becomes

$$9x'^2 + 25y'^2 = 225.$$

Dividing by 225, we find

$$\frac{x'^2}{25} + \frac{y'^2}{9} = 1,$$

which we recognize as an ellipse with $a = 5, b = 3, c^2 = a^2 - b^2 = 16, c = 4, e = \frac{4}{5}$. In the $x'y'$ system, we have: center $(0, 0)$; vertices $(\pm 5, 0)$; foci $(\pm 4, 0)$.

In the xy system, we use the relations $x = x' - 1, y = y' + 2$ to obtain: center $(-1, 2)$; vertices $(4, 2), (-6, 2)$; foci $(3, 2), (-5, 2)$. The curve is sketched in Fig. 8–30. ◀

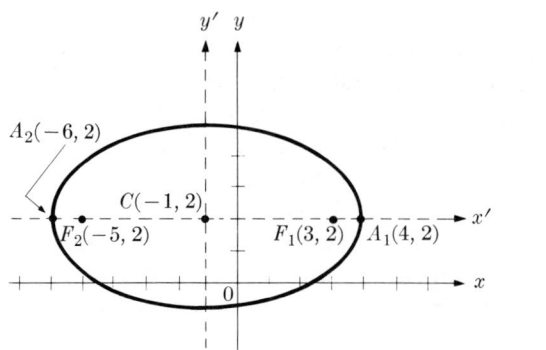

Figure 8-30

Example 2. Discuss the properties of the graph of the equation

$$x^2 + 4x + 4y - 4 = 0.$$

Solution. Here we have $A = 1$, $C = 0$, $D = 4$, $E = 4$, $F = -4$. There is only one second-degree term, and we complete the square to obtain

$$(x^2 + 4x + 4) = -4y + 4 + 4 \quad \Leftrightarrow \quad (x + 2)^2 = -4(y - 2).$$

We read off the appropriate translation of axes. It is

$$x' = x + 2, \quad y' = y - 2,$$

and we have

$$x'^2 = -4y',$$

which we recognize as a parabola. In the $x'y'$ system, the vertex is at $(0, 0)$, focus at $(0, -1)$ (since $p = 2$); directrix, $y' = 1$. In the xy system (since $h = -2$, $k = 2$), the vertex is at $(-2, 2)$, focus at $(-2, 1)$, and the directrix is the line $y = 3$. The curve is sketched in Fig. 8-31. ◂

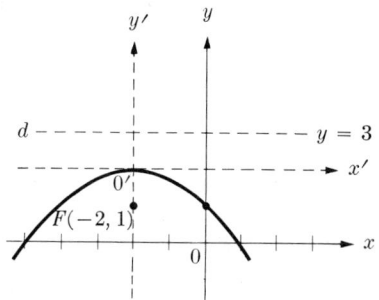

Figure 8-31

Example 3. Find the equation of the ellipse with eccentricity $\frac{1}{2}$ and foci at $(4, 2)$ and $(2, 2)$.

Solution. The center of the ellipse is halfway between the foci and therefore is at the point $(3, 2)$. The major axis of the ellipse is along the line $y = 2$. To move the origin to the center of the ellipse by a translation of coordinates, we let

$$x' = x - 3, \quad y' = y - 2.$$

In this system the foci are at $(\pm 1, 0)$; $e = c/a = \frac{1}{2}$ and, since $c = 1$, we have $a = 2$. Also $c^2 = a^2 - b^2$ gives us $b^2 = 3$. In the $x'y'$ system, the equation is

$$\frac{x'^2}{4} + \frac{y'^2}{3} = 1.$$

In the xy system, the equation is

$$\frac{(x-3)^2}{4} + \frac{(y-2)^2}{3} = 1,$$

or

$$3(x-3)^2 + 4(y-2)^2 = 12 \quad \Leftrightarrow \quad 3x^2 + 4y^2 - 18x - 16y + 31 = 0. \quad \blacktriangleleft$$

We are now in a position to discuss all the curves that an equation

$$Ax^2 + Cy^2 + Dx + Ey + F = 0$$

can possibly represent, presented here in the form of a theorem:

Theorem 2. (a) *If A and C are both positive or both negative, then the graph is an ellipse, a circle (if $A = C$), a point, or nothing.* (b) *If A and C are of opposite signs, the graph is a hyperbola or two intersecting lines.* (c) *If either A or C is zero, the graph is a parabola, two parallel lines, one line, or nothing.*

We shall not prove this theorem, but some of its possible uses will be illustrated by examples.

The equation $3x^2 + y^2 + 5 = 0$ has no graph, since the sum of positive quantities can never add up to zero. This exhibits the last case under (a) of Theorem 2. Under (c), the equation $x^2 - 2x = 0$ ($A = 1$, $C = 0$, $D = -2$, $E = 0$, $F = 0$) is an example of two parallel lines, while $x^2 - 2x + 1 = 0$ ($A = 1$, $C = 0$, $D = -2$, $E = 0$, $F = 1$) is an example of one line.

A proof of this theorem is within the scope of any curious student.

PROBLEMS

1. A Cartesian system of coordinates is translated to a new $x'y'$ system, whose origin is at $(-3, 4)$. The points P, Q, R, and S have coordinates $(4, 3)$, $(-2, 7)$, $(-6, \frac{3}{2})$, and $(-5, -2)$, respectively, in the original system. Find the coordinates of these points in the new system.

2. A translation of coordinates moves the origin to the point of intersection of the lines $2x + 3y - 4 = 0$ and $x + 4y - 1 = 0$. Find the translation of coordinates and the equations of these lines in the new system.

3. A translation of coordinates moves the origin to the point of intersection of the lines $3x - 7y + 1 = 0$ and $2x - 5y - 6 = 0$. Find the equations of these lines in the new coordinate system.

In Problems 4 through 17, translate the coordinates so as to eliminate the first-degree terms (or one first-degree term in the case of parabolas), describe the principal properties (as in Examples 1 and 2) and sketch the curves.

4. $y^2 + 8x - 6y + 1 = 0$
5. $25x^2 + 16y^2 + 50x - 64y - 311 = 0$
6. $9x^2 + 16y^2 - 36x - 32y - 92 = 0$
7. $9x^2 - 4y^2 + 18x + 16y + 29 = 0$

8. $x^2 - 4x + 6y + 16 = 0$
9. $4x^2 + 9y^2 + 8x - 36y + 4 = 0$
10. $y^2 - 4x + 10y + 5 = 0$
11. $9x^2 + 4y^2 - 18x + 8y + 4 = 0$
12. $4x^2 - 9y^2 + 8x + 18y + 4 = 0$
13. $3x^2 + 4y^2 - 12x + 8y + 4 = 0$
14. $x^2 + 6x - 8y + 1 = 0$
15. $3x^2 - 2y^2 + 6x - 8y - 17 = 0$
16. $2x^2 + 3y^2 - 8x - 6y + 11 = 0$
17. $4x^2 - 3y^2 + 8x + 12y - 8 = 0$

In Problems 18 through 23, find the equation of the graph indicated; sketch the curve.

18. Parabola: vertex at $(2, 1)$; directrix $y = 3$
19. Ellipse: vertices at $(2, -3)$ and $(2, 5)$; foci at $(2, -2)$ and $(2, 4)$
20. Ellipse: foci at $(-2, 3)$ and $(4, 3)$; length of major axis is 10
21. Hyperbola: vertices at $(-3, 1)$ and $(5, 1)$; passing through $(-5, 3)$
22. Hyperbola: vertices at $(2, -5)$ and $(2, 7)$; foci at $(2, -6)$ and $(2, 8)$
23. Parabola: axis $y = -1$; directrix $x = 2$; focus $(5, -1)$

In Problems 24 through 28, find the equation of the given graph. Identify the curve.

24. The graph of all points whose distances from $(4, 3)$ equal their distances from the line $x = 6$
25. The graph of all points whose distances from $(2, 1)$ are equal to one-half their distances from the line $y = -2$
26. The graph of all points whose distances from $(-1, 2)$ are equal to twice their distances from the line $x = -4$
27. The graph of all points whose distances from $(2, -3)$ are equal to twice their distances from $(-4, 3)$
28. The graph of all points whose distances from $(1, 2)$ are equal to their distances from the line $3x - 4y - 5 = 0$
29. By referring to Fig. 8–29, establish formulas (1) on page 278. Use directed distances and verify that the result is the same if P in Fig. 8–29 is located in the second, third, or fourth quadrants.
30. Prove Theorem 2.

9. ROTATION OF AXES. THE GENERAL EQUATION OF THE SECOND DEGREE

Suppose we make a transformation of coordinates from an xy system to an $x'y'$ system in the following way. The origin is kept fixed, and the x' and y' axes are obtained rotating the x and y axes counterclockwise an amount θ, as shown in Fig. 8–32. Every point P will have coordinates (x, y) with respect to the original system and coordinates (x', y') with respect to the new system. We now find the relationship between (x, y) and (x', y'). Draw the lines OP, AP, and BP, as shown in Fig. 8–33, and note that $x = \overline{OA}$, $y = \overline{AP}$, $x' = \overline{OB}$, $y' = \overline{BP}$. We have

$$x = \overline{OA} = \overline{OP}\cos(\theta + \alpha), \qquad y = \overline{AP} = \overline{OP}\sin(\theta + \alpha).$$

Recalling from trigonometry* the formula for the sine and cosine of the sum of

---

*A review of all the basic formulas in trigonometry may be found in Appendix 1 at the end of the book.

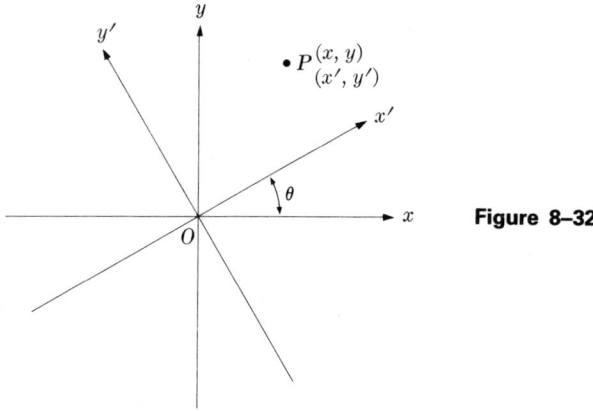

Figure 8–32

two angles, we obtain

$$x = \overline{OP} \cos \theta \cos \alpha - \overline{OP} \sin \theta \sin \alpha,$$
$$y = \overline{OP} \sin \theta \cos \alpha + \overline{OP} \cos \theta \sin \alpha.$$

But $\overline{OP} \cos \alpha = \overline{OB} = x'$ and $\overline{OP} \sin \alpha = \overline{BP} = y'$. Therefore

$$x = x' \cos \theta - y' \sin \theta, \qquad y = x' \sin \theta + y' \cos \theta.$$

When two coordinate systems, xy and $x'y'$, satisfy these equations, we say that they are related by a **rotation of axes;** more specifically, we can say that the x' and y' axes are obtained by rotating the x and y axes through the angle θ.

To express x' and y' in terms of x and y, we merely solve the above equations for x' and y'. Multiplying the first equation by $\cos \theta$, the second by $\sin \theta$, and adding, we find

$$x \cos \theta + y \sin \theta = x'(\cos^2 \theta + \sin^2 \theta) \qquad \text{or} \qquad x' = x \cos \theta + y \sin \theta.$$

Similarly, we get

$$y' = -x \sin \theta + y \cos \theta.$$

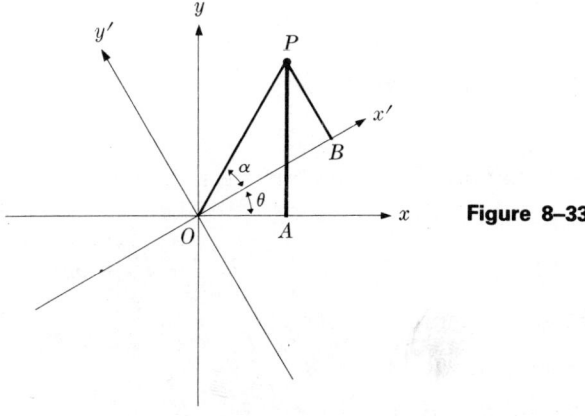

Figure 8–33

Example 1. A Cartesian coordinate system is rotated 60°. Find the coordinates of the point $P(3, -1)$ in the new system. What is the equation of the line $2x - 3y + 1 = 0$ in the rotated system?

Solution. We have $\sin 60° = \frac{1}{2}\sqrt{3}$, $\cos 60° = \frac{1}{2}$. The equations relating the xy system and the $x'y'$ system are

$$x' = \tfrac{1}{2}x + \tfrac{1}{2}\sqrt{3}\,y, \qquad y' = -\tfrac{1}{2}\sqrt{3}\,x + \tfrac{1}{2}y.$$

The coordinates of P are $x' = \tfrac{3}{2} - \tfrac{1}{2}\sqrt{3}$, $y' = -\tfrac{3}{2}\sqrt{3} - \tfrac{1}{2}$. The relationships giving the xy system in terms of the $x'y'$ system are

$$x = \tfrac{1}{2}x' - \tfrac{1}{2}\sqrt{3}\,y', \qquad y = \tfrac{1}{2}\sqrt{3}\,x' + \tfrac{1}{2}y';$$

and so the equation of the line in the new system is

$$2(\tfrac{1}{2}x' - \tfrac{1}{2}\sqrt{3}\,y') - 3(\tfrac{1}{2}\sqrt{3}\,x' + \tfrac{1}{2}y') + 1 = 0,$$

$$\Leftrightarrow (1 - \tfrac{3}{2}\sqrt{3})x' - (\tfrac{3}{2} + \sqrt{3})y' + 1 = 0. \qquad \blacktriangleleft$$

We recall that the most general equation of the second degree has the form

$$\boxed{Ax^2 + Bxy + Cy^2 + Dx + Ey + F = 0 \qquad (A, B, C \text{ not all zero}).}$$

In Section 8 we showed (when $B = 0$) how the axes could be translated so that in the new system $D = E = 0$ (except in the case of parabolas, when we could make only one of them zero).

Now we shall show that *it is always possible to rotate the coordinates in such a way that in the new system there is no $x'y'$ term*. To do this we take the equations

$$x = x' \cos \theta - y' \sin \theta, \qquad y = x' \sin \theta + y' \cos \theta,$$

and we substitute in the general equation of the second degree. Then we have

$$Ax^2 = A(x' \cos \theta - y' \sin \theta)^2,$$
$$Bxy = B(x' \cos \theta - y' \sin \theta)(x' \sin \theta + y' \cos \theta),$$
$$Cy^2 = C(x' \sin \theta + y' \cos \theta)^2,$$
$$Dx = D(x' \cos \theta - y' \sin \theta),$$
$$Ey = E(x' \sin \theta + y' \cos \theta),$$
$$F = F.$$

We add these equations and obtain (after multiplying out the right side)

$$Ax^2 + Bxy + Cy^2 + Dx + Ey + F$$
$$= A'x'^2 + B'x'y' + C'y'^2 + D'x' + E'y' + F' = 0,$$

where A', B', C', D', E', and F' are the abbreviations of

$$A' = A \cos^2 \theta + B \sin \theta \cos \theta + C \sin^2 \theta,$$
$$B' = 2(C - A) \sin \theta \cos \theta + B(\cos^2 \theta - \sin^2 \theta),$$

$$C' = A \sin^2 \theta - B \sin \theta \cos \theta + C \cos^2 \theta,$$
$$D' = D \cos \theta + E \sin \theta,$$
$$E' = -D \sin \theta + E \cos \theta,$$
$$F' = F.$$

Our purpose is to select θ so that the $x'y'$ term is missing, in order that B' will be zero. Let us set B' equal to zero and see what happens:

$$2(C - A) \sin \theta \cos \theta + B(\cos^2 \theta - \sin^2 \theta) = 0.$$

We recall from trigonometry the double-angle formulas

$$\sin 2\theta = 2 \sin \theta \cos \theta, \qquad \cos 2\theta = \cos^2 \theta - \sin^2 \theta,$$

and we write

$$(C - A) \sin 2\theta + B \cos 2\theta = 0$$

or

$$\boxed{\cot 2\theta = \frac{A - C}{B}.}$$

In other words, if we select θ so that $\cot 2\theta = (A - C)/B$ we will obtain $B' = 0$. There is always an angle 2θ between 0 and π which solves this equation.

The next example will illustrate the process.

Example 2. Given the equation $8x^2 - 4xy + 5y^2 = 36$. Choose new axes by rotation so as to eliminate the $x'y'$ term; sketch the curve and locate the principal quantities.

Solution. We have $A = 8$, $B = -4$, $C = 5$, $D = E = 0$, $F = -36$. We select $\cot 2\theta = (A - C)/B = (8 - 5)/(-4) = -\frac{3}{4}$. This means that 2θ is in the second quadrant and $\cos 2\theta = -\frac{3}{5}$.

We remember from trigonometry the half-angle formulas

$$\sin \theta = \sqrt{(1 - \cos 2\theta)/2}, \qquad \cos \theta = \sqrt{(1 + \cos 2\theta)/2},$$

and we use these to get

$$\sin \theta = \sqrt{\tfrac{4}{5}}, \qquad \cos \theta = \sqrt{\tfrac{1}{5}},$$

which gives the rotation

$$x = \frac{1}{\sqrt{5}} x' - \frac{2}{\sqrt{5}} y' = \frac{x' - 2y'}{\sqrt{5}}, \qquad y = \frac{2}{\sqrt{5}} x' + \frac{1}{\sqrt{5}} y' = \frac{2x' + y'}{\sqrt{5}}.$$

Substituting in the given equation, we obtain

$$\tfrac{8}{5}(x' - 2y')^2 - \tfrac{4}{5}(x' - 2y')(2x' + y') + \tfrac{5}{5}(2x' + y')^2 = 36;$$

286 LINES, CIRCLES, CONICS

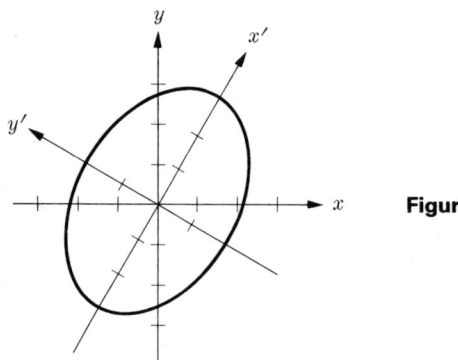

Figure 8-34

when we multiply out, we find that

$$4x'^2 + 9y'^2 = 36 \quad \Leftrightarrow \quad \frac{x'^2}{9} + \frac{y'^2}{4} = 1.$$

The graph is an ellipse, which is sketched in Fig. 8-34. The $x'y'$ coordinates of the vertices are at $(\pm 3, 0)$, and the foci are at $(\pm\sqrt{5}, 0)$. The eccentricity is $\sqrt{5}/3$. In the original system, the vertices are at

$$(3/\sqrt{5}, 6/\sqrt{5}), (-3/\sqrt{5}, -6/\sqrt{5}).$$

The foci are at $(1, 2)$, $(-1, -2)$. ◀

In Example 2 the coefficients D and E are zero. When a problem arises with B, D, and E all different from zero, we may eliminate them by performing in succession a rotation (eliminating B) and a translation (eliminating the D and E terms). Here we have an example of the way processes develop in mathematics: a complicated structure is erected by combining simple building blocks. The technique of performing several transformations in succession occurs frequently in mathematical problems.

Example 3. Given the equation

$$4x^2 - 12xy + 9y^2 - 52x + 26y + 81 = 0,$$

reduce it to standard form by eliminating B, D, and E. Identify the curve and locate the principal quantities.

Solution. We first rotate and choose θ so that

$$\cot 2\theta = \frac{A - C}{B} = \frac{4 - 9}{-12} = \frac{5}{12}.$$

Then 2θ is in the first quadrant and $\cos 2\theta = \frac{5}{13}$. From the half-angle formulas (as in Example 2),

$$\cos \theta = \sqrt{\tfrac{9}{13}}, \quad \sin \theta = \sqrt{\tfrac{4}{13}}.$$

The desired rotation of coordinates is

$$x = \frac{1}{\sqrt{13}}(3x' - 2y'), \qquad y = \frac{1}{\sqrt{13}}(2x' + 3y').$$

Substituting in the given equation, we obtain

$$\tfrac{4}{13}(9x'^2 - 12x'y' + 4y'^2) - \tfrac{12}{13}(6x'^2 + 5x'y' - 6y'^2)$$
$$+ \tfrac{9}{13}(4x'^2 + 12x'y' + 9y'^2) - 4\sqrt{13}(3x' - 2y')$$
$$+ 2\sqrt{13}(2x' + 3y') + 81 = 0,$$

and, after simplification,

$$13y'^2 - 8\sqrt{13}\,x' + 14\sqrt{13}\,y' + 81 = 0.$$

Note that in the process of eliminating the $x'y'$ term we also eliminated the x'^2 term. We now realize that the curve must be a parabola. To translate the coordinates properly, we complete the square in y. This yields

$$13(y'^2 + \tfrac{14}{13}\sqrt{13}\,y' \qquad) = 8\sqrt{13}\,x' - 81$$
$$\Leftrightarrow \quad 13\left(y' + \frac{7}{\sqrt{13}}\right)^2 = 8\sqrt{13}\left(x' - \frac{4}{\sqrt{13}}\right).$$

The translation of axes,

$$x'' = x' - \frac{4}{\sqrt{13}}, \qquad y'' = y' + \frac{7}{\sqrt{13}},$$

leads to the equation

$$y''^2 = \frac{8}{\sqrt{13}}\,x''.$$

In the $x''y''$ system, $p = 4/\sqrt{13}$, the vertex is at the origin, the focus is at $(2/\sqrt{13}, 0)$, the directrix is the line $x'' = -2/\sqrt{13}$, and the x'' axis is the axis of the parabola. In the $x'y'$ system, the focus is at $(6/\sqrt{13}, -7/\sqrt{13})$, and the directrix is the line $x' = 2/\sqrt{13}$. As for the original xy system, we find that the focus is at $(\tfrac{32}{13}, -\tfrac{9}{13})$. The directrix is the line $3x + 2y = 2$. The parabola and all sets of axes are sketched in Fig. 8–35; the points $(3, 2)$ and $(5, 1)$ in the sketch are merely aids in plotting the axes. ◀

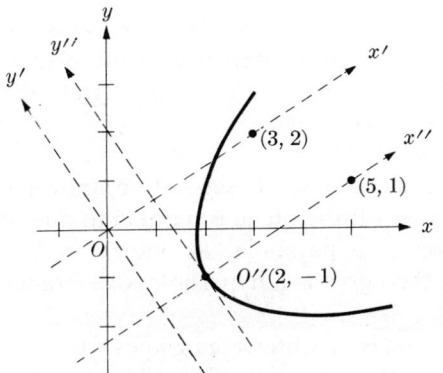

Figure 8–35

In a rotation of the coordinates, the general equation of the second degree,

$$Ax^2 + Bxy + Cy^2 + Dx + Ey + F = 0,$$

goes into the equation

$$A'x^2 + B'xy + C'y^2 + D'x + E'y + F = 0,$$

with

$$A' = A\cos^2\theta + B\sin\theta\cos\theta + C\sin^2\theta,$$
$$B' = 2(C - A)\sin\theta\cos\theta + B(\cos^2\theta - \sin^2\theta),$$
$$C' = A\sin^2\theta - B\sin\theta\cos\theta + C\cos^2\theta.$$

The quantity $A' + C'$, when calculated, is

$$A' + C' = A(\cos^2\theta + \sin^2\theta) + C(\sin^2\theta + \cos^2\theta) = A + C.$$

In other words, even though A changes to A' and C changes to C' when a rotation through *any* angle is made, the quantity $A + C$ does not change at all. We say that $A + C$ is **invariant** under a rotation of coordinates.

If we compute the expression $B'^2 - 4A'C'$ (a tedious computation) and use some trigonometry, we find that

$$B'^2 - 4A'C' = B^2 - 4AC.$$

In other words, $B^2 - 4AC$ is also an *invariant* under rotation of axes.

It can readily be checked that $A + C$ and $B^2 - 4AC$ are invariant under translation of axes, and therefore we can formulate the following theorem for *general equations of the second degree.*

Theorem 3. (a) *If $B^2 - 4AC < 0$, the curve is an ellipse, a circle, a point, or there is no curve.* (b) *If $B^2 - 4AC > 0$, the curve is a hyperbola or two intersecting straight lines.* (c) *If $B^2 - 4AC = 0$, the curve is a parabola, two parallel lines, one line, or there is no curve.*

The circle, ellipse, parabola, and hyperbola are often called **conic sections,** for all of them can be obtained as sections cut from a right circular cone by planes. The cone is thought of as extending indefinitely on both sides of its vertex; the part of the cone on one side of the vertex is called a **nappe.**

If the plane intersects only one nappe, as in Fig. 8–36(a), the curve of the intersection is an ellipse. (A circle is a special case of the ellipse, and occurs when the plane is perpendicular to the axis of the cone.) If the plane is parallel to one of the generators of the cone, the intersection is a parabola, as shown in Fig. 8–36(b). If the plane intersects both nappes, the curve is a hyperbola, one branch coming from each nappe, as in Fig. 8–36(c).

Certain degenerate cases also occur; the graph is two intersecting lines when the plane intersects both nappes and also passes through the vertex. If the plane

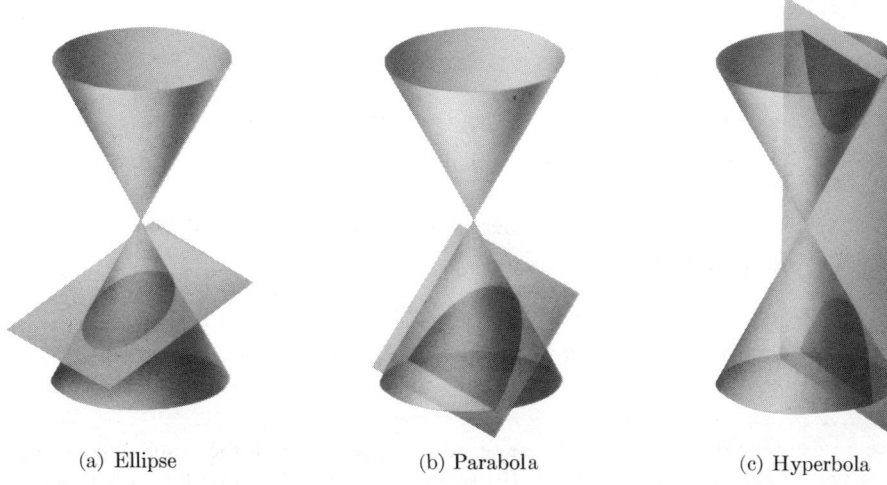

(a) Ellipse (b) Parabola (c) Hyperbola

Figure 8–36

contains one of the generators, the graph of the intersection is a single straight line. Finally, the graph is a single point if the plane contains the vertex and does not intersect either nappe of the cone. The degenerate graph of two parallel lines cannot be obtained as a plane section of a cone.

PROBLEMS

In Problems 1 through 12, in each case change from an xy system to an $x'y'$ system such that the $x'y'$ term is missing. For the ellipses and hyperbolas, find the coordinates of the vertices. For each parabola find the xy coordinates of the focus and the xy equation of the directrix. If the graph consists of lines, find their xy equations.

1. $x^2 + 4xy + 4y^2 = 9$
2. $7x^2 - 4xy + 4y^2 = 240$
3. $2x^2 + 3xy - 2y^2 = 25$
4. $7x^2 - 6xy - y^2 = 0$
5. $x^2 + 2xy + y^2 - 8x + 8y = 0$
6. $x^2 + 4xy + y^2 + 32 = 0$
7. $8x^2 + 12xy + 13y^2 = 884$
8. $x^2 - 4xy + 4y^2 - 40x - 20y = 0$
9. $x^2 - 2xy = 10$
10. $xy = 4$
11. $2xy - 3y^2 = 5$
12. $xy = -3$

In Problems 13 through 22, in each case rotate the axes to an $x'y'$ system such that the $x'y'$ term is missing. Then translate the axes so that the first-degree terms are absent. Sketch the graphs and identify the principal quantities.

13. $3x^2 + 10xy + 3y^2 - 2x - 14y - 5 = 0$
14. $4x^2 - 8xy - 2y^2 + 20x - 4y + 15 = 0$
15. $4x^2 + 4xy + y^2 - 24x + 38y - 139 = 0$
16. $16x^2 - 24xy + 9y^2 + 56x - 42y + 49 = 0$
17. $12x^2 + 24xy + 19y^2 - 12x - 40y + 31 = 0$
18. $4x^2 - 4\sqrt{5}\,xy + 5y^2 + 8x - 4\sqrt{5}\,y - 21 = 0$
19. $31x^2 + 10\sqrt{3}\,xy + 21y^2 - (124 - 40\sqrt{3})x + (168 - 20\sqrt{3})y + 316 - 80\sqrt{3} = 0$

20. $x^2 + 2xy + y^2 - 4x - 4y + 4 = 0$
21. $x^2 - \sqrt{3}\, xy + 2\sqrt{3}\, x - 3y - 3 = 0$
22. $3xy - 4y^2 + x - 2y + 1 = 0$
23. Prove that a second-degree equation with an xy term in it can never represent a circle.
24. Prove that a second-degree equation with D and E absent (i.e., no x and y terms) cannot be a parabola.
25. Prove that $B^2 - 4AC$ is invariant under rotation and translation of axes.
26. Given the transformation of coordinates

$$x' = ax + by, \qquad y' = cx + dy,$$

with a, b, c, and d numbers such that $ad - bc$ is positive. If the general equation of the second degree undergoes such a transformation, what can be said about $B^2 - 4AC$?

27. Give an example of an equation of the second degree such that the graph degenerates to (a) two intersecting straight lines; (b) two parallel lines; (c) one line; (d) a point.
28. Prove Theorem 3.

9

THE TRIGONOMETRIC AND EXPONENTIAL FUNCTIONS

1. SOME SPECIAL LIMITS

In Chapter 4 we learned several techniques for evaluating the limits of various algebraic expressions. We shall now employ a new device for finding the limit in a special case. We consider the function

$$f(\theta) = \frac{\sin \theta}{\theta}$$

and suppose that θ is measured in *radians*. This function is defined for all values of θ except $\theta = 0$. Since $\sin 0 = 0$, straight substitution gives $f(0) = 0/0$, a meaningless expression. Nevertheless, we can find the value of

$$\lim_{\theta \to 0} \frac{\sin \theta}{\theta}.$$

We shall start by supposing that θ is positive and less than $\pi/2$. Figure 9-1 shows a typical value of θ. We draw an arc $\widehat{IP}$ of a circle of radius 1 and construct the perpendiculars PQ and TI, as shown. From this figure, it is clear that

$$\text{area } \triangle IOP < \text{area sector } IOP < \text{area } \triangle IOT.$$

As usual, we use the symbol $|AB|$ to denote the length (always positive) of a line segment AB. Since $|OI| = |OP| = 1$, we have

$$\text{area } \triangle IOP = \tfrac{1}{2}|PQ| \cdot |OI| = \tfrac{1}{2}|PQ|;$$
$$\text{area sector } IOP = \tfrac{1}{2}\theta r^2 = \tfrac{1}{2}\theta \ (\theta \text{ in radians});$$
$$\text{area } \triangle IOT = \tfrac{1}{2}|IT| \cdot |OI| = \tfrac{1}{2}|IT|.$$

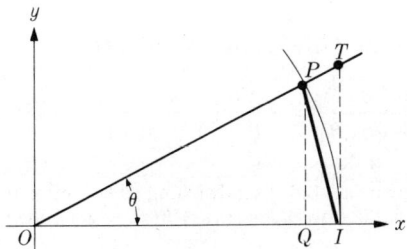

Figure 9-1

On the other hand, we know that

$$\sin\theta = \frac{|PQ|}{|OP|} = |PQ|; \quad \tan\theta = \frac{|IT|}{|OI|} = |IT|.$$

Therefore the above area inequalities are the same as the inequalities

$$\tfrac{1}{2}\sin\theta < \tfrac{1}{2}\theta < \tfrac{1}{2}\tan\theta.$$

We multiply by 2 and divide by $\sin\theta$. (This keeps the same direction for the inequalities, since everything is positive.) We obtain

$$1 < \frac{\theta}{\sin\theta} < \frac{1}{\cos\theta}.$$

Now we take reciprocals; this has the effect of reversing the direction. (Recall that if a and b are any positive numbers and $a > b$, then $1/a < 1/b$.) We conclude that

$$1 > \frac{\sin\theta}{\theta} > \cos\theta.$$

As θ tends to zero, we know that $\cos\theta \to 1$. Since $\sin\theta/\theta$ is always between 1 and a number tending to 1, it also must approach 1 as θ approaches zero.* In the proof we considered θ positive, but now we note that

$$f(\theta) = \frac{\sin\theta}{\theta}$$

is an *even* function of θ. For, if we replace θ by $-\theta$, we have

$$f(-\theta) = \frac{\sin(-\theta)}{-\theta} = \frac{-\sin\theta}{-\theta} = \frac{\sin\theta}{\theta} = f(\theta).$$

Therefore, as $\theta \to 0$ through negative values, the result must be identical with the one obtained when $\theta \to 0$ through positive values. The above development gives the following result.

Theorem 1. *If θ is measured in radians, then*

$$\lim_{\theta \to 0} \frac{\sin\theta}{\theta} = 1.$$

Corollary

$$\lim_{\theta \to 0} \frac{1 - \cos\theta}{\theta} = 0.$$

Proof. This result may be obtained from Theorem 1 by using the identity

$$\frac{1 - \cos\theta}{\theta} = \frac{(1 - \cos\theta)(1 + \cos\theta)}{\theta(1 + \cos\theta)} = \frac{\sin^2\theta}{\theta(1 + \cos\theta)} = \frac{\sin\theta}{\theta} \cdot \frac{\sin\theta}{1 + \cos\theta}.$$

---

* This fact, intuitively clear, follows from a result known as the "sandwiching theorem": if $\lim_{x \to a} f(x) = L$, $\lim_{x \to a} g(x) = L$, and if for all x near the value a we have $f(x) \leq h(x) \leq g(x)$, then $\lim_{x \to a} h(x) = L$.

We know that
$$\lim_{\theta \to 0} \frac{\sin \theta}{\theta} = 1 \quad \text{and} \quad \lim_{\theta \to 0} \frac{\sin \theta}{1 + \cos \theta} = 0,$$
since the second limit may be obtained by straight substitution. From the fact that the limit of a product is the product of the limits, we obtain the result given by the corollary.

Example 1. Find the value of
$$\lim_{\theta \to 0} \frac{\sin 2\theta}{\theta},$$
θ measured in radians.

Solution. We write
$$\frac{\sin 2\theta}{\theta} = \frac{2 \sin (2\theta)}{(2\theta)}.$$
Therefore
$$\lim_{\theta \to 0} \frac{\sin 2\theta}{\theta} = 2 \lim_{(2\theta) \to 0} \frac{\sin (2\theta)}{(2\theta)} = 2 \cdot 1.$$
We made use of the fact that when $\theta \to 0$, so does 2θ. ◀

Example 2. Find the value of
$$\lim_{\alpha \to 0} \frac{\sin \alpha}{\alpha},$$
α measured in degrees.

Solution. The functions $\sin \alpha°$ and $\sin (\pi\alpha/180)$ have the same values, with α the measure in degrees. We have
$$\lim_{\alpha \to 0} \frac{\sin \alpha}{\alpha} = \lim_{\alpha \to 0} \frac{\pi}{180} \frac{\sin \frac{\pi\alpha}{180}}{\frac{\pi\alpha}{180}} = \frac{\pi}{180} \lim_{\frac{\pi\alpha}{180} \to 0} \frac{\sin \frac{\pi\alpha}{180}}{\frac{\pi\alpha}{180}} = \frac{\pi}{180}.$$
◀

Example 2 exhibits one of the reasons why radian measure is used throughout calculus. The limit of $\sin \theta/\theta$ is 1 if θ is measured in radians and $\pi/180$ if θ is measured in degrees. In the next section we shall see why the first is highly preferable to the second.

2. THE DIFFERENTIATION OF TRIGONOMETRIC FUNCTIONS

With the help of the limits obtained in the preceding section, we can find the derivatives of all the trigonometric functions.

Theorem 2. *If $f(x) = \sin x$, then $f'(x) = \cos x$.*

Proof. We apply the three-step rule.

Step 1: $f(x + h) - f(x) = \sin(x + h) - \sin x.$

Step 2: $\dfrac{f(x + h) - f(x)}{h} = \dfrac{\sin(x + h) - \sin x}{h}.$

As usual, it is at this point that we must employ ingenuity. We first use the fact that $\sin(x + h) = \sin x \cos h + \cos x \sin h$ and write

$$\frac{f(x + h) - f(x)}{h} = \frac{\sin x \cos h + \cos x \sin h - \sin x}{h}$$

$$= \cos x \frac{\sin h}{h} + \sin x \frac{\cos h - 1}{h}.$$

Then, taking the limit as $h \to 0$, we use Theorem 1 and the Corollary of the preceding section to perform Step 3 and conclude that

$$f'(x) = \lim_{h \to 0} \frac{f(x + h) - f(x)}{h} = \cos x.$$

The simplicity of the result—that the derivative of $\sin x$ is $\cos x$—is due to the use of radians. If, for example, degrees were used, then $\lim_{h \to 0} (\sin h/h)$ would be $\pi/180$, and all the derivative formulas for trigonometric functions would have an extra, inconvenient factor in them. Later we shall see that a similar situation occurs in the study of logarithms.

Suppose that u is a function of x; that is, $u = u(x)$. Then, by using the Chain Rule, we easily find the formula for the derivative of $\sin u$. If we write

$$y = \sin u, \quad \text{then} \quad \frac{dy}{dx} = \cos u \frac{du}{dx}.$$

In differentials this becomes

$$d \sin u = \cos u \, du.$$

Example 1. Find the derivative of $y = \sin(3x^2 - 6x + 1)$.

Solution. Taking $u = 3x^2 - 6x + 1$, we obtain

$$\frac{dy}{dx} = \cos(3x^2 - 6x + 1) \cdot (6x - 6) = 6(x - 1) \cos(3x^2 - 6x + 1). \quad \blacktriangleleft$$

Theorem 3. *If $f(x) = \cos x$, then $f'(x) = -\sin x$.*

Proof. We can obtain this result by the three-step rule. However, by writing

$$f(x) = \cos x = \sin\left(\frac{\pi}{2} - x\right),$$

we may apply the Chain Rule and obtain the result more simply. We find

$$f'(x) = \frac{d}{dx} \sin\left(\frac{\pi}{2} - x\right) = \cos\left(\frac{\pi}{2} - x\right)(-1), \quad \text{since} \quad \frac{d}{dx}\left(\frac{\pi}{2} - x\right) = -1.$$

Therefore
$$f'(x) = -\cos\left(\frac{\pi}{2} - x\right) = -\sin x.$$

The Chain Rule again shows that if

$$y = \cos u, \quad \text{then} \quad \frac{dy}{dx} = -\sin u \frac{du}{dx}$$

and, in differentials,

$$d\cos u = -\sin u\, du.$$

With the help of the derivatives of the sine and cosine functions, we may find the derivatives of the remaining trigonometric functions. The following example shows the method.

Example 2. Find the derivative of $y = \tan x$.

Solution. We write
$$y = \frac{\sin x}{\cos x}$$
and use the formula for the derivative of a quotient. We have
$$\frac{dy}{dx} = \frac{\cos x\,(\cos x) - \sin x\,(-\sin x)}{\cos^2 x}.$$

This yields
$$\frac{dy}{dx} = \frac{1}{\cos^2 x} = \sec^2 x. \quad \blacktriangleleft$$

The formulas for differentiating trigonometric functions are summarized in the following table (expressed in terms of differentials):

$d\sin u = \cos u\, du$	$d\cot u = -\csc^2 u\, du$
$d\cos u = -\sin u\, du$	$d\sec u = \sec u \tan u\, du$
$d\tan u = \sec^2 u\, du$	$d\csc u = -\csc u \cot u\, du$

The general rules for differentiating polynomials and rational functions, together with the formulas for products and quotients, open up vast possibilities for differentiating complicated expressions. The next examples show some of the ramifications.

Example 3. Given that $f(x) = \sin^3 x$, find $f'(x)$.

Solution. $f(x) = (\sin x)^3$, and so
$$f'(x) = 3(\sin x)^2 \cdot \frac{d(\sin x)}{dx} = 3\sin^2 x \cos x. \quad \blacktriangleleft$$

Example 4. Given that $y = \tan^4(2x + 1)$, find dy/dx.

Solution. $y = [\tan(2x + 1)]^4$;

$$\frac{dy}{dx} = 4[\tan(2x + 1)]^3 \cdot \frac{d}{dx}\tan(2x + 1).$$

But

$$\frac{d}{dx}\tan(2x + 1) = \sec^2(2x + 1) \cdot \frac{d}{dx}(2x + 1) = \sec^2(2x + 1) \cdot (2).$$

Therefore

$$\frac{dy}{dx} = 4\tan^3(2x + 1) \cdot 2\sec^2(2x + 1) = 8\tan^3(2x + 1)\sec^2(2x + 1). \quad \blacktriangleleft$$

Example 5. Given that $f(x) = x^3 \sec^2 3x$, find $f'(x)$.

Solution. $f(x) = x^3[\sec(3x)]^2$. We see that this is in the form of a product, and write

$$f'(x) = x^3 \cdot 2[\sec(3x)]\frac{d}{dx}(\sec 3x) + [\sec(3x)]^2 \cdot 3x^2$$

$$= 2x^3 \sec 3x \cdot \left[\sec 3x \tan 3x \cdot \frac{d(3x)}{dx}\right] + 3x^2 \sec^2 3x$$

$$= 3x^2(\sec^2 3x)(2x \tan 3x + 1). \quad \blacktriangleleft$$

PROBLEMS

In each of Problems 1 through 12, evaluate the limit (all angles in radians).

1. $\lim\limits_{h \to 0} \dfrac{\sin 5h}{h}$
2. $\lim\limits_{\theta \to 1} \dfrac{\sin 2(\theta - 1)}{\theta - 1}$
3. $\lim\limits_{h \to 0} \dfrac{\sin 2h}{\sin 5h}$
4. $\lim\limits_{h \to 0} \dfrac{h}{\tan h}$
5. $\lim\limits_{\theta \to 0} \dfrac{1 - \cos \theta}{4\theta^2}$
6. $\lim\limits_{\theta \to \pi/2} \dfrac{\theta - \dfrac{\pi}{2}}{\cos\left(\theta - \dfrac{\pi}{2}\right)}$
7. $\lim\limits_{\phi \to 0} \dfrac{1}{\phi \cot \phi}$
8. $\lim\limits_{\theta \to \pi/2} \dfrac{\cos \theta}{(\pi/2) - \theta}$
9. $\lim\limits_{x \to 0} \dfrac{x}{\csc^2 x}$
10. $\lim\limits_{x \to 0} \dfrac{\cot^2 x}{4x^2}$
11. $\lim\limits_{x \to 0} \dfrac{x^3}{\tan^3 2x}$
12. $\lim\limits_{x \to 0} \dfrac{4x^3}{1 - \cos^2 \frac{1}{2}x}$

In Problems 13 through 39, differentiate the trigonometric functions.

13. $f(x) = \tan 3x$
14. $g(y) = 2 \sin y \cos y$
15. $f(s) = \sec(s^2)$
16. $\phi(t) = \sin^2 3t$
17. $f(y) = \csc^3 \frac{1}{3}y$
18. $g(x) = \sin 2x - \frac{1}{3}(\sin^3 2x)$
19. $f(x) = 2x + \cot 2x$
20. $h(u) = u \sin u + \cos u$
21. $x(s) = \sqrt{\tan 2s}$
22. $h(t) = \sec^2 2t - \tan^2 2t$
23. $f(x) = x^2 \tan^2(x/2)$
24. $g(x) = (\tan^3 4x)/3 + \tan 4x$
25. $f(x) = \frac{1}{3}(\sec^3 2x) - \sec 2x$
26. $f(x) = \sqrt{1 + \cos x}$

27. $f(x) = (\sin 2x)/x^2$
28. $f(x) = (\sin 2x)/x^5$
29. $f(x) = (\cos^2 3x)/(1 + x^2)$
30. $f(x) = \sin^3 2x \tan^2 3x$
31. $f(x) = \cos^2 3x \csc^3 2x$
32. $f(x) = (\sin 2x)/(1 + \cos 2x)$
33. $f(x) = (\cot^2 ax)/(1 + x^2)$
34. $f(x) = (1 + \tan 3x)^{2/3}$
35. $f(x) = (1 + \sin^2 2x)^{1/2}$
36. $f(x) = \sin nx \sin^n x$
37. $f: x \to \sin(\cos x)$
38. $f: x \to \dfrac{\cot(x/2)}{\sqrt{1 - \cot^2(x/2)}}$
39. $f(x) = \dfrac{(1 + \cos^2 3x)^{1/3}}{(1 + x^2)^{1/2}}$

40. If $f(x) = \cos x$ prove, by using the three-step rule, that $f'(x) = -\sin x$.
41. Prove that if $f(x) = \cot x$, then $f'(x) = -\csc^2 x$.
42. Prove that if $f(x) = \sec x$, then $f'(x) = \sec x \tan x$.
43. Prove that if $f(x) = \csc x$, then $f'(x) = -\csc x \cot x$.
44. If $f(x) = \sin \lambda x$, show that $f''(x) + \lambda^2 f(x) = 0$.

In each of Problems 45 through 48 find the relative maximum and minimum points of the graph.

45. $y = \sin x + \cos x$
46. $y = \sin 3x - 3 \sin x$
47. $y = \sqrt{3} \sin 2x - \cos 2x$
48. $y = 4 \sin^3 x - 3 \sin x$

49. The function $f(x) = (\tan x)^{1/2}$ is defined for $0 \le x < \pi/2$. Show that $f'(x) > 0$ for $0 < x < \pi/2$. Is it true that $f''(x) > 0$ for $0 < x < \pi/2$?

50. Find all the maximum and minimum points of the function

$$f(x) = \sin \frac{\pi}{x}$$

defined for $0 < x \le 1$. (There are infinitely many.)

51. Find all the relative maximum and minimum points of the function

$$f(x) = x \sin \frac{\pi}{x}$$

defined for $0 < x \le 1$. [Answer in terms of x_0, a solution of $\tan u = u$.]

52. Given

$$f(x) = x^p \sin \frac{\pi}{x}$$

for $0 < x \le 1$. Show that if p is any positive number, there are always infinitely many relative maxima and minima. If these points are labeled $(x_1, y_1), (x_2, y_2), \ldots$ in order of decreasing x_i, then it follows that $x_i \to 0$, $y_i \to 0$ as $i \to \infty$.

3. INTEGRATION OF TRIGONOMETRIC FUNCTIONS

Every differentiation formula automatically gives us a corresponding integration formula. From our knowledge of the derivatives of the trigonometric functions, we obtain at once the following indefinite integral formulas:

$$\int \cos u \, du = \sin u + c,$$

$$\int \sin u \, du = -\cos u + c,$$

$$\int \sec^2 u \, du = \tan u + c,$$

$$\int \csc^2 u \, du = -\cot u + c,$$

$$\int \sec u \tan u \, du = \sec u + c,$$

$$\int \csc u \cot u \, du = -\csc u + c.$$

In the above formulas u may be any function of x, and du is the differential of u. These formulas may be combined with the methods of integration of Chapter 7, Section 7, to help us integrate a wide variety of functions. We shall illustrate the technique with some examples.

Example 1. Find $\int \sin 4x \, dx$.

Solution. We let $u = 4x$ and $du = 4 \, dx$. Then the above integral becomes

$$\int \sin u \cdot \tfrac{1}{4} du = \tfrac{1}{4} \int \sin u \, du = -\tfrac{1}{4} \cos u + c = -\tfrac{1}{4} \cos 4x + c. \quad \blacktriangleleft$$

Example 2. Find $\int \tan^2 (2x - 3) \, dx$.

Solution. This does not seem to fit any of the formulas for integration until we recall* the trigonometric identity $1 + \tan^2 \theta = \sec^2 \theta$. Then the above integral becomes

$$\int [\sec^2 (2x - 3) - 1] \, dx = \int \sec^2 (2x - 3) \, dx - \int dx.$$

We now let $u = 2x - 3$, $du = 2 \, dx$, and the first integral on the right becomes

$$\int \sec^2 u \cdot \tfrac{1}{2} du = \tfrac{1}{2} \int \sec^2 u \, du = \tfrac{1}{2} \tan u + c = \tfrac{1}{2} \tan (2x - 3) + c.$$

The second integral gives us merely $x + c$ and so, combining the constants of integration, we have

$$\int \tan^2 (2x - 3) \, dx = \tfrac{1}{2} \tan (2x - 3) - x + c. \quad \blacktriangleleft$$

---

* The basic trigonometric formulas may be found in Appendix 1 at the end of the book.

9-3 INTEGRATION OF TRIGONOMETRIC FUNCTIONS

Example 3. Find $\int \sin^3 x \, dx$.

Solution. We again make use of trigonometric identities and write
$$\sin^3 x = \sin x \sin^2 x = \sin x (1 - \cos^2 x) = \sin x - \sin x \cos^2 x.$$
Therefore
$$\int \sin^3 x \, dx = \int \sin x \, dx - \int \sin x \cos^2 x \, dx.$$

The first integral is easy and gives us $-\cos x + c$. The second integral is a little harder but still manageable, if we let $u = \cos x$. Then $du = -\sin x \, dx$, and we obtain
$$\int \sin x \cos^2 x \, dx = -\int u^2 \, du = -\tfrac{1}{3} u^3 + c = -\tfrac{1}{3} \cos^3 x + c.$$

The final result is
$$\int \sin^3 x \, dx = -\cos x + \tfrac{1}{3} \cos^3 x + c. \quad \blacktriangleleft$$

PROBLEMS

In each of the following problems, perform the integration.

1. $\int \sin 6x \, dx$
2. $\int \cos 4v \, dv$
3. $\int \csc^2 \left(\dfrac{2x + 1}{3} \right) dx$
4. $\int \sec (x - 1) \tan (x - 1) \, dx$
5. $\int \dfrac{\sin 3x}{\cos^3 3x} \, dx$
6. $\int \dfrac{\sec^2 4x}{\tan^7 4x} \, dx$
7. $\int \cot^2 (2x - 6) \, dx$
8. $\int \cos^3 2x \, dx$
9. $\int \sin^5 x \, dx$
10. $\int \cos^5 x \, dx$
11. $\int x \sin (x^2) \, dx$
12. $\int x^2 \sec^2 (4x^3) \, dx$
13. $\int \dfrac{\sin x \, dx}{\cos^5 x}$
14. $\int \cos^6 2x \sin 2x \, dx$
15. $\int \tan^5 x \sec^2 x \, dx$
16. $\int \csc^7 2x \cot 2x \, dx$
17. $\int \dfrac{\csc^2 (3 - 2x)}{\cot^4 (3 - 2x)} \, dx$
18. $\int \cos^2 \tfrac{1}{2} x \, dx$ [*Hint:* Use half-angle formula.]
19. $\int \sin^2 3x \, dx$
20. $\int \sin^4 x \, dx$

21. $\int \sin 3x \cos 5x\, dx$ [See *Hint* in Problem 29 below.]

22. $\int \cos 6x \cos 9x\, dx$ [Same hint as for Problem 21.]

In each of Problems 23 through 28 find the area of the region R as described.

23. $R = \left\{(x, y): 0 \le x \le \dfrac{\pi}{2},\ 0 \le y \le \sin x\right\}$

24. $R = \left\{(x, y): 0 \le x \le \dfrac{\pi}{3},\ 0 \le y \le \tan^2 x\right\}$

25. $R = \left\{(x, y): \dfrac{\pi}{6} \le x \le \dfrac{\pi}{3},\ 0 \le y \le \sec^2 x\right\}$

26. $R = \left\{(x, y): 0 \le x \le \dfrac{\pi}{6},\ \sin x \le y \le \cos x\right\}$

27. $R = \left\{(x, y): 0 \le x \le \dfrac{\pi}{3},\ \sin^3 x \le y \le \sin x\right\}$

28. $R = \left\{(x, y): \dfrac{\pi}{2} \le x \le \pi,\ 0 \le y \le \sin^2 x\right\}$

29. If m and n are positive integers, show that

a) $\displaystyle\int_{-\pi}^{\pi} \sin mx \sin nx\, dx = \int_{-\pi}^{\pi} \cos mx \cos nx\, dx = \begin{cases} 0 & \text{if } m \ne n, \\ \pi & \text{if } m = n; \end{cases}$

b) $\displaystyle\int_{-\pi}^{\pi} \sin mx \cos nx\, dx = 0$ for all m, n.

[*Hint:* Use the formulas

$$\sin A \sin B = \tfrac{1}{2}[\cos(A - B) - \cos(A + B)]$$
$$\cos A \cos B = \tfrac{1}{2}[\cos(A - B) + \cos(A + B)]$$
$$\sin A \cos B = \tfrac{1}{2}[\sin(A - B) + \sin(A + B)].]$$

4. RELATIONS AND INVERSE FUNCTIONS*

So far we have been concerned with functions from R_1 to R_1, although the conic sections discussed in Chapter 8 provided examples of relations in R_2. Before discussing inverse functions, the principal topic of this section, let us repeat several of the definitions concerning relations which were given in Chapter 2, Section 4.

Definitions. *Any set in the number plane R_2 is called a **relation**. More specifically, a set in R_2 is called a **relation from R_1 to R_1**. The **domain** of a relation is the set of all x such that (x, y) is in the set of points comprising the relation. The set of all y such that (x, y) is in the relation is called the **range** of the relation.*

---

* Except for the definition of inverse, the material in this Section may be omitted without loss of continuity.

The distinction between a function and a relation is simply that for a function, to each element of the domain there corresponds exactly one element of the range. No such requirement is imposed for a relation. However, as is the case for functions, the *domain* of a relation is the projection of the relation on the x axis and the *range* of the relation is the projection on the y axis.

Definition. *Suppose that S is a relation. That is, S is made up of points (x, y) in R_2. We define the* **inverse relation of** *S to consist of all points (x, y) such that (y, x) is a point of S.*

The inverse relation interchanges the role of domain and range. For example, if $(3, 5)$ is in the relation S, then $(5, 3)$ is in the inverse relation. More generally, the domain of the inverse relation is the range of the original relation and the range of the inverse is the domain of the given relation. Also, the inverse of the inverse of a given relation is just the given relation.

A relation is usually defined by an equation involving x and y. According to the definition above, the inverse relation is the solution set of the equation we get when we interchange x and y in the equation. We give four examples.

i) If a relation is defined by
$$x^2 - y^2 = 1,$$
the inverse relation is the solution set of
$$y^2 - x^2 = 1.$$

ii) If a relation is defined by
$$y = x^2,$$
the inverse relation is the solution set of
$$y^2 = x.$$

iii) If a relation is defined by
$$y = (x - 1)^3,$$
the inverse relation is the solution set of
$$(y - 1)^3 = x \quad \Leftrightarrow \quad y = 1 + x^{1/3}.$$

iv) If a relation is defined by
$$y^2 = x^3,$$
the inverse relation is the solution set of
$$y^3 = x^2 \quad \Leftrightarrow \quad y = x^{2/3}.$$

In example (i), neither the original nor the inverse relation is a function. In example (ii), the given relation is a function and the inverse relation is not. In example (iii), both the given relation and its inverse are functions. Finally, in example (iv), the given relation is not a function, but its inverse is. Therefore we see that there is no simple interconnection between inverse relations and functions.

Since a function is a special case of a relation, no additional effort is needed to define the **inverse** of a function. As we saw in example (ii) above, the inverse of a function need not be a function, although it may be one, as example (iii) shows. In general, a function f which has the equation $y = f(x)$ has its inverse given by

$$x = f(y).$$

The inverse of a function is a relation which may consist of several functions. These functions are called the **branches** of the inverse relation. If we can solve the equation $x = f(y)$ for y in terms of x, then we can obtain explicit expressions for the branches of the inverse of f.

Example 1. Given the function $f(x) = x^2 - 2x + 2$, plot the inverse relation of f. Show that the inverse consists of two branches and find explicit formulas for these branches.

Solution. We write $y = x^2 - 2x + 2$ and then the inverse relation is the solution set of the equation

$$x = y^2 - 2y + 2.$$

The graph is the parabola sketched in Fig. 9–2. Solving the above quadratic equation for y, we get the branches

$$y = 1 - \sqrt{x - 1} \equiv g_1(x), \qquad 1 \leq x < \infty,$$
$$y = 1 + \sqrt{x - 1} \equiv g_2(x), \qquad 1 \leq x < \infty. \qquad \blacktriangleleft$$

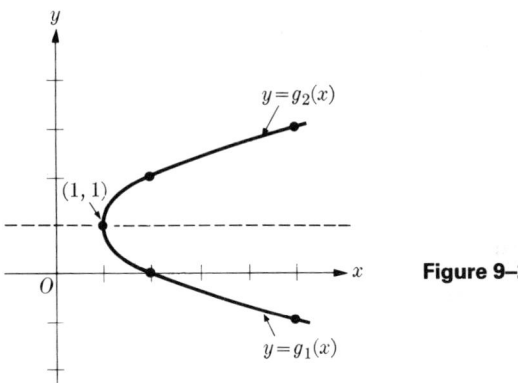

Figure 9–2

Suppose we are given a function $y = f(x)$. There are many times when it is impossible to solve the equation $x = f(y)$ explicitly for y in terms of x. However, it is not always necessary to do this in order to determine whether or not the inverse relation is a function. The next theorem, stated without proof, gives a procedure for deciding when the inverse of a function is also a function.

Theorem 4 (Inverse Function Theorem). *Let f be a continuous, increasing function which has an interval I for domain and has range J (Fig. 9–3). Then f has an*

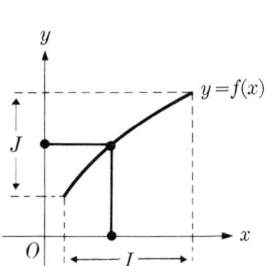
Figure 9–3

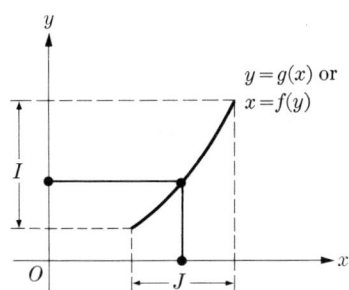
Figure 9–4

inverse g which is a continuous, increasing function with domain J and range I (Fig. 9–4). We have, moreover,

$$g[f(x)] = x \quad \text{for} \quad x \text{ on } I, \tag{1}$$

$$f[g(x)] = x \quad \text{for} \quad x \text{ on } J. \tag{2}$$

If f is decreasing on I instead of increasing, the same result holds with g decreasing on J. If $f'(x) \neq 0$ on the interior of I, then g is differentiable on the interior of J.

Theorem 4 can be given a graphic interpretation. We may think of a function f as an **operator** which carries one object (an element of the domain) into another (the image element of the range). The function f "operates" in that it carries elements of R_1 into elements of R_1. The term **mapping** is also used as a synonym for operation. We also say that f **maps** elements of R_1 (the domain) into elements of R_1 (the range). We may interpret formulas (1) and (2) of Theorem 4 by saying that g is the operator which reverses the action of f. It is the inverse operator or inverse mapping. Whatever action f performs, g undoes it.

If a function f has a derivative, then Theorem 4 may be used to decide whether or not the inverse of f is a function. We merely compute f' and if this quantity is always positive or always negative on the domain of f, then the inverse of f must be a function. The next examples show how this idea may be applied to analyze the inverse of a function.

Example 2. Given the function

$$y = f(x) = \frac{x}{1 + x^2} \quad \text{for} \quad 1 < x < \infty.$$

Determine the range of this function and decide whether or not the inverse of f is a function.

Solution. We first compute the derivative of f:

$$y' = \frac{1 - x^2}{(1 + x^2)^2}.$$

Since the derivative is negative for $1 < x < \infty$, the function is decreasing (see Fig. 9–5). We have $f(1) = \frac{1}{2}$ and $\lim_{x \to \infty} f(x) = 0$. Therefore the range of the function is the interval $(0, \frac{1}{2})$. By the Inverse Function Theorem we know that f has a

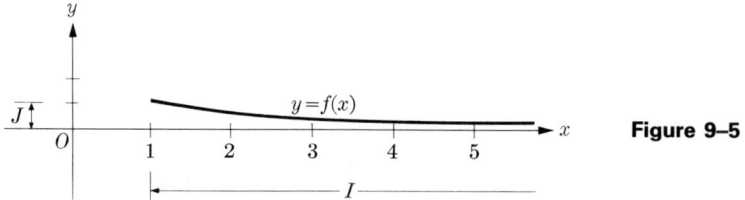

Figure 9–5

continuous inverse function g which has domain $(0, \frac{1}{2})$ and range $(1, \infty)$ and is decreasing; g is determined by the equation

$$x = \frac{y}{1 + y^2} \quad \text{or} \quad xy^2 - y + x = 0.$$

Solving this quadratic equation for y in terms of x and choosing the appropriate branch, we get

$$y = g(x) = \frac{1 + \sqrt{1 - 4x^2}}{2x}, \quad 0 < x < \tfrac{1}{2}. \quad \blacktriangleleft$$

In case f is not always increasing or always decreasing, its inverse may be analyzed as follows: First find the intervals $I_1, I_2, \ldots$ on which f is increasing or decreasing. If $f_1, f_2, \ldots$ denote the parts of f for x on $I_1, I_2, \ldots$, respectively, then the inverse of each of these functions is a function. The various inverse functions which we call $g_1, g_2, \ldots$ may be identified and plotted provided the equation $x = f(y)$ can be solved for y. The next example shows a systematic method for obtaining the branches of the inverse relation.

Example 3. Given the function

$$f(x) = x^3 - 3x + 2,$$

discuss and plot the inverse relation of f, indicating the various functions $g_1, g_2, \ldots$.

Solution. The graph of f is shown in Fig. 9–6. The graph of the inverse is determined by the equation

$$x = f(y) = y^3 - 3y + 2.$$

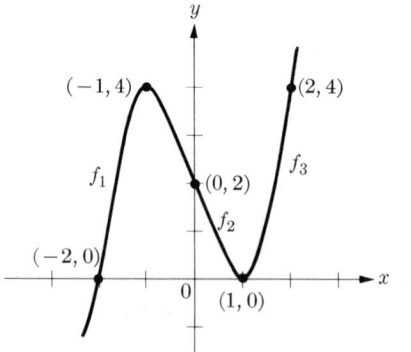

Figure 9–6

We find that
$$f'(y) = 3y^2 - 3.$$
We construct the table

y	-2	-1	0	1	2
$f'(y)$	$+$	0	$-$	0	$+$
$f(y)$	0	4	2	0	4

and conclude that

f is increasing on $I_1 = (-\infty, -1]$ and $J_1 = (-\infty, 4]$,
f is decreasing on $I_2 = [-1, 1]$ and $J_2 = [0, 4]$,
f is increasing on $I_3 = [1, \infty)$ and $J_3 = [0, \infty)$.

The inverse relation of f is plotted in Fig. 9–7. ◀

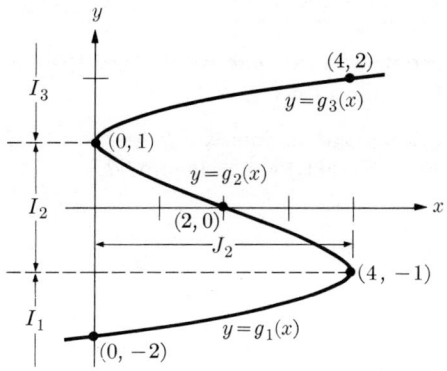

Figure 9–7

PROBLEMS

For each of the functions f given in Problems 1 through 10, determine whether or not there is an inverse function g. If there is, determine the domain of g and find an expression for $g(x)$, if possible. Draw a graph.

1. $f(x) = 3x + 2,\ -\infty < x < \infty$
2. $f(x) = x^2 + 2x - 3,\ 0 \le x < \infty$
3. $f(x) = x^3 + 4x - 5,\ -\infty < x < \infty$
4. $f(x) = \dfrac{x}{x+1},\ -1 < x < \infty$
5. $f(x) = 3 - 2x - x^2,\ -\infty < x < 0$
6. $f(x) = 3 - 2x - x^2,\ -5 < x < 1$
7. $f: x \to 3 - 2x - x^2,\ 0 < x < \infty$
8. $f: x \to \dfrac{x^2}{(x+1)^2},\ -2 < x < 2,\ x \ne -1$
9. $f: x \to \dfrac{1}{x^2},\ 0 < x \le 1$
10. $f: x \to \dfrac{2+x}{2-x},\ 0 \le x < 2$

For each of the functions f given in Problems 11 through 22, find the domains for which the Inverse Function Theorem applies. Find the inverse functions g and give their domains, if possible.

11. $f(x) = x^2 + 4x - 1$
12. $f(x) = 6x - x^2$
13. $f(x) = 2x^2 + 3x + 4$
14. $f(x) = 5x + 6 - 2x^2$
15. $f(x) = \dfrac{x}{x+1}$
16. $f(x) = \dfrac{2x}{1+x^2}$
17. $f(x) = \dfrac{2x-1}{x+2}$
18. $f(x) = x + \dfrac{1}{x}$
19. $f: x \to x^3 + 3x^2 - 9x + 7$
20. $f: x \to 2x^3 + 3x^2 - 3x$
21. $f: x \to x^3 + 3x^2 + 6x - 3$
22. $f: x \to x^4 + \tfrac{4}{3}x^3 - 4x^2 + \tfrac{2}{3}$

23. Assume that f and g in Theorem 4 are differentiable. By using the Chain Rule and implicit differentiation in the equation $f[g(x)] = x$, show that at each point of the domain of f, the derivative of f is the reciprocal of the derivative of g.

24. Assume that f and g in Theorem 4 have second derivatives. If f and g are increasing, show that f'' and g'' have opposite signs at each point. If f and g are decreasing, show that f'' and g'' have the same sign at each point.

25. a) Suppose that f is a differentiable function on an interval I and that g is the inverse function of f. Show that if f' is zero for some value x_0 in I, then g cannot be differentiable at x_0.

    b) Give an example of a differentiable function f on an interval I which has an inverse on I and is such that $f'(x_0) = 0$ for some x_0 in I.

26. Suppose that f_1 and f_2 are increasing functions on an interval I. (a) Is $f_1 + f_2$ increasing on I? (b) Is $f_1 \cdot f_2$ increasing on I? (c) Is $f_1[f_2]$ increasing on I?

27. Define the function f on $[0, \infty)$ by the formulas

$$f(x) = \begin{cases} x & \text{for } 0 \le x \le 1 \\ f(n) + \dfrac{1}{n}(x - n) & \text{for } n < x \le n+1, \quad n = 1, 2, \ldots . \end{cases}$$

Find the inverse function g. Describe its domain and range. Is g differentiable?

5. THE INVERSE TRIGONOMETRIC FUNCTIONS

A procedure for constructing graphs of the elementary trigonometric functions is given in Section 2 of Appendix 1. We reproduce in Fig. 9–8 the graph of $f(x) = \sin x$. The inverse of the function $y = \sin x$ is the *relation* defined by the equation

$$x = \sin y.$$

A graph of this relation is sketched in Fig. 9–9, with the curve repeating indefinitely in both upward and downward directions. Since the *range* of the sine function is the interval $[-1, 1]$, this interval is the *domain* of the inverse of the

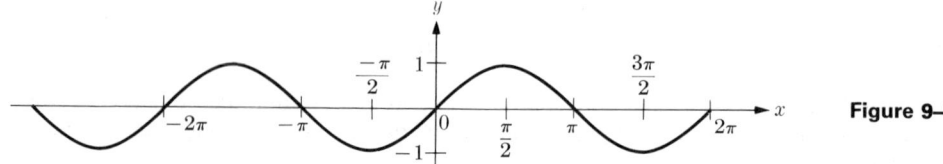

Figure 9–8

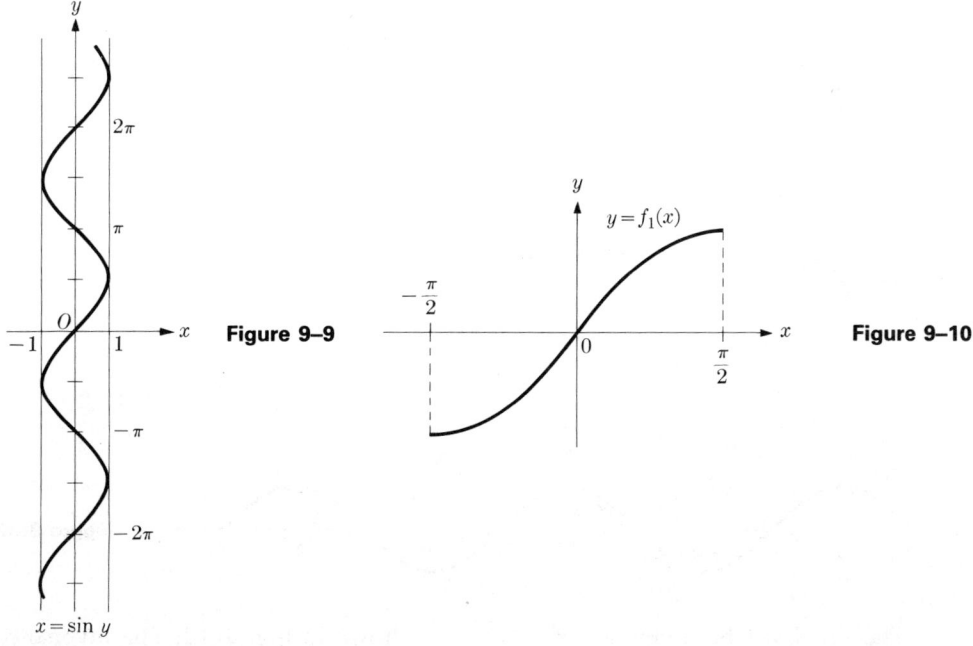

Figure 9-9

Figure 9-10

sine function. For each number x on $[-1, 1]$, there are infinitely many values of y such that $\sin y = x$. For example, if $x = \frac{1}{2}$, then

$$y = \frac{\pi}{6} + 2n\pi, \qquad y = \frac{5\pi}{6} + 2n\pi, \qquad \text{where} \qquad n = 0, \pm 1, \pm 2, \ldots.$$

Referring to Fig. 9-8, we see that $\sin x$ increases as x goes from $-\pi/2$ to $\pi/2$, decreases from $\pi/2$ to $3\pi/2$, increases from $3\pi/2$ to $5\pi/2$, decreases from $5\pi/2$ to $7\pi/2$, and so on. We define the *increasing* function f_1 by restricting the domain of the sine function to the interval $[-\pi/2, \pi/2]$:

$$f_1(x) = \sin x, \qquad -\frac{\pi}{2} \le x \le \frac{\pi}{2}.$$

The graph of f_1 is shown in Fig. 9-10. We define the **arcsin function** as the inverse of f_1; that is,*

$$y = \arcsin x \quad \Leftrightarrow \quad x = \sin y \quad \text{and} \quad -\frac{\pi}{2} \le y \le \frac{\pi}{2}.$$

The graph of this function is indicated by the solid line in Fig. 9-11. Since arcsin is the inverse of f_1, formulas (1) and (2) of Section 4 (page 303) become

$$\arcsin (\sin x) = x \qquad \text{for} \quad -\frac{\pi}{2} \le x \le \frac{\pi}{2},$$

$$\sin (\arcsin x) = x \qquad \text{for} \quad -1 \le x \le 1.$$

---

* The common notation $\sin^{-1} x$ will sometimes be used for arcsin x. This alternate notation will also be used for the other inverse trigonometric functions. When this symbol is used, *the -1 is not an exponent.* It is a synonym for the "arc" notation.

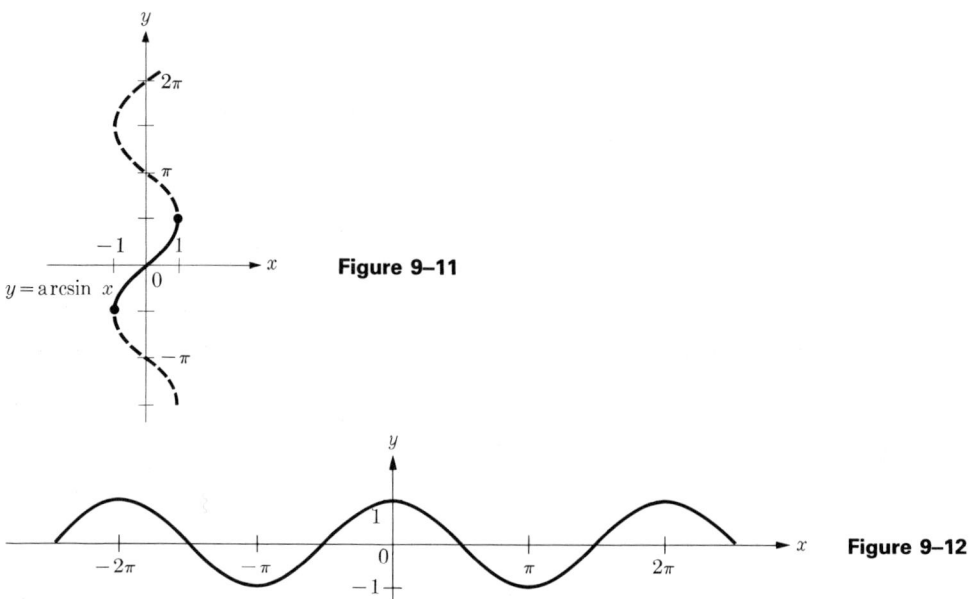

Figure 9–11

Figure 9–12

The graph of the function $f(x) = \cos x$ is shown in Fig. 9–12. The inverse of the function $y = \cos x$ is the relation defined by the equation

$$x = \cos y.$$

The graph of this relation is shown in Fig. 9–13, with part of the curve drawn solid and the remainder dashed. By restricting $\cos x$ to the interval $[0, \pi]$, we obtain a decreasing function of x. We define (see Fig. 9–14)

$$f_2(x) = \cos x, \quad 0 \leq x \leq \pi,$$

and so obtain the **arccos function** as the inverse of f_2; that is,

$$y = \arccos x \quad \text{if} \quad x = \cos y \quad \text{and} \quad 0 \leq y \leq \pi.$$

The graph of this function is indicated by the solid line in Fig. 9–13.

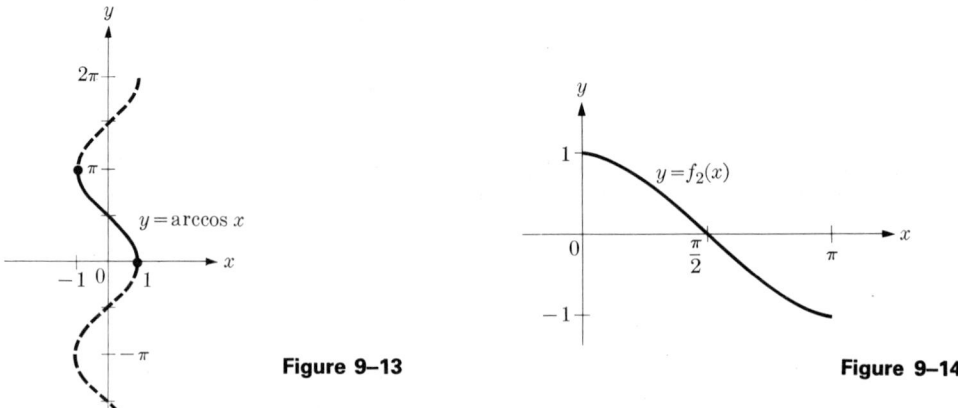

Figure 9–13

Figure 9–14

Example 1. Sketch the graph of
$$y = 2 \arccos\left(\frac{x}{2}\right).$$

Solution. We have

$$y = 2 \arccos\left(\frac{x}{2}\right) \Leftrightarrow \left(\frac{y}{2}\right) = \arccos\left(\frac{x}{2}\right)$$

$$\Leftrightarrow \left(\frac{x}{2}\right) = \cos\left(\frac{y}{2}\right) \quad \text{and} \quad 0 \leq \frac{y}{2} \leq \pi$$

$$\Leftrightarrow x = 2\cos\left(\frac{y}{2}\right) \quad \text{and} \quad 0 \leq y \leq 2\pi.$$

The graph is shown in Fig. 9–15. ◀

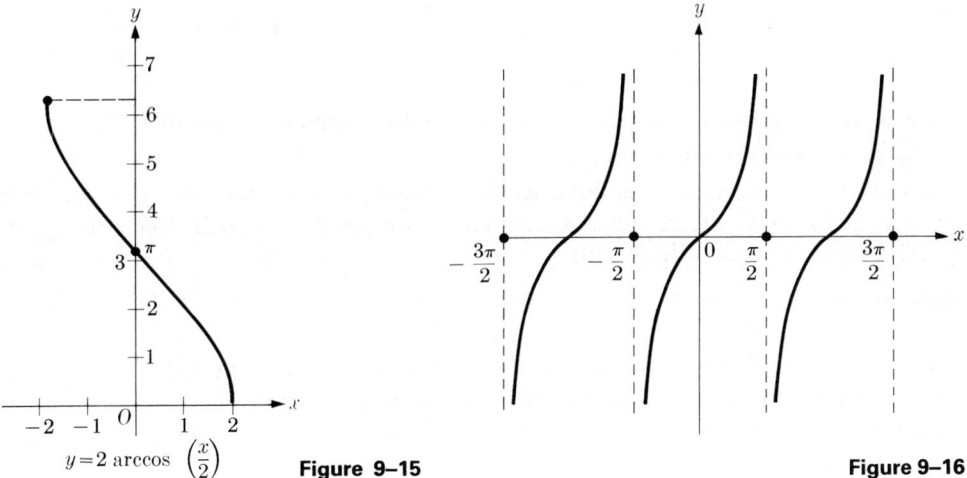

Figure 9–15   Figure 9–16

The graph of the function $f(x) = \tan x$ is shown in Fig. 9–16. The inverse of the function $y = \tan x$ is the relation defined by the equation

$$x = \tan y.$$

The graph of this relation is shown in Fig. 9–17. Repeating the method used for the sine and cosine functions, we observe that $y = \tan x$ is an increasing function in the interval $(-\pi/2, \pi/2)$. Therefore the **arctan function** is defined by

$$y = \arctan x \quad \text{if} \quad x = \tan y \quad \text{and} \quad -\frac{\pi}{2} < y < \frac{\pi}{2}.$$

This function is shown as the solid curve in Fig. 9–17.

Using exactly the same procedure as for the sine, cosine, and tangent functions, we defined arccot x as the number y for which $0 < y < \pi$ and cot $y = x$ (Fig. 9–18).

For $x \geq 1$, arcsec x is defined as the number y such that $0 \leq y < \pi/2$ and

310   THE TRIGONOMETRIC AND EXPONENTIAL FUNCTIONS

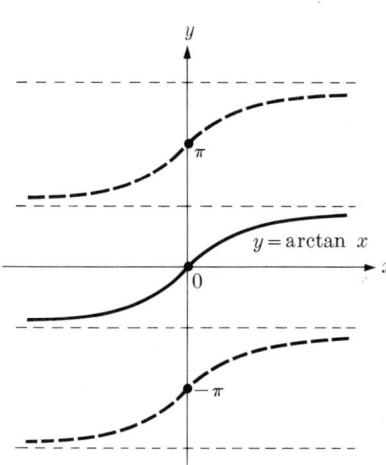

Figure 9–17

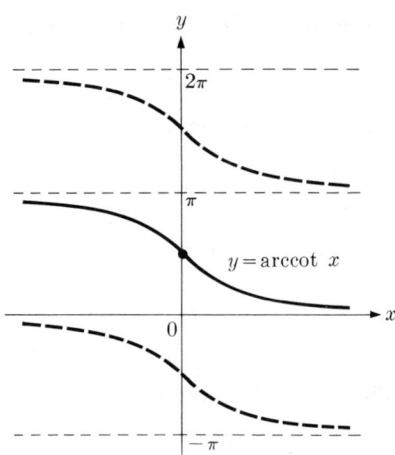

Figure 9–18

$\sec y = x$; for $x \le -1$, arcsec x is defined as the number y such that $-\pi \le y < -\pi/2$ and $\sec y = x$ (Fig. 9–19).

For $x \ge 1$, arccsc x is defined as the number y such that $0 < y \le \pi/2$ and $\csc y = x$; for $x \le -1$, arccsc x is defined as the number y such that $-\pi < y \le -\pi/2$ and $\csc y = x$ (Fig. 9–20).

To sum up:

a) the values of arcsin x and arctan x are between $-\pi/2$ and $\pi/2$;
b) the values of arccos x and arccot x are between 0 and π;
c) the values of arcsec x and arccsc x are between 0 and $\pi/2$ if $x \ge 1$ and between $-\pi$ and $-\pi/2$ if $x \le -1$.

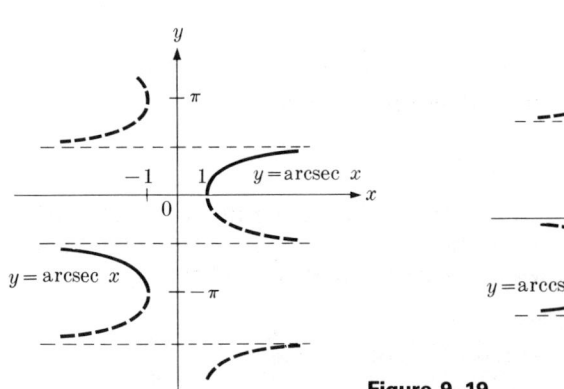

Figure 9–19

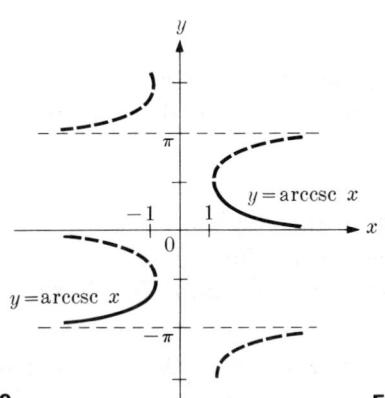

Figure 9–20

Example 2. Evaluate (a) $\arcsin \frac{1}{2}$, (b) $\operatorname{arccot}(-1)$, (c) $\arctan(-\sqrt{3})$, (d) $\operatorname{arcsec}(-2)$.

Solution. (a) We set $y = \arcsin \frac{1}{2}$, and this gives us $\sin y = \frac{1}{2}$ or $y = \pi/6$. In the same way we have (b) $\text{arccot}(-1) = 3\pi/4$ and (c) $\arctan(-\sqrt{3}) = -\pi/3$. As for (d), by setting $y = \text{arcsec}(-2)$, we obtain $\sec y = -2$, and y must be in the third quadrant. Therefore $y = -2\pi/3$ ($\cos y = -\frac{1}{2}$, $y = -120°$). ◀

The following theorems tell us how to find the derivatives of the inverse trigonometric functions.

Theorem 5. *Let $y = \arcsin x$. Then*

$$\frac{dy}{dx} = \frac{1}{\sqrt{1-x^2}} \quad \text{if } -1 < x < 1.$$

Proof. From the definition of inverse trigonometric functions, we have

$$\sin y = x, \quad -\frac{\pi}{2} \leq y \leq \frac{\pi}{2}.$$

From the Inverse Function Theorem we know that dy/dx exists for $-1 < x < 1$. By the Chain Rule for derivatives, we get

$$\cos y \cdot \frac{dy}{dx} = 1,$$

and we note that $\cos y > 0$ so long as y is in the interval $(-\pi/2, \pi/2)$. Therefore, from the relation

$$\sin^2 y + \cos^2 y = 1,$$

we find that

$$\cos y = +\sqrt{1 - \sin^2 y}, \quad -\frac{\pi}{2} < y < \frac{\pi}{2},$$

or

$$\cos y = \sqrt{1 - x^2}.$$

We conclude that

$$\frac{dy}{dx} = \frac{1}{\cos y} = \frac{1}{\sqrt{1-x^2}}.$$

Theorem 6. *If $y = \arccos x$, then*

$$\frac{dy}{dx} = -\frac{1}{\sqrt{1-x^2}}, \quad -1 < x < 1.$$

Proof. We have

$$\cos y = x, \quad 0 \leq y \leq \pi,$$

and by the Chain Rule for derivatives and the Inverse Function Theorem,

$$-\sin y \frac{dy}{dx} = 1.$$

We see that $\sin y > 0$ for $0 < y < \pi$, and so

$$\sin y = +\sqrt{1 - \cos^2 y} = \sqrt{1 - x^2}, \quad -1 < x < 1,$$

from which we conclude that

$$\frac{dy}{dx} = -\frac{1}{\sqrt{1-x^2}}, \quad -1 < x < 1.$$ ◀

The next two formulas for derivatives are established in a similar way.

Theorem 7. *If* $y = \arctan x$, *then*

$$\frac{dy}{dx} = \frac{1}{1+x^2}.$$

Theorem 8. *If* $y = \text{arccot } x$, *then*

$$\frac{dy}{dx} = -\frac{1}{1+x^2}.$$

We prove the following theorem.

Theorem 9. *If* $y = \text{arcsec } x$, *then*

$$\frac{dy}{dx} = \frac{1}{x\sqrt{x^2-1}}, \quad \text{if } |x| > 1.$$

Proof. From the definition of the inverse secant function, we may write

$$\sec y = x, \quad 0 \le y < \frac{\pi}{2} \quad \text{or} \quad -\pi \le y < -\frac{\pi}{2}.$$

We note that $\tan y > 0$ in the first and third quadrants, and therefore from $\tan^2 y + 1 = \sec^2 y$ we obtain

$$\tan y = +\sqrt{\sec^2 y - 1} = \sqrt{x^2 - 1}, \quad |x| > 1.$$

By the Chain Rule for derivatives and the Inverse Function Theorem, we see that

$$\sec^2 y \frac{dy}{dx} = \tfrac{1}{2}(x^2 - 1)^{-1/2}(2x) = \frac{x}{\sqrt{x^2 - 1}}.$$

Now we solve for dy/dx and set $\sec^2 y = x^2$ to obtain

$$\frac{dy}{dx} = \frac{1}{x\sqrt{x^2-1}}.$$

Theorem 10. *If* $y = \text{arccsc } x$, *then*

$$\frac{dy}{dx} = \frac{-1}{x\sqrt{x^2-1}}, \quad |x| > 1.$$

The proof of this result follows the same lines as the proof of Theorem 9.

If u is any function of x, we may use the Chain Rule to find derivatives of the inverse trigonometric functions of u. We state the basic formulas in terms of differentials:

$$d \arcsin u = \frac{du}{\sqrt{1-u^2}}, \qquad d \arccos u = \frac{-du}{\sqrt{1-u^2}},$$

$$d \arctan u = \frac{du}{1+u^2}, \qquad d \operatorname{arccot} u = \frac{-du}{1+u^2},$$

$$d \operatorname{arcsec} u = \frac{du}{u\sqrt{u^2-1}}, \qquad d \operatorname{arccsc} u = \frac{-du}{u\sqrt{u^2-1}}.$$

Example 3. Given that $y = \arcsin(\frac{5}{3}x)$, find dy/dx.

Solution. We let
$$u = \tfrac{5}{3}x \quad \text{and} \quad du = \tfrac{5}{3}dx.$$
Therefore
$$\frac{dy}{dx} = \frac{\tfrac{5}{3}}{\sqrt{1-(\tfrac{5}{3}x)^2}} = \frac{5}{\sqrt{9-25x^2}}. \qquad \blacktriangleleft$$

Example 4. Given that $f(x) = x \arctan(x^2)$, find $f'(x)$.

Solution. This is in the form of the product of x and $\arctan(x^2)$. From the product formula we have
$$f'(x) = x \cdot \frac{d}{dx}(\arctan(x^2)) + \arctan(x^2) \cdot 1.$$
Since
$$\frac{d}{dx}(\arctan(x^2)) = \frac{2x}{1+(x^2)^2} = \frac{2x}{1+x^4},$$
we conclude that
$$f'(x) = \frac{2x^2}{1+x^4} + \arctan(x^2). \qquad \blacktriangleleft$$

PROBLEMS

In Problems 1 through 4, in each case plot the graph of the equation.

1. $y = 2 \arccos x$
2. $y = 3 \arcsin(x/2)$
3. $y = \tfrac{1}{2} \arcsin 2x$
4. $y = \tfrac{1}{2} \arctan 2x$

In Problems 5 through 8, find the value of each of the expressions.

5. (a) $\arcsin(\sqrt{3}/2)$ (b) $\arccos(-1/\sqrt{2})$ (c) $\arctan(-1)$
6. (a) $\arccos(-\tfrac{1}{2})$ (b) $\operatorname{arcsec}(-2/\sqrt{3})$ (c) $\operatorname{arccsc}(-2)$
7. (a) $\operatorname{arccot}\sqrt{3}$ (b) $\operatorname{arcsec} 2$ (c) $\arctan(-1/\sqrt{3})$
8. (a) $\arcsin(-\tfrac{1}{2})$ (b) $\operatorname{arccot}(-\sqrt{3})$ (c) $\operatorname{arccsc}(-2/\sqrt{3})$

In Problems 9 through 25, in each case find the derivative.

9. $f(x) = \arctan 2x$
10. $g(s) = \operatorname{arccot}(s^2)$

11. $\phi(y) = \arccos\sqrt{y}$
12. $f(t) = \text{arcsec}\,(1/t)$
13. $y = x \arccos x$
14. $f(u) = u^2 \arcsin 2u$
15. $f(x) = (\text{arccot}\,3x)/(1+x^2)$
16. $f(x) = \arctan(x/\sqrt{1-x^2})$
17. $f : x \to \arcsin(x/\sqrt{1+x^2})$
18. $g : x \to \text{arccot}\,[(1-x)/(1+x)]$
19. $f : x \to \arctan[(x-3)/(1+3x)]$
20. $f(x) = (\arccos 2x)/\sqrt{1+4x^2}$
21. $f(x) = x \arcsin 2x + \tfrac{1}{2}\sqrt{1-4x^2}$
22. $f(x) = \arctan(3 \tan x)$
23. $f(x) = \sqrt{x^2-4} - 2\arctan[\tfrac{1}{2}\sqrt{x^2-4}]$
24. $f(x) = \sec^{-1}(\sqrt{1+x^2}/x)$
25. $f(x) = \arctan[(3\sin x)/(4 + 5\cos x)]$
26. Given that $f(x) = \arcsin x + \arccos x$, find $f'(x)$. What can you conclude about $f(x)$?
27. Given that $f(x) = \text{arcsec}\,x + \text{arccsc}\,x$, find $f'(x)$. What can you conclude about $f(x)$?
28. Prove Theorem 7.
29. Prove Theorem 8.
30. Prove Theorem 10, showing the required restrictions on the domain of the arccsc function.
31. Given $f(x) = \arcsin(\cos x)$ for $0 \leq x < \pi$, show that $f''(x) = 0$ for all x. Hence f must be a linear function of the form $ax + b$. Find the values of a and b.
32. Same as Problem 31 for $g(x) = \text{arccot}\,(\tan x)$, $-(\pi/2) < x < (\pi/2)$.
33. Prove that

$$\arctan x + \arctan c = \arctan\left(\frac{x+c}{1-xc}\right) \quad (1)$$

provided that $0 < c < 1$ and $0 \leq x \leq 1$ where c is a constant. [*Hint:* Show that the derivative of the left side equals the derivative of the right side and that the two sides of (1) are equal for one particular value of x.]

34. Prove that if $x \geq 1$, then $\text{arcsec}\,(1/x) = \arccos x$. [Same hint as for Problem 33.]

6. INTEGRATIONS YIELDING INVERSE TRIGONOMETRIC FUNCTIONS

Each new derivative formula carries with it a new integration formula. Therefore the theorems of Section 5 may be rephrased to give the following integration formulas:

$$\int \frac{du}{\sqrt{1-u^2}} = \arcsin u + c,$$

$$\int \frac{du}{1+u^2} = \arctan u + c,$$

$$\int \frac{du}{u\sqrt{u^2-1}} = \text{arcsec}\,u + c.$$

Example 1. Find

$$\int \frac{dx}{\sqrt{1-4x^2}}.$$

Solution. Let $u = 2x$ and $du = 2\,dx$. Then

$$\int \frac{dx}{\sqrt{1-4x^2}} = \int \frac{\tfrac{1}{2}\,du}{\sqrt{1-u^2}} = \tfrac{1}{2}\arcsin u + c = \tfrac{1}{2}\arcsin 2x + c. \quad \blacktriangleleft$$

Example 2. Find

$$\int \frac{dx}{9+x^2}.$$

Solution. We write

$$\int \frac{dx}{9+x^2} = \int \frac{dx}{9[1+(x/3)^2]} = \frac{1}{9}\int \frac{dx}{1+(x/3)^2}.$$

Making the substitution $u = \tfrac{1}{3}x$, $du = \tfrac{1}{3}\,dx$, we obtain

$$\frac{1}{9}\int \frac{3\,du}{1+u^2} = \frac{1}{3}\arctan u + c = \frac{1}{3}\arctan \frac{1}{3}x + c. \quad \blacktriangleleft$$

The reader may wonder why three of the formulas for integration were omitted. For example, we did not include the formula

$$\int \frac{du}{\sqrt{1-u^2}} = -\arccos u + c.$$

This is not a new formula because it can be verified that for all values of u between -1 and $+1$ we have

$$\arccos u = \frac{\pi}{2} - \arcsin u.$$

(See Problem 26 of Section 5.) Therefore the above integration formula is the same as the one leading to $\arcsin u$. The other two formulas present similar situations.

PROBLEMS

In Problems 1 through 12, in each case perform the integration.

1. $\int \dfrac{dx}{\sqrt{1-16x^2}}$

2. $\int \dfrac{dx}{1+25x^2}$

3. $\int \dfrac{dx}{x\sqrt{9x^2-1}}$

4. $\int \dfrac{dx}{\sqrt{9-16x^2}}$

5. $\int \dfrac{dx}{1+7x^2}$

6. $\int \dfrac{dx}{16+9x^2}$

7. $\int \dfrac{dx}{x\sqrt{x^2-4}}$

8. $\int \dfrac{dx}{\sqrt{12-5x^2}}$

9. $\int \dfrac{(x^2+1)\,dx}{4+x^2}$

10. $\int \dfrac{(2x-3)\,dx}{\sqrt{1-4x^2}}$

11. $\int \dfrac{dx}{x\sqrt{2x^2-10}}$

12. $\int \dfrac{dx}{7+(\tfrac{1}{3}x)^2}$

In Problems 13 through 16, in each case find the value of the integral.

13. $\displaystyle\int_{-1/2}^{(1/2)\sqrt{3}} \frac{dx}{\sqrt{1-x^2}}$

14. $\displaystyle\int_{-1}^{\sqrt{3}} \frac{dx}{1+x^2}$

15. $\displaystyle\int_{(2/3)\sqrt{3}}^{2} \frac{dx}{x\sqrt{x^2-1}}$

16. $\displaystyle\int_{-2}^{-\sqrt{2}} \frac{dx}{x\sqrt{x^2-1}}$

17. Show that arctan x + arccot x = constant. Find the value of the constant.

*18. Find the area of the region
$$R = \{(x, y) : 0 \le x \le 1, x^2 \le y \le \sqrt{4 - 3x^2}\}.$$
$\left[\text{Hint: Let } x = \dfrac{2}{\sqrt{3}} \sin\theta \text{ and see Problem 19 of Sec. 3.}\right]$

19. Find the area of the region
$$R = \{(x, y) : 0 \le x \le \tfrac{1}{2}\sqrt{2}, x^2 \le y \le (1 + 2x^2)^{-1}\}.$$

20. Define the function
$$F(x; a, b) = \int_0^x \frac{1}{\sqrt{a - bu^2}} \, du, \qquad a > 0, b > 0.$$

Show that
$$F(x; a, b) = \frac{1}{\sqrt{b}} F\left(\sqrt{\frac{b}{a}} x; 1, 1\right)$$
and observe that $F(x; 1, 1)$ = arcsin x.

21. Define the function
$$G(x; a, b) = \int_0^x \frac{1}{a + bu^2} \, du, \qquad a > 0, b > 0.$$

Show that
$$G(x; a, b) = \frac{1}{\sqrt{ab}} G\left(\sqrt{\frac{b}{a}} x; 1, 1\right)$$
and observe that $G(x; 1, 1)$ = arctan x.

7. THE LOGARITHM FUNCTION

The student has undoubtedly learned the basic ideas of logarithms in elementary courses in algebra and trigonometry. The underlying principle occurs in the law of exponents, which states that
$$b^\alpha \cdot b^\beta = b^{\alpha+\beta},$$
where α, β, and b are any numbers. If the exponents α and β are integers, the meaning of the above equation is clear from the very definition of an exponent as a symbol for repeated multiplication. Similarly, if the exponents are rational and b is positive, the notion of the nth root of a number allows us to derive the above law of exponents. However, it is not clear how we would go about defining the number
$$3^{\sqrt{2}}.$$

9-7 THE LOGARITHM FUNCTION

The definition of numbers with irrational exponents requires an additional effort. The customary introduction of logarithms proceeds on the *assumption* that all such numbers are known. In particular, if b and N are any numbers, with $b > 0$, $b \neq 1$, it is assumed that the equation

$$b^x = N$$

may always be solved uniquely for x. The value of x which solves this equation is written as

$$\log_b N.$$

The usual laws of logarithms, such as

$$\log_b M + \log_b N = \log_b (MN),$$

$$\log_b M - \log_b N = \log_b \left(\frac{M}{N}\right),$$

follow immediately from the law of exponents.

Rather than follow the traditional method of elementary mathematics to define logarithms, we shall use the help of the calculus. First we recall the formula

$$\int t^n \, dt = \frac{t^{n+1}}{n+1} + c, \qquad n \neq -1.$$

This integration formula fails for $n = -1$. Nevertheless, if we plot the graph (Fig. 9-21) of $y = 1/t$ for positive values of t, the expression

$$\int_1^x \frac{1}{t} \, dt, \qquad x > 0,$$

has meaning, as it is simply the area under the curve between the points 1 and x. Its value will depend on x and, in fact, it is a function of x so long as x is positive. We define a new function in the following way.

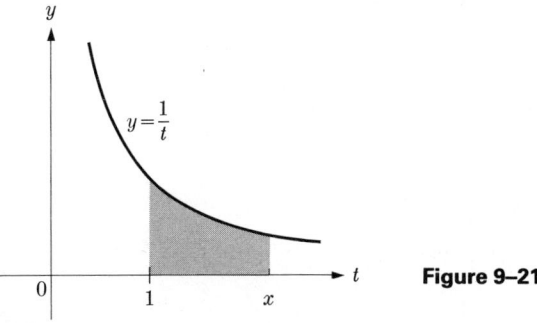

Figure 9-21

Definition. *For values of $x > 0$, we define*

$$\ln x = \int_1^x \frac{1}{t} \, dt.$$

Remark. For the present, read ln x as "ell-en x" or "ell-en of x."

318   THE TRIGONOMETRIC AND EXPONENTIAL FUNCTIONS

The integral of the function $1/t$ has a useful property which we shall state in the form of a lemma.

Lemma. *If a and b are any positive numbers, then*

$$\int_a^{a \cdot b} \frac{1}{t}\, dt = \int_1^b \frac{1}{t}\, dt.$$

Proof. We let $u = (1/a)t$ in the integral on the left; then $du = (1/a)\, dt$. After making this change, we have for the limits of integration: $t = a$ corresponds to $u = 1$, and $t = ab$ corresponds to $u = b$. Therefore

$$\int_a^{a \cdot b} \frac{dt}{t} = \int_1^b \frac{a\, du}{au} = \int_1^b \frac{du}{u} = \int_1^b \frac{dt}{t},$$

the last equality being valid because the letter used for the variable of integration is irrelevant.

Geometrically, the lemma states that the shaded areas shown in Figs. 9–22 and 9–23 are equal. With the aid of this lemma the following basic theorem is easy to establish.

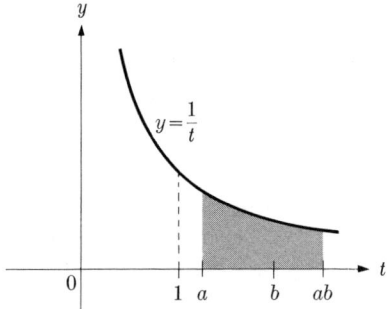

Figure 9–22

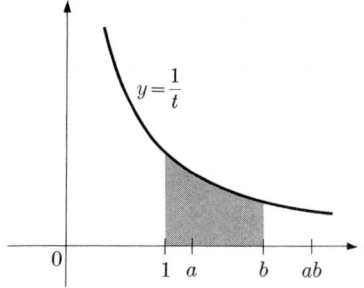

Figure 9–23

Theorem 11. *If a and b are any positive numbers, then*

i) $\ln (a \cdot b) = \ln a + \ln b$,
ii) $\ln (a/b) = \ln a - \ln b$,
iii) $\ln 1 = 0$,
iv) $\ln a^r = r \ln a$, *if r is any rational number.*

Proof. To prove (i), we have from the definition of the ln function that

$$\ln (a \cdot b) = \int_1^{a \cdot b} \frac{dt}{t}.$$

This integral may be written in the form

$$\int_1^{a \cdot b} \frac{dt}{t} = \int_1^a \frac{dt}{t} + \int_a^{a \cdot b} \frac{dt}{t},$$

and the equality holds whether or not a falls between 1 and $a \cdot b$. At this point we use the lemma to replace the second integral on the right and get

$$\ln(a \cdot b) = \int_1^a \frac{dt}{t} + \int_1^b \frac{dt}{t} = \ln a + \ln b.$$

To prove (ii), we write $a = b \cdot (a/b)$ and apply (i) to the product of b and a/b. We then have

$$\ln a = \ln\left(b \cdot \frac{a}{b}\right) = \ln b + \ln\left(\frac{a}{b}\right),$$

and when we transfer $\ln b$ to the left side, we see that

$$\ln \frac{a}{b} = \ln a - \ln b.$$

As for (iii), we apply (ii) with $b = a$. This says that

$$\ln 1 = \ln \frac{a}{a} = \ln a - \ln a = 0.$$

To establish (iv) we proceed in stages. If the exponent is a positive integer, we use (i) and mathematical induction. First, if $b = a$, we observe that (i) states that

$$\ln a^2 = \ln(a \cdot a) = \ln a + \ln a = 2 \ln a.$$

Then, using a step-by-step method, we obtain

$$\ln a^n = n \ln a, \quad \text{if } n \text{ is a positive integer.}$$

To obtain the result for negative integers, we write $a^{-n} = 1/a^n$ and use (ii) to get

$$\ln(a^{-n}) = \ln\left(\frac{1}{a^n}\right) = \ln 1 - \ln a^n.$$

Since $\ln 1 = 0$ and $\ln a^n = n \ln a$, we find that

$$\ln(a^{-n}) = -n \ln a.$$

If r is any rational number, then $r = p/q$, where p and q are integers (we take q as positive). We define $u = a^{1/q}$; then $u^q = a$ and $\ln a = q \ln u$, since q is an integer. Also we have

$$u^p = a^{p/q} = a^r.$$

Therefore

$$\ln(a^r) = \ln(u^p) = p \ln u = \frac{p}{q} q \ln u = \frac{p}{q} \ln a;$$

and so

$$\ln(a^r) = r \ln a.$$

The **domain** of the ln function consists of all positive numbers. The next theorem establishes some of the basic properties of this function and also tells us that the **range** of the ln function consists of all real numbers.

Theorem 12. *If $f(x) = \ln x$, then*

i) $f'(x) = 1/x$, $x > 0$,
ii) f is increasing,
iii) $\frac{1}{2} \leq \ln 2 \leq 1$,
iv) $\ln x \to +\infty$ as $x \to +\infty$,
v) $\ln x \to -\infty$ as $x \to 0^+$,
vi) the range of f consists of all real numbers.

Proof. Item (i) is simply the statement that differentiation and integration are inverse processes (Fundamental Theorem of Calculus). The derivative is positive, and part (ii) states that a function with a positive derivative is increasing, a fact which we have already learned. To establish (iii), we make use of our knowledge of integrals as areas. The value of

$$\ln 2 = \int_1^2 \frac{1}{t}\, dt$$

is shown as the shaded area in Fig. 9–24. From the theorem on integrals which gives upper and lower bounds, we have

$$\text{area } ADFE \leq \int_1^2 \frac{1}{t}\, dt \leq \text{area } BCFE \quad \text{or} \quad \tfrac{1}{2} \leq \ln 2 \leq 1.$$

To prove (iv), we must show that $\ln x$ increases without bound as x does. Let n be any positive integer. If $x > 2^n$, then

$$\ln x > \ln (2^n),$$

since we know from (ii) that $\ln x$ is an increasing function. But

$$\ln x > \ln 2^n = n \ln 2 \geq \tfrac{1}{2} n \quad \text{(by (iii))}.$$

Since $\tfrac{1}{2} n \to +\infty$ as $n \to +\infty$, we have

$$\ln x \to +\infty \quad \text{as} \quad x \to +\infty.$$

To obtain (v), we use a device similar to that used for (iv). If n is a positive integer, $(1/2^n) \to 0$ as $n \to +\infty$. If $0 < x < 1/2^n$, we know (by (ii)) that

$$\ln x < \ln\left(\frac{1}{2^n}\right) = -n \ln 2.$$

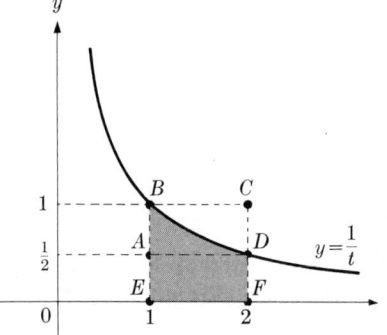

Figure 9–24

Since $\ln 2 > \frac{1}{2}$, we can say that
$$\ln x < -\tfrac{1}{2}n.$$
Since $x < 1/2^n$, we know that $x \to 0$ as $n \to +\infty$. As $n \to +\infty$, the quantity $-\frac{1}{2}n$ tends to $-\infty$, and so $\ln x \to -\infty$ as $x \to 0$. Note that x must tend to zero through positive values, since $\ln x$ is not defined for $x \leq 0$.

Items (iv) and (v) show that the range of $\ln x$ extends from $-\infty$ to $+\infty$. The Intermediate Value Theorem then allows us to conclude that no numbers in the range are omitted, thus proving (vi).

A graph of the function $\ln x$ is given in Fig. 9–25.

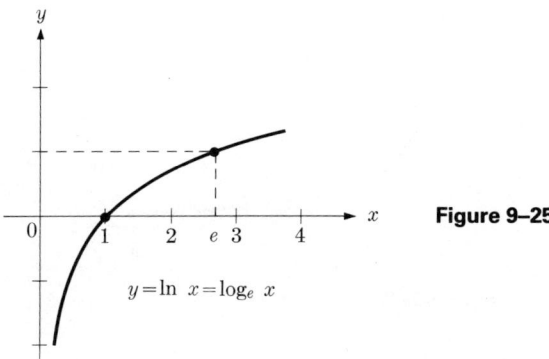

Figure 9–25

Remark. Suppose that u is any positive function of x. Then from the Chain Rule, we get the important formula

$$\frac{d}{dx} \ln u = \frac{1}{u} \frac{du}{dx}.$$

Example 1. Given that $\ln 2 = 0.69315$ and $\ln 3 = 1.09861$, find $\ln 24$ and $\ln 0.1875$.

Solution. We write $24 = 2^3 \cdot 3$ and use the properties of Theorem 11 to obtain
$$\ln 24 = 3 \ln 2 + \ln 3 = 3(0.69315) + 1.09861 = 3.17806.$$
Observe that $0.1875 = \frac{3}{16}$, and so
$$\ln 0.1875 = \ln 3 - 4 \ln 2 = -1.67399. \qquad \blacktriangleleft$$

Example 2. Use the theorem giving upper and lower bounds for integrals to obtain an upper and lower bound for $\ln 5$.

Solution. From Fig. 9–26, we see that a lower bound is given by the shaded area $ABCD$; it is $\frac{4}{5}$. An upper bound is given by the area $ABEF$; it is 4. In other words,
$$\tfrac{4}{5} \leq \ln 5 \leq 4.$$

A refined upper bound is given by the area of the trapezoid $ABCF$; it is
$$\tfrac{1}{2}(4)(\tfrac{1}{5} + 1) = \tfrac{12}{5}.$$
A more precise lower bound may be obtained by dividing the interval from 1 to 5 into sections and computing the areas separately, as shown in Fig. 9–27:
$$\text{Area} = 1 \cdot (\tfrac{1}{2}) + 1 \cdot (\tfrac{1}{3}) + 1 \cdot (\tfrac{1}{4}) + 1 \cdot (\tfrac{1}{5}) = \tfrac{77}{60}.$$
The estimates for $\ln 5$ are
$$\tfrac{77}{60} \leq \ln 5 \leq \tfrac{12}{5}. \qquad \blacktriangleleft$$

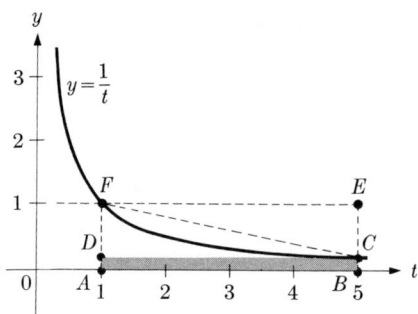

Figure 9–26 **Figure 9–27**

Example 3. Given that $F(x) = \ln(x^2 + 5)$, find $F'(x)$.

Solution. The Chain Rule tells us that
$$\frac{d}{dx} \ln u = \frac{1}{u} \cdot \frac{du}{dx}.$$
Therefore, letting $u = x^2 + 5$, we obtain
$$F'(x) = \frac{1}{x^2 + 5} \cdot 2x = \frac{2x}{x^2 + 5}. \qquad \blacktriangleleft$$

Example 4. Given that $G(x) = \ln \sqrt{(1+x)/(1-x)}$ for $-1 < x < 1$, find $G'(x)$.

Solution. The use of the properties of logarithms simplifies the work of differentiation. We first write
$$G(x) = \tfrac{1}{2} \ln \frac{1+x}{1-x} = \tfrac{1}{2} \ln(1+x) - \tfrac{1}{2} \ln(1-x).$$
Then
$$G'(x) = \frac{1}{2} \frac{1}{1+x} - \frac{1}{2} \frac{1}{1-x} \cdot (-1) = \frac{1}{2}\left(\frac{1}{1+x} + \frac{1}{1-x}\right) = \frac{1}{1-x^2}. \qquad \blacktriangleleft$$

PROBLEMS

In Problems 1 through 5, compute the required quantities, using the fact that $\ln 2 = 0.69315$, $\ln 3 = 1.09861$, and $\ln 10 = 2.30259$.

1. ln 4, ln 5, ln 6, ln 8, ln 9
2. ln 12, ln 15, ln 16, ln 18, ln 20
3. ln 24, ln 25, ln 27, ln 30, ln 32
4. ln 1.5, ln 2.7, ln 0.25, ln 1.25
5. ln $\sqrt{30}$, ln $\sqrt[3]{270}$, ln $\sqrt{2/3}$, ln $\sqrt[4]{7.2}$

In Problems 6 through 15, use the definition of the function ln x and the properties of the area of elementary figures to obtain in each case an upper and lower bound for the given quantity (method of Example 2).

6. ln 4.5
7. ln 7
8. ln 14
9. ln 3.2
10. ln 1.05
11. ln $\frac{1}{2}$
12. ln 0.15
13. ln 0.001
14. ln 0.95
15. ln 0.995

In Problems 16 through 31, in each case perform the differentiation.

16. $f(x) = \ln(2x + 3)$
17. $f(x) = \ln x^4$
18. $f(x) = \ln(x^2 + 2x + 3)$
19. $f(x) = \ln(x^3 - 3x + 1)$
20. $g(x) = 2 \ln \sin x$
21. $h(x) = \ln \sec 2x$
22. $F(x) = \ln(\sec x + \tan x)$
23. $F(x) = (\ln x)^4$
24. $G(x) = \ln(\ln x), (x > 1)$
25. $G(x) = x \ln x$
26. $H: x \to \ln \sqrt{1 - x^2}$
27. $f: x \to \ln(x + \sqrt{x^2 + a^2})$
28. $H: x \to \ln(x^2 \sqrt{x^2 + 1})$
29. $H: x \to \ln[(x^2 + 1)/(x^2 - 1)]$
30. $T(x) = \ln \sqrt{(1 + \cos x)/(1 - \cos x)}$
31. $f(x) = (\ln x)^2/(1 + x^2)$

In Problems 32 through 40, use the definition of the function ln x and the values of ln 2, ln 3, and ln 10 which are given at the beginning of the exercises to evaluate the following integrals.

32. $\int_1^5 \frac{1}{x} dx$
33. $\int_4^{10} \frac{dx}{x}$
34. $\int_3^{15} \frac{du}{u}$
35. $\int_5^8 \frac{du}{u}$
36. $\int_1^{1/4} \frac{du}{u}$
37. $\int_{1/3}^1 \frac{du}{u}$
38. $\int_{1/2}^{12} \frac{du}{u}$
39. $\int_{1/8}^{1/5} \frac{du}{u}$
40. $\int_{0.01}^{10} \frac{du}{u}$

41. Using a method similar to that of the proof of the lemma of this section, show that
$$\int_a^{a \cdot b} \frac{du}{u^2} = \frac{1}{a} \int_1^b \frac{du}{u^2}, \quad a, b \text{ positive.}$$

42. Extend the result given in Problem 41 to show that if n is any integer, then
$$\int_a^{a \cdot b} \frac{du}{u^n} = \frac{1}{a^{n-1}} \int_1^b \frac{du}{u^n}.$$

43. Find the equation of the line tangent to the curve $y = 2 \ln(x^2 - 1)$ at the point where $x = 2$.

44. Find the area of the region bounded by the curve $y = x/(x^2 + 1)$ and the lines $x = 1$, $x = 3$, $y = 0$.

45. Find the area of the region
$$R = \{(x, y): 1 \le x \le 2, x - 1 \le y \le 5x^2/(2x^3 - 1)\}.$$

324 THE TRIGONOMETRIC AND EXPONENTIAL FUNCTIONS

46. Show that if k is any integer larger than 1, then
$$\frac{1}{2} + \frac{1}{3} + \cdots + \frac{1}{k} < \ln k < 1 + \frac{1}{2} + \cdots + \frac{1}{k-1}.$$
[*Hint:* See Example 2.]

47. Find the domain and range of the function $f(x) = \ln (\ln x)$. Show that
$$x \frac{d}{dx} [xf'(x)] = -(\ln x)^{-2}.$$

48. a) Show that $\ln (1 + x) = \int_0^x 1/(1 + u)\, du$ for $x > -1$.
    b) If $x > 0$, show that $\ln (1 + x) < x$.
    c) If $x > 0$, show that $x - \frac{1}{2}x^2 < \ln (1 + x)$.
    [*Hint:* Use the inequality $1 - u < 1/(1 + u)$ for $u > 0$.]

8. THE EXPONENTIAL FUNCTION

We saw in the previous section that the domain of the ln function consists of all positive numbers, while the range consists of all real numbers (Theorem 12). Since the derivative of ln x is always positive, we may apply the Inverse Function Theorem (Theorem 4), and so define the *inverse function* of ln x.

Definition. **The exponential function,** *denoted by* exp, *is defined to be the inverse of the function* ln.

Remark. From the paragraph above it is clear that the domain of exp x consists of all real numbers, and the range consists of all positive numbers.

The basic properties of exp x are given in the next theorem.

Theorem 13. *If* $f(x) = \exp x$, *then*

i) f *is increasing for all values of* x,
ii) $\exp (x_1 + x_2) = (\exp x_1)(\exp x_2)$,
iii) $\exp (x_1 - x_2) = (\exp x_1)/(\exp x_2)$,
iv) $\exp rx = (\exp x)^r$ *if* r *is rational*,
v) $f(x) \to +\infty$ *as* $x \to +\infty$,
vi) $f(x) \to 0$ *as* $x \to -\infty$.

Proof. To establish (i), let $y_1 = \exp x_1$ and $y_2 = \exp x_2$, with $x_2 > x_1$. We wish to show that $y_2 > y_1$. From the definition of inverse functions we know that
$$x_1 = \ln y_1 \quad \text{and} \quad x_2 = \ln y_2.$$
Since ln is an increasing function, the only way that the inequality $x_2 > x_1$ can hold is if $y_2 > y_1$.

Item (ii) is proved by the same technique—going back to the ln function. Let
$$y_1 = \exp x_1 \quad \text{and} \quad y_2 = \exp x_2.$$

Then $x_1 = \ln y_1$, $x_2 = \ln y_2$, and $x_1 + x_2 = \ln y_1 + \ln y_2 = \ln(y_1 y_2)$. This last statement asserts that
$$y_1 y_2 = \exp(x_1 + x_2)$$
or
$$(\exp x_1)(\exp x_2) = \exp(x_1 + x_2).$$

The argument for (iii) is analogous to that for (ii).

We prove part (iv) by mathematical induction. We know that
$$\exp(2x) = \exp(x + x) = (\exp x)(\exp x) = (\exp x)^2,$$
and therefore that
$$\exp(nx) = (\exp x)^n, \quad \text{for } n \text{ a positive integer.}$$

The remaining argument parallels the proof of (iv) in Theorem 11, Section 7. Parts (v) and (vi) are derived by appealing to the analogous result for the function ln. To state that ln and exp are inverse functions is equivalent to writing the compact formulas (see Eqs. (1) and (2) in Section 4, page 303)

$$\ln(\exp x) = x \quad \text{for all } x \quad \text{and} \quad \exp(\ln x) = x, \quad \text{if } x > 0.$$

Figure 9–28 shows the functions ln and exp. They are the reflections of each other in the line $y = x$, since an interchange of the roles of x and y sends one function into the other.

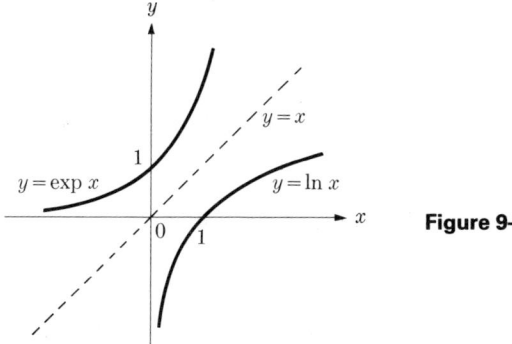

Figure 9–28

If a is any positive number and r is any rational number, the formula
$$\exp(\ln x) = x$$
becomes for $x = a^r$ the formula
$$\exp(\ln a^r) = a^r.$$
But $\ln a^r = r \ln a$, and so we have
$$\exp(r \ln a) = a^r, \quad \text{if } a > 0 \text{ and } r \text{ is rational.}$$

The right side of this expression has meaning if r is rational. The left side has meaning if r is any real number. This gives us a clue for defining a^r when r is any real number.

Definition. *If $a > 0$, we define*

$$a^x = \exp(x \ln a), \quad \text{for any } x.$$

Remark. The procedure used in defining a^x for any real number x is typical of a technique that is used in many branches of mathematics. The process is one that starts with an equation which holds under certain restrictions. If one side of the equation has meaning without these restrictions while the other side is *undefined* if the restrictions are relaxed, the equation itself may be used to define the hitherto meaningless side.* ◄

Now that numbers of the form a^x have meaning for any real number x,† we are in a position to define logarithm in the familiar way.

Definition. *If $b > 0$, $b \neq 1$, and $N > 0$, we define $\log_b N$ as that number y such that $b^y = N$. In words:*

The logarithm of a positive number N to the positive base b is that power to which b must be raised to obtain N.

The familiar properties of exponents and logarithms to any base may now be established from the definition of a^x and from the properties of the ln and exp functions as given in Theorems 11, 12, and 13. Some of these properties are:

i) $a^x \cdot a^y = a^{x+y}$; $a^x/a^y = a^{x-y}$; $(a^x)^y = a^{xy}$; $(ab)^x = a^x \cdot b^x$;
ii) $\log_b (x \cdot y) = \log_b x + \log_b y$;
iii) $\log_b (x/y) = \log_b x - \log_b y$;
iv) $\log_b (x^y) = y \log_b x$, if $x > 0$.

Definition. *We define a number which we designate e by the formula*

$$e = \exp 1.$$

The importance of this number is exhibited in the next two theorems.

Theorem 14. *If $x > 0$, then $\ln x = \log_e x$.*

Proof. Since, by definition, $e = \exp 1$, we have $\ln e = \ln \exp 1$. However, because ln and exp are inverse functions, we know that $\ln \exp 1 = 1$. Therefore $\ln e = 1$. If we set $y = \log_e x$, then $x = e^y$; further, $\ln x = \ln e^y = y \ln e = y$. ◄

Theorem 14 states that the function we have been calling "ell-en of x" is simply a logarithm function to the base e.

Definitions. *Logarithms to the base e are called* **natural logarithms** *(also Napierian logarithms). Logarithms to the base 10 are called* **common logarithms.**

---

*An example of this is given by the equation $x^2 = b$, where b is a number. The equation determines the square root of a number b if it is nonnegative; but the right side has meaning if b is *any real number*. This equation may be used as the *definition* of imaginary numbers.
† Moreover since $a^x = \exp(x \ln a)$, we see that the range of a^x is all positive numbers if $a \neq 1$ (Theorem 13).

Theorem 15. *We have*

i) $\exp x = e^x$ *for all x,*
ii) *e lies between 2 and 4.*

Proof. For the first part, we know by definition of a number to an exponent, as given on page 326, that
$$e^x = \exp(x \ln e).$$
But we saw in the proof of Theorem 14 that $\ln e = 1$. This yields $e^x = \exp x$. To establish (ii), we first recall that in Theorem 12, part (iii), we found that
$$\tfrac{1}{2} \leq \ln 2 \leq 1.$$
The inequality
$$\ln 2 = \log_e 2 \leq 1$$
is interpreted from the definition of logarithm as $2 \leq e^1 = e$. From the relation $\ln 4 = \ln 2 + \ln 2 \geq \tfrac{1}{2} + \tfrac{1}{2} = 1$, we see that $\log_e 4 \geq 1$, and therefore
$$4 \geq e^1 = e.$$

The actual numerical value of e is 2.71828$^+$, correct to five decimal places. Methods for computing e to any desired number of places will be given later. (See Section 10 of this chapter.)

Notation. The notations $\ln x$ and $\log_e x$ are sometimes replaced by $\log x$. Therefore, whenever no base is indicated in the logarithm function, it will always be understood that the base is e.

Example 1. Find the value of $\log_2 \tfrac{1}{8}$.

Solution. Letting $x = \log_2 \tfrac{1}{8}$, we see from the definition of logarithm that
$$2^x = \tfrac{1}{8}.$$
Writing $\tfrac{1}{8}$ as 2^{-3}, we obtain $x = -3$. ◄

Example 2. Sketch the graph of $y = 2^x$.

Solution. We construct a table of values, as shown. Observe that $y \to +\infty$ as $x \to +\infty$, and $y \to 0$ as $x \to -\infty$. The curve is sketched in Fig. 9–29.

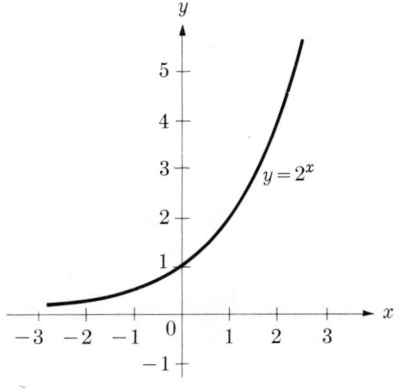

Figure 9–29

x	0	1	2	3	4	−1	−2	−3	−4
y	1	2	4	8	16	$\frac{1}{2}$	$\frac{1}{4}$	$\frac{1}{8}$	$\frac{1}{16}$

Example 3. Sketch the graph of $y = \log_3 x$.

Solution. One way to get a table of values is to note that the equation $y = \log_3 x$ is equivalent to $3^y = x$. Letting y take on a succession of values, we construct the table and then sketch the curve, as shown in Fig. 9–30.

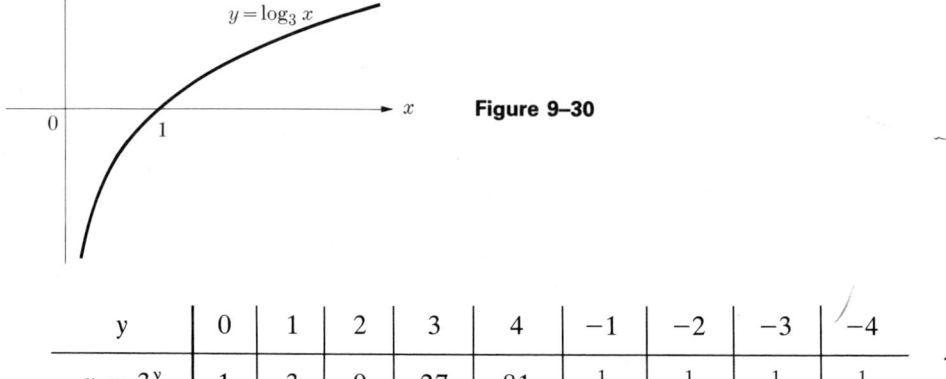

Figure 9–30

y	0	1	2	3	4	−1	−2	−3	−4
$x = 3^y$	1	3	9	27	81	$\frac{1}{3}$	$\frac{1}{9}$	$\frac{1}{27}$	$\frac{1}{81}$

Example 4. Prove that if $x > 0$, $b > 0$, and $b \neq 1$, *then*

$$\log_b x = \frac{\ln x}{\ln b}.$$

Solution. We set $y = \log_b x$ and write the equivalent expression $b^y = x$. This yields

$$\ln (b^y) = \ln x$$

and, consequently,

$$y \ln b = \ln x \quad \Leftrightarrow \quad y = \frac{\ln x}{\ln b}.$$

PROBLEMS

In Problems 1 through 14, sketch the curves.

1. $y = 4^x$
2. $y = 3^x$
3. $y = (\frac{1}{2})^x$
4. $y = (\frac{1}{3})^{2x}$
5. $y = 2^{-x}$
6. $y = 3^{-2x}$
7. $y = 2^{x+1}$
8. $y = 2^{3-x}$
9. $y = \log_2 x$
10. $y = \log_4 x$
11. $y = \log_{1/2} x$
12. $y = \log_{1/5} x$
13. $y = \log_3 2x$
14. $y = \log_2 (x + 1)$

In Problems 15 through 18, find in each case the value of the given expression.

15. (a) $\log_3 81$ (b) $\log_4 16$ (c) $\log_2 \frac{1}{32}$
16. (a) $\log_3 \frac{1}{27}$ (b) $\log_5 125$ (c) $\log_4 \frac{1}{64}$
17. (a) $\log_2 1$ (b) $\log_7 \frac{1}{49}$ (c) $\log_a a$
18. (a) $\log_{1/2} 8$ (b) $\log_{1/6} 216$ (c) $\log_{1/4} \frac{1}{16}$
19. From the definition of the function a^x, prove that $a^x \cdot a^y = a^{x+y}$ for any real numbers x and y and $a > 0$.
20. (a) Prove that $(2^{\sqrt{3}})^{\sqrt{3}} = 8$.
    (b) Given that x and y are any real numbers, prove that $(a^x)^y = a^{xy}$.
21. From the definition of the ln function, show that $\ln 2.4 \leq 1$ and, consequently, that $2.4 \leq e$.
22. From the definition of the ln function, show that $\ln 3.5 \geq 1$ and, consequently, that $3.5 \geq e$.

Using the results of Example 4, express the following quantities in terms of natural logarithms.

23. $\log_3 8$, $\log_7 12$, $\log_2 15$ 24. $\log_{1/3} 28$, $\log_{1/3} x^2$, $\log_8 12$

Without using tables, find the value of the following, given that $\ln 2 = 0.693$, $\ln 3 = 1.099$, $\ln 10 = 2.303$.

25. $\log_3 16$ 26. $\log_2 30$ 27. $\log_{1/5} 6$
28. $\log_{10} 24$ 29. $\log_6 18$ 30. $\log_{15} 9$
31. Suppose that a is a number such that $0 < a < 1$. Show that $\lim_{x \to +\infty} a^x = 0$.
32. Find the value of $\lim_{x \to 0+} x \log x$.
33. Find the value of $\lim_{n \to \infty} \frac{\ln n}{n}$.
34. Suppose that f is a continuous function for $-\infty < x < \infty$ and $f(a + b) = f(a)f(b)$ for all a, b. Show that either $f(0) = 0$ or $f(0) = 1$. If $f(0) = 1$ show that f can never vanish. If also $f(1) \neq 1$, show that $\lim_{x \to +\infty} f(x) = 0$ or $+\infty$, in which case $\lim_{x \to -\infty} f(x) = +\infty$ or 0.
35. Determine the domain and range of the function

$$f: x \to \ln\left(\frac{x-1}{x+1}\right).$$

Find the inverse function (or functions) and determine its domain and range.

9. DIFFERENTIATION OF EXPONENTIAL FUNCTIONS; LOGARITHMIC DIFFERENTIATION

In Section 7 we saw that the derivative of $\ln x$ is $1/x$. In Section 8 we showed that $\ln x$ is $\log_e x$ and, therefore, we have the formula

$$\frac{d}{dx}(\log_e x) = \frac{1}{x}, \quad x > 0.$$

The following theorem is convenient when x may have negative values.

Theorem 16. *If $f(x) = \log_e |x|$, then $f'(x) = 1/x$, $x \neq 0$.*

Proof. If x is positive, $\log_e |x| = \log_e x$, and the result is nothing new. If x is negative, then $-x$ is positive and

$$\frac{d}{dx} \log_e (-x) = \frac{1}{-x} \frac{d(-x)}{dx} = \frac{1}{x}.$$

Therefore, whatever x is, we have

$$\frac{d}{dx} \log_e |x| = \frac{1}{x}. \quad \blacktriangleleft$$

The derivative of the exponential function is obtained by using the fact that it is the inverse of the natural logarithm.

Theorem 17. *If $f(x) = e^x$, then $f'(x) = e^x$.*

Proof. By letting $y = e^x = \exp x$, we may write $\ln y = x$. Then, by the Chain Rule and the Inverse Function Theorem, we have

$$\frac{1}{y} \frac{dy}{dx} = 1 \quad \text{or} \quad f'(x) = \frac{dy}{dx} = y = e^x. \quad \blacktriangleleft$$

We note the interesting fact that the exponential function is its own derivative, which is the content of Theorem 17.

The problem of finding the derivative of a^x can be reduced to that of obtaining the derivative of the exponential function, as the proof of the next theorem shows.

Theorem 18. *If $f(x) = a^x$, then $f'(x) = a^x \log_e a$.*

Proof. We set $y = a^x$. Then $\log_e y = x \log_e a$ and, by differentiating, we find

$$\frac{1}{y} \frac{dy}{dx} = \log_e a \quad \text{and} \quad f'(x) = \frac{dy}{dx} = y \log_e a = a^x \log_e a.$$

We summarize the formulas in the following table:

$$\ln u = \log_e u, \qquad a^u = e^{u \ln a},$$
$$\exp u = e^u, \qquad \log_b x = \frac{\ln x}{\ln b},$$
$$d \ln u = \frac{du}{u}, \qquad d \log_a u = \frac{du}{u \ln a},$$
$$de^u = e^u \, du, \qquad da^u = a^u (\ln a) \, du.$$

Example 1. Given that $f(x) = \ln |\sin x|$, find $f'(x)$.

Solution. We have

$$d \ln |\sin x| = \frac{1}{\sin x} d(\sin x) = \frac{\cos x \, dx}{\sin x}.$$

Therefore
$$f'(x) = \cot x. \blacktriangleleft$$

Example 2. Given that $f(x) = x^2 e^{-x^2}$, find $f'(x)$.

Solution. We have
$$d(x^2 e^{-x^2}) = x^2 d(e^{-x^2}) + e^{-x^2} d(x^2)$$
$$= x^2 e^{-x^2} d(-x^2) + 2xe^{-x^2} dx = e^{-x^2}(-2x^3 + 2x) dx.$$
Therefore
$$f'(x) = 2xe^{-x^2}(1 - x^2). \blacktriangleleft$$

Example 3. Given that $f(x) = \ln |\sec x + \tan x|$, find $f'(x)$.

Solution. We have
$$f'(x) = \frac{1}{\sec x + \tan x} \cdot \frac{d}{dx}(\sec x + \tan x)$$
$$= \frac{1}{\sec x + \tan x}(\sec x \tan x + \sec^2 x) = \sec x. \blacktriangleleft$$

Remark. Since each differentiation formula carries with it a corresponding integration formula, we see that the result
$$\frac{d}{dx} \ln |\sec x + \tan x| = \sec x$$
gives the interesting integration formula
$$\int \sec x \, dx = \ln |\sec x + \tan x| + C.$$

This follows, of course, from the fact that differentiation and integration are inverse processes. In a similar way, we can easily deduce that
$$\int \csc x \, dx = -\ln |\csc x + \cot x| + C. \blacktriangleleft$$

The function x^n is now defined for n any *real* number (see definition on page 326). The next theorem shows how we can extend to the case for any real number the differentiation formula for x^n stated previously for the case n a rational number on page 127.

Theorem 19. *If n is any real number and f is defined by $f(x) = x^n$, then $f'(x) = nx^{n-1}$, $x > 0$.*

Proof. We let $y = x^n$. From the definition on page 326, we have
$$y = x^n = \exp(n \log x) = e^{n \log x}.$$

332 THE TRIGONOMETRIC AND EXPONENTIAL FUNCTIONS

From the derivative formulas, we get

$$\frac{dy}{dx} = e^{n \log x} \frac{d}{dx}(n \log x) = e^{n \log x} n\left(\frac{1}{x}\right).$$

Therefore

$$\frac{dy}{dx} = x^n \cdot n\frac{1}{x} = nx^{n-1}. \qquad \blacktriangleleft$$

We learned in Theorem 18 how to differentiate functions of the form a^u, where a is constant and u is variable, and Theorem 19 yields the formula for u^a, where u is a variable and a is any constant. Sometimes we get functions of the form $u(x)^{v(x)}$, where u and v are both functions of x. If u and v can each be differentiated by known methods it is possible to develop a formula for the derivative of u^v. The following examples illustrate the procedure, called **logarithmic differentiation**.

Example 4. Given that $f(x) = x^x$, find $f'(x)$.

Solution. Write $y = x^x$, and take logarithms of both sides of this equation. Then $\log y = x \log x$. Differentiating this last equation implicitly, we find

$$\frac{1}{y}\frac{dy}{dx} = x \cdot \frac{1}{x} + \log x \cdot 1 = 1 + \log x.$$

Therefore

$$f'(x) = \frac{dy}{dx} = y(1 + \log x) = x^x(1 + \log x). \qquad \blacktriangleleft$$

Example 5. Given that

$$f(x) = \left|\frac{x^2 \sqrt[3]{3x + 2}}{(2x - 3)^3}\right|, \qquad x \neq \tfrac{3}{2}, 0, -\tfrac{2}{3},$$

find $f'(x)$.

Solution. Taking logs, we have

$$\log f(x) = 2 \log |x| + \tfrac{1}{3} \log |3x + 2| - 3 \log |2x - 3|.$$

Therefore

$$\frac{1}{f}f' = \frac{2}{x} + \frac{1}{3x + 2} - \frac{6}{2x - 3}$$

and

$$f'(x) = \left|\frac{x^2 \sqrt[3]{3x + 2}}{(2x - 3)^3}\right|\left(\frac{2}{x} + \frac{1}{3x + 2} - \frac{6}{2x - 3}\right). \qquad \blacktriangleleft$$

Example 6. Perform the integration:

$$\int (x - 1)e^{-x^2 + 2x} \, dx.$$

Solution. Set $u = -x^2 + 2x$. Then $du = (-2x + 2)\, dx$ and

$$\int (x - 1)e^{-x^2+2x}\, dx = -\frac{1}{2}\int e^u\, du = -\frac{1}{2}e^u = -\frac{1}{2}e^{-x^2+2x} + c.$$

PROBLEMS

In each of Problems 1 through 24, find the derivative.

1. $f(x) = e^{-3x}$
2. $f(x) = e^{x^3+2x^2-6}$
3. $f(x) = x^3 e^{4x}$
4. $f(x) = x^7 e^{-2x^2}$
5. $H(x) = e^{3x} \ln |x|$
6. $f(x) = xe^{-x} \ln(x^3)$
7. $g(x) = e^{\tan x}$
8. $F(x) = \exp(\sin x)$
9. $f(x) = 3^{5x}$
10. $f(x) = 4^{-2x}$
11. $f(x) = (x^3 + 3)2^{-7x}$
12. $F(x) = 2^{5x} \cdot 3^{4x^2}$
13. $F(x) = x^{\sin x}$
14. $F(x) = x^{\ln x}$
15. $G(x) = x^{\sqrt{x}}$
16. $G(x) = (\sin x)^x$
17. $H(x) = (\ln x)^x$
18. $H(x) = (3/x)^x$
19. $f(x) = \dfrac{x^2\sqrt{2x + 3}}{(x^2 + 1)^4}$
20. $f(x) = |x\sqrt{2x - 1}\sqrt[3]{3x + 3}|$
21. $g: x \to \sqrt{(x + 2)(x + 3)/(x + 1)}$
22. $g: x \to \left|\dfrac{(x + 2)^2(2x - 3)^{1/2}}{\sqrt[3]{3x - 2}}\right|$
23. $g(x) = \left|\dfrac{x^2 \arctan x}{1 + x^2}\right|$
24. $f(x) = \dfrac{\exp(x^2)}{\sqrt{x^2 + 1}\sqrt[4]{2x^2 + 3}}$

25. By differentiating the right side, verify the integration formula

$$\int \csc x\, dx = -\log|\cot x + \csc x| + C.$$

26. Integrate

$$\int x \sec(3x^2)\, dx.$$

27. Integrate

$$\int \frac{dx}{x^2 \cos(1/x)}.$$

28. Integrate

$$\int \frac{\cos^2 x\, dx}{\sin x}.$$

In each of Problems 29 through 32, perform the integration.

29. $\int e^{-3x}\, dx$
30. $\int xe^{-1+x^2}\, dx$
31. $\int e^{-\sin x}(\cos x)\, dx$
32. $\int \dfrac{1}{x} e^{-2\log_3 x}\, dx$

33. If n is any positive integer and $y = \ln x$, show that

$$\frac{d^n y}{dx^n} = (-1)^{n-1}\frac{(n - 1)!}{x^n}.$$

34. If n is any positive integer and $y = xe^x$, show that
$$\frac{d^n y}{dx^n} = y\left(1 + \frac{n}{x}\right).$$

35. Suppose that $F(x) = [f(x)]^{g(x)}$ where f and g are differentiable functions and $f(x) > 0$ for all x. Show that
$$F'(x) = gf^g\left[\frac{f'}{f} + \frac{g'}{g}\ln f\right].$$

36. The symbol x^{x^x} is interpreted as $x^{(x^x)}$.
   a) Find $f'(x)$ if $f(x) = x^{x^x}$.
   b) Find $g'(x)$ if $g(x) = x^{x^{x^x}}$.

10. THE NUMBER e

The function $f(x) = \log_e x$ has as its derivative the function $1/x$. We shall use this fact to gain additional insight into the number e. Suppose that we wish to find the derivative of $\log_e x$ from the very definition of derivative. We proceed in the following way (three-step rule).

Step 1: $\Delta f = f(x + h) - f(x) = \log_e (x + h) - \log_e x.$

Step 2: $\dfrac{1}{h}\Delta f = \dfrac{\log_e (x + h) - \log_e x}{h}.$

We use the properties of logarithms to write this last expression in the form
$$\frac{1}{h}\Delta f = \frac{1}{h}\log_e \frac{x + h}{x} = \frac{1}{h}\log_e\left(1 + \frac{h}{x}\right).$$

Before letting $h \to 0$ we multiply and divide by x, getting
$$\frac{1}{h}\Delta f = \frac{1}{x}\cdot\frac{x}{h}\log_e\left(1 + \frac{h}{x}\right) = \frac{1}{x}\cdot\log_e\left(1 + \frac{h}{x}\right)^{x/h}.$$

Now we are ready to let h tend to zero. We have (Step 3)
$$\lim_{h\to 0}\frac{1}{h}\Delta f = f'(x) = \frac{1}{x}\lim_{h\to 0}\log_e\left(1 + \frac{h}{x}\right)^{x/h}.$$

Since the answer must be $1/x$, we conclude that
$$\lim_{h\to 0}\log_e\left(1 + \frac{h}{x}\right)^{x/h} = 1.$$

Since $\log_e e = 1$, this answer is possible only if
$$\lim_{h\to 0}\left(1 + \frac{h}{x}\right)^{x/h} = e.$$

Letting $t = h/x$, we find that e is obtainable in the limiting form:
$$e = \lim_{t\to 0}(1 + t)^{1/t}.$$

By simple substitution, we can get an idea of how the function
$$G(t) = (1 + t)^{1/t}$$
behaves as $t \to 0$. We have
$$G(1) = 2, \quad G(\tfrac{1}{2}) = \tfrac{9}{4}, \quad G(\tfrac{1}{3}) = \tfrac{64}{27}, \quad G(\tfrac{1}{4}) = \tfrac{625}{256} = 2.4^+, \quad G(\tfrac{1}{10}) = 2.59^+.$$

PROBLEMS

1. Compute values of the function $G(t) = (1 + t)^{1/t}$ for $t = \tfrac{1}{5}, \tfrac{1}{6}, \tfrac{1}{7}, \tfrac{1}{8}$. Use logarithms to find $G(\tfrac{1}{20})$, $G(\tfrac{1}{50})$.
2. Compute $G(2), G(3), G(4), G(5)$. Sketch the graph.
3. Show that the function $H(t) = [1 + (1/t)]^t$ is an increasing function of t.
4. Let $H(t) = (1 + 1/t)^t$ and consider $H(n)$ for large integer values of n. The binomial formula yields
$$H(n) = \left(1 + \frac{1}{n}\right)^n = 1 + \frac{n}{1!}\left(\frac{1}{n}\right) + \frac{n(n-1)}{2!}\left(\frac{1}{n}\right)^2 + \cdots.$$
Use this result to show that
$$H(n) \leq 1 + 1 + \frac{1}{2!} + \cdots + \frac{1}{n!}.$$
Since $\lim_{n \to \infty} H(n) = e$, get an approximate value for e to three decimal places.
5. For $0 \leq u \leq 1$, integrate the obvious inequality $0 \leq e^u \leq e$ between 0 and x to obtain the inequality $0 \leq e^x - 1 \leq ex$. Repeat this n times and obtain the inequality
$$0 \leq e^x - 1 - x - \frac{x^2}{2!} - \cdots - \frac{x^n}{n!} \leq \frac{x^{n+1}}{(n+1)!} e.$$

11. APPLICATIONS

In Chapter 6 we discussed various applications of differentiation. Now we are able to differentiate many more types of functions than we could at that time—including logarithmic, exponential, trigonometric, and inverse trigonometric functions. Although the general methods we shall now discuss are the same as those described in Chapter 6, the degree of permissible complication has expanded greatly. The following examples illustrate some of the techniques which we can use.

Example 1. Find the equation of the line tangent to the curve $y = xe^{2x}$ at the point where $x = 2$.

Solution. We have
$$\frac{dy}{dx} = 2xe^{2x} + e^{2x}$$
and, at $x = 2$,
$$\frac{dy}{dx} = 4e^4 + e^4 = 5e^4.$$

Further, if $x = 2$, then $y = 2e^4$. The tangent line has the equation
$$y - 2e^4 = 5e^4(x - 2) \quad \Leftrightarrow \quad 5e^4 x - y = 8e^4. \quad \blacktriangleleft$$

Example 2. A gutter is to be made out of a long sheet of metal 12 cm wide by turning up strips of width 4 cm along each side so that they make equal angles θ with the vertical, as in Fig. 9–31. For what value of θ will the carrying capacity be greatest?

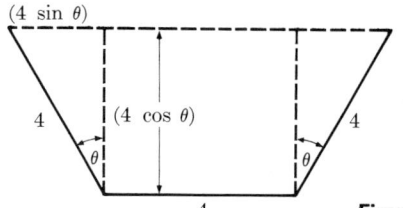

Figure 9–31

Solution. The carrying capacity is proportional to the area A of a cross section. If we find the maximum value for A, the capacity will also be a maximum. From Fig. 9–31 we see that the area of each of the triangles is $8 \sin \theta \cos \theta$.

Therefore, for $A = A(\theta)$, we have the formula

$$A = 16 \cos \theta + 16 \sin \theta \cos \theta, \quad 0 \leq \theta < \frac{\pi}{2},$$

$$\frac{dA}{d\theta} = -16 \sin \theta - 16 \sin^2 \theta + 16 \cos^2 \theta = 16(1 - \sin \theta - 2 \sin^2 \theta).$$

Setting the derivative equal to zero, we get

$$0 = 1 - \sin \theta - 2 \sin^2 \theta = (1 + \sin \theta)(1 - 2 \sin \theta).$$

The equation $\sin \theta + 1 = 0$ gives no root for $0 \leq \theta < (\pi/2)$, while

$$1 - 2 \sin \theta = 0 \quad \text{yields} \quad \theta = \frac{\pi}{6}.$$

The corresponding value of $\cos \theta$ is $\frac{1}{2}\sqrt{3}$, and we find that

$$A\left(\frac{\pi}{6}\right) = 12\sqrt{3}.$$

Since $A(0) = 16$ and $A(\theta) \to 0$ as $\theta \to (\pi/2)$, the maximum carrying capacity occurs at $\pi/6$ (rather than at an endpoint), and $A = 12\sqrt{3} \approx 20.8$. $\blacktriangleleft$

Example 3. A balloon leaving the ground 1200 meters from an observer rises at the rate of 200 meters per minute. How fast is the angle of elevation of the observer's line of sight increasing when the balloon is at an altitude of 1600 meters?

Solution. As shown in Fig. 9–32, the angle of elevation is θ and the rate of change of this angle is $d\theta/dt$, where t denotes the time in minutes. We have the

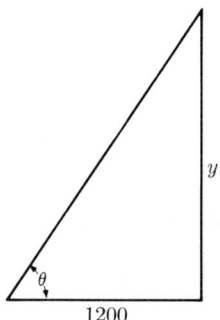

Figure 9–32

formula

$$\theta = \arctan\left(\frac{y}{1200}\right),$$

where $y = y(t)$ is the height of the balloon at time t. The rate at which the balloon is rising is $dy/dt = 200$. Differentiating with respect to t, we obtain

$$\frac{d\theta}{dt} = \frac{1}{1 + (y/1200)^2} \frac{d}{dt}\left(\frac{y}{1200}\right) = \frac{1200}{(1200)^2 + y^2} \frac{dy}{dt}.$$

Inserting $y = 1600$ and $dy/dt = 200$, we find $d\theta/dt = 0.06$ rad/min. ◀

PROBLEMS

In Problems 1 through 8, find in each case the equations of the lines tangent and normal to the given curve at the point on the curve corresponding to the given value of x.

1. $y = \arctan x$, $x = 1$
2. $y = 2^x$, $x = -1$
3. $y = \text{arcsec } x$, $x = -2$
4. $y = \arcsin x$, $x = -\frac{1}{2}$
5. $y = \log x$, $x = e$
6. $y = x^2 e^{-x}$, $x = 1$
7. $y = x \log x$, $x = e$
8. $y = x^{-2} e^{2x}$, $x = -1$

In Problems 9 through 15, find in each case the point or points of intersection of the curves given; then find the tangent of the acute angle between the curves at these intersection points.

9. $y = \arcsin x$, $y = \arccos x$
10. $y = \arctan x$, $y = \arcsin (x/2)$
11. $y = e^{2x}$, $y = 2e^x$
12. $y = \log x$, $y = \log x^2$
13. $y = \text{arccot } x$, $y = \frac{1}{2} \arcsin x$
14. $y = xe^x$, $y = x^2 e^x$
15. $y = xe^{-x}$, $y = x^2 e^{-x}$

In Problems 16 through 27, find in each case the relative maxima and minima, the points of inflection, the intervals in which f is increasing, those in which f is decreasing, those in which the graph is concave upward, and those in which it is concave downward. Sketch the graph.

16. $f(x) = e^{-x}$
17. $f(x) = xe^{-x}$
18. $f(x) = x^2 e^{-x}$
19. $f(x) = e^{-x^2}$
20. $f(x) = 2xe^{-(1/2)x^2}$
21. $f(x) = x^2 e^{-x^2}$

22. $f(x) = \log(1 + x)$
23. $f(x) = x \log x$
24. $f(x) = 2x^2 \log x$
25. $f(x) = 2x (\log x)^2$
26. $f(x) = e^{-x} \sin x$
27. $f(x) = e^{-x} \cos x$
28. Find the minimum value of x^x for $x > 0$.
29. A revolving light 3 km from a straight shoreline makes 2 revolutions/min. Find the speed of the spot of light along the shore when it is 2 km away from the point on the shore nearest the light.
30. A wall 8 m high is $\frac{27}{8}$ m from a building. Find the length of the shortest ladder which will clear the wall and rest with one end on the ground and the other end on the building. Also find the angle which this ladder makes with the horizontal (see Fig. 9–33).

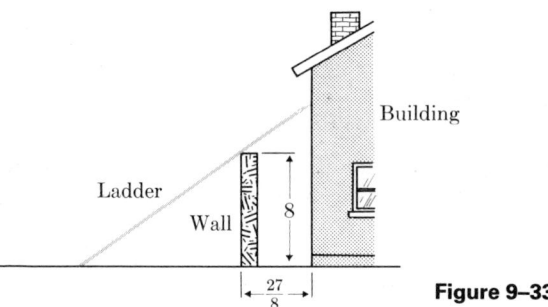

Figure 9–33

31. An airplane at an altitude of 4400 m is flying horizontally directly away from an observer. At the instant when the angle of elevation is 45°, the angle is decreasing at the rate of 0.05 radian/sec. How fast is the airplane flying at that instant?
32. A cone whose generators make an angle θ with its axis is inscribed in a sphere of radius R. For what value of θ will the lateral area of the cone be greatest?
33. A man is running along a sidewalk at the rate of 5 m/sec. A searchlight on the ground 30 m from the walk is kept trained on him. At what rate is the searchlight revolving when the man is 20 m away from the point on the sidewalk nearest the light?
34. A tablet 7 m high is placed on a wall with its base 9 m above the level of an observer's eye. How far from the wall should an observer stand in order that the angle subtended at his eye by the tablet be a maximum?
35. A steel girder 27 m long is moved horizontally along a corridor 8 m wide and around a corner into a hall at right angles to the corridor. How wide must the hall be to permit this? Neglect the width of the girder (see Fig. 9–34).

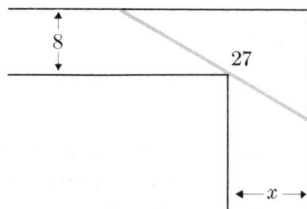

Figure 9–34

36. A man on a wharf is pulling in a small boat. His hands are 20 m above the level of the point on the boat where the rope is tied. If he is pulling in the rope at 2 m/sec, how

fast is the angle that the rope makes with the horizontal increasing when there are 52 m of rope out?

37. Find the angle of the sector which should be removed from a circular piece of canvas of radius 12 m so that the conical tent made from the remaining piece will have the greatest volume. Repeat the problem for a circular piece of canvas of radius R m.

38. The end B of a piston rod moves back and forth on a line through O; B is attached to a rod AB of length 10 cm which, in turn, is attached to a crank OA of length 4 cm, which revolves about O. Find an expression for the velocity of B if OA makes 10 rev/sec (see Fig. 9–35).

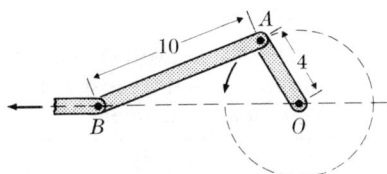

Figure 9–35

39. A hemispherical dome is 50 m in diameter. Just at sunset a ball is dropped from a point on the line joining the top of the dome with the sun. Assuming that the ball drops according to the law $u = 10t^2$ (u in m, t in sec), obtain an expression for the speed of its shadow along the dome in terms of t. Find its speed when $t = 1$ and the limit of its speed as $t \to 0$.

40. A weight is drawn along a level table by means of a rope, attached to a point P, which passes over a pulley whose axle is fixed horizontally at the level of the table and perpendicular to the line of motion of P. If the pulley, which has a radius of 1 m, turns at the rate of $\frac{1}{2}$ rev/sec, find an expression for the rate of change of the angle which the rope makes with the table. Assume that P is at the level of the table.

41. The hour hand of a clock in a tower is $\frac{3}{4}$ meter long and the minute hand is 1 meter long. At what rate in meters per hour are the ends of the hands approaching each other at 3 o'clock?

42. A radioactive substance disintegrates at a rate proportional to the amount of substance present. Let $F(t)$ be the amount at time t. Then $F(t)$ is determined by the relation
$$F'(t) = -cF(t)$$
where c is a positive constant depending on the substance.

a) Show that $F(t)$ is given by
$$F(t) = Ae^{-ct}$$
where A is the amount present at time $t = 0$.

b) Find the half-life of the substance; that is, find the value of t such that F has the value $\frac{1}{2}A$. (Answer in terms of c.)

43. Bacteria grow at a rate proportional to the amount present. Using Problem 42 as a guide, show that the number of bacteria $B(t)$ follows the law
$$B(t) = Ae^{ct}$$
where c is a positive constant and A is the number at time $t = 0$. Find the length of time required for the population to double. (Answer in terms of c.)

44. When a bank provides interest at 6% annually "compounded daily," then it provides each day 0.06/360 of the principal P as interest. (Banks consider a year to be 360 days.) In n days the total amount A becomes $A = P(1 + 0.06/360)^n$. Actually banks compound interest "continuously" by approximating the above formula with

$$A = Pe^{0.06n/360}.$$

a) Justify the approximation by finding the value of A by the two methods for $n = 1, 2, 3$ days.

b) At 6% interest compounded continuously, how long does it take for an initial deposit to double?

c) If an amount of money is deposited so that it doubles in 15 years, what must be the interest rate if the compounding is continuous?

45. The current I (in amperes) in a battery (and other electrical networks) obeys the law $LI'(t) = -RI(t)$ where L is the inductance (henrys) and R is the resistance (ohms). Express I as a function of t if L and R are constant and $I = 2$ when $t = 0$.

12. THE HYPERBOLIC FUNCTIONS

Exponential functions occur often in many applications, especially those in physics and engineering. Certain combinations of them occur so often that they have been tabulated and given special names: the **hyperbolic functions.** These functions satisfy relations which bear a great similarity to the basic formulas which the trigonometric functions satisfy. Furthermore, the differentiation and integration properties of hyperbolic functions are quite similar to those of the trigonometric functions. However, one of the most important properties of the trigonometric functions—that of periodicity—is not shared by any of the hyperbolic functions.

We provide in Appendix 3 at the end of the book two sections giving the definitions, elementary properties, and differentiation and integration formulas of hyperbolic and inverse hyperbolic functions. Graphs of the various functions may be found there. Readers not interested in hyperbolic functions may skip this subject without loss of continuity in the development of either the theory or the techniques of calculus.

10

PARAMETRIC EQUATIONS. ARC LENGTH

1. PARAMETRIC EQUATIONS

We consider the two equations
$$x = t^2 + 2t, \quad y = t - 2,$$
involving the quantities x, y, and t. Each value of t determines a value of x and one of y and therefore a point (x, y) in the plane. The totality of these points obtained by letting t take on all possible values is a relation in R_2. The graph of this relation may be exhibited by the following method. We set up a table of values for t, x, and y by letting t take on various values and then computing x and y from the given equations. The points (x, y) are plotted in the usual way on a Cartesian coordinate system. The table is shown below and the graph is sketched in Fig. 10–1. The t scale is completely separate and does not appear in the graph. The quantity t is called a **parameter** and the above equations are called the parametric equations of the graph shown in Fig. 10–1.

t	0	1	2	3	4	−1	−2	−3	−4
x	0	3	8	15	24	−1	0	3	8
y	−2	−1	0	1	2	−3	−4	−5	−6

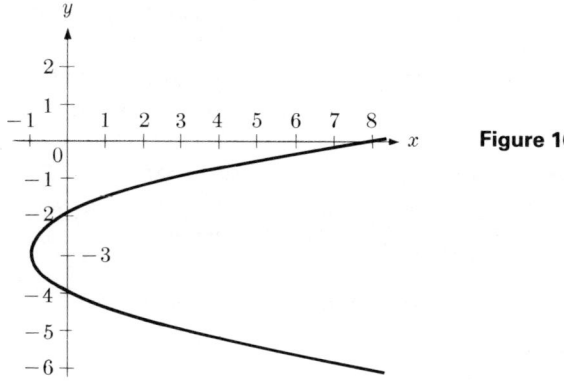

Figure 10–1

In general, if f and g are functions with the same domain S in R_1, and if we write
$$x = f(t), \quad y = g(t),$$
then we say that the two equations form a set of **parametric equations**. The **graph** of the parametric equations is the set of points in the xy plane which we get when t takes on all possible values in the domain S.

Let us now consider the particularly simple case in which both f and g are *linear* functions. That is, we suppose that
$$f(t) = x_0 + at, \quad g(t) = y_0 + bt,$$
where x_0, y_0, a, and b are numbers. We establish the following result.

Theorem 1. *The set of parametric equations*
$$x = x_0 + at, \quad y = y_0 + bt, \tag{1}$$
so long as a and b are not both zero, has as its graph a straight line in the xy plane. This line (denoted by L in Fig. 10–2) *passes through the point (x_0, y_0). If $a \neq 0$, then L has slope b/a. If $a = 0$, the line is vertical.*

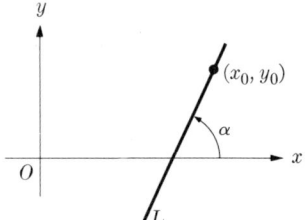

Figure 10–2

Proof. Suppose that $a \neq 0$. Solving $x = x_0 + at$ for t and substituting in the equation $y = y_0 + bt$, we find that each point on the graph of (1) satisfies the equation
$$y - y_0 = \frac{b}{a}(x - x_0), \tag{1'}$$
which we recognize as the equation of a straight line through the point (x_0, y_0) with slope b/a. Moreover, if (x, y) is any point on the line $(1')$ and we define $t = (x - x_0)/a$, then x and y satisfy (1).

If $a = 0$, we see that $x = x_0$ always and, since $b \neq 0$, y may have any value. Thus L must be a vertical line through (x_0, y_0), and the theorem is proved.

Remark. The converse of Theorem 1 is not difficult to establish. That is, *if a line L is given by an equation of the form $y - y_0 = m(x - x_0)$, or if the line is vertical, then it is possible to represent L parametrically by Eqs.* (1). (See Problem 22 at the end of this section.)

If k is any number different from zero, the equations
$$x = x_0 + kat, \quad y = y_0 + kbt \tag{2}$$

represent the same line as Eqs. (1). To see this we observe that if $a \neq 0$, then the slope of the line determined by (2) is $kb/ka = b/a$ and the line goes through the point (x_0, y_0). Also, if $a = 0$, the line must be vertical and pass through (x_0, y_0). Denoting the inclination of L by α (see Fig. 10–2), we may select k so that the parametric equations of L have the form

$$x = x_0 + (\cos \alpha)t, \qquad y = y_0 + (\sin \alpha)t. \qquad (3)$$

Once a and b are given, we verify easily that the selections

$$k = 1/\sqrt{a^2 + b^2} \qquad \text{for } b > 0,$$
$$k = -1/\sqrt{a^2 + b^2} \qquad \text{for } b < 0,$$
$$k = 1/a \qquad \text{for } b = 0,$$

transform the parametric equations (2) into the form (3). Thus we see (by the Remark after Theorem 1) that *every straight line in the plane may be represented parametrically by Eqs. (3)*. We call them the **parametric equations of a line in standard form.**

Example 1. Find parametric equations of the line L through the points $(2, -1)$ and $(4, 7)$. Also determine the parametric equations in standard form.

Solution. We may select $(2, -1)$ as (x_0, y_0). The slope of L is

$$\frac{7 - (-1)}{4 - 2} = 4.$$

Selecting $b = 4$ and $a = 1$, we substitute in Eqs. (1) to get the parametric equations

$$x = 2 + t, \qquad y = -1 + 4t.$$

Since $b > 0$, we choose $k = 1/\sqrt{1 + 16} = 1/\sqrt{17}$ and the standard form of the parametric equations is

$$x = 2 + \frac{1}{\sqrt{17}} t, \qquad y = -1 + \frac{4}{\sqrt{17}} t.$$

We note that since $\cos \alpha = 1/\sqrt{17}$, $\sin \alpha = 4/\sqrt{17}$, the inclination α of L is an acute angle whose value is $\arccos (1/\sqrt{17})$. ◀

When we are given a pair of parametric equations it is sometimes possible to *eliminate the parameter* by solving for t in one of the equations and then substituting in the other. In the equations

$$x = t^2 + 2t, \qquad y = t - 2,$$

which we considered at the beginning of the section, we may solve the second equation for t, getting $t = y + 2$, and then substitute in the first equation. We obtain

$$x = (y + 2)^2 + 2(y + 2) = y^2 + 6y + 8,$$

and we recognize the equation $x = y^2 + 6y + 8$ as that of a parabola, as indeed Fig. 10–1 shows. We have shown that any point on the graph of the parametric equations satisfies this equation. Moreover, if (x, y) is any point satisfying this equation, we may define $t = y + 2$ and find that x and y satisfy the parametric equations. It can happen, however that the graph of the parametric equations is not the same as the graph of the equation in x and y obtained by "eliminating the parameter t." For example, the graph of the parametric equations

$$x = \cos t, \qquad y = \cos^2 t \tag{4}$$

is just part of the graph of the parabola $y = x^2$ obtained by eliminating t. In fact, since $-1 \leq \cos t \leq 1$ for all t, the graph of (4) is that part of the parabola $y = x^2$ for which $-1 \leq x \leq 1$. See Problem 23 at the end of this section.

The equations

$$x = t^7 + 3t^2 - 1,$$
$$y = e^t + 2t^2 - 3\sqrt{t},$$

are a pair of parametric equations. Because of the square-root sign, it is clear that the domain for t is restricted to nonnegative real numbers. It is not possible to eliminate the parameter in any simple way, since solving either equation for t gives the appearance of being a herculean task.

Example 2. Plot the graph of the equations

$$x = t^2 + 2t - 1, \qquad y = t^2 + t - 2.$$

Eliminate the parameter, if possible.

Solution. We construct the following table:

t	-4	-3	-2	-1	0	1	2	3
x	7	2	-1	-2	-1	2	7	14
y	10	4	0	-2	-2	0	4	10

By plotting these points, we obtain the curve shown in Fig. 10–3. To eliminate the parameter, we first subtract the equations, obtaining

$$x - y = t + 1 \quad \Leftrightarrow \quad t = x - y - 1.$$

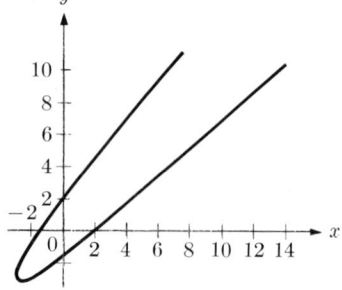

Figure 10–3

Substituting for t in the second of the equations, we find that
$$y = (x - y - 1)^2 + (x - y - 1) - 2,$$
and, simplifying,
$$x^2 - 2xy + y^2 - x - 2 = 0. \qquad \blacktriangleleft$$

Example 3. A projectile moves approximately according to the law
$$x = (v_0 \cos \alpha)t, \qquad y = (v_0 \sin \alpha)t - 5t^2,$$
where v_0 and α are constants and t is the time, in seconds, after the projectile is fired. The equations give the Cartesian coordinates x, y (in meters) of the center of the projectile in the vertical plane of motion with the muzzle of the gun at the origin of the coordinate system, the x axis horizontal, and the y axis vertical; v_0 is the muzzle velocity, i.e., the velocity of the projectile at the instant it leaves the gun; α is the angle of inclination of the projectile as it leaves the gun (Fig. 10–4). If $v_0 = 100$ m/sec, $\cos \alpha = \frac{3}{5}$, and $\sin \alpha = \frac{4}{5}$, what are the coordinates of the center of the projectile at times $t = 1, 2, 3, 4,$ and 5? Find the time T when the projectile hits the ground, and find the distance R from the muzzle of the gun to the place where the projectile strikes the ground.

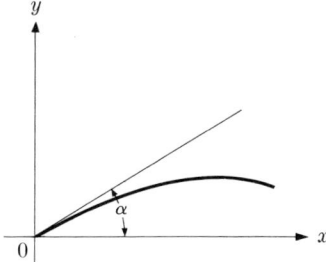

Figure 10–4

Solution. We have $v_0 \cos \alpha = 60$, $v_0 \sin \alpha = 80$, and so
$$x = 60t, \qquad y = 80t - 5t^2.$$
The projectile hits the ground when $y = 0$. Therefore $0 = 80t - 5t^2$ and $t = 0$, 16. Consequently, $T = 16$. To find R, we insert $t = 16$ in $x = 60t$, obtaining $R = 960$ m. We construct the table:

t	0	1	2	3	4	5	6	7	8	16
x	0	60	120	180	240	300	360	420	480	960
y	0	75	140	195	240	275	300	315	320	0

Plotting the points yields the curve shown in Fig. 10–5; t is restricted to the interval $0 \le t \le 16$. $\qquad \blacktriangleleft$

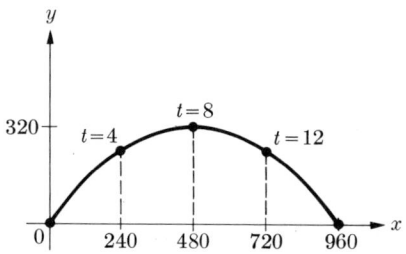

Figure 10–5

A circular hoop starts rolling along a stretch of level ground. A point on the rim of the hoop has a mark on it. We wish to find the path traced out by the marked point. Figure 10–6 shows the hoop in a number of different positions as it rolls along. The curve traced out by the marked point, which we denote by P, may be expressed in terms of parametric equations. Let a be the radius of the hoop, and suppose that when the hoop begins rolling the point P is on the ground at the point labeled O in Fig. 10–7. Figure 10–7 shows the position of the point P after the hoop has turned through an angle θ. Since the hoop is assumed to roll without slipping, we have, from the diagram,

$$|OM| = \text{arc } \widehat{MP} = a\theta, \quad \theta \text{ in radians.}$$

From $\triangle CPQ$ we read off

$$|PQ| = a \sin \theta, \quad 0 \leq \theta \leq \frac{\pi}{2},$$

$$|QC| = a \cos \theta, \quad 0 \leq \theta \leq \frac{\pi}{2}.$$

Denoting the coordinates of P by (x, y), we see that

$$x = |ON| = |OM| - |NM| = \text{arc } \widehat{MP} - |PQ| = a\theta - a \sin \theta,$$
$$y = |NP| = |MC| - |QC| = a - a \cos \theta.$$

Even though these equations were derived for θ between 0 and $\pi/2$, it can be shown that for all values of θ the parametric equations

$$x = a(\theta - \sin \theta), \quad y = a(1 - \cos \theta)$$

represent the path of the marked point on the rim. This curve is called a **cycloid**.

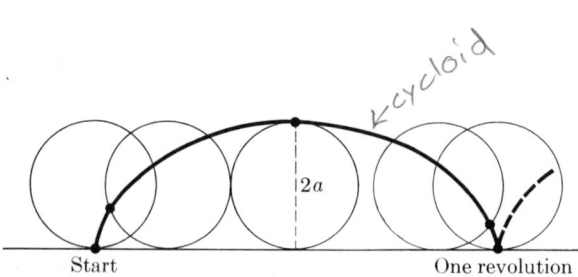

Figure 10–6

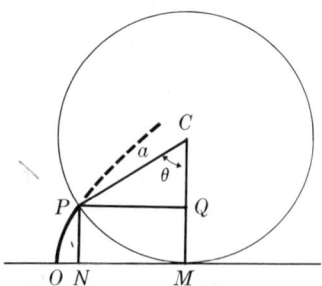

Figure 10–7

The cycloid is a particularly good example of a curve which is obtained without too much difficulty by use of parametric equations, while any attempt to find the relation between x and y without restoring to a parameter would lead to an almost insurmountable problem.

PROBLEMS

In Problems 1 through 6, in each case find the set of parametric equations in standard form for the line L determined by the given conditions.

1. L passes through the points $(3, -7)$ and $(-4, 6)$.
2. L passes through the points $(-3, 1)$ and $(-3, 8)$.
3. L passes through $(4, -2)$ and is parallel to the line with equation $2x + 5y = 1$.
4. L passes through $(-1, -3)$ and is perpendicular to the line with equation $x + 3y = 2$.
5. L has slope $\frac{2}{3}$ and passes through the intersection of the lines $2x - 3y - 3 = 0$, $x + 2y = 5$.
6. L passes through $(2, -6)$ and is perpendicular to the line $y = 7$.

In Problems 7 through 21, plot each curve as in Example 2; in each case eliminate the parameter and get an equation relating x and y (which is satisfied by every point on the graph).

7. $x = 2t,\ y = -5t$
8. $x = t - 1,\ y = t^2$
9. $x = t,\ y = 1/t$
10. $x = 2 + (s^2/2),\ y = -1 + \frac{1}{8}s^3$
11. $x = -1 + \cos\theta$, $y = 2 + 2\sin\theta$
12. $x = 3\cos\theta$, $y = 2\sin\theta$
13. $x = 2\cos^3\theta,\ y = 2\sin^3\theta$
14. $x = 3\sec t,\ y = 2\tan t$
15. $x = t^2 + 2t + 3$, $y = t^2 + t - 1$
16. $x = t^2 + t + 1$, $y = \frac{1}{2}t^2 + t - 1$
17. $x = \dfrac{20t}{4 + t^2}$, $y = \dfrac{5(4 - t^2)}{4 + t^2}$
18. $x = \dfrac{3(2 - t)^2(2 + t)}{6t^2 + 8}$, $y = \dfrac{3(2 - t)(2 + t)^2}{6t^2 + 8}$
19. $x = e^t,\ y = e^{-t}$
20. $x = \ln s,\ y = e^{2s}$
21. $x = e^t + e^{-t},\ y = e^t - e^{-t}$

22. Prove that every line in the plane has a parametric representation in the form
$$x = x_0 + at, \qquad y = y_0 + bt.$$
[Hint: Consider two cases: one in which the line is vertical and one in which the line is not vertical.]

23. Plot the graph of each of the following pairs of parametric equations.
    - i) $x = \cos t,\quad y = \cos^2 t$
    - ii) $x = \cos^2 t,\quad y = \cos^4 t$
    - iii) $x = t^2,\quad y = t^4$
    - iv) $x = -\sqrt{-t},\quad y = -t$

    Draw the graph of $y = x^2$ and compare each of the above graphs with this graph. Also find two more sets of parametric equations which yield a part of the parabola $y = x^2$ different from (i) through (iv) above.

24. Assuming that the equations of Example 3 hold, and given that $v_0 = 500$ m/sec,

$\cos\alpha = \frac{4}{5}$, $\sin\alpha = \frac{3}{5}$, find x and y for $t = 2, 4, 6, 8$, and 10. Also find T and R and plot the trajectory.

25. Assuming that the equations of Example 3 hold, find T and R in terms of v_0 and α. Find y in terms of x (i.e., eliminate the parameter).

26. A wheel of radius a rolls without slipping along a level stretch of ground. A point on one of the spokes of the wheel is marked. Find the path traced by this point, given that it is at a distance b from the center.

27. Find the locus of a point P on a circle of radius a which rolls without slipping on the inside of a circle of radius $4a$. Choose the center of the large circle as origin, choose the positive x axis through a point where P touches the large circle, and choose the parameter θ as the angle xOC, where C is the center of the small circle (Fig. 10–8). [*Hint:* Note that arc $\widehat{BHP}$ = arc $\widehat{MB}$.]

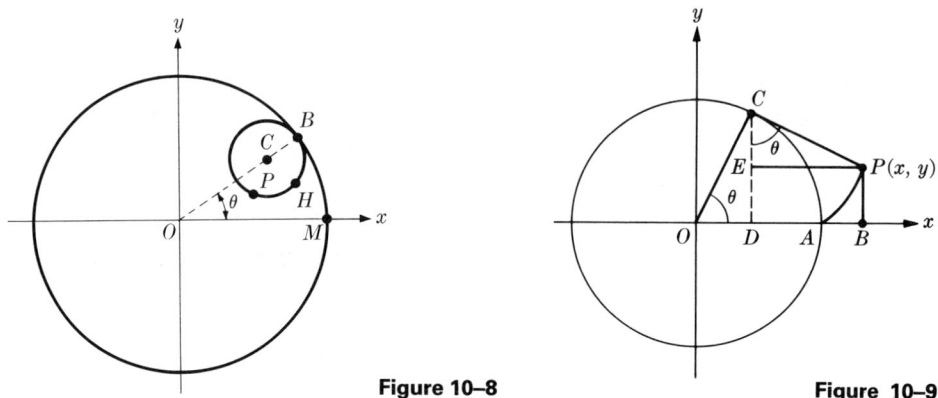

Figure 10–8 Figure 10–9

28. A string is wound about a circle of radius a. The path traced by the end of the string as it is unwound is called the **involute of the circle.** Refer to Fig. 10–9 and show that the equations of the involute are

$$x = a\cos\theta + a\theta\sin\theta$$
$$y = a\sin\theta - a\theta\cos\theta.$$

29. Plot the graph of the equations

$$x = a\sin\theta, \quad y = a\tan\theta(1 + \sin\theta).$$

Eliminate the parameter and show that the line $x = a$ is an asymptote. The curve is called a **strophoid.**

30. A circle of radius a is drawn tangent to the x axis as shown in Fig. 10–10. The line OA intersects the circle at point B. Then the projection of AB on the vertical line through A is the segment AP. The graph of the point P as B moves around the circle is called the **Witch of Agnesi.** Show that the curve is given by the parametric equations

$$x = 2a\cot\theta, \quad y = 2a\sin^2\theta.$$

Eliminate the parameter and show that the x axis is an asymptote.

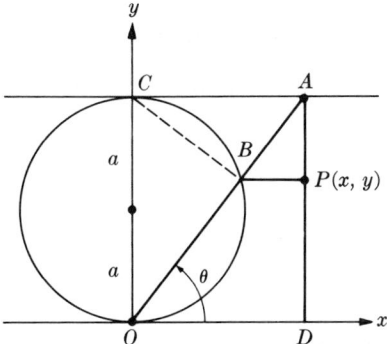

Figure 10–10

2. DERIVATIVES AND PARAMETRIC EQUATIONS

In previous chapters we learned how to use the first and second derivatives of functions as aids in drawing graphs. Such aids can also be used with parametric equations, even when the parameter cannot be eliminated. Suppose we have the equations

$$x = f(t), \quad y = g(t)$$

which represent a relation in the xy plane. Usually the graph of this relation is a curve in the plane, but it is only rarely that the graph will represent a function. No matter how complicated the shape of the curve may be, if it is "smooth," then a tangent line may be drawn at each point (see Fig. 10–11). The slope of the curve can be found if we can compute dy/dx. When x and y are given parametrically, this derivative may be found by the Chain Rule. We have

$$\frac{dy}{dx} = \frac{dy}{dt} \cdot \frac{dt}{dx} = \frac{dy/dt}{dx/dt}.$$

This derivative will be given in terms of t. If t cannot be expressed in terms of x or y—as is frequently the case—the process of getting the second derivative requires some explanation. The idea is to use the Chain Rule again. We write

$$\frac{d^2y}{dx^2} = \frac{d}{dx}\left(\frac{dy}{dx}\right) = \frac{d}{dt}\left(\frac{dy}{dx}\right)\frac{dt}{dx} = \frac{d}{dt}\left(\frac{dy}{dx}\right) \div \frac{dx}{dt}.$$

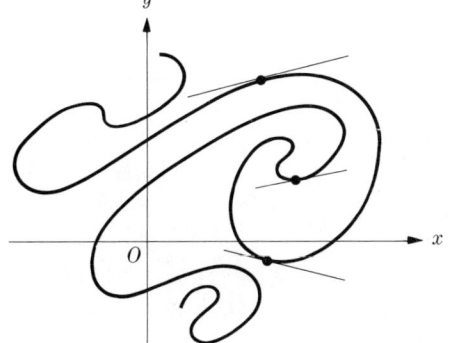

Figure 10–11

Since dy/dx is given in terms of t, finding $\dfrac{d}{dt}\left(\dfrac{dy}{dx}\right)$ is a routine matter. Furthermore, we calculated dx/dt previously, when we obtained dy/dx. We illustrate the technique with three examples.

Example 1. Find dy/dx, d^2y/dx^2, and d^3y/dx^3, given that $x = t^2 + 3t - 2$, $y = 2 - t - t^2$.

Solution. We have
$$\frac{dx}{dt} = 2t + 3, \qquad \frac{dy}{dt} = -1 - 2t.$$

Therefore
$$\frac{dy}{dx} = \frac{dy/dt}{dx/dt} = \frac{-(2t+1)}{2t+3}.$$

The second derivative is given by
$$\frac{d^2y}{dx^2} = \frac{\dfrac{d}{dt}\left(\dfrac{dy}{dx}\right)}{dx/dt},$$

and since
$$\frac{d}{dt}\left(\frac{dy}{dx}\right) = \frac{d}{dt}\left(-\frac{2t+1}{2t+3}\right) = -\frac{(2t+3)(2) - (2t+1)(2)}{(2t+3)^2} = \frac{-4}{(2t+3)^2},$$

we get
$$\frac{d^2y}{dx^2} = -\frac{4/(2t+3)^2}{2t+3} = \frac{-4}{(2t+3)^3}.$$

We have not derived a formula for the third derivative, but the process is analogous. We write
$$\frac{d^3y}{dx^3} = \frac{d}{dx}\left(\frac{d^2y}{dx^2}\right) = \frac{d}{dt}\left(\frac{d^2y}{dx^2}\right) \cdot \frac{dt}{dx} = \frac{\dfrac{d}{dt}\left(\dfrac{d^2y}{dx^2}\right)}{dx/dt}.$$

For the numerator, we obtain
$$\frac{d}{dt}\left(\frac{d^2y}{dx^2}\right) = \frac{d}{dt}\left(\frac{-4}{(2t+3)^3}\right) = \frac{24}{(2t+3)^4},$$

and so
$$\frac{d^3y}{dx^3} = \frac{24/(2t+3)^4}{2t+3} = \frac{24}{(2t+3)^5}.$$

Example 2. Given that $x = x(t) = t^2 - t$, $y = y(t) = t^3 - 3t$. Find dy/dx and d^2y/dx^2. Plot the graph.

Solution. We compute

$$\frac{dx}{dt} = 2t - 1, \qquad \frac{dy}{dt} = 3(t^2 - 1);$$

$$\frac{dy}{dx} = \frac{3(t^2 - 1)}{2t - 1}, \qquad \frac{d^2y}{dx^2} = \frac{\frac{d}{dt}\left(\frac{dy}{dx}\right)}{dx/dt} = 6\frac{t^2 - t + 1}{(2t - 1)^3}.$$

We construct the table of values:

t	-3	-2	$-\sqrt{3}$	-1	0	$\tfrac{1}{2}$	1	$\sqrt{3}$	2	3
x	12	6	$3+\sqrt{3}$	2	0	$-\tfrac{1}{4}$	0	$3-\sqrt{3}$	2	6
x'	$-$	$-$	$-$	$-$	$-$	0	$+$	$+$	$+$	$+$
y	-18	-2	0	2	0	$-\tfrac{11}{8}$	-2	0	2	18
y'	$+$	$+$	$+$	0	$-$	$-$	0	$+$	$+$	$+$

We draw the graph, as in Fig. 10–12.

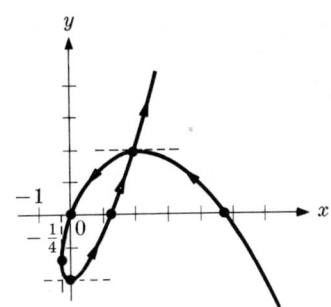

Figure 10–12

Example 3. Find the equation of the line tangent to the curve

$$x = t^2 - 2, \qquad y = t^3 - 2t + 1,$$

at the point where $t = 2$.

Solution. When $t = 2$, we have $x = 2$, $y = 5$. Further, $dx/dt = 2t$, $dy/dt = 3t^2 - 2$, and $dy/dx = (3t^2 - 2)/2t$. The slope of the tangent line is

$$\frac{3(2)^2 - 2}{2(2)} = \frac{5}{2}.$$

The equation of the desired line is

$$y - 5 = \tfrac{5}{2}(x - 2) \quad \Leftrightarrow \quad 5x - 2y = 0.$$

PROBLEMS

In Problems 1 through 10, find dy/dx and d^2y/dx^2. Give results in terms of the parameter.

1. $x = 4t + 6,\ y = 2 - 5t$
2. $x = 2t^2 - 3,\ y = t^2 + 4t - 1$
3. $x = 2 - t^2,\ y = t^3 + 4t^2 + 2t$
4. $x = 3 \cos t,\ y = 5 \sin t$
5. $x = e^{-2t},\ y = 1 + 3t^2$
6. $x = 2t + 1,\ y = 2e^{3t}$
7. $x = 4 \cos t,\ y = 2 \sin^2 t$
8. $x = 2 \sin 2\theta,\ y = 2 \sin \theta$
9. $x = a \cos^3 \theta,\ y = a \sin^3 \theta$
10. $x = a\theta - a \sin \theta,\ y = a - a \cos \theta$ (cycloid)
11. Find $dy/dx,\ d^2y/dx^2,\ d^3y/dx^3,\ d^4y/dx^4$, given that $x = t^3 + t,\ y = \tfrac{3}{2}t^4 + t^2$.
12. Same as Problem 11, given that $x = e^{2t},\ y = e^t + e^{-t}$.

In Problems 13 through 20, find in each case the equations of the lines tangent and normal to the specified curve at the point corresponding to the given value of the parameter.

13. $x = t^2 + 1,\ y = t^3 + 2t,\ t = -2$
14. $x = e^{2t},\ y = 2 + t^2,\ t = 1$
15. $x = 4 \cos t,\ y = 2 \sin^2 t,\ t = \pi/3$
16. $x = 5 \cos t,\ y = 4 \sin t,\ t = \pi/3$
17. $x = t^2 - 1,\ y = 2e^t,\ t = -1$
18. $x = 2 \sin 2\theta,\ y = 2 \sin \theta,\ \theta = \pi/4$
19. $x = 2 \cos^3 \theta,\ y = 2 \sin^3 \theta,\ \theta = \pi/4$
20. $x = 3\theta - 3 \sin \theta,\ y = 3 - 3 \cos \theta,\ \theta = \pi/2$

In Problems 21 through 26, find in each case the intervals of values of the parameter when x and y are increasing and decreasing. Plot the curve. Solve for y in terms of x.

21. $x = 2 \tan \phi,\ y = \sec \phi,\ -\pi/2 < \phi < 3\pi/2$
22. $x = 2 \tan \theta,\ y = 4 \cos^2 \theta,\ -\pi/2 < \theta < \pi/2$
23. $x = t^2 + 2t,\ y = t^2 + t,\ -\infty < t < \infty$
24. $x = \cos \theta,\ y = \sin 2\theta,\ 0 \le \theta \le 2\pi$
25. $x = e^{2t} + 1,\ y = 1 - e^{-t},\ -\infty < t < \infty$
26. $x = t^2(t - 2),\ y = t(t - 2)^2,\ -\infty < t < \infty$. (Do not try to solve for y in terms of x.)
27. Given that
$$x = 1 + f(t), \qquad y = \frac{1 - f(t)}{1 + f(t)}, \qquad \text{find}\ \frac{dy}{dx}$$
in terms of f and f'.
28. Given that $x = a \cos g(t),\ y = b \sin g(t)$. Prove that
$$xy^2 \frac{d^2y}{dx^2} = b^2 \frac{dy}{dx}.$$
29. A projectile moves so that its parametric equations are
$$x = (v_0 \cos \alpha)t, \qquad y = (v_0 \sin \alpha)t - 5t^2$$
where v_0 is the initial speed and α is the angle the direction of motion makes with the horizontal at $t = 0$.
a) Eliminate the parameter t and show that the path of the projectile is parabolic.
b) The speed s at any time t is given by
$$s = \left[\left(\frac{dx}{dt}\right)^2 + \left(\frac{dy}{dt}\right)^2\right]^{1/2}.$$
Find s when $t = 1$ and $t = 2$.
c) Show that the projectile hits the ground when $t = \tfrac{1}{5}v_0 \sin \alpha$. Find how far the projectile has travelled horizontally at that time.

30. A particle moves so that its parametric equations are given by

$$x = \frac{4t}{t^2 + 4}, \quad y = \frac{t^2 - 4}{t^2 + 4}.$$

a) Show that the point moves on the circle of radius 1 with center at the origin.
b) The speed s is given by

$$s = \left[\left(\frac{dx}{dt}\right)^2 + \left(\frac{dy}{dt}\right)^2\right]^{1/2}.$$

Find the speed in terms of t.

c) The magnitude of the acceleration, denoted $|\mathbf{a}|$ is given by

$$\left[\left(\frac{d^2x}{dt^2}\right)^2 + \left(\frac{d^2y}{dt^2}\right)^2\right]^{1/2}.$$

Find $|\mathbf{a}|$ when $t = 1$ and $t = 2$.

---

3. ARC LENGTH

We frequently draw the graph of a function $y = f(x)$ and refer to it as the curve representing the function. When picturing such a graph, we automatically associate a length with any part of it (e.g., the portion going from P_1 to P_2 in Fig. 10–13). We now raise three questions:

1. What are the kinds of curves with which we shall associate a length?
2. How do we define length?
3. Once we have defined length, how do we measure it?

These questions are easily answered for straight lines. Every straight line segment has a length given by the distance formula

$$d = \sqrt{(x_1 - x_2)^2 + (y_1 - y_2)^2},$$

where (x_1, y_1) and (x_2, y_2) are the coordinates of the endpoints of the segment. We are familiar with the formula for the length of a circular arc, but for many students the derivation of this formula is but a vague memory from high-school geometry or trigonometry. A first step in a precise discussion of the length of curves, of which a circular arc is a special case, is the definition of what we call an *arc*.

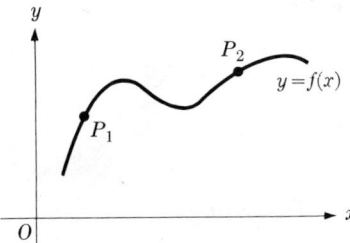

Figure 10–13

Definition. *If a graph is given by a function such as $y = f(x)$, $a \leq x \leq b$, and if f is continuous on this interval, then the graph of f is called an* **arc.** *When the graph is given by the parametric equations*

$$x = F(t), \qquad y = G(t), \qquad c \leq t \leq d,$$

it is called an **arc** *if F and G are continuous on the interval $[c, d]$ and if for two different values of the parameter, t_1 and t_2, it can never happen that both $F(t_1) = F(t_2)$ and $G(t_1) = G(t_2)$.*

Remark. This last condition, which guarantees that the graph does not intersect itself, may be written more compactly as

$$[F(t_1) - F(t_2)]^2 + [G(t_1) - G(t_2)]^2 > 0 \qquad \text{if} \qquad t_1 \neq t_2.$$

The first of the three questions we raised may now be answered by stating that we shall discuss the lengths of only those curves which are arcs.

Turning to the second question—that of defining length—we shall proceed by making use of the one type of arc whose length we know how to measure: the line segment. Let C be an arc in the plane, as shown in Fig. 10–14. Suppose that we wish to define the length of such an arc between the points A and B. *The definition of length requires a limiting process.* As a first step we mark off a number of points on the arc (in order) between A and B and label them $P_1, P_2, \ldots, P_{n-1}$. Setting $P_0 = A$ and $P_n = B$, we draw a straight line segment from each point P_i to the next point P_{i+1}, as in Fig. 10–15. From the distance formula for the length of a line segment, we compute the lengths

$$|P_0 P_1|, \qquad |P_1 P_2|, \qquad |P_2 P_3|, \qquad \ldots, \qquad |P_{n-1} P_n|,$$

and add them. We can write the sum concisely in the form

$$\sum_{i=1}^{n} |P_{i-1} P_i|.$$

If the points P_i are "close together," we feel intuitively that the total length of the line segments will be "close to" the as-yet-undefined length of the curve. Therefore, if a sequence of subdivisions is made with an increasing number of points in succeeding subdivisions, we would expect that in the limit the length of the arc would be attained. This is indeed the case. In any subdivision of C we

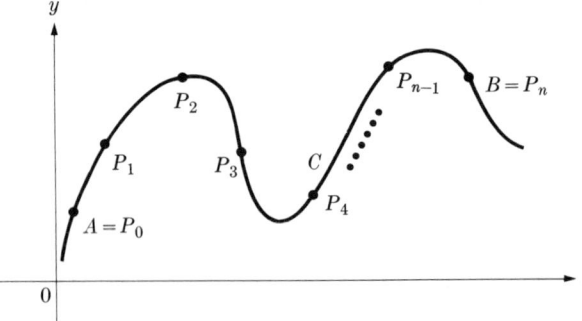

Figure 10–14

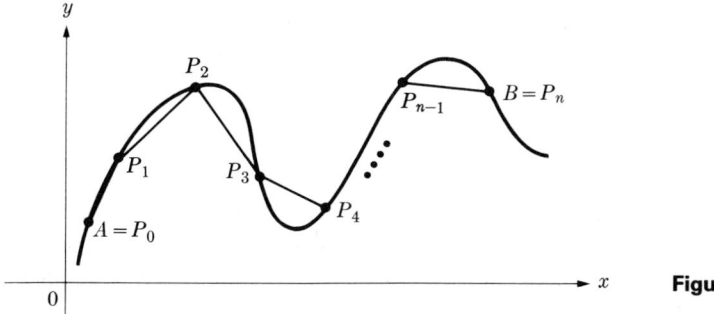

Figure 10–15

denote the length of the longest line segment connecting successive points by $\|\Delta\|$ and call this the **norm of the subdivision.** We recall that the norm of a subdivision of a section of the x axis was encountered in the definition of integral.

Definition. **An arc C from A to B has length** *if there is a number L with the following property: For each $\epsilon > 0$ there is a $\delta > 0$ such that*

$$\left| \sum_{i=1}^{n} |P_{i-1}P_i| - L \right| < \epsilon$$

for every subdivision $A = P_0, P_1, P_2, \ldots, P_{n-1}, P_n = B$ with $\|\Delta\| < \delta$. The number L is called the **length** *of the arc C.*

This number L is unique if it exists at all.

Any arc which has a length is called **rectifiable.** According to the definition, *the length L is the limit of the lengths of inscribed polygons, as the maximum distance between successive points of these inscribed polygons tends to zero.*

We now turn to the third question—that of actually finding ways of computing the length of an arc. Suppose that the arc C which has length s is given in the form $y = f(x)$, with f possessing a continuous derivative and with the endpoints of the arc at $A(a, f(a))$, $B(b, f(b))$, as shown in Fig. 10–16. We introduce subdivision points $P_1, P_2, \ldots, P_{n-1}$, with coordinates $P_i(x_i, y_i)$. The length of the inscribed polygon is given by the formula

$$\sum_{i=1}^{n} |P_{i-1}P_i| = \sum_{i=1}^{n} \sqrt{(x_i - x_{i-1})^2 + (y_i - y_{i-1})^2},$$

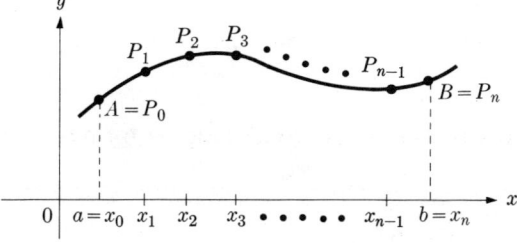

Figure 10–16

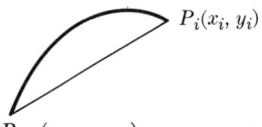

Figure 10–17

where we have denoted $a = x_0$, $f(a) = y_0$, $b = x_n$, and $f(b) = y_n$. At this point we recall the Theorem of the Mean (Chapter 6, Section 2), which states that if a function f has a continuous derivative in an interval (c, d), then there is a value ξ between c and d such that

$$\frac{f(d) - f(c)}{d - c} = f'(\xi), \quad c < \xi < d.$$

This theorem can be used in the present situation. A typical line segment is shown in Fig. 10–17, and we remember that $y_{i-1} = f(x_{i-1})$, $y_i = f(x_i)$, and therefore there is a value ξ_i between x_{i-1} and x_i such that

$$\frac{f(x_i) - f(x_{i-1})}{x_i - x_{i-1}} = f'(\xi_i), \quad x_{i-1} < \xi_i < x_i,$$

or

$$y_i - y_{i-1} = f'(\xi_i)(x_i - x_{i-1}).$$

We find such an equation for every i from 1 to n, and we substitute in the formula for the length of the inscribed polygons, getting

$$\sum_{i=1}^{n} |P_{i-1}P_i| = \sum_{i=1}^{n} \sqrt{1 + [f'(\xi_i)]^2} \, (x_i - x_{i-1}).$$

We now set $\Delta_i x = x_i - x_{i-1}$ and, recalling the definition of an integral, we see that the limit of

$$\sum_{i=1}^{n} \sqrt{1 + [f'(\xi_i)]^2} \, \Delta_i x$$

is nothing but

$$s = \int_a^b \sqrt{1 + [f'(x)]^2} \, dx. \tag{1}$$

This is the desired formula for the length of an arc, and we have the answer to the third question raised at the beginning of the section.

Example 1. Find the length of the arc $y = x^{3/2}$ from the point $A(1, 1)$ to $B(2, 2\sqrt{2})$.

Solution. We have $dy/dx = \frac{3}{2}x^{1/2}$ and $(dy/dx)^2 = \frac{9}{4}x$. According to formula (1), the length of arc, s, is given by

$$s = \int_1^2 \sqrt{1 + \tfrac{9}{4}x} \, dx.$$

To integrate this we set $u = 1 + \frac{9}{4}x$, $du = \frac{9}{4} dx$, and obtain

$$s = \int_{13/4}^{11/2} u^{1/2} \cdot \tfrac{4}{9} \, du,$$

new limits having been inserted because $u = \frac{13}{4}$ when $x = 1$ and $u = \frac{11}{2}$ when $x = 2$. Therefore

$$s = [\tfrac{4}{9} \cdot \tfrac{2}{3} u^{3/2}]_{13/4}^{11/2} = \tfrac{2}{27}(11\sqrt{22} - \tfrac{13}{2}\sqrt{13}). \quad \blacktriangleleft$$

Suppose that an arc C is given in the parametric form

$$x = x(t), \qquad y = y(t), \qquad a \le t \le b.$$

We proceed as before, by introducing a number of subdivision points between the endpoints, $A(x(a), y(a))$ and $B(x(b), y(b))$. The sum of the lengths of the inscribed polygon is given by the formula

$$\sum_{i=1}^{n} |P_{i-1}P_i| = \sum_{i=1}^{n} \sqrt{[x(t_i) - x(t_{i-1})]^2 + [y(t_i) - y(t_{i-1})]^2}.$$

Assuming that $x(t)$ and $y(t)$ have continuous first derivatives on the interval $[a, b]$, we may apply the Theorem of the Mean to both functions, getting

$$x(t_i) - x(t_{i-1}) = x'(\xi_i)(t_i - t_{i-1}), \qquad t_{i-1} < \xi_i < t_i,$$
$$y(t_i) - y(t_{i-1}) = y'(\eta_i)(t_i - t_{i-1}), \qquad t_{i-1} < \eta_i < t_i.$$

We find that

$$\sum_{i=1}^{n} |P_{i-1}P_i| = \sum_{i=1}^{n} \sqrt{[x'(\xi_i)]^2 + [y'(\eta_i)]^2} (t_i - t_{i-1}).$$

Proceeding to the limit as the norm of the subdivision tends to zero and taking into account the definition of integral,* we obtain the formula for arc length:

$$s = \int_a^b \sqrt{[x'(t)]^2 + [y'(t)]^2} \, dt.$$

From the Fundamental Theorem of Calculus, which states that differentiation and integration are inverse processes, the above integral formula for s yields

$$s'(t) = \frac{ds}{dt} = \sqrt{\left(\frac{dx}{dt}\right)^2 + \left(\frac{dy}{dt}\right)^2}.\dagger$$

Or we may express this formula quite simply by using differentials:

$$ds^2 = dx^2 + dy^2.$$

---

* The argument here is not quite precise. To apply the definition of integral directly, it is necessary that ξ_i and η_i be the same. However, the fact that the result is valid when $\xi_i \ne \eta_i$ is not difficult to establish.
† Here $s(T)$ is thought of as that function of T which gives the length of the part of the arc for t between a and T.

If the arc is in the form $y = f(x)$ or $x = g(y)$ we obtain, respectively,

$$\frac{ds}{dx} = \sqrt{1 + (dy/dx)^2} \quad \text{and} \quad \frac{ds}{dy} = \sqrt{1 + (dx/dy)^2}.$$

Example 2. Given that $x = t^3 + 1$, $y = 2t^{9/2} - 4$. Find the length of the arc from the point where $t = 1$ to the point where $t = 3$.

Solution. We have $x'(t) = 3t^2$, $y'(t) = 9t^{7/2}$. Therefore

$$s = \int_1^3 \sqrt{9t^4 + 81t^7}\, dt = 9\int_1^3 \sqrt{\tfrac{1}{9} + t^3}\, t^2\, dt.$$

Letting $u = \tfrac{1}{9} + t^3$, $du = 3t^2\, dt$, and integrating, we get

$$s = [2(\tfrac{1}{9} + t^3)^{3/2}]_1^3 = \tfrac{2}{27}(244\sqrt{244} - 10\sqrt{10}). \quad \blacktriangleleft$$

Example 3. Find the length of one arch of a cycloid, given that

$$x = a(\theta - \sin\theta), \quad y = a(1 - \cos\theta), \quad 0 \le \theta \le 2\pi.$$

Solution. We have $dx/d\theta = a(1 - \cos\theta)$, $dy/d\theta = a\sin\theta$, and therefore

$$s = \int_0^{2\pi} \sqrt{a^2(1 - \cos\theta)^2 + a^2\sin^2\theta}\, d\theta = a\int_0^{2\pi} \sqrt{2 - 2\cos\theta}\, d\theta.$$

Making use of the formula $\sin(\theta/2) = \sqrt{(1 - \cos\theta)/2}$, we obtain

$$s = 2a\int_0^{2\pi} \sin\frac{\theta}{2}\, d\theta = \left[-4a\cos\frac{\theta}{2}\right]_0^{2\pi} = 8a. \quad \blacktriangleleft$$

PROBLEMS

In Problems 1 through 9, find the length of arc in each case.

1. $y = \tfrac{1}{4}x^2 - \tfrac{1}{2}\ln x$, $1 \le x \le 2$
2. $y = 2x^{3/2}$, $0 \le x \le 2$
3. $y = \tfrac{1}{6}x^3 + 1/(2x)$, $1 \le x \le 3$
4. $y = \tfrac{1}{2}(e^x + e^{-x})$, $0 \le x \le 2$
5. $y = \sqrt{36 - x^2}$, $0 \le x \le 4$
6. $x = 6\cos t$, $y = 6\sin t$, $\dfrac{\pi}{3} \le t \le \dfrac{\pi}{2}$
7. $f: x \to (9 - x^{2/3})^{3/2}$, $1 \le x \le 2$
8. $y = x^{2/3}$, $-8 \le x \le -2$
9. $x = e^{-t}\cos t$, $y = e^{-t}\sin t$, $0 \le t \le \pi/2$

In Problems 10 through 18, in each case set up the integral for arc length but do not attempt to evaluate the integral.

10. $y = \sqrt{x}$, $1 \le x \le 4$
11. $y = x^3$, $1 \le x \le 2$
12. $x = y^4 - 2y^2 + 3$, $1 \le y \le 2$
13. $y = \tfrac{1}{3}(x - 3)\sqrt{x}$, $1 \le x \le 2$
14. $x = \tfrac{1}{6}t^3$, $1/2t$, $1 \le t \le 3$
15. $x = t^3 + 2t - 1$, $y = t^2 - t + 5$, $1 \le t \le 5$
16. $x = 5\cos t$, $y = 4\sin t$, $0 \le t \le \pi/2$
17. $y = \tfrac{4}{5}\sqrt{25 - x^2}$, $0 \le x \le 4$
18. $x = t\cos t$, $y = t\sin t$, $0 \le t \le \pi$

In Problems 19 through 22, in each case subdivide the given interval into the number of equal subintervals indicated by the integer n. Compute the quantity

$$\sum_{i=1}^{n} \sqrt{1 + [f'(\xi_i)]^2}\,(x_i - x_{i-1}),$$

by taking the value ξ_i at the midpoint of the ith subinterval, and in this way approximate the length of the arc.

19. $y = x^3/3$, $0 \le x \le 2$, $n = 4$
20. $y = \frac{2}{5}x^{5/2}$, $0 \le x \le 2$, $n = 4$
21. $y = \frac{4}{5}\sqrt{25 - x^2}$, $0 \le x \le 2$, $n = 4$
22. $y = \dfrac{2}{\sqrt{x^2 + 1}}$, $0 \le x \le 2$, $n = 4$

In Problems 23 through 25, follow the directions for Problems 19 through 22, except that the quantity

$$\sum_{i=1}^{n} \sqrt{[x'(\xi_i)]^2 + [y'(\xi_i)]^2}\,(t_i - t_{i-1})$$

is to be computed, with ξ_i the midpoint of $[t_{i-1}, t_i]$.

23. $x = t^3 + 1$, $y = t + 2$, $0 \le t \le 1$, $n = 5$
24. $x = 2\sqrt{1 + t}$, $y = \dfrac{2}{\sqrt{1 + t}}$, $0 \le t \le 1$, $n = 5$
25. $x = \dfrac{t}{t + 1}$, $y = \dfrac{2}{\sqrt{t + 1}}$, $0 \le t \le 2$, $n = 4$

26. Find the length of the circumference of a circle of radius 1 by inscribing regular polygons in the circle, finding their lengths and then allowing the number of sides of the polygons to tend to infinity. [*Hint:* The parametric equations of the circle may be taken as $x = \cos t$, $y = \sin t$, $0 \le t < 2\pi$.]

27. Find the total length of the loop of the curve $6y^2 = x(x - 2)^2$ between $x = 0$ and $x = 2$.

*28. Find the total length of the loop of the curve $3y^2 = x^2(2x + 1)$ between $x = 0$ and $x = -\frac{1}{2}$.

29. Given the curve $ay^2 = x(x - b)^2$ which is similar to that in Problem 27. Find three pairs of values for a and b so that the length of the loop from $x = 0$ to $x = b$ can be found by evaluating an elementary integral.

30. Let f be any function which has arc length on $[a, b]$. Let $s(x)$ be the length of the arc from the point a to x. Show that s is an increasing function of x for $x \in [a, b]$.

31. Suppose that f is continuous on $[a, b]$ and that f has a derivative except at one point where the derivative has a jump. Show how to find the length of the arc from a to b.

4. CURVATURE

Our aim in this section is to obtain a measure of the rapidity with which curves change direction. Suppose that a curve is given by the equation $y = f(x)$ and that f has a continuous second derivative. At a particular point $P_0(x_0, y_0)$ the tangent to the curve makes an angle with the positive x direction which we call ϕ (Fig. 10–18). From the definition of derivative we know that $\tan \phi = f'(x_0)$ or that at

360 PARAMETRIC EQUATIONS. ARC LENGTH

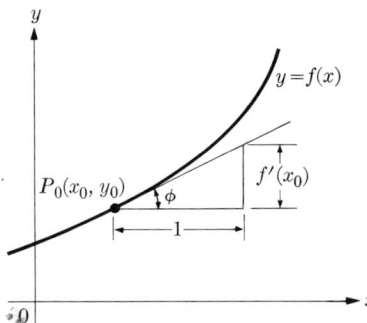

Figure 10–18

a point $P(x, y)$,

$$\phi(x) = \arctan f'(x).$$

The way ϕ changes as we move along the curve is a measure of the sharpness of the curve. Note that for a straight line ϕ doesn't change at all, and it changes very little even after traversing a long section of the arc of a gradual curve.

Definition. *The **curvature** κ of an arc* given in the form $y = f(x)$ is the rate of change of the angle ϕ with respect to the arc length s. That is,*

$$\kappa = \frac{d\phi}{ds}.$$

The following theorem tells us how to compute the curvature when we are given the function f.

Theorem 2. *For an arc of the form $y = f(x)$, the curvature κ is given by*

$$\kappa(x) = \frac{f''(x)}{\{1 + [f'(x)]^2\}^{3/2}}. \tag{1}$$

Proof. From the Chain Rule, we know that

$$\kappa = \frac{d\phi}{ds} = \frac{d\phi}{dx} \cdot \frac{dx}{ds}.$$

We first find $d\phi/dx$:

$$\frac{d\phi}{dx} = \frac{d}{dx}(\arctan f'(x)) = \frac{1}{1 + [f'(x)]^2} \cdot f''(x).$$

To obtain dx/ds, we recall that since differentiation and integration are inverse processes, the formula

$$s = \int \sqrt{1 + [f'(x)]^2}\, dx$$

---

* The Greek letter kappa (κ) is customarily used to denote curvature.

may be differentiated to give

$$ds = \sqrt{1 + [f'(x)]^2}\, dx,$$

or

$$\frac{dx}{ds} = \frac{1}{\sqrt{1 + [f'(x)]^2}}.$$

Therefore we get

$$\kappa = \frac{d\phi}{dx} \cdot \frac{dx}{ds} = \frac{f''(x)}{1 + [f'(x)]^2} \cdot \frac{1}{\sqrt{1 + [f'(x)]^2}} = \frac{f''(x)}{\{1 + [f'(x)]^2\}^{3/2}},$$

and Theorem 2 is established.

If a curve is given in the form $x = g(y)$, or in the parametric form $x = x(t)$, $y = y(t)$, the angle ϕ which the tangent line forms with the positive x direction may still be defined. It can be shown that for an arc of the form $x = g(y)$ the curvature is given by the formula

$$\kappa(y) = -\frac{g''(y)}{\{1 + [g'(y)]^2\}^{3/2}}, \tag{2}$$

while for arcs in parametric form $x = x(t)$, $y = y(t)$, κ is given by the formula

$$\kappa(t) = \frac{x'(t)y''(t) - x''(t)y'(t)}{(x'^2 + y'^2)^{3/2}}. \tag{3}$$

It can be shown that the magnitude of the curvature κ of a curve C is an intrinsic property of the curve and does not depend on the coordinate system. However, the sign of κ is determined by the parameter we select to describe C. For example, if C is given in the form $y = f(x)$, then the curvature has the same sign as $f''(x)$, as can be seen from formula (1) on page 360. This fact reflects the method we used to define arc length s. When defining s, we stated that s is positive as we traverse C from smaller values of x to larger values of x. This definition of arc length amounts to the selection of a **direction** or an **orientation** of every arc C. It is clear that each arc has two possible orientations, according as we go along it in one direction or the opposite direction. However, once we choose an orientation, the direction of positive arc length is decided and, along with it, the sign of the curvature. For curves described by parametric equations $x = x(t)$, $y = y(t)$, the orientation of the curve and therefore the sign of κ is determined by the direction along C which corresponds to an increase in the parameter t.

If ϕ is increasing so that the curve is "turning to the left" as the parameter increases, then κ is positive; κ is negative if ϕ is decreasing. This is equivalent to saying that the concave side of the curve is on the left if $\kappa > 0$ and is on the right if $\kappa < 0$ (Fig. 10–19).

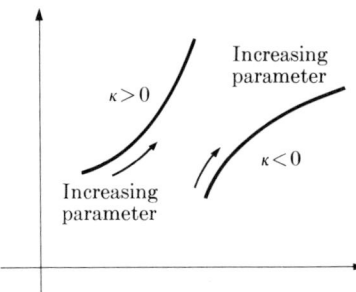

Figure 10–19

Example 1. Find the curvature of the circle $x = a \cos t$, $y = a \sin t$.

Solution. Differentiation yields

$$x' = -a \sin t, \quad x'' = -a \cos t, \quad y' = a \cos t, \quad y'' = -a \sin t.$$

Therefore

$$\kappa = \frac{a^2 \sin^2 t + a^2 \cos^2 t}{(a^2 \sin^2 t + a^2 \cos^2 t)^{3/2}} = \frac{a^2}{a^3} = \frac{1}{a}.$$ ◂

We conclude that a circle of radius a has constant curvature equal to $1/a$.

Definitions. *The* **radius of curvature** *R* **of an arc at a point** *is defined as the reciprocal of the absolute value of the curvature at that point; that is,*

$$R = \frac{1}{|\kappa|}.$$

The **circle of curvature of an arc at a point** *P is that circle passing through P which has radius equal to R, the radius of curvature, and whose center C lies on the concave side of the curve along the normal through P (see Fig. 10–20).*

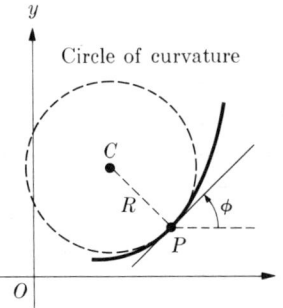

Figure 10–20

Example 2. Find the curvature κ and the radius of curvature R for the parabola $y = x^2$. Find the center of the circle of curvature at the point $(1, 1)$.

Solution. We have $dy/dx = 2x$, $d^2y/dx^2 = 2$, and therefore

$$\kappa = \frac{2}{(1 + 4x^2)^{3/2}}; \quad R = \tfrac{1}{2}(1 + 4x^2)^{3/2}.$$

At (1, 1) the slope is 2 and the equation of the normal line is

$$y - 1 = -\tfrac{1}{2}(x - 1) \quad \Leftrightarrow \quad x + 2y - 3 = 0.$$

The radius of curvature R is $\tfrac{5}{2}\sqrt{5}$. (See Fig. 10–21.) To find the center C of the circle of curvature we first write the equation of the circle with center at (1, 1) and radius $\tfrac{5}{2}\sqrt{5}$. It is

$$(x - 1)^2 + (y - 1)^2 = \tfrac{125}{4}.$$

Solving this simultaneously with $x + 2y - 3 = 0$, the equation of the normal line, we obtain for the coordinates x_c, y_c of C, the values

$$x_c = -4, \quad y_c = \tfrac{7}{2}.$$

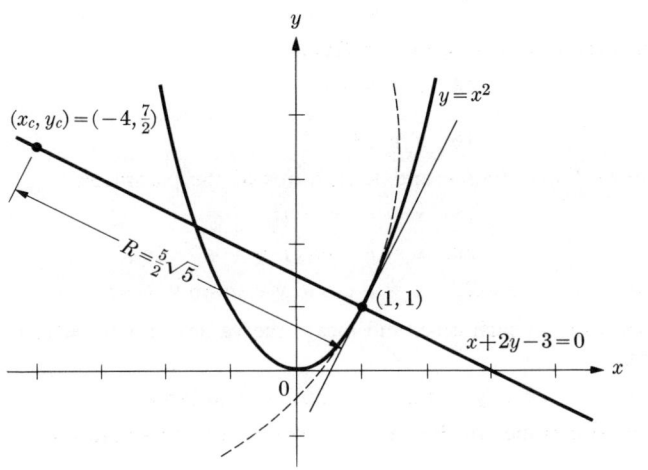

Figure 10–21

Example 3. Find the curvature and radius of curvature of the cycloid

$$x = a(\theta - \sin \theta), \quad y = a(1 - \cos \theta), \quad 0 < \theta < 2\pi.$$

Solution. We have

$$x'(\theta) = a(1 - \cos \theta), \quad x''(\theta) = a \sin \theta,$$
$$y'(\theta) = a \sin \theta, \quad y''(\theta) = a \cos \theta.$$

Therefore, using formula (3) on page 361, we find

$$\kappa(\theta) = \frac{a(1 - \cos \theta)a \cos \theta - a \sin \theta \, a \sin \theta}{[a^2(1 - \cos \theta)^2 + a^2 \sin^2 \theta]^{3/2}}$$

$$= \frac{1}{a} \frac{\cos \theta - 1}{[2(1 - \cos \theta)]^{3/2}} = -\frac{1}{a2\sqrt{2}\sqrt{1 - \cos \theta}} = -\frac{1}{4a \sin (\theta/2)}.$$

$$R(\theta) = 4a \sin \frac{\theta}{2}, \quad 0 < \theta < 2\pi.$$

PROBLEMS

In Problems 1 through 12, find in each case the curvature $\kappa(x)$ and the radius of curvature $R(x)$.

1. $y = 2x^2 + 1$
2. $y = e^{-2x}$
3. $y = 1 + 3\sqrt{x}$
4. $y = x + \dfrac{2}{x}, \quad x > 0$
5. $y = \sin x$
6. $y = \tfrac{1}{2}\cos 3x$
7. $y = ae^{bx}$, a, b constants
8. $y = \tan x$
9. $y = \arcsin x$
10. $y = (a^{2/3} - x^{2/3})^{3/2}$
11. $y = \dfrac{b}{a}\sqrt{a^2 - x^2}, \quad |x| < a$
12. $y = \dfrac{b}{a}\sqrt{x^2 - a^2}, \quad |x| > a$

In Problems 13 through 16, in each case find $\kappa(y)$ and $R(y)$.

13. $x = \cot y$
14. $x = \ln y$
15. $x = -\sqrt{a^2 - y^2}$
16. $x = \tfrac{1}{3}y^3$

In Problems 17 through 22, in each case find κ and R in terms of the parameter.

17. $x = \tfrac{1}{2}t^2, \ y = \tfrac{1}{3}t^3$
18. $x = e^t, \ y = t^2$
19. $x = e^t \sin t, \ y = e^t \cos t$
20. $x = e^{-2t}\cos t, \ y = e^{-2t}\sin t$
21. $x = a\cos^3\theta, \ y = a\sin^3\theta$
22. $x = a\cos\theta, \ y = b\sin\theta$

In Problems 23 through 26, find $\kappa(x)$ in each case, and locate the values of x for which κ has relative maxima and minima.

23. $y = \tfrac{1}{2}x^2$
24. $y = \ln x$
25. $y = 1/x$
26. $y = \ln(\sin x)$

27. Prove that for any arc the coordinates of the center of the circle of curvature C are given by

$$x_c = x - \frac{\sin\phi}{\kappa}, \qquad y_c = y + \frac{\cos\phi}{\kappa}.$$

In Problems 28 through 31, in each case find κ and R, and locate the center of the circle of curvature at the point indicated.

28. $y = x^2, \ (-1, 1)$
29. $y = \cos x, \ (\pi/3, \tfrac{1}{2})$
30. $y = x^3 - 2x^2 + 3x - 5, \ (1, -1)$
31. $y = e^x, \ (0, 1)$

In Problems 32 and 33, find in terms of t the coordinates of the center of the circle of curvature.

32. $x = a\cos t, \ y = b\sin t$
33. $x = a\sec t, \ y = b\tan t$

34. Establish formula (2) on page 361 for the curvature $\kappa(y)$ of an arc given by $x = g(y)$.

35. Establish formula (3) on page 361 for the curvature $\kappa(t)$ of an arc given parametrically by $x = x(t), \ y = y(t)$.

36. a) Show that if $x = x(t) = t$ and $y = y(t)$, then formula (3) on page 361 for κ reduces to formula (1) on page 360. Observe that when a curve is given by an equation $y = f(x)$, this is equivalent to the parametric representation $x = t, \ y = f(t)$.
    b) Show the result analogous to that in part (a) when $x = x(t)$ and $y = y(t) = t$.

37. Find the point on the graph of $y = 3\ln(2x + 1)$ where the curvature has its minimum value. Show that there is no point where the curvature is a maximum.

38. Given the cycloid

$$x = a(\theta - \sin \theta), \quad y = a(1 - \cos \theta),$$

show that the curvature is a maximum at the point on the curve where the tangent to the curve is horizontal. Explain why there is no point on the curve where the curvature is a minimum.

11
POLAR COORDINATES

1. POLAR COORDINATES

A coordinate system in the plane allows us to associate an ordered pair of numbers with each point in the plane. Until now we have considered Cartesian coordinate systems exclusively. However, there are many problems for which other systems of coordinates are more advantageous than are Cartesian coordinates. With this in mind we now describe the system known as **polar coordinates.**

We begin by selecting a point in the plane which we call the **pole** or **origin** and label it O. From this point we draw a half-line starting at the pole and extending indefinitely in one direction. This line is usually drawn horizontally and to the right of the pole, as shown in Fig. 11-1. It is called the **initial line** or **polar axis.**

Let P be any point in the plane. Its position will be determined by its distance from the pole and by the angle that the line OP makes with the initial line. We measure angles θ from the initial line as in trigonometry—positive in a counterclockwise direction and negative in a clockwise direction. The distance r from the origin to the point P will be taken as positive. The coordinates of P (Fig. 11-2) in the polar coordinate system are (r, θ). There is a sharp distinction between Cartesian and polar coordinates, in that a point P may be represented in just one way by a pair of Cartesian coordinates, but it may be represented in many ways by polar coordinates. For example, the point Q with polar coordinates $(2, \pi/6)$ also has polar coordinates

$$\left(2, 2\pi + \frac{\pi}{6}\right), \quad \left(2, 4\pi + \frac{\pi}{6}\right), \quad \left(2, 6\pi + \frac{\pi}{6}\right),$$

$$\left(2, -2\pi + \frac{\pi}{6}\right), \quad \left(2, -4\pi + \frac{\pi}{6}\right), \quad \text{etc.}$$

In other words, there are infinitely many representations of the same point. Furthermore, it is convenient to allow r, the distance from the origin, to take on negative values. We establish the convention that a pair of coordinates such as

Figure 11-1

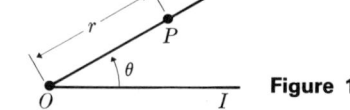

Figure 11-2

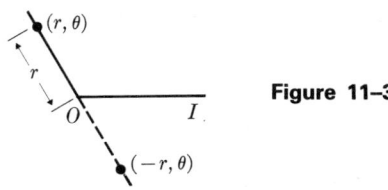

Figure 11-3

$(-3, \theta)$ is simply another representation of the point with coordinates $(3, \theta + \pi)$. Figure 11-3 shows the relationship of the points (r, θ) and $(-r, \theta)$.

Example 1. Plot the points whose polar coordinates are

$$P\left(3, \frac{\pi}{3}\right), \quad Q\left(-2, \frac{2\pi}{3}\right), \quad R\left(-2, \frac{\pi}{4}\right), \quad S\left(2, \frac{3\pi}{4}\right), \quad T\left(3, -\frac{\pi}{6}\right).$$

Solution. The points are plotted in Fig. 11-4. ◀

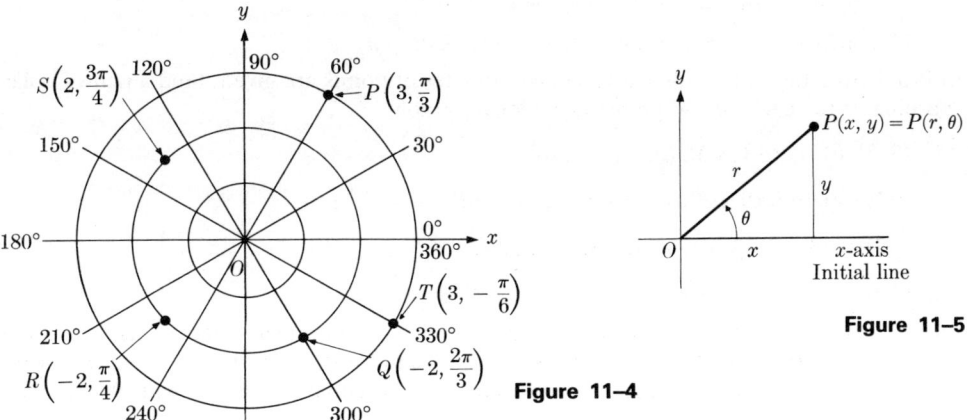

It is important to know the connection between Cartesian and polar coordinate systems. To find this relationship, let us consider a plane with one system superimposed on the other in such a way that the origin of the Cartesian system is at the pole and the positive x axis coincides with the initial line (Fig. 11-5). The relationship between the Cartesian coordinates (x, y) and the polar coordinates (r, θ) of a point P is given by the equations

$$x = r \cos \theta, \quad y = r \sin \theta.$$

When we are given r and θ, these equations tell us how to find x and y. We also have the formulas

$$r = \pm\sqrt{x^2 + y^2}, \quad \tan \theta = \frac{y}{x},$$

which give us r and θ when the Cartesian coordinates are known.

Example 2. The Cartesian coordinates of a point are $(\sqrt{3}, -1)$. Find a set of polar coordinates for this point.

Solution. We have

$$r = \sqrt{3 + 1} = 2 \quad \text{and} \quad \tan \theta = -1/\sqrt{3}.$$

Since the point is in the fourth quadrant, we select for θ the value $-\pi/6$ (or $11\pi/6$). The answer is $(2, -\pi/6)$. ◂

PROBLEMS

In Problems 1 through 4, the polar coordinates of points are given. Find the Cartesian coordinates of the same points and plot them on graphs.

1. $(4, \pi/6), (3, 3\pi/4), (2, \pi), (1, 0), (-2, \pi)$
2. $(2, \pi/4), (1, \pi/3), (3, \pi/2), (4, 3\pi/2), (-1, 7\pi/6)$
3. $(-1, 0), (2, -\pi/6), (4, -\pi/3), (-3, 3\pi/4), (0, \pi/2)$
4. $(2, -\pi/2), (-1, -3\pi/2), (2, 4\pi/3), (-1, -\pi/4), (0, -\pi)$

In Problems 5 through 8, the Cartesian coordinates of points are given. Find a pair of polar coordinates for each of the points and plot on graphs.

5. $(3, 3), (0, 4), (-1, \sqrt{3}), (0, -1), (2, 0)$
6. $(-2, -2), (-4, 0), (\sqrt{3}, 1), (-\sqrt{3}, -1), (0, -2)$
7. $(-2, 2), (3, -3), (-\sqrt{3}, 1), (2\sqrt{3}, 2), (2, 2\sqrt{3})$
8. $(4, 0), (0, 0), (6, 6), (\sqrt{6}, \sqrt{2}), (\frac{3}{2}, -\frac{3}{2})$
9. Describe the graph of all points which, in polar coordinates, satisfy the condition $r = 5$; do the same for the condition $\theta = \pi/3$ and for the condition $\theta = -5\pi/6$. What can be said about the angle of intersection of the curve $r = $ const, with $\theta = $ const?
10. Find the distance between the points with polar coordinates $(3, \pi/4), (2, \pi/3)$.
11. Find the distance between the points with polar coordinates $(1, \pi/2), (4, 5\pi/6)$.
12. Find a formula for the distance between the points (r_1, θ_1) and (r_2, θ_2).
13. Find the polar coordinates of the midpoint of the line segment connecting the points P and Q whose polar coordinates are $P(3, \pi/4), Q(6, \pi/6)$.
14. Find a formula for the coordinates $Q(\bar{r}, \bar{\theta})$ of the midpoint of the line segment connecting the points $P_1(r_1, \theta_1)$ and $P_2(r_2, \theta_2)$.

2. GRAPHS IN POLAR COORDINATES

Suppose that r and θ are connected by some equation such as $r = 3 \cos 2\theta$ or $r^2 = 4 \sin 3\theta$. We define the **graph of an equation in polar coordinates** (r, θ) as the set of all points P each of which has at least one pair of polar coordinates (r, θ) which satisfies the given equation. To plot the graph of such an equation we must find all ordered pairs (r, θ) which satisfy the given equation and then plot the points obtained. We can obtain a good approximation of the graph in polar coordinates, as we can in Cartesian coordinates, by making a sufficiently complete table of values, plotting the points, and connecting them by a smooth curve.

Example 1. Draw the graph of the equation $r = 3 \cos \theta$.

Solution. We construct the table:

θ	0	$\pi/6$	$\pi/3$	$\pi/2$	$2\pi/3$	$5\pi/6$
$\theta°$	0	30	60	90	120	150
r	3	$\frac{3}{2}\sqrt{3}$	$\frac{3}{2}$	0	$-\frac{3}{2}$	$-\frac{3}{2}\sqrt{3}$
r approx.	3.00	2.60	1.50	0	-1.50	-2.60

θ	π	$7\pi/6$	$4\pi/3$	$3\pi/2$	$5\pi/3$	$11\pi/6$
$\theta°$	180	210	240	270	300	330
r	-3	$-\frac{3}{2}\sqrt{3}$	$-\frac{3}{2}$	0	$\frac{3}{2}$	$\frac{3}{2}\sqrt{3}$
r approx.	-3.00	-2.60	-1.50	0	1.50	2.60

The graph is shown in Fig. 11-6. It can be shown to be a circle with center at $(\frac{3}{2}, 0)$ and radius $\frac{3}{2}$. It is worth noting that even though there are 12 entries in the table, only 6 points are plotted. The curve is symmetric with respect to the initial line, and we could have saved effort if this fact had been taken into account. Since $\cos(-\theta) = \cos \theta$ for all values of θ, we could have obtained the points on the graph for values of θ between 0 and $-\pi$ without extra computation. ◀

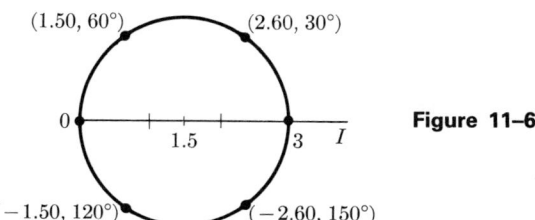

Figure 11-6

We shall set forth the rules of symmetry, as we did in Chapter 2, Section 4. These rules, which are useful as aids in graphing, are described in terms of symmetries with respect to x and y axes in Cartesian coordinates. That is, we suppose that the positive x axis coincides with the initial line of the polar coordinate system.

Rule I. *If the substitution of $(r, -\theta)$ for (r, θ) yields the same equation, the graph is symmetric with respect to the x axis.*

Rule II. *If the substitution of $(r, \pi - \theta)$ for (r, θ) yields the same equation, the graph is symmetric with respect to the y axis.*

Rule III. *If the substitution of $(-r, \theta)$ or of $(r, \theta + \pi)$ for (r, θ) yields the same equation, the graph is symmetric with respect to the pole.*

It is true, as we have already seen, that if any two of the three **symmetries** hold, the remaining one holds automatically. However, *it is possible for a graph to have certain symmetry properties which the rules above will fail to exhibit.*

Example 2. Discuss for symmetry and plot the graph of $r = 3 + 2\cos\theta$.

Solution. The graph is symmetric with respect to the x axis, since $\cos(-\theta) = \cos\theta$ and Rule I applies. We construct the table:

θ	0	$\pm \pi/6$	$\pm \pi/3$	$\pm \pi/2$	$\pm 2\pi/3$	$\pm 5\pi/6$	$\pm \pi$
$\theta°$	0	$\pm 30°$	$\pm 60°$	$\pm 90°$	$\pm 120°$	$\pm 150°$	$\pm 180°$
r	5	$3 + \sqrt{3}$	4	3	2	$3 - \sqrt{3}$	1
r approx.	5.00	4.73	4.00	3.00	2.00	1.27	1.00

We use values of θ from $-\pi$ to $+\pi$ only, since $\cos(\theta + 2\pi) = \cos\theta$ and no new points could possibly be obtained with higher values of θ. The graph, called a **limaçon,** is shown in Fig. 11–7. ◀

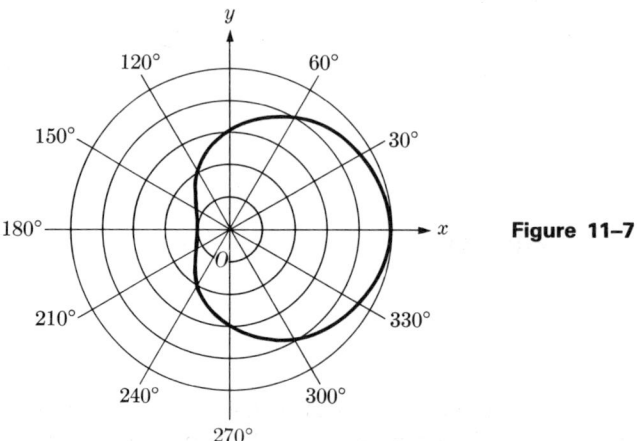

Figure 11–7

Example 3. Test for symmetry and plot the graph of the equation

$$r = 2\cos 2\theta.$$

Solution. If we replace θ by $-\theta$ the equation is unchanged and, therefore, by Rule I, we have symmetry with respect to the x axis. When we replace θ by $\pi - \theta$

we get
$$\cos 2(\pi - \theta) = \cos 2\pi \cos 2\theta + \sin 2\pi \sin 2\theta = \cos 2\theta$$
and, by Rule II, the graph is symmetric with respect to the y axis. Consequently, the graph is symmetric with respect to the pole. We construct the table:

θ	0	$\pi/12$	$\pi/6$	$\pi/4$	$\pi/3$	$5\pi/12$	$\pi/2$
2θ	0	$\pi/6$	$\pi/3$	$\pi/2$	$2\pi/3$	$5\pi/6$	π
$2\theta°$	0°	30°	60°	90°	120°	150°	180°
r	2	$\sqrt{3}$	1	0	-1	$-\sqrt{3}$	-2
r approx.	2.00	1.73	1.00	0	-1.00	-1.73	-2.00

We plot the points as shown in Fig. 11–8. Now, making use of the symmetries, we can easily complete the graph (Fig. 11–9). This curve is called a **four-leaved rose.**

◂

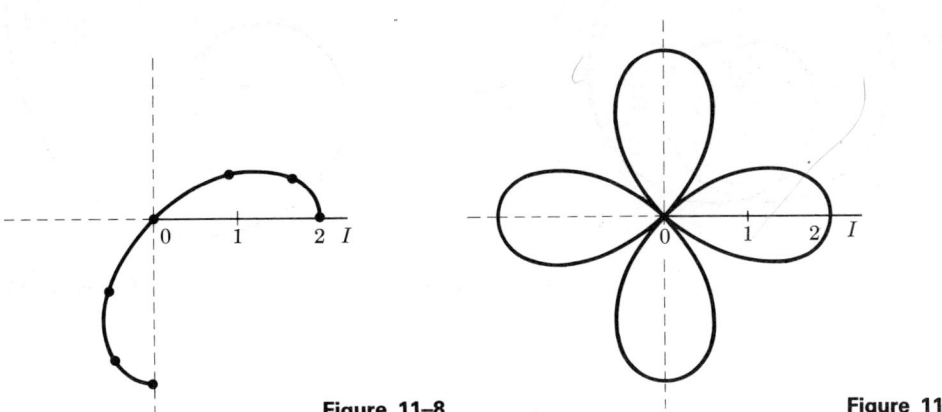

Figure 11–8 **Figure 11–9**

Equations of the form
$$r = a \sin n\theta, \qquad r = a \cos n\theta,$$
where n is a positive integer, have graphs which are called **rose** or **petal** curves. The number of petals is equal to n if n is an odd integer and is equal to $2n$ if n is an even integer. If $n = 1$, there is one petal and it is circular.

The graphs of equations of the form
$$r = a \pm b \cos \theta \qquad \text{or} \qquad r = a \pm b \sin \theta$$
are called **limaçons.** In the special cases in which $a = b$, the graphs are called **cardioids** (see Problems 6, 7, and 8 below). The graphs of equations of the form
$$r^2 = a^2 \cos 2\theta \qquad \text{or} \qquad r^2 = a^2 \sin 2\theta$$

are called **lemniscates;** these graphs have the appearance of "figure eights" (see Problems 15 and 16 below). Polar coordinates are particularly well suited to the study of certain curves called **spirals.** The so-called **spiral of Archimedes** has an equation of the form

$$r = k\theta$$

and its graph is drawn in Fig. 11–10 (the dashed portion arises from negative values of θ). The **logarithmic spiral** has an equation of the form

$$\log_b r = \log_b a + k\theta$$

or

$$r = a \cdot b^{k\theta}$$

and its graph is drawn in Fig. 11–11.

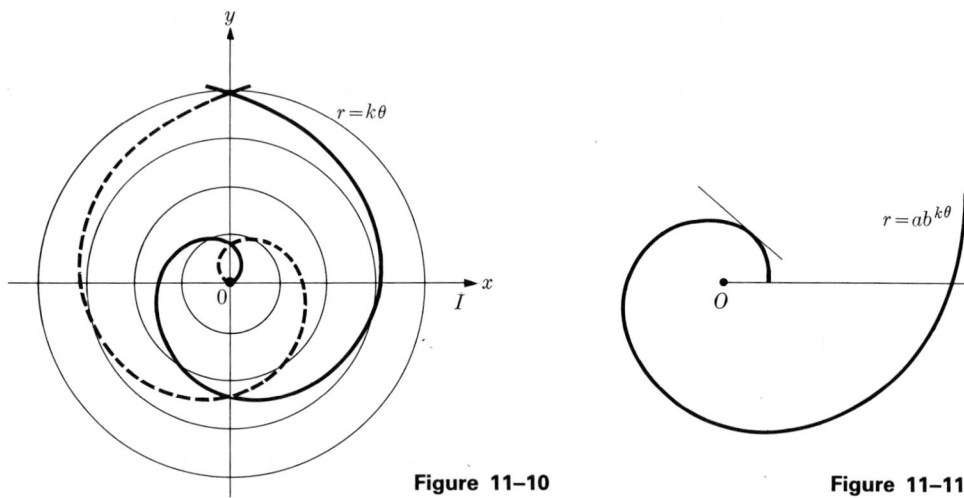

Figure 11–10

Figure 11–11

PROBLEMS

In the following problems, in each case discuss for symmetry and plot the graph of the equation.

1. $r = 4 \cos \theta$
2. $r = -4 \cos \theta$
3. $r = -3 \sin \theta$
4. $r = 2 \cos (\theta + \pi/6)$
5. $r = 3 \sin \left(\theta + \dfrac{\pi}{4} \right)$
6. $r = 3(1 - \cos \theta)$
7. $r = 2(1 + \cos \theta)$
8. $r = 2(1 + \sin \theta)$
9. $r = 4 - 2 \sin \theta$
10. $r \cos \theta = 2$
11. $r = 2 + 4 \cos \theta$
12. $r = 2 - 4 \sin \theta$
13. $r \sin \theta = 1$
14. $r^2 = 2 \cos \theta$
15. $r^2 = \cos 2\theta$
16. $r^2 = 4 \sin 2\theta$

17. $r = 5 \cos 2\theta$
18. $r = 5 \sin 2\theta$
19. $r = 5 \cos 3\theta$
20. $r = 5 \sin 3\theta$
21. $r = 4 \sin^2 \tfrac{1}{2}\theta$
22. $r = 2 \cos \tfrac{1}{2}\theta$
23. $r = \tan \theta$
24. $r = 5 \cos 4\theta$
25. $r = 5 \sin 4\theta$
26. $r = \cot \theta$
27. $r = 2 \csc \theta$
28. $r(2 - \cos \theta) = 4$
29. $r(1 - \cos \theta) = 2$
30. $r(1 + 2\cos \theta) = 4$
31. $r(1 + 2 \sin \theta) = 2$
32. $r(1 + \cos \theta) = -1$
33. $r = 2 + \theta$
34. $r = 3 - 2\theta$
35. $r = 2e^\theta$
36. $r = 3e^{-2\theta}$

37. Prove that if the symmetries described by Rules I and III hold, then the symmetry described by Rule II must hold. Give an example in which the symmetry of Rule II holds, but the rule itself does not.

In each of Problems 38 through 45, sketch the curves and find the coordinates of the point (or points) of intersection, if any.

38. $r = 3 \cos \theta,\ r = 2 \sin \theta$
39. $r = \sin \theta,\ r = \sin 2\theta$
40. $r = 2 \cos \theta,\ r \cos \theta = 1$
41. $r = \cos \theta,\ r^2 = \cos 2\theta$
42. $r = \tan \theta,\ r = 2 \sin \theta$
43. $r = 2 + \cos \theta,\ r = 6 \cos \theta$
44. $r = \sin 2\theta,\ r = \sin 4\theta$
45. $r = 1 + \cos \theta,\ r = 1 + \sin \theta$.

46. Sketch the graphs of the curves $r = \cos^n \theta$ for $n = 1, 2, 3, 4$. Find the limiting curve as $n \to \infty$.

3. EQUATIONS IN CARTESIAN AND POLAR COORDINATES

The curves discussed in Section 2 find their natural setting in polar coordinates. However, the equations of some curves may be simpler in appearance or their properties more transparent in one coordinate system than in another. For this reason it is useful to be able to transform an equation given in one coordinate system into the corresponding equation in another system.

Suppose that we are given the equation of a curve in the form $y = f(x)$ in a Cartesian system. If we simply make the substitution

$$x = r \cos \theta, \qquad y = r \sin \theta,$$

we have the same equation in a polar coordinate system. We can also go the other way, so that if we are given $r = g(\theta)$ in polar coordinates, the substitution

$$r^2 = x^2 + y^2, \qquad \theta = \arctan \frac{y}{x}, \qquad \left(\text{i.e., } \tan \theta = \frac{y}{x}\right)$$

transforms the relation into one in Cartesian coordinates. In this connection, it is frequently useful to make the substitutions

$$\sin \theta = \frac{y}{\sqrt{x^2 + y^2}}, \qquad \cos \theta = \frac{x}{\sqrt{x^2 + y^2}}, \qquad \tan \theta = \frac{y}{x},$$

rather than use the formula for θ itself.

Example 1. Find the graph of the equation in polar coordinates which corresponds to the graph of the equation

$$x^2 + y^2 - 3x = 0$$

in Cartesian coordinates.

Solution. Substituting $x = r \cos \theta$, $y = r \sin \theta$, we obtain

$$r^2 - 3r \cos \theta = 0 \quad \text{or} \quad r(r - 3 \cos \theta) = 0.$$

Therefore the graph is $r = 0$ or $r - 3 \cos \theta = 0$. Because the pole is part of the graph of $r - 3 \cos \theta = 0$ (since $r = 0$ when $\theta = \pi/2$), the result is

$$r = 3 \cos \theta.$$

We recognize this graph as the circle of Example 1, Section 2. [In Cartesian coordinates, the equation is $(x - \frac{3}{2})^2 + y^2 = \frac{9}{4}$.] ◀

Example 2. Given the polar coordinate equation

$$r = \frac{1}{1 - \cos \theta},$$

find the corresponding equation in Cartesian coordinates.

Solution. We write

$$r - r \cos \theta = 1 \quad \text{or} \quad r = 1 + r \cos \theta.$$

Substituting for r and $\cos \theta$, we get

$$\pm \sqrt{x^2 + y^2} = 1 + x.$$

Squaring both sides (a dangerous operation, since this may introduce extraneous solutions), we obtain

$$x^2 + y^2 = 1 + 2x + x^2 \quad \text{or} \quad y^2 = 2x + 1.$$

When we squared both sides we introduced the extraneous solution

$$r = -(1 + r \cos \theta).$$

The graph of this equation is the same as the graph of the equation

$$r = \frac{-1}{1 + \cos \theta}$$

as is seen by solving for r. But this last equation can be put in the form

$$(-r) = \frac{1}{1 - \cos (\theta + \pi)},$$

which is obtained from the original by replacing (r, θ) by $(-r, \theta + \pi)$. Since we know that (r, θ) and $(-r, \theta + \pi)$ are polar coordinates of the same point, the extraneous graph coincides with the original.

11-3 EQUATIONS IN CARTESIAN AND POLAR COORDINATES

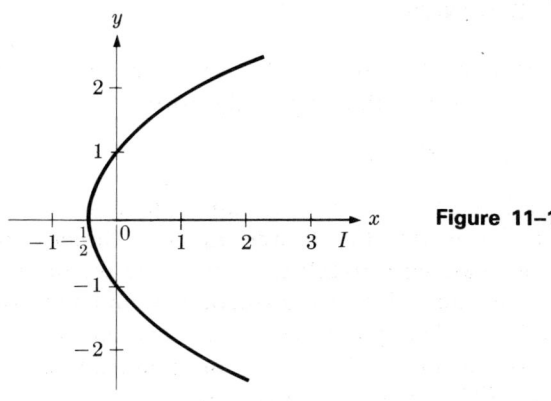

Figure 11-12

Therefore the correct result is the parabola
$$y^2 = 2x + 1.$$
The graph is drawn in Fig. 11-12. ◀

PROBLEMS

In Problems 1 through 12 an equation is given in Cartesian coordinates. In each case find an equation in polar coordinates which describes the same graph.

1. $x = -2$
2. $y = 3$
3. $x - y = 0$
4. $2x + 3y = 0$
5. $2x + y\sqrt{2} = 4$
6. $3x + 3y = 7$
7. $xy = 4$
8. $y^2 = 2x$
9. $x^2 + y^2 + 2x - 4y = 0$
10. $x^2 + y^2 - 4x = 0$
11. $x^2 + y^2 + 2x + 6y = 0$
12. $x^2 + y^2 - 4x + 4y + 4 = 0$

In Problems 13 through 32, in each case find a polynomial equation in Cartesian coordinates whose graph contains that of the equation given in polar coordinates. Discuss possible extraneous solutions.

13. $r = 7$
14. $\theta = \pi/3$
15. $r = 3 \cos \theta$
16. $r = 4 \sin \theta$
17. $r \cos \theta = 5$
18. $r \sin \theta = -2$
19. $r \cos (\theta - \pi/4) = 2$
20. $r \cos (\theta + \pi/6) = 1$
21. $r^2 \cos 2\theta = 4$
22. $r^2 \sin 2\theta = 2$
23. $r^2 = \sin 2\theta$
24. $r^2 = \cos 2\theta$
25. $r = 2 \sec \theta \tan \theta$
26. $r(2 - \cos \theta) = 2$
27. $r(1 - 2 \cos \theta) = 2$
28. $r = a \cos 3\theta$
29. $r = a \sin 3\theta$
30. $r = 1 - \cos \theta$
31. $r = 1 - 2 \sin \theta$
32. $r = a \cos 2\theta$

33. The equation $r^2 = \cos 4\theta$ corresponds to a polynomial equation in x and y. What is the degree of this polynomial?

34. The equation $r^2 = \cos 2n\theta$, with n a positive integer, corresponds to a polynomial equation in x and y. What is the degree of this polynomial?

4. STRAIGHT LINES, CIRCLES, AND CONICS

The straight line was discussed in detail in Chapters 3 and 8. We found that every straight line has an equation (in Cartesian coordinates) of the form

$$Ax + By + C = 0.$$

To obtain the equation in polar coordinates we can simply substitute $x = r \cos \theta$, $y = r \sin \theta$. However, it is also possible to find the desired equation directly. In Fig. 11–13, we consider a line L (not passing through the pole), and we draw a perpendicular from the pole to L. Call the point of intersection N and suppose that the polar coordinates of N are (p, α). The quantities p and α are sufficient to determine the line L completely. To see this let $P(r, \theta)$ be any point on L, and note, according to Fig. 11–13, that $\angle PON = \theta - \alpha$. Therefore

$$r \cos (\theta - \alpha) = p.$$

When α and p are given, this is *the equation of a straight line*. The student may verify that the same equation results regardless of the positions of L and N. The above equation can also be written as

$$r \cos \theta \cos \alpha + r \sin \theta \sin \alpha = p,$$

and changing to Cartesian coordinates yields

$$x \cos \alpha + y \sin \alpha = p,$$

which is known as the **normal form of the equation of a straight line.** Whenever the coefficients of x and y are the cosine and sine of an angle, respectively, the constant term gives the distance from the origin. Any linear equation may be put in normal form. For example,

$$3x + 4y = 7$$

may be written (upon division by $\sqrt{3^2 + 4^2} = 5$) as

$$\tfrac{3}{5}x + \tfrac{4}{5}y = \tfrac{7}{5}.$$

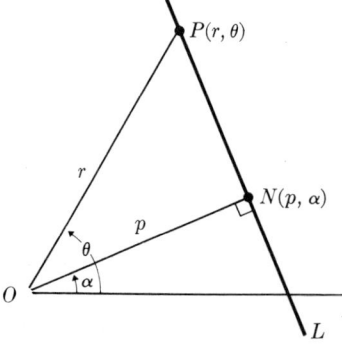

Figure 11–13

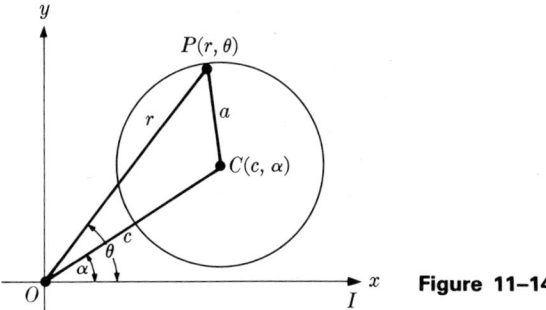

Figure 11–14

This equation is now in normal form, and the distance from the origin to the line is $\frac{7}{5}$ units. (Note that this procedure is the same as the one using the formula for the distance from a point to a line, as given in Chapter 8, Section 1.)

A line through the pole has $p = 0$, and its polar coordinate equation is

$$\cos(\theta - \alpha) = 0.$$

This is equivalent to the statement

$$\theta = \text{const},$$

a condition which we already know represents straight lines passing through the pole.

We shall employ a direct method for obtaining the equation of a circle in polar coordinates. Let the circle have radius a and center at the point C with coordinates (c, α), as shown in Fig. 11–14. If $P(r, \theta)$ is any point on the circle, then in $\triangle POC$, side $|OP| = r$, side $|OC| = c$, side $|PC| = a$, and $\angle POC = \theta - \alpha$. We use the law of cosines in triangle POC, obtaining

$$\boxed{r^2 - 2cr\cos(\theta - \alpha) + c^2 = a^2.}$$

If c, α, and a are given, that is, if the center and radius of the circle are known, then the equation of the circle is completely determined. It is instructive to draw a figure similar to 11–14 but with $a > c$ and to repeat the proof with $P(r, \theta)$ in various positions.

Example 1. A circle has center at $(5, \pi/3)$ and radius 2. Find its equation in polar coordinates.

Solution. We have $c = 5$, $\alpha = \pi/3$, $a = 2$. Therefore

$$r^2 - 10r\cos\left(\theta - \frac{\pi}{3}\right) + 25 = 4 \quad \text{or} \quad r^2 - 10r\cos\left(\theta - \frac{\pi}{3}\right) + 21 = 0. \blacktriangleleft$$

If a circle passes through the origin, so that $c = \pm a$, the above equation for a circle becomes particularly simple. The graph is represented by

$$r = 2c\cos(\theta - \alpha).$$

An illustration is given in Example 1, Section 2, where we have $c = \frac{3}{2}$ and $\alpha = 0$.

378 POLAR COORDINATES

We now take up the study of parabolas, ellipses, and hyperbolas, which we discussed previously in Chapter 8. Polar coordinates are particularly well suited to the discussion of these conics, since the equations of all three types have the same form in this coordinate system. To derive the equations of conics in polar coordinates, we shall solve a problem which differs from the ones we employed in Cartesian coordinates. We shall show that *the graph of a point which moves so that the ratio of its distance from a fixed point to its distance from a fixed line remains constant is*

- *a parabola*   if the ratio is 1,
- *an ellipse*   if the ratio is between 0 and 1,
- *a hyperbola* if the ratio is larger than 1.

To derive the equation of the above graph, we place the fixed point at the pole and let the fixed line L be a line perpendicular to the initial line and p units to the left of the pole (Fig. 11–15). If $P(r, \theta)$ is a point on the graph (and $r > 0$), then the conditions say that $|OP|/|PD|$ = a constant which we call the **eccentricity,** e; and we write (since $|OP| = r$),

$$r = e|PD|.$$

We can see from Fig. 11–15 that

$$|PD| = |PE| + |ED| = r \cos \theta + p.$$

Therefore we have

$$r = er \cos \theta + ep$$

or

$$r = \frac{ep}{1 - e \cos \theta}. \qquad (1)$$

Equation (1) represents an ellipse if $0 < e < 1$, a parabola if $e = 1$, and a hyperbola if $e > 1$.

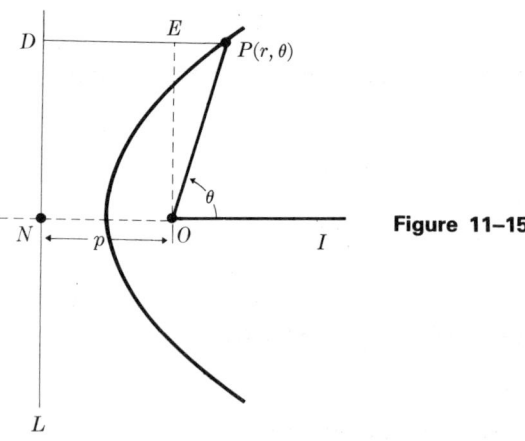

Figure 11–15

We obtain the result most easily by transforming Eq. (1) to Cartesian coordinates and then employing the information on conics in Chapter 8. See Problem 30 at the end of this section. In the above definition of conic, the fixed point is a **focus** of the conic, and we call the fixed line a **directrix**.

Example 2. Find the equation of the conic with focus (fixed point) at the origin, with directrix (fixed line) the line $r \cos \theta = -3$, and with eccentricity $\frac{1}{2}$.

Solution. The equation $r \cos \theta = -3$ tells us that $p = 3$ in Eq. (1) above. Also, we have $e = \frac{1}{2}$, and so

$$r = \frac{\frac{1}{2}(3)}{1 - \frac{1}{2} \cos \theta} = \frac{3}{2 - \cos \theta}.$$

The curve is an ellipse. ◀

Example 3. Given the conic with equation

$$r = \frac{7}{4 - 5 \cos \theta},$$

find the eccentricity and locate the directrix. Identify the conic.

Solution. In order to get the equation in the *exact* form of Eq. (1), we divide numerator and denominator by 4, getting

$$r = \frac{\frac{7}{4}}{1 - \frac{5}{4} \cos \theta}.$$

Then we read off the eccentricity $e = \frac{5}{4}$ and $ep = \frac{7}{4}$. This gives $p = \frac{7}{5}$, and the directrix is the line $r \cos \theta = -\frac{7}{5}$. The curve is a hyperbola. ◀

It can be shown that the equation

$$r = \frac{ep}{1 - e \cos (\theta - \alpha)}$$

is a conic of eccentricity e with focus at the pole and directrix the line

$$r \cos (\theta - \alpha) = -p.$$

PROBLEMS

In Problems 1 through 6, in each case the line L passes through the point N and is perpendicular to the line going through N and the pole O. Find the equation of L in polar coordinates when the coordinates of N are as given.

1. $N(2, \pi/3)$
2. $N(3, 5\pi/6)$
3. $N(-1, \pi/2)$
4. $N(2, 0)$
5. $N(-3, -\pi)$
6. $N(-3, -\pi/2)$

In Problems 7 through 13, in each case find in polar coordinates the equation of the circle satisfying the given conditions.

7. Center $(3, \pi/6)$, radius 2
8. Center $(-2, 5\pi/6)$, radius 3
9. Center $(4, 0)$, radius 4
10. Center $(5, \pi/2)$, radius 5

11. Center $(4, \pi/3)$, passing through $(7, \pi/4)$

12. Center $(5, \pi/4)$ passing through $(6, 0)$

13. Center on the line $\theta = \pi/3$ and passing through $(5, \pi/2)$ and $(0, 0)$

In Problems 14 through 19 find, in polar coordinates, the equation of the conic with focus at the pole and having the given eccentricity and directrix.

14. $e = 1$, directrix $r \cos \theta = -4$
15. $e = 2$, directrix $r \cos \theta = -5$
16. $e = \frac{1}{2}$, directrix $r \cos \theta = 2$
17. $e = \frac{1}{2}$, directrix $r \cos(\theta - \pi/2) = 1$
18. $e = 1$, directrix $r \cos(\theta - \pi/4) = 2\sqrt{2}$
19. $e = 2$, directrix $r \cos(\theta + \pi/3) = 3$

20. Given the line with equation $r \cos(\theta - \pi/3) = 2$, find the distance from the point $(2, \pi/2)$ to this line.

21. Given the line with equation $r \cos(\theta - \alpha) = p$, find a formula for the distance d from a point (r_1, θ_1) to this line.

In Problems 22 through 29, conics are given. Find the eccentricity and directrix of each and sketch the curve.

22. $r = \dfrac{6}{1 - \cos \theta}$

23. $r = \dfrac{9}{2 - \cos \theta}$

24. $r = \dfrac{4}{2 - 3 \cos(\theta - \pi/4)}$

25. $r = \dfrac{2}{1 - \sin \theta}$ $\left[\text{Hint: } \sin \theta = \cos\left(\theta - \dfrac{\pi}{2}\right).\right]$

26. $r = \dfrac{4}{2 - 4 \sin \theta}$

27. $r = \dfrac{5}{1 - 2 \cos(\theta - \pi/3)}$

28. $r = \dfrac{2}{2 + \cos \theta}$

29. $r = \dfrac{4}{6 + \sin \theta}$

30. Transform the equation
$$r = \frac{ep}{1 - e \cos \theta}$$
into Cartesian coordinates. Make use of the results on conics in Chapter 8 to show that the above equation is an ellipse if $0 < e < 1$, a hyperbola if $e > 1$, and a parabola if $e = 1$.

31. Consider the line $L: r \sin \theta = p$, where p is a constant. Find, in polar coordinates, the equation of the graph of all points such that the ratio of their distances from the pole to their distances from L is a constant $(= e)$. Use the method in Problem 30 to identify the conic for various values of e.

32. Transform the equation
$$r = \frac{ep}{1 - e \cos(\theta - \alpha)}$$
into Cartesian coordinates (p, e, α are constants). Use the results of Chapter 8, Section 9 to identify the conic for various values of e.

33. If $a^2 + b^2 > 0$ and $c \neq 0$, show that the graph of
$$r = \frac{c}{a \cos \theta + b \sin \theta}$$
is always a straight line.

34. Given the hyperbola $r = 3/(1 - 3 \cos \theta)$, show that the asymptotes have inclinations
$$\theta = \arccos(1/3), \qquad \theta = \arccos(-1/3).$$

35. Given the equation $r = ep/(1 - e \cos \theta)$, $e > 1$, show that the asymptotes have inclinations
$$\theta = \arccos(1/e), \qquad \theta = \arccos(-1/e).$$

36. Show that the graphs of the equations $r = a \sec^2(\theta/2)$ and $r = a \csc^2(\theta/2)$ are parabolas.

5. DERIVATIVES IN POLAR COORDINATES

Consider a curve given in polar coordinates by an equation of the form $r = f(\theta)$. We wish to investigate the meaning of the derivative. Before doing so, however, we shall examine the graph of one of the simplest equations in polar coordinates. The equation

$$r = \theta$$

represents a **spiral of Archimedes,** which we discussed on page 372. We saw there that the curve is an ever-increasing spiral around the origin. The dashed portion in Fig. 11–16 corresponds to negative values of θ.

We know that for a function given in Cartesian coordinates, the derivative gives the slope of the tangent line. If we plunge ahead blindly and take derivatives in polar coordinates with the intention of obtaining the slope, trouble arises at once. In the example of the spiral of Archimedes we have $dr/d\theta = 1$, and the derivative is clearly not the slope. We need more information in order to learn about slopes in polar coordinates.

If $r = f(\theta)$, then the equations in Cartesian coordinates,

$$x = r \cos \theta, \qquad y = r \sin \theta,$$

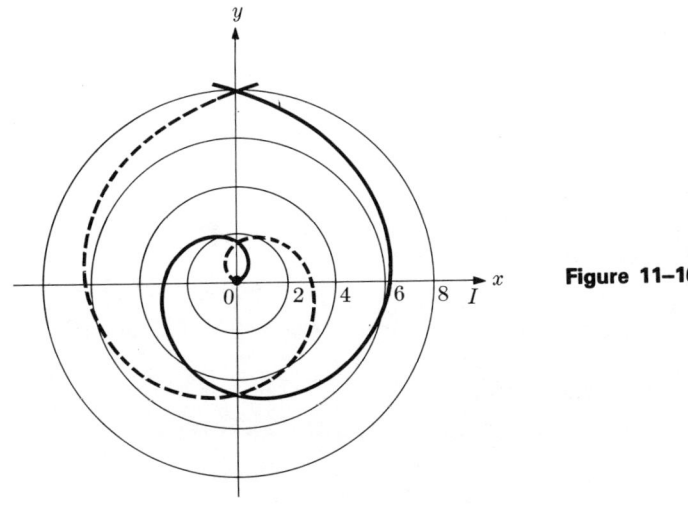

Figure 11–16

when we substitute $f(\theta)$ for r, may be considered as the parametric equations of a curve with θ as parameter. We then have

$$x = f(\theta)\cos\theta, \qquad y = f(\theta)\sin\theta.$$

Differentiating, we find that

$$\frac{dx}{d\theta} = f'(\theta)\cos\theta - f(\theta)\sin\theta, \qquad \frac{dy}{d\theta} = f'(\theta)\sin\theta + f(\theta)\cos\theta.$$

The slope is

$$\frac{dy}{dx} = \frac{dy/d\theta}{dx/d\theta}.$$

Suppose that the curve $r = f(\theta)$ has the appearance of the one in Fig. 11–17. At a point P the tangent line is drawn, and we recognize the slope of this line to be $\tan\phi$. The slope is not particularly convenient in polar coordinates, but the angle ψ between the tangent line and the line from P passing through the pole turns out to be convenient. As seen in Fig. 11–17, ψ and ϕ bear the simple relationship

$$\psi = \phi - \theta,$$

and so

$$\tan\psi = \tan(\phi - \theta) = \frac{\tan\phi - \tan\theta}{1 + \tan\phi\tan\theta}.$$

We know that

$$\tan\phi = \frac{dy}{dx} = \frac{dy/d\theta}{dx/d\theta} = \frac{f'(\theta)\sin\theta + f(\theta)\cos\theta}{f'(\theta)\cos\theta - f(\theta)\sin\theta}.$$

It is a good exercise in algebraic manipulation to substitute the above expression for $\tan\phi$ into the formula for $\tan\psi$ and obtain the simple relation

$$\tan\psi = \frac{f(\theta)}{f'(\theta)}.$$

We can also write

$$\cot\psi = \frac{f'(\theta)}{f(\theta)} = \frac{1}{r}\frac{dr}{d\theta}, \qquad r \neq 0.$$

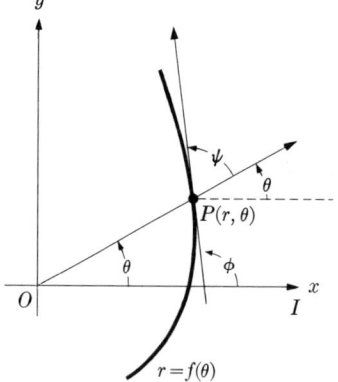

Figure 11–17

11–5 DERIVATIVES IN POLAR COORDINATES

The significance of the derivative in polar coordinates is now becoming clear. The derivative at a point P is related to the angle that the tangent line forms with the line through the point P and the pole, according to the above formula for $\cot \psi$.

Example 1. Given the circle $r = 4 \cos \theta$, find the angle between the tangent line and the line from the pole through the point of tangency. Evaluate this angle at $(2, \pi/3)$.

Solution. We have

$$\cot \psi = \frac{1}{r} \frac{dr}{d\theta} = \frac{1}{4 \cos \theta} (-4 \sin \theta) = -\tan \theta,$$

and so

$$\psi = \theta + \frac{\pi}{2}.$$

For $\theta = \pi/3$, we obtain $\psi = 5\pi/6$. ◀

Example 2. Given the curve $r = 3e^{2\theta}$. Find $\cot \psi$ at any point, and sketch the curve.

Solution. We have

$$\cot \psi = \frac{1}{r} \frac{dr}{d\theta} = \frac{1}{3e^{2\theta}} \cdot 3 \cdot 2e^{2\theta} = 2.$$

In other words, the angle which the tangent makes with the line from the pole through the point of tangency is always the same. The curve, called a **logarithmic spiral,** is shown in Fig. 11–18. (See the discussion on page 372). ◀

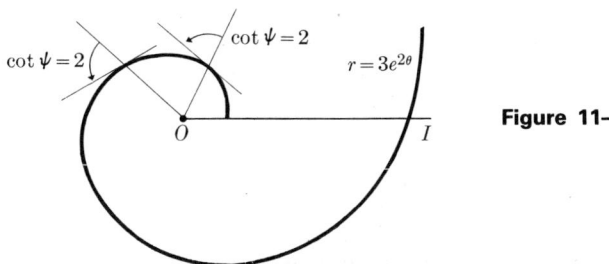

Figure 11–18

To obtain a formula for **arc length** in polar coordinates, we start with the formula

$$ds^2 = dx^2 + dy^2,$$

which we obtained in Chapter 10, Section 3. If $r = f(\theta)$ has a continuous first derivative, we then write, as before,

$$\frac{dx}{d\theta} = f'(\theta) \cos \theta - f(\theta) \sin \theta, \qquad \frac{dy}{d\theta} = f'(\theta) \sin \theta + f(\theta) \cos \theta.$$

Substitution of these expressions in the formula

$$\frac{ds}{d\theta} = \sqrt{(dx/d\theta)^2 + (dy/d\theta)^2}$$

yields (after some algebraic manipulation)

$$\frac{ds}{d\theta} = \sqrt{[f'(\theta)]^2 + [f(\theta)]^2}$$

and, upon integration,

$$s = \int_{\theta_0}^{\theta_1} \sqrt{(dr/d\theta)^2 + r^2}\, d\theta.$$

We provide an example showing how this formula may be used to compute arc length when the equation of the arc is given in polar coordinates.

Example 3. Find the arc length of the curve $r = 3e^{2\theta}$ from $\theta = 0$ to $\theta = \pi/6$.

Solution. We have $dr/d\theta = 6e^{2\theta}$, and so

$$s = \int_0^{\pi/6} \sqrt{36e^{4\theta} + 9e^{4\theta}}\, d\theta = 3\sqrt{5} \int_0^{\pi/6} e^{2\theta}\, d\theta = [\tfrac{3}{2}\sqrt{5}\, e^{2\theta}]_0^{\pi/6}.$$

Therefore

$$s = \tfrac{3}{2}\sqrt{5}[e^{\pi/3} - 1] = 6.2 \text{ approx.} \quad \blacktriangleleft$$

PROBLEMS

In Problems 1 through 14, find $ds/d\theta$ and $\cot \psi$ for each equation.

1. $r = 2a \sin \theta$
2. $r = 3 \cos \theta + 4 \sin \theta$
3. $r = 3(1 - \cos \theta)$
4. $r = 2(1 + \sin \theta)$
5. $r = -5\theta$
6. $r = 3\theta^2$
7. $r = 3 + 2 \cos \theta$
8. $r = 1 - 2 \cos \theta$
9. $r = e^{5\theta}$
10. $r = e^{-4\theta}$
11. $r(2 - \sin \theta) = 2$
12. $r(1 + \cos \theta) = 3$
13. $r(1 + 2 \cos \theta) = 2$
14. $r(5 - 6 \sin \theta) = -1$

In Problems 15 through 18, find the length of arc as indicated.

15. $r = 3\theta^2$ from $\theta = 1$ to $\theta = 2$
16. $r = 2e^{3\theta}$ from $\theta = 0$ to $\theta = 3$
17. $r = 3 \cos \theta$ from $\theta = 0$ to $\theta = \pi/4$
18. $r = 3(1 + \cos \theta)$ from $\theta = 0$ to $\theta = \pi/2$

19. Suppose that $r = f(\theta)$ and $x = f(\theta) \cos \theta$, $y = f(\theta) \sin \theta$. Find an expression for d^2y/dx^2 in terms of θ.

$$\left[\text{Hint:}\ \frac{d^2y}{dx^2} = \frac{\frac{d}{d\theta}\left(\frac{dy}{dx}\right)}{dx/d\theta}. \right]$$

20. Using the result of Problem 19, show that the curvature κ is given by the formula

$$\kappa = \frac{r^2 + 2(dr/d\theta)^2 - r(d^2r/d\theta^2)}{[r^2 + (dr/d\theta)^2]^{3/2}}.$$

21. Using the formula in Problem 20, find κ, given that $r = 2a \sin \theta$.
22. Using the formula of Problem 20, find κ, given that $r = 1 - \cos \theta$.
23. Draw the curve of $r = 1 - \cos \theta$ and construct the tangent at the point where $\theta = \pi/3$. Find the equation of the tangent line at this point in Cartesian coordinates.
24. Draw the curve of $r = 2 + \sqrt{2} \cos \theta$ and construct the tangent at the point where $\theta = 3\pi/4$. Find the equation of this tangent in Cartesian coordinates.
25. Given that $r = f(\theta)$. Find, in polar coordinates, the formula for the equation of the line tangent to the curve at the point (r_1, θ_1) on the curve.
26. Given that $r = f(\theta)$. Find, in polar coordinates, the formula for the line normal to the curve at the point (r_1, θ_1) on the curve.

6. AREA IN POLAR COORDINATES

Two basic concepts are needed for the development of area in Cartesian coordinate systems. They are (1) the idea of the limit of a sum, and (2) the formula for the area of a rectangle: length times width.

To find areas enclosed by curves given in polar coordinates, again two concepts are necessary. They are, first, the idea of the limit of a sum (as in Cartesian coordinates) and second, the formula for the area of a sector of a circle. We recall that *the area of a sector of a circle of radius a with angle opening θ (measured in radians) is*

$$\tfrac{1}{2}\theta a^2.$$

Suppose that $r = f(\theta)$ is a continuous, positive function defined for values of θ between $\theta = \alpha$ and $\theta = \beta$, with $0 \le \alpha < \beta \le 2\pi$. We construct the lines $\theta = \alpha$ and $\theta = \beta$, and we pose the problem of determining the area of the region bounded by these straight lines and the curve with equation $r = f(\theta)$. (See Fig. 11-19.) We subdivide the θ scale into n parts between α and β by introducing the values $\alpha = \theta_0 < \theta_1 < \theta_2 < \cdots < \theta_{n-1} < \theta_n = \beta$. In this way we obtain n subintervals; in each subinterval we select a value of θ which we call ξ_i. We then

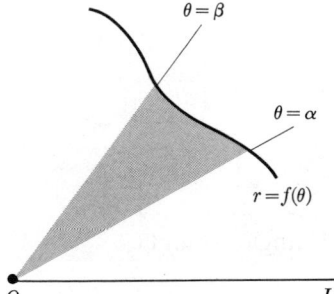

Figure 11-19

386 POLAR COORDINATES

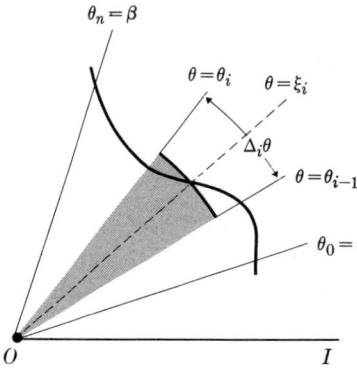

Figure 11–20

compute the area of the circular sector of radius $f(\xi_i)$ and angle opening $\Delta_i\theta = \theta_i - \theta_{i-1}$, as shown in Fig. 11–20. According to the formula for the area of a sector of a circle, the area is

$$\tfrac{1}{2}[f(\xi_i)]^2\,\Delta_i\theta.$$

We add these areas for i between 1 and n, getting

$$\tfrac{1}{2}[f(\xi_1)]^2\,\Delta_1\theta + \tfrac{1}{2}[f(\xi_2)]^2\,\Delta_2\theta + \tfrac{1}{2}[f(\xi_3)]^2\,\Delta_3\theta + \cdots + \tfrac{1}{2}[f(\xi_n)]^2\,\Delta_n\theta,$$

or, concisely,

$$\tfrac{1}{2}\sum_{i=1}^{n}[f(\xi_i)]^2\,\Delta_i\theta.$$

It can be shown (although we shall not do so) that as the number of subdivision points increases without bound and as the norm of the subdivision (length of the largest $\Delta_i\theta$) tends to zero, the area A, bounded by the lines $\theta = \alpha$, $\theta = \beta$, and by the curve $r = f(\theta)$, is given by

$$A = \tfrac{1}{2}\int_{\alpha}^{\beta}[f(\theta)]^2\,d\theta.$$

We have just obtained the desired formula for area in polar coordinates.

Example. 1. Find the area bounded by the curve

$$r = 2 + \cos\theta$$

and the lines $\theta = 0$, $\theta = \pi/2$ (see Fig. 11–21).

Solution. We have

$$A = \tfrac{1}{2}\int_0^{\pi/2}[2 + \cos\theta]^2\,d\theta = \tfrac{1}{2}\int_0^{\pi/2}[4 + 4\cos\theta + \cos^2\theta]\,d\theta$$

$$= [2\theta + 2\sin\theta]_0^{\pi/2} + \tfrac{1}{2}\int_0^{\pi/2}\cos^2\theta\,d\theta = \pi + 2 + \tfrac{1}{2}\int_0^{\pi/2}\cos^2\theta\,d\theta.$$

To perform the remaining integration we use the half-angle formula

$$\cos^2\theta = (1 + \cos 2\theta)/2,$$

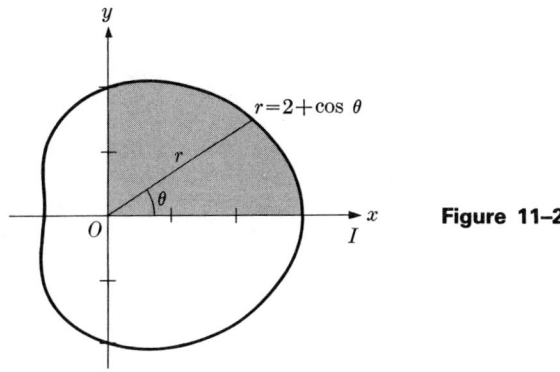

Figure 11–21

and obtain

$$\tfrac{1}{2}\int_0^{\pi/2} \cos^2\theta\, d\theta = \tfrac{1}{4}\int_0^{\pi/2}(1+\cos 2\theta)\, d\theta = [\tfrac{1}{4}(\theta + \tfrac{1}{2}\sin 2\theta)]_0^{\pi/2}.$$

Therefore

$$A = \pi + 2 + \frac{\pi}{8} = \frac{9\pi}{8} + 2. \qquad \blacktriangleleft$$

Example 2. Find the entire area enclosed by the curve $r = 2\sin 3\theta$.

Solution. We first draw the curve, as shown in Fig. 11–22. In fact, in all area problems in polar coordinates (as well as in Cartesian coordinates), a sketch should be made. The curve is a Rose Curve with three congruent petals. *The main difficulty in this problem is determining the limits of integration.* To find the area of the loop in the first quadrant, we observe that $r = 0$ when $\theta = 0$, and that the value of r increases steadily to a maximum when $\theta = \pi/6$ and $r = 2$; then r decreases steadily until $\theta = \pi/3$ when $r = 0$. In other words, the petal is described completely as θ goes from 0 to $\pi/3$. The total area A is three times the

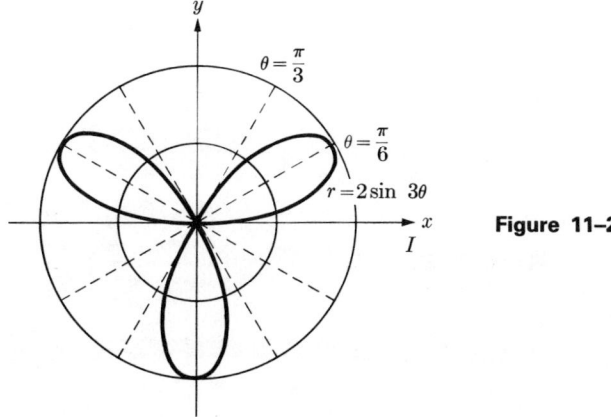

Figure 11–22

area of this loop. Therefore we have

$$\tfrac{1}{3}A = \tfrac{1}{2}\int_0^{\pi/3} (2\sin 3\theta)^2\, d\theta \quad \text{or} \quad A = 6\int_0^{\pi/3} \frac{1 - \cos 6\theta}{2}\, d\theta,$$

where we have used the half-angle formula, $\sin^2\phi = (1 - \cos 2\phi)/2$. Integration yields

$$A = 6\left[\frac{\theta}{2} - \frac{1}{12}\sin 6\theta\right]_0^{\pi/3} = \pi. \quad \blacktriangleleft$$

Example 3. Find the area inside the circle $r = 5\cos\theta$ and outside the curve $r = 2 + \cos\theta$.

Solution. We make a sketch, as shown in Fig. 11–23. The two curves intersect when

$$5\cos\theta = 2 + \cos\theta, \quad \cos\theta = \tfrac{1}{2},$$

or

$$\theta = \frac{\pi}{3},\ -\frac{\pi}{3}.$$

The limits of integration are $-\pi/3$ and $+\pi/3$. The area A inside $5\cos\theta$ and outside $2 + \cos\theta$ is given by

$$A = \tfrac{1}{2}\int_{-\pi/3}^{+\pi/3} (5\cos\theta)^2\, d\theta - \tfrac{1}{2}\int_{-\pi/3}^{+\pi/3} (2 + \cos\theta)^2\, d\theta.$$

We can shorten the work in two ways: (1) we observe that both curves are symmetric with respect to the x axis, and (2) the integrals may be combined, since the limits of integration are the same. In this way we obtain

$$A = \int_0^{\pi/3} [25\cos^2\theta - (4 + 4\cos\theta + \cos^2\theta)]\, d\theta$$

$$= \int_0^{\pi/3} (8 + 12\cos 2\theta - 4\cos\theta)\, d\theta$$

$$= [8\theta + 6\sin 2\theta - 4\sin\theta]_0^{\pi/3} = \frac{8\pi}{3} + \sqrt{3}. \quad \blacktriangleleft$$

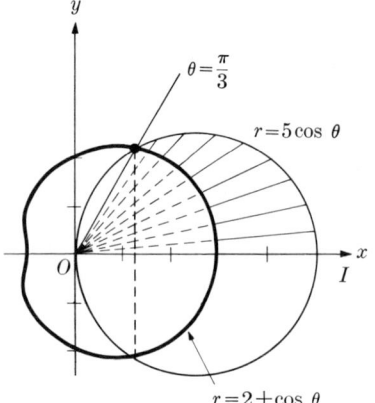

Figure 11–23

PROBLEMS

In Problems 1 through 7, in each case find the area enclosed by the curve and the two lines.

1. $r = 2\theta, \theta = 0, \theta = \pi$
2. $r = 2\tan\theta, \theta = 0, \theta = \pi/6$
3. $r = e^\theta, \theta = 0, \theta = 3\pi/2$
4. $r = 2\cos 2\theta, \theta = -\pi/4, \theta = \pi/4$
5. $r = \theta^2, \theta = 0, \theta = \pi$
6. $r = -3\theta^3, \theta = 0, \theta = \pi/2$
7. $r = 1/\theta, \theta = \pi/4, \theta = \pi/2$

In Problems 8 through 16, sketch each curve and find the entire area bounded by it.

8. $r = 3 + 2\cos\theta$
9. $r = 2\cos 3\theta$
10. $r^2 = \sin\theta$
11. $r^2 = 2\sin 2\theta$
12. $r = 2\sin 2\theta$
13. $r = 2(1 + \cos\theta)$
14. $r = \cos n\theta$, n a positive integer
15. $r = \sin n\theta$, n a positive integer
16. $r = 4 + 2\sin\theta$

In Problems 17 through 23, find the indicated area.

17. Inside the circle $r = 4\cos\theta$ and outside $r = 2$
18. Inside $r = 2 + 2\cos\theta$ and outside $r = 1$
19. Inside $r^2 = 8\cos 2\theta$ and outside $r = 2$
20. Inside $r = \sqrt{2}\cos\theta$ and outside $r = 2(1 - \cos\theta)$
21. Inside the curves $r = 3\cos\theta$ and $r = 2 - \cos\theta$
22. Inside the curves $r = 2\sin\theta$ and $r^2 = 2\cos 2\theta$
23. Inside the small loop of the curve $r = 1 + 2\cos\theta$
24. Find the area bounded by $r = a\cos\theta$ and the lines $\theta = \alpha$, $\theta = \pi/2$. By elementary methods, verify that the result is the formula for the area of a segment of a circle.

12

SOLID ANALYTIC GEOMETRY

1. THE NUMBER SPACE R_3. COORDINATES. THE DISTANCE FORMULA

In the study of analytic geometry of the plane we were careful to distinguish the geometric plane from the number plane R_2. As we saw in Chapter 2, page 18, the number plane consists of the collection of ordered pairs of real numbers. In the study of three-dimensional geometry it is useful to introduce the set of all ordered *triples* of real numbers. Calling this set the **three-dimensional number space,** we denote it by R_3. Each individual number triple is a **point** in R_3. The three elements in each number triple are called its **coordinates.** We now show how three-dimensional number space may be represented on a geometric or Euclidean three-dimensional space.

In three-dimensional space, consider three mutually perpendicular lines which intersect in a point O. We designate these lines the **coordinate axes** and, starting from O, set up identical number scales on each of them. If the positive directions of the x, y, and z axes are labeled x, y, and z, as shown in Fig. 12–1, we say the axes form a **right-handed system.** Figure 12–2 illustrates the axes in a **left-handed system.** We shall use a right-handed coordinate system throughout.

Any two intersecting lines in space determine a plane. A plane containing two of the coordinate axes is called a **coordinate plane.** Clearly, there are three such planes.

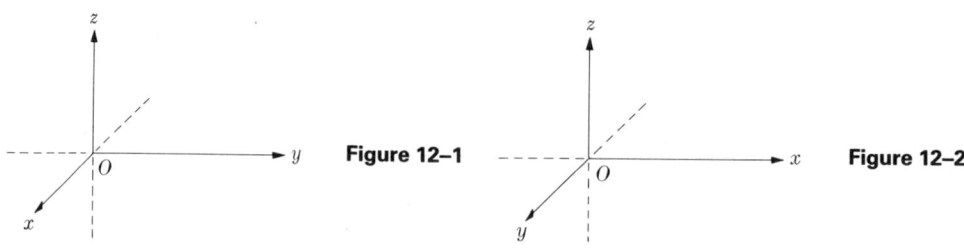

Figure 12–1   Figure 12–2

To each point P in three-dimensional space we can assign a point in R_3 in the following way. Through P construct three planes, each parallel to one of the coordinate planes as shown in Fig. 12–3. We label the intersections of the planes through P with the coordinate axes Q, R, and S, as shown. Then, If Q is x_0 units from the origin O, R is y_0 units from O, and S is z_0 units from O, we assign to P

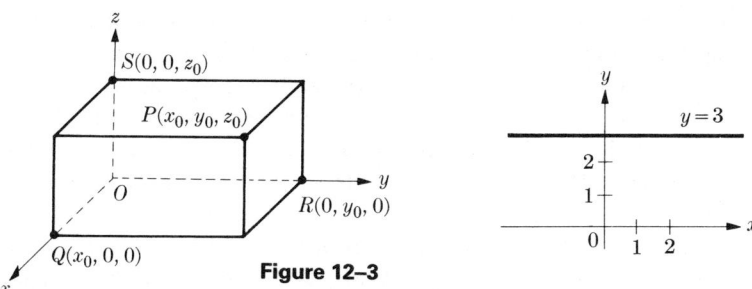

Figure 12–3

Figure 12–4

the number triple (x_0, y_0, z_0) and say that the point P has **Cartesian coordinates** (x_0, y_0, z_0). To each point in space there corresponds exactly one ordered number triple and, conversely, to each ordered number triple there is associated exactly one point in three-dimensional space. We have just described a **Cartesian** or **rectangular** coordinate system. In Section 7 we shall discuss other coordinate systems.

In studying analytic geometry in the plane, we saw that an equation such as

$$y = 3$$

represents all points lying on a line parallel to the x axis and three units above it (Fig. 12–4). In set notation, we write

$$\{(x, y): y = 3\}$$

to denote all the points on this line. The equation

$$y = 3$$

in the context of three-dimensional geometry represents something entirely different. The graph of points satisfying this equation is a plane parallel to the xz plane (the xz plane is the coordinate plane determined by the x and the z axis) and three units from it (Fig. 12–5). In set notation, we represent this plane by writing

$$\{(x, y, z): y = 3\}.$$

We see that set notation, by use of the symbols (x, y) or (x, y, z), indicates clearly when we are dealing with two- or three-dimensional geometry. The equation $y = 3$ by itself is ambiguous unless we know in advance the dimension of the geometry.

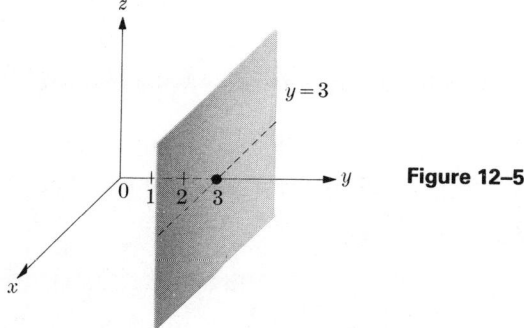

Figure 12–5

392 SOLID ANALYTIC GEOMETRY

In three dimensions the plane represented by $y = 3$ is perpendicular to the y axis and passes through the point $(0, 3, 0)$. Since there is exactly one plane which is perpendicular to a given line and which passes through a given point, we see that the graph of the equation $y = 3$ consists of one and only one such plane. Conversely, from the very definition of a Cartesian coordinate system every point with y coordinate 3 must lie in this plane. Equations such as $x = a$ or $y = b$ or $z = c$ always represent planes parallel to the coordinate planes.

We recall from Euclidean solid geometry that *any two nonparallel planes intersect in a straight line*. Therefore, the graph of all points which simultaneously satisfy the equations

$$x = a \quad \text{and} \quad y = b$$

is a line parallel to (or coincident with) the z axis. Conversely, any such line is the graph of a pair of equations of the above form. Since the plane $x = a$ is parallel to the z axis, and the plane $y = b$ is parallel to the z axis, the line of intersection must be parallel to the z axis also. (Corresponding statements hold with the axes interchanged.)

Theorem 1. *The distance d between the points $P_1(x_1, y_1, z_1)$ and $P_2(x_2, y_2, z_2)$ is*

$$d = \sqrt{(x_2 - x_1)^2 + (y_2 - y_1)^2 + (z_2 - z_1)^2}.$$

Proof. We make the construction shown in Fig. 12–6. By the Pythagorean theorem we have

$$d^2 = |P_1Q|^2 + |QP_2|^2.$$

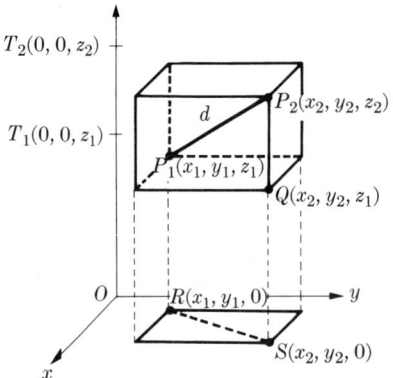

Figure 12–6

Noting that $|P_1Q| = |RS|$, we use the formula for distance in the xy plane to get

$$|P_1Q|^2 = |RS|^2 = (x_2 - x_1)^2 + (y_2 - y_1)^2.$$

Furthermore, since P_2 and Q are on a line parallel to the z axis, we see that

$$|QP_2|^2 = |T_1T_2|^2 = (z_2 - z_1)^2.$$

Therefore

$$d^2 = (x_2 - x_1)^2 + (y_2 - y_1)^2 + (z_2 - z_1)^2. \quad \blacktriangleleft$$

The midpoint P of the line segment connecting the point $P_1(x_1, y_1, z_1)$ and $P_2(x_2, y_2, z_2)$ has coordinates $P(\bar{x}, \bar{y}, \bar{z})$ given by the formulas

$$\bar{x} = \frac{x_1 + x_2}{2}, \quad \bar{y} = \frac{y_1 + y_2}{2}, \quad \bar{z} = \frac{z_1 + z_2}{2}.$$

If P_1 and P_2 lie in the xy plane—that is, if $z_1 = 0$ and $z_2 = 0$—then so does the midpoint P, and the formula is the one we learned in plane analytic geometry. The above formula for $\bar{x}$ is proved as in Chapter 3, Section 1, by passing planes through P_1, P, and P_2 perpendicular to the x axis. The formulas for $\bar{y}$ and $\bar{z}$ are established by analogy. (See Problem 24 at the end of this section.)

Example 1. Find the coordinates of the point Q which divides the line segment from $P_1(1, 4, -2)$ to $P_2(-3, 6, 7)$ in the proportion 3 to 1.

Solution. The midpoint P of the segment P_1P_2 has coordinates $P(-1, 5, \frac{5}{2})$. When we find the midpoint of PP_2 we get $Q(-2, \frac{11}{2}, \frac{19}{4})$. ◀

Example 2. One endpoint of a segment P_1P_2 has coordinates $P_1(-1, 2, 5)$. The midpoint P is known to lie in the xz plane, while the other endpoint is known to lie on the intersection of the planes $x = 5$ and $z = 8$. Find the coordinates of P and P_2.

Solution. For $P(\bar{x}, \bar{y}, \bar{z})$ we note that $\bar{y} = 0$, since P is in the xz plane. Similarly, for $P_2(x_2, y_2, z_2)$ we have $x_2 = 5$ and $z_2 = 8$. From the midpoint formula we get

$$\bar{x} = \frac{-1 + 5}{2}, \quad 0 = \bar{y} = \frac{2 + y_2}{2}, \quad \bar{z} = \frac{5 + 8}{2}.$$

Therefore the points have coordinates $P(2, 0, \frac{13}{2})$, $P_2(5, -2, 8)$. ◀

PROBLEMS

In Problems 1 through 5, find the lengths of the sides of triangle ABC and state whether the triangle is a right triangle, an isosceles triangle, or both.

1. $A(2, 1, 3)$, $B(3, -1, -2)$, $C(0, 2, -1)$
2. $A(4, 3, 1)$, $B(2, 1, 2)$, $C(0, 2, 4)$
3. $A(3, -1, -1)$, $B(1, 2, 1)$, $C(6, -1, 2)$
4. $A(1, 2, -3)$, $B(4, 3, -1)$, $C(3, 1, 2)$
5. $A(0, 0, 0)$, $B(4, 1, 2)$, $C(-5, -5, -1)$

In Problems 6 and 7, find the midpoint of the segment joining the given points A, B.

6. $A(4, -2, 6)$, $B(-2, 8, 1)$
7. $A(-2, 3, 5)$, $B(-6, 0, 4)$

In Problems 8 and 9, in each case find the coordinates of the three points which divide the given segment AB into four equal parts.

8. $A(3, 4, -1)$, $B(7, -2, 5)$
9. $A(1, -6, 0)$, $B(6, 12, 7)$

In Problems 10 through 13, find the lengths of the medians of the given triangles ABC.

10. $A(2, 1, 3)$, $B(3, -1, -2)$, $C(0, 2, -1)$
11. $A(4, 3, 1)$, $B(2, 1, 2)$, $C(0, 2, 4)$
12. $A(3, -1, -1)$, $B(1, 2, 1)$, $C(6, -1, 2)$
13. $A(1, 2, -3)$, $B(4, 3, -1)$, $C(3, 1, 2)$
14. One endpoint of a line segment is at $P(4, 6, -3)$ and the midpoint is at $Q(2, 1, 6)$. Find the other endpoint.

15. One endpoint of a line segment is at $P_1(-2, 1, 6)$ and the midpoint Q lies in the plane $y = 3$. The other endpoint, P_2, lies on the intersection of the planes $x = 4$ and $z = -6$. Find the coordinates of P_2 and Q.

In Problems 16 through 19, determine whether or not the three given points lie on a line.

16. $A(1, -1, 2)$, $B(-1, -4, 3)$, $C(3, 2, 1)$
17. $A(2, 3, 1)$, $B(4, 6, 5)$, $C(-2, -2, -7)$
18. $A(1, -1, 2)$, $B(3, 3, 4)$, $C(-2, -6, -1)$
19. $A(-4, 5, -6)$, $B(-1, 2, -1)$, $C(3, -3, 6)$
20. Describe the set of points in space given by $\{(x, y, z) : x = 2, y = -4\}$
21. Describe the set of points in space given by $\{(x, y, z) : -2 \leq y \leq 5\}$.
22. Describe the set of points in space given by $\{(x, y, z) : x \geq 0, y \geq 0, y \geq 0\}$.
23. Describe the set of points in space given by $\{(x, y, z) : x^2 + y^2 + z^2 < 1\}$.
24. Derive the formula for determining the midpoint of a line segment.
25. The formula for the coordinates of a point $Q(x_0, y_0, z_0)$ which divides the line segment from $P_1(x_1, y_1, z_1)$ to $P_2(x_2, y_2, z_2)$ in the ratio p to q is

$$x_0 = \frac{px_2 + qx_1}{p + q}, \quad y_0 = \frac{py_2 + qy_1}{p + q}, \quad z_0 = \frac{pz_2 + qz_1}{p + q}.$$

Derive this formula.

26. Find the equation of the graph of all points equidistant from the points $(2, -1, 3)$ and $(3, 1, -1)$. Can you describe the graph?
27. Find the equation of the graph of all points equidistant from the points $(5, 1, 0)$ and $(2, -1, 4)$. Can you describe the graph?
28. Find the equation of the graph of all points such that the sum of the distances from $(1, 0, 0)$ and $(-1, 0, 0)$ is always equal to 4. Describe the graph.
29. The points $A(0, 0, 0)$, $B(1, 0, 0)$, $C(\frac{1}{2}, \frac{1}{2}, 1/\sqrt{2})$, $D(0, 1, 0)$ are the vertices of a four-sided figure. Show that $|AB| = |BC| = |CD| = |DA| = 1$. Prove that the figure is not a rhombus.
30. Prove that the diagonals joining opposite vertices of a rectangular parallelepiped (there are four of them which are interior to the parallelepiped) bisect each other.

2. DIRECTION COSINES AND NUMBERS

In three-dimensional space we consider a line passing through the origin O and place an arrow on it so that one of the two possible directions on the line is distinguished (Fig. 12–7). We call such a line a **directed line.** If no arrow is placed, then L is called an **undirected line.** We use the symbol $\vec{L}$ to indicate a directed

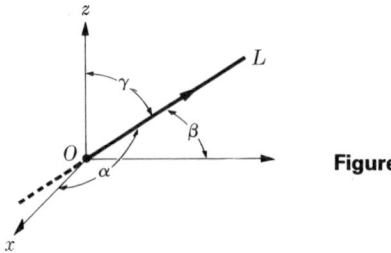

Figure 12–7

line while the letter L without the arrow over it indicates an undirected line. We denote by α, β, and γ the angles made by the directed line $\vec{L}$ and the positive directions of the x, y, and z axes, respectively. We define these angles to be the **direction angles** of the directed line $\vec{L}$. The *undirected line L* will have two possible sets of direction angles according to the ordering chosen. The two sets are

$$\alpha, \beta, \gamma \quad \text{and} \quad 180° - \alpha, \quad 180° - \beta, \quad 180° - \gamma.$$

The term "line" without further specification shall mean undirected line.

Definition. *If α, β, γ are direction angles of a directed line $\vec{L}$, then $\cos \alpha$, $\cos \beta$, $\cos \gamma$ are called the **direction cosines** of $\vec{L}$.*

Since $\cos(180° - \theta) = -\cos \theta$, we see that if λ, μ, ν are direction cosines of a directed line $\vec{L}$, then λ, μ, ν and $-\lambda, -\mu, -\nu$ are the two sets of direction cosines of the undirected line L.

We shall show that the direction cosines of any line L satisfy the relation

$$\cos^2 \alpha + \cos^2 \beta + \cos^2 \gamma = 1.$$

Let $P(x_0, y_0, z_0)$ be a point on a line L which goes through the origin. Then the distance d or P from the origin is

$$d = \sqrt{x_0^2 + y_0^2 + z_0^2},$$

and (see Fig. 12–8) we have

$$\cos \alpha = \frac{x_0}{d}, \quad \cos \beta = \frac{y_0}{d}, \quad \cos \gamma = \frac{z_0}{d}.$$

Squaring and adding, we get the desired result.

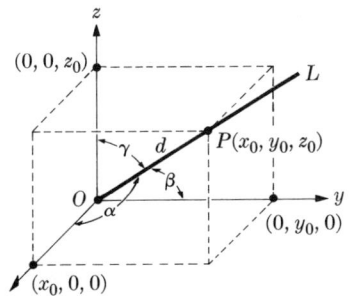

Figure 12–8

To define the direction cosines of any line L in space, we simply consider the line L' parallel to L which passes through the origin, and assert that *by definition L has the same direction cosines as L'*. Thus *all parallel lines in space have the same direction cosines.*

Definition. *Two sets of number triples, a, b, c and a', b', c', neither all zero, are said to be **proportional** if there is a number k such that*

$$a' = ka, \quad b' = kb, \quad c' = kc.$$

396 SOLID ANALYTIC GEOMETRY

Remark. The number k may be positive or negative but not zero, since by hypothesis neither of the number triples is 0, 0, 0. If none of the numbers a, b, and c is zero, we may write the proportionality relations as

$$\frac{a'}{a} = k, \qquad \frac{b'}{b} = k, \qquad \frac{c'}{c} = k$$

or, more simply,

$$\frac{a'}{a} = \frac{b'}{b} = \frac{c'}{c}.$$

Definition. *Suppose that a line L has direction cosines λ, μ, ν. Then a set of numbers a, b, c is called a **set of direction numbers** for L if a, b, c and λ, μ, ν are proportional.*

A line L has unlimited sets of direction numbers.

Theorem 2. *If $P_1(x_1, y_1, z_1)$ and $P_2(x_2, y_2, z_2)$ are two points on a line L, then*

$$\lambda = \frac{x_2 - x_1}{d}, \qquad \mu = \frac{y_2 - y_1}{d}, \qquad \nu = \frac{z_2 - z_1}{d}$$

is a set of direction cosines of L where d is the distance from P_1 to P_2.

Proof. In Fig. 12–9 the angles α, β, and γ are equal to the direction angles, since the lines P_1A, P_1B, P_1C are parallel to the coordinate axes. We read off from the figure that

$$\cos \alpha = \frac{x_2 - x_1}{d}, \qquad \cos \beta = \frac{y_2 - y_1}{d}, \qquad \cos \gamma = \frac{z_2 - z_1}{d},$$

which is the desired result.

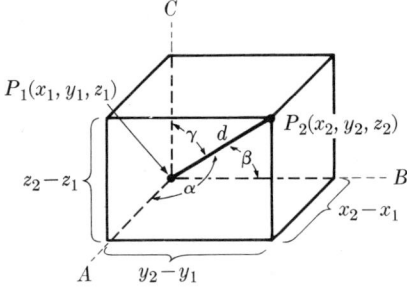

Figure 12–9

Corollary 1. *If $P_1(x_1, y_1, z_1)$ and $P_2(x_2, y_2, z_2)$ are two points on a line L, then*

$$x_2 - x_1, \qquad y_2 - y_1, \qquad z_2 - z_1$$

constitute a set of direction numbers for L.

Multiplying λ, μ, ν of Theorem 2 by the constant d, we obtain the result of the Corollary.

Example 1. Find direction numbers and direction cosines for the line L passing through the points $P_1(1, 5, 2)$ and $P_2(3, 7, -4)$.

Solution. From the Corollary, it is clear that 2, 2, -6 form a set of direction numbers. We compute

$$d = |P_1P_2| = \sqrt{4 + 4 + 36} = \sqrt{44} = 2\sqrt{11},$$

and so

$$\frac{1}{\sqrt{11}}, \quad \frac{1}{\sqrt{11}}, \quad -\frac{3}{\sqrt{11}}$$

form a set of direction cosines. Since L is undirected, it has two such sets, the other being $-1/\sqrt{11}$, $-1/\sqrt{11}$, $3/\sqrt{11}$. ◄

Example 2. Do the three points $P_1(3, -1, 4)$, $P_2(1, 6, 8)$, and $P_3(9, -22, -8)$ lie on the same straight line?

Solution. A set of direction numbers for the line L_1 through P_1 and P_2 is $-2, 7, 4$. A set of direction numbers for the line L_2 through P_2 and P_3 is $8, -28, -16$. Since the second set is proportional to the first (with $k = -4$), we conclude that L_1 and L_2 have the same direction cosines. Therefore the two lines are parallel. However, they have the point P_2 in common and so must coincide. ◄

From Theorem 2 and the statements in Example 2, we easily obtain the next result.

Corollary 2. *A line L_1 is parallel to a line L_2 if and only if a set of direction numbers of L_1 is proportional to a set of direction numbers of L_2.*

The angle between two intersecting lines in space is defined in the same way as the angle between two lines in the plane. It may happen that two lines L_1 and L_2 in space are neither parallel nor intersecting. Such lines are said to be **skew** to each other. Nevertheless, the angle between L_1 and L_2 can still be defined. Denote by L_1' and L_2' the lines passing through the origin and parallel to L_1 and L_2, respectively. The **angle between L_1 and L_2 is defined to be the angle between the intersecting lines L_1' and L_2'.**

Theorem 3. *If L_1 and L_2 have direction cosines λ_1, μ_1, ν_1 and λ_2, μ_2, ν_2, respectively, and if θ is the angle between L_1 and L_2, then*

$$\cos \theta = \lambda_1\lambda_2 + \mu_1\mu_2 + \nu_1\nu_2,$$

Proof. From the way the angle between two lines is defined we may consider L_1 and L_2 as lines passing through the origin. Let $P_1(x_1, y_1, z_1)$ be a point on L_1 and

398 SOLID ANALYTIC GEOMETRY

$P_2(x_2, y_2, z_2)$ a point on L_2, neither O (see Fig. 12–10). Denote by d_1 the distance of P_1 from O, by d_2 the distance of P_2 from O; let $d = |P_1P_2|$. We apply the Law of Cosines to triangle OP_1P_2, getting

$$d^2 = d_1^2 + d_2^2 - 2d_1d_2 \cos\theta \quad \text{or} \quad \cos\theta = \frac{d_1^2 + d_2^2 - d^2}{2d_1d_2}$$

and $\cos\theta =$

$$\frac{x_1^2 + y_1^2 + z_1^2 + x_2^2 + y_2^2 + z_2^2 - (x_2 - x_1)^2 - (y_2 - y_1)^2 - (z_2 - z_1)^2}{2d_1d_2}.$$

After simplification we find

$$\cos\theta = \frac{x_1x_2 + y_1y_2 + z_1z_2}{d_1d_2} = \frac{x_1}{d_1} \cdot \frac{x_2}{d_2} + \frac{y_1}{d_1} \cdot \frac{y_2}{d_2} + \frac{z_1}{d_1} \cdot \frac{z_2}{d_2}$$

$$= \lambda_1\lambda_2 + \mu_1\mu_2 + \nu_1\nu_2.$$

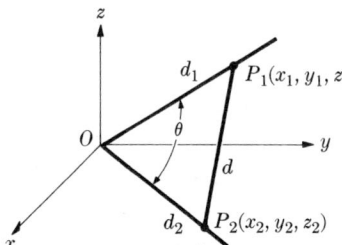

Figure 12–10

Corollary. *Two lines L_1 and L_2 with direction numbers a_1, b_1, c_1 and a_2, b_2, c_2, respectively, are perpendicular if and only if*

$$a_1a_2 + b_1b_2 + c_1c_2 = 0.$$

Example 3. Find the cosine of the angle between the line L_1, passing through the points $P_1(1, 4, 2)$ and $P_2(3, -1, 3)$, and the line L_2, passing through the points $Q_1(3, 1, 2)$ and $Q_2(2, 1, 3)$.

Solution. A set of direction numbers for L_1 is 2, -5, 1. A set for L_2 is -1, 0, 1. Therefore direction cosines for the two lines are

$$L_1: \frac{2}{\sqrt{30}}, \quad \frac{-5}{\sqrt{30}}, \quad \frac{1}{\sqrt{30}}; \quad L_2: \frac{-1}{\sqrt{2}}, \quad 0, \quad \frac{1}{\sqrt{2}}.$$

We obtain

$$\cos\theta = -\frac{1}{\sqrt{15}} + 0 + \frac{1}{2\sqrt{15}} = -\frac{1}{2\sqrt{15}}. \quad \blacktriangleleft$$

We observe that two lines always have two possible supplementary angles of intersection. If $\cos\theta$ is negative, we have obtained the obtuse angle and, if it is positive, the acute angle of intersection.

PROBLEMS

In Problems 1 through 4, find a set of direction numbers and a set of direction cosines for the line passing through the given points.

1. $A(2, 1, 4), B(3, 5, -2)$
2. $A(4, 2, -3), B(1, 0, 5)$
3. $A(-2, 1, -4), B(0, -5, -7)$
4. $A(6, 7, -2), B(8, -5, 1)$

In each of Problems 5 through 8, a point P_1 and a set of direction numbers are given. Find the coordinates of another point on the line L determined by P_1 and the given direction numbers.

5. $P_1(1, 4, -2)$,   direction numbers 2, 1, 4
6. $P_1(3, 5, -1)$,   direction numbers 2, 0, 4
7. $P_1(0, 4, -3)$,   direction numbers 0, 0, 5
8. $P_1(1, 2, 0)$,   direction numbers 4, 0, 0

In each of Problems 9 through 12, determine whether or not the three given points lie on a line.

9. $A(3, 1, 0), B(2, 2, 2), C(0, 4, 6)$
10. $A(2, -1, 1), B(4, 1, -3), C(7, 4, -9)$
11. $A(4, 2, -1), B(2, 1, 1), C(0, 0, 2)$
12. $A(5, 8, 6), B(-2, -3, 1), C(4, 2, 8)$

In each of Problems 13 through 15, determine whether or not the line through the points P_1, P_2 is parallel to the line through the points Q_1, Q_2.

13. $P_1(4, 8, 0), P_2(1, 2, 3); Q_1(0, 5, 0), Q_2(-3, -1, 3)$
14. $P_1(2, 1, 1), P_2(3, 2, -1); Q_1(0, 1, 4), Q_2(2, 3, 0)$
15. $P_1(3, 1, 4), P_2(-3, 2, 5); Q_1(4, 6, 1), Q_2(0, 5, 8)$

In each of Problems 16 through 18, determine whether or not the line through the points P_1, P_2 is perpendicular to the line through the points Q_1, Q_2.

16. $P_1(2, 1, 3), P_2(4, 0, 5); Q_1(3, 1, 2), Q_2(2, 1, 6)$
17. $P_1(2, -1, 0), P_2(3, 1, 2); Q_1(2, 1, 4), Q_2(4, 0, 4)$
18. $P_1(0, -4, 2), P_2(5, -1, 0); Q_1(3, 0, 2), Q_2(2, 1, 1)$

In each of Problems 19 through 21, find $\cos \theta$ where θ is the angle between the line L_1 passing through P_1, P_2 and L_2 passing through Q_1, Q_2.

19. $P_1(2, 1, 4), P_2(-1, 4, 1); Q_1(0, 5, 1), Q_2(3, -1, -2)$
20. $P_1(4, 0, 5), P_2(-1, -3, -2); Q_1(2, 1, 4), Q_2(2, -5, 1)$
21. $P_1(0, 0, 5), P_2(4, -2, 0); Q_1(0, 0, 6), Q_2(3, -2, 1)$

22. A **regular tetrahedron** is a 4-sided figure each side of which is an equilateral triangle. Find 4 points in space which are the vertices of a regular tetrahedron with each edge of length 2 units.

23. A **regular pyramid** is a 5-sided figure with a square base and sides consisting of 4 congruent isosceles triangles. If the base has a side of length 4 and if the height of the pyramid is 6 units, find the area of each of the triangular faces.

24. The points $P_1(1, 2, 3), P_2(2, 1, 2), P_3(3, 0, 1), P_4(5, 2, 7)$ are the vertices of a plane quadrilateral. Find the coordinates of the midpoints of the sides. What kind of quadrilateral do these four midpoints form?

25. Prove that the four interior diagonals of a parallelepiped bisect each other.

26. Given the set $S = \{(x, y, z) : x^2 + y^2 + z^2 = 1\}$. Find the points of $S \cap L$ where L is the line passing through the origin and having direction numbers 2, 1, 3.

27. Let P, Q, R, S be the vertices of any quadrilateral in three-space. Prove that the lines joining the midpoints of the sides form a parallelogram.

28. The vertices of a tetrahedron are at $(3, 0, 0)$, $(6, 0, 0)$, $(0, 9, 0)$, and $(6, 12, 15)$. Show that the three lines joining the midpoints of opposite sides bisect each other. How general is this result?

3. EQUATIONS OF A LINE

In plane analytic geometry a single equation of the first degree,

$$Ax + By + C = 0,$$

is the equation of a line (so long as A and B are not both zero). *In the geometry of three dimensions such an equation represents a plane.* Although we shall postpone a systematic study of planes until the next section, we assert now that in three-dimensional geometry it is not possible to represent a line by a single first-degree equation.

A line in space is determined by two points. If $P_0(x_0, y_0, z_0)$ and $P_1(x_1, y_1, z_1)$ are given points, we seek an analytic method of representing the line L determined by these points. The result is obtained by solving a geometric problem. A point $P(x, y, z)$, different from P_0, is on L if and only if the direction numbers determined by P and P_0 are proportional to those determined by P_1 and P_0 (Fig. 12–11). Calling the proportionality constant t, we see that the conditions are

$$x - x_0 = t(x_1 - x_0), \qquad y - y_0 = t(y_1 - y_0), \qquad z - z_0 = t(z_1 - z_0).$$

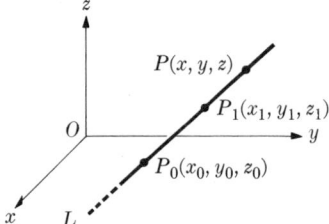

Figure 12–11

Thus we obtain the **two-point form of the parametric equations of a line:**

$$\begin{aligned} x &= x_0 + (x_1 - x_0)t, \\ y &= y_0 + (y_1 - y_0)t, \\ z &= z_0 + (z_1 - z_0)t. \end{aligned} \tag{1}$$

Definition. Given two points $P_0(x_0, y_0, z_0)$ and $P_1(x_1, y_1, z_1)$, the point which is t **of the way from** P_0 **to** P_1 is the point $P(x, y, z)$ whose coordinates are given by equations (1).

Example 1. Find the parametric equations of the line through the points $A(3, 2, -1)$ and $B(4, 4, 6)$. Locate three additional points on the line.

Solution. Substituting in (1) we obtain

$$x = 3 + t, \quad y = 2 + 2t, \quad z = -1 + 7t.$$

To get an additional point on the line we let $t = 2$ and obtain $P_1(5, 6, 13)$; $t = -1$ yields $P_2(2, 0, -8)$ and $t = 3$ gives $P_3(6, 8, 20)$. ◀

Theorem 4. *The parametric equations of a line L through the point $P_0(x_0, y_0, z_0)$ with direction numbers a, b, c are given by*

$$x = x_0 + at, \quad y = y_0 + bt, \quad z = z_0 + ct. \tag{2}$$

Proof. The point $P_1(x_0 + a, y_0 + b, z_0 + c)$ must be on L, since the direction numbers formed by P_0 and P_1 are just a, b, c. Using the two-point form (1) for the equations of a line through P_0 and P_1, we get (2) precisely.

If $\cos \alpha$, $\cos \beta$, $\cos \gamma$ are the direction cosines of a line L passing through the point (x_0, y_0, z_0), the equations of the line are

$$x = x_0 + t \cos \alpha, \quad y = y_0 + t \cos \beta, \quad z = z_0 + t \cos \gamma.$$

Example 2. Find the parametric equations of the line L through the point $A(3, -2, 5)$ with direction numbers $4, 0, -2$. What is the relation of L to the coordinate planes?

Solution. Substituting in (2), we obtain

$$x = 3 + 4t, \quad y = -2, \quad z = 5 - 2t.$$

Since all points on the line must satisfy all three of the above equations, L must lie in the plane $y = -2$. This plane is parallel to the xz plane. Therefore L is parallel to the xz plane. ◀

If none of the direction numbers is zero, the parameter t may be eliminated from the system of equations (2). We may write

$$\frac{x - x_0}{a} = \frac{y - y_0}{b} = \frac{z - z_0}{c} \tag{3}$$

for the equations of a line. For any value of t in (2) the ratios in (3) are equal. Conversely, if the ratios in (3) are all equal we may set the common value equal to t and (2) is satisfied.

If one of the direction numbers is zero, the form (3) may still be used if the zero in the denominator is interpreted properly. The equations

$$\frac{x - x_0}{a} = \frac{y - y_0}{b} = \frac{z - z_0}{0}$$

are understood to stand for the equations

$$\frac{x - x_0}{a} = \frac{y - y_0}{b} \quad \text{and} \quad z = z_0.$$

The system
$$\frac{x - x_0}{0} = \frac{y - y_0}{b} = \frac{z - z_0}{0}$$
stands for
$$x = x_0 \quad \text{and} \quad z = z_0.$$

We recognize these last two equations as those of planes parallel to coordinate planes. In other words, *a line is represented as the intersection of two planes.* This point will be discussed further in the next section.

The two-point form for the equations of a line also may be written *symmetrically.* The equations

$$\frac{x - x_0}{x_1 - x_0} = \frac{y - y_0}{y_1 - y_0} = \frac{z - z_0}{z_1 - z_0}$$

are called the **symmetric form for the equations of a line.**

Example 3. Find the point of intersection of the line
$$L = \{(x, y, z) : x = 3 - t, \quad y = 2 + 3t, \quad z = -1 + 2t\}$$
with the plane $S = \{(x, y, z) : z = 5\}$.

Solution. Denoting by $P(x_0, y_0, z_0)$ the point of intersection of L and S, we see that z_0 must have the value 5. From the equation $z = -1 + 2t$, we conclude that at the point of intersection, $5 = -1 + 2t$ or $t = 3$. Then $x_0 = 3 - 3 = 0$ and $y_0 = 2 + 3(3) = 11$. Hence $P = L \cap S$ has coordinates $(0, 11, 5)$. ◂

Example 4. Find the equations of the line through the point $P(2, -1, 3)$ and parallel to the line through the points $Q(1, 4, -6)$ and $R(-2, -1, 5)$.

Solution. The line through Q and R has direction numbers: $-2 - 1 = -3$, $-1 - 4 = -5$, $5 - (-6) = 11$. Therefore the parallel line through P is given by the equations
$$x = 2 - 3t, \quad y = -1 - 5t, \quad z = 3 + 11t. \quad ◂$$

PROBLEMS

In each of Problems 1 through 4, find the equations of the line going through the given points.

1. $A(1, 3, 2), B(2, -1, 4)$
2. $A(1, 0, 5), B(-2, 0, 1)$
3. $A(4, -2, 0), B(3, 2, -1)$
4. $A(5, 5, -2), B(6, 4, -4)$

In each of Problems 5 through 9, find the equations of the line passing through the given point with the given direction numbers.

5. $P_1(1, 0, -1)$, direction numbers 2, 1, -3
6. $P_1(-2, 1, 3)$, direction numbers 3, -1, -2
7. $P_1(4, 0, 0)$, direction numbers 2, -1, -3
8. $P_1(1, 2, 0)$, direction numbers 0, 1, 3
9. $P_1(3, -1, -2)$, direction numbers 2, 0, 0

In each of Problems 10 through 14, decide whether or not L_1 and L_2 are perpendicular.

10. $L_1: \dfrac{x-2}{2} = \dfrac{y+1}{-3} = \dfrac{z-1}{4}$; $L_2: \dfrac{x-2}{-3} = \dfrac{y+1}{2} = \dfrac{z-1}{3}$

11. $L_1: \dfrac{x}{1} = \dfrac{y+1}{2} = \dfrac{z+1}{3}$; $L_2: \dfrac{x-3}{0} = \dfrac{y+1}{-3} = \dfrac{z+4}{2}$

12. $L_1: \dfrac{x+2}{-1} = \dfrac{y-2}{2} = \dfrac{z+3}{3}$; $L_2: \dfrac{x+2}{1} = \dfrac{y-2}{2} = \dfrac{z+3}{-1}$

13. $L_1: \dfrac{x+5}{4} = \dfrac{y-1}{3} = \dfrac{z+8}{5}$; $L_2: \dfrac{x-4}{3} = \dfrac{y+7}{2} = \dfrac{z+4}{1}$

14. $L_1: \dfrac{x+1}{0} = \dfrac{y-2}{1} = \dfrac{z+8}{0}$; $L_2: \dfrac{x-3}{1} = \dfrac{y+2}{0} = \dfrac{z-1}{0}$

15. Find the equations of the medians of the triangle with vertices at $A(4, 0, 2)$, $B(3, 1, 4)$, $C(2, 5, 0)$.

16. Find the equations of any line through $P_1(2, 1, 4)$ and perpendicular to any line having direction numbers 4, 1, 3.

17. Find the equations of any line through $P_1(2, -1, 5)$ and perpendicular to any line having direction numbers 2, -3, 1.

18. Find the points of intersection of the line
$$L = \{(x, y, z): x = 3 + 2t, \quad y = 7 + 8t, \quad z = -2 + t\}$$
with each of the coordinate planes.

19. Find the points of intersection of the line
$$L = \left\{(x, y, z): \dfrac{x+1}{-2} = \dfrac{y+1}{3} = \dfrac{z-1}{7}\right\}$$
with each of the coordinate planes.

20. Show that the following lines are coincident:
$$\dfrac{x-1}{2} = \dfrac{y+1}{-3} = \dfrac{z}{4}; \quad \dfrac{x-5}{2} = \dfrac{y+7}{-3} = \dfrac{z-8}{4}.$$

21. Find the equations of the line through $(3, 1, 5)$ which is parallel to the line
$$L = \{(x, y, z): x = 4 - t, \quad y = 2 + 3t, \quad z = -4 + t\}.$$

22. Find the equations of the line through $(3, 1, -2)$ which is perpendicular and intersects the line
$$\dfrac{x+1}{1} = \dfrac{y+2}{1} = \dfrac{z+1}{1}.$$

[*Hint*: Let (x_0, y_0, z_0) be the point of intersection and determine its coordinates.]

23. A triangle has vertices at $A(2, 1, 6)$, $B(-3, 2, 4)$ and $C(5, 8, 7)$. Perpendiculars are drawn from these vertices to the xz plane. Locate the points A', B', and C' which are

the intersections of the perpendiculars through A, B, C and the xz plane. Find the equations of the sides of the triangle $A'B'C'$.

The point $(\bar{x}, \bar{y}, \bar{z})$ which is h of the way from (x_1, y_1, z_1) to (x_2, y_2, z_2) has its coordinates given by

$$\bar{x} = x_1 + h(x_2 - x_1), \quad \bar{y} = y_1 + h(y_2 - y_1), \quad \bar{z} = z_1 + h(z_2 - z_1). \tag{4}$$

Equations (4) are known as the **point of division formula**.

24. Use Equations (4) to find the point $\frac{1}{3}$ of the way from $P_1(2, -1, 3)$ to $P_2(6, 2, -5)$.

25. Show that the medians of any triangle in three-space intersect at a point which divides each median in the proportion $2:1$.

4. THE PLANE

Any three points not on a straight line determine a plane. While this characterization of a plane is quite simple, it is not convenient for beginning the study of planes. Instead we use the fact that *there is exactly one plane which passes through a given point and is perpendicular to a given line.*

Let $P_0(x_0, y_0, z_0)$ be a given point, and suppose that a given line L goes through the point $P_1(x_1, y_1, z_1)$ and has direction numbers A, B, C.

Theorem 5. *The equation of the plane passing through P_0 and perpendicular to L is*

$$A(x - x_0) + B(y - y_0) + C(z - z_0) = 0.$$

Proof. We establish the result by using the above characterization of a plane. Let $P(x, y, z)$ be a point on the plane (Fig. 12–12). From three-dimensional Euclidean geometry we recall that if a line L_1 through P_0 and P is perpendicular to L, then P must be in the desired plane. A set of direction numbers for the line L_1 is

$$x - x_0, \quad y - y_0, \quad z - z_0.$$

Since L has direction numbers A, B, C, we conclude that the two lines L and L_1 are perpendicular if and only if their direction numbers satisfy the relation

$$A(x - x_0) + B(y - y_0) + C(z - z_0) = 0,$$

which is the equation we seek.

Remark. Note that only the direction of L—and not the coordinates of P_1—enters the above equation. We obtain the same plane and the same equation if any line parallel to L is used in its stead.

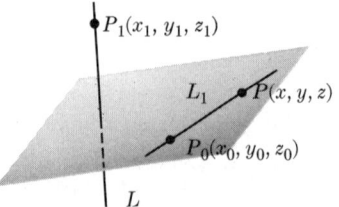

Figure 12–12

Example 1. Find the equation of the plane through the point $P_0(5, 2, -3)$ which is perpendicular to the line through the points $P_1(5, 4, 3)$ and $P_2(-6, 1, 7)$.

Solution. The line through the points P_1 and P_2 has direction numbers $-11, -3, 4$. The equation of the plane is

$$-11(x - 5) - 3(y - 2) + 4(z + 3) = 0$$
$$\Leftrightarrow \quad 11x + 3y - 4z - 73 = 0. \quad \blacktriangleleft$$

All lines perpendicular to the same plane are parallel and therefore have proportional direction numbers.

Definition. *A set of **attitude numbers** of a plane is any set of direction numbers of a line perpendicular to the plane.*

In Example 1 above, $11, 3, -4$ form a set of attitude numbers of the plane.

Example 2. What are sets of attitude numbers for planes parallel to the coordinate planes?

Solution. A plane parallel to the yz plane has an equation of the form $x - c = 0$, where c is a constant. A set of attitude numbers for this plane is $1, 0, 0$. A plane parallel to the xz plane has attitude numbers $0, 1, 0$, and any plane parallel to the xy plane has attitude numbers $0, 0, 1$. $\quad \blacktriangleleft$

Since lines perpendicular to the same or parallel planes are themselves parallel, we get at once the next theorem.

Theorem 6. *Two planes are parallel if and only if their attitude numbers are proportional.*

Theorem 7. *If A, B, and C are not all zero, the graph of an equation of the form*

$$Ax + By + Cz + D = 0 \tag{1}$$

is a plane.

Proof. Suppose that $C \neq 0$, for example. Then the point $P_0(0, 0, -D/C)$ is on the graph, as its coordinates satisfy the above equation. Therefore we may write

$$A(x - 0) + B(y - 0) + C\left(z + \frac{D}{C}\right) = 0,$$

and the graph is the plane passing through P_0 perpendicular to any line with direction numbers A, B, C. $\quad \blacktriangleleft$

An equation of the plane through three points not on a line can be found by assuming that the plane has an equation of the form (1), substituting in turn the coordinates of the three points, and solving simultaneously the three resulting equations. The fact that there are four constants, A, B, C, D, and only three equations is illusory, since we may divide through by one of them (say D) and obtain three equations in the unknowns A/D, B/D, C/D. This is equivalent to setting D (or one of the other constants) equal to some convenient value. An example illustrates the procedure.

Example 3. Find an equation of the plane passing through the points $(2, 1, 3)$, $(1, 3, 2)$, $(-1, 2, 4)$.

Solution. Since the three points lie in the plane, each of them satisfies equation (1). We have

$$(2, 1, 3): \quad 2A + B + 3C + D = 0,$$
$$(1, 3, 2): \quad A + 3B + 2C + D = 0,$$
$$(-1, 2, 4): \quad -A + 2B + 4C + D = 0.$$

Solving for A, B, C in terms of D, we obtain

$$A = -\tfrac{3}{25}D, \qquad B = -\tfrac{4}{25}D, \qquad C = -\tfrac{5}{25}D.$$

Setting $D = -25$, we get the equation

$$3x + 4y + 5z - 25 = 0. \qquad \blacktriangleleft$$

PROBLEMS

In each of Problems 1 through 4, find the equation of the plane which passes through the given point P_0 and has the given attitude numbers.

1. $P_0(1, 4, 2);\ 3, 1, -4$
2. $P_0(2, 1, -5);\ 3, 0, 2$
3. $P_0(4, -2, -5);\ 0, 3, -2$
   $P_0(-1, -2, -3);\ 4, 0, 0$

In each of Problems 5 through 8, find the equation of the plane which passes through the three points.

5. $(1, -2, 1), (2, 0, 3), (0, 1, -1)$
6. $(2, 2, 1), (-1, 2, 3), (3, -5, -2)$
7. $(3, -1, 2), (1, 2, -1), (2, 3, 1)$
8. $(-1, 3, 1), (2, 1, 2), (4, 2, -1)$

In each of Problems 9 through 12, find the equation of the plane passing through P_1 and perpendicular to the line L_1.

9. $P_1(2, -1, 3);\qquad L_1: x = -1 + 2t,\ y = 1 + 3t,\ z = -4t$
10. $P_1(1, 2, -3);\qquad L_1: x = t,\ y = -2 - 2t,\ z = 1 + 3t$
11. $P_1(2, -1, -2);\qquad L_1: x = 2 + 3t,\ y = 0,\ z = -1 - 2t$
12. $P_1(-1, 2, -3);\qquad L_1: x = -1 + 5t,\ y = 1 + 2t,\ z = -1 + 3t$

In each of Problems 13 through 16, find the equations of the line through P_1 and perpendicular to the given plane M_1.

13. $P_1(-2, 3, 1);\qquad M_1: 2x + 3y + z - 3 = 0$
14. $P_1(1, -2, -3);\qquad M_1: 3x - y - 2z + 4 = 0$
15. $P_1(-1, 0, -2);\qquad M_1: x + 2z + 3 = 0$
16. $P_1(2, -1, -3);\qquad M_1: x = 4$

In each of Problems 17 through 20, find an equation of the plane through P_1 and parallel to the plane Φ.

17. $P_1(1, -2, -1)$; $\Phi: 3x + 2y - z + 4 = 0$
18. $P_1(-1, 3, 2)$; $\Phi: 2x + y - 3z + 5 = 0$
19. $P_1(2, -1, 3)$; $\Phi: x - 2y - 3z + 6 = 0$
20. $P_1(3, 0, 2)$; $\Phi: x + 2y + 1 = 0$

In each of Problems 21 through 23, find the equations of the line through P_1 parallel to the given line L.

21. $P_1(2, -1, 3)$; $L: \dfrac{x-1}{3} = \dfrac{y+2}{-2} = \dfrac{z-2}{4}$

22. $P_1(0, 0, 1)$; $L: \dfrac{x+2}{1} = \dfrac{y-1}{3} = \dfrac{z+1}{-2}$

23. $P_1(1, -2, 0)$; $L: \dfrac{x-2}{2} = \dfrac{y+2}{-1} = \dfrac{z-3}{4}$

In each of Problems 24 through 28, find the equation of the plane containing L_1 and L_2.

24. $L_1: \dfrac{x+1}{2} = \dfrac{y-2}{3} = \dfrac{z-1}{1}$; $L_2: \dfrac{x+1}{1} = \dfrac{y-2}{-1} = \dfrac{z-1}{2}$

25. $L_1: \dfrac{x-1}{3} = \dfrac{y+2}{2} = \dfrac{z-2}{2}$; $L_2: \dfrac{x-1}{1} = \dfrac{y+2}{1} = \dfrac{z-2}{0}$

26. $L_1: \dfrac{x+2}{1} = \dfrac{y}{0} = \dfrac{z+1}{2}$; $L_2: \dfrac{x+2}{2} = \dfrac{y}{3} = \dfrac{z+1}{1}$

27. $L_1: \dfrac{x}{2} = \dfrac{y-1}{3} = \dfrac{z+2}{-1}$; $L_2: \dfrac{x-2}{2} = \dfrac{y+1}{3} = \dfrac{z}{-1}$ $(L_1 \parallel L_2)$

28. $L_1: \dfrac{x-2}{2} = \dfrac{y+1}{-1} = \dfrac{z}{3}$; $L_2: \dfrac{x+1}{2} = \dfrac{y}{-1} = \dfrac{z+2}{3}$ $(L_1 \parallel L_2)$

In Problems 29 and 30, find the equation of the plane through P_1 and the given line L.

29. $P_1(3, -1, 2)$; $L = \left\{(x, y, z): \dfrac{x-2}{2} = \dfrac{y+1}{3} = \dfrac{z}{-2}\right\}$

30. $P_1(1, -2, 3)$; $L = \{(x, y, z): x = -1 + t,\ y = 2 + 2t,\ z = 2 - 2t\}$

31. Show that the plane $2x - 3y + z - 2 = 0$ is parallel to the line
$$\dfrac{x-2}{1} = \dfrac{y+2}{1} = \dfrac{z+1}{1}.$$

32. Show that the plane $5x - 3y - z - 6 = 0$ contains the line
$$x = 1 + 2t, \qquad y = -1 + 3t, \qquad z = 2 + t.$$

33. A plane has attitude numbers A, B, C, and a line has direction numbers a, b, c. What condition must be satisfied in order that the plane and line be parallel?

34. Show that the three planes
$$7x - 2y - 2z - 5 = 0,$$
$$3x + 2y - 3z - 10 = 0,$$
$$7x + 2y - 5z - 16 = 0$$
all contain a common line. Find the coordinates of two points on this line.

408　SOLID ANALYTIC GEOMETRY

*35. Find a condition that three planes

$$A_1x + B_1y + C_1z + D_1 = 0,$$
$$A_2x + B_2y + C_2z + D_2 = 0,$$
$$A_3x + B_3y + C_3z + D_3 = 0$$

either have a line in common or have no point in common.

5. ANGLES. DISTANCE FROM A POINT TO A PLANE

The angle between two lines was defined in Section 2. We recall that if line L_1 has direction cosines λ_1, μ_1, ν_1 and line L_2 has direction cosines λ_2, μ_2, ν_2, then

$$\cos \theta = \lambda_1\lambda_2 + \mu_1\mu_2 + \nu_1\nu_2.$$

where θ is the angle between L_1 and L_2.

Definition. *Let Φ_1 and Φ_2 be two planes, and let L_1 and L_2 be two lines which are perpendicular to Φ_1 and Φ_2, respectively. Then the* **angle between Φ_1 and Φ_2** *is, by definition, the angle between L_1 and L_2.* (See Fig. 12–13.) *Furthermore, we make the convention that we always select the acute angle between these lines as the angle between Φ_1 and Φ_2.*

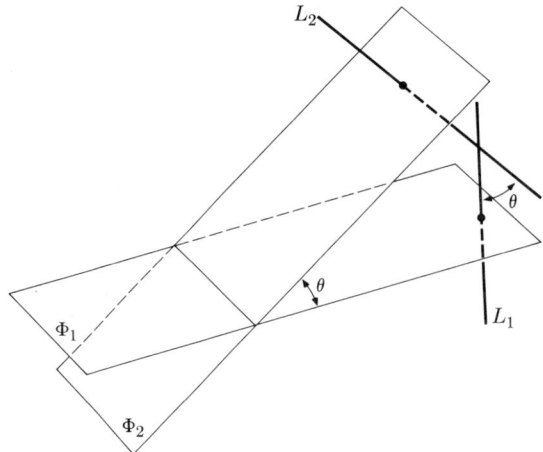

Figure 12–13

Theorem 8. *The angle θ between the planes $A_1x + B_1y + C_1z + D_1 = 0$ and $A_2x + B_2y + C_2z + D_2 = 0$ is given by*

$$\cos \theta = \frac{|A_1A_2 + B_1B_2 + C_1C_2|}{\sqrt{A_1^2 + B_1^2 + C_1^2}\sqrt{A_2^2 + B_2^2 + C_2^2}}.$$

Proof. From the definition of attitude numbers for a plane, we know that they are direction numbers of any line perpendicular to the plane. Converting to direction cosines, we get the above formula.

Corollary. *Two planes with attitude numbers A_1, B_1, C_1 and A_2, B_2, C_2 are perpendicular if and only if*

$$A_1A_2 + B_1B_2 + C_1C_2 = 0.$$

Example 1. Find $\cos \theta$ where θ is the angle between the planes $3x - 2y + z = 4$ and $x + 4y - 3z - 2 = 0$.

Solution. Substituting in the formula of Theorem 8, we have

$$\cos \theta = \frac{|3 - 8 - 3|}{\sqrt{9 + 4 + 1}\sqrt{1 + 16 + 9}} = \frac{4}{\sqrt{91}}. \quad \blacktriangleleft$$

Two nonparallel planes intersect in a line. Every point on the line satisfies the equations of both planes and, conversely, every point which satisfies the equations of both planes must be on the line. *Therefore we may characterize any line in space by finding two planes which contain it.* Since every line has an unlimited number of planes which pass through it and since *any* two of them are sufficient to determine the line uniquely, we see that there is an unlimited number of ways of writing the equations of a line. The next example shows how to transform one representation into another.

Example 2. The two planes

$$2x + 3y - 4z - 6 = 0 \quad \text{and} \quad 3x - y + 2z + 4 = 0$$

intersect in a line. (That is, the points which satisfy *both* equations constitute the line.) Find a set of parametric equations of the line of intersection.

Solution. We solve the above equations for x and y in terms of z, getting

$$x = -\tfrac{2}{11}z - \tfrac{6}{11}, \qquad y = \tfrac{16}{11}z + \tfrac{26}{11} \quad \text{and} \quad \frac{x + \tfrac{6}{11}}{-\tfrac{2}{11}} = \frac{y - \tfrac{26}{11}}{\tfrac{16}{11}} = \frac{z}{1}.$$

We can therefore write

$$x = -\tfrac{6}{11} - \tfrac{2}{11}t, \qquad y = \tfrac{26}{11} + \tfrac{16}{11}t, \qquad z = t,$$

which are the desired parametric equations. $\blacktriangleleft$

Three planes may be parallel, may pass through a common line, may have no common points, or may have a unique point of intersection. If they have a unique point of intersection, the intersection point may be found by solving simultaneously the three equations of the planes. If they have no common point, an attempt to solve simultaneously will fail. A further examination will show whether or not two or more of the planes are parallel.

Example 3. Determine whether or not the planes $\Phi_1: 3x - y + z - 2 = 0$; $\Phi_2: x + 2y - z + 1 = 0$; $\Phi_3: 2x + 2y + z - 4 = 0$ intersect. If so, find the point of intersection.

Solution. Eliminating z between Φ_1 and Φ_2, we have
$$4x + y - 1 = 0. \tag{1}$$
Eliminating z between Φ_2 and Φ_3, we find
$$3x + 4y - 3 = 0. \tag{2}$$
We solve equations (1) and (2) simultaneously to get
$$x = \tfrac{1}{13}, \qquad y = \tfrac{9}{13}.$$
Substituting in the equation for Φ_1, we obtain $z = \tfrac{32}{13}$. Therefore the single point of intersection of the three planes is $(\tfrac{1}{13}, \tfrac{9}{13}, \tfrac{32}{13})$. ◂

Example 4. Find the point of intersection of the plane
$$3x - y + 2z - 3 = 0$$
and the line
$$\frac{x+1}{3} = \frac{y+1}{2} = \frac{z-1}{-2}.$$

Solution. We write the equations of the line in parametric form:
$$x = -1 + 3t, \qquad y = -1 + 2t, \qquad z = 1 - 2t.$$
The point of intersection is given by a value of t; call it t_0. This point must satisfy the equation of the plane. We have
$$3(-1 + 3t_0) - (-1 + 2t_0) + 2(1 - 2t_0) - 3 = 0 \quad \Leftrightarrow \quad t_0 = 1.$$
The desired point is $(2, 1, -1)$. ◂

We now derive an important formula which tells us how to find the perpendicular distance from a point in space to a plane.

Theorem 9. *The distance d from the point $P_1(x_1, y_1, z_1)$ to the plane*
$$Ax + By + Cz + D = 0$$
is given by
$$d = \frac{|Ax_1 + By_1 + Cz_1 + D|}{\sqrt{A^2 + B^2 + C^2}}.$$

Proof. We write the equations of the line L through P_1 which is perpendicular to the plane. They are
$$L : x = x_1 + At, \qquad y = y_1 + Bt, \qquad z = z_1 + Ct.$$
Denote by (x_0, y_0, z_0) the point of intersection of L and the plane. Then
$$d^2 = (x_1 - x_0)^2 + (y_1 - y_0)^2 + (z_1 - z_0)^2. \tag{3}$$
Also (x_0, y_0, z_0) is on both the line and the plane. Therefore, for some value t_0

we have
$$x_0 = x_1 + At_0,$$
$$y_0 = y_1 + Bt_0, \quad (4)$$
$$z_0 = z_1 + Ct_0$$

and
$$Ax_0 + By_0 + Cz_0 + D = 0 = A(x_1 + At_0) + B(y_1 + Bt_0) + C(z_1 + Ct_0) + D.$$

Thus, from (3) and (4), we write
$$d = \sqrt{A^2 + B^2 + C^2}\,|t_0|,$$

and now, inserting the relation
$$t_0 = \frac{-(Ax_1 + By_1 + Cz_1 + D)}{A^2 + B^2 + C^2}$$

in the preceding expression for d, we obtain the desired formula.

Example 5. Find the distance from the point $(2, -1, 5)$ to the plane
$$3x + 2y - 2z - 7 = 0.$$

Solution
$$d = \frac{|6 - 2 - 10 - 7|}{\sqrt{9 + 4 + 4}} = \frac{13}{\sqrt{17}}. \quad \blacktriangleleft$$

PROBLEMS

In each of Problems 1 through 4, find $\cos\theta$ where θ is the angle between the given planes.
1. $2x - y + 2z - 3 = 0, \quad 3x + 2y - 6z - 11 = 0$
2. $x + 2y - 3z + 6 = 0, \quad x + y + z - 4 = 0$
3. $2x - y + 3z - 5 = 0, \quad 3x - 2y + 2z - 7 = 0$
4. $x + 4z - 2 = 0, \quad y + 2z - 6 = 0$

In each of Problems 5 through 8, find the equations in parametric form of the line of intersection of the given planes.
5. $3x + 2y - z + 5 = 0, \quad 2x + y + 2z - 3 = 0$
6. $x + 2y + 2z - 4 = 0, \quad 2x + y - 3z + 5 = 0$
7. $x + 2y - z + 4 = 0, \quad 2x + 4y + 3z - 7 = 0$
8. $2x + 3y - 4z + 7 = 0, \quad 3x - 2y + 3z - 6 = 0$

In each of Problems 9 through 12, find the point of intersection of the given line and the given plane.
9. $3x - y + 2z - 5 = 0, \quad \dfrac{x-1}{2} = \dfrac{y+1}{3} = \dfrac{z-1}{-2}$
10. $2x + 3y - 4z + 15 = 0, \quad \dfrac{x+3}{2} = \dfrac{y-1}{-2} = \dfrac{z+4}{3}$

11. $x + 2z + 3 = 0$, $\quad \dfrac{x+1}{1} = \dfrac{y}{0} = \dfrac{z+2}{2}$

12. $2x + 3y + z - 3 = 0$, $\quad \dfrac{x+2}{2} = \dfrac{y-3}{3} = \dfrac{z-1}{1}$

In each of Problems 13 through 16, find the distance from the given point to the given plane.

13. $(2, 1, -1)$, $\quad x - 2y + 2z + 5 = 0$
14. $(3, -1, 2)$, $\quad 3x + 2y - 6z - 9 = 0$
15. $(-1, 3, 2)$, $\quad 2x - 3y + 4z - 5 = 0$
16. $(0, 4, -3)$, $\quad 3y + 2z - 7 = 0$
17. Find the equation of the plane through the line
$$\dfrac{x+1}{3} = \dfrac{y-1}{2} = \dfrac{z-2}{4}$$
which is perpendicular to the plane
$$2x + y - 3z + 4 = 0.$$

18. Find the equation of the plane through the line
$$\dfrac{x-2}{2} = \dfrac{y-2}{3} = \dfrac{z-1}{-2}$$
which is parallel to the line
$$\dfrac{x+1}{3} = \dfrac{y-1}{2} = \dfrac{z+1}{1}.$$

19. Find the equation of the plane through the line
$$\dfrac{x+2}{3} = \dfrac{y}{-2} = \dfrac{z+1}{2}$$
which is parallel to the line
$$\dfrac{x-1}{2} = \dfrac{y+1}{3} = \dfrac{z-1}{4}.$$

20. Find the equations of any line through the point $(1, 4, 2)$ which is parallel to the plane
$$2x + y + z - 4 = 0.$$

21. Find the equation of the plane through $(3, 2, -1)$ and $(1, -1, 2)$ which is parallel to the line
$$\dfrac{x-1}{3} = \dfrac{y+1}{2} = \dfrac{z}{-2}.$$

In each of Problems 22 through 26, find all the points of intersection of the three given planes. If the three planes pass through a line, find the equations of the line in parametric form.

22. $2x + y - 2z - 1 = 0$, $\quad 3x + 2y + z - 10 = 0$, $\quad x + 2y - 3z + 2 = 0$
23. $x + 2y + 3z - 4 = 0$, $\quad 2x - 3y + z - 2 = 0$, $\quad 3x + 2y - 2z - 5 = 0$
24. $3x - y + 2z - 4 = 0$, $\quad x + 2y - z - 3 = 0$, $\quad 3x - 8y + 7z + 1 = 0$

25. $2x + y - 2z - 3 = 0$, $x - y + z + 1 = 0$, $x + 5y - 7z - 3 = 0$
26. $x + 2y + 3z - 5 = 0$, $2x - y - 2z - 2 = 0$, $x - 8y - 13z + 11 = 0$

In each of Problems 27 through 29, find the equations in parametric form of the line through the given point P_1 which intersects and is perpendicular to the given line L.

27. $P_1(3, -1, 2)$; $L: \dfrac{x-1}{2} = \dfrac{y+1}{-1} = \dfrac{z}{3}$

28. $P_1(-1, 2, 3)$; $L: \dfrac{x}{2} = \dfrac{y-2}{0} = \dfrac{z+3}{-3}$

29. $P_1(0, 2, 4)$; $L: \dfrac{x-1}{3} = \dfrac{y-2}{1} = \dfrac{z-3}{4}$

30. If $A_1x + B_1y + C_1z + D_1 = 0$ and $A_2x + B_2y + C_2z + D_2 = 0$ are two intersecting planes, what is the graph of all points which satisfy

$$A_1x + B_1y + C_1z + D_1 + k(A_2x + B_2y + C_2z + D_2) = 0,$$

where k is a constant?

31. Find the equation of the plane passing through the point $(2, 1, -3)$ and the intersection of the planes $3x + y - z - 2 = 0$, $2x + y + 4z - 1 = 0$. [*Hint*: See Problem 30.]

32. Given a regular tetrahedron with each edge a units in length. (See Problem 22 at the end of Section 2.) Find the distance from a vertex to the opposite face.

33. Given a regular pyramid with each edge of the base 2 units in length and each lateral edge 4 units in length. (See Problem 23 at the end of Section 2.) Find the distance from the top of the pyramid to the base. Also find the distance from one of the vertices of the base to a face opposite to that vertex.

34. A slice is made in a cube by cutting through a diagonal of one face and proceeding through one of the vertices of the opposite face as shown in Fig. 12–14. Find the angle between the planar slice and the first face which is cut.

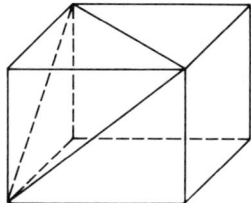

Figure 12–14

6. THE SPHERE. CYLINDERS

A **sphere** is the graph of all points at a given distance from a fixed point. The fixed point is called the **center** and the fixed distance is called the **radius**.

If the center is at the point (h, k, l), the radius is r, and (x, y, z) is any point on the sphere, then, from the formula for the distance between two points, we obtain the relation

$$(x - h)^2 + (y - k)^2 + (z - l)^2 = r^2. \tag{1}$$

Equation (1) is the **equation of a sphere.** If it is multiplied out and the terms collected we have the equivalent form

$$x^2 + y^2 + z^2 + Dx + Ey + Fz + G = 0. \qquad (2)$$

Example 1. Find the center and the radius of the sphere with equation

$$x^2 + y^2 + z^2 + 4x - 6y + 9z - 6 = 0.$$

Solution. We complete the square by first writing

$$x^2 + 4x \quad + y^2 - 6y \quad + z^2 + 9z \quad = 6;$$

then, adding the appropriate quantities to both sides, we have

$$(x^2 + 4x + 4) + (y^2 - 6y + 9) + (z^2 + 9z + \tfrac{81}{4}) = 6 + 4 + 9 + \tfrac{81}{4}$$
$$\Leftrightarrow \quad (x + 2)^2 + (y - 3)^2 + (z + \tfrac{9}{2})^2 = \tfrac{157}{4}.$$

The center is at $(-2, 3, -\tfrac{9}{2})$ and the radius is $\tfrac{1}{2}\sqrt{157}$. ◀

Example 2. Find the equation of the sphere which passes through $(2, 1, 3)$, $(3, 2, 1)$, $(1, -2, -3)$, $(-1, 1, 2)$.

Solution. Substituting these points in the form (2) above for the equation of a sphere, we obtain

$$(2, 1, 3): \quad 2D + E + 3F + G = -14,$$
$$(3, 2, 1): \quad 3D + 2E + F + G = -14,$$
$$(1, -2, -3): \quad D - 2E - 3F + G = -14,$$
$$(-1, 1, 2): \quad -D + E + 2F + G = -6.$$

Solving these by elimination (first G, then D, then F) we obtain, successively,

$$D + E - 2F = 0, \quad D + 3E + 6F = 0, \quad 3D + F = -8,$$

and

$$2E + 8F = 0, \quad 3E - 7F = 8; \quad \text{and so} \quad -38E = -64.$$

Therefore

$$E = \tfrac{32}{19}, \quad F = -\tfrac{8}{19}, \quad D = -\tfrac{48}{19}, \quad G = -\tfrac{178}{19}.$$

The desired equation is

$$x^2 + y^2 + z^2 - \tfrac{48}{19}x + \tfrac{32}{19}y - \tfrac{8}{19}z - \tfrac{178}{19} = 0. \quad ◀$$

A **cylindrical surface** is a surface which consists of a collection of parallel lines. Each of the parallel lines is called a **generator** of the **cylinder** or cylindrical surface.

The customary right circular cylinder of elementary geometry is clearly a special case of the type of cylinder we are considering. Figure 12–15 shows some examples of cylindrical surfaces. Note that a plane is a cylindrical surface.

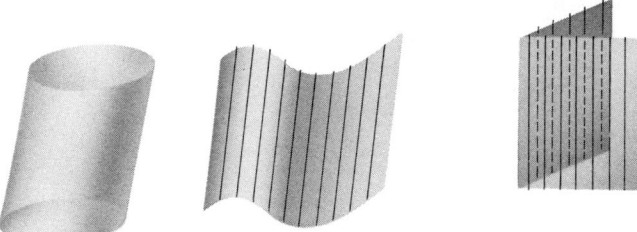

Figure 12–15

Theorem 10. *An equation of the form*

$$f(x, y) = 0$$

is a cylindrical surface with generators all parallel to the z axis. The surface intersects the xy plane in the curve

$$f(x, y) = 0, \quad z = 0.$$

A similar result holds with axes interchanged.

Proof. Suppose that x_0, y_0 satisfies $f(x_0, y_0) = 0$. Then any point (x_0, y_0, z) for $-\infty < z < \infty$ satisfies the same equation, since z is absent. Therefore the line parallel to the z axis through $(x_0, y_0, 0)$ is a generator.

Example 3. Describe and sketch the graph of the equation $x^2 + y^2 = 9$.

Solution. The graph is a right circular cylinder with generators parallel to the z axis (Theorem 10). It is sketched in Fig. 12–16. ◄

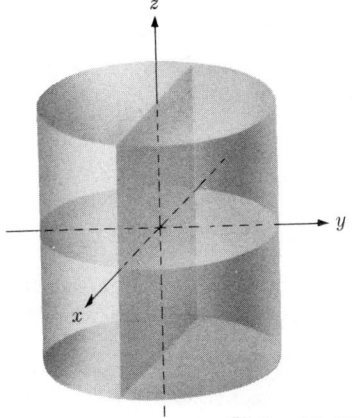

Figure 12–16

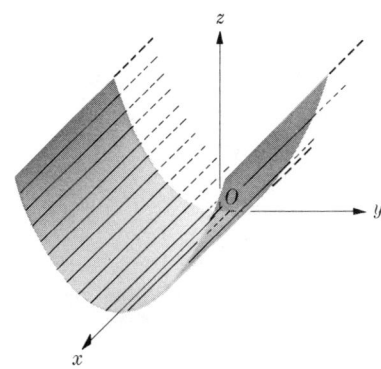

Figure 12–17

Example 4. Describe and sketch the graph of the equation $y^2 = 4z$.

Solution. According to Theorem 10 the graph is a cylindrical surface with generators parallel to the x axis. The intersection with the yz plane is a parabola. The graph, called a **parabolic cylinder,** is sketched in Fig. 12–17. ◄

PROBLEMS

In each of Problems 1 through 4, find the equation of the sphere with center at C and given radius r.

1. $C(1, 4, -2)$, $r = 3$
2. $C(2, 0, -3)$, $r = 5$
3. $C(0, 1, 4)$, $r = 6$
4. $C(-2, 1, -3)$, $r = 1$

In each of Problems 5 through 9, determine the graph of the equation. If it is a sphere, find its center and radius.

5. $x^2 + y^2 + z^2 + 2x - 4z + 1 = 0$
6. $x^2 + y^2 + z^2 - 4x + 2y + 6z - 2 = 0$
7. $x^2 + y^2 + z^2 - 2x + 4y - 2z + 7 = 0$
8. $x^2 + y^2 + z^2 + 4x - 2y + 4z + 9 = 0$
9. $x^2 + y^2 + z^2 - 6x + 4y + 2z + 10 = 0$
10. Find the equation of the graph of all points which are twice as far from $A(3, -1, 2)$ as from $B(0, 2, -1)$.
11. Find the equation of the graph of all points which are three times as far from $A(2, 1, -3)$ as from $B(-2, -3, 5)$.
12. Find the equation of the graph of all points whose distances from the point $(0, 0, 4)$ are equal to their perpendicular distances from the xy plane.

In each of Problems 13 through 24, describe and sketch the graph of the given equation.

13. $x = 3$
14. $x^2 + y^2 = 16$
15. $2x + y = 3$
16. $x + 2z = 4$
17. $x^2 = 4z$
18. $z = 2 - y^2$
19. $4x^2 + y^2 = 16$
20. $4x^2 - y^2 = 16$
21. $y^2 + x^2 = 9$
22. $z^2 = 2 - 2x$
23. $x^2 + y^2 - 2x = 0$
24. $z^2 = y^2 + 4$

In each of Problems 25 through 28, describe the curve of intersection, if any, of the given surface S and the given plane Φ. That is, describe the set $S \cap \Phi$.

25. $S: x^2 + y^2 + z^2 = 25$; $\quad \Phi: z = 3$
26. $S: 4x = y^2 + z^2$; $\quad \Phi: x = 4$
27. $S: x^2 + 2y^2 + 3z^2 = 12$; $\quad \Phi: y = 4$
28. $S: x^2 + y^2 = z^2$; $\quad \Phi: 2x + z = 4$

29. Verify that the graph of the equation $(x - 2)^2 + (y - 1)^2 = 0$ is a straight line. Show that every straight line parallel to one of the coordinate axes can be represented by a *single* equation of the *second* degree.

30. If
$$S_1 = \{(x, y, z): x^2 + y^2 + z^2 + A_1 x + B_1 y + C_1 z + D_1 = 0\}$$
and
$$S_2 = \{(x, y, z): x^2 + y^2 + z^2 + A_2 x + B_2 y + C_2 z + D_2 = 0\}$$
are two spheres, then the **radical plane** is obtained by subtraction of the equations for S_1 and S_2. It is
$$\Phi = \{(x, y, z): (A_1 - A_2)x + (B_1 - B_2)y + (C_1 - C_2)z + (D_1 - D_2) = 0\}.$$

Show that the radical plane is perpendicular to the line joining the centers of the spheres S_1 and S_2. Develop the properties of Φ in analogy with the properties of the radical axis given in Appendix 2, Section 1.

31. Show that the three radical planes (see Problem 30) of three spheres intersect in a common line or are parallel.

7. OTHER COORDINATE SYSTEMS

In plane analytic geometry we employed a Cartesian coordinate system for certain types of problems and a polar coordinate system for others. We saw that there are circumstances in which one system is more convenient than the other. A similar situation prevails in three-dimensional geometry, and we now take up systems of coordinates other than the Cartesian one which we have studied exclusively so far. One such system, known as **cylindrical coordinates,** is described in the following way. A point P in space with Cartesian coordinates (x, y, z) may also be located by replacing the x and y values with the corresponding polar coordinates r, θ and by allowing the z value to remain unchanged. In other words, to each ordered number triple of the form (r, θ, z), there is associated a point in space. The transformation from cylindrical to Cartesian coordinates is given by the equations

$$x = r \cos \theta, \qquad y = r \sin \theta, \qquad z = z.$$

The transformation from Cartesian to cylindrical coordinates is given by

$$r^2 = x^2 + y^2, \qquad \tan \theta = y/x, \qquad z = z.$$

If the coordinates of a point are given in one system, the above equations show how to get the coordinates in the other. Figure 12–18 exhibits the relation between the two systems. It is always assumed that the origins of the systems coincide and that $\theta = 0$ corresponds to the xz plane. We see that the graph $\theta = $ const consists of all points in a plane containing the z axis. The graph of $r = $ const consists of all points on a right circular cylinder with the z axis as its central axis. (The term "cylindrical coordinates" comes from this fact.) The graph of $z = $ const consists of all points in a plane parallel to the xy plane.

Example 1. Find cylindrical coordinates of the points whose Cartesian coordinates are $P(3, 3, 5)$, $Q(2, 0, -1)$, $R(0, 4, 4)$, $S(0, 0, 5)$, $T(2, 2\sqrt{3}, 1)$.

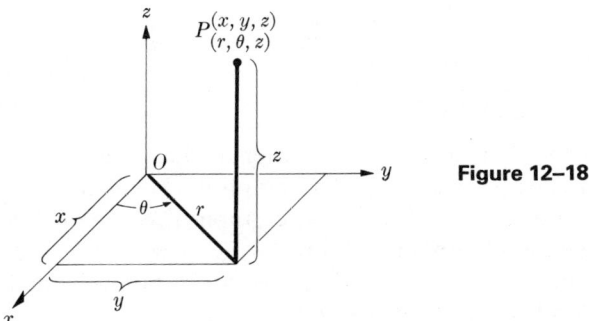

Figure 12–18

Solution. For the point P we have $r = \sqrt{9+9} = 3\sqrt{2}$, $\tan\theta = 1$, $\theta = \pi/4$, $z = 5$. Therefore one set of coordinates is $(3\sqrt{2}, \pi/4, 5)$. For Q we have $r = 2$, $\theta = 0$, $z = -1$. The coordinates are $(2, 0, -1)$. For R we get $r = 4$, $\theta = \pi/2$, $z = 4$. The result is $(4, \pi/2, 4)$. For S we see at once that the coordinates are $(0, \theta, 5)$ for any θ. For T we get $r = \sqrt{4+12} = 4$, $\tan\theta = \sqrt{3}$, $\theta = \pi/3$. The answer is $(4, \pi/3, 1)$. ◀

Remark. Just as polar coordinates do not give a one-to-one correspondence between ordered number pairs and points in the plane, so cylindrical coordinates do not give a one-to-one correspondence between ordered number triples and points in space.

A **spherical coordinate system** is defined in the following way. A point P with Cartesian coordinates (x, y, z) has spherical coordinates (ρ, θ, ϕ) where ρ is the distance of the point P from the origin, θ is the same quantity as in cylindrical coordinates, and ϕ is the angle that the directed line $\overrightarrow{OP}$ makes with the positive z direction. Figure 12–19 exhibits the relation between Cartesian and spherical coordinates. The transformation from spherical to Cartesian coordinates is given by the equations

$$x = \rho \sin\phi \cos\theta, \qquad y = \rho \sin\phi \sin\theta, \qquad z = \rho \cos\phi.$$

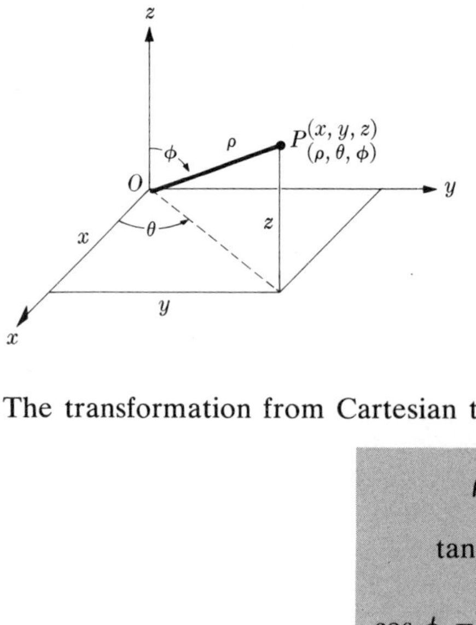

Figure 12–19

The transformation from Cartesian to spherical coordinates is given by

$$\rho^2 = x^2 + y^2 + z^2,$$
$$\tan\theta = \frac{y}{x},$$
$$\cos\phi = \frac{z}{\sqrt{x^2 + y^2 + z^2}}.$$

We note that the graph of $\rho = $ const is a sphere with center at the origin (from which is derived the term "spherical coordinates"). The graph of $\theta = $ const is a plane through the z axis, as in cylindrical coordinates. The graph of $\phi = $ const is a **cone** with vertex at the origin and angle opening 2ϕ if $0 < \phi < \pi/2$. (See Fig. 12–20.) The lower portion of the cone in Fig. 12–20 is or is not included according as negative values of ρ are or are not allowed.

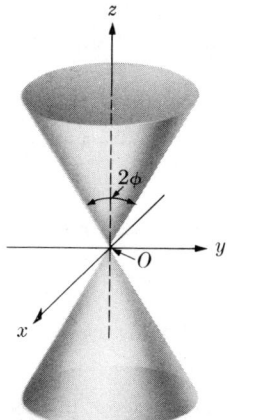

Figure 12-20

Example 2. Find an equation in spherical coordinates of the sphere

$$x^2 + y^2 + z^2 - 2z = 0.$$

Sketch the graph.

Solution. Since $\rho^2 = x^2 + y^2 + z^2$ and $z = \rho \cos \phi$, we have

$$\rho^2 - 2\rho \cos \phi = 0 \quad \Leftrightarrow \quad \rho(\rho - 2 \cos \phi) = 0.$$

The graph of this equation is the graph of $\rho = 0$ and $\rho - 2 \cos \phi = 0$. The graph of $\rho = 0$ is on the graph of $\rho - 2 \cos \phi = 0$ (with $\phi = \pi/2$). Plotting the surface

$$\rho = 2 \cos \phi,$$

we get the surface shown in Fig. 12-21. ◀

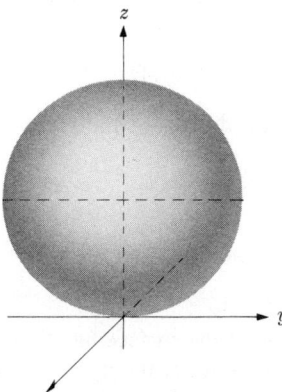

Figure 12-21

If ρ is constant, then the quantities (θ, ϕ) form a coordinate system on the surface of a sphere. Latitude and longitude on the surface of the earth also form a coordinate system. If we restrict θ so that $-\pi < \theta \leq \pi$, then θ is called the **longitude** of the point in spherical coordinates. If ϕ is restricted so that $0 \leq \phi \leq$

π, then ϕ is called the **colatitude** of the point. That is, ϕ is $(\pi/2)$ − latitude, where latitude is taken in the ordinary sense—i.e., positive north of the equator and negative south of it.

PROBLEMS

1. Find a set of cylindrical coordinates for each of the points whose Cartesian coordinates are
   a) $(3, 3, 7)$
   b) $(4, 8, 2)$
   c) $(-2, 3, 1)$.

2. Find the Cartesian coordinates of the points whose cylindrical coordinates are
   a) $(2, \pi/3, 1)$,
   b) $(3, -\pi/4, 2)$,
   c) $(7, 2\pi/3, -4)$.

3. Find a set of spherical coordinates for each of the points whose Cartesian coordinates are
   a) $(2, 2, 2)$,
   b) $(2, -2, -2)$,
   c) $(-1, \sqrt{3}, 2)$.

4. Find the Cartesian coordinates of the points whose spherical coordinates are
   a) $(4, \pi/6, \pi/4)$,
   b) $(6, 2\pi/3, \pi/3)$,
   c) $(8, \pi/3, 2\pi/3)$.

5. Find a set of cylindrical coordinates for each of the points whose spherical coordinates are
   a) $(4, \pi/3, \pi/2)$,
   b) $(2, 2\pi/3, 5\pi/6)$,
   c) $(7, \pi/2, \pi/6)$.

6. Find a set of spherical coordinates for each of the points whose cylindrical coordinates are
   a) $(2, \pi/4, 1)$,
   b) $(3, \pi/2, 2)$,
   c) $(1, 5\pi/6, -2)$.

In each of Problems 7 through 16, find an equation in cylindrical coordinates of the graph whose (x, y, z) equation is given. Sketch

7. $x^2 + y^2 + z^2 = 9$
8. $x^2 + y^2 + 2z^2 = 8$
9. $x^2 + y^2 = 4z$
10. $x^2 + y^2 - 2x = 0$
11. $x^2 + y^2 = z^2$
12. $x^2 + y^2 + 2z^2 + 2z = 0$
13. $x^2 - y^2 = 4$
14. $xy + z^2 = 5$
15. $x^2 + y^2 - 4y = 0$
16. $x^2 + y^2 + z^2 - 2x + 3y - 4z = 0$

In each of Problems 17 through 22, find an equation in spherical coordinates of the graph whose (x, y, z) equation is given. Sketch.

17. $x^2 + y^2 + z^2 - 4z = 0$
18. $x^2 + y^2 + z^2 + 2z = 0$
19. $x^2 + y^2 = z^2$
20. $x^2 + y^2 = 4$
21. $x^2 + y^2 = 4z + 4$ (Solve for ρ in terms of ϕ.)
22. $x^2 + y^2 - z^2 + z - y = 0$

23. Theorem 10 on page 415 describes a surface when the equation of the surface has one of the Cartesian coordinates absent. Describe, in the form of a theorem, the nature of a surface with equation $f(r, z) = 0$ where r, θ, z are cylindrical coordinates. Do the same when the equation is of the form $f(\theta, z) = 0$.

24. Same as Problem 23 for spherical coordinates when the equation is of the form $f(\rho, \phi) = 0$.

13

VECTORS IN THE PLANE AND IN THREE SPACE

1. DIRECTED LINE SEGMENTS AND VECTORS IN THE PLANE

Let A and B be two points in a plane. We recall that the length of the line segment joining A and B is denoted $|AB|$.

Definition. *The **directed segment fron** A **to** B is defined as the line segment AB which is ordered so that A precedes B. We use the symbol $\overrightarrow{AB}$ to denote such a directed segment. We call A its **base** and B its **head**.*

If we select the opposite ordering of the line segment AB, then we get the directed segment $\overrightarrow{BA}$. In this case B precedes A and we call B the base and A the head. To distinguish geometrically the base and head of a directed line segment, we usually draw an arrow at the head, as shown in Fig. 13–1.

Directed line segment $\overrightarrow{AB}$ **Figure 13–1**

Definitions. *The **magnitude** of a directed line segment $\overrightarrow{AB}$ is its length $|AB|$. Two directed line segments $\overrightarrow{AB}$ and $\overrightarrow{CD}$ are said to **have the same magnitude and direction** if and only if either one of the following two conditions holds:*

i) $\overrightarrow{AB}$ *and* $\overrightarrow{CD}$ *are both on the same line L, their magnitudes are equal, and the heads B and D are pointing in the same direction, as shown in* Fig. 13–2.
ii) *The points A, C, D, and B are the vertices of a parallelogram, as shown in* Fig. 13–3.

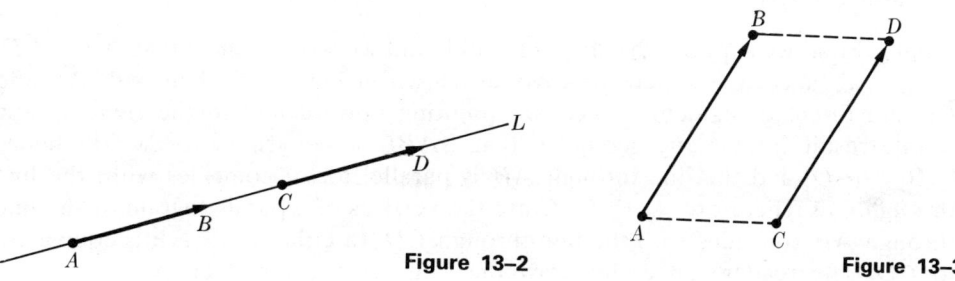

Figure 13–2   Figure 13–3

421

422 VECTORS IN THE PLANE AND IN THREE SPACE

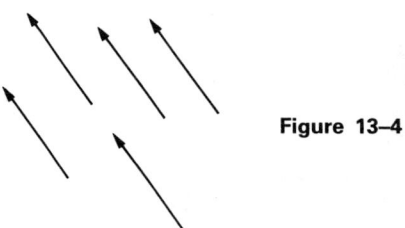

Figure 13–4

Figure 13–4 shows several line segments having the same magnitude and direction. Whenever two directed line segments $\overrightarrow{AB}$ and $\overrightarrow{CD}$ have the same magnitude and direction, we say that they are **equivalent** and write

$$\overrightarrow{AB} \approx \overrightarrow{CD}.$$

We now prove a theorem which expresses this equivalence relationship in terms of coordinates.

Theorem 1. *Suppose that A, B, C, and D are points in a plane. Denote the coordinates of A, B, C, and D by (x_A, y_A), (x_B, y_B), (x_C, y_C), (x_D, y_D), respectively. (See Fig. 13–5.)*

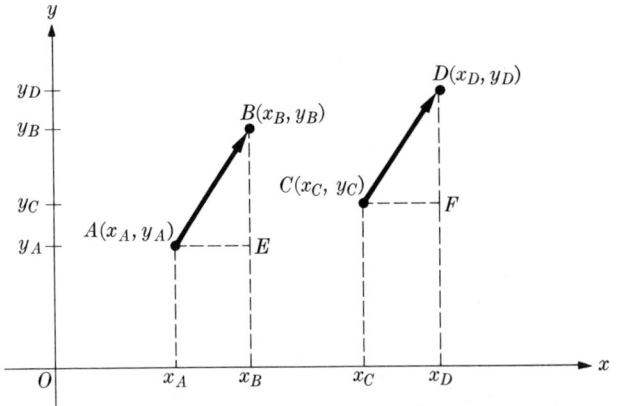

Figure 13–5

i) *If the coordinates above satisfy the equations*

$$x_B - x_A = x_D - x_C \quad \text{and} \quad y_B - y_A = y_D - y_C, \tag{1}$$

then

$$\overrightarrow{AB} \approx \overrightarrow{CD}.$$

ii) *Conversely, if $\overrightarrow{AB}$ is equivalent to $\overrightarrow{CD}$, then the coordinates of the four points satisfy Eqs. (1).*

Proof. First, we suppose that Eqs. (1) hold, and we wish to show that $\overrightarrow{AB} \approx \overrightarrow{CD}$. If $x_B - x_A$ and $y_B - y_A$ are positive, as shown in Fig. 13–5, then both $\overrightarrow{AB}$ and $\overrightarrow{CD}$ are directed segments which are pointing upward and to the right. It is a simple result from plane geometry that $\triangle ABE$ is congruent to $\triangle CDF$; hence $|AB| = |CD|$ and the line through AB is parallel to (or coincides with) the line through CD. Therefore A, B, D, C are the vertices of a parallelogram or the line through AB coincides with the line through CD. In either case, $\overrightarrow{AB}$ is equivalent to $\overrightarrow{CD}$. The reader can easily draw the same conclusion if either $x_B - x_A$ or

$y_B - y_A$ is negative or if both quantities are negative. See Problem 27 at the end of Section 2.

To prove the converse, we assume that $\overrightarrow{AB} \approx \overrightarrow{CD}$. That is, we suppose that $|AB| = |CD|$ and that the line through AB is parallel to (or coincides with) the line through CD. If the directed line segments appear as in Fig. 13–5, we conclude from plane geometry that $\triangle ABE$ is congruent to $\triangle CDF$. Consequently $x_B - x_A = x_D - x_C$ and $y_B - y_A = y_D - y_C$. The reader may verify that the result holds generally if $\overrightarrow{AB}$ and $\overrightarrow{CD}$ are pointing in directions other than upward and to the right. See Problem 28 at the end of Section 2.

If we are given a directed line segment $\overrightarrow{AB}$, we see at once that there is an unlimited number of equivalent ones. In fact, if C is any given point in the plane, we can use Eqs. (1) of Theorem 1 to find the coordinates of the unique point D such that $\overrightarrow{CD} \approx \overrightarrow{AB}$. Theorem 1 also yields various simple properties of the relation $\approx$. For example, if $\overrightarrow{AB} \approx \overrightarrow{CD}$, then $\overrightarrow{CD} \approx \overrightarrow{AB}$; also if $\overrightarrow{AB} \approx \overrightarrow{CD}$ and $\overrightarrow{CD} \approx \overrightarrow{EF}$, then $\overrightarrow{AB} \approx \overrightarrow{EF}$.

The definition of a vector involves an abstract concept—that of a collection of directed line segments.

Definition. *A **vector** is a collection of all directed line segments having a given magnitude and a given direction. We shall use boldface letters to denote vectors; thus when we write **v** for a vector, it stands for an entire collection of directed line segments. A particular directed line segment in the collection is called a **representative** of the vector **v**. Any member of the collection may be used as a representative.*

Figure 13–4 shows five representatives of the same vector. Since any two representative directed line segments of the same vector are equivalent, the collection used to define a vector is called an **equivalence class.**

[The vector as we have defined it is sometimes called a **free vector.** There are other ways of introducing vectors; one is to call a directed line segment a vector. We then would make the convention that directed line segments with the same magnitude and direction (i.e., equivalent) are equal vectors. This leads to certain logical difficulties which we wish to avoid.]

Vectors occur with great frequency in various branches of physics and engineering. Problems in mechanics, especially those involving forces, are concerned with "lines of action," i.e., the lines along which forces act. In such problems it is convenient (but not necessary) to define a vector as the equivalence class of all directed line segments which lie along a given straight line and which have a given magnitude.

Definition. *The **length** of a vector is the common length of all its representative segments. A **unit vector** is a vector of length one. Two vectors are said to be **orthogonal** (or **perpendicular**) if any representative of one vector is perpendicular to any representative of the other (i.e., the representatives lie along perpendicular lines).*

For convenience, we consider directed line segments of zero length; these are simply points. The **zero vector,** denoted by **0**, is the class of directed line segments of zero length.* We make the convention that the zero vector is orthogonal to every vector.

---

* We could develop a theory of vectors using ordered pairs of *points* instead of directed line segments. Then the zero vector would be the collection of all ordered pairs of points (A, B) in which $B = A$.

2. OPERATIONS WITH VECTORS

Vectors may be added to yield other vectors. Suppose **u** and **v** are vectors, i.e., each is a collection of directed line segments. To add **u** and **v**, first select a representative of **u**, say $\overrightarrow{AB}$, as shown in Fig. 13–6(a). Next take the particular representative of **v** which has its base at the point B, and label it $\overrightarrow{BC}$. Then draw the directed line segment $\overrightarrow{AC}$. The sum **w** of **u** and **v** is the equivalence class of directed line segments of which $\overrightarrow{AC}$ is a representative. We write

$$\mathbf{u} + \mathbf{v} = \mathbf{w}.$$

It is important to note that we could have started with any representative of **u**, say $\overrightarrow{A'B'}$ in Fig. 13–6(b). Then we could have selected the representative of **v** with base at B'. The directed line segments $\overrightarrow{A'C'}$ and $\overrightarrow{AC}$ are representatives of the same vector, as is easily seen from Theorem 1. (See Problem 22 at the end of this section.)

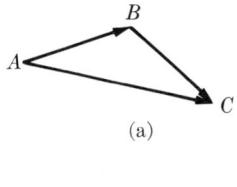

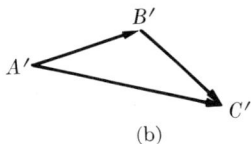

Figure 13–6

Vectors may be multiplied by numbers to yield new vectors. If **v** is a vector and c is a *positive* real number, then $c\mathbf{v}$ is a vector with all representatives having the same direction as those of **v** but with magnitudes c times as long as the representatives of **v**. If c is a negative number, then all representatives of $c\mathbf{v}$ have the opposite direction to those of **v** and their magnitudes are $|c|$ times as long as those of **v**. If c is zero, we get the vector **0**. Figure 13–7 shows various multiples of a representative line segment $\overrightarrow{AB}$ of a vector **v**. We write $-\mathbf{v}$ for the vector $(-1)\mathbf{v}$.

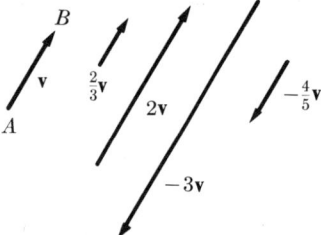

Figure 13–7

Definitions. *Suppose we are given a Cartesian coordinate system in the plane. We call I the point with coordinates $(1, 0)$ and J the point with coordinates $(0, 1)$ as shown in Fig. 13–8. The **unit vector i** is defined as the vector which has $\overrightarrow{OI}$ as one*

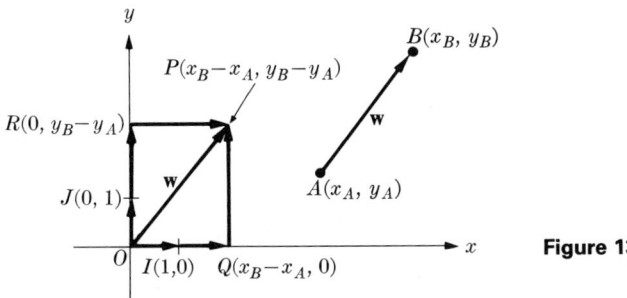

Figure 13-8

of its representatives. The **unit vector j** is defined as the vector which has $\overrightarrow{OJ}$ as a representative.

Theorem 2. Suppose a vector **w** has $\overrightarrow{AB}$ as a representative. Denote the coordinates of A and B by (x_A, y_A) and (x_B, y_B), respectively. Then **w** may be expressed in the form

$$\mathbf{w} = (x_B - x_A)\mathbf{i} + (y_B - y_A)\mathbf{j}.$$

Proof. From Eqs. (1) in Theorem 1, we know that **w** has the representative $\overrightarrow{OP}$ where P has coordinates $(x_B - x_A, y_B - y_A)$. (See Fig. 13-8.) Let $Q(x_B - x_A, 0)$ and $R(0, y_B - y_A)$ be the points on the coordinate axes as shown in Fig. 13-8. It is clear geometrically that $\overrightarrow{OQ}$ is $(x_B - x_A)$ times as long as $\overrightarrow{OI}$ and that $\overrightarrow{OR}$ is $(y_B - y_A)$ times as long as $\overrightarrow{OJ}$. We denote by **u** the vector which has $\overrightarrow{OQ}$ as representative and by **v** the vector which has $\overrightarrow{OR}$ as representative. Using the rule for addition of vectors, we obtain

$$\mathbf{w} = \mathbf{u} + \mathbf{v},$$

since $\overrightarrow{QP}$ as well as $\overrightarrow{OR}$ is a representative of **v**. Since $\mathbf{u} = (x_B - x_A)\mathbf{i}$ and $\mathbf{v} = (y_B - y_A)\mathbf{j}$, the result of the theorem is established.

Example 1. A vector **v** has $\overrightarrow{AB}$ as a representative. Given that A has coordinates $(3, -2)$ and B has coordinates $(1, 1)$, express **v** in terms of **i** and **j**. Draw a figure.

Solution. Using the formula in Theorem 2, we have (see Fig. 13-9)

$$\mathbf{v} = (1 - 3)\mathbf{i} + (1 + 2)\mathbf{j} = -2\mathbf{i} + 3\mathbf{j}. \qquad \blacktriangleleft$$

Notation. The length of a vector **v** will be denoted by $|\mathbf{v}|$.

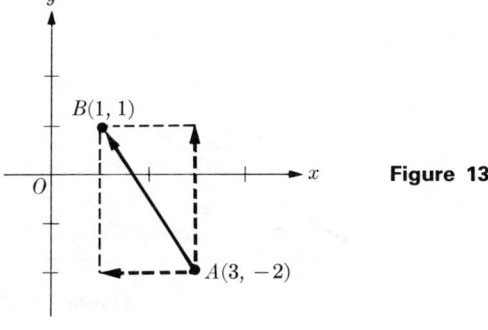

Figure 13-9

426 VECTORS IN THE PLANE AND IN THREE SPACE

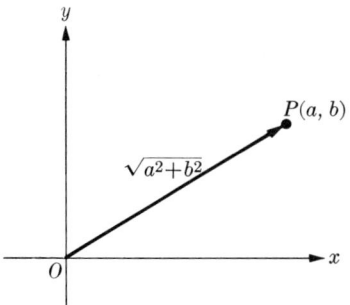

Figure 13–10

Theorem 3. *If* $\mathbf{v} = a\mathbf{i} + b\mathbf{j}$, *then*

$$|\mathbf{v}| = \sqrt{a^2 + b^2}.$$

Therefore $\mathbf{v} = \mathbf{0}$ *if and only if* $a = b = 0$.

Proof. By means of Theorem 2, we know that $\mathbf{v}$ has as one of its representatives the directed segment $\overrightarrow{OP}$ where P has coordinates (a, b). (See Fig. 13–10.) Then $|\overrightarrow{OP}| = \sqrt{a^2 + b^2}$; since, by definition, the length of a vector is the length of any of its representatives, the result follows.

The next theorem is useful for problems concerned with the addition of vectors and the multiplication of vectors by numbers.

Theorem 4. *If* $\mathbf{v} = a\mathbf{i} + b\mathbf{j}$ *and* $\mathbf{w} = c\mathbf{i} + d\mathbf{j}$, *then*

$$\mathbf{v} + \mathbf{w} = (a + c)\mathbf{i} + (b + d)\mathbf{j}.$$

Further, if h is any number, then

$$h\mathbf{v} = (ha)\mathbf{i} + (hb)\mathbf{j}.$$

Proof. Let P, Q, and S have coordinates as shown in Fig. 13–11. Then $\overrightarrow{OP}$ and $\overrightarrow{OQ}$ are representatives of $\mathbf{v}$ and $\mathbf{w}$, respectively. Since $\overrightarrow{PS} \approx \overrightarrow{OQ}$, we use the rule for addition of vectors to find that $\overrightarrow{OS}$ is a representative of $\mathbf{v} + \mathbf{w}$. The point R (see Fig. 13–12) is h of the way from O to P. Hence $\overrightarrow{OR}$ is a representative of $h\mathbf{v}$.

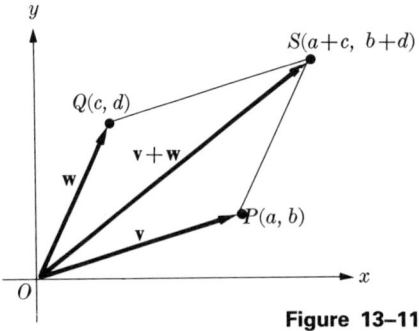

Figure 13–11

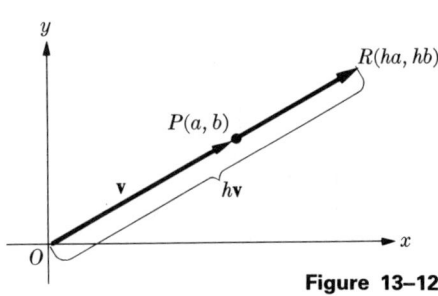

Figure 13–12

We conclude from the theorems above that the addition of vectors and their multiplication by numbers satisfy the following laws:

$$\left.\begin{array}{r}\mathbf{u} + (\mathbf{v} + \mathbf{w}) = (\mathbf{u} + \mathbf{v}) + \mathbf{w} \\ c(d\mathbf{v}) = (cd)\mathbf{v}\end{array}\right\} \text{Associative laws}$$

$$\mathbf{u} + \mathbf{v} = \mathbf{v} + \mathbf{u} \qquad \text{Commutative law}$$

$$\left.\begin{array}{r}(c + d)\mathbf{v} = c\mathbf{v} + d\mathbf{v} \\ c(\mathbf{u} + \mathbf{v}) = c\mathbf{u} + c\mathbf{v}\end{array}\right\} \text{Distributive laws}$$

$$1 \cdot \mathbf{u} = \mathbf{u}, \qquad 0 \cdot \mathbf{u} = \mathbf{0}, \qquad (-1)\mathbf{u} = -\mathbf{u}$$

where $-\mathbf{u}$ denotes that vector such that $\mathbf{u} + (-\mathbf{u}) = \mathbf{0}$. These laws hold for all $\mathbf{u}$, $\mathbf{v}$, $\mathbf{w}$ and all numbers c and d. It is important to note that multiplication and division of vectors in the ordinary sense is not (and will not be) defined. However, subtraction is defined. If $\overrightarrow{AB}$ is a representative of $\mathbf{v}$ and $\overrightarrow{AC}$ is one of $\mathbf{w}$, then $\overrightarrow{CB}$ is a representative of $\mathbf{v} - \mathbf{w}$ (see Fig. 13–13).

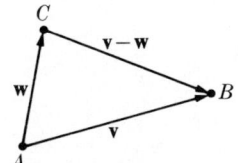

Figure 13–13

Example 2. Given the vectors $\mathbf{u} = 2\mathbf{i} - 3\mathbf{j}$, $\mathbf{v} = -4\mathbf{i} + \mathbf{j}$. Express the vector $2\mathbf{u} - 3\mathbf{v}$ in terms of $\mathbf{i}$ and $\mathbf{j}$.

Solution. $2\mathbf{u} = 4\mathbf{i} - 6\mathbf{j}$ and $-3\mathbf{v} = 12\mathbf{i} - 3\mathbf{j}$. Adding these vectors, we get $2\mathbf{u} - 3\mathbf{v} = 16\mathbf{i} - 9\mathbf{j}$. ◀

Definition. *Let $\mathbf{v}$ be any vector except $\mathbf{0}$. The unit vector $\mathbf{u}$ in the direction of $\mathbf{v}$ is defined by*

$$\mathbf{u} = \left(\frac{1}{|\mathbf{v}|}\right)\mathbf{v}.$$

Example 3. Given the vector $\mathbf{v} = -2\mathbf{i} + 3\mathbf{j}$, find a unit vector in the direction of $\mathbf{v}$.

Solution. We have $|\mathbf{v}| = \sqrt{4 + 9} = \sqrt{13}$. The desired vector $\mathbf{u}$ is

$$\mathbf{u} = \frac{1}{\sqrt{13}}\mathbf{v} = -\frac{2}{\sqrt{13}}\mathbf{i} + \frac{3}{\sqrt{13}}\mathbf{j}. \qquad ◀$$

Example 4. Given the vector $\mathbf{v} = 2\mathbf{i} - 4\mathbf{j}$. Find the representative $\overrightarrow{AB}$ of $\mathbf{v}$, given that A has coordinates $(3, -5)$.

Solution. Denote the coordinates of B by x_B, y_B. Then we have (by Theorem 2)

$$x_B - 3 = 2 \qquad \text{and} \qquad y_B + 5 = -4.$$

Therefore $x_B = 5$, $y_B = -9$. ◀

PROBLEMS

In Problems 1 through 5, express **v** in terms of **i** and **j**, given that the endpoints A and B of the representative $\overrightarrow{AB}$ of **v** have the given coordinates. Draw a figure.

1. $A(3, -2), B(1, 5)$
2. $A(-4, 1), B(2, -1)$
3. $A(5, -6), B(0, -2)$
4. $A(4, 4), B(8, -3)$
5. $A(7, -2), B(-5, -6)$

In Problems 6 through 9, in each case find a unit vector **u** in the direction of **v**. Express **u** in terms of **i** and **j**.

6. $\mathbf{v} = 4\mathbf{i} + 3\mathbf{j}$
7. $\mathbf{v} = -5\mathbf{i} - 12\mathbf{j}$
8. $\mathbf{v} = 2\mathbf{i} - 2\sqrt{3}\mathbf{j}$
9. $\mathbf{v} = -2\mathbf{i} + 5\mathbf{j}$

In Problems 10 through 15, find the representative $\overrightarrow{AB}$ of the vector **v** from the information given. Draw a figure.

10. $\mathbf{v} = 7\mathbf{i} - 3\mathbf{j}, A(2, -1)$
11. $\mathbf{v} = -2\mathbf{i} + 4\mathbf{j}, A(6, 2)$
12. $\mathbf{v} = 3\mathbf{i} + 2\mathbf{j}, B(-2, 1)$
13. $\mathbf{v} = -4\mathbf{i} - 2\mathbf{j}, B(0, 5)$
14. $\mathbf{v} = 3\mathbf{i} + 2\mathbf{j}$, midpoint of segment AB has coordinates $(3, 1)$
15. $\mathbf{v} = -2\mathbf{i} + 3\mathbf{j}$, midpoint of segment AB has coordinates $(-4, 2)$
16. Find a representative of the vector **v** of unit length making an angle of 30° with the positive x direction. Express **v** in terms of **i** and **j**.
17. Find the vector **v** (in terms of **i** and **j**) which has length $2\sqrt{2}$ and makes an angle of 45° with the positive y axis (two solutions).
18. Given that $\mathbf{u} = 3\mathbf{i} - 2\mathbf{j}, \mathbf{v} = 4\mathbf{i} + 3\mathbf{j}$. Find $\mathbf{u} + \mathbf{v}$ in terms of **i** and **j**. Draw a figure.
19. Given that $\mathbf{u} = -2\mathbf{i} + 3\mathbf{j}, \mathbf{v} = \mathbf{i} - 2\mathbf{j}$. Find $\mathbf{u} + \mathbf{v}$ in terms of **i** and **j**. Draw a figure.
20. Given that $\mathbf{u} = -3\mathbf{i} - 2\mathbf{j}, \mathbf{v} = 2\mathbf{i} + \mathbf{j}$. Find $3\mathbf{u} - 2\mathbf{v}$ in terms of **i** and **j**. Draw a figure.
21. Show that if $\overrightarrow{AB} \approx \overrightarrow{CD}$ and $\overrightarrow{CD} \approx \overrightarrow{EF}$, then $\overrightarrow{AB} \approx \overrightarrow{EF}$.
22. Show that if $\overrightarrow{AB} \approx \overrightarrow{DE}$ and $\overrightarrow{BC} \approx \overrightarrow{EF}$, then $\overrightarrow{AC} \approx \overrightarrow{DF}$. Draw a figure.
23. Show that if $\overrightarrow{AB} \approx \overrightarrow{DE}$, c is any real number, C is the point c of the way from A to B, and F is the point c of the way from D to E, then $\overrightarrow{AC} \approx \overrightarrow{DF}$. Draw a figure.
24. Show that the vectors $\mathbf{v} = 2\mathbf{i} + 4\mathbf{j}$ and $\mathbf{w} = 10\mathbf{i} - 5\mathbf{j}$ are orthogonal.
25. Show that the vectors $\mathbf{v} = -3\mathbf{i} + \sqrt{2}\mathbf{j}$ and $\mathbf{w} = 4\sqrt{2}\mathbf{i} + 12\mathbf{j}$ are orthogonal.
26. Let **u** and **v** be two nonzero vectors. Show that they are orthogonal if and only if the equation
$$|\mathbf{u} + \mathbf{v}|^2 = |\mathbf{u}|^2 + |\mathbf{v}|^2$$
holds.
27. Vectors in the plane may be divided into categories as follows: pointing upward to the right, upward to the left, downward to the right, downward to the left; also those pointing vertically and those horizontally. Theorem 1, part (i) was established for vectors pointing upward to the right (Fig. 13–5). Write a proof of Theorem 1(i) for three of the remaining five possibilities. Draw a figure and devise, if you can, a general proof.
28. The same as Problem 27 for part (ii) of Theorem 1.
29. Write out a proof establishing the associative, commutative, and distributive laws for vectors. (See page 427.)

3. OPERATIONS WITH PLANE VECTORS, CONTINUED. THE SCALAR PRODUCT

Two vectors **v** and **w** are said to be **parallel** or **proportional** when each is a scalar multiple of the other (and neither is zero). The representatives of parallel vectors are all parallel directed line segments.

By the **angle between two vectors** **v** and **w** (neither = **0**), we mean the measure of the angle between any representatives of **v** and **w** having the same base (see Fig. 13–14). Two parallel vectors make an angle either of 0 or of π, depending on whether they are pointing in the same or opposite directions (i.e., whether the scalar multiple is positive or negative).

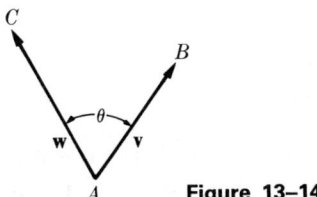

Figure 13–14

Suppose that **i** and **j** are the usual unit vectors pointing in the direction of the x and y axes, respectively. Then we have the following theorem.

Theorem 5. *If θ is the angle between the vectors*

$$\mathbf{v} = a\mathbf{i} + b\mathbf{j} \quad \text{and} \quad \mathbf{w} = c\mathbf{i} + d\mathbf{j},$$

then

$$\cos\theta = \frac{ac + bd}{|\mathbf{v}||\mathbf{w}|}.$$

Proof. We draw the representatives of **v** and **w** with base at the origin of the coordinate system, as shown in Fig. 13–15. Using Theorem 2 of Section 2, we see that the coordinates of P are (a, b) and those of Q are (c, d). The length of **v** is $|OP|$ and the length of **w** is $|OQ|$. We apply the law of cosines to $\triangle OPQ$, obtaining

$$\cos\theta = \frac{|OP|^2 + |OQ|^2 - |QP|^2}{2|OP||OQ|}.$$

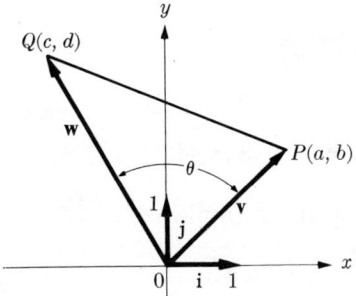

Figure 13–15

Therefore

$$\cos \theta = \frac{a^2 + b^2 + c^2 + d^2 - (a-c)^2 - (b-d)^2}{2|OP||OQ|} = \frac{ac + bd}{|\mathbf{v}||\mathbf{w}|}.$$

Example 1. Given the vectors $\mathbf{v} = 2\mathbf{i} - 3\mathbf{j}$ and $\mathbf{w} = \mathbf{i} - 4\mathbf{j}$, compute the cosine of the angle between $\mathbf{v}$ and $\mathbf{w}$.

Solution. We have

$$|\mathbf{v}| = \sqrt{4 + 9} = \sqrt{13}, \qquad |\mathbf{w}| = \sqrt{1 + 16} = \sqrt{17}.$$

Therefore

$$\cos \theta = \frac{2 \cdot 1 + (-3)(-4)}{\sqrt{17}\sqrt{13}} = \frac{14}{\sqrt{221}}. \qquad \blacktriangleleft$$

Suppose that we have two directed line segments $\overrightarrow{AB}$ and $\overrightarrow{CD}$, as shown in Fig. 13–16. The **projection of** $\overrightarrow{AB}$ **in the direction** $\overrightarrow{CD}$ is defined as the directed length $\overrightarrow{EF}$ of the line segment EF obtained by dropping perpendiculars from A and B to the line containing $\overrightarrow{CD}$. We can find the projection of a vector $\mathbf{v}$ along a vector $\mathbf{w}$ by first taking a representative of $\mathbf{v}$ and finding its projection (call it $\overrightarrow{EF}$) in the direction of a representative of $\mathbf{w}$.

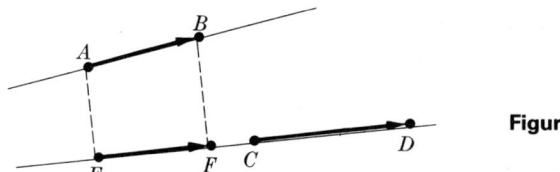

Figure 13–16

If θ is the angle between two vectors $\mathbf{v}$ and $\mathbf{w}$, the quantity

$$|\mathbf{v}| \cos \theta$$

is positive if θ is an acute angle and the projection of $\mathbf{v}$ on $\mathbf{w}$ is positive. If θ is an obtuse angle, then $|\mathbf{v}| \cos \theta$ is negative, and the projection of $\mathbf{v}$ on $\mathbf{w}$ is negative.

Definition. *The quantity* $|\mathbf{v}| \cos \theta$ *is called* **the projection of v on w.** *We write* $\text{Proj}_\mathbf{w} \mathbf{v}$ *for this quantity. If* $\mathbf{v} = a\mathbf{i} + b\mathbf{j}$ *and* $\mathbf{w} = c\mathbf{i} + d\mathbf{j}$, *then*

$$\boxed{\text{Proj}_\mathbf{w} \mathbf{v} = |\mathbf{v}| \cos \theta = \frac{ac + bd}{|\mathbf{w}|}.}$$

Example 2. Find the projection of $\mathbf{v} = 3\mathbf{i} - 2\mathbf{j}$ on $\mathbf{w} = -2\mathbf{i} - 4\mathbf{j}$.

Solution. We obtain the desired projection by using the formula

$$\text{Proj}_\mathbf{w}\mathbf{v} = \frac{-6 + 8}{\sqrt{20}} = \frac{1}{\sqrt{5}}. \qquad \blacktriangleleft$$

The **scalar product** of two nonzero vectors **v** and **w**, written **v** · **w**, is defined by the formula

$$\mathbf{v} \cdot \mathbf{w} = |\mathbf{v}| |\mathbf{w}| \cos \theta,$$

where θ is the angle between **v** and **w**. If one of the vectors is **0**, the scalar product is defined to be 0. The terms **dot product** and **inner product** are also used to designate scalar product. It is evident from the definition that scalar product satisfies the relations

$$\mathbf{v} \cdot \mathbf{w} = \mathbf{w} \cdot \mathbf{v}, \qquad \mathbf{v} \cdot \mathbf{v} = |\mathbf{v}|^2.$$

Furthermore, if **v** and **w** are orthogonal, then

$$\mathbf{v} \cdot \mathbf{w} = 0,$$

and conversely. If **v** and **w** are parallel, we have $\mathbf{v} \cdot \mathbf{w} = \pm |\mathbf{v}| |\mathbf{w}|$, and conversely. In terms of the orthogonal unit vectors **i** and **j**, vectors $\mathbf{v} = a\mathbf{i} + b\mathbf{j}$ and $\mathbf{w} = c\mathbf{i} + d\mathbf{j}$ have as their scalar product (see Theorem 5)

$$\mathbf{v} \cdot \mathbf{w} = ac + bd.$$

In addition, it can be verified that the **distributive law**

$$\mathbf{u} \cdot (\mathbf{v} + \mathbf{w}) = \mathbf{u} \cdot \mathbf{v} + \mathbf{u} \cdot \mathbf{w}$$

holds for any three vectors.

Example 3. Given the vectors $\mathbf{u} = 3\mathbf{i} + 2\mathbf{j}$ and $\mathbf{v} = 2\mathbf{i} + a\mathbf{j}$. Determine the number a so that **u** and **v** are orthogonal. Determine a so that **u** and **v** are parallel. For what value of a will **u** and **v** make an angle of $\pi/4$?

Solution. If **u** and **v** are orthogonal, we have

$$3 \cdot 2 + 2 \cdot a = 0 \qquad \text{and} \qquad a = -3.$$

For **u** and **v** to be parallel, we must have

$$\mathbf{u} \cdot \mathbf{v} = \pm |\mathbf{u}| |\mathbf{v}|$$

or

$$6 + 2a = \pm \sqrt{13} \cdot \sqrt{4 + a^2}.$$

Solving, we obtain $a = \frac{4}{3}$.

Employing the formula in Theorem 5, we find

$$\tfrac{1}{2}\sqrt{2} = \cos \frac{\pi}{4} = \frac{6 + 2a}{\sqrt{13} \cdot \sqrt{4 + a^2}},$$

so that **u** and **v** make an angle of $\pi/4$ when

$$a = 10, -\tfrac{2}{5}.$$

◀

Because they are geometric quantities which are independent of the coordinate system, vectors are well suited for establishing certain types of theorems in plane geometry. We give two examples to exhibit the technique.

Example 4. Let $\overrightarrow{OA}$ be a representative of **u** and $\overrightarrow{OB}$ a representative of **v**. Let C be the point on line AB which is $\frac{2}{3}$ of the way from A to B (Fig. 13–17). Express in terms of **u** and **v** the vector **w** which has $\overrightarrow{OC}$ as representative.

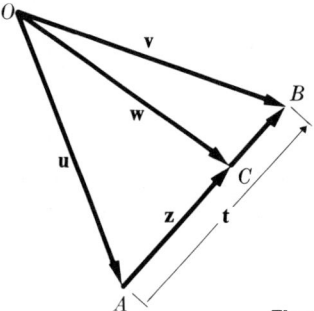

Figure 13–17

Solution. Let **z** be the vector with $\overrightarrow{AC}$ as representative, and **t** the vector with $\overrightarrow{AB}$ as representative. We have

$$\mathbf{w} = \mathbf{u} + \mathbf{z} = \mathbf{u} + \tfrac{2}{3}\mathbf{t}.$$

Also we know that $\mathbf{t} = \mathbf{v} - \mathbf{u}$, and so

$$\mathbf{w} = \mathbf{u} + \tfrac{2}{3}(\mathbf{v} - \mathbf{u}) = \tfrac{1}{3}\mathbf{u} + \tfrac{2}{3}\mathbf{v}. \qquad \blacktriangleleft$$

Let $\overrightarrow{AB}$ be a directed line segment. We introduce a convenient symbol for the vector which has $\overrightarrow{AB}$ as a representative.

Notation. The symbol $\mathbf{v}[\overrightarrow{AB}]$ denotes the vector which has $\overrightarrow{AB}$ as a representative.

Example 5. Let $ABDC$ be a parallelogram, as shown in Fig. 13–18. Suppose that E is the midpoint of CD and F is $\frac{2}{3}$ of the way from A to E on AE. Show that F is $\frac{2}{3}$ of the way from B to C.

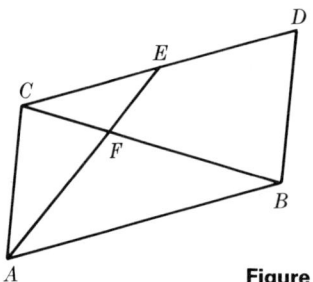

Figure 13–18

Solution. Let $\overrightarrow{AB}$, $\overrightarrow{AC}$, $\overrightarrow{AF}$, $\overrightarrow{AE}$, $\overrightarrow{BF}$, $\overrightarrow{BC}$, and $\overrightarrow{CE}$ all be representatives of vectors. Then $\mathbf{v}[\overrightarrow{AB}]$, $\mathbf{v}[\overrightarrow{AC}]$, $\mathbf{v}[\overrightarrow{AF}]$, etc., are the vectors which have the directed line segments shown in brackets as representatives. By hypothesis, we have

$$\mathbf{v}[\overrightarrow{AB}] = \mathbf{v}[\overrightarrow{CD}] \quad \text{and} \quad \mathbf{v}[\overrightarrow{CE}] = \tfrac{1}{2}\mathbf{v}[\overrightarrow{CD}].$$

The rule for addition of vectors gives us

$$\mathbf{v}[\overrightarrow{AE}] = \mathbf{v}[\overrightarrow{AC}] + \mathbf{v}[\overrightarrow{CE}] = \mathbf{v}[\overrightarrow{AC}] + \tfrac{1}{2}\mathbf{v}[\overrightarrow{AB}].$$

Also, since $\mathbf{v}[\overrightarrow{AF}] = \tfrac{2}{3}\mathbf{v}[\overrightarrow{AE}]$, we obtain

$$\mathbf{v}[\overrightarrow{AF}] = \tfrac{2}{3}\mathbf{v}[\overrightarrow{AC}] + \tfrac{1}{3}\mathbf{v}[\overrightarrow{AB}].$$

The rule for subtraction of vectors yields

$$\mathbf{v}[\overrightarrow{BF}] = \mathbf{v}[\overrightarrow{AF}] - \mathbf{v}[\overrightarrow{AB}] = \tfrac{2}{3}\mathbf{v}[\overrightarrow{AC}] - \tfrac{2}{3}\mathbf{v}[\overrightarrow{AB}] = \tfrac{2}{3}(\mathbf{v}[\overrightarrow{AC}] - \mathbf{v}[\overrightarrow{AB}]).$$

Since $\mathbf{v}[\overrightarrow{BC}] = \mathbf{v}[\overrightarrow{AC}] - \mathbf{v}[\overrightarrow{AB}]$, we conclude that

$$\mathbf{v}[\overrightarrow{BF}] = \tfrac{2}{3}\mathbf{v}[\overrightarrow{BC}],$$

which is the desired result. ◄

PROBLEMS

In Problems 1 through 6, given that θ is the angle between $\mathbf{v}$ and $\mathbf{w}$, find $|\mathbf{v}|$, $|\mathbf{w}|$, $\cos \theta$, and the projection of $\mathbf{v}$ on $\mathbf{w}$.

1. $\mathbf{v} = 4\mathbf{i} - 3\mathbf{j}$, $\mathbf{w} = -4\mathbf{i} - 3\mathbf{j}$
2. $\mathbf{v} = 2\mathbf{i} + \mathbf{j}$, $\mathbf{w} = \mathbf{i} - \mathbf{j}$
3. $\mathbf{v} = 3\mathbf{i} - 4\mathbf{j}$, $\mathbf{w} = 5\mathbf{i} + 12\mathbf{j}$
4. $\mathbf{v} = 4\mathbf{i} + \mathbf{j}$, $\mathbf{w} = 6\mathbf{i} - 8\mathbf{j}$
5. $\mathbf{v} = 3\mathbf{i} + 2\mathbf{j}$, $\mathbf{w} = 2\mathbf{i} - 3\mathbf{j}$
6. $\mathbf{v} = 6\mathbf{i} + 5\mathbf{j}$, $\mathbf{w} = 2\mathbf{i} + 5\mathbf{j}$

In Problems 7 through 12, find the projection of the vector with representative $\overrightarrow{AB}$ on the vector with representative $\overrightarrow{CD}$. Draw a figure in each case.

7. $A(1, 0)$, $B(2, 3)$, $C(1, 1)$, $D(-1, 1)$
8. $A(0, 0)$, $B(1, 4)$, $C(0, 0)$, $D(2, 7)$
9. $A(3, 1)$, $B(5, 2)$, $C(-2, -1)$, $D(-1, 3)$
10. $A(2, 4)$, $B(4, 7)$, $C(6, -1)$, $D(2, 2)$
11. $A(2, -1)$, $B(1, 3)$, $C(5, 2)$, $D(9, 3)$
12. $A(1, 6)$, $B(2, 5)$, $C(5, 2)$, $D(9, 3)$

In Problems 13 through 17, find $\cos \theta$ and $\cos \alpha$, given that $\theta = \angle ABC$ and $\alpha = \angle BAC$. Use vector methods and draw figures.

13. $A(-1, 1)$, $B(3, -1)$, $C(3, 4)$
14. $A(2, 1)$, $B(-1, 2)$, $C(1, 3)$
15. $A(3, 4)$, $B(5, 1)$, $C(4, 1)$
16. $A(4, 1)$, $B(1, -1)$, $C(3, 3)$
17. $A(0, 0)$, $B(3, -5)$, $C(6, -10)$

In Problems 18 through 24, determine the number a (if possible) such that the given condition for $\mathbf{v}$ and $\mathbf{w}$ is satisfied.

18. $\mathbf{v} = 2\mathbf{i} + a\mathbf{j}$, $\mathbf{w} = \mathbf{i} + 3\mathbf{j}$, $\mathbf{v}$ and $\mathbf{w}$ orthogonal
19. $\mathbf{v} = \mathbf{i} - 3\mathbf{j}$, $\mathbf{w} = 2a\mathbf{i} + \mathbf{j}$, $\mathbf{v}$ and $\mathbf{w}$ orthogonal
20. $\mathbf{v} = 3\mathbf{i} - 4\mathbf{j}$, $\mathbf{w} = 2\mathbf{i} + a\mathbf{j}$, $\mathbf{v}$ and $\mathbf{w}$ parallel

21. $\mathbf{v} = a\mathbf{i} + 2\mathbf{j}, \mathbf{w} = 2\mathbf{i} - a\mathbf{j}, \mathbf{v}$ and $\mathbf{w}$ parallel
22. $\mathbf{v} = a\mathbf{i}, \mathbf{w} = 2\mathbf{i} - 3\mathbf{j}, \mathbf{v}$ and $\mathbf{w}$ parallel
23. $\mathbf{v} = 5\mathbf{i} + 12\mathbf{j}, \mathbf{w} = \mathbf{i} + a\mathbf{j}, \mathbf{v}$ and $\mathbf{w}$ make an angle of $\pi/3$
24. $\mathbf{v} = 4\mathbf{i} - 3\mathbf{j}, \mathbf{w} = 2\mathbf{i} + a\mathbf{j}, \mathbf{v}$ and $\mathbf{w}$ make an angle of $\pi/6$
25. Prove the distributive law for the scalar product, as stated on page 432.
26. Let $\mathbf{i}$ and $\mathbf{j}$ be the usual unit vectors of one coordinate system, and let $\mathbf{i}_1$ and $\mathbf{j}_1$ be the unit orthogonal vectors corresponding to another Cartesian system of coordinates. Given that
$$\mathbf{v} = a\mathbf{i} + b\mathbf{j}, \quad \mathbf{w} = c\mathbf{i} + d\mathbf{j}, \quad \mathbf{v} = a_1\mathbf{i}_1 + b_1\mathbf{j}_1, \quad \mathbf{w} = c_1\mathbf{i}_1 + d_1\mathbf{j}_1,$$
show that
$$ac + bd = a_1c_1 + b_1d_1.$$

In Problems 27 through 30, the quantity $|AB|$ denotes (as is customary) the length of the line segment AB, the quantity $|AC|$, the length of AC, etc.

27. Given $\triangle ABC$, in which $\angle A = 120°$, $|AB| = 4$, and $|AC| = 7$. Find $|BC|$ and the projections of $\overrightarrow{AB}$ and $\overrightarrow{AC}$ on $\overrightarrow{BC}$. Draw a figure.
28. Given $\triangle ABC$, with $\angle A = 45°$, $|AB| = 8$, $|AC| = 6\sqrt{2}$. Find $|BC|$ and the projections of $\overrightarrow{AB}$ and $\overrightarrow{AC}$ on $\overrightarrow{BC}$. Draw a figure.
29. Given $\triangle ABC$, with $|AB| = 10$, $|AC| = 9$, $|BC| = 7$. Find the projections of $\overrightarrow{AC}$ and $\overrightarrow{BC}$ on $\overrightarrow{AB}$. Draw a figure.
30. Given $\triangle ABC$, with $|AB| = 5$, $|AC| = 7$, $|BC| = 9$. Find the projections of $\overrightarrow{AB}$ and $\overrightarrow{AC}$ on $\overrightarrow{CB}$. Draw a figure.
31. Given the line segments AB and AC, with D on AB $\frac{2}{3}$ of the way from A to B. Let E be the midpoint of AC. Express $\mathbf{v}[\overrightarrow{DE}]$ in terms of $\mathbf{v}[\overrightarrow{AB}]$ and $\mathbf{v}[\overrightarrow{AC}]$. Draw a figure.
32. Suppose that $\mathbf{v}[\overrightarrow{AD}] = \frac{1}{4}\mathbf{v}[\overrightarrow{AB}]$ and $\mathbf{v}[\overrightarrow{BE}] = \frac{1}{2}\mathbf{v}[\overrightarrow{BC}]$. Find $\mathbf{v}[\overrightarrow{DE}]$ in terms of $\mathbf{v}[\overrightarrow{AB}]$ and $\mathbf{v}[\overrightarrow{BC}]$. Draw a figure.
33. Given $\square ABDC$, a parallelogram, with E $\frac{2}{3}$ of the way from B to D, and F as the midpoint of segment CD. Find $\mathbf{v}[\overrightarrow{EF}]$ in terms of $\mathbf{v}[\overrightarrow{AB}]$ and $\mathbf{v}[\overrightarrow{AC}]$.
34. Given parallelogram $ABDC$, with E $\frac{1}{4}$ of the way from B to C, and F $\frac{1}{4}$ of the way from A to D. Find $\mathbf{v}[\overrightarrow{EF}]$ in terms of $\mathbf{v}[\overrightarrow{AB}]$ and $\mathbf{v}[\overrightarrow{AC}]$.
35. Given parallelogram $ABDC$, with E $\frac{1}{3}$ of the way from B to D, and F $\frac{1}{4}$ of the way from B to C. Show that F is $\frac{3}{4}$ of the way from A to E.
36. Suppose that on the sides of $\triangle ABC$, $\mathbf{v}[\overrightarrow{BD}] = \frac{2}{3}\mathbf{v}[\overrightarrow{BC}]$, $\mathbf{v}[\overrightarrow{CE}] = \frac{2}{3}\mathbf{v}[\overrightarrow{CA}]$, and $\mathbf{v}[\overrightarrow{AF}] = \frac{2}{3}\mathbf{v}[\overrightarrow{AB}]$. Draw a figure and show that
$$\mathbf{v}[\overrightarrow{AD}] + \mathbf{v}[\overrightarrow{BE}] + \mathbf{v}[\overrightarrow{CF}] = \mathbf{0}.$$
37. Show that the conclusion of Problem 36 holds when the fraction $\frac{2}{3}$ is replaced by any real number h.
38. Let $\mathbf{a} = \mathbf{v}[\overrightarrow{OA}]$, $\mathbf{b} = \mathbf{v}[\overrightarrow{OB}]$, and $\mathbf{c} = \mathbf{v}[\overrightarrow{OC}]$. Show that the medians of $\triangle ABC$ meet at a point P, and express $\mathbf{v}[\overrightarrow{OP}]$ in terms of $\mathbf{a}$, $\mathbf{b}$, and $\mathbf{c}$. Draw a figure.

4. VECTORS IN THREE DIMENSIONS

The development of vectors in three-dimensional space is a direct extension of the theory of vectors in the plane as given in Sections 1–3. This section, which is completely analogous to Section 1, may be read quickly. The essential distinction between two- and three-dimensional vectors appears in Sections 5, 6, and 7.

A **directed line segment** $\overrightarrow{AB}$ is defined as before, except that now the **base** A and the **head** B may be situated anywhere in three-space. The **magnitude** of a directed line segment is its length. Two directed line segments $\overrightarrow{AB}$ and $\overrightarrow{CD}$ are said to **have the same magnitude and direction** if and only if either one of the following two conditions holds:

i) $\overrightarrow{AB}$ and $\overrightarrow{CD}$ are both on the same directed line $\vec{L}$ and their directed lengths are equal; or
ii) the points $A, C, D,$ and B are the vertices of a parallelogram as shown in Fig. 13–19.

We note that the above definition is the same as that given in Section 1.

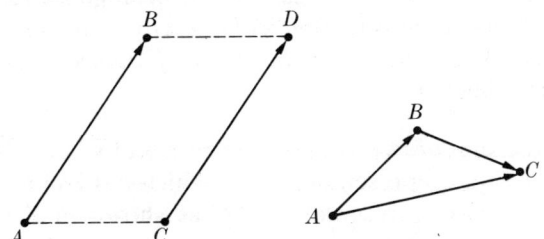

Figure 13–19

Whenever two directed line segments $\overrightarrow{AB}$ and $\overrightarrow{CD}$ have the same magnitude and direction, we say they are **equivalent** and write

$$\overrightarrow{AB} \approx \overrightarrow{CD}.$$

We shall next state a theorem which is a direct extension of Theorem 1 in Section 1.

Theorem 6. *Suppose that $A, B, C,$ and D are points in space. Denote the coordinates of $A, B, C,$ and D by $(x_A, y_A, z_A), (x_B, y_B, z_B),$ and so forth. (i) If the coordinates satisfy the equations*

$$x_B - x_A = x_D - x_C, \quad y_B - y_A = y_D - y_C, \quad z_B - z_A = z_D - z_C, \quad (1)$$

then $\overrightarrow{AB} \approx \overrightarrow{CD}$. (ii) Conversely, if $\overrightarrow{AB} \approx \overrightarrow{CD}$, the coordinates satisfy the equations in (1).

Sketch of proof. (i) We assume that the equations in (1) hold. Then also,

$$x_C - x_A = x_D - x_B, \quad y_C - y_A = y_D - y_B, \quad z_C - z_A = z_D - z_B. \quad (2)$$

As in the proof of Theorem 1 in Section 1, we can conclude that the lines AB and CD are parallel because their direction numbers are equal. Similarly, AC and BD are parallel. Therefore $\overrightarrow{AB} \approx \overrightarrow{CD}$ or all the points are on a directed line $\vec{L}$. The proof that $\overrightarrow{AB} \approx \overrightarrow{CD}$ in this latter case is given in Appendix 6. (ii) To prove the converse, assume that $\overrightarrow{AB} \approx \overrightarrow{CD}$. Then either $ACDB$ is a parallelogram or all the points are on a directed line $\vec{L}$. When $ACDB$ is a parallelogram there is a unique point E such that

$$x_E - x_C = x_B - x_A, \quad y_E - y_C = y_B - y_A, \quad z_E - z_C = z_B - z_A.$$

As in the proof of Theorem 1 in Section 1, we conclude from part (i) that $ACEB$ is a parallelogram and $D = E$; hence equations (1) hold. The proof that equations (1) hold when the points are on a directed line $\vec{L}$ is given in Appendix 6. ◂

If we are given a directed line segment $\overrightarrow{AB}$, it is clear that there is an unlimited number of equivalent ones. In fact, if C is any given point in three-space, we can use equations (1) of Theorem 6 to find the coordinates of the unique point D such that $\overrightarrow{CD} \approx \overrightarrow{AB}$.

Definitions. *A* **vector** *is the collection of all directed line segments having a given magnitude and direction. We shall use boldface letters to denote vectors. A particular directed line segment in a collection* **v** *is called a* **representative** *of the vector* **v**. *The* **length** *of a vector is the common length of all its representatives. A* **unit vector** *is a vector of length one. Two vectors are said to be* **orthogonal** (*or* **perpendicular**) *if any representative of one vector is perpendicular to any representative of the other. The* **zero vector**, *denoted by* **0**, *is the class of directed line "segments" of zero length (i.e., simply points).*

As in Section 2, we can define the sum of two vectors. Given **u** and **v**, let $\overrightarrow{AB}$ be a representative of **u** and let $\overrightarrow{BC}$ be that representative of **v** which has its base at B. Then $\mathbf{u} + \mathbf{v}$ is the vector which has representative $\overrightarrow{AC}$ as shown in Fig. 13–20. If $\overrightarrow{A'B'}$ and $\overrightarrow{B'C'}$ are other representatives of **u** and **v**, respectively, it follows from Theorem 6 that $\overrightarrow{A'C'} \approx \overrightarrow{AC}$. Therefore $\overrightarrow{A'C'}$ is also a representative of $\mathbf{u} + \mathbf{v}$. In other words, the rule for forming the sum of two vectors does not depend on the particular representatives we select in making the calculation.

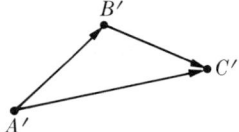

Figure 13–20

Vectors may be multiplied by numbers (scalars). Given a vector **u** and a number c, let $\overrightarrow{AB}$ be a representative of **u** and let C be the point c of the way from A to B. Then $\overrightarrow{AC}$ is a representative of $c\mathbf{u}$. It follows easily from Theorem 6 that if $\overrightarrow{A'B'}$ is another representative of **u** and C' is c of the way from A' to B', then $\overrightarrow{A'C'} \approx \overrightarrow{AC}$ and so $\overrightarrow{A'C'}$ is another representative of $c\mathbf{u}$.

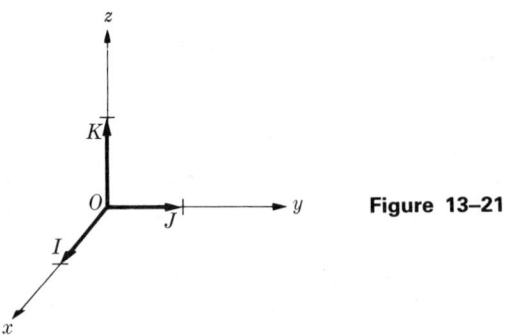

Figure 13–21

Definitions. *Suppose that a Cartesian coordinate system is given. Figure 13–21 shows such a system with the points $I(1, 0, 0)$, $J(0, 1, 0)$, and $K(0, 0, 1)$ identified.*

The **unit vector i** is defined as the vector which has $\overrightarrow{OI}$ as one of its representatives. The **unit vector j** is defined as the vector which has $\overrightarrow{OJ}$ as one of its representatives. The **unit vector k** is defined as the vector which has $\overrightarrow{OK}$ as one of its representatives.

We now establish a direct extension of Theorem 2 in Section 2.

Theorem 7. *Suppose a vector* **w** *has* $\overrightarrow{AB}$ *as a representative. Denote the coordinates of A and B by* (x_A, y_A, z_A) *and* (x_B, y_B, z_B), *respectively. Then* **w** *may be expressed in the form*

$$\mathbf{w} = (x_B - x_A)\mathbf{i} + (y_B - y_A)\mathbf{j} + (z_B - z_A)\mathbf{k}.$$

Proof. From equations (1) of Theorem 6, we know that **w** has the representative $\overrightarrow{OP}$ where P has coordinates $(x_B - x_A, y_B - y_A, z_B - z_A)$. Let

$$Q(x_B - x_A, 0, 0), \quad R(0, y_B - y_A, 0),$$
$$S(0, 0, z_B - z_A), \quad T(x_B - x_A, y_B - y_A, 0)$$

be as shown in Fig. 13-22. Then Q is $x_B - x_A$ of the way from O to $I(1, 0, 0)$, and similarly for R and S with regard to J and K. Therefore,

$$\mathbf{v}(\overrightarrow{OQ}) = (x_B - x_A)\mathbf{i},$$
$$\mathbf{v}(\overrightarrow{QT}) = \mathbf{v}(\overrightarrow{OR}) = (y_B - y_A)\mathbf{j},$$
$$\mathbf{v}(\overrightarrow{TP}) = \mathbf{v}(\overrightarrow{OS}) = (z_B - z_A)\mathbf{k}.$$

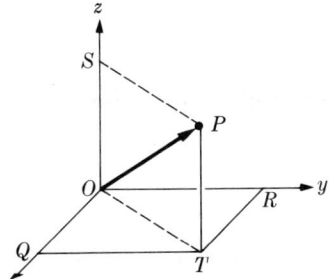

Figure 13-22

Using the rule for addition of vectors, we find

$$\mathbf{v}(\overrightarrow{OP}) = \mathbf{v}(\overrightarrow{OQ}) + \mathbf{v}(\overrightarrow{QT}) + \mathbf{v}(\overrightarrow{TP}),$$

and the proof is complete.

Example 1. A vector **v** has $\overrightarrow{AB}$ as a representative. If A and B have coordinates $(3, -2, 4)$ and $(2, 1, 5)$, respectively, express **v** in terms of **i**, **j**, and **k**.

Solution. From Theorem 7, we obtain

$$\mathbf{v}(\overrightarrow{AB}) = (2 - 3)\mathbf{i} + (1 + 2)\mathbf{j} + (5 - 4)\mathbf{k} = -\mathbf{i} + 3\mathbf{j} + \mathbf{k}. \quad \blacktriangleleft$$

The next two theorems are direct extensions of Theorems 3 and 4 in Section 2. The proofs are left to the student.

The **length** of a vector $\mathbf{v}$ is denoted by $|\mathbf{v}|$.

Theorem 8. *If* $\mathbf{v} = a\mathbf{i} + b\mathbf{j} + c\mathbf{k}$, *then*

$$|\mathbf{v}| = \sqrt{a^2 + b^2 + c^2}.$$

Theorem 9. *If* $\mathbf{v} = a_1\mathbf{i} + b_1\mathbf{j} + c_1\mathbf{k}$, $\mathbf{w} = a_2\mathbf{i} + b_2\mathbf{j} + c_2\mathbf{k}$, *then*

$$\mathbf{v} + \mathbf{w} = (a_1 + a_2)\mathbf{i} + (b_1 + b_2)\mathbf{j} + (c_1 + c_2)\mathbf{k}.$$

If h is any number, then

$$h\mathbf{v} = ha_1\mathbf{i} + hb_1\mathbf{j} + hc_1\mathbf{k}.$$ ◂

In complete analogy with vectors in the plane, we conclude from the theorems above that the addition of vectors and their multiplication by numbers satisfy the following laws:

$$\left.\begin{array}{c}\mathbf{u} + (\mathbf{v} + \mathbf{w}) = (\mathbf{u} + \mathbf{v}) + \mathbf{w} \\ c(d\mathbf{v}) = (cd)\mathbf{v}\end{array}\right\} \text{Associative laws}$$

$$\mathbf{u} + \mathbf{v} = \mathbf{v} + \mathbf{u} \qquad \text{Commutative law}$$

$$\left.\begin{array}{c}(c + d)\mathbf{v} = c\mathbf{v} + d\mathbf{v} \\ c(\mathbf{u} + \mathbf{v}) = c\mathbf{u} + c\mathbf{v}\end{array}\right\} \text{Distributive laws}$$

$$1 \cdot \mathbf{u} = \mathbf{u}, \qquad 0 \cdot \mathbf{u} = \mathbf{0}, \qquad (-1)\mathbf{u} = -\mathbf{u}.$$

Definition. *Let $\mathbf{v}$ be any vector except $\mathbf{0}$. The **unit vector $\mathbf{u}$ in the direction of $\mathbf{v}$** is defined by*

$$\mathbf{u} = \frac{1}{|\mathbf{v}|}\mathbf{v}.$$

Example 2. Given the vectors $\mathbf{u} = 3\mathbf{i} - 2\mathbf{j} + 4\mathbf{k}$ and $\mathbf{v} = 6\mathbf{i} - 4\mathbf{j} - 2\mathbf{k}$, express the vector $3\mathbf{u} - 2\mathbf{v}$ in terms of $\mathbf{i}$, $\mathbf{j}$, and $\mathbf{k}$.

Solution. $3\mathbf{u} = 9\mathbf{i} - 6\mathbf{j} + 12\mathbf{k}$ and $-2\mathbf{v} = -12\mathbf{i} + 8\mathbf{j} + 4\mathbf{k}$. Adding these vectors, we get $3\mathbf{u} - 2\mathbf{v} = -3\mathbf{i} + 2\mathbf{j} + 16\mathbf{k}$. ◂

Example 3. Given the vector $\mathbf{v} = 2\mathbf{i} - 3\mathbf{j} + \mathbf{k}$, find a unit vector in the direction of $\mathbf{v}$.

Solution. We have $|\mathbf{v}| = \sqrt{4 + 9 + 1} = \sqrt{14}$. The desired vector $\mathbf{u}$ is

$$\mathbf{u} = \frac{1}{\sqrt{14}}\mathbf{v} = \frac{2}{\sqrt{14}}\mathbf{i} - \frac{3}{\sqrt{14}}\mathbf{j} + \frac{1}{\sqrt{14}}\mathbf{k}.$$ ◂

Example 4. Given the vector $\mathbf{v} = 2\mathbf{i} + 4\mathbf{j} - 3\mathbf{k}$, find the representative $\overrightarrow{AB}$ of $\mathbf{v}$ if the point A has coordinates $(2, 1, -5)$.

Solution. Denote the coordinates of B by x_B, y_B, z_B. Then we have

$$x_B - 2 = 2, \qquad y_B - 1 = 4, \qquad z_B + 5 = -3.$$

Therefore, $x_B = 4$, $y_B = 5$, $z_B = -8$. ◀

PROBLEMS

In Problems 1 through 6, express $\mathbf{v}$ in terms of $\mathbf{i}$, $\mathbf{j}$, and $\mathbf{k}$, given that the endpoints P and Q of the representative $\overrightarrow{PQ}$ of $\mathbf{v}$ have the given coordinates. Also find another representative of the same vector $\mathbf{v}$.

1. $P(2, 0, 3)$, $\quad Q(1, 4, -3)$
2. $P(1, 1, 0)$, $\quad Q(-1, 2, 0)$
3. $P(-4, -2, 1)$, $\quad Q(1, -3, 4)$
4. $P(3, 2, 1)$, $\quad Q(3, 3, 3)$
5. $P(2, 0, 0)$, $\quad Q(0, 0, -3)$
6. $P(4, -5, -1)$, $\quad Q(-2, 1, -3)$

In Problems 7 through 10, in each case find a unit vector $\mathbf{u}$ in the direction of $\mathbf{v}$. Express $\mathbf{u}$ in terms of $\mathbf{i}$, $\mathbf{j}$, and $\mathbf{k}$.

7. $\mathbf{v} = 3\mathbf{i} + 2\mathbf{j} - 4\mathbf{k}$
8. $\mathbf{v} = \mathbf{i} - \mathbf{j} + \mathbf{k}$
9. $\mathbf{v} = 2\mathbf{i} - 4\mathbf{j} - \mathbf{k}$
10. $\mathbf{v} = -2\mathbf{i} + 3\mathbf{j} + 5\mathbf{k}$

In Problems 11 through 17, find the representative $\overrightarrow{AB}$ of the vector $\mathbf{v}$ from the information given.

11. $\mathbf{v} = 2\mathbf{i} + \mathbf{j} - 3\mathbf{k}$, $\quad A(1, 2, -1)$
12. $\mathbf{v} = -\mathbf{i} + 3\mathbf{j} - 2\mathbf{k}$, $\quad A(2, 0, 4)$
13. $\mathbf{v} = 3\mathbf{i} + 2\mathbf{j} - 4\mathbf{k}$, $\quad B(2, 0, -4)$
14. $\mathbf{v} = -2\mathbf{i} + 4\mathbf{j} + \mathbf{k}$, $\quad B(0, 0, -5)$
15. $\mathbf{v} = \mathbf{i} - 2\mathbf{j} + 2\mathbf{k}$; the midpoint of the segment AB has coordinates $(2, -1, 4)$.
16. $\mathbf{v} = 3\mathbf{i} + 4\mathbf{k}$; the midpoint of the segment AB has coordinates $(1, 2, -5)$.
17. $\mathbf{v} = -\mathbf{i} + \mathbf{j} - 2\mathbf{k}$; the point three-fourths of the distance from A to B has coordinates $(1, 0, 2)$.
18. Find a vector $\mathbf{u}$ in the direction of $\mathbf{v} = -\mathbf{i} + \mathbf{j} - \mathbf{k}$ and having half the length of $\mathbf{v}$.
19. Given $\mathbf{u} = \mathbf{i} + 2\mathbf{j} - 4\mathbf{k}$, $\mathbf{v} = 3\mathbf{i} - 7\mathbf{j} + 5\mathbf{k}$, find $\mathbf{u} + \mathbf{v}$ in terms of $\mathbf{i}$, $\mathbf{j}$, and $\mathbf{k}$. Sketch a figure.
20. Given that $\mathbf{u} = -3\mathbf{i} + 7\mathbf{j} - 4\mathbf{k}$, $\mathbf{v} = 2\mathbf{i} + \mathbf{j} - 6\mathbf{k}$, find $3\mathbf{u} - 7\mathbf{v}$ in terms of $\mathbf{i}$, $\mathbf{j}$, and $\mathbf{k}$.
21. Let a and b be any real numbers. Show that the vector $\mathbf{k}$ is orthogonal to $a\mathbf{i} + b\mathbf{j}$.

22. Show that the vector $\mathbf{u} = \mathbf{i} + 2\mathbf{j} - 3\mathbf{k}$ is orthogonal to $a\mathbf{v} + b\mathbf{w}$ where $\mathbf{v} = 2\mathbf{i} + 2\mathbf{j} + 2\mathbf{k}$, $\mathbf{w} = -\mathbf{i} + 2\mathbf{j} + \mathbf{k}$ and a, b are any real numbers. Interpret this statement geometrically.

23. Discuss the relationship between the direction numbers of a line and the representation of a vector $\mathbf{v}$ in terms of the vectors $\mathbf{i}$, $\mathbf{j}$, and $\mathbf{k}$.

24. Suppose two representatives of $\mathbf{u}$ and $\mathbf{v}$ determine a plane. Discuss the relationship between the attitude numbers of this plane and the representations of $\mathbf{u}$ and $\mathbf{v}$ in terms of $\mathbf{i}$, $\mathbf{j}$, and $\mathbf{k}$.

5. LINEAR DEPENDENCE AND INDEPENDENCE*

Two vectors $\mathbf{u}$ and $\mathbf{v}$, neither zero, are said to be **proportional** if and only if there is a number c such that $\mathbf{u} = c\mathbf{v}$; that is, each vector is a scalar multiple of the other. If $\mathbf{v}_1, \mathbf{v}_2, \ldots, \mathbf{v}_k$ are any vectors and $c_1, c_2, \ldots, c_k$ are numbers, we call an expression of the form

$$c_1\mathbf{v}_1 + c_2\mathbf{v}_2 + \cdots + c_k\mathbf{v}_k$$

a **linear combination** of the vectors $\mathbf{v}_1, \mathbf{v}_2, \ldots, \mathbf{v}_k$. If two vectors $\mathbf{u}$ and $\mathbf{v}$ are proportional, the definition shows that a linear combination of them is the zero vector. In fact, $\mathbf{u} - c\mathbf{v} = \mathbf{0}$. A set of vectors $\{\mathbf{v}_1, \mathbf{v}_2, \ldots, \mathbf{v}_k\}$ is **linearly dependent** if and only if there is a set of constants $\{c_1, c_2, \ldots, c_k\}$, *not all zero*, such that

$$c_1\mathbf{v}_1 + c_2\mathbf{v}_2 + \cdots + c_k\mathbf{v}_k = \mathbf{0}. \tag{1}$$

If no such set of constants exists, then the set $\{\mathbf{v}_1, \mathbf{v}_2, \ldots, \mathbf{v}_k\}$ is said to be **linearly independent.**

It is clear that any two proportional vectors are linearly dependent. As another example, the vectors

$$\mathbf{v}_1 = 2\mathbf{i} + 3\mathbf{j} - \mathbf{k}, \quad \mathbf{v}_2 = -2\mathbf{i} - \mathbf{j} + \mathbf{k}, \quad \mathbf{v}_3 = 2\mathbf{i} + 7\mathbf{j} - \mathbf{k}$$

form a linearly dependent set since the selection $c_1 = 3$, $c_2 = 2$, $c_3 = -1$ shows that

$$c_1\mathbf{v}_1 + c_2\mathbf{v}_2 + c_3\mathbf{v}_3 = 3(2\mathbf{i} + 3\mathbf{j} - \mathbf{k}) + 2(-2\mathbf{i} - \mathbf{j} + \mathbf{k}) - (2\mathbf{i} + 7\mathbf{j} - \mathbf{k}) = \mathbf{0}.$$

A set $\{\mathbf{v}_1, \mathbf{v}_2, \ldots, \mathbf{v}_k\}$ is linearly dependent if and only if one member of the set can be expressed as a linear combination of the remaining members. To see this, we observe that in Eq. (1) one of the terms on the left-hand side, say $\mathbf{v}_i$, must have a nonzero coefficient and so may be transferred to the right-hand side. Dividing by the coefficient $-c_i$, we express this particular $\mathbf{v}_i$ as a linear combination of the remaining $\mathbf{v}$'s. If some $\mathbf{v}_i$ is expressible in terms of the others, it follows by transposing $\mathbf{v}_i$ that $\mathbf{v}_1, \mathbf{v}_2, \ldots, \mathbf{v}_k$ are linearly dependent.

The following statement, a direct consequence of the definition of linear dependence, is often useful in proofs of theorems. If $\{\mathbf{v}_1, \mathbf{v}_2, \ldots, \mathbf{v}_k\}$ is a linearly

---

*For an understanding of Sections 5, 6, and 7, it is assumed that the reader is acquainted with determinants of the second and third order. For those unfamiliar with the subject a brief discussion is provided in Appendix 5.

independent set and if
$$c_1\mathbf{v}_1 + c_2\mathbf{v}_2 + \cdots + c_k\mathbf{v}_k = \mathbf{0},$$
then it follows that $c_1 = c_2 = \cdots = c_k = 0$.

The set $\{\mathbf{i}, \mathbf{j}, \mathbf{k}\}$ is linearly independent. To show this observe that the equation
$$c_1\mathbf{i} + c_2\mathbf{j} + c_3\mathbf{k} = \mathbf{0} \tag{2}$$
holds if and only if $|c_1\mathbf{i} + c_2\mathbf{j} + c_3\mathbf{k}| = 0$. But
$$|c_1\mathbf{i} + c_2\mathbf{j} + c_3\mathbf{k}| = \sqrt{c_1^2 + c_2^2 + c_3^2},$$
and this last expression is zero if and only if $c_1 = c_2 = c_3 = 0$. Thus no nonzero constants satisfying (2) exist and $\{\mathbf{i}, \mathbf{j}, \mathbf{k}\}$ is a linearly independent set.

The proof of the next theorem employs the tools on determinants given in Appendix 5. The details of the proof of Theorem 10 are carried out in Appendix 6.

Theorem 10. *Let*
$$\mathbf{u} = a_{11}\mathbf{i} + a_{12}\mathbf{j} + a_{13}\mathbf{k},$$
$$\mathbf{v} = a_{21}\mathbf{i} + a_{22}\mathbf{j} + a_{23}\mathbf{k},$$
$$\mathbf{w} = a_{31}\mathbf{i} + a_{32}\mathbf{j} + a_{33}\mathbf{k},$$
and denote by D the determinant
$$D = \begin{vmatrix} a_{11} & a_{12} & a_{13} \\ a_{21} & a_{22} & a_{23} \\ a_{31} & a_{32} & a_{33} \end{vmatrix}.$$
Then the set $\{\mathbf{u}, \mathbf{v}, \mathbf{w}\}$ is linearly independent if and only if $D \neq 0$.

Example 1. Determine whether or not the vectors
$$\mathbf{u} = 2\mathbf{i} - \mathbf{j} + \mathbf{k}, \quad \mathbf{v} = \mathbf{i} + 2\mathbf{j} + \mathbf{k}, \quad \mathbf{w} = -\mathbf{i} + \mathbf{j} + 3\mathbf{k}$$
form a linearly independent set.

Solution. Expanding D by its first row, we have
$$D = \begin{vmatrix} 2 & -1 & 1 \\ 1 & 2 & 1 \\ -1 & 1 & 3 \end{vmatrix} = 2\begin{vmatrix} 2 & 1 \\ 1 & 3 \end{vmatrix} + \begin{vmatrix} 1 & 1 \\ -1 & 3 \end{vmatrix} + \begin{vmatrix} 1 & 2 \\ -1 & 1 \end{vmatrix}.$$
Therefore $D = 2(5) + 4 + 3 = 17 \neq 0$. The set is linearly independent. ◀

Theorem 11. *If $\{\mathbf{u}, \mathbf{v}, \mathbf{w}\}$ is a linearly independent set and $\mathbf{r}$ is any vector, then there are constants A_1, A_2, and A_3 such that*
$$\mathbf{r} = A_1\mathbf{u} + A_2\mathbf{v} + A_3\mathbf{w}. \tag{3}$$

Proof. According to Theorem 7, *every* vector can be expressed as a linear combination of **i**, **j**, and **k**. Therefore

$$\mathbf{u} = a_{11}\mathbf{i} + a_{12}\mathbf{j} + a_{13}\mathbf{k},$$
$$\mathbf{v} = a_{21}\mathbf{i} + a_{22}\mathbf{j} + a_{23}\mathbf{k},$$
$$\mathbf{w} = a_{31}\mathbf{i} + a_{32}\mathbf{j} + a_{33}\mathbf{k},$$
$$\mathbf{r} = b_1\mathbf{i} + b_2\mathbf{j} + b_3\mathbf{k}.$$

When we insert all these expressions in (3) and collect all terms on one side, we get a linear combination of **i**, **j**, and **k** equal to zero. Since {**i**, **j**, **k**} is a linearly independent set, the coefficients of **i**, **j**, and **k** are equal to zero separately. Computing these coefficients, we get the equations

$$a_{11}A_1 + a_{21}A_2 + a_{31}A_3 = b_1,$$
$$a_{12}A_1 + a_{22}A_2 + a_{32}A_3 = b_2, \qquad (4)$$
$$a_{13}A_1 + a_{23}A_2 + a_{33}A_3 = b_3.$$

We have here three equations in the three unknowns A_1, A_2, A_3. The determinant D' of the coefficients in (4) differs from the determinant D of Theorem 10 in that the rows and columns are interchanged. Since {**u**, **v**, **w**} is an independent set, we know that $D \neq 0$; also Theorem 3 of Appendix 5 proves that $D = D'$, and so $D' \neq 0$. We now use Cramer's rule (Theorem 11, Appendix 5) to solve for A_1, A_2, A_3. ◀

Note that the proof of Theorem 11 gives the method for finding A_1, A_2, A_3. We work an example.

Example 2. Given the vectors

$$\mathbf{u} = 2\mathbf{i} + 3\mathbf{j} + \mathbf{k}, \qquad \mathbf{v} = -\mathbf{i} + \mathbf{j} + 2\mathbf{k},$$
$$\mathbf{w} = 3\mathbf{i} - \mathbf{j} + 3\mathbf{k}, \qquad \mathbf{r} = \mathbf{i} + 2\mathbf{j} - 6\mathbf{k},$$

show that **u**, **v**, and **w** are linearly independent and express **r** as a linear combination of **u**, **v**, and **w**.

Solution. Expanding D by its first row, we obtain

$$D = \begin{vmatrix} 2 & 3 & 1 \\ -1 & 1 & 2 \\ 3 & -1 & 3 \end{vmatrix} = 2\begin{vmatrix} 1 & 2 \\ -1 & 3 \end{vmatrix} - 3\begin{vmatrix} -1 & 2 \\ 3 & 3 \end{vmatrix} + \begin{vmatrix} -1 & 1 \\ 3 & -1 \end{vmatrix}$$
$$= 2(5) - 3(-9) + (-2) = 35.$$

Hence $D \neq 0$ and so {**u**, **v**, **w**} is a linearly independent set. Using equations (4), we now obtain the equations

$$2A_1 - A_2 + 3A_3 = 1,$$
$$3A_1 + A_2 - A_3 = 2,$$
$$A_1 + 2A_2 + 3A_3 = -6.$$

Solving these, we find that $A_1 = 1$, $A_2 = -2$, $A_3 = -1$. Finally,

$$\mathbf{r} = \mathbf{u} - 2\mathbf{v} - \mathbf{w}.$$

PROBLEMS

In Problems 1 through 5 state whether or not the given vectors are linearly independent.

1. $\mathbf{u} = 2\mathbf{i} + \mathbf{j} - \mathbf{k}$, $\quad \mathbf{v} = \mathbf{i} - 2\mathbf{j} + 5\mathbf{k}$, $\quad \mathbf{w} = 2\mathbf{i} - 7\mathbf{j} + \mathbf{k}$
2. $\mathbf{u} = \mathbf{i} + 2\mathbf{j} + 3\mathbf{k}$, $\quad \mathbf{v} = 2\mathbf{i} + \mathbf{j} + 4\mathbf{k}$, $\quad \mathbf{w} = 3\mathbf{j} + 2\mathbf{k}$
3. $\mathbf{u} = 2\mathbf{i} + 3\mathbf{j}$, $\quad \mathbf{v} = \mathbf{i} - 4\mathbf{j}$, $\quad \mathbf{w} = \mathbf{i} + 2\mathbf{j}$
4. $\mathbf{u} = -\mathbf{i} + 2\mathbf{j}$, $\quad \mathbf{v} = \mathbf{i} + \mathbf{j} + \mathbf{k}$, $\quad \mathbf{w} = -2\mathbf{j} + 6\mathbf{k}$
5. $\mathbf{u} = \mathbf{i} + \mathbf{j}$, $\quad \mathbf{v} = 2\mathbf{i} - 6\mathbf{j} + 3\mathbf{k}$, $\quad \mathbf{w} = -\mathbf{i} + \mathbf{j}$, $\quad \mathbf{r} = 4\mathbf{k}$

In Problems 6 through 11, show that $\mathbf{u}$, $\mathbf{v}$, and $\mathbf{w}$ are linearly independent and express $\mathbf{r}$ in terms of $\mathbf{u}$, $\mathbf{v}$, and $\mathbf{w}$.

6. $\mathbf{u} = 2\mathbf{i} - \mathbf{j} + \mathbf{k}$, $\quad \mathbf{v} = -\mathbf{i} + \mathbf{j} - 2\mathbf{k}$, $\quad \mathbf{w} = 2\mathbf{i} - \mathbf{j} + 2\mathbf{k}$, $\quad \mathbf{r} = 3\mathbf{i} - \mathbf{j} + 2\mathbf{k}$
7. $\mathbf{u} = \mathbf{i} - \mathbf{j} + \mathbf{k}$, $\quad \mathbf{v} = -\mathbf{i} + 2\mathbf{j} - \mathbf{k}$, $\quad \mathbf{w} = 2\mathbf{i} - \mathbf{j} + \mathbf{k}$, $\quad \mathbf{r} = 2\mathbf{i} + 3\mathbf{j} + 4\mathbf{k}$
8. $\mathbf{u} = 3\mathbf{i} + \mathbf{j} - 2\mathbf{k}$, $\quad \mathbf{v} = 2\mathbf{i} - \mathbf{k}$, $\quad \mathbf{w} = -\mathbf{i} + 2\mathbf{j} + \mathbf{k}$, $\quad \mathbf{r} = \mathbf{i} + 2\mathbf{j} - 3\mathbf{k}$
9. $\mathbf{u} = 2\mathbf{i} - \mathbf{j} + \mathbf{k}$, $\quad \mathbf{v} = \mathbf{i} + \mathbf{j}$, $\quad \mathbf{w} = -\mathbf{i} + \mathbf{j} + 2\mathbf{k}$, $\quad \mathbf{r} = 2\mathbf{i} - \mathbf{j} - 2\mathbf{k}$
10. $\mathbf{u} = \mathbf{i} - 2\mathbf{j} - 3\mathbf{k}$, $\quad \mathbf{v} = 2\mathbf{i} - \mathbf{j} - 2\mathbf{k}$, $\quad \mathbf{w} = -\mathbf{i} + \mathbf{j} + \mathbf{k}$, $\quad \mathbf{r} = 2\mathbf{i} + 3\mathbf{j} + 4\mathbf{k}$
11. $\mathbf{u} = 2\mathbf{i} - 3\mathbf{k}$, $\quad \mathbf{v} = \mathbf{i} + 4\mathbf{j} - \mathbf{k}$, $\quad \mathbf{w} = -2\mathbf{i} + 5\mathbf{j} + 3\mathbf{k}$, $\quad \mathbf{r} = -\mathbf{i} + 20\mathbf{j} + 3\mathbf{k}$
12. Prove Theorem 8.
13. Prove Theorem 9.
14. Show that in three dimensions any set of four vectors must be linearly dependent.
15. Show that if $\overrightarrow{OA}$, $\overrightarrow{OB}$, and $\overrightarrow{OC}$ are representatives of $\mathbf{u}$, $\mathbf{v}$, and $\mathbf{w}$, respectively, and if $\{\mathbf{u}, \mathbf{v}, \mathbf{w}\}$ is a linearly dependent set, then the three representatives lie in one plane.
16. Find the equations of the line passing through the point $P(2, 1, -3)$ and parallel to any representative of $\mathbf{v} = 3\mathbf{i} - 2\mathbf{j} + 7\mathbf{k}$.
17. Find the equations of the line through the point $A(1, -4, 0)$ and perpendicular to the plane determined by two (intersecting) representatives of $\mathbf{v} = \mathbf{i} + 2\mathbf{j} - \mathbf{k}$ and $\mathbf{w} = 3\mathbf{i} - \mathbf{j} + \mathbf{k}$.

6. THE SCALAR (INNER OR DOT) PRODUCT

Two vectors are said to be **parallel** or **proportional** when each is a scalar multiple of the other (and neither is zero). The representatives of parallel vectors are all parallel directed line segments.

By the **angle between two vectors v and w** (neither $= \mathbf{0}$), we mean the measure of the angle between any directed line containing a representative of $\mathbf{v}$ and an intersecting directed line containing a representative of $\mathbf{w}$ (Fig. 13–23). Two parallel vectors make an angle of 0 or π, depending on whether they are pointing in the same or in the opposite direction.

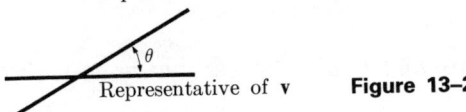

Figure 13–23

Theorem 12. *If θ is the angle between the vectors*
$$\mathbf{v} = a_1\mathbf{i} + a_2\mathbf{j} + a_3\mathbf{k} \quad \text{and} \quad \mathbf{w} = b_1\mathbf{i} + b_2\mathbf{j} + b_3\mathbf{k},$$
then
$$\cos\theta = \frac{a_1b_1 + a_2b_2 + a_3b_3}{|\mathbf{v}| \cdot |\mathbf{w}|}.$$

The proof is a straightforward extension of the proof of the analogous theorem in the plane (Theorem 5 of Section 3) and will therefore be omitted.

Example 1. Given the vectors $\mathbf{v} = 2\mathbf{i} + \mathbf{j} - 3\mathbf{k}$ and $\mathbf{w} = -\mathbf{i} + 4\mathbf{j} - 2\mathbf{k}$, find the cosine of the angle between $\mathbf{v}$ and $\mathbf{w}$.

Solution. We have
$$|\mathbf{v}| = \sqrt{4 + 1 + 9} = \sqrt{14}, \quad |\mathbf{w}| = \sqrt{1 + 16 + 4} = \sqrt{21}.$$
Therefore
$$\cos\theta = \frac{-2 + 4 + 6}{\sqrt{14} \cdot \sqrt{21}} = \frac{8}{7\sqrt{6}}. \quad \blacktriangleleft$$

Definitions. *Given the vectors $\mathbf{u}$ and $\mathbf{v}$, the* **scalar** **(inner** *or* **dot) product** $\mathbf{u} \cdot \mathbf{v}$ *is defined by the formula*
$$\mathbf{u} \cdot \mathbf{v} = |\mathbf{u}|\,|\mathbf{v}|\cos\theta,$$
where θ is the angle between the vectors. If either $\mathbf{u}$ or $\mathbf{v}$ is $\mathbf{0}$, we define $\mathbf{u} \cdot \mathbf{v} = 0$. Two vectors $\mathbf{u}$ and $\mathbf{v}$ are **orthogonal** *if and only if $\mathbf{u} \cdot \mathbf{v} = 0$.*

Thus we see that $\mathbf{0}$ is orthogonal to every vector. These definitions are identical with those for plane vectors.

Theorem 13. *The scalar product satisfies the laws*

a) $\mathbf{u} \cdot \mathbf{v} = \mathbf{v} \cdot \mathbf{u}$;   b) $\mathbf{u} \cdot \mathbf{u} = |\mathbf{u}|^2$.

c) *If* $\mathbf{u} = a_1\mathbf{i} + b_1\mathbf{j} + c_1\mathbf{k}$ *and* $\mathbf{v} = a_2\mathbf{i} + b_2\mathbf{j} + c_2\mathbf{k}$, *then*
$$\mathbf{u} \cdot \mathbf{v} = a_1a_2 + b_1b_2 + c_1c_2.$$

Proof. Parts (a) and (b) are direct consequences of the definition; part (c) follows from Theorem 12 since
$$\mathbf{u} \cdot \mathbf{v} = |\mathbf{u}|\,|\mathbf{v}|\cos\theta = |\mathbf{u}|\,|\mathbf{v}|\frac{a_1a_2 + b_1b_2 + c_1c_2}{|\mathbf{u}|\,|\mathbf{v}|}.$$

Corollary. (a) *If c and d are any numbers and if $\mathbf{u}, \mathbf{v}, \mathbf{w}$ are any vectors, then*
$$\mathbf{u} \cdot (c\mathbf{v} + d\mathbf{w}) = c(\mathbf{u} \cdot \mathbf{v}) + d(\mathbf{u} \cdot \mathbf{w}).$$

b) We have
$$\mathbf{i} \cdot \mathbf{i} = \mathbf{j} \cdot \mathbf{j} = \mathbf{k} \cdot \mathbf{k} = 1, \qquad \mathbf{i} \cdot \mathbf{j} = \mathbf{i} \cdot \mathbf{k} = \mathbf{j} \cdot \mathbf{k} = 0.$$

Example 2. Find the scalar product of the vectors
$$\mathbf{u} = 3\mathbf{i} + 2\mathbf{j} - 4\mathbf{k} \qquad \text{and} \qquad \mathbf{v} = -2\mathbf{i} + \mathbf{j} + 5\mathbf{k}.$$

Solution. $\mathbf{u} \cdot \mathbf{v} = 3(-2) + 2 \cdot 1 + (-4)(5) = -24.$ ◀

Example 3. Express $|3\mathbf{u} + 5\mathbf{v}|^2$ in terms of $|\mathbf{u}|^2$, $|\mathbf{v}|^2$, and $\mathbf{u} \cdot \mathbf{v}$.

Solution.
$$\begin{aligned}|3\mathbf{u} + 5\mathbf{v}|^2 &= (3\mathbf{u} + 5\mathbf{v}) \cdot (3\mathbf{u} + 5\mathbf{v}) \\ &= 9(\mathbf{u} \cdot \mathbf{u}) + 15(\mathbf{u} \cdot \mathbf{v}) + 15(\mathbf{v} \cdot \mathbf{u}) + 25(\mathbf{v} \cdot \mathbf{v}) \\ &= 9|\mathbf{u}|^2 + 30(\mathbf{u} \cdot \mathbf{v}) + 25|\mathbf{v}|^2.\end{aligned}$$ ◀

Definition. *Let $\mathbf{v}$ and $\mathbf{w}$ be two vectors which make an angle θ. We denote by $|\mathbf{v}| \cos \theta$ the **projection of $\mathbf{v}$ on $\mathbf{w}$**. We also call this quantity the **component of $\mathbf{v}$ along $\mathbf{w}$**. As before we denote this quantity by $\text{Proj}_\mathbf{w} \mathbf{v}$.*

From the formula for $\cos \theta$, we may write

$$\text{Proj}_\mathbf{w} \mathbf{v} = |\mathbf{v}| \cos \theta = |\mathbf{v}| \frac{\mathbf{v} \cdot \mathbf{w}}{|\mathbf{v}| |\mathbf{w}|} = \frac{\mathbf{v} \cdot \mathbf{w}}{|\mathbf{w}|}.$$

Example 4. Find the projection of $\mathbf{v} = -\mathbf{i} + 2\mathbf{j} + 3\mathbf{k}$ on $\mathbf{w} = 2\mathbf{i} - \mathbf{j} - 4\mathbf{k}$.

Solution. We have $\mathbf{v} \cdot \mathbf{w} = (-1)(2) + (2)(-1) + (3)(-4) = -16$; $|\mathbf{w}| = \sqrt{21}$. Therefore, the projection of $\mathbf{v}$ on $\mathbf{w}$ is $|\mathbf{v}| \cos \theta = -16/\sqrt{21}$. ◀

An application* of scalar product to mechanics occurs in the calculation of work done by a constant force $\mathbf{F}$ when its point of application moves along a segment from A to B. The **work done** in this case is defined as the product of the distance from A to B and the projection of $\mathbf{F}$ on $\mathbf{v}$ ($\overrightarrow{AB}$). We have

$$\text{Proj}_\mathbf{w} \mathbf{F} = \text{Projection of } \mathbf{F} \text{ on } \mathbf{v} = \frac{\mathbf{F} \cdot \mathbf{v}}{|\mathbf{v}|};$$

since the distance from A to B is exactly $|\mathbf{v}|$, we conclude that

$$\text{Work done by } \mathbf{F} = \mathbf{F} \cdot \mathbf{v}.$$

Example 5. Find the work done by the force
$$\mathbf{F} = 5\mathbf{i} - 3\mathbf{j} + 2\mathbf{k}$$
as its point of application moves from the point $A(2, 1, 3)$ to $B(4, -1, 5)$.

---

* This paragraph and Problems 11–14 at the end of this section may be omitted by readers who have not studied the previous applications of calculus to problems on work.

Solution. We have
$$\mathbf{v}(\overrightarrow{AB}) = (4-2)\mathbf{i} + (-1-1)\mathbf{j} + (5-3)\mathbf{k} = 2\mathbf{i} - 2\mathbf{j} + 2\mathbf{k}.$$
Therefore, work done $= 5 \cdot 2 + 3 \cdot 2 + 2 \cdot 2 = 20$. ◀

Theorem 14. *If $\mathbf{u}$ and $\mathbf{v}$ are not $\mathbf{0}$, there is a unique number k such that $\mathbf{v} - k\mathbf{u}$ is orthogonal to $\mathbf{u}$. In fact, k can be found from the formula*
$$k = \frac{\mathbf{u} \cdot \mathbf{v}}{|\mathbf{u}|^2}.$$

Proof. $(\mathbf{v} - k\mathbf{u})$ is orthogonal to $\mathbf{u}$ if and only if $\mathbf{u} \cdot (\mathbf{v} - k\mathbf{u}) = 0$. But
$$\mathbf{u} \cdot (\mathbf{v} - k\mathbf{u}) = \mathbf{u} \cdot \mathbf{v} - k|\mathbf{u}|^2 = 0.$$
Therefore, selection of $k = \mathbf{u} \cdot \mathbf{v}/|\mathbf{u}|^2$ yields the result.

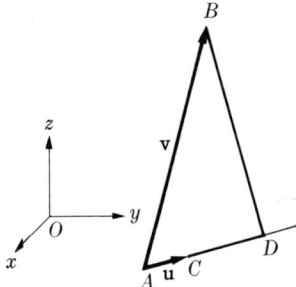

Figure 13–24

Figure 13–24 shows geometrically how k is to be selected. We drop a perpendicular from the head of $\mathbf{v}$ (point B) to the line containing $\mathbf{u}$ (point D). The directed segment $\overrightarrow{AD}$ gives the proper multiple of $\mathbf{u}$ ($\overrightarrow{AC}$), and the directed segment $\overrightarrow{DB}$ represents the orthogonal vector.

Example 6. Find a linear combination of
$$\mathbf{u} = 2\mathbf{i} + 3\mathbf{j} - \mathbf{k} \quad \text{and} \quad \mathbf{v} = \mathbf{i} + 2\mathbf{j} + \mathbf{k}$$
which is orthogonal to $\mathbf{u}$.

Solution. We select $k = (2 + 6 - 1)/14 = \frac{1}{2}$, and the desired vector is $\frac{1}{2}\mathbf{j} + \frac{3}{2}\mathbf{k}$. ◀

PROBLEMS

In each of Problems 1 through 5, find $\cos \theta$ where θ is the angle between $\mathbf{v}$ and $\mathbf{w}$.
1. $\mathbf{v} = \mathbf{i} + 3\mathbf{j} - 2\mathbf{k},$ $\quad \mathbf{w} = 2\mathbf{i} + 4\mathbf{j} - \mathbf{k}$
2. $\mathbf{v} = -\mathbf{i} + 2\mathbf{j} + 3\mathbf{k},$ $\quad \mathbf{w} = 3\mathbf{i} - 2\mathbf{j} - 2\mathbf{k}$
3. $\mathbf{v} = 4\mathbf{i} - 3\mathbf{j} + 5\mathbf{k},$ $\quad \mathbf{w} = 2\mathbf{i} + \mathbf{j} + \mathbf{k}$
4. $\mathbf{v} = 2\mathbf{i} + 3\mathbf{k},$ $\quad \mathbf{w} = \mathbf{j} + 4\mathbf{k}$
5. $\mathbf{v} = -2\mathbf{i} - 3\mathbf{j} - 4\mathbf{k},$ $\quad \mathbf{w} = 2\mathbf{i} - 3\mathbf{j} + 4\mathbf{k}$

In each of Problems 6 through 10, find the projection of the vector **v** on **u**.

6. $\mathbf{u} = 2\mathbf{i} - 6\mathbf{j} + 3\mathbf{k}$, $\quad \mathbf{v} = \mathbf{i} + 2\mathbf{j} - 2\mathbf{k}$
7. $\mathbf{u} = 6\mathbf{i} + 2\mathbf{j} - 3\mathbf{k}$, $\quad \mathbf{v} = -\mathbf{i} + 8\mathbf{j} + 4\mathbf{k}$
8. $\mathbf{u} = 12\mathbf{i} + 3\mathbf{j} + 4\mathbf{k}$, $\quad \mathbf{v} = 4\mathbf{i} + 8\mathbf{j} + \mathbf{k}$
9. $\mathbf{u} = 3\mathbf{i} + 5\mathbf{j} - 4\mathbf{k}$, $\quad \mathbf{v} = 4\mathbf{i} - 3\mathbf{j} + 5\mathbf{k}$
10. $\mathbf{u} = 2\mathbf{i} - 5\mathbf{j} + 3\mathbf{k}$, $\quad \mathbf{v} = -\mathbf{i} + 2\mathbf{j} + 7\mathbf{k}$

In each of Problems 11 through 14, find the work done by the force **F** when its point of application moves from A to B.

11. $\mathbf{F} = -32\mathbf{k}$, $\quad A:(-1, 1, 2)$, $\quad B:(3, 2, -1)$
12. $\mathbf{F} = 5\mathbf{i} - 2\mathbf{j} + 3\mathbf{k}$, $\quad A:(1, -2, 2)$, $\quad B:(3, 1, -1)$
13. $\mathbf{F} = -2\mathbf{i} + 3\mathbf{j} + 4\mathbf{k}$, $\quad A:(2, -1, -2)$, $\quad B:(-1, 2, 3)$
14. $\mathbf{F} = 3\mathbf{i} - 2\mathbf{j} - 3\mathbf{k}$, $\quad A:(-1, 2, 3)$, $\quad B:(2, 1, -1)$

In each of Problems 15 through 17, find a unit vector in the direction of **u**.

15. $\mathbf{u} = 2\mathbf{i} - 6\mathbf{j} + 3\mathbf{k}$
16. $\mathbf{u} = -\mathbf{i} + 2\mathbf{k}$
17. $\mathbf{u} = 3\mathbf{i} - 2\mathbf{j} + 7\mathbf{k}$

In each of Problems 18 through 21, find the value of k so that $\mathbf{v} - k\mathbf{u}$ is orthogonal to **u**. Also, find the value h so that $\mathbf{u} - h\mathbf{v}$ is orthogonal to **v**.

18. $\mathbf{u} = 2\mathbf{i} - \mathbf{j} + 2\mathbf{k}$, $\quad \mathbf{v} = 3\mathbf{i} + \mathbf{j} + 2\mathbf{k}$
19. $\mathbf{u} = 2\mathbf{i} - 3\mathbf{j} + 6\mathbf{k}$, $\quad \mathbf{v} = 7\mathbf{i} + 14\mathbf{k}$
20. $\mathbf{u} = 3\mathbf{i} + 4\mathbf{j} - 5\mathbf{k}$, $\quad \mathbf{v} = 9\mathbf{i} + 12\mathbf{j} - 5\mathbf{k}$
21. $\mathbf{u} = \mathbf{i} + 3\mathbf{j} - 2\mathbf{k}$, $\quad \mathbf{v} = 6\mathbf{i} + 10\mathbf{j} - 3\mathbf{k}$
22. Write a detailed proof of Theorem 12.
23. Show that if **u** and **v** are any vectors ($\neq \mathbf{0}$), then **u** and **v** make equal angles with **w** if

$$\mathbf{w} = \left(\frac{|\mathbf{v}|}{|\mathbf{u}| + |\mathbf{v}|}\right)\mathbf{u} + \left(\frac{|\mathbf{u}|}{|\mathbf{u}| + |\mathbf{v}|}\right)\mathbf{v}.$$

24. Show that if **u** and **v** are any vectors, the vectors $|\mathbf{v}|\mathbf{u} + |\mathbf{u}|\mathbf{v}$ and $|\mathbf{v}|\mathbf{u} - |\mathbf{u}|\mathbf{v}$ are orthogonal.

In each of Problems 25 through 27, determine the relation between g and h so that $g\mathbf{u} + h\mathbf{v}$ is orthogonal to **w**.

25. $\mathbf{u} = 3\mathbf{i} - 2\mathbf{j} + \mathbf{k}$, $\quad \mathbf{v} = \mathbf{i} + 2\mathbf{j} - 3\mathbf{k}$, $\quad \mathbf{w} = -\mathbf{i} + \mathbf{j} + 2\mathbf{k}$
26. $\mathbf{u} = 2\mathbf{i} + \mathbf{j} - 2\mathbf{k}$, $\quad \mathbf{v} = \mathbf{i} - \mathbf{j} + \mathbf{k}$, $\quad \mathbf{w} = -\mathbf{i} + 2\mathbf{j} + 3\mathbf{k}$
27. $\mathbf{u} = \mathbf{i} + 2\mathbf{j} - 3\mathbf{k}$, $\quad \mathbf{v} = 3\mathbf{i} + \mathbf{j} - \mathbf{k}$, $\quad \mathbf{w} = 4\mathbf{i} - \mathbf{j} + 2\mathbf{k}$

In each of Problems 28 through 30, determine g and h so that $\mathbf{w} - g\mathbf{u} - h\mathbf{v}$ is orthogonal to both **u** and **v**.

28. $\mathbf{u} = 2\mathbf{i} - \mathbf{j} + \mathbf{k}$, $\quad \mathbf{v} = \mathbf{i} + \mathbf{j} + 2\mathbf{k}$, $\quad \mathbf{w} = 2\mathbf{i} - \mathbf{j} + 4\mathbf{k}$
29. $\mathbf{u} = \mathbf{i} + \mathbf{j} - 2\mathbf{k}$, $\quad \mathbf{v} = -\mathbf{i} + 2\mathbf{j} + 3\mathbf{k}$, $\quad \mathbf{w} = 5\mathbf{i} + 8\mathbf{k}$
30. $\mathbf{u} = 3\mathbf{i} - 2\mathbf{j}$, $\quad \mathbf{v} = 2\mathbf{i} - \mathbf{k}$, $\quad \mathbf{w} = 4\mathbf{i} - 2\mathbf{k}$
31. If **u** and **v** are nonzero vectors, under what conditions is it true that

$$|\mathbf{u} + \mathbf{v}| = |\mathbf{u}| + |\mathbf{v}|?$$

32. Suppose that $\overrightarrow{AB}$, $\overrightarrow{AC}$, and $\overrightarrow{AD}$ are representatives of **u**, **v**, and $\mathbf{u} + \mathbf{v}$, respectively, with $|\mathbf{u}| = |\mathbf{v}|$. Show that $\overrightarrow{AD}$ bisects the angle between $\overrightarrow{AB}$ and $\overrightarrow{AC}$.
33. Let P be a vertex of a cube. Draw a diagonal of the cube from P and a diagonal of

one of the faces from P. Use vectors to find the cosine of the angle between these two diagonals.

34. Use vectors to find the cosine of the angle between two faces of a regular tetrahedron. (See Problem 22, Chapter 12, Section 2.)

7. THE VECTOR OR CROSS PRODUCT

We saw in Section 6 that the scalar product of two vectors **u** and **v** associates an ordinary number, i.e., a scalar, with each pair of vectors. The vector or cross product, on the other hand, associates a *vector* with each ordered pair of vectors. However, before defining the cross product, we shall discuss the notion of "right-handed" and "left-handed" triples of vectors.

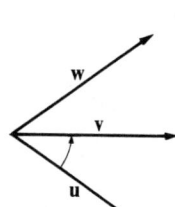

Figure 13–25

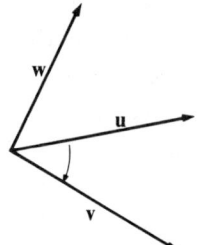

Figure 13–26

An **ordered triple** $\{\mathbf{u}, \mathbf{v}, \mathbf{w}\}$ of linearly independent vectors is said to be **right-handed** if the vectors are situated as in Fig. 13–25. If the ordered triple is situated as in Fig. 13–26, the vectors are said to form a **left-handed triple.** The notion of left-handed and right-handed triple is not defined if the vectors form a linearly dependent set.

Definition. *Two sets of ordered triples of vectors are said to be* **similarly oriented** *if and only if both sets are right-handed or both are left-handed. Otherwise they are* **oppositely oriented.**

Suppose $\{\mathbf{u}_1, \mathbf{v}_1, \mathbf{w}_1\}$ and $\{\mathbf{u}_2, \mathbf{v}_2, \mathbf{w}_2\}$ are ordered linearly independent sets of triples. From Theorem 11, it follows that we may express $\mathbf{u}_2, \mathbf{v}_2,$ and $\mathbf{w}_2$ in terms of $\mathbf{u}_1, \mathbf{v}_1,$ and $\mathbf{w}_1$ by equations of the form

$$\mathbf{u}_2 = a_{11}\mathbf{u}_1 + a_{12}\mathbf{v}_1 + a_{13}\mathbf{w}_1,$$
$$\mathbf{v}_2 = a_{21}\mathbf{u}_1 + a_{22}\mathbf{v}_1 + a_{23}\mathbf{w}_1,$$
$$\mathbf{w}_2 = a_{31}\mathbf{u}_1 + a_{32}\mathbf{v}_1 + a_{33}\mathbf{w}_1.$$

We denote by D the determinant

$$D = \begin{vmatrix} a_{11} & a_{12} & a_{13} \\ a_{21} & a_{22} & a_{23} \\ a_{31} & a_{32} & a_{33} \end{vmatrix}.$$

Although the proof is beyond the scope of this book, it is a fact that the two triples above are similarly oriented if and only if $D > 0$; they are oppositely oriented if and only if $D < 0$. Note that the determinant cannot be zero, for then

$\mathbf{u}_2$, $\mathbf{v}_2$, and $\mathbf{w}_2$ would not be linearly independent. (See the proof of Theorem 5' in Appendix 6.)

It is also true that if $\{\mathbf{u}_1, \mathbf{v}_1, \mathbf{w}_1\}$ and $\{\mathbf{u}_2, \mathbf{v}_2, \mathbf{w}_2\}$ are similarly oriented and if $\{\mathbf{u}_2, \mathbf{v}_2, \mathbf{w}_2\}$ and $\{\mathbf{u}_3, \mathbf{v}_3, \mathbf{w}_3\}$ are similarly oriented, then $\{\mathbf{u}_1, \mathbf{v}_1, \mathbf{w}_1\}$ and $\{\mathbf{u}_3, \mathbf{v}_3, \mathbf{w}_3\}$ are similarly oriented.

The facts above lead to the following result.

Theorem 15. *If $\{\mathbf{u}, \mathbf{v}, \mathbf{w}\}$ is a right-handed triple, then (i) $\{\mathbf{v}, \mathbf{u}, -\mathbf{w}\}$ is a right-handed triple, and (ii) $\{c_1\mathbf{u}, c_2\mathbf{v}, c_3\mathbf{w}\}$ is a right-handed triple provided that $c_1 c_2 c_3 > 0$.*

To prove (i) we apply the above determinant condition on similar orientation by regarding $\{\mathbf{u}, \mathbf{v}, \mathbf{w}\}$ as $\{\mathbf{u}_1, \mathbf{v}_1, \mathbf{w}_1\}$ and $\{\mathbf{v}, \mathbf{u}, -\mathbf{w}\}$ as $\{\mathbf{u}_2, \mathbf{v}_2, \mathbf{w}_2\}$. To establish (ii) we regard $\{c_1\mathbf{u}, c_2\mathbf{v}, c_3\mathbf{w}\}$ as $\{\mathbf{u}_2, \mathbf{v}_2, \mathbf{w}_2\}$. The details are left to the student.

Definition. *Given the vectors $\mathbf{u}$ and $\mathbf{v}$, the **vector** or **cross product** $\mathbf{u} \times \mathbf{v}$ is defined as follows:*

i) *if either $\mathbf{u}$ or $\mathbf{v}$ is $\mathbf{0}$, then*
$$\mathbf{u} \times \mathbf{v} = \mathbf{0};$$

ii) *if $\mathbf{u}$ is proportional to $\mathbf{v}$, then*
$$\mathbf{u} \times \mathbf{v} = \mathbf{0};$$

iii) *otherwise,*
$$\mathbf{u} \times \mathbf{v} = \mathbf{w}$$

where $\mathbf{w}$ has the three properties: (a) it is orthogonal to both $\mathbf{u}$ and $\mathbf{v}$; (b) it has magnitude $|\mathbf{w}| = |\mathbf{u}| |\mathbf{v}| \sin \theta$, where θ is the angle between $\mathbf{u}$ and $\mathbf{v}$, and (c) it is directed so that $\{\mathbf{u}, \mathbf{v}, \mathbf{w}\}$ is a right-handed triple.

Remark. We shall always assume that any coordinate triple $\{\mathbf{i}, \mathbf{j}, \mathbf{k}\}$ is right-handed. (We have assumed this up to now without pointing out this fact specifically.)

The proofs of the next two theorems are given in Appendix 6.

Theorem 16. *Suppose that $\mathbf{u}$ and $\mathbf{v}$ are any vectors, that $\{\mathbf{i}, \mathbf{j}, \mathbf{k}\}$ is a right-handed triple, and that t is any number. Then*

i) $\mathbf{v} \times \mathbf{u} = -(\mathbf{u} \times \mathbf{v})$,
ii) $(t\mathbf{u}) \times \mathbf{v} = t(\mathbf{u} \times \mathbf{v}) = \mathbf{u} \times (t\mathbf{v})$,
iii) $\mathbf{i} \times \mathbf{j} = -\mathbf{j} \times \mathbf{i} = \mathbf{k}$,
    $\mathbf{j} \times \mathbf{k} = -\mathbf{k} \times \mathbf{j} = \mathbf{i}$,
    $\mathbf{k} \times \mathbf{i} = -\mathbf{i} \times \mathbf{k} = \mathbf{j}$,
iv) $\mathbf{i} \times \mathbf{i} = \mathbf{j} \times \mathbf{j} = \mathbf{k} \times \mathbf{k} = \mathbf{0}$.

450   VECTORS IN THE PLANE AND IN THREE SPACE

Theorem 17. *If* $\mathbf{u}$, $\mathbf{v}$, $\mathbf{w}$ *are any vectors, then*

i) $\mathbf{u} \times (\mathbf{v} + \mathbf{w}) = (\mathbf{u} \times \mathbf{v}) + (\mathbf{u} \times \mathbf{w})$ *and*
ii) $(\mathbf{v} + \mathbf{w}) \times \mathbf{u} = (\mathbf{v} \times \mathbf{u}) + (\mathbf{w} \times \mathbf{u})$.

With the aid of Theorems 16 and 17, the next theorem, an extremely useful one, is easily established.

Theorem 18. *If*

$$\mathbf{u} = a_1\mathbf{i} + a_2\mathbf{j} + a_3\mathbf{k} \quad \text{and} \quad \mathbf{v} = b_1\mathbf{i} + b_2\mathbf{j} + b_3\mathbf{k},$$

then

$$\mathbf{u} \times \mathbf{v} = (a_2 b_3 - a_3 b_2)\mathbf{i} + (a_3 b_1 - a_1 b_3)\mathbf{j} + (a_1 b_2 - a_2 b_1)\mathbf{k}. \tag{1}$$

Proof. By using the laws in Theorems 16 and 17 we obtain (being careful to keep the order of the factors)

$$\begin{aligned}\mathbf{u} \times \mathbf{v} = &\; a_1 b_1 (\mathbf{i} \times \mathbf{i}) + a_1 b_2 (\mathbf{i} \times \mathbf{j}) + a_1 b_3 (\mathbf{i} \times \mathbf{k}) \\ &+ a_2 b_1 (\mathbf{j} \times \mathbf{i}) + a_2 b_2 (\mathbf{j} \times \mathbf{j}) + a_2 b_3 (\mathbf{j} \times \mathbf{k}) \\ &+ a_3 b_1 (\mathbf{k} \times \mathbf{i}) + a_3 b_2 (\mathbf{k} \times \mathbf{j}) + a_3 b_3 (\mathbf{k} \times \mathbf{k}).\end{aligned}$$

The result follows from Theorem 16, parts (iii) and (iv) by collecting terms. ◀

The formula (1) above is useful in calculating the cross product. The following *symbolic form* is a great aid in remembering the formula. We write

$$\mathbf{u} \times \mathbf{v} = \begin{vmatrix} \mathbf{i} & \mathbf{j} & \mathbf{k} \\ a_1 & a_2 & a_3 \\ b_1 & b_2 & b_3 \end{vmatrix},$$

where it is understood that this "determinant" is to be expanded formally according to its first row. The student may easily verify that when the above expression is expanded, it is equal to (1).

Example 1. Find $\mathbf{u} \times \mathbf{v}$ if $\mathbf{u} = 2\mathbf{i} - 3\mathbf{j} + \mathbf{k}$, $\mathbf{v} = \mathbf{i} + \mathbf{j} - 2\mathbf{k}$.

Solution. Carrying out the formal expansion, we obtain

$$\begin{vmatrix} \mathbf{i} & \mathbf{j} & \mathbf{k} \\ 2 & -3 & 1 \\ 1 & 1 & -2 \end{vmatrix} = \begin{vmatrix} -3 & 1 \\ 1 & -2 \end{vmatrix} \mathbf{i} - \begin{vmatrix} 2 & 1 \\ 1 & -2 \end{vmatrix} \mathbf{j} + \begin{vmatrix} 2 & -3 \\ 1 & 1 \end{vmatrix} \mathbf{k} = 5\mathbf{i} + 5\mathbf{j} + 5\mathbf{k}. \;\blacktriangleleft$$

Remarks. In mechanics the cross product is used for the computation of the vector moment of a force $\mathbf{F}$ applied at a point B, about a point A. There are also applications of cross product to problems in electricity and magnetism. However, we shall confine our attention to applications in geometry.

Theorem 19. *The area of a parallelogram with adjacent sides AB and AC is given by**

$$|\mathbf{v}(\overrightarrow{AB}) \times \mathbf{v}(\overrightarrow{AC})|.$$

The area of $\triangle ABC$ *is then* $\frac{1}{2}|\mathbf{v}(\overrightarrow{AB}) \times \mathbf{v}(\overrightarrow{AC})|.$

Proof. From Fig. 13–27, we see that the area of the parallelogram is

$$|AB|\, h = |AB|\,|AC|\sin\theta.$$

The result then follows from the definition of cross product.

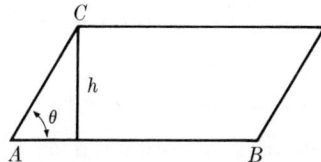

Figure 13–27

Example 2. Find the area of $\triangle ABC$ with $A(-2, 1, 3)$, $B(1, -1, 1)$, $C(3, -2, 4)$.

Solution. We have $\mathbf{v}(\overrightarrow{AB}) = 3\mathbf{i} - 2\mathbf{j} - 2\mathbf{k}$, $\mathbf{v}(\overrightarrow{AC}) = 5\mathbf{i} - 3\mathbf{j} + \mathbf{k}$. From Theorem 19 we obtain

$$\mathbf{v}(\overrightarrow{AB}) \times \mathbf{v}(\overrightarrow{AC}) = -8\mathbf{i} - 13\mathbf{j} + \mathbf{k}$$

and

$$\tfrac{1}{2}|-8\mathbf{i} - 13\mathbf{j} + \mathbf{k}| = \tfrac{1}{2}\sqrt{64 + 169 + 1} = \tfrac{3}{2}\sqrt{26}. \qquad \blacktriangleleft$$

The vector product may be used to find the equation of a plane through three points. The next example illustrates the technique.

Example 3. Find the equation of the plane through the points $A(-1, 1, 2)$, $B(1, -2, 1)$, $C(2, 2, 4)$.

Solution. A vector normal to the plane will be perpendicular to both the vectors

$$\mathbf{v}(\overrightarrow{AB}) = 2\mathbf{i} - 3\mathbf{j} - \mathbf{k} \quad \text{and} \quad \mathbf{v}(\overrightarrow{AC}) = 3\mathbf{i} + \mathbf{j} + 2\mathbf{k}.$$

One such vector is the cross product

$$\mathbf{v}(\overrightarrow{AB}) \times \mathbf{v}(\overrightarrow{AC}) = -5\mathbf{i} - 7\mathbf{j} + 11\mathbf{k}.$$

Therefore the numbers $-5, -7, 11$ form a set of *attitude numbers* (see Chapter 12, Section 4) of the desired plane. Using $A(-1, 1, 2)$ as a point on the plane, we get for the equation

$$-5(x + 1) - 7(y - 1) + 11(z - 2) = 0$$

or

$$5x + 7y - 11z + 20 = 0. \qquad \blacktriangleleft$$

---

* The notation $\mathbf{v}(\overrightarrow{AB})$ to indicate a vector $\mathbf{v}$ with representative $\overrightarrow{AB}$ was introduced in Section 3.

Example 4. Find the perpendicular distance between the skew lines
$$L_1: \frac{x+2}{2} = \frac{y-1}{3} = \frac{z+1}{-1}, \quad L_2: \frac{x-1}{-1} = \frac{y+1}{2} = \frac{z-2}{4}.$$

Solution. The vector
$$\mathbf{v}_1 = 2\mathbf{i} + 3\mathbf{j} - \mathbf{k}$$
is a vector along L_1. The vector
$$\mathbf{v}_2 = -\mathbf{i} + 2\mathbf{j} + 4\mathbf{k}$$
is a vector along L_2. A vector perpendicular to both $\mathbf{v}_1$ and $\mathbf{v}_2$ (i.e., to both L_1 and L_2) is
$$\mathbf{v}_1 \times \mathbf{v}_2 = 14\mathbf{i} - 7\mathbf{j} + 7\mathbf{k}.$$

Call this common perpendicular $\mathbf{w}$. The desired length may be obtained as a *projection*. Select any point on L_1 (call it P_1) and any point on L_2 (call it P_2). Then the desired length is the projection of the vector $\mathbf{v}(\overrightarrow{P_1P_2})$ on $\mathbf{w}$. To get this, we select $P_1(-2, 1, -1)$ on L_1 and $P_2(1, -1, 2)$ on L_2; and so
$$\mathbf{v}(\overrightarrow{P_1P_2}) = 3\mathbf{i} - 2\mathbf{j} + 3\mathbf{k}.$$

Therefore,
$$\text{Projection of } \mathbf{v}(\overrightarrow{P_1P_2}) \text{ on } \mathbf{w} = \frac{\mathbf{v}(\overrightarrow{P_1P_2}) \cdot \mathbf{w}}{|\mathbf{w}|}$$
$$= \frac{3 \cdot 14 + (-2)(-7) + 3(7)}{7\sqrt{6}} = \frac{11}{\sqrt{6}}. \quad \blacktriangleleft$$

PROBLEMS

In each of Problems 1 through 6, find the cross product $\mathbf{u} \times \mathbf{v}$.
1. $\mathbf{u} = \mathbf{i} + 3\mathbf{j} - \mathbf{k}$, $\quad \mathbf{v} = 2\mathbf{i} - \mathbf{j} + \mathbf{k}$
2. $\mathbf{u} = 4\mathbf{i} - 2\mathbf{j} + 3\mathbf{k}$, $\quad \mathbf{v} = -\mathbf{i} - 2\mathbf{j} - \mathbf{k}$
3. $\mathbf{u} = -\mathbf{i} + 2\mathbf{j}$, $\quad \mathbf{v} = \mathbf{i} + 3\mathbf{j} - 2\mathbf{k}$
4. $\mathbf{u} = 3\mathbf{j} + 2\mathbf{k}$, $\quad \mathbf{v} = 2\mathbf{i} - 3\mathbf{j}$
5. $\mathbf{u} = -2\mathbf{i} + 4\mathbf{j} + 5\mathbf{k}$, $\quad \mathbf{v} = 4\mathbf{i} + 5\mathbf{k}$
6. $\mathbf{u} = 2\mathbf{i} - 3\mathbf{j} + \mathbf{k}$, $\quad \mathbf{v} = 4\mathbf{k}$

In Problems 7 through 11, find in each case the area of $\triangle ABC$ and the equation of the plane through A, B, and C. Use vector methods.
7. $A(1, -2, 3)$, $\quad B(3, 1, 2)$, $\quad C(2, 3, -1)$
8. $A(3, 2, -2)$, $\quad B(4, 1, 2)$, $\quad C(1, 2, 3)$
9. $A(2, -1, 1)$, $\quad B(3, 2, -1)$, $\quad C(-1, 3, 2)$
10. $A(1, -2, 3)$, $\quad B(2, -1, 1)$, $\quad C(4, 2, -1)$
11. $A(-2, 3, 1)$, $\quad B(4, 2, -2)$, $\quad C(2, 0, 1)$

In Problems 12 through 14, find in each case the perpendicular distance between the given lines.

12. $\dfrac{x+1}{2} = \dfrac{y-3}{-3} = \dfrac{z+2}{4}$; $\dfrac{x-2}{3} = \dfrac{y+1}{2} = \dfrac{z-1}{5}$

13. $\dfrac{x-1}{3} = \dfrac{y+1}{2} = \dfrac{z-1}{5}$; $\dfrac{x+2}{4} = \dfrac{y-1}{3} = \dfrac{z+1}{-2}$

14. $\dfrac{x+1}{2} = \dfrac{y-1}{-4} = \dfrac{z+2}{3}$; $\dfrac{x}{3} = \dfrac{y}{5} = \dfrac{z-2}{-2}$

In Problems 15 through 19, use vector methods to find, in each case, the equations in symmetric form of the line through the given point P and parallel to the two given planes.

15. $P(-1, 3, 2)$, $3x - 2y + 4z + 2 = 0$, $2x + y - z = 0$
16. $P(2, 3, -1)$, $x + 2y + 2z - 4 = 0$, $2x + y - 3z + 5 = 0$
17. $P(1, -2, 3)$, $3x + y - 2z + 3 = 0$, $2x + 3y + z - 6 = 0$
18. $P(-1, 0, -2)$, $2x + 3y - z + 4 = 0$, $3x - 2y + 2z - 5 = 0$
19. $P(3, 0, 1)$, $x + 2y = 0$, $3y - z = 0$

In Problems 20 and 21, find in each case equations in symmetric form of the line of intersection of the given planes. Use the method of vector products.

20. $2(x - 1) + 3(y + 1) - 4(z - 2) = 0$
    $3(x - 1) - 4(y + 1) + 2(z - 2) = 0$
21. $3(x + 2) - 2(y - 1) + 2(z + 1) = 0$
    $(x + 2) + 2(y - 1) - 3(z + 1) = 0$

In each of Problems 22 through 26, find an equation of the plane through the given point or points and parallel to the given line or lines.

22. $(1, 3, 2)$; $\dfrac{x+1}{2} = \dfrac{y-2}{-1} = \dfrac{z+3}{3}$; $\dfrac{x-2}{1} = \dfrac{y+1}{-2} = \dfrac{z+2}{2}$

23. $(2, -1, -3)$; $\dfrac{x-1}{3} = \dfrac{y+2}{2} = \dfrac{z}{-4}$; $\dfrac{x}{2} = \dfrac{y-1}{-3} = \dfrac{z-2}{2}$

24. $(2, 1, -2)$; $(1, -1, 3)$; $\dfrac{x+1}{3} = \dfrac{y-1}{2} = \dfrac{z-2}{2}$

25. $(1, -2, 3)$; $(-1, 2, -1)$; $\dfrac{x-2}{2} = \dfrac{y+1}{3} = \dfrac{z-1}{4}$

26. $(0, 1, 2)$; $(2, 0, 1)$; $\dfrac{x-1}{3} = \dfrac{y+1}{0} = \dfrac{z+1}{1}$

In Problems 27 through 29, find in each case the equation of the plane through the line L_1 which also satisfies the additional condition.

27. $L_1: \dfrac{x-1}{2} = \dfrac{y+1}{3} = \dfrac{z-2}{1}$;   through $(2, 1, 1)$

28. $L_1: \dfrac{x-2}{2} = \dfrac{y-2}{3} = \dfrac{z-1}{-2}$;   parallel to $\dfrac{x+1}{3} = \dfrac{y-1}{2} = \dfrac{z+1}{1}$

29. $L_1: \dfrac{x+1}{1} = \dfrac{y-1}{2} = \dfrac{z-2}{-2}$;   perpendicular to $2x + 3y - z + 4 = 0$

In Problems 30 and 31, find the equation of the plane through the given points and perpendicular to the given planes.

30. $(1, 2, -1)$; $2x - 3y + 5z - 1 = 0$; $3x + 2y + 4z + 6 = 0$
31. $(-1, 3, 2)$; $(1, 6, 1)$; $3x - y + 4z - 7 = 0$

In Problems 32 and 33, find equations in symmetric form of the line through the given point P, which is perpendicular to and intersects the given line. Use the cross product.

32. $P(3, 3, -1)$; $\dfrac{x}{-1} = \dfrac{y-3}{1} = \dfrac{z+1}{1}$

33. $P(3, -2, 0)$; $\dfrac{x-4}{3} = \dfrac{y-4}{-4} = \dfrac{z-5}{-1}$

34. Suppose that $\mathbf{a}$ is a nonzero vector. If $\mathbf{a} \cdot \mathbf{x} = \mathbf{a} \cdot \mathbf{y}$ and $\mathbf{a} \times \mathbf{x} = \mathbf{a} \times \mathbf{y}$, is it true that $\mathbf{x} = \mathbf{y}$? Justify your answer.

35. Given that $\mathbf{u} + \mathbf{v} + \mathbf{w} - \mathbf{r} = \mathbf{0}$ and $\mathbf{u} - \mathbf{v} + \mathbf{w} + 2\mathbf{r} = \mathbf{0}$. Prove that $\mathbf{r} \times \mathbf{v} = \mathbf{0}$, that $\mathbf{v} \times \mathbf{w} = \frac{3}{2}\mathbf{r} \times \mathbf{w}$, and that

$$2\mathbf{u} \times \mathbf{w} = \mathbf{w} \times \mathbf{r} = \mathbf{r} \times \mathbf{u}.$$

8. PRODUCTS OF THREE VECTORS

Since two types of multiplication, the scalar product and the cross product, may be performed on vectors, we can combine three vectors in several ways. For example, we can form the product

$$(\mathbf{u} \times \mathbf{v}) \cdot \mathbf{w}$$

and the product

$$(\mathbf{u} \times \mathbf{v}) \times \mathbf{w}.$$

Also, we can consider the combinations

$$\mathbf{u} \cdot (\mathbf{v} \times \mathbf{w}) \quad \text{and} \quad \mathbf{u} \times (\mathbf{v} \times \mathbf{w}).$$

The next theorem gives a simple rule for computing $(\mathbf{u} \times \mathbf{v}) \cdot \mathbf{w}$ and also an elegant geometric interpretation of the quantity $|(\mathbf{u} \times \mathbf{v}) \cdot \mathbf{w}|$.

Theorem 20. *Suppose that $\mathbf{u}_1, \mathbf{u}_2, \mathbf{u}_3$, are vectors and that the points A, B, C, D are chosen so that*

$$\mathbf{v}(\overrightarrow{AB}) = \mathbf{u}_1, \qquad \mathbf{v}(\overrightarrow{AC}) = \mathbf{u}_2, \qquad \mathbf{v}(\overrightarrow{AD}) = \mathbf{u}_3.$$

Then

i) *the quantity $|(\mathbf{u}_1 \times \mathbf{u}_2) \cdot \mathbf{u}_3|$ is the volume of the parallelepiped with one vertex at A and adjacent vertices at $B, C,$ and D. (See Fig. 13–28.) This volume is zero if and only if the four points A, B, C, D lie in a plane;*

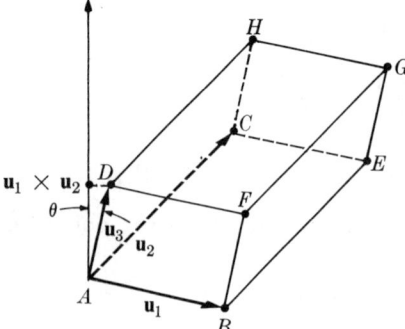

Figure 13–28

13-8 PRODUCTS OF THREE VECTORS

ii) *if* $\{\mathbf{i}, \mathbf{j}, \mathbf{k}\}$ *is a right-handed coordinate triple and if*

$$\mathbf{u}_1 = a_1\mathbf{i} + b_1\mathbf{j} + c_1\mathbf{k}, \qquad \mathbf{u}_2 = a_2\mathbf{i} + b_2\mathbf{j} + c_2\mathbf{k},$$
$$\mathbf{u}_3 = a_3\mathbf{i} + b_3\mathbf{j} + c_3\mathbf{k},$$

then

$$(\mathbf{u}_1 \times \mathbf{u}_2) \cdot \mathbf{u}_3 = \begin{vmatrix} a_1 & b_1 & c_1 \\ a_2 & b_2 & c_2 \\ a_3 & b_3 & c_3 \end{vmatrix};$$

iii) $(\mathbf{u}_1 \times \mathbf{u}_2) \cdot \mathbf{u}_3 = \mathbf{u}_1 \cdot (\mathbf{u}_2 \times \mathbf{u}_3)$.

Proof. To prove (i), note that $|\mathbf{u}_1 \times \mathbf{u}_2|$ is the area of the parallelogram $ABEC$ and that

$$|(\mathbf{u}_1 \times \mathbf{u}_2) \cdot \mathbf{u}_3| = |\mathbf{u}_1 \times \mathbf{u}_2| \, |\mathbf{u}_3| \, |\cos \theta|,$$

where θ is the angle between the two vectors $\mathbf{u}_3$ and $\mathbf{u}_1 \times \mathbf{u}_2$. The quantity $|\mathbf{u}_3| \, |\cos \theta|$ is the length of the projection of $\mathbf{u}_3$ on the normal to the plane of $ABEC$. Clearly, $(\mathbf{u}_1 \times \mathbf{u}_2) \cdot \mathbf{u}_3 = 0 \Leftrightarrow \mathbf{u}_1 \times \mathbf{u}_2 = \mathbf{0}$ or $\mathbf{u}_3 = \mathbf{0}$ or $\cos \theta = 0$. If $\cos \theta = 0$, then $\mathbf{u}_3$ is parallel to the plane of $\mathbf{u}_1$ and $\mathbf{u}_2$ and all four points lie in a plane. The proof of parts (ii) and (iii) follow from Theorems 13 and 18 and are left to the student.

Theorem 21. *If* $\mathbf{u}$, $\mathbf{v}$, *and* $\mathbf{w}$ *are any vectors, then*

i) $(\mathbf{u} \times \mathbf{v}) \times \mathbf{w} = (\mathbf{u} \cdot \mathbf{w})\mathbf{v} - (\mathbf{v} \cdot \mathbf{w})\mathbf{u}$,
ii) $\mathbf{u} \times (\mathbf{v} \times \mathbf{w}) = (\mathbf{u} \cdot \mathbf{w})\mathbf{v} - (\mathbf{u} \cdot \mathbf{v})\mathbf{w}$.

Proof. If $\mathbf{u}$ and $\mathbf{v}$ are proportional or if $\mathbf{w}$ is orthogonal to both $\mathbf{u}$ and $\mathbf{v}$, then both sides of (i) are zero. Otherwise, we see that $(\mathbf{u} \times \mathbf{v}) \times \mathbf{w}$ is orthogonal to the perpendicular to the plane determined by $\mathbf{u}$ and $\mathbf{v}$. Hence $(\mathbf{u} \times \mathbf{v}) \times \mathbf{w}$ is in the plane of $\mathbf{u}$ and $\mathbf{v}$. We choose a right-handed coordinate triple $\{\mathbf{i}, \mathbf{j}, \mathbf{k}\}$ so that $\mathbf{i}$ is in the direction of $\mathbf{u}$ and $\mathbf{j}$ is in the plane of $\mathbf{u}$ and $\mathbf{v}$. Then there are numbers a_1, a_2, b_2, etc., so that

$$\mathbf{u} = a_1\mathbf{i}, \qquad \mathbf{v} = a_2\mathbf{i} + b_2\mathbf{j}, \qquad \mathbf{w} = a_3\mathbf{i} + b_3\mathbf{j} + c_3\mathbf{k}.$$

The student may now compute both sides of (i) to see that they are equal. The proof of (ii) is left to the student.

Example 1. Given $A(3, -1, 2)$, $B(1, 2, -2)$, $C(2, 1, -2)$, and $D(-1, 3, 2)$, find the volume of the parallelepiped having AB, AC, and AD as edges.

Solution. We have

$$\mathbf{u}_1 = \mathbf{v}(\overrightarrow{AB}) = -2\mathbf{i} + 3\mathbf{j} - 4\mathbf{k},$$
$$\mathbf{u}_2 = \mathbf{v}(\overrightarrow{AC}) = -\mathbf{i} + 2\mathbf{j} - 4\mathbf{k},$$
$$\mathbf{u}_3 = \mathbf{v}(\overrightarrow{AD}) = -4\mathbf{i} + 4\mathbf{j}.$$

We compute $\mathbf{u}_2 \times \mathbf{u}_3 = 16\mathbf{i} + 16\mathbf{j} + 4\mathbf{k}$. Therefore
$$|\mathbf{u}_1 \cdot (\mathbf{u}_2 \times \mathbf{u}_3)| = |-32 + 48 - 16| = 0.$$
Hence the four points are in a plane. The volume is zero. ◄

Example 2. Find the equations of the line through the point $(3, -2, 1)$ perpendicular to the line L (and intersecting it) given by
$$L = \left\{(x, y, z): \frac{x-2}{2} = \frac{y+1}{-2} = \frac{z}{1}\right\}.$$

Solution. Let $P_0(3, -2, 1)$ and $P_1(2, -1, 0)$,
$$\mathbf{u} = 2\mathbf{i} - 2\mathbf{j} + \mathbf{k},$$
$$\mathbf{v} = \mathbf{v}(\overrightarrow{P_0 P_1}) = -\mathbf{i} + \mathbf{j} - \mathbf{k}.$$
The plane containing L and P_0 has a normal perpendicular to $\mathbf{u}$ and $\mathbf{v}$. Hence this normal is proportional to $\mathbf{u} \times \mathbf{v}$. The desired line is in this plane and perpendicular to L. Therefore it has a direction $\mathbf{w}$ perpendicular to $\mathbf{u}$ and $\mathbf{u} \times \mathbf{v}$. Thus for some number c, we have
$$c\mathbf{w} = \mathbf{u} \times (\mathbf{u} \times \mathbf{v}) = (\mathbf{u} \cdot \mathbf{v})\mathbf{u} - (\mathbf{u} \cdot \mathbf{u})\mathbf{v}$$
$$= -5(2\mathbf{i} - 2\mathbf{j} + \mathbf{k}) - 9(-\mathbf{i} + \mathbf{j} - \mathbf{k}) = -\mathbf{i} + \mathbf{j} + 4\mathbf{k}.$$
Consequently, the desired line has equations
$$\frac{x-3}{-1} = \frac{y+2}{1} = \frac{z-1}{4}.$$ ◄

PROBLEMS

In Problems 1 through 4, find the volume of the parallelepiped having edges AB, AC, and AD, or else show that A, B, C, and D lie on a plane or on a line. If they lie on a plane, find its equation; if they lie on a line, find its equations.

1. $A = (2, -1, 3)$, $B = (-1, 2, 2)$, $C = (1, 0, 1)$, $D = (4, 1, -1)$
2. $A = (3, 1, -2)$, $B = (1, 2, 1)$, $C = (2, -1, 3)$, $D = (4, 3, -7)$
3. $A = (1, 2, -3)$, $B = (3, 1, -2)$, $C = (-1, 3, 1)$, $D = (-3, 4, 3)$
4. $A = (-1, -2, 2)$, $B = (2, -1, 1)$, $C = (0, 1, 3)$, $D = (3, 2, -1)$
5. Prove Theorem 20, parts (ii) and (iii).
6. Complete the proof of Theorem 21.

In Problems 7 through 10, compute $(\mathbf{u} \times \mathbf{v}) \times \mathbf{w}$ directly and by using Theorem 21.

7. $\mathbf{u} = 2\mathbf{i} + 3\mathbf{j} - \mathbf{k}$, $\mathbf{v} = \mathbf{i} - 2\mathbf{j} + \mathbf{k}$, $\mathbf{w} = -\mathbf{i} + \mathbf{j} + 2\mathbf{k}$
8. $\mathbf{u} = 3\mathbf{i} - 2\mathbf{j} + \mathbf{k}$, $\mathbf{v} = \mathbf{i} + \mathbf{j} + 2\mathbf{k}$, $\mathbf{w} = 2\mathbf{i} - \mathbf{j} + 3\mathbf{k}$
9. $\mathbf{u} = \mathbf{i} + 2\mathbf{j} - 3\mathbf{k}$, $\mathbf{v} = -\mathbf{i} + \mathbf{j} - 2\mathbf{k}$, $\mathbf{w} = 3\mathbf{i} - \mathbf{j} + \mathbf{k}$
10. $\mathbf{u} = 2\mathbf{i} - \mathbf{j} + 3\mathbf{k}$, $\mathbf{v} = \mathbf{i} + 2\mathbf{j} + \mathbf{k}$, $\mathbf{w} = 3\mathbf{i} - 2\mathbf{j} - \mathbf{k}$
11. Show that every vector $\mathbf{v}$ satisfies the identity
$$\mathbf{i} \times (\mathbf{v} \times \mathbf{i}) + \mathbf{j} \times (\mathbf{v} \times \mathbf{j}) + \mathbf{k} \times (\mathbf{v} \times \mathbf{k}) = 2\mathbf{v}.$$

In Problems 12 through 14, find, in each case, the equations of the line through the given point and perpendicular to the given line and intersecting it.

12. $(1, 3, -2)$, $\dfrac{x-2}{3} = \dfrac{y+1}{-2} = \dfrac{z-1}{4}$

13. $(2, -1, 3)$, $\dfrac{x+1}{2} = \dfrac{y-2}{3} = \dfrac{z+1}{-5}$

14. $(-1, 2, 4)$, $\dfrac{x-1}{4} = \dfrac{y+2}{-3} = \dfrac{z-1}{2}$

In Problems 15 and 16, express $(\mathbf{t} \times \mathbf{u}) \times (\mathbf{v} \times \mathbf{w})$ in terms of $\mathbf{v}$ and $\mathbf{w}$.

15. $\mathbf{t} = \mathbf{i} + \mathbf{j} - 2\mathbf{k}$, $\mathbf{u} = 3\mathbf{i} - \mathbf{j} + 2\mathbf{k}$, $\mathbf{v} = 2\mathbf{i} + 2\mathbf{j} - \mathbf{k}$, $\mathbf{w} = -\mathbf{i} + \mathbf{j} + 2\mathbf{k}$

16. $\mathbf{t} = 2\mathbf{i} - \mathbf{j} + \mathbf{k}$, $\mathbf{u} = \mathbf{i} + 2\mathbf{j} - 3\mathbf{k}$, $\mathbf{v} = 3\mathbf{i} + \mathbf{j} + 2\mathbf{k}$, $\mathbf{w} = -\mathbf{i} + 2\mathbf{j} - 2\mathbf{k}$

17. Derive a formula expressing $(\mathbf{t} \times \mathbf{u}) \times (\mathbf{v} \times \mathbf{w})$ in terms of $\mathbf{v}$ and $\mathbf{w}$.

18. Given $\mathbf{a} = \mathbf{i} - \mathbf{j} + \mathbf{k}$, $\mathbf{b} = 2\mathbf{i} + 3\mathbf{j} + \mathbf{k}$, $p = 1$. Solve the equations $\mathbf{a} \cdot \mathbf{v} = p$, $\mathbf{a} \times \mathbf{v} = \mathbf{b}$ for $\mathbf{v}$.

19. Given that $\mathbf{a} \cdot \mathbf{b} = 0$, $\mathbf{a} \neq \mathbf{0}$, $\mathbf{b} \neq \mathbf{0}$, find a formula for the solution $\mathbf{v}$ of the equations
$$\mathbf{a} \cdot \mathbf{v} = p, \qquad \mathbf{a} \times \mathbf{v} = \mathbf{b}.$$
[*Hint:* Note that $\mathbf{a}$, $\mathbf{b}$, and $\mathbf{a} \times \mathbf{b}$ are mutually orthogonal.]

20. Show that for any vectors $\mathbf{u}, \mathbf{v}, \mathbf{w}$, we have

   i) $(\mathbf{u} \pm \mathbf{v}) \cdot [(\mathbf{u} + \mathbf{v}) \times (\mathbf{u} - \mathbf{v})] = 0$
   ii) $(\mathbf{u} + \mathbf{v}) \cdot [(\mathbf{u} \times \mathbf{w}) \times (\mathbf{u} + \mathbf{v})] = 0$.

21. If $\mathbf{u}, \mathbf{v}, \mathbf{w}$ are any vectors show that
$$[\mathbf{u} \times (\mathbf{v} \times \mathbf{w})] + [\mathbf{v} \times (\mathbf{w} \times \mathbf{u})] + [\mathbf{w} \times (\mathbf{u} \times \mathbf{v})] = \mathbf{0}.$$

*22. Let $\mathscr{F}$ be a collection of objects with $\mathbf{A}, \mathbf{B}, \mathbf{C}$ members of $\mathscr{F}$. Let $\oplus$ be an operation between members of $\mathscr{F}$ satisfying the relation
$$(\mathbf{A} \oplus \mathbf{B}) \oplus \mathbf{C} = \alpha \mathbf{B} - \beta \mathbf{A}.$$
State general conditions on the numbers α and β such that the formula
$$[(\mathbf{A} \oplus \mathbf{B}) \oplus \mathbf{C}] + [(\mathbf{B} \oplus \mathbf{C}) \oplus \mathbf{A}] + [(\mathbf{C} \oplus \mathbf{A}) \oplus \mathbf{B}] = \mathbf{0}$$
should hold.

9. VECTOR FUNCTIONS IN THE PLANE AND THEIR DERIVATIVES

We recall that a function is defined as an ordered pair in which the second member (an element of the range) is a number. We now extend the definition of function to the case where the range may be a vector. In this section we restrict ourselves to vectors in the plane. The more general situation in which the range consists of vectors in three-space is discussed in Section 10.

We consider the collection of all vectors in the plane and we denote this set by V_2. That is, any vector in the plane is a member of V_2.

Definition. *A **vector function** is the collection of ordered pairs $(t, \mathbf{v})$ in which t is a real number (an element of R_1) and $\mathbf{v}$ is a vector in V_2; this collection of ordered* *pairs must have the property that no two pairs have the same first element. The* **domain** *consists of all possible values of t in the collection, and the* **range** *consists of all vectors which occur.*

We see that in terms of mappings, a vector function is a mapping from a set in R_1 into a set in V_2.

Vector functions are more complicated than ordinary functions, since the elements of the range, namely vectors, are themselves equivalence classes of directed line segments. However, if we concentrate on the representation—the directed line segments—the concept becomes more concrete. Also, many of the properties of ordinary functions extend easily to vector functions when suitably interpreted. We shall use boldface letters such as **f**, **g**, **v**, **F**, **G** to represent vector functions. If the dependence on the independent variable is to be indicated, we shall write $\mathbf{f}(t)$, $\mathbf{g}(s)$, $\mathbf{F}(x)$, and so forth for the function values.

Definition. *A vector function* **f** *is continuous at* $t = a$ *if* $\mathbf{f}(a)$ *is defined and if for each* $\epsilon > 0$ *there is a* $\delta > 0$ *such that*

$$|\mathbf{f}(t) - \mathbf{f}(a)| < \epsilon \qquad \text{for all } t \text{ such that} \qquad 0 < |t - a| < \delta.$$

We note that the form of the definition of continuity for vector functions is identical with that for ordinary functions. (See page 95.) We must realize, however, that $\mathbf{f}(t) - \mathbf{f}(a)$ is a *vector*, and that the symbol $|\mathbf{f}(t) - \mathbf{f}(a)|$ stands for the length of a vector, whereas in the case of ordinary functions, $|f(t) - f(a)|$ is the absolute value of a number. In words, the continuity of a vector function asserts that as $t \to a$ the vector $\mathbf{f}(t)$ approaches $\mathbf{f}(a)$, in *both* length and direction. When **f** is continuous at a, we also write

$$\lim_{t \to a} \mathbf{f}(t) = \mathbf{f}(a).$$

If **i** and **j** are the customary unit vectors associated with a Cartesian coordinate system in the plane, a vector function **f** can be written in the form

$$\mathbf{f}(t) = f_1(t)\mathbf{i} + f_2(t)\mathbf{j},$$

where f_1 and f_2 are functions in the ordinary sense. Statements about vector functions **f** may always be interpreted as statements about a pair of functions (f_1, f_2).

Theorem 22. *A function* **f** *is continuous at* $t = a$ *if and only if* f_1 *and* f_2 *are continuous at* $t = a$.

Proof. We have $\mathbf{f}(a) = f_1(a)\mathbf{i} + f_2(a)\mathbf{j}$ and

$$|\mathbf{f}(t) - \mathbf{f}(a)| = |(f_1(t) - f_1(a))\mathbf{i} + (f_2(t) - f_2(a))\mathbf{j}|.$$

Also, (see Fig. 13-29)

$$|\mathbf{f}(t) - \mathbf{f}(a)| = \sqrt{|f_1(t) - f_1(a)|^2 + |f_2(t) - f_2(a)|^2}.$$

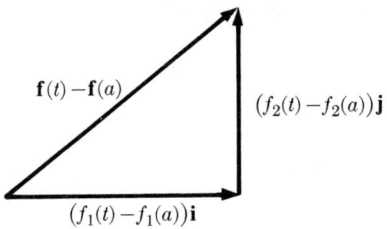

Figure 13-29

If
$$\lim_{t \to a} f_1(t) = f_1(a) \quad \text{and} \quad \lim_{t \to a} f_2(t) = f_2(a),$$
it follows that
$$\lim_{t \to a} \mathbf{f}(t) = \mathbf{f}(a).$$
On the other hand, the inequalities
$$|f_1(t) - f_1(a)| \le |\mathbf{f}(t) - \mathbf{f}(a)|, \quad |f_2(t) - f_2(a)| \le |\mathbf{f}(t) - \mathbf{f}(a)|$$
show that if
$$\lim_{t \to a} \mathbf{f}(t) = \mathbf{f}(a),$$
then *both*
$$\lim_{t \to a} f_1(t) = f_1(a) \quad \text{and} \quad \lim_{t \to a} f_2(t) = f_2(a).$$

Definition. *If $\mathbf{f}$ is a vector function, we define the **derivative $\mathbf{f}'$** as*
$$\mathbf{f}'(t) = \lim_{h \to 0} \frac{\mathbf{f}(t + h) - \mathbf{f}(t)}{h},$$
whenever the limit exists. If $\mathbf{f}$ is given in terms of functions f_1 and f_2 by
$$\mathbf{f}(t) = f_1(t)\mathbf{i} + f_2(t)\mathbf{j},$$
then the derivative may be computed by the simple formula
$$\mathbf{f}'(t) = f_1'(t)\mathbf{i} + f_2'(t)\mathbf{j}. \tag{1}$$

The quantities $f_1'(t)$ and $f_2'(t)$ are derivatives in the ordinary sense. Formula (1) follows directly from the definition of derivative and the theorem on the limit of a sum.

Example 1. Find the derivative $\mathbf{f}'(t)$ if $\mathbf{f}(t) = (t^2 + 2t - 1)\mathbf{i} + (3t^3 - 2)\mathbf{j}$.

Solution. $\mathbf{f}'(t) = (2t + 2)\mathbf{i} + 9t^2\mathbf{j}.$ ◄

Example 2. Given that $\mathbf{f}(t) = (\sin t)\mathbf{i} + (3 - 2\cos t)\mathbf{j}$, find $\mathbf{f}''(t)$.

Solution. $\mathbf{f}'(t) = \cos t\mathbf{i} + 2\sin t\mathbf{j}.$ Hence $\mathbf{f}''(t) = -\sin t\mathbf{i} + 2\cos t\mathbf{j}.$ ◄

Example 3. Given $\mathbf{f}(t) = (3t - 2)\mathbf{i} + (2t^2 + 1)\mathbf{j}$. Find the value of
$$\mathbf{f}'(t) \cdot \mathbf{f}''(t).$$

Solution. $\mathbf{f}'(t) = 3\mathbf{i} + 4t\mathbf{j}, \mathbf{f}''(t) = 4\mathbf{j}$, and
$$\mathbf{f}'(t) \cdot \mathbf{f}''(t) = (3\mathbf{i} + 4t\mathbf{j}) \cdot (4\mathbf{j}) = 16t. \blacktriangleleft$$

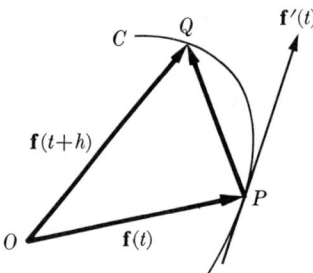

Figure 13–30

The derivative of a vector function has a simple geometric interpretation in terms of representatives. Draw the particular representative of $\mathbf{f}(t)$ which has its base at the origin of the coordinate system. Then the head of this representative will trace out a curve C as t takes on all possible values in its domain (Fig. 13–30). The directed line segment $\overrightarrow{OP}$ represents $\mathbf{f}(t)$. Let $\overrightarrow{OQ}$ represent $\mathbf{f}(t + h)$. Then $\mathbf{f}(t + h) - \mathbf{f}(t)$ has as one of its representatives the directed line segment $\overrightarrow{PQ}$. Multiplying $\mathbf{f}(t + h) - \mathbf{f}(t)$ by $1/h$ gives a vector in the direction of $\overrightarrow{PQ}$, but $1/h$ times as long. As h tends to zero, the quantity

$$\frac{\mathbf{f}(t + h) - \mathbf{f}(t)}{h}$$

tends to a vector, with one of its representatives tangent to the curve C at the point P. Using vector terminology instead of directed line segments, we conclude that $\mathbf{f}'(t)$ is the vector tangent to the curve $\mathbf{f}(t)$.

PROBLEMS

In Problems 1 through 6, calculate $\mathbf{f}'(t)$ and $\mathbf{f}''(t)$.

1. $\mathbf{f}(t) = (t^2 + 1)\mathbf{i} + (t^3 - 3t)\mathbf{j}$

2. $\mathbf{f}(t) = \left(\dfrac{t - 3}{t + 1}\right)\mathbf{i} - \dfrac{t^2 + 1}{t^2 + t + 1}\mathbf{j}$

3. $\mathbf{f}(t) = (\tan 3t)\mathbf{i} + (\cos \pi t)\mathbf{j}$

4. $\mathbf{f}(t) = \dfrac{1}{t}\mathbf{i} - \dfrac{1}{t^2 + 1}\mathbf{j}$

5. $\mathbf{f}(t) = e^{2t}\mathbf{i} + e^{-2t}\mathbf{j}$

6. $\mathbf{f}(t) = (e^t + e^{-t})\mathbf{i} + (e^t - e^{-t})\mathbf{j}$

7. Find $\mathbf{f}(t) \cdot \mathbf{f}'(t)$ if $\mathbf{f}(t) = \dfrac{t^2 + 1}{t^2 + 2}\mathbf{i} + 2t\mathbf{j}$.

8. Find $\dfrac{d}{dt}(\mathbf{f}(t) \cdot \mathbf{f}'(t))$ if $\mathbf{f}(t) = 2t\mathbf{i} + \dfrac{1}{t + 1}\mathbf{j}$.

9. Find $\dfrac{d}{dt}|\mathbf{f}(t)|$ if $\mathbf{f}(t) = \sin 2t\mathbf{i} + \cos 3t\mathbf{j}$.

10. Find $\dfrac{d}{dt}|\mathbf{f}(t)|$ if $\mathbf{f}(t) = (t^2 + 1)\mathbf{i} + (3 + 2t^2)\mathbf{j}$.

11. Find $\dfrac{d}{dt}|\mathbf{f}(t)|$ if $\mathbf{f}(t) = \cos\dfrac{1}{t}\mathbf{i} + \sin\dfrac{1}{t}\mathbf{j}$.

12. Find $\dfrac{d}{dt}(\mathbf{f}'(t) \cdot \mathbf{f}''(t))$ if $\mathbf{f}(t) = (3t + 1)\mathbf{i} + (2t^2 - t^3)\mathbf{j}$.

13. Find $\dfrac{d}{dt}(\mathbf{f}'(t) \cdot \mathbf{f}''(t))$ if $\mathbf{f}(t) = (\log t)\mathbf{i} + \dfrac{2}{t}\mathbf{j}$.

14. Find $\dfrac{d}{dt}(\mathbf{f}''(t) \cdot \mathbf{f}'''(t))$ if $\mathbf{f}(t) = e^{3t}\mathbf{i} + e^{-3t}\mathbf{j}$.

15. Given that $\mathbf{f}(t) = (2t + 1)\mathbf{i} + 3t\mathbf{j}$, $\mathbf{g}(t) = 4t\mathbf{j}$. Find $d\theta/dt$, where $\theta = \theta(t)$ is the angle between $\mathbf{f}$ and $\mathbf{g}$.

16. Write out a proof establishing Formula (1) on page 459.

17. Prove the formula
$$\frac{d}{dt}(F(t)\mathbf{f}(t)) = F(t)\mathbf{f}'(t) + F'(t)\mathbf{f}(t).$$

18. Show that the following Chain Rule holds:
$$\frac{d}{dt}\{\mathbf{f}[g(t)]\} = \mathbf{f}'[g(t)]g'(t).$$

19. Given the vector $\mathbf{f}(t) = (2t + 1)\mathbf{i} + 2t\mathbf{j}$. Describe the curve traced out by the tip of the representative which has its base at the origin.

20. Given the vector $\mathbf{g}(t) = \cos t\mathbf{i} + \sin t\mathbf{j}$. Describe the curve traced out by the tip of the representative which has its base at the origin.

21. Prove that if $\mathbf{f}(t) = \sin 2t\mathbf{i} + \cos 2t\mathbf{j}$, then $\mathbf{f}(t) \cdot \mathbf{f}'(t) = 0$. What is the geometric interpretation of this result?

22. Prove that if
$$\mathbf{f}(t) = \frac{\mathbf{g}(t)}{h(t)}, \quad \text{then} \quad \mathbf{f}'(t) = \frac{h(t)\mathbf{g}'(t) - \mathbf{g}(t)h'(t)}{h^2(t)}$$

23. Show that if
$$\mathbf{f}(t) = f_1(t)\mathbf{i} + f_2(t)\mathbf{j}, \quad \text{then} \quad \frac{d}{dt}(\mathbf{f}(t) \cdot \mathbf{f}(t)) = 2\mathbf{f}(t) \cdot \mathbf{f}'(t).$$

24. Show that
$$\frac{d}{dt}(\mathbf{f}(t) \cdot \mathbf{g}(t)) = \mathbf{f}(t) \cdot \mathbf{g}'(t) + \mathbf{f}'(t) \cdot \mathbf{g}(t).$$

[*Hint:* Write $\mathbf{f}(t) = f_1(t)\mathbf{i} + f_2(t)\mathbf{j}$ and $\mathbf{g}(t) = g_1(t)\mathbf{i} + g_2(t)\mathbf{j}$.]

10. VECTOR VELOCITY AND ACCELERATION IN THE PLANE

The vector function
$$\mathbf{f}(t) = x(t)\mathbf{i} + y(t)\mathbf{j}$$

is equivalent to the pair of parametric equations

$$x = x(t), \quad y = y(t),$$

since the head of the directed line segment with base at the origin, which represents $\mathbf{f}$, traces out a curve which is identical with the curve given in parametric form by these equations.

We recall (Chapter 10, Section 3, page 357) that the arc length $s(t)$ of a curve in parametric form satisfies the relation

$$\frac{ds}{dt} = \sqrt{(dx/dt)^2 + (dy/dt)^2},$$

and therefore

$$\frac{ds}{dt} = |\mathbf{f}'(t)|.$$

In our earlier study of the motion of a particle (Chapter 4, Section 5), we were concerned primarily with motion along a straight line. Now we shall take up the motion of a particle along a curve C in the plane given by the parametric equations

$$C: x = x(t), \quad y = y(t).$$

Letting t denote the time, we define the **velocity vector v** at the time t as

$$\mathbf{v}(t) \equiv \frac{d\mathbf{f}}{dt} \equiv \mathbf{f}'(t) = x'(t)\mathbf{i} + y'(t)\mathbf{j}.$$

According to the geometrical interpretation of the derivative of a vector function given in the preceding section, the velocity vector is always tangent to the path describing the motion. We define the **speed** of the particle to be *the magnitude of the velocity vector*. The speed is

$$|\mathbf{v}(t)| = |\mathbf{f}'(t)| = \sqrt{(dx/dt)^2 + (dy/dt)^2},$$

which tells us that the speed is identical with the quantity ds/dt; in other words, the speed measures the rate of change in arc length s with respect to time t.

The **acceleration vector** $\mathbf{a}(t)$ is defined as the derivative of the velocity vector, or

$$\mathbf{a}(t) = \mathbf{v}'(t) = \mathbf{f}''(t).$$

Example 1. Suppose that a particle P moves according to the law

$$\mathbf{f}(t) = (3\cos 2t)\mathbf{i} + (3\sin 2t)\mathbf{j}.$$

Find $\mathbf{v}(t)$, $\mathbf{a}(t)$, $s'(t)$, $s''(t)$, $|\mathbf{a}(t)|$, and $\mathbf{v}(t) \cdot \mathbf{a}(t)$.

13-10 VECTOR VELOCITY AND ACCELERATION IN THE PLANE

Solution. We have

$$\mathbf{v}(t) = \mathbf{f}'(t) = (-6 \sin 2t)\mathbf{i} + (6 \cos 2t)\mathbf{j},$$
$$\mathbf{a}(t) = (-12 \cos 2t)\mathbf{i} - (12 \sin 2t)\mathbf{j},$$
$$s'(t) = [(-6 \sin 2t)^2 + (6 \cos 2t)^2]^{1/2} = 6,$$
$$s''(t) = 0,$$
$$|\mathbf{a}(t)| = [(-12 \cos 2t)^2 + (12 \sin 2t)^2]^{1/2} = 12,$$
$$\mathbf{v}(t) \cdot \mathbf{a}(t) = 0. \quad \blacktriangleleft$$

Remark. We note that P is moving around a circle with center at O and radius 3, with a constant speed but a changing velocity vector! Since

$$\mathbf{v}(t) \cdot \mathbf{a}(t) = 0,$$

and since $\mathbf{v}(t)$ is always tangent to the circle, we conclude that the acceleration vector is always pointing toward the center of the circle.

Example 2. Suppose that a particle P moves according to the law

$$x(t) = t \cos t, \qquad y(t) = t \sin t,$$

or, equivalently,

$$\mathbf{f}(t) = (t \cos t)\mathbf{i} + (t \sin t)\mathbf{j}.$$

Find $\mathbf{v}(t)$, $\mathbf{a}(t)$, $s'(t)$, $s''(t)$, and $|\mathbf{a}(t)|$.

Solution
$$\mathbf{v}(t) = (-t \sin t + \cos t)\mathbf{i} + (t \cos t + \sin t)\mathbf{j},$$
$$\mathbf{a}(t) = (-t \cos t - 2 \sin t)\mathbf{i} + (-t \sin t + 2 \cos t)\mathbf{j}.$$

Therefore

$$s'(t) = |\mathbf{v}(t)| = \sqrt{1 + t^2},$$
$$s''(t) = \tfrac{1}{2}(1 + t^2)^{-1/2}(2t) = \frac{t}{\sqrt{1 + t^2}},$$
$$|\mathbf{a}(t)| = \sqrt{4 + t^2}. \quad \blacktriangleleft$$

PROBLEMS

In Problems 1 through 8, assume that a particle P moves according to the given law, t denoting the time. Compute $\mathbf{v}(t), |\mathbf{v}(t)|, \mathbf{a}(t), |\mathbf{a}(t)|, s'(t)$, and $s''(t)$.

1. $\mathbf{f}(t) = t^2\mathbf{i} - 3t\mathbf{j}$
2. $\mathbf{f}(t) = \tfrac{1}{2}t^2\mathbf{i} + \tfrac{1}{3}t^3\mathbf{j}$
3. $\mathbf{f}(t) = 3t\mathbf{i} + (1 - t^{-1})\mathbf{j}$
4. $\mathbf{f}(t) = (t - \tfrac{1}{2}t^2)\mathbf{i} + 2t^{1/2}\mathbf{j}$
5. $\mathbf{f}(t) = (\log \sec t)\mathbf{i} + t\mathbf{j}$
6. $\mathbf{f}(t) = (3 \cos t)\mathbf{i} + (2 \sin t)\mathbf{j}$
7. $x = 2e^t, y = 3e^{-t}$
8. $x = e^{-t} \cos t, y = e^{-t} \sin t$

In Problems 9 through 13, assume that a particle P moves according to the given law. Find $\mathbf{v}(t)$, $\mathbf{a}(t)$, $s'(t)$, and $s''(t)$ at the given time t.

9. $x = t^2 - t - 1, y = t^2 - 2t, t = 1$
10. $\mathbf{f}(t) = (5 \cos t)\mathbf{i} + (4 \sin t)\mathbf{j}, t = 2\pi/3$
11. $x = 3 \sec t, y = 2 \tan t, t = \pi/6$
12. $\mathbf{f}(t) = 4(\pi t - \sin \pi t)\mathbf{i} + 4(1 - \cos \pi t)\mathbf{j}, t = \frac{3}{4}$
13. $x = \log(1 + t), y = 3/t, t = 2$
14. Suppose that P moves according to the law
$$\mathbf{f}(t) = (R \cos wt)\mathbf{i} + (R \sin wt)\mathbf{j},$$
where $R > 0$ and w are constants. Find $\mathbf{v}(t), \mathbf{a}(t), s'(t)$, and $s''(t)$. Show that
$$|\mathbf{a}(t)| = \frac{1}{R}|\mathbf{v}(t)|^2.$$
15. Let $\mathbf{T}(t)$ be a vector one unit long and parallel to the *velocity vector*. Show that
$$\mathbf{T}(t) = \frac{\mathbf{v}(t)}{ds/dt} = \frac{x'(t)}{s'(t)}\mathbf{i} + \frac{y'(t)}{s'(t)}\mathbf{j}.$$
16. Using the formula of Problem 15, compute $\mathbf{T}(t)$ if the law of motion is
$$\mathbf{f}(t) = (3t - 1)\mathbf{i} + (t^2 + 2)\mathbf{j}.$$
17. Using the formula of Problem 15, compute $\mathbf{T}(t)$ if the law of motion is
$$\mathbf{f}(t) = (4e^{2t})\mathbf{i} + (3e^{-2t})\mathbf{j}.$$
18. A formula for $\mathbf{T}(t)$ is given in Problem 15. Compute $\mathbf{T}'(t)$ and show that $\mathbf{T}(t) \cdot \mathbf{T}'(t) = 0$. [*Hint:* Use the fact that $x'^2 + y'^2 = s'^2$.]

11. VECTOR FUNCTIONS IN SPACE. SPACE CURVES. TANGENTS AND ARC LENGTH

The collection of all vectors in three-space is devoted by V_3. We now extend the definition of vector function given in Section 9 to the case where the range is an element of V_3.

Definition. A **vector function in three-space,** or simply a *vector function,* is the collection of ordered pairs $(t, \mathbf{v})$ in which t is a real number (an element of R_1) and $\mathbf{v}$ is a vector in V_3. As usual, no two pairs have the same first element.

The definition of continuity for a vector function in three-space is identical with that given in Section 9 for vector functions in the plane. Part (b) of the next theorem is proved exactly as is Theorem 22 in Section 9. The proofs of the remaining parts are left to the reader.

Theorem 23. *Suppose that*
$$\mathbf{f}(t) = f_1(t)\mathbf{i} + f_2(t)\mathbf{j} + f_3(t)\mathbf{k}$$
is a vector function and that the vector $\mathbf{c} = c_1\mathbf{i} + c_2\mathbf{j} + c_3\mathbf{k}$ *is a constant. Then*

a) $\mathbf{f}(t) \to \mathbf{c}$ *as* $t \to a$ *if and only if*
$$f_1(t) \to c_1 \quad \text{and} \quad f_2(t) \to c_2 \quad \text{and} \quad f_3(t) \to c_3.$$

b) **f** is continuous at a if and only if f_1, f_2, and f_3 are.
c) $\mathbf{f}'(t)$ exists if and only if $f_1'(t)$, $f_2'(t)$, and $f_3'(t)$ do.
d) We have the formula

$$\mathbf{f}'(t) = f_1'(t)\mathbf{i} + f_2'(t)\mathbf{j} + f_3'(t)\mathbf{k}.$$

e) If $\mathbf{v}(t) = a\mathbf{w}(t)$, then $\mathbf{v}'(t) = a\mathbf{w}'(t)$ where a is a constant.
f) If $\mathbf{v}(t) = c(t)\mathbf{w}(t)$, then $\mathbf{v}'(t) = c(t)\mathbf{w}'(t) + c'(t)\mathbf{w}(t)$.
g) If $\mathbf{v}(t) = \mathbf{w}(t)/c(t)$, then

$$\mathbf{v}'(t) = \frac{c(t)\mathbf{w}'(t) - c'(t)\mathbf{w}(t)}{[c(t)]^2}.$$

Example 1. Given $\mathbf{f}(t) = 3t^2\mathbf{i} - 2t^3\mathbf{j} + (t^2 + 3)\mathbf{k}$, find $\mathbf{f}'(t)$, $\mathbf{f}''(t)$, $\mathbf{f}'''(t)$.

Solution
$$\mathbf{f}'(t) = 6t\mathbf{i} - 6t^2\mathbf{j} + 2t\mathbf{k},$$
$$\mathbf{f}''(t) = 6\mathbf{i} - 12t\mathbf{j} + 2\mathbf{k},$$
$$\mathbf{f}'''(t) = -12\mathbf{j}. \quad \blacktriangleleft$$

The next theorem shows how to differentiate functions involving scalar and vector products. The proofs are left to the reader.

Theorem 24. *If $\mathbf{u}(t)$ and $\mathbf{v}(t)$ are differentiable, then the derivative of $f(t) = \mathbf{u}(t) \cdot \mathbf{v}(t)$ is given by the formula*

$$f'(t) = \mathbf{u}(t) \cdot \mathbf{v}'(t) + \mathbf{u}'(t) \cdot \mathbf{v}(t). \tag{1}$$

The derivative of $\mathbf{w}(t) = \mathbf{u}(t) \times \mathbf{v}(t)$ is given by the formula

$$\mathbf{w}'(t) = \mathbf{u}(t) \times \mathbf{v}'(t) + \mathbf{u}'(t) \times \mathbf{v}(t). \tag{2}$$

The proofs of formulas (1) and (2) follow from the corresponding differentiation formulas for ordinary functions. Note that in formula (2) it is essential to retain the order of the factors in each vector product.

Example 2. Find the derivative $f'(t)$ and $\mathbf{w}'(t)$ of

$$f(t) = \mathbf{u}(t) \cdot \mathbf{v}(t) \quad \text{and} \quad \mathbf{w}(t) = \mathbf{u}(t) \times \mathbf{v}(t)$$

when

$$\mathbf{u}(t) = (t + 3)\mathbf{i} + t^2\mathbf{j} + (t^3 - 1)\mathbf{k}$$

and

$$\mathbf{v}(t) = 2t\mathbf{i} + (t^4 - 1)\mathbf{j} + (2t + 3)\mathbf{k}.$$

Solution. By formula (1) we have

$$f(t) = [(t + 3)\mathbf{i} + t^2\mathbf{j} + (t^3 - 1)\mathbf{k}] \cdot (2\mathbf{i} + 4t^3\mathbf{j} + 2\mathbf{k})$$
$$+ (\mathbf{i} + 2t\mathbf{j} + 3t^2\mathbf{k}) \cdot [2t\mathbf{i} + (t^4 - 1)\mathbf{j} + (2t + 3)\mathbf{k}]$$
$$= (2t + 6) + 4t^5 + 2t^3 - 2 + 2t + 2t^5 - 2t + 6t^3 + 9t^2$$
$$= 6t^5 + 8t^3 + 9t^2 + 2t + 4.$$

According to (2), we have
$$\mathbf{w}'(t) = [(t+3)\mathbf{i} + t^2\mathbf{j} + (t^3 - 1)\mathbf{k}] \times (2\mathbf{i} + 4t^3\mathbf{j} + 2\mathbf{k})$$
$$+ (\mathbf{i} + 2t\mathbf{j} + 3t^2\mathbf{k}) \times [2t\mathbf{i} + (t^4 - 1)\mathbf{j} + (2t + 3)\mathbf{k}].$$

Computing the two cross products on the right, we get
$$\mathbf{w}'(t) = (-7t^6 + 4t^3 + 9t^2 + 6t)\mathbf{i} + (8t^3 - 4t - 11)\mathbf{j}$$
$$+ (5t^4 + 12t^3 - 6t^2 - 1)\mathbf{k}. \quad \triangleleft$$

Consider a Cartesian coordinate system and a directed line segment from the origin O to a point P in space. As in two dimensions, we denote the vector $\mathbf{v}(\overrightarrow{OP})$ by $\mathbf{r}$. We define an **arc** C in space in a way completely analogous to that in which an arc in the plane was defined. (See page 353.) The vector equation
$$\mathbf{r}(t) = x(t)\mathbf{i} + y(t)\mathbf{j} + z(t)\mathbf{k}$$
is considered to be equivalent to the parametric equations
$$x = x(t), \quad y = y(t), \quad z = z(t),$$
which represent an arc C in space.

The definition of **length** of an arc C in three-space is identical with its definition in the plane. If we denote the length of an arc by $l(C)$ we get, in a way that is similar to the two-dimensional analysis, the formulas

$$l(C) = \int_a^b \sqrt{[x'(t)]^2 + [y'(t)]^2 + [z'(t)]^2} \, dt \qquad (3)$$

and

$$s'(t) = \sqrt{[x'(t)]^2 + [y'(t)]^2 + [z'(t)]^2} \qquad (4)$$

where $s(t)$ is an arc length function. We could also write the above formulas (3) and (4) in the vector form

$$l(C) = \int_a^b |\mathbf{r}'(t)| \, dt, \qquad s'(t) = |\mathbf{r}'(t)|.$$

Definitions. If $\mathbf{r}'(t) \neq \mathbf{0}$, we define the vector $\mathbf{T}(t) = \mathbf{r}'(t)/|\mathbf{r}'(t)|$ as the **unit tangent vector** to the path corresponding to the value t. The line through the point P_0 corresponding to $\mathbf{r}(t_0)$ and parallel to $\mathbf{T}(t_0)$ is called the **tangent line to the arc** at t_0; the line directed in the same way as $\mathbf{T}(t_0)$ is called the **directed tangent line at** t_0 (Fig. 13–31).

Example 3. The graph of the equations
$$x = a \cos t, \quad y = a \sin t, \quad z = bt \qquad (5)$$
is called a **helix**.

i) Find $s'(t)$.
ii) Find the length of that part of the helix for which $0 \leq t \leq 2\pi$.
iii) Show that the unit tangent vector makes a constant angle with the z axis.

13-11 VECTOR FUNCTIONS IN SPACE. SPACE CURVES. TANGENTS AND ARC LENGTH

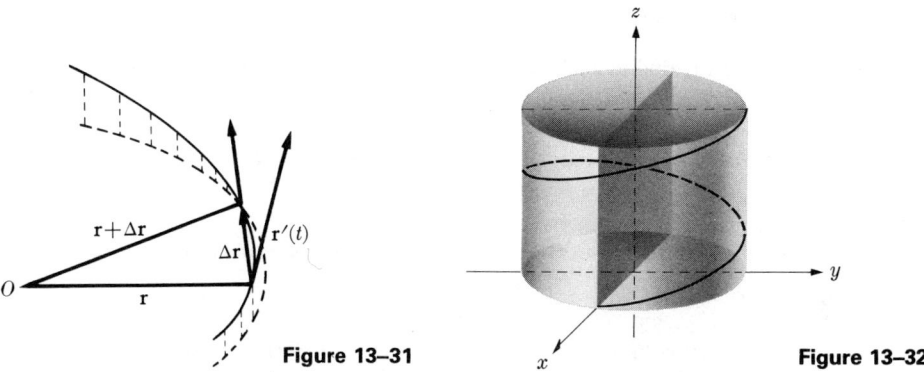

Figure 13-31   Figure 13-32

Solution. (See Fig. 13-32 for the graph.)

i) $x'(t) = -a \sin t$, $y'(t) = a \cos t$, $z'(t) = b$. Therefore,
$$s'(t) = \sqrt{a^2 + b^2}.$$

ii) $l(C) = 2\pi\sqrt{a^2 + b^2}$.

iii) $\mathbf{T}(t) = \dfrac{-y\mathbf{i} + x\mathbf{j} + b\mathbf{k}}{\sqrt{a^2 + b^2}}$.

Letting ϕ be the angle between $\mathbf{T}$ and $\mathbf{k}$, we get
$$\cos \phi = b/\sqrt{a^2 + b^2}. \qquad \blacktriangleleft$$

Remark. We note that the helix (5) winds around the cylinder $x^2 + y^2 = a^2$.

Definitions. *If t denotes time in the parametric equations of an arc $\mathbf{r}(t)$, then $\mathbf{r}'(t)$ is the **velocity vector** $\mathbf{v}(t)$, and $\mathbf{r}''(t) = \mathbf{v}'(t)$ is called the **acceleration vector**. The quantity $s'(t)$ (a scalar) is called the **speed** of a particle moving according to the law*
$$\mathbf{r}(t) = x(t)\mathbf{i} + y(t)\mathbf{j} + z(t)\mathbf{k}.$$

PROBLEMS

In each of Problems 1 through 8, find the derivatives $\mathbf{f}'(t)$ and $\mathbf{f}''(t)$.

1. $\mathbf{f}(t) = t^2\mathbf{i} + (t^2 + 1)\mathbf{j} + (2 - 3t)\mathbf{k}$
2. $\mathbf{f}(t) = (2t^3 + t - 1)\mathbf{i} + (t^{-1} + 1)\mathbf{j} + (t^2 + t^{-2})\mathbf{k}$
3. $\mathbf{f}(t) = (\cos 2t)\mathbf{i} + (\sin 2t)\mathbf{j} + 2t\mathbf{k}$
4. $\mathbf{f}(t) = e^{2t}\mathbf{i} + t^2\mathbf{j} + e^{-2t}\mathbf{k}$
5. $\mathbf{f}(t) = \left(\dfrac{t^2}{t^2 + 1}\right)\mathbf{i} + \dfrac{1}{t^2 + 1}\mathbf{j} + (t^2 + 1)\mathbf{k}$
6. $\mathbf{f}(t) = (\log 2t)\mathbf{i} + e^{3t}\mathbf{j} + (t \log t)\mathbf{k}$
7. $\mathbf{f}: t \to (\cos t)\mathbf{i} + (\tan t)\mathbf{j} + (\sin t)\mathbf{k}$

8. $\mathbf{f}: t \to [(t^2 + 1)(t^2 - 2)]\mathbf{i} + (t^3 + 2t^{-3})\mathbf{j} + [\ln(t^2 + 1)]\mathbf{k}$

In Problems 9 through 11, find in each case either $f'(t)$ or $\mathbf{f}'(t)$, whichever is appropriate.

9. $f(t) = \mathbf{u}(t) \cdot \mathbf{v}(t)$, where
$$\mathbf{u}(t) = 3t\mathbf{i} + 2t^2\mathbf{j} + \frac{1}{t}\mathbf{k}, \quad \mathbf{v}(t) = t^2\mathbf{i} + \frac{1}{t}\mathbf{j} + t^3\mathbf{k}.$$

10. $\mathbf{f}(t) = \mathbf{u}(t) \times \mathbf{v}(t)$, where
$$\mathbf{u}(t) = (\cos t)\mathbf{i} + (\sin t)\mathbf{j} + t\mathbf{k}, \quad \mathbf{v}(t) = (\sin t)\mathbf{i} + (\cos t)\mathbf{j} + t^2\mathbf{k}.$$

11. $f(t) = \mathbf{u}(t) \cdot [\mathbf{v}(t) \times \mathbf{w}(t)]$, where
$$\mathbf{u}(t) = t\mathbf{i} + (t+1)\mathbf{k}, \quad \mathbf{v}(t) = t^2\mathbf{j}, \quad \mathbf{w}(t) = \frac{1}{t^2}\mathbf{k}.$$

In Problems 12 through 15, find $l(C)$.

12. $C: x = t, \ y = t^2/\sqrt{2}, \ z = t^3/3; \quad 0 \leq t \leq 2$
13. $C: x = t, \ y = 3t^2/2, \ z = 3t^3/2; \quad 0 \leq t \leq 2$
14. $C: x = t, \ y = \ln(\sec t + \tan t), \ z = \ln \sec t; \quad 0 \leq t \leq \pi/4$
15. $C: x = t \cos t, \ y = t \sin t, \ z = t; \quad 0 \leq t \leq \pi/2$

In Problems 16 through 18, the parameter t is the time in seconds. Taking s to be the length in meters, in each case find the velocity, speed, and acceleration of a particle moving according to the given law.

16. $\mathbf{r}(t) = t^2\mathbf{i} + 2t\mathbf{j} + (t^3 - 1)\mathbf{k}$
17. $\mathbf{r}(t) = (t \sin t)\mathbf{i} + (t \cos t)\mathbf{j} + t\mathbf{k}$
18. $\mathbf{r}(t) = e^{3t}\mathbf{i} + e^{-3t}\mathbf{j} + te^{3t}\mathbf{k}$

19. a) Given $\mathbf{f}(t) = (3 \cos t)\mathbf{i} + (4 \cos t)\mathbf{j} + (5 \sin t)\mathbf{k}$, show that
$$\mathbf{f}(t) \cdot \mathbf{f}'(t) = 0$$
for all t.

b) Let $\mathbf{f}(t)$ be a vector function such that $|\mathbf{f}(t)| = 1$ for all t. Show that $\mathbf{f}(t)$ and $\mathbf{f}'(t)$ are orthogonal for all t.

20. Suppose that $\mathbf{f}: t \to \alpha(t)\mathbf{i} + \beta(t)\mathbf{j} + \gamma(t)\mathbf{k}$ has the properties: $\mathbf{f}(t) = \mathbf{f}'(t)$ for all t and $\alpha(0) = \beta(0) = \gamma(0) = 2$. Find $\mathbf{f}(t)$.

21. Show that every differentiable vector function $\mathbf{v}(t)$ satisfies the identity
$$2\mathbf{v}'(t) = \mathbf{i} \times (\mathbf{v}' \times \mathbf{i}) + \mathbf{j} \times (\mathbf{v}' \times \mathbf{j}) + \mathbf{k} \times (\mathbf{v}' \times \mathbf{k}).$$

22. Write out a complete proof of Theorem 23.
23. Write out a complete proof of Theorem 24.

12. TANGENTIAL AND NORMAL COMPONENTS. THE MOVING TRIHEDRAL*

We consider the graph of the equation
$$\mathbf{r} = \mathbf{r}(t) \quad \Leftrightarrow \quad \mathbf{r}(t) = x(t)\mathbf{i} + y(t)\mathbf{j} + z(t)\mathbf{k}.$$

---

* This optional section may be omitted without loss of continuity.

13–12 TANGENTIAL AND NORMAL COMPONENTS. THE MOVING TRIHEDRAL

In the last section, we defined the *unit tangent vector* $\mathbf{T}(t)$ by the relation

$$\mathbf{T}(t) = \frac{\mathbf{r}'(t)}{|\mathbf{r}'(t)|},$$

which we could also write

$$\mathbf{T}(t) = \frac{\mathbf{r}'(t)}{s'(t)} = \frac{d\mathbf{r}}{ds}.$$

Now, since $\mathbf{T} \cdot \mathbf{T} = 1$ for all t, we differentiate this relation with respect to t to obtain

$$\mathbf{T}(t) \cdot \mathbf{T}'(t) + \mathbf{T}'(t) \cdot \mathbf{T}(t) = 2\mathbf{T}(t) \cdot \mathbf{T}'(t) = 0.$$

Therefore, the vector $\mathbf{T}'(t)$ is orthogonal to $\mathbf{T}(t)$. We define the vector $\boldsymbol{\kappa}(t)$ by the relation

$$\boldsymbol{\kappa}(t) = \frac{d\mathbf{T}}{ds} = \frac{\mathbf{T}'(t)}{s'(t)}.$$

Taking the scalar product of $\boldsymbol{\kappa}$ and $\mathbf{T}$, we see that

$$\boldsymbol{\kappa}(t) \cdot \mathbf{T}(t) = \frac{1}{s'(t)} \mathbf{T}'(t) \cdot \mathbf{T}(t) = 0,$$

and so $\boldsymbol{\kappa}$ is orthogonal to $\mathbf{T}$. The vector $\boldsymbol{\kappa}$ is the vector rate of change of direction of the curve with respect to arc length. Bearing in mind the definition of curvature of a plane curve (see Chapter 10, Section 4), we make the following definitions.

Definitions. *The vector $\boldsymbol{\kappa}(t)$, defined above, is called the* **curvature vector** *of the curve $\mathbf{r} = \mathbf{r}(t)$. The magnitude $|\boldsymbol{\kappa}(t)| = \kappa(t)$ is called the* **curvature.** *If $\boldsymbol{\kappa}(t) \neq \mathbf{0}$, we define the* **principal normal vector** $\mathbf{N}(t)$ *and the* **binormal vector** $\mathbf{B}(t)$ *by the relations*

$$\mathbf{N}(t) = \frac{\boldsymbol{\kappa}(t)}{|\boldsymbol{\kappa}(t)|}, \qquad \mathbf{B}(t) = \mathbf{T}(t) \times \mathbf{N}(t).$$

The **center of curvature** $C(t)$ *is defined by the equation*

$$\mathbf{v}[\overrightarrow{OC}(t)] = \mathbf{r}(t) + \frac{1}{\kappa(t)} \mathbf{N}(t).$$

The **radius of curvature** *is $R(t) = 1/\kappa(t)$. The* **osculating plane** *corresponding to the value t is the plane which contains the tangent line to the path and the center of curvature at t; it is defined only for those values of t for which $\kappa(t) \neq 0$.*

Example 1. Given the curve

$$x = t, \qquad y = t^2, \qquad z = 1 + t^2,$$

find $\mathbf{T}(t)$ and $\boldsymbol{\kappa}(t)$.

470 VECTORS IN THE PLANE AND IN THREE SPACE

Solution. We may write
$$\mathbf{r}(t) = t\mathbf{i} + t^2\mathbf{j} + (1 + t^2)\mathbf{k}.$$
Therefore
$$\mathbf{r}'(t) = \mathbf{i} + 2t\mathbf{j} + 2t\mathbf{k} \quad \text{and} \quad |\mathbf{r}'(t)| = \sqrt{1 + 8t^2} = s'(t).$$
Since $\mathbf{T} = \mathbf{r}'(t)/|\mathbf{r}'(t)|$, we have
$$\mathbf{T}(t) = \frac{1}{\sqrt{1 + 8t^2}}(\mathbf{i} + 2t\mathbf{j} + 2t\mathbf{k}).$$
We may differentiate to get
$$\mathbf{T}'(t) = \frac{-8t}{(1 + 8t^2)^{3/2}}(\mathbf{i} + 2t\mathbf{j} + 2t\mathbf{k}) + \frac{1}{\sqrt{1 + 8t^2}}(2\mathbf{j} + 2\mathbf{k}).$$
Hence
$$\kappa(t) = \frac{\mathbf{T}'(t)}{s'(t)} = \frac{-8t}{(1 + 8t^2)^2}(\mathbf{i} + 2t\mathbf{j} + 2t\mathbf{k}) + \frac{1}{1 + 8t^2}(2\mathbf{j} + 2\mathbf{k})$$
$$= \frac{-8t}{(1 + 8t^2)^2}\mathbf{i} + \frac{2}{(1 + 8t^2)^2}\mathbf{j} + \frac{2}{(1 + 8t^2)^2}\mathbf{k}. \quad \blacktriangleleft$$

It is clear from the definitions of $\mathbf{T}$ and $\mathbf{N}$ that they are unit vectors which are orthogonal. From the definition of cross product, we see at once that $\mathbf{B}$ is a unit vector orthogonal to both $\mathbf{T}$ and $\mathbf{N}$. The triple $\mathbf{T}(t)$, $\mathbf{N}(t)$, $\mathbf{B}(t)$ forms a mutually orthogonal triple of unit vectors at each point, called the **trihedral** at the point. See Fig. 13–33 where the representatives of $\mathbf{T}(t)$, $\mathbf{N}(t)$, and $\mathbf{B}(t)$ are drawn with their bases at the point P on the curve corresponding to the value t of the parameter. (It is assumed that $\kappa(t) \neq 0$.) We may also write the formulas

$$\frac{d\mathbf{T}}{ds} = \kappa \mathbf{N} \quad \text{or} \quad \mathbf{T}'(t) = \kappa(t)s'(t)\mathbf{N}(t).$$

The equation of the osculating plane at $t = t_0$ is given by
$$b_1(x - x_0) + b_2(y - y_0) + b_3(z - z_0) = 0,$$
where (x_0, y_0, z_0) is the point on the curve corresponding to
$$t = t_0 \quad \text{and} \quad \mathbf{B}(t_0) = b_1\mathbf{i} + b_2\mathbf{j} + b_3\mathbf{k}.$$

Example 2. Given the helix
$$x = 4 \cos t, \quad y = 4 \sin t, \quad z = 2t$$
or, equivalently,
$$\mathbf{r}(t) = (4 \cos t)\mathbf{i} + (4 \sin t)\mathbf{j} + (2t)\mathbf{k},$$
find $\mathbf{T}$, $\mathbf{N}$, and $\mathbf{B}$ for $t = 2\pi/3$. Also find κ, the equation of the osculating plane, and the equations of the tangent line.

13-12 TANGENTIAL AND NORMAL COMPONENTS. THE MOVING TRIHEDRAL

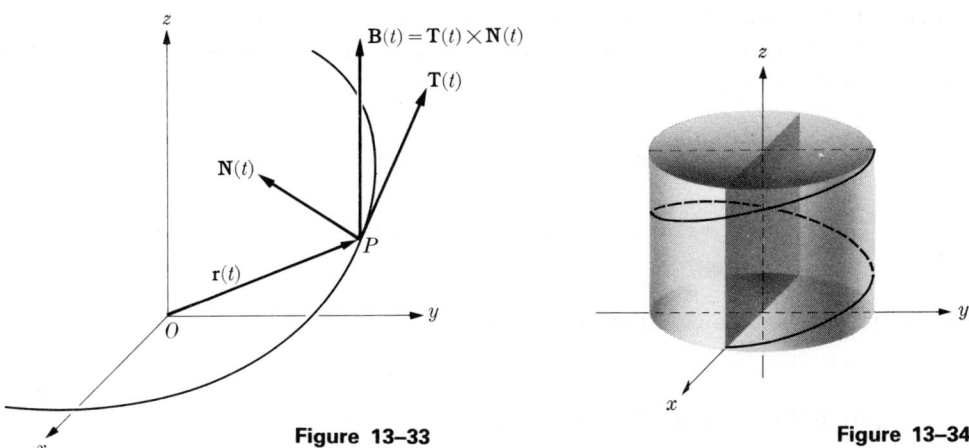

Figure 13-33   **Figure 13-34**

Solution. The curve is shown in Fig. 13-34. We have

$$x' = -4 \sin t, \quad y' = 4 \cos t, \quad z' = 2, \quad s' = 2\sqrt{5},$$

$$\mathbf{T}(t) = \frac{1}{2\sqrt{5}}(-y\mathbf{i} + x\mathbf{j} + 2\mathbf{k}), \quad \mathbf{T}'(t) = \frac{1}{2\sqrt{5}}(-x\mathbf{i} - y\mathbf{j}).$$

We compute

$$\kappa(t) = \frac{|\mathbf{T}'(t)|}{s'(t)} = \frac{1}{5}$$

and, using the formula $\mathbf{T}'(t) = \kappa(t)s'(t)\mathbf{N}(t)$, we get

$$\mathbf{N}(t) = -(\cos t)\mathbf{i} - (\sin t)\mathbf{j}.$$

For $t = 2\pi/3$, $x_0 = -2$, $y_0 = 2\sqrt{3}$, $z_0 = 4\pi/3$, we obtain

$$\mathbf{T} = \frac{1}{\sqrt{5}}(-\sqrt{3}\mathbf{i} - \mathbf{j} + \mathbf{k}), \quad \mathbf{N} = \frac{1}{2}\mathbf{i} - \frac{\sqrt{3}}{2}\mathbf{j},$$

$$\mathbf{B} = \frac{1}{2\sqrt{5}}(\sqrt{3}\mathbf{i} + \mathbf{j} + 4\mathbf{k}), \quad \kappa\left(\frac{2\pi}{3}\right) = \frac{1}{5}.$$

The tangent line is

$$\frac{x + 2}{\sqrt{3}} = \frac{y - 2\sqrt{3}}{1} = \frac{z - 4\pi/3}{-1},$$

and the equation of the osculating plane is

$$\sqrt{3}(x + 2) + (y - 2\sqrt{3}) + 4(z - 4\pi/3) = 0. \quad \blacktriangleleft$$

Definitions. *If a particle moves according to the law* $\mathbf{r} = \mathbf{r}(t)$, *the* **tangential** *and* **normal components** *of any vector are its components along* $\mathbf{T}$ *and* $\mathbf{N}$, *respectively.*

Theorem 25. *If a particle moves according to the law* $\mathbf{r} = \mathbf{r}(t)$, *with t the time and* $\mathbf{T}(t)$, $\mathbf{N}(t)$, $R(t)$ *defined as above, then the acceleration vector* $\mathbf{a}(t) = \mathbf{v}'(t)$ *satisfies the equation*

$$\mathbf{a}(t) = s''(t)\mathbf{T}(t) + \frac{|\mathbf{v}(t)|^2}{R(t)}\mathbf{N}(t). \tag{1}$$

Proof. Since $\mathbf{v}(t) = \mathbf{r}'(t)$, we have, from the definition of $\mathbf{T}(t)$, that

$$\mathbf{v}(t) = s'(t)\mathbf{T}(t).$$

We differentiate this equation and obtain

$$\mathbf{a}(t) = s''(t)\mathbf{T}(t) + s'(t)\mathbf{T}'(t).$$

The relation $\mathbf{T}'(t) = \kappa(t)s'(t)\mathbf{N}(t)$ yields

$$\mathbf{a}(t) = s''(t)\mathbf{T}(t) + \kappa(t)[s'(t)]^2\mathbf{N}(t).$$

We observe that $|\mathbf{v}(t)| = s'(t)$ and $\kappa(t) = 1/R(t)$, and the proof is complete.

Remark. Since $\mathbf{T}$, $\mathbf{N}$, and $\mathbf{B}$ are mutually orthogonal vectors, formula (1) shows that $s''(t)$ and $|\mathbf{v}(t)|^2/R(t)$ are the tangential and normal components, respectively, of $\mathbf{a}(t)$.

Example 3. A particle moves according to the law

$$x = t, \quad y = t^2, \quad z = t^3.$$

Find (a) its vector velocity, (b) its acceleration, (c) its speed. Also find (d) the unit vector $\mathbf{T}$, and (e) the unit vector $\mathbf{N}$; find (f) the normal and (g) the tangential components of the acceleration vector. Evaluate all these quantities at $t = 1$.

Solution. Since $\mathbf{r}(t) = t\mathbf{i} + t^2\mathbf{j} + t^3\mathbf{k}$, we have

$$\mathbf{v}(t) = \mathbf{r}'(t) = \mathbf{i} + 2t\mathbf{j} + 3t^2\mathbf{k}, \qquad \mathbf{a}(t) = \mathbf{v}'(t) = 2\mathbf{j} + 6t\mathbf{k}.$$

Also,

$$s'(t) = (1 + 4t^2 + 9t^4)^{1/2}; \qquad s''(t) = \frac{4t + 18t^3}{(1 + 4t^2 + 9t^4)^{1/2}}.$$

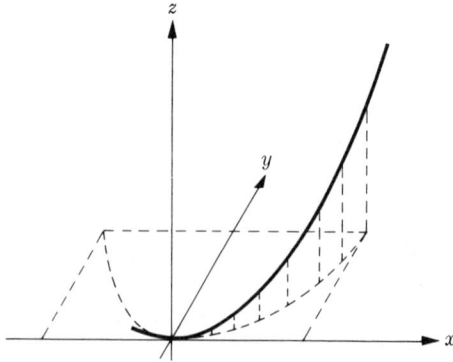

Figure 13-35

The path of the particle is sketched in Fig. 13–35. For $t = 1$,

a) $\qquad\qquad\mathbf{v} = \mathbf{i} + 2\mathbf{j} + 3\mathbf{k},$

b) $\qquad\qquad\mathbf{a} = 2\mathbf{j} + 6\mathbf{k}.$

c) The speed is $s'(1) = \sqrt{14}$. We obtain at $t = 1$

$$s''(1) = \frac{22}{\sqrt{14}};$$

d) $$\mathbf{T} = \frac{1}{\sqrt{14}}(\mathbf{i} + 2\mathbf{j} + 3\mathbf{k});$$

$$\mathbf{a} - s''\mathbf{T} = \frac{(s')^2}{R}\mathbf{N} = 2\mathbf{j} + 6\mathbf{k} - \frac{11}{7}(\mathbf{i} + 2\mathbf{j} + 3\mathbf{k})$$

$$= \frac{1}{7}(-11\mathbf{i} - 8\mathbf{j} + 9\mathbf{k}).$$

Therefore

$$|\mathbf{a} - s''\mathbf{T}| = \frac{1}{7}\sqrt{266} = \frac{14}{R};$$

e) $$\mathbf{N} = \frac{1}{\sqrt{266}}(-11\mathbf{i} - 8\mathbf{j} + 9\mathbf{k});$$

$$R = \frac{98}{\sqrt{266}}.$$

At $t = 1$, the tangential component is $s''(1)$, and so for (g) we find $22/\sqrt{14}$. To find the normal component (f), we compute $|\mathbf{v}(1)|^2/R(1)$ and get the value $\sqrt{266}/7$. ◄

PROBLEMS

In each of Problems 1 through 6, find the unit tangent vector $\mathbf{T}(t)$.

1. $\mathbf{r}(t) = t^3\mathbf{i} + (1 - t)\mathbf{j} + (2t + 1)\mathbf{k}$
2. $\mathbf{r}(t) = (\sin t)\mathbf{i} + (\cos t)\mathbf{j} + 3t^4\mathbf{k}$
3. $x(t) = e^{-2t},\qquad y(t) = e^{2t},\qquad z(t) = 1 + t^2$
4. $\mathbf{r}(t) = \dfrac{t}{1 + t}\mathbf{i} + \dfrac{t^2}{1 + t}\mathbf{j} + \dfrac{1 - t}{1 + t}\mathbf{k}$
5. $x(t) = e^t \sin t,\qquad y(t) = e^{2t} \cos t,\qquad z(t) = e^{-t}$
6. $\mathbf{r}(t) = \ln(1 + t)\mathbf{i} + \dfrac{t}{1 + t^2}\mathbf{j} - 2t^3\mathbf{k}$

In each of Problems 7 through 11, for the particular value of t given, find the vectors $\mathbf{T}$, $\mathbf{N}$, and $\mathbf{B}$; the curvature κ; the equations of the tangent line; and the equation of the osculating plane to the curves given.

7. $x = 1 + t,\qquad y = 3 - t,\qquad z = 2t + 4,\qquad t = 3$
8. $\mathbf{r}(t) = t\mathbf{i} + t^2\mathbf{j} + \frac{1}{3}t^3\mathbf{k},\qquad t = 0$
9. $x = e^t \cos t,\qquad y = e^t \sin t,\qquad z = e^t,\qquad t = 0$
10. $x = \frac{1}{3}t^3,\qquad y = 2t,\qquad z = 2/t,\qquad t = 2$

11. $x = (e^{t/2} + e^{-t/2}),\quad y = (e^{t/2} - e^{-t/2}),\quad z = 2t,\quad t = 0$

12. Given the path $x = e^t \cos t$, $y = e^t \sin t$, $z = e^t$, show that the path lies on the upper half of the cone $z^2 = x^2 + y^2$, and that the tangent vector $\mathbf{T}(t)$ cuts the generators of the cone at a constant angle and makes a constant angle with the z axis for all t. Also, find $\mathbf{N}$ and κ in terms of t.

In Problems 13 through 16, a particle moves according to the law given. In each case, find its vector velocity and acceleration, its speed, the radius of curvature of its path, the unit vectors $\mathbf{T}$ and $\mathbf{N}$, and the tangential and normal components of acceleration at the given time.

13. $x = t,\quad y = \frac{3}{2}t^2,\quad z = \frac{3}{2}t^3,\quad t = 2$

14. $x = t,\quad y = \ln(\sec t + \tan t),\quad z = \ln \sec t,\quad t = \pi/3$

15. $x = t^3/3,\quad y = 2t,\quad z = 2/t,\quad t = 1$

16. $x = t \cos t,\quad y = t \sin t,\quad z = t,\quad t = 0$

14

FORMULAS AND METHODS OF INTEGRATION

1. INTEGRATION BY SUBSTITUTION

We saw in Chapter 7 that any function continuous on a closed interval can be integrated. This fact, while it is of theoretical value, gives no hint whatsoever as to ways of actually finding the integral of a function. For example, we know that

$$\int u^n \, du = \frac{u^{n+1}}{n+1} + c, \qquad n \neq -1,$$

but the integral

$$\int (4 - 7x)^{15} \, dx$$

is not quite in the same form. In Section 7 of Chapter 7 we saw that if we let

$$u = 4 - 7x, \qquad du = -7 \, dx,$$

the integral becomes

$$-\tfrac{1}{7} \int u^{15} \, du = -\tfrac{1}{112} u^{16} + c = -\tfrac{1}{112}(4 - 7x)^{16} + c.$$

By making a change of variable we are able to change the integral into a known form, and therefore we can perform the integration. This technique is known as the **method of substitution.**

It is easy to find functions for which there is no way of performing the integration. For example, it is possible to show that the integrals

$$\int \frac{dx}{\sqrt{1 + x^3}}, \qquad \int e^{-x^2} \, dx, \qquad \int \frac{\sin x}{x} \, dx$$

cannot be integrated in terms of any of the functions we have studied so far (polynomials, rational functions, trigonometric functions, logarithms, etc.).

Appearances may be deceiving, for while

$$\int e^{-x^2} \, dx$$

cannot be integrated, the slightly more "complicated" integral

$$\int x e^{-x^2} \, dx$$

easily can be.

In this chapter we shall learn a number of methods for performing the integrations in many feasible cases. For ready reference, we collect the formulas for integration which we have already learned:

1. $\int u^n \, du = \dfrac{u^{n+1}}{n+1} + C, \; n \ne -1$ 2. $\int \dfrac{du}{u} = \ln |u| + C$

3. $\int e^u \, du = e^u + C$ 4. $\int a^u \, du = \dfrac{a^u}{\ln a} + C, \; a > 0$

The trigonometric formulas are:

5. $\int \sin u \, du = -\cos u + C$ 6. $\int \cos u \, du = \sin u + C$

7. $\int \sec^2 u \, du = \tan u + C$ 8. $\int \csc^2 u \, du = -\cot u + C$

9. $\int \sec u \tan u \, du = \sec u + C$ 10. $\int \csc u \cot u \, du = -\csc u + C$

11. $\int \sec u \, du = \ln |\sec u + \tan u| + C$

12. $\int \csc u \, du = -\ln |\csc u + \cot u| + C$

The inverse trigonometric formulas ($a > 0$) are:

13. $\int \dfrac{du}{\sqrt{a^2 - u^2}} = \arcsin \dfrac{u}{a} + C$ 14. $\int \dfrac{du}{a^2 + u^2} = \dfrac{1}{a} \arctan \dfrac{u}{a} + C$

15. $\int \dfrac{du}{u\sqrt{u^2 - a^2}} = \dfrac{1}{a} \operatorname{arcsec} \dfrac{u}{a} + C, \quad |u| > a$

Formulas 11 and 12 are obtained from the Remark following Example 3 in Chapter 9, Section 9, page 331. We recall that every integration formula may be verified by differentiation of the result. Therefore the reader does not have to refer to earlier material to assure himself of the validity of Formulas 11 and 12; differentiation of $\ln |\sec u + \tan u|$ and $-\ln |\csc u + \cot u|$ suffices.

It is evident that Formulas 13 through 15 are derived by using the substitution $v = u/a$ in the simpler differentiation formulas

$$d \arcsin v = \frac{dv}{\sqrt{1 - v^2}}, \quad d \arctan v = \frac{dv}{1 + v^2}, \quad d \operatorname{arcsec} v = \frac{dv}{v\sqrt{v^2 - 1}}.$$

We illustrate the method of substitution with seven examples.

Example 1. Find $\int x e^{x^2} \, dx.$

Solution. Try $u = x^2$. Then $du = 2x\,dx$, and therefore

$$\int xe^{x^2}\,dx = \tfrac{1}{2}\int e^u\,du = \tfrac{1}{2}e^u + C = \tfrac{1}{2}e^{x^2} + C. \quad \blacktriangleleft$$

Example 2. Find

$$\int \frac{\cos 3x\,dx}{1 + \sin 3x}.$$

Solution. Try $u = 1 + \sin 3x$. Then $du = 3\cos 3x\,dx$, and

$$\int \frac{\cos 3x\,dx}{1 + \sin 3x} = \frac{1}{3}\int \frac{du}{u} = \frac{1}{3}\ln|u| + C = \frac{1}{3}\ln|1 + \sin 3x| + C. \quad \blacktriangleleft$$

Example 3. Find

$$\int \frac{dt}{\sqrt{4 - 9t^2}}.$$

Solution. Try $u = 3t$, $du = 3\,dt$. Then

$$\int \frac{dt}{\sqrt{4 - 9t^2}} = \frac{1}{3}\int \frac{du}{\sqrt{4 - u^2}}.$$

This expression is Formula 13 with $a = 2$. Therefore

$$\int \frac{dt}{\sqrt{4 - 9t^2}} = \frac{1}{3}\arcsin\frac{u}{2} + C = \frac{1}{3}\arcsin\frac{3t}{2} + C. \quad \blacktriangleleft$$

Example 4. Find

$$\int \frac{t\,dt}{\sqrt{4 - 9t^2}}.$$

Solution. Try $u = 4 - 9t^2$. Then $du = -18t\,dt$, and therefore

$$\int \frac{t\,dt}{\sqrt{4 - 9t^2}} = -\frac{1}{18}\int \frac{du}{u^{1/2}} = -\frac{1}{9}u^{1/2} + C = -\frac{1}{9}\sqrt{4 - 9t^2} + C. \quad \blacktriangleleft$$

Remark. The method of substitution proceeds by trial and error. If a particular substitution does not reduce the integral to a known formula, we should feel no hesitation about abandoning that substitution and trying one of an entirely different nature. Example 4, although it is similar in appearance to Example 3, integrates according to a different substitution.

Example 5. Find

$$\int \frac{e^y\,dy}{(1 + e^y)^3}.$$

Solution. Try $u = 1 + e^y$. Then $du = e^y\,dy$, and

$$\int \frac{e^y\,dy}{(1 + e^y)^3} = \int u^{-3}\,du = -\frac{1}{2u^2} + C = -\frac{1}{2(1 + e^y)^2} + C. \quad \blacktriangleleft$$

Remarks. When an expression such as e^{x^2} or $\sin(\ln x)$ occurs, it is generally worth while to set $u = x^2$ in the first case or $u = \ln x$ in the second case. Such a substitution may reduce the integral to a more recognizable form. It is also fairly easy to spot integrals which are reducible to the forms given in Formulas 13 through 15. But, as Examples 2, 4, and 5 show, it may happen that Formulas 1 and 2 are applicable. Sometimes we must make several substitutions in succession, as the next example shows.

Example 6. Find
$$\int \frac{dx}{x[9 + 4(\ln x)^2]}.$$

Solution. Try $u = \ln x$. Then $du = (1/x)\, dx$, and
$$\int \frac{dx}{x[9 + 4(\ln x)^2]} = \int \frac{du}{9 + 4u^2}.$$

Let $v = 2u$. Then $dv = 2\, du$, and therefore the integral becomes
$$\frac{1}{2}\int \frac{dv}{9 + v^2} = \frac{1}{2} \cdot \frac{1}{3} \arctan \frac{v}{3} + C$$
$$= \frac{1}{6} \arctan \frac{2u}{3} + C = \frac{1}{6} \arctan \frac{2\ln x}{3} + C. \quad \blacktriangleleft$$

Substitution provides us with a short method for evaluating definite integrals. We illustrate the process in the following example.

Example 7. Evaluate
$$\int_0^{\pi/4} \tan^2 x \sec^2 x\, dx.$$

Solution. Try $u = \tan x$, $du = \sec^2 x\, dx$. Note that $u(0) = 0$, $u(\pi/4) = 1$. Therefore
$$\int_0^{\pi/4} \tan^2 x \sec^2 x\, dx = \int_0^1 u^2\, du = \tfrac{1}{3}u^3\big]_0^1 = \tfrac{1}{3}. \quad \blacktriangleleft$$

There are handbooks that contain extensive tables of integrals. Therefore, when we are confronted with a problem in integration, we should first look through a table to seek the answer. However, the integrals that usually occur in practice are not precisely in a form in which the answer can be read directly from a table. The seven examples above show how an integral may be manipulated and reduced so that the answer can be obtained from one of the Formulas 1 through 15, formulas which constitute our "table of integrals."

We now enlarge the table by adding several formulas. These are integrals of algebraic expressions in which the result may be expressed in terms of the logarithm function. The verification of the formulas below follows the customary method of differentiation of the result. (See Example 8 which follows, and Problems 17 through 20 at the end of this section.) We assume that $a > 0$ in the following formulas.

16. $\int \dfrac{du}{\sqrt{u^2 + a^2}} = \ln(u + \sqrt{u^2 + a^2}) + C$

17. $\int \dfrac{du}{\sqrt{u^2 - a^2}} = \ln|u + \sqrt{u^2 - a^2}| + C$ if $|u| > a$

18. $\int \dfrac{du}{a^2 - u^2} = \begin{cases} \dfrac{1}{2a} \ln\left(\dfrac{u+a}{u-a}\right) + C \text{ if } |u| > a \\ \dfrac{1}{2a} \ln\left(\dfrac{a+u}{a-u}\right) + C \text{ if } |u| < a \end{cases}$

19. $\int \dfrac{du}{u\sqrt{a^2 - u^2}} = -\dfrac{1}{a} \ln\left|\dfrac{a + \sqrt{a^2 - u^2}}{u}\right| + C$ for $0 < |u| < a$

20. $\int \dfrac{du}{u\sqrt{a^2 + u^2}} = -\dfrac{1}{a} \ln\left|\dfrac{a + \sqrt{a^2 + u^2}}{u}\right| + C$ for $u \neq 0$

Example 8. Use differentiation to verify Formula 16 above.

Solution. We have

$$\dfrac{d}{du} \ln(u + \sqrt{u^2 + a^2}) = \dfrac{1}{u + \sqrt{u^2 + a^2}} [1 + \tfrac{1}{2}(u^2 + a^2)^{-1/2} \cdot 2u]$$

$$= \dfrac{1}{u + \sqrt{u^2 + a^2}} \left[1 + \dfrac{u}{\sqrt{u^2 + a^2}}\right]$$

$$= \dfrac{1}{u + \sqrt{u^2 + a^2}} \cdot \dfrac{\sqrt{u^2 + a^2} + u}{\sqrt{u^2 + a^2}}$$

$$= \dfrac{1}{\sqrt{u^2 + a^2}}.$$

PROBLEMS

In Problems 1 through 16, find the indefinite integrals.

1. $\int \dfrac{dx}{\sqrt{3 - 2x}}$

2. $\int \dfrac{x^3\, dx}{x^4 + 1}$

3. $\int \sin 5x\, dx$

4. $\int x^2 \sin(x^3)\, dx$

5. $\int \sec^2 2y\, dy$

6. $\int \csc^2 4x\, dx$

7. $\int \tan 2x\, dx$

8. $\int \cot x\, dx$

9. $\int e^{x/2}\, dx$

10. $\int \dfrac{dx}{\sqrt{9 - 4x^2}}$

11. $\int \dfrac{dx}{16 + 9x^2}$

12. $\int \dfrac{dx}{x\sqrt{9x^2 - 4}}$

13. $\displaystyle\int \frac{\sin\sqrt{x}\,dx}{\sqrt{x}}$
14. $\displaystyle\int x\sec^2(x^2)\,dx$
15. $\displaystyle\int xe^{x^2+1}\,dx$
16. $\displaystyle\int x^2 e^{2x^3}\,dx$

17. Verify by differentiation Formula 17.
18. Verify by differentiation Formula 18.
19. Verify by differentiation Formula 19.
20. Verify by differentiation Formula 20.

In Problems 21 through 38, in each case find the indefinite integral.

21. $\displaystyle\int \sin^2 x \cos x\,dx$
22. $\displaystyle\int \tan 2x \sec 2x\,dx$
23. $\displaystyle\int \tan 2x \sec^2 2x\,dx$
24. $\displaystyle\int \frac{\cos(\ln x)\,dx}{x}$
25. $\displaystyle\int \frac{\ln x\,dx}{x[1+(\ln x)^2]}$
26. $\displaystyle\int \frac{2\,dx}{x\sqrt{4x^2-9}}$
27. $\displaystyle\int \frac{\cos 2x\,dx}{\sin 2x}$
28. $\displaystyle\int \left(x-\frac{1}{x}\right)^2 dx$
29. $\displaystyle\int (\ln x)^3\,\frac{dx}{x}$
30. $\displaystyle\int \frac{dx}{x\ln x}$
31. $\displaystyle\int \frac{e^x\,dx}{(e^x+2)^2}$
32. $\displaystyle\int \sec^2 3x \sec 3x \tan 3x\,dx$
33. $\displaystyle\int \sec^7 2x \tan 2x\,dx$
34. $\displaystyle\int \frac{x^2}{x^2+1}\,dx$
35. $\displaystyle\int \frac{e^x\,dx}{e^x+1}$
36. $\displaystyle\int \frac{\sqrt{x}\,dx}{4+x^3}$
37. $\displaystyle\int \frac{dy}{1+e^y}$
38. $\displaystyle\int \frac{x^5\,dx}{1+x^2}$

In Problems 39 through 50, evaluate the definite integrals.

39. $\displaystyle\int_0^2 \frac{(2x+1)\,dx}{\sqrt{x^2+x+1}}$
40. $\displaystyle\int_1^3 \frac{\sqrt[3]{\ln x}\,dx}{x}$
41. $\displaystyle\int_0^{\pi/3} \sec^3 x \tan x\,dx$
42. $\displaystyle\int_0^{\pi/4} \tan^3 x \sec^2 x\,dx$
43. $\displaystyle\int_0^{1/\sqrt{2}} \frac{x\,dx}{\sqrt{1-x^4}}$
44. $\displaystyle\int_0^{\sqrt{2}} \frac{3x\,dx}{4+x^4}$
45. $\displaystyle\int_0^{\pi/2} \sin^3 x \cos x\,dx$
46. $\displaystyle\int_0^{\pi/2} \frac{\cos x\,dx}{1+\sin^2 x}$
47. $\displaystyle\int_0^{\pi/3} \frac{\sin x\,dx}{\cos^3 x}$
48. $\displaystyle\int_{\pi/8}^{\pi/4} \cot 2x \csc^2 2x\,dx$
49. $\displaystyle\int_0^1 \frac{\sqrt{1+e^{-2x}}}{e^{-3x}}\,dx$
50. $\displaystyle\int_0^{1/2} \frac{3\arcsin x}{\sqrt{1-x^2}}\,dx$

2. INTEGRATION BY SUBSTITUTION. MORE TECHNIQUES

If the indicated substitution is simple enough, the reader may shorten the integration process by rewriting the integral so that it is expressed in one of the standard forms. The following examples exhibit the method.

Example 1. Find
$$\int \frac{\sin \theta \, d\theta}{\sqrt{1 + \cos \theta}}.$$

Solution. We see that
$$d(1 + \cos \theta) = -\sin \theta \, d\theta,$$
and so
$$\int \frac{\sin \theta \, d\theta}{\sqrt{1 + \cos \theta}} = -\int (1 + \cos \theta)^{-1/2} \, d(1 + \cos \theta).$$

This is Formula 1, and we obtain
$$\int \frac{\sin \theta \, d\theta}{\sqrt{1 + \cos \theta}} = -2(1 + \cos \theta)^{1/2} + C. \quad \blacktriangleleft$$

Example 2. Find
$$\int \frac{dx}{x \ln x}.$$

Solution. Since $d(\ln x) = (1/x) \, dx$, we obtain
$$\int \frac{dx}{x \ln x} = \int \frac{d(\ln x)}{\ln x} = \ln |\ln x| + C. \quad \blacktriangleleft$$

When the integrand consists of a rational function (one polynomial divided by another polynomial) a simplification can be achieved when *the degree of the numerator is greater than or equal to the degree of the denominator,* simply by performing the division. The next two examples illustrate this technique.

Example 3. Find
$$\int \frac{x^2 - 2x - 1}{x + 2} \, dx.$$

Solution. We first divide $x^2 - 2x - 1$ by $x + 2$, using ordinary division:

$$\begin{array}{r}
x - 4 \\
x + 2 \overline{)x^2 - 2x - 1} \\
\underline{x^2 + 2x} \\
-4x - 1 \\
\underline{-4x - 8} \\
+ 7
\end{array}$$

Then
$$\int \frac{x^2 - 2x - 1}{x + 2} \, dx = \int \left(x - 4 + \frac{7}{x + 2} \right) dx$$
$$= \tfrac{1}{2}x^2 - 4x + 7 \int \frac{d(x + 2)}{x + 2}$$
$$= \tfrac{1}{2}x^2 - 4x + 7 \ln |x + 2| + C. \quad \blacktriangleleft$$

Example 4. Find
$$\int \frac{x+5}{x-1} dx.$$

Solution. Since the degree of the numerator is equal to the degree of the denominator, we may use ordinary division to obtain
$$\frac{x+5}{x-1} = 1 + \frac{6}{x-1}.$$
Therefore
$$\int \frac{x+5}{x-1} dx = \int \left(1 + \frac{6}{x-1}\right) dx = x + 6 \ln|x-1| + C. \blacktriangleleft$$

PROBLEMS

In Problems 1 through 35, find the indefinite integrals.

1. $\int \frac{(x+1)\,dx}{x^2+2x+2}$
2. $\int \csc^2\left(\frac{x}{2}\right) dx$
3. $\int \frac{\sec^2 x\,dx}{1+\tan x}$
4. $\int \frac{e^x\,dx}{1+e^{2x}}$
5. $\int \cot 2x\,dx$
6. $\int \frac{dx}{x\sqrt{\ln x}}$
7. $\int \frac{x^2+3x-2}{x+1}\,dx$
8. $\int \frac{x^3+2x^2-x+1}{x+2}\,dx$
9. $\int \frac{x}{2x-1}\,dx$
10. $\int \frac{x+1}{2x+3}\,dx$
11. $\int \frac{dx}{4+3x^2}$
12. $\int xe^{-x^2}\,dx$
13. $\int \frac{x\,dx}{(x^2+1)^2}$
14. $\int \cos 2x e^{\sin 2x}\,dx$
15. $\int x 2^{x^2}\,dx$
16. $\int \cot^2 \frac{1}{3}x \csc^2 \frac{1}{3}x\,dx$
17. $\int \frac{x^2+2x-1}{3-x}\,dx$
18. $\int \frac{x^4+x^3-2}{x+5}\,dx$
19. $\int \frac{x^2+2x+1}{x^2+1}\,dx$
20. $\int \frac{3x-2}{4x+7}\,dx$
21. $\int \frac{2x^2+6x-2}{x^2+2x-2}\,dx$
22. $\int \frac{4-5x}{2-3x}\,dx$
23. $\int e^{\tan x}\sec^2 x\,dx$
24. $\int \frac{e^{\arctan x}}{1+x^2}\,dx$
25. $\int \frac{e^x\,dx}{\sqrt{3-e^{2x}}}$
26. $\int \cot 2x \csc^3 2x\,dx$
27. $\int x^{-1/2}\sec\sqrt{x}\tan\sqrt{x}\,dx$
28. $\int \frac{\arctan 3x}{1+9x^2}\,dx$
29. $\int \frac{\sin 2x\,dx}{2+\cos 2x}$
30. $\int \frac{2x+1}{x^2+9}\,dx$
31. $\int \frac{x-2}{\sqrt{5-x^2}}\,dx$
32. $\int \frac{x^2+4}{x\sqrt{x^2-7}}\,dx$
33. $\int \frac{(2-\sin x)\,dx}{(2x+\cos x)^3}$
34. $\int \frac{1+e^{2x}}{e^x}\,dx$
35. $\int \frac{\sin 2x\,dx}{3+\cos^2 2x}$

In Problems 36 through 45, in each case evaluate the definite integral.

36. $\int_0^2 \frac{2x+3}{x^2+3x+5}\,dx$
37. $\int_{\pi/3}^{\pi} \frac{\cos(x/2)}{\sin(x/2)}\,dx$

38. $\int_1^2 \dfrac{x^2 + 1}{x + 1}\, dx$

39. $\int_{-1}^1 xe^{-3x^2}\, dx$

40. $\int_0^1 x^2 3^{(x^3)}\, dx$

41. $\int_{\pi/6}^{\pi/4} \tan(3x/2) \sec^2(3x/2)\, dx$

42. $\int_{1/2}^1 \dfrac{\arctan(2x)}{1 + 4x^2}\, dx$

43. $\int_0^{\pi^3/27} \dfrac{\sin(x^{1/3})}{x^{2/3} \cos(x^{1/3})}\, dx$

44. $\int_0^1 \dfrac{3x + 2}{x^2 + 4}\, dx$

45. $\int_{-1}^2 \dfrac{e^x}{2 + e^x}\, dx$

3. CERTAIN TRIGONOMETRIC INTEGRALS

We consider the problem of finding the indefinite integral of expressions of the following three types:

(a) $\displaystyle\int \sin^m u \cos^n u\, du,$ (b) $\displaystyle\int \tan^m u \sec^n u\, du,$ (c) $\displaystyle\int \cot^m u \csc^n u\, du.$

TYPE (a)

Under Type (a) there are two cases:

CASE 1. **Either m or n is an odd, positive integer.** If m is odd, we factor out $\sin u\, du$ and change the remaining even power of sine to powers of cosine by the trigonometric identity

$$\sin^2 u + \cos^2 u = 1.$$

If n is odd, we factor out $\cos u\, du$ and change the remaining even power of cosine to powers of sine by the same identity. Two examples illustrate the technique.

Example 1. Find $\int \sin^3 x \cos^{-5} x\, dx$.

Solution. Since $m\ (= 3)$ is odd and positive, we have

$$\int \sin^3 x \cos^{-5} x\, dx = \int \sin^2 x \cos^{-5} x \sin x\, dx$$

$$= \int (1 - \cos^2 x) \cos^{-5} x \sin x\, dx$$

$$= -\int \cos^{-5} x\, d(\cos x) + \int \cos^{-3} x\, d(\cos x)$$

$$= \dfrac{1}{4 \cos^4 x} - \dfrac{1}{2 \cos^2 x} + C. \qquad \blacktriangleleft$$

Example 2. Find $\int \sin^4 2x \cos^5 2x\, dx$.

Solution. Since $n \,(= 5)$ is odd and positive, we write

$$\int \sin^4 2x \cos^5 2x \, dx = \int \sin^4 2x \cos^4 2x \cos 2x \, dx$$

$$= \int \sin^4 2x (1 - \sin^2 2x)^2 \cos 2x \, dx$$

$$= \tfrac{1}{2} \int (\sin^4 2x - 2 \sin^6 2x + \sin^8 2x) \, d(\sin 2x)$$

$$= \tfrac{1}{10} \sin^5 2x - \tfrac{1}{7} \sin^7 2x + \tfrac{1}{18} \sin^9 2x + C. \quad \blacktriangleleft$$

CASE 2. **Both m and n are even and positive integers or zero.** In this case, the half-angle formulas are used to lower the degree of the expression. These formulas (which the student should memorize) are

$$\sin^2 u = \frac{1 - \cos 2u}{2}, \quad \cos^2 u = \frac{1 + \cos 2u}{2}.$$

The method of reduction is shown in the next two examples.

Example 3. Find $\int \sin^2 x \cos^2 x \, dx$.

Solution. By the half-angle formulas, we have

$$\int \sin^2 x \cos^2 x \, dx = \frac{1}{4} \int (1 - \cos^2 2x) \, dx = \frac{x}{4} - \frac{1}{4} \int \cos^2 2x \, dx.$$

To this last integral we again apply the half-angle formula, and get

$$\int \sin^2 x \cos^2 x \, dx = \frac{x}{4} - \frac{1}{4} \int \frac{1 + \cos 4x}{2} \, dx$$

$$= \frac{x}{4} - \frac{x}{8} - \frac{1}{32} \sin 4x + C = \frac{1}{8}\left(x - \frac{1}{4} \sin 4x\right) + C.$$

Example 4. Find $\int \sin^4 3u \, du$.

Solution. This is Case 2, with $m = 4$, $n = 0$. We have

$$\int \sin^4 3u \, du = \frac{1}{4} \int (1 - \cos 6u)^2 \, du$$

$$= \frac{1}{4} \int (1 - 2\cos 6u + \cos^2 6u) \, du$$

$$= \frac{u}{4} - \frac{1}{12} \sin 6u + \frac{1}{4} \int \frac{1 + \cos 12u}{2} \, du$$

$$= \frac{3u}{8} - \frac{1}{12} \sin 6u + \frac{1}{96} \sin 12u + C. \quad \blacktriangleleft$$

We recognize that Cases 1 and 2 do not cover all possible integrals of Type (a). For example, if both m and n are negative, the methods just described are not applicable, and other techniques must be employed. However, when both m and n are negative, an integral of Type (a) can often be changed to one of Type (b) or Type (c) with different values of m and n. Then the methods described below for Types (b) and (c) may be tried.

TYPE (b)

Under Type (b) there are also two cases:

CASE 1. **n is an even positive integer.** We factor out $\sec^2 u\, du$ and change the remaining secants to tangents, using the trigonometric identity

$$\sec^2 u = 1 + \tan^2 u.$$

Example 5. Find

$$\int \frac{\sec^4 u\, du}{\sqrt{\tan u}}.$$

Solution. We write

$$\int (\tan u)^{-1/2} \sec^4 u\, du = \int (\tan u)^{-1/2} (1 + \tan^2 u) \sec^2 u\, du$$

$$= \int (\tan u)^{-1/2} d(\tan u) + \int (\tan u)^{3/2} d(\tan u)$$

$$= 2(\tan u)^{1/2} + \tfrac{2}{5}(\tan u)^{5/2} + C. \quad \blacktriangleleft$$

CASE 2. **m is an odd positive integer.** We factor out $\sec u \tan u\, du$ and change the remaining even power of the tangents to secants, again using the identity $\tan^2 u = \sec^2 u - 1$.

Example 6. Find

$$\int \frac{\tan^3 x\, dx}{\sqrt[3]{\sec x}}.$$

Solution. We write

$$\int (\sec x)^{-1/3} \tan^3 x\, dx = \int (\sec x)^{-4/3} \tan^2 x (\sec x \tan x\, dx)$$

$$= \int (\sec x)^{-4/3} (\sec^2 x - 1)\, d(\sec x)$$

$$= \int [(\sec x)^{2/3} - (\sec x)^{-4/3}]\, d(\sec x)$$

$$= \tfrac{3}{5}(\sec x)^{5/3} + 3(\sec x)^{-1/3} + C. \quad \blacktriangleleft$$

The breakdown into cases of integrals of Type (c) is identical with that of Type (b). We simply employ the formula $\csc^2 u = 1 + \cot^2 u$ whenever necessary.

Additional methods for integrating trigonometric integrals are developed in

486 FORMULAS AND METHODS OF INTEGRATION

Section 6 where the technique known as integration by parts is described. Example 6 in that Section gives an alternate way of integrating an expression of the form $\int \sin^n x \, dx$ where n is any positive integer. Also, Example 8 shows how to integrate an expression of Type (b) where n is an odd integer, a case not covered by the above discussion.

PROBLEMS

In Problems 1 through 10, in each case evaluate the definite integral.

1. $\displaystyle\int_{-\pi/3}^{\pi/6} \cos^2 x \sin x \, dx$
2. $\displaystyle\int_0^{\pi/6} \cos^3 3x \, dx$
3. $\displaystyle\int_0^{\pi/4} \frac{\sin^3 x}{\sqrt{\cos x}} \, dx$
4. $\displaystyle\int_{-\pi/6}^{\pi/6} \tan^2 x \sec^2 x \, dx$
5. $\displaystyle\int_{\pi/6}^{\pi/3} \cot^2 2x \csc^2 2x \, dx$
6. $\displaystyle\int_0^{\pi/3} \tan^3 x \sec x \, dx$
7. $\displaystyle\int_0^{\pi/2} \cos^2 x \, dx$
8. $\displaystyle\int_0^{\pi/4} \sin^2 2x \cos^2 2x \, dx$
9. $\displaystyle\int_0^{\pi/2} \cos^4 x \, dx$
10. $\displaystyle\int_0^{\pi/3} \sin^6 x \, dx$

In Problems 11 through 32, find the indefinite integrals.

11. $\displaystyle\int \sin^5 (x/2) \, dx$
12. $\displaystyle\int \sin^3 2x \cos^2 2x \, dx$
13. $\displaystyle\int \frac{\cos^3 x}{\sin x} \, dx$
14. $\displaystyle\int \frac{\sin^5 3x}{\cos 3x} \, dx$
15. $\displaystyle\int \tan 2x \sec^2 2x \, dx$
16. $\displaystyle\int \tan (x/3) \sec^3 (x/3) \, dx$
17. $\displaystyle\int \sec^4 2x \, dx$
18. $\displaystyle\int \csc^6 x \, dx$
19. $\displaystyle\int \cot^5 x \csc^3 x \, dx$
20. $\displaystyle\int \tan^2 2x \sec^4 2x \, dx$
21. $\displaystyle\int \tan^2 x \, dx$
22. $\displaystyle\int \cot^3 (x/2) \csc^2 (x/2) \, dx$
23. $\displaystyle\int \cot^4 x \, dx$
24. $\displaystyle\int \cot^3 4x \csc^4 4x \, dx$
25. $\displaystyle\int x \sin^3 (x^2) \, dx$
26. $\displaystyle\int \frac{\sec^4 x}{\tan^2 x} \, dx$
27. $\displaystyle\int \frac{\cot x}{\csc^3 x} \, dx$
28. $\displaystyle\int \frac{\sec^2 \sqrt{x}}{\sqrt{x} \tan \sqrt{x}} \, dx$
29. $\displaystyle\int \frac{\sin^2 x}{\cos^4 x} \, dx \quad \tfrac{1}{3}\tan^3 x + C$
30. $\displaystyle\int \tan^2 2x \cos^2 2x \, dx$
31. $\displaystyle\int \tan x \cos^4 x \, dx$
32. $\displaystyle\int \frac{\sin^4 x}{\cos^2 x} \, dx$

33. Verify the formula

$$\int_0^{\pi/2} \sin^{2m+1}\theta \cos^{2n+1}\theta \, d\theta = \frac{1}{2}\frac{m!\,n!}{(m+n+1)!}$$

for $m = n = 0$, $m = n = 1$, and $m = 2$, $n = 1$. (Recall that $0! = 1$.)

34. Find the area bounded by the curve $y = \sin^3 x$, $0 \le x \le \pi$, and the x axis.

35. Find the area of the region

$$R = \{(x, y) : 0 \le x \le \pi/4, 0 \le y \le (\cos^3 x)(\sin x)^{5/2}\}.$$

36. Use the formula $\sin 2A = 2 \sin A \cos A$ to integrate an expression of Type (a) with $m = n = -1$. Do the same for $m = n = -2$. How general is this method?

37. We have

a) $\int \sin x \cos x \, dx = \int \sin x \, d(\sin x) = \frac{1}{2}\sin^2 x + C$,

b) $\int \sin x \cos x \, dx = -\int \cos x \, d(\cos x) = -\frac{1}{2}\cos^2 x + C$.

Why are both results correct?

4. TRIGONOMETRIC SUBSTITUTION

If the integrand contains expressions of the form $\sqrt{a^2 - x^2}, \sqrt{a^2 + x^2}$, or $\sqrt{x^2 - a^2}$, it is frequently possible to transform the integral into one of the forms discussed in Section 3 by means of a **trigonometric substitution.** The way the method works is shown in the following three examples.

A) If an expression of the form $\sqrt{a^2 - x^2}$ occurs, make the substitution $x = a \sin \theta$ (where we take $\theta = \arcsin(x/a)$).

In performing the substitution, the student should sketch a right triangle, as shown in Fig. 14–1. The figure contains all the essential ingredients of the process, and it is evident that $\sqrt{a^2 - x^2} = a \cos \theta$.

Example 1. Evaluate

$$\int_{-1}^{\sqrt{3}} \sqrt{4 - x^2} \, dx.$$

Solution. Let $x = 2 \sin \theta$. Then $dx = 2 \cos \theta \, d\theta$ and, according to Fig. 14–1 with $a = 2$, we have $\sqrt{4 - x^2} = 2 \cos \theta$. Therefore

$$\int_{-1}^{\sqrt{3}} \sqrt{4 - x^2} \, dx = 4 \int_{-\pi/6}^{\pi/3} \cos^2 \theta \, d\theta = 2 \int_{-\pi/6}^{\pi/3} (1 + \cos 2\theta) \, d\theta$$

$$= 2\left[\theta + \tfrac{1}{2}\sin 2\theta\right]_{-\pi/6}^{\pi/3} = \pi + \sqrt{3}. \quad \blacktriangleleft$$

We note that Fig. 14–1 is only an aid in determining the substitution. In making the change $x = 2 \sin \theta$ or, equivalently, $\theta = \arcsin(x/2)$, the range allowed for θ is $-\pi/2 \le \theta \le \pi/2$.

B) If an expression of the form $\sqrt{a^2 + x^2}$ occurs, make the substitution $x = a \tan \theta$ (or $\theta = \arctan(x/a)$).

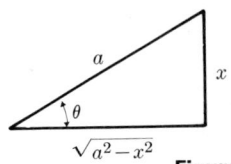
Figure 14-1

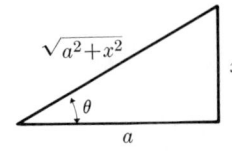
Figure 14-2

A sketch of this substitution is shown in Fig. 14–2, from which we see that $\sqrt{a^2 + x^2} = a \sec \theta$.

Example 2. Find $\int x^3 \sqrt{7 + x^2} \, dx$.

Solution. Let $x = \sqrt{7} \tan \theta$. According to Fig. 14–2, with $a = \sqrt{7}$, we have $\sqrt{7 + x^2} = \sqrt{7} \sec \theta$. Also $dx = \sqrt{7} \sec^2 \theta \, d\theta$. Therefore

$$\int x^3 \sqrt{7 + x^2} \, dx = \int 7\sqrt{7} \tan^3 \theta \sqrt{7} \sec \theta \cdot \sqrt{7} \sec^2 \theta \, d\theta$$

$$= 49\sqrt{7} \int \tan^3 \theta \sec^3 \theta \, d\theta.$$

This last integral is of Type (b) of Section 3, with m odd and positive. We obtain

$$\int x^3 \sqrt{7 + x^2} \, dx = 49\sqrt{7} \int (\sec^2 \theta - 1) \sec^2 \theta \cdot (\tan \theta \sec \theta) \, d\theta$$

$$= 49\sqrt{7} (\tfrac{1}{5} \sec^5 \theta - \tfrac{1}{3} \sec^3 \theta) + C.$$

Referring to Fig. 14–2 again, we now find that $\sec \theta = (1/\sqrt{7})\sqrt{7 + x^2}$, and so

$$\int x^3 \sqrt{7 + x^2} \, dx = \tfrac{1}{5}(7 + x^2)^{5/2} - \tfrac{7}{3}(7 + x^2)^{3/2} + C. \quad \blacktriangleleft$$

C) If an expression of the form $\sqrt{x^2 - a^2}$ occurs, make the substitution $x = a \sec \theta$. A sketch of this substitution is shown in Fig. 14–3. It follows that $\sqrt{x^2 - a^2} = a \tan \theta$ when we take $\theta = \operatorname{arcsec} (x/a)$.

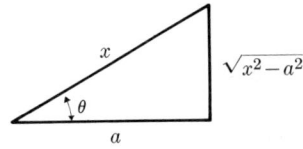
Figure 14-3

Example 3. Evaluate

$$\int_3^6 \frac{\sqrt{x^2 - 9}}{x} \, dx.$$

Solution. We make the substitution $x = 3 \sec \theta$, $dx = 3 \tan \theta \sec \theta \, d\theta$. Further, $\sqrt{x^2 - 9} = 3 \tan \theta$. When $x = 3$, we have $\sec \theta = 1$ and $\theta = 0$; when $x = 6$, we have $\sec\theta = 2$ and $\theta = \pi/3$. Therefore

$$\int_3^6 \frac{\sqrt{x^2-9}}{x}\,dx = \int_0^{\pi/3} \frac{3\tan\theta}{3\sec\theta} \cdot 3\tan\theta\sec\theta\,d\theta$$

$$= 3\int_0^{\pi/3} \tan^2\theta\,d\theta$$

$$= 3\int_0^{\pi/3} (\sec^2\theta - 1)\,d\theta$$

$$= 3[\tan\theta - \theta]_0^{\pi/3}$$

$$= 3\left(\sqrt{3} - \frac{\pi}{3}\right). \blacktriangleleft$$

Remarks. When there is an expression of the form $(a^2 - x^2)$, $(a^2 + x^2)$, or $(x^2 - a^2)$, it sometimes helps to make a substitution of Type (A), (B), or (C), respectively. We see that we can frequently perform an integration when the integrand contains the square root of a quadratic expression. If a third, fourth, or higher-degree polynomial occurs under the radical sign, it can be shown that there is no general method for carrying out the integration.

PROBLEMS

In Problems 1 through 6, evaluate the definite integrals.

1. $\displaystyle\int_0^1 \frac{dx}{\sqrt{4-x^2}}$

2. $\displaystyle\int_0^2 \frac{x^2\,dx}{x^2+4}$

3. $\displaystyle\int_{-6}^{-2\sqrt{3}} \frac{dx}{x\sqrt{x^2-9}}$

4. $\displaystyle\int_0^2 \frac{x^3\,dx}{\sqrt{16-x^2}}$

5. $\displaystyle\int_0^{2\sqrt{3}} \frac{x^3\,dx}{\sqrt{x^2+4}}$

6. $\displaystyle\int_0^1 \frac{x^2\,dx}{\sqrt{4-x^2}}$

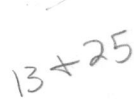

In Problems 7 through 20, find the indefinite integrals.

7. $\displaystyle\int \frac{2\,dx}{x\sqrt{x^2-5}}$

8. $\displaystyle\int \frac{\sqrt{2x^2-5}}{x}\,dx$

9. $\displaystyle\int \frac{x^3\,dx}{\sqrt{9-x^2}}$

10. $\displaystyle\int \frac{x^3\,dx}{\sqrt{3x^2-5}}$

11. $\displaystyle\int \frac{x^3\,dx}{\sqrt{2x^2+7}}$

12. $\displaystyle\int t^3\sqrt{a^2t^2-b^2}\,dt$

13. $\displaystyle\int u^3\sqrt{a^2u^2+b^2}\,du$

14. $\displaystyle\int u^3\sqrt{a^2-b^2u^2}\,du$

15. $\displaystyle\int \frac{\sqrt{x^2-a^2}}{x}\,dx$

16. $\displaystyle\int \frac{dx}{x^2\sqrt{a^2-x^2}}$

17. $\displaystyle\int \frac{dx}{x^2\sqrt{x^2+a^2}}$

18. $\displaystyle\int \frac{dx}{(a^2-x^2)^{3/2}}$

19. $\displaystyle\int \frac{dx}{x^4\sqrt{a^2-x^2}}$

20. $\displaystyle\int \frac{dx}{(x^2+a^2)^2}$

In Problems 21 through 28, evaluate the definite integrals.

21. $\displaystyle\int_0^4 \frac{dx}{(16+x^2)^{3/2}}$

22. $\displaystyle\int_{2/\sqrt{3}}^2 \frac{dx}{(x^2-1)^{3/2}}$

23. $\int_0^1 \dfrac{x^2\,dx}{(4-x^2)^{3/2}}$

24. $\int_{2\sqrt{3}}^{6} \dfrac{x^3\,dx}{(x^2-1)^{3/2}}$

25. $\int_{\sqrt{3}}^{3\sqrt{3}} \dfrac{dx}{x^2\sqrt{x^2+9}}$

26. $\int_{-2}^{2\sqrt{3}} x^3\sqrt{x^2+4}\,dx$

27. $\int_0^{\sqrt{5}} x^2\sqrt{5-x^2}\,dx$

28. $\int_1^3 \dfrac{dx}{x^4\sqrt{x^2+3}}$

29. State the domain for θ in substitutions of Types (B) and (C).

30. Suppose that an integrand contains an expression of the form
$$\sqrt{bx^2+cx+d}.$$
Complete the square and describe the conditions on b, c, and d which are required to reduce the expression to one of the three forms:
$$A\sqrt{a^2-u^2},\quad A\sqrt{a^2+u^2},\quad A\sqrt{u^2-a^2},$$
where A is some positive constant.

31. Given the integral
$$\int \dfrac{\sqrt{1-x^2}}{x^4}\,dx,$$
evaluate it two ways: (a) by trigonometric substitution and (b) by setting $x=1/t$. Compare results.

32. Evaluate
$$\int \dfrac{1}{\sqrt{1-e^{-x}}}\,dx.$$
[*Hint:* Let $u = -e^{(1/2)x}$.]

33. Given the integral
$$\int \dfrac{dx}{x^n\sqrt{1-x^2}}.$$
For what values of n can the integral be evaluated by trigonometric substitution and the methods of Section 3 applied? For what values of n can the integral be evaluated by the substitution $x = 1/t^{\alpha}$?

5. INTEGRANDS INVOLVING QUADRATIC FUNCTIONS

When an integrand involves a quadratic function of the form ax^2+bx+c, the integration is usually simplified by completing the square; we write
$$ax^2+bx+c = a\left(x^2+\dfrac{b}{a}x\right)+c = a\left(x+\dfrac{b}{2a}\right)^2+c-\dfrac{b^2}{4a^2}.$$
Then the substitution
$$y = x + \dfrac{b}{2a}$$
changes the quadratic term in the intergrand to a form resembling those considered in the previous sections. The next three examples show this procedure.

14–5 INTEGRANDS INVOLVING QUADRATIC FUNCTIONS

Example 1. Find
$$\int \frac{(2x-3)\,dx}{x^2+2x+2}.$$

Solution. We complete the square, obtaining
$$x^2+2x+2 = (x^2+2x+1)+1 = (x+1)^2+1.$$
Let $y = x+1$, $dy = dx$, and write
$$\int \frac{(2x-3)\,dx}{x^2+2x+2} = \int \frac{2y-5}{y^2+1}\,dy = \int \frac{2y\,dy}{y^2+1} - 5\int \frac{dy}{y^2+1}$$
$$= \log(y^2+1) - 5\arctan y + C$$
$$= \log(x^2+2x+2) - 5\arctan(x+1) + C. \quad \blacktriangleleft$$

Example 2. Find
$$\int \frac{dx}{\sqrt{2x-x^2}}.$$

Solution. We write
$$2x - x^2 = -(x^2-2x+1)+1 = -(x-1)^2 + 1.$$
We let $y = x-1$, $dy = dx$, and so
$$\int \frac{dx}{\sqrt{2x-x^2}} = \int \frac{dy}{\sqrt{1-y^2}} = \arcsin y + C = \arcsin(x-1) + C. \quad \blacktriangleleft$$

Example 3. Find
$$\int \frac{2x^3\,dx}{2x^2-4x+3}.$$

Solution. We perform the ordinary long division, getting
$$\int \frac{2x^3\,dx}{2x^2-4x+3} = \int \left(x+2 + \frac{5x-6}{2x^2-4x+3}\right)dx$$
$$= \tfrac{1}{2}x^2 + 2x + \int \frac{5x-6}{2x^2-4x+3}\,dx.$$
We complete the square to obtain
$$2x^2 - 4x + 3 = 2(x-1)^2 + 1;$$
therefore the substitution $y = x-1$ yields
$$\int \frac{2x^3\,dx}{2x^2-4x+3}$$
$$= \tfrac{1}{2}x^2 + 2x + \int \frac{5y-1}{2y^2+1}\,dy$$
$$= \tfrac{1}{2}x^2 + 2x + \tfrac{5}{4}\int \frac{d(2y^2+1)}{2y^2+1} - \tfrac{1}{2}\int \frac{dy}{y^2+\tfrac{1}{2}}$$
$$= \tfrac{1}{2}x^2 + 2x + \tfrac{5}{4}\ln(2y^2+1) - \tfrac{1}{2}\sqrt{2}\arctan(y\sqrt{2}) + C$$
$$= \tfrac{1}{2}x^2 + 2x + \tfrac{5}{4}\ln(2x^2-4x+3) - \tfrac{1}{2}\sqrt{2}\arctan\sqrt{2}(x-1) + C. \quad \blacktriangleleft$$

PROBLEMS

In Problems 1 through 22, find the indefinite integrals.

1. $\displaystyle\int \frac{dx}{x^2 + 4x + 5}$

2. $\displaystyle\int \frac{dx}{\sqrt{5 + 4x - x^2}}$

3. $\displaystyle\int \frac{x^2\, dx}{x^2 + 2x + 5}$

4. $\displaystyle\int \frac{x^3 - 2x}{x^2 + x + 3}\, dx$

5. $\displaystyle\int \frac{dx}{(x + 2)\sqrt{x^2 + 4x + 3}}$

6. $\displaystyle\int \frac{dx}{x^2 - x + 1}$

7. $\displaystyle\int \frac{dx}{\sqrt{5 - 2x + x^2}}$

8. $\displaystyle\int \frac{dx}{(x - 1)\sqrt{x^2 - 2x - 3}}$

9. $\displaystyle\int \frac{x + 3}{x^2 + 2x + 5}\, dx$

10. $\displaystyle\int \frac{(2x - 5)\, dx}{\sqrt{4x - x^2}}$

11. $\displaystyle\int \frac{x^4\, dx}{x^2 + 3x + 1}$

12. $\displaystyle\int \frac{4x^3}{2x^2 + 6x + 1}\, dx$

13. $\displaystyle\int \frac{(3x + 4)\, dx}{\sqrt{2x + x^2}}$

14. $\displaystyle\int \frac{\sqrt{x^2 + 2x}}{x + 1}\, dx$

15. $\displaystyle\int \frac{4x + 7}{(x^2 - 2x + 3)^2}\, dx$

16. $\displaystyle\int \frac{(5x - 3)\, dx}{(x^2 + 4x + 7)^2}$

17. $\displaystyle\int \frac{4x + 5}{(2x^2 + 3x + 4)^2}\, dx$

18. $\displaystyle\int \frac{(3x - 2)\, dx}{(2x^2 - 5x + 8)^2}$

19. $\displaystyle\int \frac{(4x + 5)\, dx}{(x^2 - 2x + 2)^{3/2}}$

20. $\displaystyle\int \frac{(x + 2)\, dx}{(3 + 2x - x^2)^{3/2}}$

21. $\displaystyle\int \frac{(2x - 3)\, dx}{(x^2 + 2x - 3)^{3/2}}$

22. $\displaystyle\int \frac{dx}{(x^2 - 2x + 5)^2}$

In Problems 23 through 28, in each case find the value of the definite integral.

23. $\displaystyle\int_0^2 \frac{dx}{x^2 + 3x + 4}$

24. $\displaystyle\int_1^2 \frac{x^2\, dx}{x^2 + 4x + 5}$

25. $\displaystyle\int_{-1}^1 \frac{dx}{x^2 - 4x + 5}$

26. $\displaystyle\int_2^4 \frac{dx}{\sqrt{8x - x^2}}$

27. $\displaystyle\int_0^1 \frac{2x + 3}{(x^2 - 2x + 5)^2}\, dx$

28. $\displaystyle\int_0^1 \frac{dx}{(x^2 + 4x + 13)^{3/2}}$

29. Find the indefinite integral of
$$\int \frac{dx}{x^2 - 1}.$$

Is it possible to use this formula to evaluate the definite integral from -2 to $+2$? Explain. Find all possible intervals in which the above integral can be evaluated as a definite integral.

In Problems 30 through 35, we state general integration formulas. Verify the validity of each by differentiating the right side. Specify conditions on a, b, and c which ensure the correctness of the formulas.

30. $\int \dfrac{dx}{\sqrt{ax^2 + bx + c}} = \dfrac{1}{\sqrt{a}} \log \left(\sqrt{ax^2 + bx + c} + \sqrt{a}\, x + \dfrac{b}{2\sqrt{a}} \right) + C$

31. $\int \dfrac{dx}{\sqrt{ax^2 + bx + c}} = \dfrac{1}{\sqrt{-a}} \arcsin \left(\dfrac{-2ax - b}{\sqrt{b^2 - 4ac}} \right) + C$

32. $\int \dfrac{dx}{(ax^2 + bx + c)^{3/2}} = 2\, \dfrac{2ax + b}{(4ac - b^2)\sqrt{ax^2 + bx + c}} + C$

33. $\int \dfrac{dx}{x\sqrt{ax^2 + bx + c}} = -\dfrac{1}{\sqrt{c}} \log \left(\dfrac{\sqrt{ax^2 + bx + c} + \sqrt{c}}{x} + \dfrac{b}{2\sqrt{c}} \right) + C$

34. $\int \dfrac{dx}{x\sqrt{ax^2 + bx + c}} = \dfrac{1}{\sqrt{-c}} \arcsin \left(\dfrac{bx + 2c}{x\sqrt{b^2 - 4ac}} \right) + C$

35. $\int \dfrac{dx}{(ax^2 + bx + c)^{5/2}} = \dfrac{2}{3} \cdot \dfrac{2ax + b}{(4ac - b^2)\sqrt{ax^2 + bx + c}}$
$\qquad\qquad\qquad\qquad \times \left(\dfrac{1}{\sqrt{ax^2 + bx + c}} + \dfrac{8a}{4ac - b^2} \right) + C$

6. INTEGRATION BY PARTS

The formula for the differential of a product is

$$d(uv) = u\, dv + v\, du.$$

By integrating both sides, we obtain $uv = \int u\, dv + \int v\, du$ or

$$\int u\, dv = uv - \int v\, du. \qquad (1)$$

Equation (1) is the formula for **integration by parts.** This formula is useful not only for evaluating integrals but also for investigating many theoretical questions.

Every integral can be written in the form $\int u\, dv$. If it happens that $\int v\, du$ is easily calculated, then the integration-by-parts formula allows us to calculate $\int u\, dv$. The next seven examples show how this method is applied.

Example 1. Find $\int xe^x\, dx$.

Solution. We wish to write this integral in the form $\int u\, dv$. To do so let

$$u = x \quad \text{and} \quad dv = e^x\, dx.$$

Then

$$du = dx \quad \text{and} \quad v = e^x.$$

The integration-by-parts formula (1) yields

$$\int xe^x\, dx = xe^x - \int e^x\, dx = xe^x - e^x + C = e^x(x - 1) + C. \qquad \blacktriangleleft$$

Example 2. Find $\int \ln x \, dx$.

Solution. To express this integral in the form $\int u \, dv$, let
$$u = \ln x \quad \text{and} \quad dv = dx.$$
Then
$$du = \frac{1}{x} dx \quad \text{and} \quad v = x.$$
Integrating by parts, we get
$$\int \ln x \, dx = x \ln x - \int x \cdot \frac{1}{x} dx = x \ln x - x + C. \quad \blacktriangleleft$$

Sometimes integration by parts must be performed several times in succession, as the following example shows.

Example 3. Find $\int x^2 e^x \, dx$.

Solution. We let
$$u = x^2 \quad \text{and} \quad dv = e^x \, dx.$$
Then
$$du = 2x \, dx \quad \text{and} \quad v = e^x.$$
Therefore
$$\int x^2 e^x \, dx = x^2 e^x - 2 \int x e^x \, dx.$$

The last integral on the right may be found by a second integration by parts, as we saw in Example 1. Taking the result given there, we obtain
$$\int x^2 e^x \, dx = x^2 e^x - 2 e^x (x - 1) + C = e^x (x^2 - 2x + 2) + C. \quad \blacktriangleleft$$

Integration by parts may be used to evaluate definite integrals. The appropriate formula is
$$\int_a^b u \, dv = \big[uv \big]_a^b - \int_a^b v \, du.$$

Example 4. Evaluate $\int_0^{\pi/2} x \cos x \, dx$.

Solution. We let
$$u = x \quad \text{and} \quad dv = \cos x \, dx.$$
Then
$$du = dx \quad \text{and} \quad v = \sin x.$$
We find that
$$\int_0^{\pi/2} x \cos x \, dx = \big[x \sin x \big]_0^{\pi/2} - \int_0^{\pi/2} \sin x \, dx$$
$$= \frac{\pi}{2} + \big[\cos x \big]_0^{\pi/2} = \frac{\pi}{2} - 1. \quad \blacktriangleleft$$

Remark. In the four examples above, we always begin with a selection of u and dv. Then we compute du and v and these choices work. In many problems the correct choices for u and dv are not immediately evident. To illustrate this, we see that in Example 4 we might easily have chosen $u = \cos x$ and $dv = x\, dx$. However, these selections lead to an integral which is more complicated than the original one. We advise the student to use *trial and error as astutely as possible* in selecting u and dv. There are no hard and fast rules. Moreover, the student should quickly abandon a selection which appears unpromising, and then go on to another choice.

The next example exhibits an interesting trick which uses integration by parts twice.

Example 5. Find $\int e^x \sin x\, dx$.

Solution. We try integration by parts by letting
$$u = e^x, \quad dv = \sin x\, dx, \quad du = e^x\, dx, \quad \text{and} \quad v = -\cos x.$$
Then
$$\int e^x \sin x\, dx = -e^x \cos x + \int e^x \cos x\, dx.$$

The integral on the right appears to be no simpler than the one with which we started. However, we now apply integration by parts to the integral on the right by letting
$$\bar{u} = e^x, \quad d\bar{v} = \cos x\, dx, \quad d\bar{u} = e^x\, dx, \quad \text{and} \quad \bar{v} = \sin x.$$
We then obtain
$$\int e^x \sin x\, dx = -e^x \cos x + e^x \sin x - \int e^x \sin x\, dx.$$
Transposing the integral on the right and dividing by 2, we get
$$\int e^x \sin x\, dx = \tfrac{1}{2} e^x (\sin x - \cos x) + C. \quad \blacktriangleleft$$

Remark. If, in the second integration by parts, we had let $\bar{u} = \cos x$, $d\bar{v} = e^x\, dx$, everything would have canceled out, as this step would have reversed the first integration by parts. The device of Example 5 is rather special, but it is spectacular when it works.

As another illustration of the power of integration by parts, we shall show how certain classes of integrals may be simplified.

Example 6. Find $\int \sin^n x\, dx$, where n is a positive integer.

Solution. The trick here is to write $\sin^n x = \sin^{n-1} x \sin x$. Then we let $u = \sin^{n-1} x$ and $dv = \sin x\, dx$. Therefore
$$du = (n-1) \sin^{n-2} x \cos x\, dx, \quad v = -\cos x.$$

Consequently
$$\int \sin^n x \, dx = -\cos x \sin^{n-1} x + (n-1) \int \sin^{n-2} x \cos^2 x \, dx.$$

We now let $\cos^2 x = 1 - \sin^2 x$ in the integral on the right, and we find that
$$\int \sin^n x \, dx = -\cos x \sin^{n-1} x + (n-1) \int \sin^{n-2} x \, dx - (n-1) \int \sin^n x \, dx.$$

Transposing the last integral on the right to the left side and dividing by n, we obtain
$$\int \sin^n x \, dx = -\frac{\cos x \sin^{n-1} x}{n} + \frac{n-1}{n} \int \sin^{n-2} x \, dx.$$

Applying the same method again to the integral on the right, we get
$$\int \sin^n x \, dx = -\frac{\cos x \sin^{n-1} x}{n}$$
$$+ \frac{n-1}{n}\left(-\frac{\cos x \sin^{n-3} x}{n-2} + \frac{n-3}{n-2} \int \sin^{n-4} x \, dx\right).$$

We continue this process until all of the integrals are evaluated. ◂

As an illustration, let $n = 5$. We have
$$\int \sin^5 x \, dx = -\frac{\cos x \sin^4 x}{5} + \frac{4}{5}\left(-\frac{\cos x \sin^2 x}{3} + \frac{2}{3} \int \sin x \, dx\right)$$
$$= -\frac{\cos x \sin^4 x}{5} - \frac{4 \cos x \sin^2 x}{15} - \frac{8}{15} \cos x + C$$
$$= -\frac{\cos x}{5}(\sin^4 x + \tfrac{4}{3} \sin^2 x + \tfrac{8}{3}) + C.$$

Remark. The integration shown in Example 6 may be performed by the methods of Section 3.

Example 7. Find $\int \sec^3 x \, dx$.

Solution. Writing $\sec^3 x = \sec x \cdot \sec^2 x$, we let $u = \sec x$, $dv = \sec^2 x \, dx$. Then $du = \tan x \sec x \, dx$ and $v = \tan x$. Therefore
$$\int \sec^3 x \, dx = \sec x \tan x - \int \tan^2 x \sec x \, dx$$
$$= \sec x \tan x - \int (\sec^2 x - 1) \sec x \, dx$$
$$= \sec x \tan x - \int \sec^3 x \, dx + \int \sec x \, dx.$$

Transferring the term $-\int \sec^3 x \, dx$ to the left side and dividing by 2, we find
$$\int \sec^3 x \, dx = \tfrac{1}{2} \sec x \tan x + \tfrac{1}{2} \int \sec x \, dx.$$

Now using Formula 11 on page 476 for $\int \sec x \, dx$, we obtain

$$\int \sec^3 x \, dx = \tfrac{1}{2} \sec x \tan x + \tfrac{1}{2} \ln |\sec x + \tan x| + C. \quad \blacktriangleleft$$

Example 8. Given that z and w are functions on R_1, with $z(0) = 0$, $z'(0) = 1$, $w(0) = 0$, and $w'(0) = 3$. Show that

$$\int_0^a z(x)w''(x) \, dx = z(a)w'(a) - z'(a)w(a) + \int_0^a w(x)z''(x) \, dx.$$

Solution. We let
$$u = z(x) \quad \text{and} \quad dv = w''(x) \, dx.$$
Then
$$du = z'(x) \, dx \quad \text{and} \quad v = w'(x).$$
Then we can write
$$\int_0^a zw'' \, dx = \Big[z(x)w'(x) \Big]_0^a - \int_0^a w'(x) z'(x) \, dx$$
$$= z(a)w'(a) - \int_0^a w'(x) z'(x) \, dx.$$

Continuing with the integral on the right side, we let
$$u = z'(x), \quad dv = w'(x) \, dx, \quad du = z''(x) \, dx, \quad \text{and} \quad v = w(x),$$
which yields
$$\int_0^a zw'' \, dx = z(a)w'(a) - \left\{ \Big[z'(x)w(x) \Big]_0^a - \int_0^a wz'' \, dx \right\}$$
$$= z(a)w'(a) - z'(a)w(a) + \int_0^a wz'' \, dx. \quad \blacktriangleleft$$

PROBLEMS

In Problems 1 through 31, find the indefinite integrals.

1. $\int x \ln x \, dx$
2. $\int x \sin x \, dx$
3. $\int x^2 \sin x \, dx$

4. $\int x^2 \ln x \, dx$
5. $\int (\ln x)^2 \, dx$
6. $\int x^3 e^{2x} \, dx$

7. $\int \arctan x \, dx$
8. $\int \arcsin x \, dx$
9. $\int \operatorname{arcsec} x \, dx$

10. $\int x \arctan x \, dx$
11. $\int x \arcsin x \, dx$
12. $\int x \operatorname{arcsec} x \, dx$

13. $\int x \csc^2 \tfrac{1}{2} x \, dx$
14. $\int (2x \, dx)/(\cos^2 2x)$
15. $\int 9x \tan^2 3x \, dx$

16. $\int x^m \ln x \, dx, \quad m \neq -1$
17. $\int 6x^2 \arcsin 2x \, dx$
18. $\int \dfrac{x \ln x \, dx}{\sqrt{x^2 - 4}}$

19. $\int \sin \sqrt{2x}\, dx$ [*Hint:* let $2x = z^2$.]

20. $\int \arcsin \sqrt{3x}\, dx$
21. $\int 2x^3 e^{x^2}\, dx$
22. $\int x^3 \arctan(x^2)\, dx$

23. $\int e^x \sin 2x\, dx$
24. $\int e^{3x} \sin 2x\, dx$
25. $\int e^{-x} \cos 3x\, dx$

26. $\int e^{ax} \cos bx\, dx$
27. $\int e^{ax} \sin bx\, dx$
28. $\int x^m (\ln x)^2\, dx$

29. $\int \sin 4x \sin 2x\, dx$
30. $\int \cos x \sin 3x\, dx$
31. $\int \sin^6 x\, dx$

In Problems 32 through 37, evaluate the definite integrals.

32. $\int_1^2 x^3 \ln x\, dx$
33. $\int_0^{(1/2)\pi^2} \cos \sqrt{2x}\, dx$
34. $\int_0^{1/2} x \arcsin 2x\, dx$

35. $\int_0^{\pi/3} x \arctan 2x\, dx$
36. $\int_0^{\pi/2} \cos^4 x\, dx$
37. $\int_0^{\pi/4} e^{3x} \sin 4x\, dx$

38. Find a formula for $\int \cos^n x\, dx$ in terms of $\int \cos^{n-4} x\, dx$.
39. Find a formula for $\int \sin^n x\, dx$ in terms of $\int \sin^{n-6} x\, dx$.
40. Given that z and w are functions of x with $z(0) = 1$, $z'(0) = 2$, $z''(0) = 3$, $z(1) = 2$, $z'(1) = -1$, $z''(1) = 2$, $w(0) = -1$, $w'(0) = 2$, $w''(0) = 0$, $w(1) = -1$, $w'(1) = -2$, and $w''(1) = -3$. Express $\int_0^1 zw'''\, dx$ in terms of $\int_0^1 wz'''\, dx$.
41. Given that $f''(x) = -af(x)$ and that $g''(x) = bg(x)$, where a and b are constants. Find the indefinite integral of $\int f(x)g''(x)\, dx$.
42. Derive the formula
$$\int (\log x)^n\, dx = x(\log x)^n - n\int (\log x)^{n-1}\, dx.$$
Apply the formula to compute $\int (\log x)^3\, dx$.
43. Derive the formula
$$\int x^p (\log x)^q\, dx = \frac{x^{p+1}(\log x)^q}{p+1} - \frac{q}{p+1}\int x^p (\log x)^{q-1}\, dx, \qquad p \neq -1.$$
Apply the formula to compute $\int x^5 (\log x)^2\, dx$.
44. Use integration by parts to evaluate $\int_1^2 x^{-2} \arctan x\, dx$.
45. Let $F(x)$ be a polynomial of the form
$$F(x) = b_0 + b_1 x + \cdots + b_n x^n.$$
Derive the formula
$$\int F(x) e^{ax}\, dx = e^{ax}\left[\sum_{k=0}^n (-1)^{k+1} a^{-k-1} F^{(k)}(x)\right]$$
where $F^{(0)}(x) = F(x)$ and $F^{(k)}(x)$ is the kth derivative of F for $k \geq 1$.
46. Derive the formula
$$\int \tan^n x\, dx = \frac{\tan^{n-1} x}{n-1} - \int \tan^{n-2} x\, dx.$$
Apply the formula to compute $\int \tan^6 x\, dx$.

47. Define the function F by the formula

$$F(x) = \int_2^x \frac{dy}{\ln y}, \quad 2 \le x < \infty.$$

Use repeated integration by parts to show that

$$F(x) = a_0 + a_1 \frac{x}{\ln x} + a_2 \frac{x}{\ln^2 x} + \cdots + a_n \frac{x}{\ln^n x} + n! \int_2^x \frac{dy}{\ln^{n+1} y}.$$

Find the values of the constants $a_0, a_1, \ldots, a_n$.

48. Use integration by parts to derive the formula

$$\int \sin(\log x)\, dx = \tfrac{1}{2}x \sin(\log x) - \tfrac{1}{2}x \cos(\log x) + C.$$

Also verify the result by differentiation of the right side.

7. INTEGRATION OF RATIONAL FUNCTIONS

If P and Q are polynomials, the integration of expressions in the form

$$\int \frac{P(x)}{Q(x)}\, dx$$

can, in theory, always be performed. In practice, however, the actual calculation of the integral depends on whether or not the denominator, $Q(x)$, can be factored. A theorem which is proved in more advanced courses states that every polynomial can be factored into a product of linear factors. That is, when $Q(x)$ is a polynomial of degree r, it may be written as a product of r linear factors:

$$Q(x) = a(x - \alpha_1)(x - \alpha_2) \cdots (x - \alpha_r).$$

In this decomposition some of the numbers $\alpha_1, \alpha_2, \ldots, \alpha_r$ may be complex. In order to be sure that we do not become involved with complex quantities, we shall use the following theorem, which is stated without proof.

Theorem 1. *Every polynomial (with real coefficients) may be decomposed into a product of linear and quadratic factors in such a way that each of the factors has real coefficients.*

For example, the polynomial

$$Q(x) = x^3 - 2x^2 + x - 2$$

can be decomposed into the linear factors

$$Q(x) = (x - 2)(x - i)(x + i).$$

Two of these factors are complex. However, the decomposition

$$Q(x) = (x - 2)(x^2 + 1)$$

into a linear and a quadratic factor has only real quantities.

The basic method of integrating the rational function P/Q consists of two steps: (1) factoring Q into a product of linear and quadratic factors, and (2) writing P/Q as a sum of simpler rational functions, each of which can be integrated by methods which we have already learned.

The student has studied methods of performing step (1) in high-school and college algebra. Finding the factors may be very difficult in specific cases, but we shall assume that this can always be done.

Before discussing step (2) in detail, let us look at some examples of simplifications leading to integrations which can be performed:

$$\frac{2x^2 - 3x + 5}{(x + 2)(x - 1)(x - 3)} = \frac{A}{x + 2} + \frac{B}{x - 1} + \frac{C}{x - 3},$$

$$\frac{x^4 - 2x^2 + 3x + 4}{(x - 1)^3(x^2 + 2x + 2)} = \frac{D}{x - 1} + \frac{E}{(x - 1)^2} + \frac{F}{(x - 1)^3} + \frac{Gx + H}{x^2 + 2x + 2},$$

$$\frac{2x^4 + 3x^3 - x - 1}{(x - 1)(x^2 + 2x + 2)^2} = \frac{J}{x - 1} + \frac{Kx + L}{x^2 + 2x + 2} + \frac{Mx + N}{(x^2 + 2x + 2)^2}.$$

Each of the above equations is an identity in x for properly chosen constants A, B, C, D, etc. Note that every term on the right side is one which we can integrate. For example,

$$\int \frac{A}{x + 2} \, dx = A \ln |x + 2| + C.$$

The integrals

$$\int \frac{dx}{(x - 1)^2}, \quad \int \frac{dx}{(x - 1)^3}$$

are routine. An expression of the form

$$\int \frac{Gx + H}{x^2 + 2x + 2} \, dx$$

succumbs if we complete the square in the denominator and proceed as described in Section 5, Example 1. The integration of

$$\int \frac{Mx + N}{(x^2 + 2x + 2)^2} \, dx$$

is performed by completing the square in the denominator and then making a trigonometric substitution of the type described in Section 4.

We now turn to the problem of decomposing a rational function P/Q into simpler expressions.

If the degree of P is larger than or equal to the degree of Q, apply long division. Then P/Q will equal a polynomial (quotient) plus a rational function (remainder divided by the divisor) in which the degree of the numerator is definitely less than the degree of the denominator. From now on, we shall always suppose that this simplification has already been performed.

The decomposition of a rational function into the sum of simpler expressions is known as the **method of partial fractions.** We divide the method into four cases, depending on the way the denominator factors. We state the results without proof.

CASE 1. **The denominator $Q(x)$ can be factored into linear factors, all different.** If we can make the decomposition

$$Q(x) = (x - a_1)(x - a_2) \cdots (x - a_r),$$

with no two of the a_i the same, then we can decompose P/Q so that

$$\frac{P(x)}{Q(x)} = \frac{A_1}{x - a_1} + \frac{A_2}{x - a_2} + \frac{A_3}{x - a_3} + \cdots + \frac{A_r}{x - a_r},$$

where $A_1, A_2, \ldots, A_r$ are properly chosen constants.

We shall show by example how the constants may be found.

Example 1. Decompose

$$(x^2 + 2x + 3)/(x^3 - x)$$

into partial fractions and integrate.

Solution. $Q(x) = x^3 - x = (x - 0)(x - 1)(x + 1)$. We write

$$\frac{x^2 + 2x + 3}{x(x - 1)(x + 1)} = \frac{A_1}{x} + \frac{A_2}{x - 1} + \frac{A_3}{x + 1},$$

which is an identity for all x ($x \neq 0, 1, -1$) if and only if

$$x^2 + 2x + 3 = A_1(x - 1)(x + 1) + A_2 x(x + 1) + A_3 x(x - 1). \tag{1}$$

Multiplying out the right side, we obtain

$$x^2 + 2x + 3 = (A_1 + A_2 + A_3)x^2 + (A_2 - A_3)x - A_1. \tag{2}$$

If these two polynomials are to be *identical*, every coefficient on the left must equal every coefficient on the right. Then the polynomials are the same for *all* values of x. It is better to work with (1) than with (2). We proceed as follows:

If in (1) we set $x = 0$: $\quad 3 = -A_1 \quad$ and $\quad A_1 = -3$.
If in (1) we set $x = 1$: $\quad 6 = 2A_2 \quad$ and $\quad A_2 = 3$.
If in (1) we set $x = -1$: $\quad 2 = 2A_3 \quad$ and $\quad A_3 = 1$.

We conclude that

$$\int \frac{(x^2 + 2x + 3)\,dx}{x(x - 1)(x + 1)} = \int \left(-\frac{3}{x} + \frac{3}{x - 1} + \frac{1}{x + 1}\right) dx$$

$$= -3\ln|x| + 3\ln|x - 1| + \ln|x + 1| + C$$

$$= \ln\left|\frac{(x - 1)^3(x + 1)}{x^3}\right| + C. \quad \blacktriangleleft$$

In the example above, we made use of the following theorem about polynomials.

Theorem 2. *Suppose that two polynomials*

$$S(x) = a_0 + a_1 x + a_2 x^2 + \cdots + a_n x^n,$$
$$T(x) = b_0 + b_1 x + \cdots + b_n x^n$$

are equal for all except possibly a finite number of values of x. Then $a_i = b_i$ *for all* $i = 0, 1, 2, \ldots, n$.

Proof. Since polynomial functions are continuous, $S(x)$ and $T(x)$ must be equal for *all* values of x. We form the expression

$$S(x) - T(x) = (a_0 - b_0) + (a_1 - b_1)x + \cdots + (a_n - b_n)x^n \equiv 0.$$

Setting $x = 0$, we obtain $S(0) - T(0) = a_0 - b_0 = 0$, or $a_0 = b_0$. Then

$$S(x) - T(x) = (a_1 - b_1)x + (a_2 - b_2)x^2 + \cdots + (a_n - b_n)x^n \equiv 0.$$

Dividing through by x and again setting $x = 0$, we obtain $a_1 - b_1 = 0$. Continuing in this way, we get $a_i = b_i$ for each i from 0 to n.

CASE 2. **The denominator $Q(x)$ can be factored into linear factors, some of which are repeated.** If we can make the decomposition

$$Q(x) = (x - a_1)^{s_1}(x - a_2)^{s_2} \cdots (x - a_r)^{s_r}$$

where the exponents $s_1, s_2, \ldots, s_r$ are positive integers, then the partial fraction decomposition introduces a number of different types of denominators. For example, a factor such as $(x - 1)^4$ gives rise to the four terms

$$\frac{A_1}{x - 1} + \frac{A_2}{(x - 1)^2} + \frac{A_3}{(x - 1)^3} + \frac{A_4}{(x - 1)^4},$$

where $A_1, \ldots, A_4$ are properly chosen constants. In general, a factor such as $(x - a)^q$ gives rise to the terms (q in number)

$$\frac{A_1}{x - a} + \frac{A_2}{(x - a)^2} + \frac{A_3}{(x - a)^3} + \cdots + \frac{A_q}{(x - a)^q}.$$

Example 2. Find

$$\int \frac{x + 5}{x^3 - 3x + 2}\, dx.$$

Solution. We factor the denominator, obtaining

$$x^3 - 3x + 2 = (x - 1)^2(x + 2).$$

This falls under Case 2, and we write

$$\frac{x + 5}{(x - 1)^2(x + 2)} = \frac{A_1}{x - 1} + \frac{A_2}{(x - 1)^2} + \frac{A_3}{x + 2}.$$

Multiplying through by $(x - 1)^2(x + 2)$, we obtain

$$x + 5 = A_1(x - 1)(x + 2) + A_2(x + 2) + A_3(x - 1)^2.$$

We let

$$x = 1: \quad 6 = 3A_2 \quad \Leftrightarrow \quad A_2 = 2;$$
$$x = -2: \quad 3 = 9A_3 \quad \Leftrightarrow \quad A_3 = \tfrac{1}{3};$$
$$x = 0: \quad 5 = -2A_1 + 2A_2 + A_3 \quad \Leftrightarrow \quad A_1 = -\tfrac{1}{3}.$$

Therefore

$$\int \frac{(x+5)\,dx}{(x-1)^2(x+2)} = -\frac{2}{x-1} - \frac{1}{3}\ln|x-1| + \frac{1}{3}\ln|x+2| + C. \quad \blacktriangleleft$$

CASE 3. **The denominator $Q(x)$ can be factored into linear and quadratic factors, and none of the quadratic factors is repeated.**

If, for example, the denominator is

$$Q(x) = (x - a_1)(x - a_2)(x - a_3)(x^2 + b_1x + c_1)(x^2 + b_2x + c_2),$$

then

$$\frac{P(x)}{Q(x)} = \frac{A_1}{x - a_1} + \frac{A_2}{x - a_2} + \frac{A_3}{x - a_3} + \frac{A_4x + A_5}{x^2 + b_1x + c_1} + \frac{A_6x + A_7}{x^2 + b_2x + c_2}.$$

In other words, each unrepeated quadratic factor gives rise to a term of the form

$$\frac{Ax + B}{x^2 + bx + c}.$$

Example 3. Find

$$\int \frac{3x^2 + x - 2}{(x-1)(x^2+1)}\,dx.$$

Solution. According to Case 3, we have

$$\frac{3x^2 + x - 2}{(x-1)(x^2+1)} = \frac{A_1}{x - 1} + \frac{A_2x + A_3}{x^2 + 1}$$

or

$$3x^2 + x - 2 = A_1(x^2 + 1) + (A_2x + A_3)(x - 1).$$

When

$$x = 1: \quad 2 = 2A_1 \quad \Leftrightarrow \quad A_1 = 1;$$
$$x = 0: \quad -2 = A_1 - A_3 \quad \Leftrightarrow \quad A_3 = 3;$$
$$x = 2: \quad 12 = 5A_1 + (2A_2 + A_3) \quad \Leftrightarrow \quad A_2 = 2.$$

Therefore

$$\int \frac{3x^2 + x - 2}{(x-1)(x^2+1)}\,dx = \ln|x - 1| + \ln(x^2 + 1) + 3\arctan x + C. \quad \blacktriangleleft$$

CASE 4. **The denominator $Q(x)$ can be factored into linear and quadratic factors, and some of the quadratic factors are repeated.** If the denominator contains a factor such as $(x^2 + 3x + 5)^3$, it will give rise to the three terms

$$\frac{A_1x + A_2}{x^2 + 3x + 5} + \frac{A_3x + A_4}{(x^2 + 3x + 5)^2} + \frac{A_5x + A_6}{(x^2 + 3x + 5)^3}.$$

In general, a factor of the form $(x^2 + bx + c)^q$ will give rise to the q terms

$$\frac{A_1x + A_2}{x^2 + bx + c} + \frac{A_3x + A_4}{(x^2 + bx + c)^2} + \cdots + \frac{A_{2q-1}x + A_{2q}}{(x^2 + bx + c)^q}.$$

Example 4. Find

$$\int \frac{2x^3 + 3x^2 + x - 1}{(x + 1)(x^2 + 2x + 2)^2} dx.$$

Solution. This is Case 4, and we have

$$\frac{2x^3 + 3x^2 + x - 1}{(x + 1)(x^2 + 2x + 2)^2} = \frac{A_1}{x + 1} + \frac{A_2x + A_3}{x^2 + 2x + 2} + \frac{A_4x + A_5}{(x^2 + 2x + 2)^2}$$

or

$$2x^3 + 3x^2 + x - 1 = A_1(x^2 + 2x + 2)^2$$
$$+ (A_2x + A_3)(x^2 + 2x + 2)(x + 1)$$
$$+ (A_4x + A_5)(x + 1).$$

When

$x = -1$: $-1 = A_1$ and $A_1 = -1$;
$x = 0$: $-1 = 4A_1 + 2A_3 + A_5$ and $2A_3 + A_5 = 3$;
$x = 1$: $5 = 25A_1 + (A_2 + A_3) \cdot (10) + (A_4 + A_5)(2)$ and
            $5A_2 + 5A_3 + A_4 + A_5 = 15$;
$x = 2$: $20A_2 + 10A_3 + 2A_4 + A_5 = 43$;
$x = -2$: $4A_2 - 2A_3 + 2A_4 - A_5 = -3$.

Solving the four equations for the four unknowns, A_2, A_3, A_4, and A_5, we get

$$A_2 = 1, \quad A_3 = 3, \quad A_4 = -2, \quad A_5 = -3.$$

Therefore

$$\int \frac{2x^3 + 3x^2 + x - 1}{(x + 1)(x^2 + 2x + 2)^2} dx = -\ln|x + 1| + \int \frac{(x + 3)\,dx}{x^2 + 2x + 2} + \int \frac{(-2x - 3)\,dx}{(x^2 + 2x + 2)^2}.$$

We complete the square and set $u = x + 1$, $du = dx$, obtaining

$$\int \frac{(x + 3)\,dx}{x^2 + 2x + 2} = \int \frac{u + 2}{u^2 + 1} du = \frac{1}{2}\ln|x^2 + 2x + 2| + 2\arctan(x + 1),$$

$$\int \frac{-2x - 3}{(x^2 + 2x + 2)^2} dx = \int \frac{-2u - 1}{(u^2 + 1)^2} du = \frac{1}{u^2 + 1} - \int \frac{du}{(u^2 + 1)^2}$$

$$= \frac{1}{x^2 + 2x + 2} - \frac{1}{2}\arctan(x + 1) - \frac{1}{2}\frac{x + 1}{x^2 + 2x + 2}.$$

The last two terms on the right are obtained by using a trigonometric substitution in the integral above. Combining these integrals, we conclude that

$$\int \frac{2x^3 + 3x^2 + x - 1}{(x + 1)(x^2 + 2x + 2)^2} dx = -\ln|x + 1| + \frac{1}{2}\ln(x^2 + 2x + 2)$$

$$+ \frac{3}{2}\arctan(x + 1) - \frac{1}{2}\frac{x - 1}{x^2 + 2x + 2} + C. \blacktriangleleft$$

PROBLEMS

In Problems 1 through 34, find the indefinite integrals.

1. $\displaystyle\int \frac{x^2 + 3x + 4}{x - 2}\,dx$

2. $\displaystyle\int \frac{x^3 + x^2 - x - 3}{x + 2}\,dx$

3. $\displaystyle\int \frac{x^3 - x^2 + 2x + 3}{x^2 + 3x + 2}\,dx$

4. $\displaystyle\int \frac{2x^3 + 3x^2 - 4}{x^2 - 4x + 3}\,dx$

5. $\displaystyle\int \frac{x^2 + 2x + 3}{x^2 - 3x + 2}\,dx$

6. $\displaystyle\int \frac{x^4 + 1}{x^3 - x}\,dx$

7. $\displaystyle\int \frac{x^2 - 2x - 1}{x^2 - 4x + 4}\,dx$

8. $\displaystyle\int \frac{x^2 + 2x + 3}{(x + 1)(x - 1)(x - 2)}\,dx$

9. $\displaystyle\int \frac{3x - 2}{(x + 2)(x + 1)(x - 1)}\,dx$

10. $\displaystyle\int \frac{x^3 + 2}{x^2 + 4}\,dx$

11. $\displaystyle\int \frac{2x^2 + 3x - 1}{(x + 3)(x + 2)(x - 1)}\,dx$

12. $\displaystyle\int \frac{x^2 - 2}{(x + 1)(x - 1)^2}\,dx$

13. $\displaystyle\int \frac{x^2 + 3x + 3}{(x + 1)(x^2 + 1)}\,dx$

14. $\displaystyle\int \frac{x^2 - 2x - 3}{(x - 1)(x^2 + 2x + 2)}\,dx$

15. $\displaystyle\int \frac{x - 3}{(x + 1)^2(x - 2)}\,dx$

16. $\displaystyle\int \frac{x^2 + 1}{(x - 1)^3}\,dx$

17. $\displaystyle\int \frac{2x + 3}{(x + 2)(x - 1)^2}\,dx$

18. $\displaystyle\int \frac{2x^2 - 1}{(x + 1)^2(x - 3)}\,dx$

19. $\displaystyle\int \frac{x^3 - 3x + 4}{(x + 1)(x - 1)^3}\,dx$

20. $\displaystyle\int \frac{x^3 + 1}{(x^2 - 1)^2}\,dx$

21. $\displaystyle\int \frac{x^3 + 3x^2 - 2x + 1}{x^4 + 5x^2 + 4}\,dx$

22. $\displaystyle\int \frac{x^2}{x^4 - 5x^2 + 4}\,dx$

23. $\displaystyle\int \frac{x^2 - x + 1}{x^4 - 5x^3 + 5x^2 + 5x - 6}\,dx$

24. $\displaystyle\int \frac{3x\,dx}{x^5 + 2x^4 - 10x^3 - 20x^2 + 9x + 18}$

25. $\displaystyle\int \frac{x^2 - 2x + 3}{(x - 1)^2(x^2 + 4)}\,dx$

26. $\displaystyle\int \frac{x^3 - 2x^2 + 3x - 4}{(x - 1)^2(x^2 + 2x + 2)}\,dx$

27. $\displaystyle\int \frac{x^3 + x^2 - 2x - 3}{(x + 1)^2(x - 2)^2}\,dx$

28. $\displaystyle\int \frac{x^2 - 3x + 5}{x^4 - 8x^2 + 16}\,dx$

29. $\displaystyle\int \frac{4x^3 + 8x^2 - 12}{(x^2 + 4)^2}\,dx$

30. $\displaystyle\int \frac{x^2 + 3x + 5}{x^3 + 8}\,dx$

31. $\displaystyle\int \frac{x^2 + 2x - 1}{x^3 - 27}\,dx$

32. $\displaystyle\int \frac{x^3 - x^2 + 2x + 3}{(x^2 + 2x + 2)^2}\,dx$

33. $\displaystyle\int \frac{x^4 + 1}{(x^2 + 4)^3}\,dx$

34. $\displaystyle\int \frac{2x^5 - 6}{(x^2 + 1)^4}\,dx$

35. Let f and g each be polynomials of degree n. If $f^{(k)}(0) = g^{(k)}(0)$, $k = 0, 1, 2, \ldots, n$, show that $f(x) \equiv g(x)$ for all x.

36. Verify that $x - \alpha$ is a factor of the third-degree polynomial $x^3 + ax^2 + bx - \alpha^3 - a\alpha^2 - b\alpha$, where a, b, and α are real numbers. Then find conditions on a, b, and α such that the polynomial is factorable into real, linear factors.

37. Suppose that $Q(x)$ is a polynomial with factors $(x - a_1), (x - a_2), \ldots, (x - a_r)$, and

that at least one of the factors is repeated. Show that in general it is *not* possible to obtain a partial fraction expansion

$$\frac{1}{Q(x)} = \frac{A_1}{x - a_1} + \frac{A_2}{x - a_2} + \cdots + \frac{A_r}{x - a_r}.$$

That is, show that Case 1 is not applicable.

38. Derive the formula

$$\int \frac{x\,dx}{(ax + b)(cx + d)} = \frac{1}{bc - ad}\left[\frac{b}{a}\log(ax + b) - \frac{d}{c}\log(cx + d)\right], \qquad bc - ad \neq 0.$$

What is the result if $bc - ad = 0$?

39. Derive the formula

$$\int \frac{x\,dx}{(ax + b)^2(cx + d)} = \frac{1}{bc - ad}\left[-\frac{b}{a(ax + b)} - \frac{d}{bc - ad}\log\frac{cx + d}{ax + b}\right],$$
$$bc - ad \neq 0.$$

What is the result if $bc - ad = 0$?

40. Write the partial fraction expansion formula for the most general rational function

$$\frac{P(x)}{Q(x)}$$

in which the degree of P is less than the degree of Q. That is, assume Q has r simple, distinct roots, s multiple roots of multiplicities $t_1, t_2, \ldots, t_s$; also that it has p distinct, simple quadratic factors, and q multiple quadratic factors of multiplicities $z_1, z_2, \ldots, z_q$.

8. TWO RATIONALIZING SUBSTITUTIONS

I. Whenever an integrand contains a single irrational expression of the form

$$(ax + b)^{p/q}, \qquad p \text{ and } q \text{ integers},$$

the substitution

$$u = (ax + b)^{1/q} \quad \Leftrightarrow \quad x = \frac{u^q - b}{a}, \qquad dx = \frac{q}{a}u^{q-1}\,du$$

will convert the given integrand into a rational function of u.

Example 1. Find

$$\int \frac{\sqrt[3]{x + 1}}{x}\,dx.$$

Solution. Let

$$u = (x + 1)^{1/3} \quad \Leftrightarrow \quad x = u^3 - 1, \qquad dx = 3u^2\,du.$$

Then

$$\int \frac{\sqrt[3]{x + 1}\,dx}{x} = \int \frac{u \cdot 3u^2\,du}{u^3 - 1} = \int 3\,du + \int \frac{3\,du}{u^3 - 1}.$$

To evaluate the second integral, we use partial fractions and write

$$\frac{3}{u^3 - 1} = \frac{3}{(u-1)(u^2 + u + 1)} = \frac{A_1}{u-1} + \frac{A_2 u + A_3}{u^2 + u + 1}$$
$$\Leftrightarrow \quad 3 = A_1(u^2 + u + 1) + (A_2 u + A_3)(u - 1).$$

When
$u = 1$: $\quad 3 = 3A_1 \quad$ and $\quad A_1 = 1$;
$u = 0$: $\quad 3 = A_1 - A_3 \quad$ and $\quad A_3 = -2$;
$u = -1$: $\quad 3 = A_1 + 2A_2 - 2A_3 \quad$ and $\quad A_2 = -1$.

Therefore

$$\int \frac{3\,du}{u^3 - 1} = \int \frac{du}{u - 1} - \int \frac{u+2}{u^2 + u + 1}\,du$$
$$= \ln|u - 1| - \tfrac{1}{2}\ln(u^2 + u + 1) - \sqrt{3}\arctan\left(\frac{2u+1}{\sqrt{3}}\right) + C.$$

We finally obtain

$$\int \frac{\sqrt[3]{x+1}}{x}\,dx = 3(x+1)^{1/3} + \ln|(x+1)^{1/3} - 1|$$
$$- \tfrac{1}{2}\ln|(x+1)^{2/3} + (x+1)^{1/3} + 1|$$
$$- \sqrt{3}\arctan\left[\frac{2(x+1)^{1/3} + 1}{\sqrt{3}}\right] + C. \quad \blacktriangleleft$$

II. If a *single* irrational expression of one of the forms

$$\sqrt{a^2 - x^2}, \quad \sqrt{x^2 + a^2}, \quad \sqrt{x^2 - a^2}$$

appears in the integrand, and if x^q (q being an *odd* integer, positive or negative) appears in the integrand, an appropriate substitution will transform the integrand into a rational function. The correct substitution is one of the following:

$$u = (a^2 - x^2)^{1/2} \quad \text{or} \quad u = (x^2 + a^2)^{1/2} \quad \text{or} \quad u = (x^2 - a^2)^{1/2},$$

depending on which expression is involved.

Example 2. Find

$$\int \frac{\sqrt{a^2 - x^2}}{x^3}\,dx.$$

Solution. This is of the form II, with $q = -3$. We let

$$u = (a^2 - x^2)^{1/2} \quad \Leftrightarrow \quad x^2 = a^2 - u^2, \quad x\,dx = -u\,du.$$

Therefore

$$\int \frac{\sqrt{a^2 - x^2}}{x^3}\,dx = \int \frac{\sqrt{a^2 - x^2}\,x\,dx}{x^4}$$
$$= -\int \frac{u^2\,du}{(a^2 - u^2)^2}$$
$$= -\int \frac{u^2\,du}{(u+a)^2(u-a)^2}.$$

We proceed by partial fractions and obtain

$$\frac{-u^2}{(u+a)^2(u-a)^2} = \frac{A_1}{u+a} + \frac{A_2}{(u+a)^2} + \frac{A_3}{u-a} + \frac{A_4}{(u-a)^2}$$

and

$$-u^2 = A_1(u+a)(u-a)^2 + A_2(u-a)^2 + A_3(u-a)(u+a)^2 + A_4(u+a)^2.$$

We find

$u = a$: $\quad -a^2 = A_4 4a^2 \quad$ and $\quad A_4 = -\frac{1}{4}$;

$u = -a$: $\quad -a^2 = A_2 4a^2 \quad$ and $\quad A_2 = -\frac{1}{4}$;

$u = 0$: $\quad 0 = a^3 A_1 + a^2 A_2 - a^3 A_3 + a^2 A_4 \quad$ and $\quad A_1 - A_3 = \frac{1}{2a}$.

$u = 2a$: $\quad -4a^2 = 3a^3 A_1 + a^2 A_2 + 9a^3 A_3 + 9a^2 A_4 \quad$ and

$$A_1 + 3A_3 = -\frac{1}{2a}.$$

We get

$$A_1 = \frac{1}{4a}, \quad A_3 = -\frac{1}{4a},$$

which yields

$$\int \frac{\sqrt{a^2 - x^2}}{x^3} dx = \frac{1}{4(u+a)} + \frac{1}{4(u-a)} + \frac{1}{4a} \ln\left|\frac{u+a}{u-a}\right| + C$$

$$= -\frac{\sqrt{a^2 - x^2}}{2x^2} + \frac{1}{2a} \ln(a + \sqrt{a^2 - x^2}) - \frac{1}{2a} \ln|x| + C.$$

PROBLEMS

In Problems 1 through 29, find the indefinite integrals.

1. $\int \frac{(2x + 3) dx}{\sqrt{x + 2}}$

2. $\int \frac{3x - 2}{\sqrt{2x - 3}} dx$

3. $\int x\sqrt{x + 1}\, dx$

4. $\int \frac{2x + 1}{(x + 2)^{2/3}} dx$

5. $\int \frac{x - 2}{(3x - 1)^{2/3}} dx$

6. $\int \frac{2x - 1}{(x - 2)^{1/3}} dx$

7. $\int \frac{x^2\, dx}{\sqrt[3]{2x + 1}}$

8. $\int (x + 2)\sqrt{x - 1}\, dx$

9. $\int \frac{\sqrt{x + 4}}{x} dx$

10. $\int \frac{\sqrt{2x + 3}}{x + 1} dx$

11. $\int \frac{\sqrt{x + 2}}{\sqrt{x - 1}} dx$

12. $\int \frac{2\sqrt{x + 1} - 3}{3\sqrt{x + 1} - 2} dx$

13. $\int \frac{x^3}{\sqrt{x^2 - 4}} dx$

14. $\int \frac{x^3 - x}{\sqrt{9 - x^2}} dx$

15. $\int \frac{x^5 + 2x^3}{\sqrt{x^2 + 4}} dx$

16. $\int x^3\sqrt{x^2 + 1}\, dx$

17. $\int (x^3 - x)\sqrt{16 - x^2}\, dx$

18. $\int x^3\sqrt{x^2 - a^2}\, dx$

19. $\int \frac{dx}{x\sqrt{a^2 - x^2}}$

20. $\int \frac{dx}{x\sqrt{x^2 + 4}}$

21. $\displaystyle\int \frac{\sqrt{a^2 - x^2}}{x}\, dx$ 22. $\displaystyle\int \frac{\sqrt{a^2 + x^2}}{x}\, dx$ 23. $\displaystyle\int \frac{\sqrt{x^2 - a^2}}{x}\, dx$

24. $\displaystyle\int \frac{\sqrt{4 - x}}{x}\, dx$ 25. $\displaystyle\int \frac{dx}{x^{1/2} + x^{2/3}}$ 26. $\displaystyle\int \sqrt{2 + \sqrt{x}}\, dx$

27. $\displaystyle\int \frac{dx}{x^3 \sqrt{x^2 + 4}}$ 28. $\displaystyle\int \frac{\sqrt{4 - x^2}}{x^3}\, dx$ 29. $\displaystyle\int \frac{\sqrt{x^2 + 9}}{x^3}\, dx$

*30. Derive the formula

$$\int x^{m-1}(ax + b)^{p/q}\, dx$$

$$= \frac{q}{a(qm + p)}\left[x^{m-1}(ax + b)^{(p/q)+1} - (m - 1)b \int x^{m-2}(ax + b)^{p/q}\, dx \right].$$

31. Given an integral of the form

$$\int \left(x + \frac{b}{2a} \right)^q F(\sqrt{ax^2 + bx + c})\, dx,$$

where q is an odd integer and F is a polynomial function. Show how this integral can be transformed into an integral of a rational function.

32. Use the substitution $u = \tan(\theta/2)$ to transform the integral

$$\int \frac{d\theta}{5 - 4 \cos \theta}$$

into the integral of a rational function. *Hint:* Use the schematic triangle in Fig. 14–4 to read off

$$\sin \frac{1}{2}\theta = \frac{u}{\sqrt{1 + u^2}}, \qquad \cos \frac{1}{2}\theta = \frac{1}{\sqrt{1 + u^2}},$$

and also employ the relations

$$\sin \theta = 2 \sin \frac{1}{2}\theta \cos \frac{1}{2}\theta = \frac{2u}{1 + u^2},$$

$$\cos \theta = \cos^2 \frac{\theta}{2} - \sin^2 \frac{\theta}{2} = \frac{1 - u^2}{1 + u^2}.$$

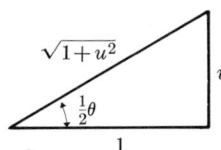

Figure 14–4

In Problems 33 through 38, use the substitution of Problem 32 to compute the integral.

33. $\displaystyle\int \frac{dx}{\tan x - \sin x}$ 34. $\displaystyle\int \frac{dx}{3 \cos x + 4 \sin x}$

35. $\displaystyle\int \frac{d\theta}{\cos \theta + \cot \theta}$ 36. $\displaystyle\int \frac{dx}{a \cos x + b \sin x}$

37. $\displaystyle\int \frac{d\theta}{4 + 5 \sin \theta}$ 38. $\displaystyle\int \cos^2 x\, dx$

9. SUMMARY

For convenient reference, we list here the methods of integration discussed in this chapter.

1) Integration by substitution in a formula. (See Sections 1 and 2.)
2) Integrations of certain integrals involving trigonometric functions. (See Section 3.)
3) Trigonometric substitutions. (See Section 4.)
4) Completing the square in quadratic functions. (See Section 5.)
5) Integration by parts. (See Section 6.)
6) Integration of rational functions. (See Section 7.)
7) Rationalizing substitutions which reduce problems to the method of (6). (See Section 8.)

The reader should become thoroughly familiar with the methods in this chapter. When one is confronted with an integral, the chances are that it will not be precisely in one of the forms given in the extensive tables of integrals found in various handbooks. The techniques developed in this chapter, however, are invaluable for transforming integrals into "handbook types."

15

APPLICATIONS OF THE INTEGRAL

1. VOLUME OF A SOLID OF REVOLUTION. DISK METHOD

In Chapter 7 we developed a method for measuring the area of plane regions. The theory rested on two ideas: (1) the formula for the area of a rectangle, and (2) a method of approximating any region by a combination of rectangular regions.

It is possible to extend the above process to the measurement of volumes of solids. First, the formula for the volume of a rectangular parallelepiped is length times width times height; next we must find a method for approximating a solid by a combination of rectangular parallelepipeds. Instead of discussing this topic in its most general form, we take up the problem of measuring the volume of solids which have special shapes.

Let R be a region in the xy plane (Fig. 15–1), and L a line which does not intersect it (although L may touch the boundary of R). If the region R is revolved about the line L, a solid results which is called a **solid of revolution.** If the region R is a circle, the resulting solid will have the shape of a doughnut (called a **torus**), as seen in Fig. 15–2. A semicircle revolved about its diameter, as shown in Fig. 15–3, generates a sphere. A rectangle revolved about one of the edges, as in Fig. 15–4, yields a right circular cylinder.

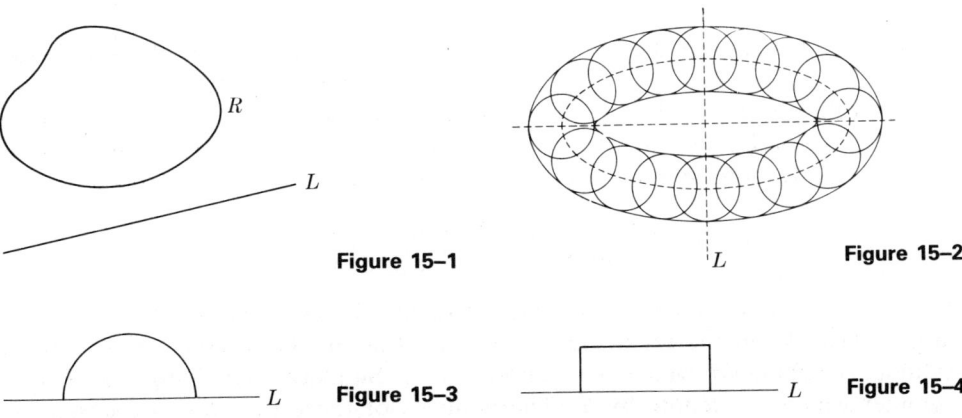

Figure 15–1   Figure 15–2

Figure 15–3   Figure 15–4

General methods for obtaining the volume of a solid by integration are developed in Chapter 18. In this section we shall discuss a method for obtaining the volume of a solid of revolution by the techniques of integration we have already studied. The basis of this method depends on two assumptions. The first is that a right circular cylinder (Fig. 15–5) with radius of base a and altitude h has a volume V given by the formula $V = \pi a^2 h$. The second assumption is that any solid of revolution may be approximated by a combination of right circular cylinders.

Let R be a region in the plane bounded by the curve $y = f(x)$, the lines $x = a$ and $x = b$, and the x axis, as shown in Fig. 15–6. The following theorem gives the formula for finding the volume of the solid obtained by revolving the region R about the x axis.

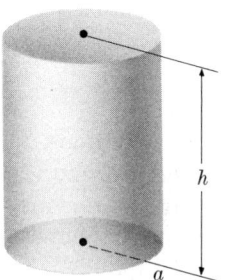

Figure 15–5

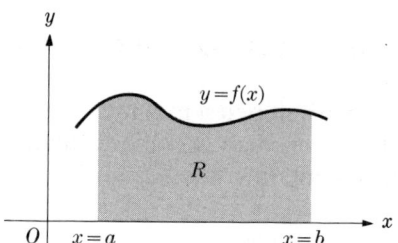

Figure 15–6

Theorem 1 (Disk Method). *Suppose that f is a continuous, nonnegative function on the interval $[a, b]$. Then the solid S obtained by revolving the region*

$$R = \{(x, y) : a \leq x \leq b, 0 \leq y \leq f(x)\}$$

about the x axis has a volume V given by the formula

$$V = \pi \int_a^b [f(x)]^2 \, dx.$$

Proof. We first divide the interval $[a, b]$ into n subintervals by the subdivision $\{a = x_0 < x_1 < \cdots < x_{n-1} < x_n = b\}$. For each integer i between 1 and n let ξ_i be the value of x at which f takes on its minimum value on the ith interval $[x_{i-1}, x_i]$. Similarly, let η_i be the value of x at which f takes on its maximum value on $[x_{i-1}, x_i]$. Denote by r_i and R_i the rectangles

$$r_i = \{(x, y) : x_{i-1} \leq x \leq x_i, 0 \leq y \leq f(\xi_i)\},$$
$$R_i = \{(x, y) : x_{i-1} \leq x \leq x_i, 0 \leq y \leq f(\eta_i)\}.$$

Figure 15–7 shows some of the smaller rectangles and Fig. 15–8 some of the larger ones. When the rectangle r_i is revolved about the x axis, a right circular cylinder is obtained which we denote by s_i. Similarly, revolving R_i yields a cylinder which we denote by S_i. The volumes of these cylinders are designated $V(s_i)$ and $V(S_i)$, respectively. (Figure 15–9 shows one quarter of a disk.) We have

$$V(s_1) + V(s_2) + \cdots + V(s_n) \leq V \leq V(S_1) + V(S_2) + \cdots + V(S_n).$$

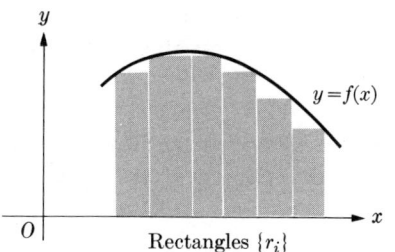
Rectangles $\{r_i\}$
Figure 15–7

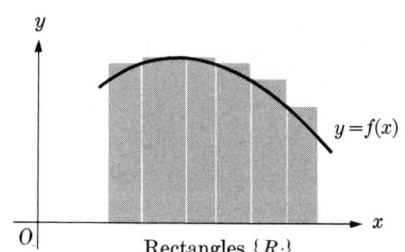
Rectangles $\{R_i\}$
Figure 15–8

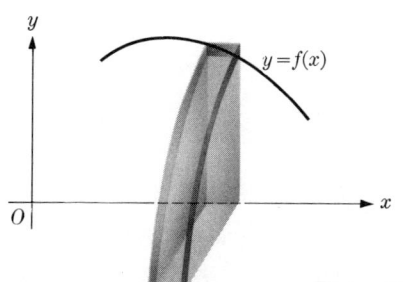
Figure 15–9

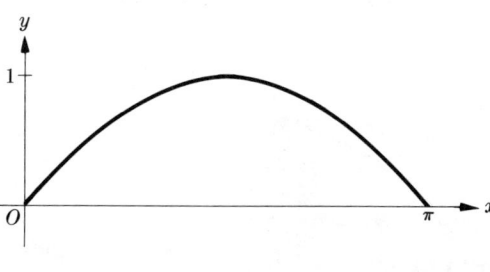
Figure 15–10

We recall that $\Delta_i x = x_i - x_{i-1}$ and from the formula for the volume of a right circular cylinder, we obtain

$$\pi[f(\xi_1)]^2 \Delta_1 x + \pi[f(\xi_2)]^2 \Delta_2 x + \cdots$$
$$+ \pi[f(\xi_n)]^2 \Delta_n x \le V \le \pi[f(\eta_1)]^2 \Delta_1 x + \cdots + \pi[f(\eta_n)]^2 \Delta_n x.$$

From the properties of integrals as described in Chapter 7, Section 3, we can conclude that the left and right sides of the above inequalities tend to

$$\pi \int_a^b [f(x)]^2 \, dx$$

as the norm of the subdivision, $\|\Delta\|$, tends to zero.*

Example 1. Find the volume of the solid generated by revolving the region bounded by the x axis and one arch of the curve $y = \sin x$ about the x axis.

Solution. The region R to be revolved is shown in Fig. 15–10. We describe it in set notation by writing $R = \{(x, y) : 0 \le x \le \pi, 0 \le y \le \sin x\}$. According to the formula of Theorem 1, we have

$$V = \pi \int_0^\pi \sin^2 x \, dx = \frac{\pi}{2} \int_0^\pi (1 - \cos 2x) \, dx = \frac{\pi}{2} \Big[x - \tfrac{1}{2} \sin 2x \Big]_0^\pi = \frac{\pi^2}{2}. \quad \blacktriangleleft$$

---

* A precise definition of volume would require a discussion analogous to the one for area. An **inner volume** V^- and an **outer volume** V^+ may be obtained in a way similar to the way in which inner and outer area are obtained. Then we must show that for regions of the type we shall consider here, we have $V^- = V^+$, and this common value, called the **volume,** is given by the integral.

Example 2. The curve $y = \sqrt{x}$, the line $x = 2$, and the x axis form the sides of a bounded region R. Find the volume of the solid generated by revolving R about the x axis.

Solution. We draw the region R which may be described in set notation by $R = \{(x, y) : 0 \leq x \leq 2, 0 \leq y \leq \sqrt{x}\}$. Figure 15–11 is a sketch of the solid of revolution. We have

$$V = \pi \int_0^2 (\sqrt{x})^2 \, dx = \frac{\pi}{2} \left[x^2 \right]_0^2 = 2\pi.$$

The solid is called a **paraboloid**. ◀

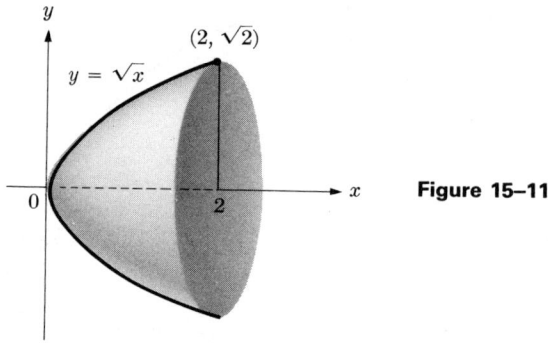

Figure 15–11

Example 3. The line $y = x + 2$ and the parabola $y = x^2$ contain a bounded region R between them. (See Fig. 15–12.) Find the volume V of the solid S generated by revolving R about the x axis.

Solution. To describe R in set notation, we find the points of intersection of the line and parabola. Solving simultaneously, we obtain $(-1, 1)$ and $(2, 4)$. Then R is described by

$$R = \{(x, y) : -1 \leq x \leq 2, x^2 \leq y \leq x + 2\}.$$

We first find the volume V_1 of the solid S_1 generated by revolving the region R_1 defined by

$$R_1 = \{(x, y) : -1 \leq x \leq 2, 0 \leq y \leq x^2\}$$

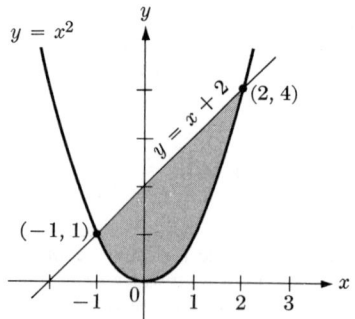

Figure 15–12

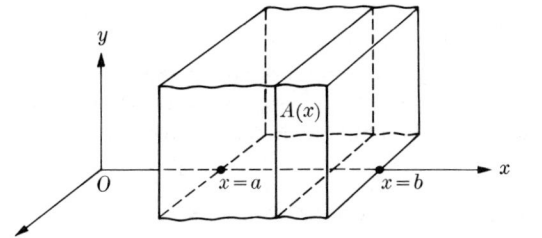

Figure 15–13

about the x axis. Then we subtract V_1 from the volume V_2 of the solid S_2 generated by revolving about the x axis the region R_2 defined by

$$R_2 = \{(x, y): -1 \le x \le 2, 0 \le y \le x + 2\}.$$

We get
$$V = V_2 - V_1 = \pi \int_{-1}^{2} (x + 2)^2 \, dx - \pi \int_{-1}^{2} x^4 \, dx$$
$$= \pi \left[\tfrac{1}{3}(x + 2)^3 - \tfrac{1}{5}x^5\right]_{-1}^{2} = \pi \left[\tfrac{64}{3} - \tfrac{32}{5} - (\tfrac{1}{3} + \tfrac{1}{5})\right] = \tfrac{72}{5}\pi. \quad \blacktriangleleft$$

Suppose that a solid S is bounded by two parallel planes $x = a$ and $x = b$ as shown in Fig. 15–13. Slice the solid with a knife parallel to these planes cutting out a cross section with area denoted by A. If A can be expressed as a function of x, then the formula for the volume V of the solid S is given by

$$V = \int_a^b A(x) \, dx.$$

This formula is a generalization of that in Theorem 1 for the volume of a solid of revolution. The proof, which we omit, follows the lines of the proof of Theorem 1. We work an example to show the technique.

Example 4. Find the volume of the tetrahedron with dimensions shown in Fig. 15–14.

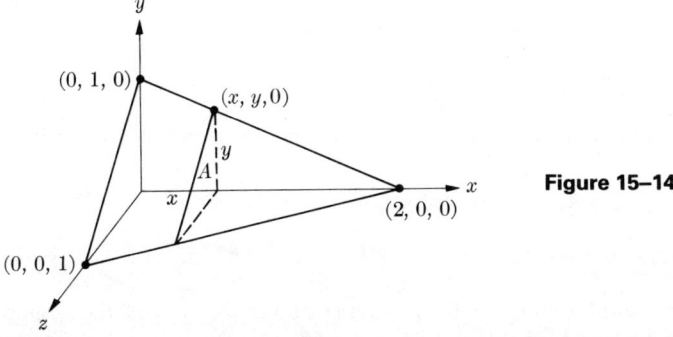

Figure 15–14

Solution. Since the typical cross section is a right isosceles triangle with side equal to y, its area A is given by $\tfrac{1}{2}y^2$. However, y and x are related by the equation $y - 0 = -\tfrac{1}{2}(x - 2)$. Therefore,

$$V = \int_0^2 A(x) \, dx = \int_0^2 \tfrac{1}{2}y^2 \, dx = \int_0^2 \tfrac{1}{4}(x^2 - 4x + 4) \, dx = \tfrac{2}{3}. \quad \blacktriangleleft$$

PROBLEMS

In Problems 1 through 24 a solid S is generated by revolving a *bounded* region R about the x axis. In each case find the volume of S. Describe the region R in set notation whenever it is not prescribed in such terms. Draw R and sketch S in each case.

1. $R = \{(x, y): 0 \le x \le 4, 0 \le y \le 2\sqrt{5x}\}$.
2. $R = \{(x, y): 0 \le x \le b, 0 \le y \le 2\sqrt{ax}\}$; a, b are positive constants.
3. R is bounded by the x axis, the y axis, and the line $2x + 3y = 1$.
4. R is bounded by the x axis, the y axis, and the line $(x/h) + (y/r) = 1$, where h and r are positive constants. Note that the solid generated is a right circular cone of altitude h and radius of base r.
5. $R = \{(x, y): -a \le x \le a, 0 \le y \le \sqrt{a^2 - x^2}\}$. Note that the result is the formula for the volume of a sphere of radius a.
6. $R = \{(x, y): a - h \le x \le a, 0 \le y \le \sqrt{a^2 - x^2}\}$, where h and a are positive constants with $h < a$.
7. R is bounded by $y = 0$ and $y = 4 - x^2$.
8. $R = \{(x, y): 0 \le x \le 4 - y^2, 0 \le y \le 2\}$.
9. R is bounded by the lines $x = \pi/3$, $y = 0$ and the curve $y = \tan x$.
10. $R = \{(x, y): 0 \le x \le 3, 0 \le y \le x/\sqrt{4 - x}\}$.
11. $R = \{(x, y): 1 \le x \le 2, 0 \le y \le \ln x\}$.
12. R is bounded by $y = xe^x$, $y = 0$, $x = 1$.
13. R is bounded by $y = 2$, $x = 0$, $x = y^2$.
14. $R = \{(x, y): y \le x \le 4/y, 1 \le y \le 2\}$.
15. $R = \{(x, y): 0 \le x \le \sqrt{y^2 + 4}, 0 \le y \le 2\}$.
16. R is bounded by $x = 0$ and $x + y^2 - 4y = 0$.
17. R is bounded below by $y = 0$ and above by the curves $x = y^2$ and $x = 8 - y^2$.
18. $R = \{(x, y): 0 \le x \le 1, x^2 \le y \le \sqrt{x}\}$.
19. R is bounded by $y = \sqrt{x}$ and $y = x^3$.
20. R is bounded by $x + y = 5$ and $xy = 4$.
21. R is bounded by the lines $x = -2$, $x = 2$, $y = 0$ and the curve
$$y = \frac{e^x + e^{-x}}{2}.$$
22. $R = \{(x, y): -b \le x \le b, 0 \le y \le (a/2)(e^{x/a} + e^{-x/a})\}$, where a and b are positive constants.
23. R is bounded by $y = x + 2$ and $y^2 - 3y = 2x$.
24. $R = \{(x, y): 1 \le y \le 2, \frac{y}{3} \le x \le \sqrt{y}\}$.

25. Find the volume of the solid generated by revolving about the x axis the region bounded by $y = 0$ and one arch of the cycloid $x = a(\theta - \sin\theta)$, $y = a(1 - \cos\theta)$. [*Hint*: $dx = a(1 - \cos\theta)\, d\theta$.]
26. Find the volume of the torus generated by revolving about the x axis the region interior to the circle $x^2 + (y - 5)^2 = 16$.
27. Find the volume of the torus generated by revolving about the x axis the region interior to the circle $x^2 + (y - b)^2 = a^2$, $b > a > 0$.

28. Suppose that f is positive and increasing on the closed interval $[a, b]$. Give a detailed proof of Theorem 1 of this section for such a function. [*Hint:* Study the proof of Theorem 1 in Chapter 7.]
29. Find the volume of the tetrahedron with vertices at $(0, 0, 0)$, $(3, 0, 0)$, $(0, 2, 0)$, $(0, 0, 4)$.
30. Find the volume of the tetrahedron with vertices at $(0, 0, 0)$, $(a, 0, 0)$, $(0, b, 0)$, $(0, 0, c)$.
31. The base of a tetrahedron is an equilateral triangle 3 cm on each side. The fourth vertex is directly over one of the vertices of the base, so that every cross section parallel to the base is an equilateral triangle. If the height of the tetrahedron is two cms, find its volume.
32. A solid bounded by the planes $x = 1$ and $x = 2$ is such that its cross sections parallel to these planes are squares. The diameters of these squares are in the xy plane with endpoints touching the curves $y = x^2$, $y = -x^2$. Find the volume of the solid.
33. Suppose that in the formula $V = \int_a^b A(x)\, dx$, the function $A(x)$ is increasing for $a \le x \le b$. Show how this formula may be derived. (See Problem 28.)

2. VOLUME OF A SOLID OF REVOLUTION. SHELL METHOD

A **cylindrical shell** is the solid contained between two concentric cylinders, as in Fig. 15–15. A cross section of such a shell is shown in Fig. 15–16. The volume V of a cylindrical shell with inner radius r_1, outer radius r_2, and height h is

$$V = \pi r_2^2 h - \pi r_1^2 h.$$

We may write the above formula in the form

$$V = \pi(r_2 + r_1)(r_2 - r_1)h = 2\pi\left(\frac{r_2 + r_1}{2}\right)(r_2 - r_1)h.$$

If we define $\bar{r}$ and Δr by the equations

$$\bar{r} = \frac{r_2 + r_1}{2}, \qquad \Delta r = r_2 - r_1,$$

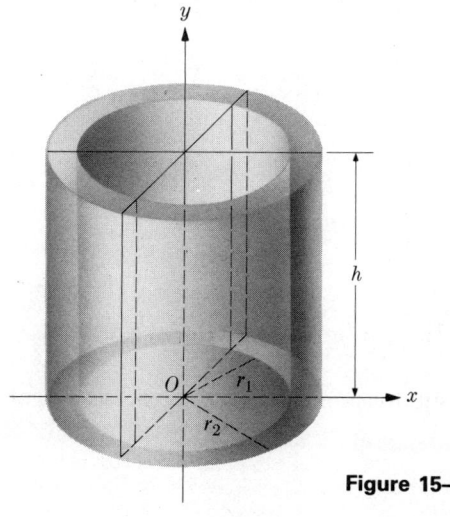

Figure 15–15

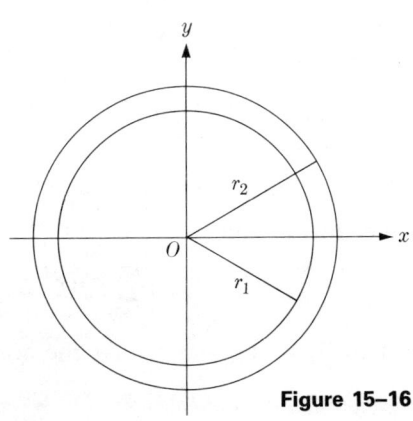

Figure 15–16

518 APPLICATIONS OF THE INTEGRAL

the formula for a cylindrical shell becomes

$$V = 2\pi \bar{r} h \, \Delta r.$$

Note that $2\pi \bar{r}$ is the length of the circumference of a circle whose radius is the *average* of r_1 and r_2, while Δr is the thickness of the shell.

A solid of revolution may be approximated by a combination of cylindrical shells. To see this, suppose that R is a region in the first quadrant bounded by the lines $x = a$, $x = b$, $y = 0$ and the curve $y = f(x)$. A solid S is formed when R is revolved about the y axis. (See Fig. 15–17.) To find the volume of S, first make a subdivision of the interval $[a, b]$:

$$\{a = x_0 < x_1 < \cdots < x_{n-1} < x_n = b\}.$$

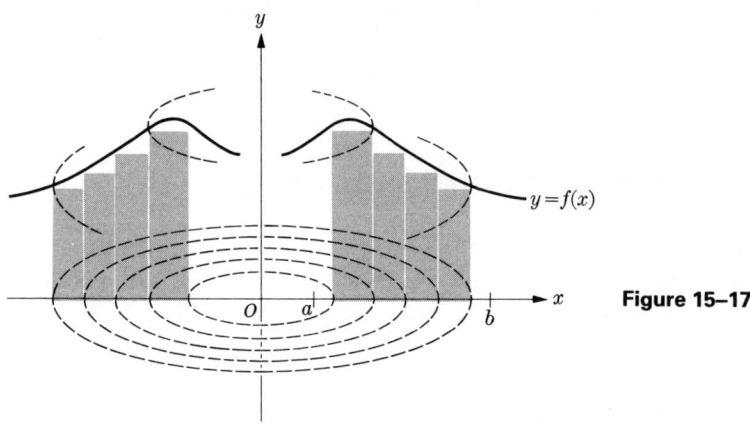

Figure 15–17

Next let $f(\xi_i)$ and $f(\eta_i)$ be the minimum and maximum values, respectively, of f on the interval $[x_{i-1}, x_i]$. As in the disk method, form the rectangles

$$r_i = \{(x, y) : x_{i-1} \leq x \leq x_i, \, 0 \leq y \leq f(\xi_i)\},$$
$$R_i = \{(x, y) : x_{i-1} \leq x \leq x_i, \, 0 \leq y \leq f(\eta_i)\}.$$

Since the region R is revolved about the y axis, the rectangles, r_i and R_i will generate cylindrical shells t_i and T_i. We have

$$V(t_i) = 2\pi \left(\frac{x_i + x_{i-1}}{2}\right) \Delta_i x f(\xi_i), \qquad V(T_i) = 2\pi \left(\frac{x_i + x_{i-1}}{2}\right) \Delta_i x f(\eta_i).$$

The volume V of the solid will be between the quantities

$$\sum_{i=1}^{n} V(t_i) \quad \text{and} \quad \sum_{i=1}^{n} V(T_i).$$

We thus obtain

$$\sum_{i=1}^{n} V(t_i) = 2\pi \sum_{i=1}^{n} \bar{x}_i f(\xi_i) \Delta_i x \leq V \leq 2\pi \sum_{i=1}^{n} \bar{x}_i f(\eta_i) \Delta_i x = \sum_{i=1}^{n} V(T_i),$$

where $\bar{x}_i = \frac{1}{2}(x_i + x_{i-1})$. As the norm of the subdivision, $\|\Delta\|$, tends to zero, we get

$$V = 2\pi \int_a^b x f(x) \, dx.$$

15-2 VOLUME OF A SOLID OF REVOLUTION. SHELL METHOD

When a region R in the first quadrant bounded by the lines $y = c$, $y = d$, $x = 0$, and the curve $x = g(y)$ is revolved about the x axis (see Fig. 15–18), the volume V of the resulting solid is given by the formula

$$V = 2\pi \int_c^d y g(y) \, dy.$$

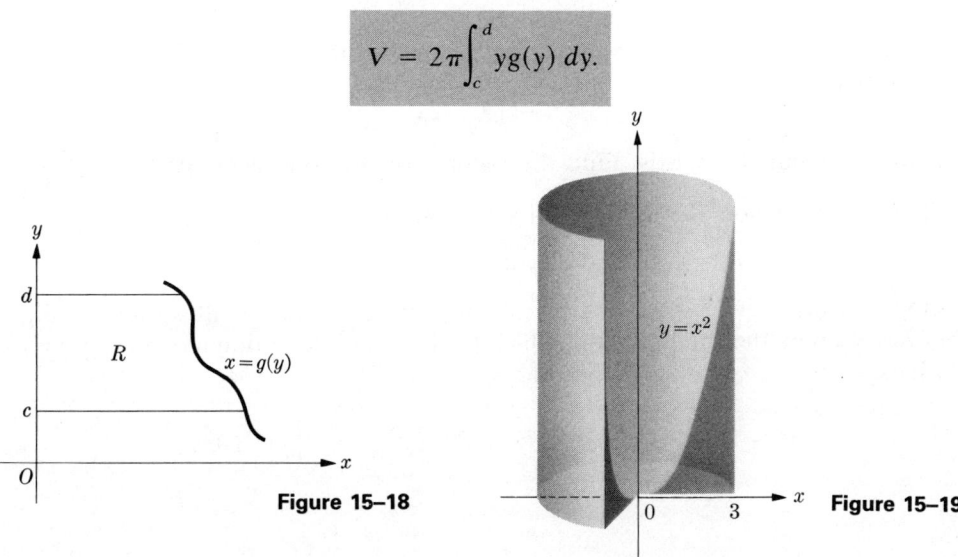

Figure 15–18

Figure 15–19

Example 1. The region R bounded by $y = x^2$ and the lines $y = 0$ and $x = 3$ is revolved about the y axis. Find the volume of the solid generated.

Solution. In set notation the region is described by writing

$$R = \{(x, y) : 0 \le x \le 3, 0 \le y \le x^2\}.$$

Figure 15–19 shows the situation. According to the shell method we have

$$V = 2\pi \int_0^3 x \cdot x^2 \, dx = \frac{2\pi}{4} \left[x^4 \right]_0^3 = \frac{81\pi}{2}. \quad \blacktriangleleft$$

Example 2. The region bounded by the positive x axis, the y axis, and the curve $y = \sqrt{a^2 - x^2}$ is revolved about the y axis. Find the volume of the solid generated.

Solution. The solid generated is a hemisphere, as shown in Fig. 15–20. Applying the method of shells, we have

$$V = 2\pi \int_0^a x \sqrt{a^2 - x^2} \, dx.$$

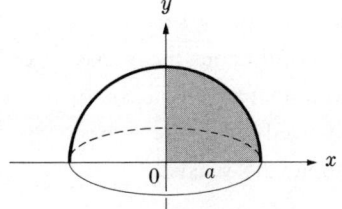

Figure 15–20

520 APPLICATIONS OF THE INTEGRAL

To integrate the above expression, we let $u = a^2 - x^2$, $du = -2x\,dx$ and obtain

$$V = -\frac{2\pi}{2}\int_{a^2}^{0}\sqrt{u}\,du = -\pi\left[\tfrac{1}{3}u^{3/2}\right]_{a^2}^{0} = \tfrac{2}{3}\pi a^3.$$ ◀

Example 3. The region R, bounded by the x axis, the y axis, the line $y = 2$, and the parabola

$$y^2 = 12 - 4x,$$

is revolved about the x axis. Find the volume of the solid generated.

Solution. The region R described in set notation as

$$R = \{(x, y) : 0 \le y \le 2,\ 0 \le x \le 3 - \tfrac{1}{4}y^2\}$$

is shown in Fig. 15–21. One-quarter of the solid generated is illustrated in Fig. 15–22. We apply the shell method, which leads to an integration along the y axis. We have $g(y) = 3 - \tfrac{1}{4}y^2$, and therefore

$$V = 2\pi\int_{0}^{2} y(3 - \tfrac{1}{4}y^2)\,dy = 2\pi\left[\tfrac{3}{2}y^2 - \tfrac{1}{16}y^4\right]_{0}^{2} = 10\pi.$$ ◀

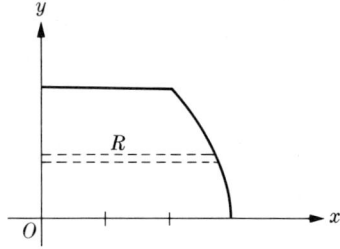

Figure 15–21

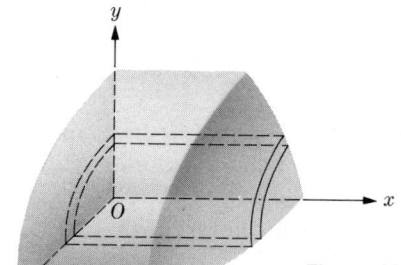

Figure 15–22

PROBLEMS

In Problems 1 through 11, the bounded region R is revolved about the given axis. Describe the region R in set notation whenever it is not prescribed in such terms. Use the method of shells to find the volume of the solid generated.

1. $R = \{(x, y) : 0 \le x \le y^2,\ 0 \le y \le 2\}$; R is revolved about the y axis.
2. R is bounded by $y = x^3$, $y = 0$, and $x = 1$; R is revolved about the y axis.
3. R is the same as in Problem 2, but the region is revolved about the x axis.
4. $R = \{(x, y) : 1 \le x \le 2,\ 0 \le y \le \log x\}$; R is revolved about the y axis.
5. R is the same as in Problem 4, but the region is revolved about the x axis.
6. R is bounded by the lines $x = 0$, $y = 0$, $y = 2$, and the curve $x = \sqrt{y^2 + 4}$; R is revolved about the y axis.
7. $R = \{(x, y) : 0 \le x \le \pi,\ 0 \le y \le \sin x\}$; R is revolved about the y axis.
8. R is bounded by $y = 0$, $x = 3$, $y = x/\sqrt{4 - x}$; R is revolved about the y axis.
9. $R = \{(x, y) : y^2 \le x \le 8 - y^2,\ 0 \le y \le 2\}$; R is revolved about the y axis.
10. R is bounded by $x = 0$ and $x + y^2 - 4y = 0$; R is revolved about the y axis.
11. R is bounded by $y = \sqrt{x}$ and $y = x^3$; R is revolved about the y axis.

12. A cylindrical hole of radius a and axis the y axis is bored through a sphere of radius b, the sphere having its center at the origin. (Of course, $b > a$.) Find the volume remaining.

13. The region R bounded by the x axis, the y axis, and the line $x + y = 1$ is revolved about the line $y = -1$. Find the volume of the solid generated.

14. The region R bounded by $y = 0$ and $y = 4 - x^2$ is revolved about the line $y = -1$. Find the volume of the solid generated.

15. Find the volume of the solid generated when the region of Problem 14 is revolved about the line $x = -2$.

16. Find the volume of the solid generated when the region R, bounded by $y = 0$ and the arch of $y = \sin x$ between 0 and π, is revolved about the line $y = -1$.

17. Find the volume of the solid generated when the region R of Problem 16 is revolved about the line $x = -1$.

18. Let a, b, c be positive numbers. Suppose that f is a positive function for $a \le x \le b$. Find a formula for the volume of the solid generated when the region $R = \{(x, y) : a \le x \le b, 0 \le y \le f(x)\}$ is revolved about the line $x = -c$.

19. Same as Problem 18, except that the region R is revolved about the line $y = -c$.

20. Same as Problem 18, except that

$$R = \{(x, y) : 0 \le x \le g(y), \alpha \le y \le \beta\}$$

and the line is $x = -\gamma$ where α, β, γ are positive numbers.

21. Same as Problem 20, except that the line is $y = -\gamma$.

22. Find by the shell method the volume of the torus generated by revolving about the x axis the region interior to the circle $x^2 + (y - b)^2 = a^2$, $b > a > 0$. (See Problem 26 in Section 1.)

3. IMPROPER INTEGRALS

The integral of $f(x) = e^{-x}$ on an interval $[0, a]$ is obtained easily. We have

$$\int_0^a e^{-x}\,dx = -e^{-x}\Big]_0^a = -e^{-a} + 1.$$

For large values of a, the quantity e^{-a} is small, and we know (see Theorem 13, Chapter 9) that e^{-a} tends to zero as a tends to infinity. (See the sketch of $f(x) = e^{-x}$ in Fig. 15-23.) Letting a tend to infinity in the above integration, we obtain

$$\lim_{a \to +\infty} \int_0^a e^{-x}\,dx = \lim_{a \to +\infty} (-e^{-a} + 1) = 0 + 1 = 1.$$

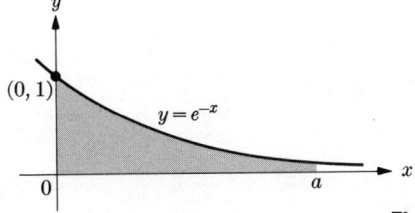

Figure 15-23

In other words, the area of the shaded region in Fig. 15–23 tends to the value 1 as a tends to infinity. It is natural to define

$$\int_0^{+\infty} e^{-x}\, dx = \lim_{a \to +\infty} \int_0^a e^{-x}\, dx$$

and assign the value 1 to the area of the entire region under the curve $y = e^{-x}$ situated to the right of the y axis.

As a second illustration, let us consider the integral

$$\int_1^a \frac{1}{x}\, dx = \Big[\ln x\Big]_1^a = \ln a,$$

where a is any number larger than 1. We recall that $\ln a$ increases without bound as a increases without bound (see Theorem 12 in Chapter 9). Since

$$\lim_{a \to +\infty} (\ln a)$$

does not exist, we conclude that

$$\lim_{a \to +\infty} \int_1^a \frac{1}{x}\, dx \qquad \text{does not exist.}$$

There is no way of assigning a value to the area of the shaded region under the curve in Fig. 15–24. We say that $\int_1^{+\infty} (1/x)\, dx$ does not exist.

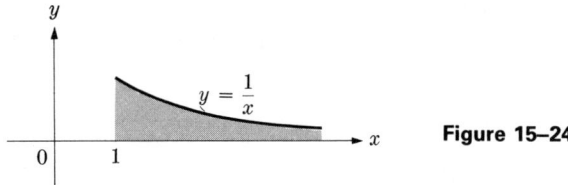

Figure 15–24

Definitions. *Suppose that f is continuous for all $x \geqq x_0$. If*

$$\lim_{a \to +\infty} \int_{x_0}^a f(x)\, dx \qquad \textit{exists, we say that} \qquad \int_{x_0}^{+\infty} f(x)\, dx$$

*is **convergent** to the value given by the above limit. If the limit does not exist, we say that $\int_{x_0}^{+\infty} f(x)\, dx$ is **divergent**, and the integral is not defined.*

Example 1. Determine whether the following integral is convergent or divergent, and if it is convergent, determine its value:

$$\int_0^{+\infty} \frac{dx}{1 + x^2}.$$

Solution

$$\int_0^a \frac{dx}{1 + x^2} = \arctan a, \qquad \lim_{a \to +\infty} (\arctan a) = \frac{\pi}{2},$$

and therefore

$$\int_0^{+\infty} \frac{dx}{1 + x^2} = \frac{\pi}{2}. \qquad \blacktriangleleft$$

Example 2. Determine whether the following integral is convergent or divergent; if it is convergent, determine its value:

$$\int_0^{+\infty} \frac{x^2}{1+x^2} \, dx.$$

Solution. By division, we have

$$\int_0^a \frac{x^2}{1+x^2} \, dx = \int_0^a \left(1 - \frac{1}{1+x^2}\right) dx$$
$$= \left[x - \arctan x\right]_0^a = a - \arctan a.$$

Therefore

$$\lim_{a \to +\infty} \int_0^a \frac{x^2}{1+x^2} \, dx = \lim_{a \to +\infty} [a - \arctan a].$$

The limit on the right does not exist and the integral is divergent. ◄

The above definitions and examples extend the definition of integral to *unbounded intervals.* That is, a function f which is continuous on an interval $[b, +\infty)$ or $(-\infty, b]$ can be integrated provided a certain limiting process can be performed. It may happen, however, that the *interval of integration is of finite length,* but that the function f to be integrated is unbounded somewhere on the interval of integration. For example, the integral

$$\int_1^3 \frac{1}{\sqrt{x-1}} \, dx$$

is not defined, since

$$\frac{1}{\sqrt{x-1}} \to +\infty \quad \text{as} \quad x \to 1.$$

(We recall Theorem 5 of Chapter 7, which states that if a function is integrable on an interval, then it is bounded on that interval.)

Since the function $1/\sqrt{x-1}$ is not defined for values of x less than 1, we clearly intend the above expression $x \to 1$ to mean that x tends to 1 through values larger than 1. We shall state this fact with more precision by writing $x \to 1^+$.

The expression

$$\int_a^3 \frac{1}{\sqrt{x-1}} \, dx$$

is integrable for every number $a > 1$, since the integrand is bounded and continuous on the interval $[a, 3]$. We can perform the integration:

$$\int_a^3 \frac{1}{\sqrt{x-1}} \, dx = 2\sqrt{x-1}\Big]_a^3 = 2\sqrt{2} - 2\sqrt{a-1}.$$

Furthermore, we can perform a limiting process:

$$\lim_{a \to 1^+} \int_a^3 \frac{1}{\sqrt{x-1}} \, dx = \lim_{a \to 1^+} \left[2\sqrt{2} - 2\sqrt{a-1}\right] = 2\sqrt{2}.$$

We assign the value $2\sqrt{2}$ to the area of the shaded region in Fig. 15–25. ◄

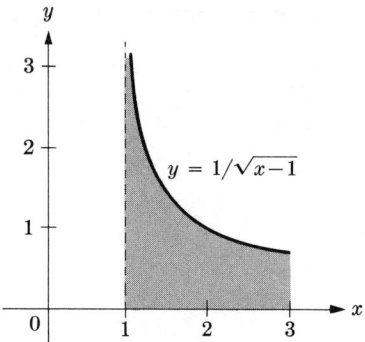

Figure 15–25

Definitions. Suppose that $f(x)$ is continuous for $a < x \leq b$ but is not continuous at a. Then $\int_a^b f(x)\, dx$ is **convergent** if

$$\lim_{\epsilon \to 0} \int_{a+\epsilon}^b f(x)\, dx$$

exists, where ϵ approaches zero through positive values. The value of $\int_a^b f(x)\, dx$ is the value of the limit. If the limit does not exist, the integral is said to be **divergent.** A similar definition is given for the case in which $f(x)$ is continuous for $a \leq x < b$ but not at b.

Example 3. Determine whether

$$\int_0^1 \frac{dx}{1-x}$$

is convergent and, if so, find its value.

Solution. The function $f(x) = 1/(1-x)$ tends to infinity as $x \to 1^-$. We let ϵ be a positive number and consider the integral

$$\int_0^{1-\epsilon} \frac{dx}{1-x}.$$

To integrate we set $u = 1 - x$ and $du = -dx$. Then we have

$$\int_0^{1-\epsilon} \frac{dx}{1-x} = -\int_1^\epsilon \frac{du}{u} = -\ln u \Big]_1^\epsilon = -\ln \epsilon.$$

However, $-\ln \epsilon \to +\infty$ as $\epsilon \to 0^+$, and the integral is divergent. ◀

Example 4. Determine whether

$$\int_0^3 \frac{dx}{(3-x)^{2/3}}$$

is convergent and, if so, find its value.

Solution. The function $f(x) = (3-x)^{-2/3}$ tends to infinity as $x \to 3^-$. We choose $\epsilon > 0$ and let $u = 3 - x$, $du = -dx$. Then we find

$$\int_0^{3-\epsilon} \frac{dx}{(3-x)^{2/3}} = -\int_3^\epsilon \frac{du}{u^{2/3}} = -3u^{1/3}\Big]_3^\epsilon = 3\sqrt[3]{3} - 3\sqrt[3]{\epsilon}.$$

The limit as $\epsilon \to 0^+$ exists. We obtain

$$\int_0^3 \frac{dx}{(3-x)^{2/3}} = 3\sqrt[3]{3}$$

for the value of the convergent integral. ◂

It may happen that the integrand $f(x)$ is unbounded somewhere in the *interior* of the interval of integration. In such cases, the next definition is employed.

Definition. *Suppose that $f(x)$ is continuous for $a \leq x \leq b$, except for $x = c$, where $a < c < b$. Then we say that $\int_a^b f(x)\, dx$ is* **convergent** *if both of the integrals $\int_a^c f(x)\, dx$ and $\int_c^b f(x)\, dx$ are convergent. We define*

$$\int_a^b f(x)\, dx = \int_a^c f(x)\, dx + \int_c^b f(x)\, dx.$$

Example 5. Determine whether

$$\int_{-1}^1 \frac{1}{x^2}\, dx$$

is convergent and, if so, find its value.

Solution. $f(x) = 1/x^2$ tends to infinity as $x \to 0$. We consider

$$\int_{-1}^{-\epsilon} \frac{1}{x^2}\, dx \quad \text{and} \quad \int_\epsilon^1 \frac{1}{x^2}\, dx, \quad \epsilon > 0.$$

The second of these yields, upon integration,

$$\int_\epsilon^1 \frac{1}{x^2}\, dx = \left[-\frac{1}{x}\right]_\epsilon^1 = \frac{1}{\epsilon} - 1.$$

The limit as $\epsilon \to 0$ does not exist, and the integral is divergent. ◂

Remark. If we had not noted that $f(x) = 1/x^2$ tends to infinity as $x \to 0$, and if we had gone blindly ahead with the integration, we would have obtained

$$\int_{-1}^1 \frac{1}{x^2}\, dx = \left[-\frac{1}{x}\right]_{-1}^1 = -1 - \left(-\frac{1}{-1}\right) = -2.$$

This result is obviously wrong, since $f(x) = 1/x^2$ is positive everywhere and could never yield a negative value for the integral.

Example 6. Determine whether

$$\int_{-2}^3 x^{-1/3}\, dx$$

is convergent and, if so, find its value.

Solution. $f(x) = (1/x^{1/3})$ is unbounded as $x \to 0$. Therefore we evaluate

$$\int_{\epsilon}^{3} x^{-1/3}\, dx = \tfrac{3}{2} x^{2/3}\Big]_{\epsilon}^{3} = \tfrac{3}{2}\sqrt[3]{9} - \tfrac{3}{2}\sqrt[3]{\epsilon^2}, \qquad \epsilon > 0,$$

getting

$$\int_{0}^{3} x^{-1/3}\, dx = \tfrac{3}{2}\sqrt[3]{9}.$$

Similarly,

$$\int_{-2}^{-\epsilon} x^{-1/3}\, dx = \tfrac{3}{2} x^{2/3}\Big]_{-2}^{-\epsilon} = \tfrac{3}{2}\sqrt[3]{\epsilon^2} - \tfrac{3}{2}\sqrt[3]{4}, \qquad \epsilon > 0,$$

and

$$\int_{-2}^{0} x^{-1/3}\, dx = -\tfrac{3}{2}\sqrt[3]{4}.$$

The integral is convergent, and

$$\int_{-2}^{3} x^{-1/3}\, dx = \tfrac{3}{2}(\sqrt[3]{9} - \sqrt[3]{4}). \qquad \blacktriangleleft$$

Remarks. The extended integrals defined in this section are called **improper integrals**. Improper integrals may be defined, in more general cases, where f may become infinite at several points in the interval of integration, which may itself be infinite. In such a case, we then divide that interval into smaller intervals of two types: In the first type, the interval is finite and f becomes infinite at one end while, in the second type, f remains continuous but the interval is infinite in one direction only. Then the original improper integral converges if and only if each of the component parts converges. The **value** is the sum of the values of the integrals taken over the smaller intervals.

Operating Method. In the evaluation of integrals the student should, as a routine procedure, check to see if the integrand becomes infinite at either of the endpoints of the interval of integration or at any interior point.

PROBLEMS

In Problems 1 through 30, determine whether or not each of the improper integrals is convergent, and compute its value if it is.

1. $\displaystyle\int_{1}^{+\infty} \frac{dx}{x^2}$
2. $\displaystyle\int_{1}^{+\infty} \frac{dx}{x^3}$
3. $\displaystyle\int_{0}^{+\infty} \frac{1}{(x+1)^{3/2}}\, dx$

4. $\displaystyle\int_{1}^{+\infty} \frac{1}{x^p}\, dx, \quad p > 1$
5. $\displaystyle\int_{1}^{+\infty} \frac{1}{x}\, dx$
6. $\displaystyle\int_{1}^{+\infty} \frac{1}{x^p}\, dx, \quad p < 1$

7. $\displaystyle\int_{0}^{1} \frac{1}{x^p}\, dx, \quad p < 1$
8. $\displaystyle\int_{0}^{1} \frac{1}{x}\, dx$
9. $\displaystyle\int_{0}^{1} \frac{1}{x^p}\, dx, \quad p > 1$

10. $\displaystyle\int_{0}^{+\infty} \frac{dx}{(x^2 + 1)^2}$
11. $\displaystyle\int_{0}^{1} \frac{dx}{\sqrt{1 - x}}$
12. $\displaystyle\int_{0}^{2} \frac{dx}{\sqrt{4 - x^2}}$

13. $\int_0^{+\infty} xe^{-x}\, dx^*$ 14. $\int_{-\infty}^0 x^2 e^x\, dx^*$ 15. $\int_{-8}^1 x^{-2/3}\, dx$

16. $\int_{-1}^1 x^{-3}\, dx$ 17. $\int_{-2}^0 \dfrac{dx}{\sqrt[3]{x+1}}$ 18. $\int_0^{+\infty} \dfrac{e^{-\sqrt{x}}}{\sqrt{x}}\, dx$

19. $\int_0^{+\infty} xe^{-x^2}\, dx$ 20. $\int_0^{+\infty} x^p e^{-x^{p+1}}\, dx,\ p > 0$ 21. $\int_0^{\pi/2} \cot\theta\, d\theta$

22. $\int_0^{\pi/4} \dfrac{\sec^2 x\, dx}{\sqrt{\tan x}}$ 23. $\int_0^4 \dfrac{x\, dx}{\sqrt{16 - x^2}}$ 24. $\int_0^{+\infty} \dfrac{dx}{\sqrt{x}(x+4)}$

25. $\int_{-1}^1 \sqrt{1 + x^{-2/3}}\, dx$ 26. $\int_0^4 \dfrac{(2-x)\, dx}{\sqrt{4x - x^2}}$ 27. $\int_0^4 \dfrac{dx}{\sqrt{4x - x^2}}$

28. $\int_{-4}^{+\infty} \dfrac{dx}{x\sqrt{x+4}}$ 29. $\int_{-\infty}^{+\infty} x^2 e^{-x^3}\, dx$ 30. $\int_{-\infty}^{+\infty} x^3 e^{-x^4}\, dx$

In Problems 31 through 34, use the substitution $t = \tan(\theta/2)$ (see Problem 32, Chapter 14, Section 8), if necessary, to evaluate each integral.

31. $\int_0^\pi \dfrac{\sin\theta\, d\theta}{\sqrt{1 + \cos\theta}}$ 32. $\int_0^\pi \dfrac{d\theta}{5 + 4\cos\theta}$

33. $\int_0^\pi \dfrac{d\theta}{2 - \sin\theta}$ 34. $\int_0^\pi \dfrac{\cos\theta\, d\theta}{5 + 4\cos\theta}$

35. The region bounded by $f(x) = 1/x$, the x axis, and the line $x = 1$, and situated to the right of $x = 1$, is revolved about the x axis. Evaluate the improper integral and assign a value to the volume of the solid generated. Note that the plane region does not have a finite area.

36. Rework Problem 35, with $f(x) = 1/x^p$, $p > 0$. For what values of p does the solid have a finite value?

37. Let $f(x) = 3^{-n}$ for $n - 1 \le x < n$, $n = 1, 2, \ldots$. Evaluate $\int_0^{+\infty} f(x)\, dx$.

*38. Let $f(x) = (x - n + 1)/n$ for $n - 1 \le x < n$, $n = 1, 2, \ldots$. Decide whether $\int_0^{+\infty} f(x)\, dx$ is convergent or divergent.

39. Let $f(x)$ be continuous and positive for $0 \le x < \infty$. Suppose there is a number B such that $\int_0^a f(x)\, dx \le B$ for all $a > 0$. Show that $\int_0^{+\infty} f(x)\, dx$ is convergent and that its value must be less than B.

40. Suppose that f and g are continuous, positive functions for $0 \le x < \infty$ and that $\int_0^{+\infty} g(x)\, dx$ is divergent. If $f(x) \ge g(x)$ for all x, show that $\int_0^{+\infty} f(x)\, dx$ is divergent. (Assume the contrary and use the result of Problem 39.)

4. ARC LENGTH

In Chapter 10, Section 3, the length of an arc is defined, and formulas for computing such lengths are obtained. Whether an arc C is given in the parametric form

$$x = x(t), \quad y = y(t), \quad a \le t \le b,$$

---

* $\lim\limits_{h \to +\infty} h^n e^{-bh} = 0$ for any n if $b > 0$.

528 APPLICATIONS OF THE INTEGRAL

with $x'(t)$ and $y'(t)$ continuous functions, then the length of C is given by the formula

$$l(C) = \int_a^b \sqrt{[x'(t)]^2 + [y'(t)]^2}\, dt.$$

If the curve is given in either of the nonparametric forms $y = f(x)$ or $x = g(y)$, the formula is

$$l = \int \sqrt{1 + (dy/dx)^2}\, dx \quad \text{or} \quad l = \int \sqrt{1 + (dx/dy)^2}\, dy.$$

By employing the powerful methods of integration which we learned in Chapter 14, we are now able to integrate many types of expressions which were beyond our grasp when we first studied arc length. Furthermore, some of the most interesting integrals for arc length lead to improper integrals. The following examples show how arc lengths may be obtained by evaluation of improper and difficult integrals.

Example 1. Find the length of the semicircle

$$C = \{(x, y): -a \le x \le a, y = \sqrt{a^2 - x^2}\}.$$

Solution. We have

$$\frac{dy}{dx} = -\frac{x}{\sqrt{a^2 - x^2}},$$

$$1 + \left(\frac{dy}{dx}\right)^2 = \frac{a^2}{a^2 - x^2},$$

and

$$l(C) = a \int_{-a}^{a} \frac{dx}{\sqrt{a^2 - x^2}}.$$

The integrand becomes infinite at $-a$ and at a. We decompose it into two parts, obtaining

$$l(C) = a \int_{-a}^{0} \frac{dx}{\sqrt{a^2 - x^2}} + a \int_{0}^{a} \frac{dx}{\sqrt{a^2 - x^2}}.$$

Evaluating each improper integral, we get

$$a \int_{-a+\epsilon}^{0} \frac{dx}{\sqrt{a^2 - x^2}} = a \left[\arcsin \frac{x}{a}\right]_{-a+\epsilon}^{0} = a\left[0 - \arcsin\left(\frac{-a+\epsilon}{a}\right)\right].$$

This integral tends to $a[0 - (-\pi/2)]$ as $\epsilon \to 0^+$. Similarly, we have

$$a \int_{0}^{a-\epsilon} \frac{dx}{\sqrt{a^2 - x^2}} = a\left[\arcsin \frac{x}{a}\right]_{0}^{a-\epsilon} \to a \cdot \arcsin 1 = \frac{\pi a}{2}.$$

Therefore $l(C) = \pi a$, a familiar result. ◂

Example 2. Find $l(C)$, where C is the arc
$$C: x = t, \quad y = \ln t, \quad 0 \le t \le 1.$$

Solution. We have $x'^2 = 1$, $y'^2 = 1/t^2$ and
$$l(C) = \int_0^1 \sqrt{1 + 1/t^2}\, dt = \int_0^1 \left\{\frac{\sqrt{1 + t^2}}{t}\right\} dt,$$

which is an improper integral. To evaluate $l(C)$ we make the substitution $u = \sqrt{1 + t^2}$, $u\, du = t\, dt$ (see Chapter 14, Section 8), and the integral becomes

$$\lim_{\epsilon \to 0^+} \int_\epsilon^1 \frac{\sqrt{1 + t^2}}{t}\, dt = \lim_{\epsilon \to 0^+} \int_{\sqrt{1+\epsilon^2}}^{\sqrt{2}} \frac{u^2\, du}{u^2 - 1}$$

$$= \lim_{\epsilon \to 0^+} \left[\int_{\sqrt{1+\epsilon^2}}^{\sqrt{2}} du + \int_{\sqrt{1+\epsilon^2}}^{\sqrt{2}} \frac{du}{u^2 - 1}\right]$$

$$= \lim_{\epsilon \to 0^+} \left[\sqrt{2} - \sqrt{1 + \epsilon^2} + \frac{1}{2}\ln\left(\frac{\sqrt{2} - 1}{\sqrt{2} + 1}\right)\right.$$

$$\left. - \frac{1}{2}\ln\left(\frac{\sqrt{1 + \epsilon^2} - 1}{\sqrt{1 + \epsilon^2} + 1}\right)\right],$$

which tends to $+\infty$ as $\epsilon \to 0^+$. The integral diverges. ◀

PROBLEMS

Find the length $l(C)$ of the arc C in each of the following problems. Note any improper integrals, and state when the length is not finite.

1. $C = \{(x, y): 0 \le x \le 1, y = x^2\}$
2. $C = \{(x, y): 1 \le x \le 2, y = \ln x\}$
3. $C = \{(x, y): x = (a^{2/3} - y^{2/3})^{3/2}, -a \le y \le a\}$
4. $C = \{(x, y): 0 \le x \le \pi/3, y = \ln \sec x\}$
5. $C = \{(x, y): x = \ln \sin y, \pi/4 \le y \le 3\pi/4\}$
6. $C = \{(x, y): x = \frac{1}{3}(y - 3)\sqrt{y}, 0 \le y \le 3\}$
7. $C = \{(x, y): -8 \le x \le 1, y = x^{2/3}\}$
8. $C = \{(x, y): x = \frac{1}{3}t^3, y = \frac{1}{2}t^2, 1 \le t \le 3\}$
9. $C = \{(x, y): x = \frac{1}{2}t^2, y = \frac{1}{4}t^4, 1 \le t \le 2\}$
10. $C = \{(x, y): 0 \le x \le 1, y = \frac{1}{4}x^2 - \frac{1}{2}\ln x\}$
11. $C = \{(x, y): 0 \le x \le 2, y = \frac{1}{6}x^3 + 1/(2x)\}$
12. $C = \{(x, y): 3 \le x \le 5, y = \ln(x + \sqrt{x^2 - 1})\}$
13. $C = \{(x, y): 1 \le x \le 2, y = \ln(x + \sqrt{x^2 - 1})\}$
14. $C = \{(x, y): x = u \cos u, y = u \sin u, 0 \le u \le \pi\}$
15. $C = \{(x, y): 0 \le x \le 1, y = \sqrt{x}\}$

16. $C = \{(x, y) : x = \cos^3 t, y = \sin^3 t, 0 \le t \le \pi/3\}$
17. $C = \{x, y) : 0 \le x \le \ln \sqrt{3}, y = \arcsin(e^{-x})\}$
18. $C : x = t^{m+1}/(m+1), y = t^{n+1}/(n+1), 1 \le t \le 2$, where $m = \frac{1}{4}(5n+1)$. Answer should be in terms of n.
19. Suppose that $f(t)$ and $g(t)$ have continuous first derivatives which are never zero for $a \le t \le b$. Let $l(T)$ be the length of the curve $C = \{(x, y) : x = f(t), y = g(t), a \le t \le b\}$ from the point a to the point T. Show that $l(T)$ is a strictly increasing function of T. Suppose that $C_1 = \{(x, y) : x = F(t), y = G(t), a \le t \le b\}$ is another curve with length $l_1(T)$. State conditions relating f, g, F, and G such that $l_1(T) \le l(T)$ for all T.
20. Let $f(x)$, $0 \le x < \infty$ be a positive function with a continuous derivative. Let $l(X)$ be the length of the curve between 0 and X. Show that $l(X) \to +\infty$ as $X \to +\infty$.

5. AREA OF A SURFACE OF REVOLUTION

When an arc situated above the x axis in the xy plane is revolved about the x axis, a **surface of revolution** is generated. We shall define the area of such a surface and show how to calculate it by integration. In defining the area of a plane region, we know that the fundamental quantity is the area of a rectangle. For the volume of a solid of revolution, the fundamental quantity is the volume of a right circular cylinder. *For determining the area of a surface of revolution, the fundamental quantity is the lateral area of the frustum of a cone.*

The lateral surface area S of a right circular cone with radius of base r and slant height l, as in Fig. 15–26, is given by the formula

$$S = \pi r l,$$

which may be seen intuitively by cutting the cone along a generator, l, and flattening the surface onto a plane. The result is a circular sector, as shown in Fig. 15–27, whose area is exactly $\pi r l$.

A frustum of a cone is shown in Fig. 15–28. The smaller radius is r_1, the larger is r_2, and the slant height is l. The surface area of such a frustum is

$$\pi r_2(l_1 + l) - \pi r_1 l_1.$$

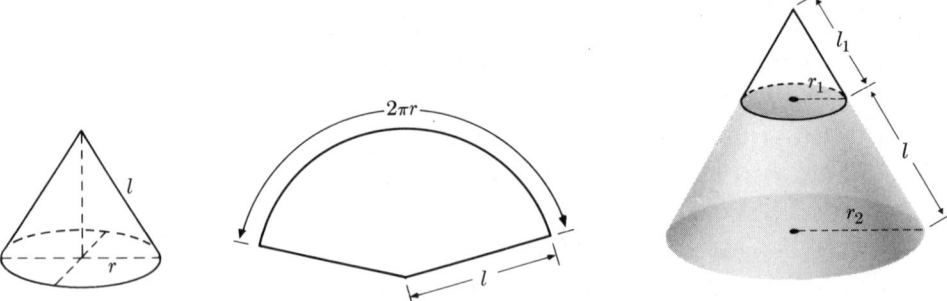

Figure 15–26 **Figure 15–27** **Figure 15–28**

We use the proportion $l_1/r_1 = l/(r_2 - r_1)$ to eliminate l_1 in the above formula, and obtain, for the **lateral surface area F of a frustum of a cone,**

$$F = \pi(r_1 + r_2)l.$$

We write this formula in the more suggestive form

$$F = 2\pi \bar{r} l,$$

where $\bar{r} = \frac{1}{2}(r_1 + r_2)$ is the average radius.

Suppose that an arc C, situated entirely above the x axis, is given in the parametric form

$$x = x(t), \quad y = y(t), \quad a \leq t \leq b.$$

We revolve C about the x axis, obtaining a surface of revolution. We let $\{a = t_0 < t_1 < t_2 < \cdots < t_n = b\}$ be a subdivision of $[a, b]$, and we construct the inscribed polygon to C exactly as we did in defining arc length. When this inscribed polygon is revolved about the x axis, we obtain a collection of frustums of cones, as in Fig. 15–29. The sum of the lateral surface areas of these frustums will be an approximate measure of the (as yet undefined) area of the surface S generated by the arc C. We argue intuitively that the "closer" the inscribed polygon is to the curve C, the more accurate will be the approximation to the surface area.

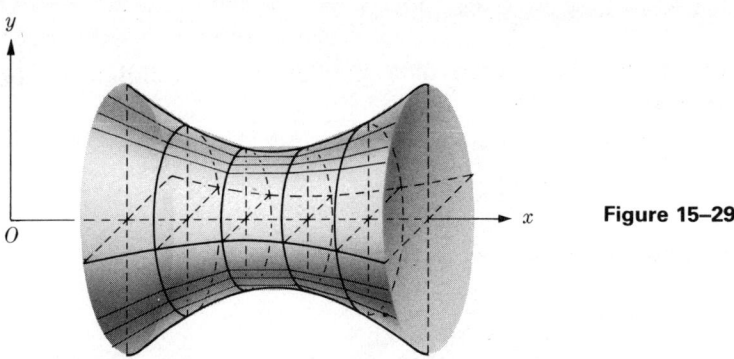

Figure 15–29

Definition. *Assume that $C: x = x(t), y = y(t), a \leq t \leq b$ is situated in the upper half-plane [that is, $y(t) \geq 0$]. Let $x_i = x(t_i), y_i = y(t_i), i = 0, 1, 2, \ldots, n$; define S as the surface of revolution obtained by revolving C about the x axis. Then the* **area** *$A(s)$ is defined as the limit approached by the sums*

$$\sum_{i=1}^{n} 2\pi \bar{y}_i \sqrt{(x_i - x_{i-1})^2 + (y_i - y_{i-1})^2},$$

where

$$\bar{y}_i = \frac{y_i + y_{i-1}}{2},$$

as the norms of the subdivisions tend to zero, provided the limit exists. Each term in the above sum is the lateral surface area of the inscribed frustum of a cone (Fig. 15–30).

532 APPLICATIONS OF THE INTEGRAL

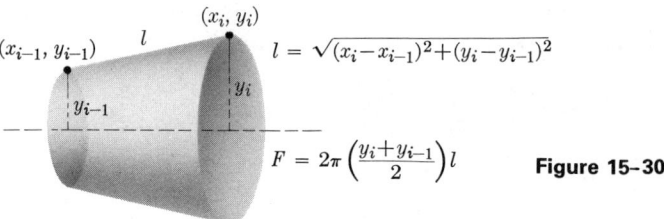

Figure 15-30

We now state without proof the formula for evaluating the area $A(s)$.

If an arc
$$C: x = x(t), \quad y = y(t), \quad a \le t \le b,$$
is in the upper half-plane, then the area $A(S)$ of the surface generated by revolving C about the x axis is given by the formula

$$A(S) = 2\pi \int_a^b y(t)\sqrt{(dx/dt)^2 + (dy/dt)^2}\, dt. \tag{1}$$

If the arc C is located to the right of the y axis and is revolved about the y axis, the area $A(S)$ of the surface generated is given by the formula

$$A(S) = 2\pi \int_a^b x(t)\sqrt{(dx/dt)^2 + (dy/dt)^2}\, dt.$$

The above formulas can be written in a simpler, more useful form. For revolution about the x axis, we write

$$A(S) = 2\pi \int_a^b y\, ds,$$

while for revolution about the y axis, we write

$$A(S) = 2\pi \int_a^b x\, ds.$$

Then any particular form for x, y, and ds is usable in these formulas. For example, if C is given in the form $y = f(x)$ and C is revolved about the x axis, we have

$$ds = \sqrt{1 + [f'(x)]^2}\, dx \quad \text{and} \quad A(S) = 2\pi \int_a^b f(x)\sqrt{1 + [f'(x)]^2}\, dx,$$

while if $y = f(x)$ and C is revolved about the y axis, we get

$$A(S) = 2\pi \int_a^b x\sqrt{1 + [f'(x)]^2}\, dx.$$

Example 1. Find the surface area of a zone of a sphere obtained by revolving about the x axis the part of the semicircle

$$\{(x, y): b \leq x \leq c, y = \sqrt{a^2 - x^2}\} \quad \text{(where } -a < b < c < a\text{).}$$

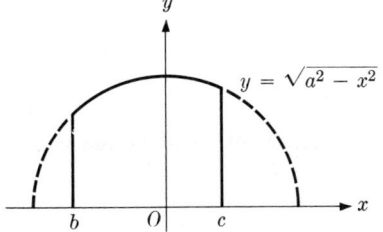

Figure 15–31

Solution. We draw the graph (Fig. 15–31) and use the formula

$$A = 2\pi \int_b^c y \, ds,$$

with $ds = \sqrt{1 + (dy/dx)^2} \, dx$ and $y = \sqrt{a^2 - x^2}$. We have

$$\frac{dy}{dx} = -\frac{x}{\sqrt{a^2 - x^2}} \quad \text{and} \quad ds = \frac{a}{\sqrt{a^2 - x^2}} \, dx.$$

Therefore

$$A = 2\pi \int_b^c \sqrt{a^2 - x^2} \cdot \frac{a}{\sqrt{a^2 - x^2}} \, dx = 2\pi a(c - b). \quad \blacktriangleleft$$

Example 2. Find the area of the surface generated by revolving about the y axis the arc C given by $C = \{(x, y): 0 \leq x \leq 1, y = x^2\}$.

Solution. We use the formula

$$A = 2\pi \int_0^1 x \, ds,$$

with $ds = \sqrt{1 + (dy/dx)^2} \, dx = \sqrt{1 + 4x^2} \, dx$. Therefore

$$A = 2\pi \int_0^1 x\sqrt{1 + 4x^2} \, dx = \frac{\pi}{6}\left[(1 + 4x^2)^{3/2}\right]_0^1 = \frac{\pi}{6}(5\sqrt{5} - 1). \quad \blacktriangleleft$$

Example 3. Find the area of the surface generated by revolving about the x axis the arc $x = t^3$, $y = \frac{3}{2}t^2$, $1 \leq t \leq 3$.

Solution. We use the formula

$$A = 2\pi \int_1^3 y \, ds,$$

with $ds = \sqrt{(dx/dt)^2 + (dy/dt)^2} \, dt$ and $y = \frac{3}{2}t^2$. Therefore

$$A = 3\pi \int_1^3 t^2\sqrt{9t^4 + 9t^2} \, dt = 9\pi \int_1^3 t^3\sqrt{1 + t^2} \, dt.$$

We make the substitution $u = 1 + t^2$, $du = 2t\, dt$, and obtain

$$A = \frac{9\pi}{2} \int_2^{10} (u-1)\sqrt{u}\, du = \frac{9}{2}\pi\left[\frac{2}{5}u^{5/2} - \frac{2}{3}u^{3/2}\right]_2^{10} = 3\pi(50\sqrt{10} - \frac{2}{5}\sqrt{2}). \blacktriangleleft$$

PROBLEMS

In Problems 1 through 21, find the area of each surface S obtained by revolving the given arc C about the x axis.

1. $C = \{(x, y): 0 \le x \le 2, y = \frac{1}{3}x^3\}$
2. $C = \{(x, y): 1 \le x \le 2, y = \frac{1}{6}x^3 + 1/2x\}$
3. $C = \{(x, y): x = t^2, y = 2t, 1 \le t \le 2\}$
4. $C = \{(x, y): 0 \le x \le 6, y^2 = 6x\}$
5. $C = \{(x, y): 0 \le x \le 2, y^2 = 2 - x\}$
6. $C = \{(x, y): 18y^2 = x(6-x)^2, 0 \le x \le 6\}$
7. $C = \{(x, y): 1 \le x \le 2, y = (x^4/8) + 1/4x^2\}$
8. $C = \{(x, y): 1 \le x \le 2, y = (x^2/4) - \frac{1}{2}\ln x\}$
9. $C = \{(x, y): 0 \le x \le 1, y = \frac{1}{2}(e^x + e^{-x})\}$
10. $C = \{(x, y): x = (y^2/4) - \frac{1}{2}\ln y, 1 \le y \le 2\}$
11. $C = \{(x, y): x = (y^3/6) + (1/2y), 1 \le y \le 2\}$
12. $C = \{(x, y): 0 \le x \le 1, y = e^x\}$
13. $C = \{(x, y): -a \le x \le a, x^{2/3} + y^{2/3} = a^{2/3}, y \ge 0\}$
14. $C = \{(x, y): 0 \le x \le \pi, y = \sin x\}$
15. $C = \{(x, y): 0 \le x \le 1, y = \frac{3}{2}x^{2/3}\}$
16. $C = \{(x, y): 0 \le x \le 1, y = e^{-x}\}$
17. $C = \{(x, y): x = a(\theta - \sin\theta), y = a(1 - \cos\theta), 0 \le \theta \le 2\pi\}$
18. $C = \{(x, y): 0 \le x \le 1, y = \frac{5}{3}x^{3/5}\}$
19. $C = \{(x, y): -a \le x \le a, (x^2/a^2) + (y^2/b^2) = 1, y \ge 0\}$
20. $C = \{(x, y): 0 \le x < \infty, y = e^{-x}\}$
21. $C = \{(x, y): 1 \le x \le 2, y = 1/x\}$
22. Decide whether or not the area of the surface generated by revolving
$$C = \{(x, y): 1 \le x < \infty, y = 1/x\}$$
about the x axis is finite.

In Problems 23 through 27, find the area of each surface S obtained by revolving the given arc C about the y axis.

23. $C = \{(x, y): x = b + a\cos t, y = a\sin t, 0 \le t \le 2\pi, b > a > 0\}$
24. $C = \{(x, y): 1 \le x \le 3, y = (x^2/4) - \frac{1}{2}\ln x\}$
25. $C = \{(x, y): 1 \le x \le 2, y = (x^3/6) + (1/2x)\}$
26. $C = \{(x, y): 0 \le x \le 1, y = \frac{1}{2}(e^x + e^{-x})\}$
27. $C = \{(x, y): 0 \le x \le 2, y = \frac{1}{3}x^3\}$
28. Sketch the arc $C = \{(x, y): 0 \le x \le 1, y = (x^2/4) - \frac{1}{2}\ln x\}$. Decide whether or not the area of the surface generated by revolving C about the y axis is finite.

29. Let C be an arc in the first quadrant given by $y = f(x)$, $a \le x \le b$. Develop a formula for the area of the surface S generated when C is revolved about the line $y = -3$. Do the same for the line $y = -k$, where $k > 0$.

Consider the transformation $x = r \cos \theta$, $y = r \sin \theta$ from Cartesian to polar coordinates. Let $r = f(\theta)$, $\theta_1 \le \theta \le \theta_2$, be an arc situated in the first quadrant which is revolved about the polar axis. Then the area A of the surface generated is given by the formula

$$A = 2\pi \int_{\theta_1}^{\theta_2} f(\theta) \sin \theta \sqrt{[f(\theta)]^2 + [f'(\theta)]^2}\, d\theta. \tag{2}$$

30. Setting $x = f(\theta) \cos \theta$, $y = f(\theta) \sin \theta$, show that the above formula is a consequence of (1) on page 532.
31. Use (2) to find the surface area generated when the arc $r = 4 \cos \theta$, $0 \le \theta \le \pi/2$ is revolved about the polar axis.
32. The portion of the cardiod $r = 2(1 + \cos \theta)$, $0 \le \theta \le \pi$, is revolved about the polar axis. Find the area of the surface generated.
33. The arc $r = f(\theta)$, $\theta_1 \le \theta \le \theta_2$ situated in the first quadrant is revolved about the line $\theta = \pi/2$. Derive the formula analogous to (2) above for the area of the surface generated. [*Hint:* Start with the appropriate formula on page 532.]

6. CENTER OF MASS

Suppose that four masses are situated on a line, as shown in Fig. 15–32. Let the distance of m_i ($i = 1, 2, 3, 4$) from the origin O be x_i (x_i is positive if m_i is to the right of O, negative if m_i is to the left). We define the **moment of** m_i **with respect to** O as the product $m_i x_i$. For the system of four masses, we define the quantity

$$\bar{x} = \frac{m_1 x_1 + m_2 x_2 + m_3 x_3 + m_4 x_4}{m_1 + m_2 + m_3 + m_4},$$

and call it the **center of mass** of the system. This definition is easily extendable to a system of n particles on a line, the center of mass being located at

$$\bar{x} = \frac{m_1 x_1 + m_2 x_2 + \cdots + m_n x_n}{m_1 + m_2 + \cdots + m_n}.$$

The center of mass is the point where the system will balance if a knife-edge is put at $\bar{x}$ and the line itself is considered weightless. Furthermore, each mass is assumed to occupy exactly one point. An important fact is that the location of x does not depend on the position of the origin O. (See Problem 5 at the end of this section.)

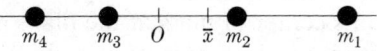

Figure 15–32

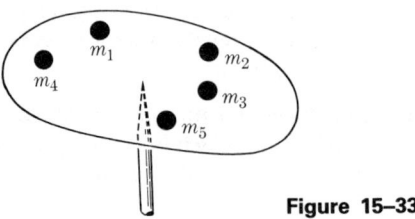

Figure 15-33

Example 1. Weights of 2, 3, 5, and 4 grams are located on the x axis at the points $(4, 0)$, $(2, 0)$, $(-6, 0)$, and $(-4, 0)$, respectively. Find the center of mass of the system.

Solution

$$\bar{x} = \frac{2(4) + 3(2) + 5(-6) + 4(-4)}{2 + 3 + 5 + 4} = -\frac{16}{7}.$$ ◀

Suppose that a number of masses, say five, are located at various points in the xy plane. We wish to find the center of mass of this system. From the point of view of mechanics, we imagine the masses supported by a weightless tray and assume that each mass occupies a single point. The center of mass is the point at which the tray will balance when supported by a sharp nail (Fig. 15–33). To locate this point, we suppose that the mass m_i is situated at the point (x_i, y_i). The **moment of** m_i **with respect to the** y **axis** is the product $m_i x_i$. The **moment of** m_i **with respect to the** x **axis** is the product $m_i y_i$. The **center of mass** $(\bar{x}, \bar{y})$ *of a system of n masses located at the points* (x_i, y_i), $i = 1, 2, \ldots, n$ is defined by the formulas

$$\bar{x} = \frac{m_1 x_1 + m_2 x_2 + \cdots + m_n x_n}{m_1 + m_2 + \cdots + m_n},$$

$$\bar{y} = \frac{m_1 y_1 + m_2 y_2 + \cdots + m_n y_n}{m_1 + m_2 + \cdots + m_n}.$$

Example 2. Find the center of mass of the system consisting of masses of 4, 2, and 3 grams located at the points $(1, 3)$, $(-2, 1)$, and $(4, -2)$, respectively.

Solution

$$\bar{x} = \frac{4(1) + 2(-2) + 3(4)}{4 + 2 + 3} = \frac{4}{3},$$

$$\bar{y} = \frac{4(3) + 2(1) + 3(-2)}{4 + 2 + 3} = \frac{8}{9}.$$ ◀

An important fact is that the center of mass is independent of the location of the coordinate axes. (See Problem 10 at the end of this section.)

A thin piece of metal of uniform density has its center of mass at the place at which it will balance horizontally when supported on the point of a nail. We idealize the situation by imagining the metal as two-dimensional and think of it as a region located in the xy plane. If the metal is uniform (and we always assume that it is) the density is constant, and the total mass of the piece of metal is proportional to the area of the region. For example, a rectangular piece of length l and width w made up of material of constant density ρ has total mass proportional to

$$\rho l w.$$

The center of mass of a rectangular region is at the center of the rectangle (Fig. 15–34). **The moment of a rectangle with respect to the y axis** is

$$\rho l w \cdot \bar{x},$$

where $\bar{x}$ is the x coordinate of the center of mass of the rectangle. Similarly, **the moment of a rectangle with respect to the x axis** is

$$\rho l w \cdot \bar{y}.$$

With these definitions, it is possible to find the center of mass of regions composed of a combination of rectangles. Each rectangle may be treated as if its mass were concentrated at the center of the rectangle. We shall illustrate with the following example.

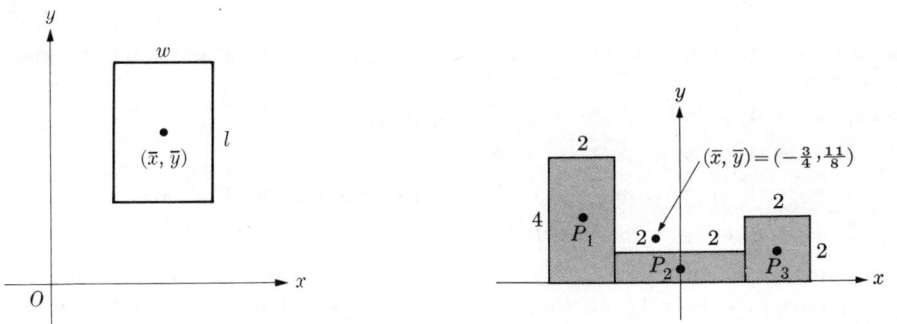

Figure 15–34 Figure 15–35

Example 3. A region is made up of a combination of rectangles of uniform density ρ, as shown in Fig. 15–35. Find the center of mass.

Solution. The centers of mass of the three rectangles are at $P_1(-3, 2)$, $P_2(0, \frac{1}{2})$, and $P_3(3, 1)$. The total masses of the rectangles are 8ρ, 4ρ, and 4ρ, respectively. We may treat the region as a system of three point masses located at the centers of mass. We obtain

$$\bar{x} = \frac{[8(-3) + 4(0) + 4(3)]\rho}{16\rho} = -\frac{3}{4},$$

$$\bar{y} = \frac{[8(2) + 4(\frac{1}{2}) + 4(1)]\rho}{16\rho} = \frac{11}{8}. \blacktriangleleft$$

Remarks. The numerator of the expression for $\bar{x}$ is the sum of the moments of the rectangles with respect to the y axis and, correspondingly, the one for $\bar{y}$ is the sum of the moments of the rectangles with respect to the x axis. In each case the denominator is the total mass.

So long as the region is of uniform density, the actual value of the density plays no part in the location of the center of mass. The above example shows how the factor ρ cancels in the computation of $\bar{x}$ and $\bar{y}$. Therefore, *we always assume that the density has the value* 1, in which case the total mass of a region is its area.

As Fig. 15–35 shows, it is possible for the center of mass of a region to lie outside the region!

PROBLEMS

In Problems 1 through 4, the masses m_i are located at the points P_i on the x axis; find the center of mass.

1. $m_1 = 3$, $m_2 = 4$, $m_3 = 1$; $P_1(2, 0)$, $P_2(6, 0)$, $P_3(-4, 0)$
2. $m_1 = 7$, $m_2 = 1$, $m_3 = 5$, $m_4 = 12$; $P_1(-4, 0)$, $P_2(1, 0)$, $P_3(5, 0)$, $P_4(-7, 0)$
3. $m_1 = 2$, $m_2 = 1$, $m_3 = 5$, $m_4 = 6$, $m_5 = 1$; $P(0, 0)$, $P_2(1, 0)$, $P_3(-1, 0)$, $P_4(2, 0)$, $P_5(-2, 0)$
4. $m_1 = 1$, $m_2 = 2$, $m_3 = 1$, $m_4 = 1$, $m_5 = 2$; $P_1(1, 0)$, $P_2(2, 0)$, $P_3(3, 0)$, $P_4(4, 0)$, $P_5(5, 0)$
5. (a) Find the center of mass $\bar{x}$ of the system $m_1 = 7$, $m_2 = 5$, $m_3 = 4$, $m_4 = 2$, located on the x axis at the points $P_1(5, 0)$, $P_2(-5, 0)$, $P_3(4, 0)$, $P_4(-2, 0)$. (b) Show that when the origin is shifted a distance h to the right, the location of the center of mass relative to the masses does not change.

In Problems 6 through 9, the masses m_i are located in the xy plane at the points P_i; find the centers of mass.

6. $m_1 = 1$, $m_2 = 3$, $m_3 = 2$; $P_1(2, 1)$, $P_2(-1, 3)$, $P_3(1, 2)$
7. $m_1 = 3$, $m_2 = 1$, $m_3 = 4$; $P_1(1, 0)$, $P_2(0, 3)$, $P_3(1, 2)$
8. $m_1 = 4$, $m_2 = 5$, $m_3 = 1$, $m_4 = 6$; $P_1(1, 1)$, $P_2(5, 0)$, $P_3(-4, 0)$, $P_4(0, 5)$
9. $m_1 = 2$, $m_2 = 8$, $m_3 = 5$, $m_4 = 2$; $P_1(0, 0)$, $P_2(0, 4)$, $P_3(5, 1)$, $P_4(-1, -1)$
10. (a) Find the center of mass $(\bar{x}, \bar{y})$ of the system $m_1 = 3$, $m_2 = 4$, $m_3 = 2$, located at the points $P_1(1, 1)$, $P_2(3, 0)$, $P_3(-1, 1)$. (b) Show that if the coordinate axes are translated so that the origin is at the point (h, k), the location of the center of mass relative to the masses does not change. Can the computation of the center of mass be simplified in this manner?

In Problems 11 through 14, introduce a convenient Cartesian coordinate system and find the coordinates of the center of mass in that coordinate system.

11. A plane region of uniform density has the shape shown in Fig. 15–36. Find the center of mass.
12. A plane region of uniform density has the shape shown in Fig. 15–37. Find the center of mass.
13. A plane region of uniform density has the shape shown in Fig. 15–38. Find the center of mass.
14. A plane region is composed of the regions R_1, R_2, and R_3, as shown in Fig. 15–39. The density of R_2 is twice that of R_1, and the density of R_3 is three times that of R_2. Find the center of mass.
15. Suppose that n masses $m_1, m_2, \ldots, m_n$ are located in the plane at the points (x_1, y_1), $(x_2, y_2), \ldots, (x_n, y_n)$, respectively. Denote the center of mass of this system by $(\bar{x}, \bar{y})$. If we translate coordinates by the customary formulas: $x' = x + h$, $y' = y + k$, show that in the new coordinate system the center of mass is located at $(\bar{x} + h, \bar{y} + k)$. That is, prove that the geometric location of the center of mass is unchanged by a translation of coordinates.

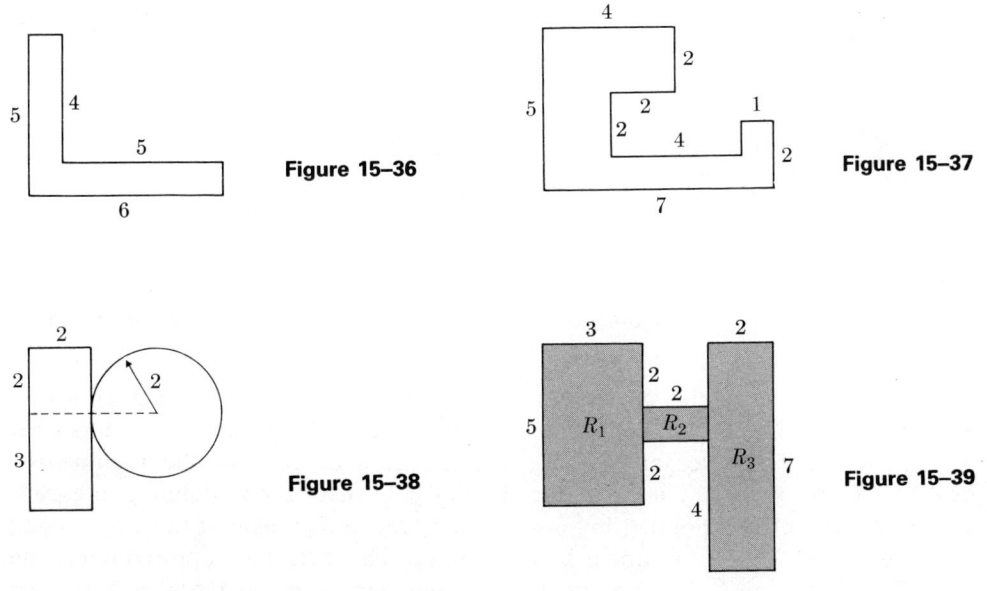

Figure 15-36

Figure 15-37

Figure 15-38

Figure 15-39

7. CENTER OF MASS OF A PLANE REGION

In the preceding section we showed how to find the center of mass of a plane region composed of a collection of rectangles, each of uniform density. We now establish methods for finding the center of mass $(\bar{x}, \bar{y})$ of more general plane figures. *We always assume that the density is constant.* It is convenient to take the density equal to 1, since then the total mass of a region is numerically equal to its area. The results rest on the following two principles, which we state without proof.

Principle 1. *The center of mass of a plane region lies on any axis of symmetry of that region.*

Principle 2. *Suppose that a plane region R with center of mass $(\bar{x}, \bar{y})$ is divided into regions $R_1, R_2, \ldots, R_n$ (no two of which overlap), having areas $A_1, A_2, \ldots, A_n$, and centers of mass at $(\bar{x}_1, \bar{y}_1), (\bar{x}_2, \bar{y}_2), \ldots, (\bar{x}_n, \bar{y}_n)$. Then*

$$\bar{x} = \frac{A_1\bar{x}_1 + A_2\bar{x}_2 + \cdots + A_n\bar{x}_n}{A_1 + A_2 + \cdots + A_n}, \quad \bar{y} = \frac{A_1\bar{y}_1 + A_2\bar{y}_2 + \cdots + A_n\bar{y}_n}{A_1 + A_2 + \cdots + A_n}.$$

Let R be the region bounded by the graph of a nonnegative function $y = f(x)$ and the lines $x = a$, $x = b$, and $y = 0$, as shown in Fig. 15-40. We make a subdivision of the interval $[a, b]$ into n parts:

$$\{a = x_0 < x_1 < x_2 < \cdots < x_n = b\}.$$

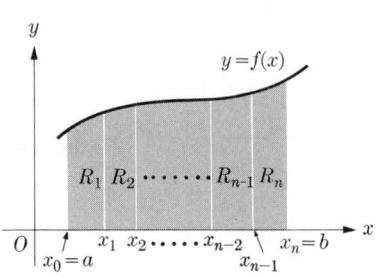
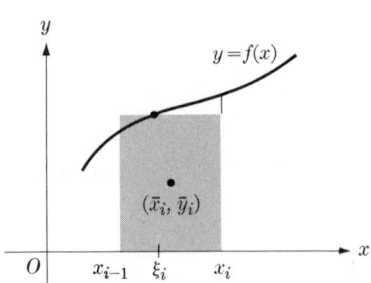

Figure 15–40   Figure 15–41

We erect ordinates at the subdivision points, thereby dividing R into n nonoverlapping regions $R_1, R_2, \ldots, R_n$. According to Principle 2, the center of mass of R can be found if the center of mass and the area of each of the regions $R_1, R_2, \ldots, R_n$ are known. The procedure is similar to that used in defining integrals. For each i we select a point ξ_i in $[x_{i-1}, x_i]$ and erect a rectangle of height $f(\xi_i)$ and width $\Delta_i x = x_i - x_{i-1}$, as shown in Fig. 15–41. This rectangle approximates the area R_i, and the center of mass of the rectangle approximates the center of mass of R_i. According to Principle 1, the center of mass $(\bar{x}_i, \bar{y}_i)$ of the rectangle is at

$$\bar{x}_i = \tfrac{1}{2}(x_{i-1} + x_i), \qquad \bar{y}_i = \tfrac{1}{2}f(\xi_i).$$

The area of the rectangle is $f(\xi_i)\,\Delta_i x$. Therefore the center of mass $(\bar{x}, \bar{y})$ of R is approximated (according to Principle 2) by the expressions

$$\frac{(f(\xi_1)\,\Delta_1 x)\bar{x}_1 + (f(\xi_2)\,\Delta_2 x)\bar{x}_2 + \cdots + (f(\xi_n)\,\Delta_n x)\bar{x}_n}{f(\xi_1)\,\Delta_1 x + f(\xi_2)\,\Delta_2 x + \cdots + f(\xi_n)\,\Delta_n x},$$

$$\frac{(f(\xi_1)\,\Delta_1 x)\bar{y}_1 + (f(\xi_2)\,\Delta_2 x)\bar{y}_2 + \cdots + (f(\xi_n)\,\Delta_n x)\bar{y}_n}{f(\xi_1)\,\Delta_1 x + f(\xi_2)\,\Delta_2 x + \cdots + f(\xi_n)\,\Delta_n x}.$$

We made the above computation for one subdivision. We now make a sequence of such subdivisions, with the norms of the subdivisions tending to zero. We can show, in the same way that we developed the integration formulas for area, that if f is continuous function the above expressions tend to the center of mass of R, and that $\bar{x}$ and $\bar{y}$ are given by

$$\bar{x} = \frac{\int_a^b xf(x)\,dx}{\int_a^b f(x)\,dx}, \qquad \bar{y} = \frac{\tfrac{1}{2}\int_a^b [f(x)]^2\,dx}{\int_a^b f(x)\,dx}.$$

Observe that the denominator in each case is the area of the region R.

Example 1. Compute the center of mass of the region

$$R = \{(x, y) : 1 \leq x \leq 3,\ 0 \leq y \leq x^2\}.$$

(See Fig. 15–42.)

Solution. The region is of the type discussed above, and therefore

$$\bar{x} = \frac{\int_1^3 x^3\, dx}{A}, \qquad \bar{y} = \frac{\frac{1}{2}\int_1^3 x^4\, dx}{A},$$

where A is the area of R. We have

$$A = \int_1^3 x^2\, dx = \tfrac{26}{3},$$

and so

$$\bar{x} = \tfrac{30}{13}, \qquad \bar{y} = \tfrac{363}{130}. \qquad \blacktriangleleft$$

Rather than memorize the above formulas for the center of mass of a region of a particular type, the student should learn the method for deriving such formulas. The steps in the method are as follows: (1) Make a subdivision of the interval on the x axis (or y axis, if that appears more convenient). The region R is approximated by erecting rectangles at the subdivision points. (2) Find the center of mass of each rectangle by Principle 1. (3) Use Principle 2 to obtain an approximate expression for $(\bar{x}, \bar{y})$, the center of mass of R. (4) Proceed to the limit by writing the appropriate integrals. (5) Evaluate the integrals.

The next example illustrates this technique.

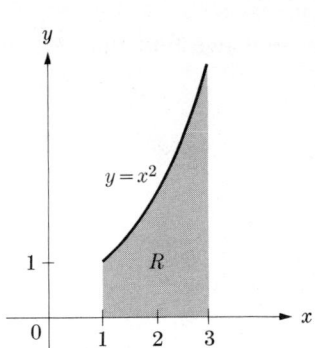

Figure 15–42

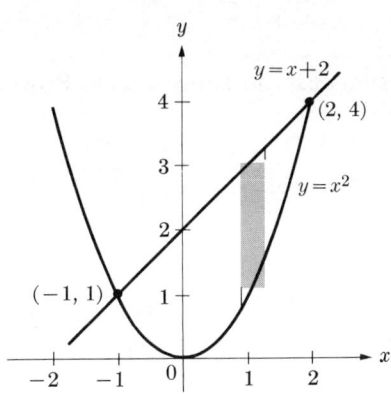

Figure 15–43

Example 2. Find the center of mass of the region R bounded by the curves $y = x^2$ and $y = x + 2$.

Solution. By solving simultaneously, we find that the curves intersect at $(-1, 1)$ and $(2, 4)$. (See Fig. 15–43.) Therefore, in set notation, we may write $R = \{(x, y)\,; -1 \leq x \leq 2,\ x^2 \leq y \leq x + 2\}$. We make a subdivision of the interval $[-1, 2]$ and erect rectangles at the subdivision points. (A typical one is shown in Fig. 15–43.) The center of mass $(\bar{x}_i, \bar{y}_i)$ of this rectangle is at

$$\bar{x}_i = \tfrac{1}{2}(x_{i-1} + x_i),$$
$$\bar{y}_i = \tfrac{1}{2}[(\xi_i + 2) + \xi_i^2],$$

where ξ_i is some number in the interval $[x_{i-1}, x_i]$. $\qquad \blacktriangleleft$

542 APPLICATIONS OF THE INTEGRAL

Remark. Since we shall always be dealing with a typical rectangle (or other typical shape), it pays to abbreviate the notation in the following way. We drop the subscript i and instead of $\frac{1}{2}(x_{i-1} + x_i)$, we write x. Since ξ_i is between x_{i-1} and x_i, we can also use x for this quantity. We then have

$$\bar{x}_i = x, \qquad \bar{y}_i = \tfrac{1}{2}[(x + 2) + x^2].$$

Although it may appear that such sloppiness will lead to errors, we need not worry, since we shall eventually proceed to the limit and replace sums by integrals. (Of course, if there is any doubt in the reader's mind in a specific case, the longer notation should be kept.)

The area of a typical rectangle (in abbreviated notation) is

$$[(x + 2) - x^2]\,\Delta x.$$

Using Principle 2, we obtain as approximate expressions for $\bar{x}$ and $\bar{y}$

$$\frac{\sum\{[(x + 2) - x^2]\,\Delta x\}x}{\sum[(x + 2) - x^2]\,\Delta x}, \qquad \frac{\sum\{[(x + 2) - x^2]\,\Delta x\}\tfrac{1}{2}[(x + 2) + x^2]}{\sum[(x + 2) - x^2]\,\Delta x}.$$

These sums extend from 1 to n. Proceeding to the limit, we find that

$$\bar{x} = \frac{\int_{-1}^{2} x[(x + 2) - x^2]\,dx}{A},$$

$$\bar{y} = \frac{\tfrac{1}{2}\int_{-1}^{2}[(x + 2) + x^2][(x + 2) - x^2]\,dx}{A},$$

where A is the area of the entire region. We get A in the usual way from the formula

$$A = \int_{-1}^{2}[(x + 2) - x^2]\,dx.$$

Evaluating the various integrals, we obtain

$$\bar{x} = \tfrac{1}{2}, \qquad \bar{y} = \tfrac{8}{5}$$

for the solution of Example 2.

Example 2 illustrates how a general formula can be obtained for the center of mass of a region R bounded by the lines $x = a$, $x = b$ and the curves $y = f(x)$, $y = g(x)$ with $f(x) \geq g(x)$. See Fig. 15-44. Following the procedure outlined in

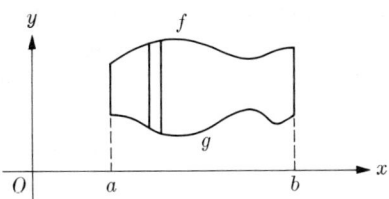

Figure 15–44

Example 2, we construct a typical rectangle, as shown, and locate its center of mass. Omitting the details, we find as a result,

$$\bar{x} = \frac{\int_a^b x[f(x) - g(x)]\,dx}{\int_a^b [f(x) - g(x)]\,dx}, \quad \bar{y} = \frac{\frac{1}{2}\int_a^b [f(x) + g(x)][f(x) - g(x)]\,dx}{\int_a^b [f(x) - g(x)]\,dx}. \quad (1)$$

Observe that the numerator of $\bar{y}$ may be written $\frac{1}{2}\int_a^b ([f(x)]^2 - [g(x)]^2)\,dx$.

Example 3. Find the center of mass of the triangular region with vertices at $(0, 0)$, $(a, 0)$, and (b, c).

Solution. We first concentrate on finding $\bar{y}$. It is convenient to subdivide the interval $[0, c]$ along the y axis. A typical rectangle is shown in Fig. 15–45. We suppose that its height is y. Then the y coordinate of its center of mass is simply y. Let l be the length and Δy the width of this rectangle. By similar triangles, we have the proportion

$$\frac{l}{a} = \frac{c - y}{c};$$

therefore the area of a typical rectangle is

$$l\,\Delta y = \frac{a}{c}(c - y)\,\Delta y.$$

According to Principle 2, $\bar{y}$ is approximated by

$$\frac{\sum (a/c)(c - y)\Delta y \cdot y}{\sum (a/c)(c - y)\Delta y}.$$

We conclude that

$$\bar{y} = \frac{(a/c)\int_0^c (c - y)y\,dy}{\frac{1}{2}ac} = \frac{\frac{1}{6}ac^2}{\frac{1}{2}ac} = \frac{1}{3}c.$$

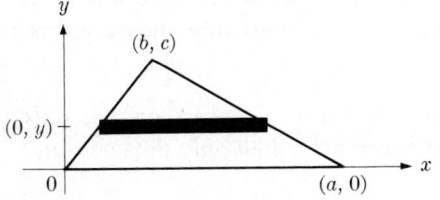

Figure 15–45

The center of mass is one-third of the distance from the base to the vertex. Since any side could have been chosen as base, we deduce that the center of mass of any triangle is at the point of intersection of its medians. ◄

PROBLEMS

In Problems 1 through 7, describe the region R in set notation. Then find the center of mass in each case.

1. R is bounded by $y = \sqrt{4x}$, the x axis, and the line $x = 6$.
2. R is bounded by $y = x^2$ and $y = x$.
3. R is bounded by one arch of $y = \sin x$ and the x axis.
4. R is bounded by $y = \sin x$, and the lines $x = 0$, $x = \pi/2$, $y = 0$.
5. R is the semicircular region bounded by $y = \sqrt{a^2 - x^2}$, $-a \leq x \leq a$, and the x axis.
6. R is the quarter-circle bounded by $y = \sqrt{a^2 - x^2}$ and the lines $x = 0$, $y = 0$.
7. R is the region to the right of $y = 3x$, and bounded by $y = 3x$, $y = x^2$, $y = 1$, and $y = 2$.

In Problems 8 through 11, find the center of mass of each of the regions R.

8. $R = \{(x, y): 0 \leq x \leq \pi/4, \ 0 \leq y \leq \sec^2 x\}$
9. $R = \{(x, y): 2 \leq x \leq 4, \ 0 \leq y \leq \ln x\}$
10. $R = \{(x, y): 0 \leq x \leq 1, \ x^3 \leq y \leq x\}$
11. $R = \{(x, y): 0 \leq x \leq \frac{1}{4}, \ x^2 \leq y \leq x - x^2\}$
12. Use the result of Example 3 of this section to find the center of mass of the trapezoid with vertices $(0, 0)$, $(a, 0)$, (b, c), (d, c), where

$$0 < b < d < a.$$

13. Complete the details of the proof of formula (1) on page 543.

In Problems 14 through 16 use formula (1) on page 543 to compute the center of mass of the region R.

14. $R = \left\{(x, y): 0 \leq x \leq \dfrac{\pi}{3}, \ \sin x \leq y \leq \sin 2x\right\}$
15. $R = \{(x, y): 0 \leq x \leq 1, \ x^3 \leq y \leq x^2\}$
16. $R = \{(x, y): 1 \leq x \leq 2, \ \frac{1}{2}(x - 1) \leq y \leq \ln x\}$

8. CENTER OF MASS OF A SOLID OF REVOLUTION

The methods we have developed enable us now to find the centers of mass of certain solids of revolution. As before, we shall assume that the object is made of uniform material of constant density. For convenience we shall select the density equal to 1, so that the total mass of an object is numerically the same as its volume.

Principle 1'. *A solid of revolution will have its center of mass on the axis of revolution. If the axis of revolution is taken as the x axis, then only the coordinate $\bar{x}$ has to be determined.*

15–8 CENTER OF MASS OF A SOLID OF REVOLUTION

The principle we use in locating this value is the analog of Principle 2 of the last section.

Principle 2'. *Suppose that a solid F is divided into solids $F_1, F_2, \ldots, F_n$ (no two of which overlap), having volumes $V_1, V_2, \ldots, V_n$. Let $\bar{x}_1, \bar{x}_2, \ldots, \bar{x}_n$ be the x values of the centers of mass of $F_1, F_2, \ldots, F_n$, respectively. Then*

$$\bar{x} = \frac{V_1\bar{x}_1 + V_2\bar{x}_2 + \cdots + V_n\bar{x}_n}{V_1 + V_2 + \cdots + V_n}.$$

Let R be a region bounded by the nonnegative function $y = f(x)$ and the lines $x = a$, $x = b$, and $y = 0$. That is,

$$R = \{(x, y): a \leq x \leq b, 0 \leq y \leq f(x)\}.$$

We revolve this region about the x axis, obtaining a solid of revolution F (Fig. 15–46). The center of mass of F will be on the x axis, according to Principle 1', and we shall show how to locate it. We make a subdivision of the interval $[a, b]: \{a = x_0 < x_1 \cdots < x_n = b\}$, and slice the solid F into domains $F_1, F_2, \ldots, F_n$, by planes through the subdivision points. We approximate each F_i by a disk of thickness $\Delta_i x = x_i - x_{i-1}$ and radius $f(\xi_i)$, where $x_{i-1} \leq \xi_i \leq x_i$. Then the volume of the disk is

$$\pi[f(\xi_i)]^2 \Delta x,$$

and the center of mass of the disk is on the x axis at $\frac{1}{2}(x_{i-1} + x_i)$. We use the abbreviated notation described in Example 2 of the preceding section and obtain (by Principle 2') the approximation to $\bar{x}$,

$$\frac{\sum (\pi[f(x)]^2 \Delta x)x}{\sum \pi[f(x)]^2 \Delta x}.$$

Proceeding to the limit, we then find

$$\bar{x} = \frac{\pi \int_a^b x[f(x)]^2\, dx}{\pi \int_a^b [f(x)]^2\, dx}.$$

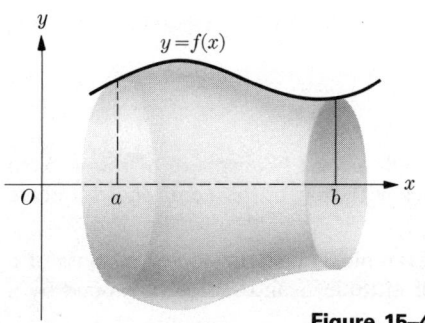

Figure 15–46

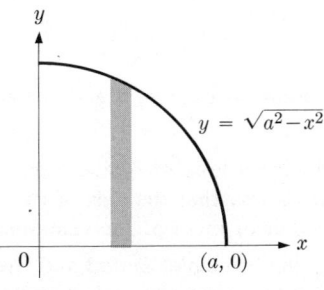

Figure 15–47

Example. Find the center of mass of the hemisphere of radius a, center at the origin, and axis along the positive x axis.

Solution. The solid is generated by revolving about the x axis the quadrant Q of a disk given by $Q = \{(x, y) : 0 \leq x \leq a,\ 0 \leq y \leq \sqrt{a^2 - x^2}\}$ (Fig. 15-47). We divide the solid into thin slabs by planes through subdivision points. According to the above description of the method, we then have

$$\bar{x} = \frac{\int_0^a x(a^2 - x^2)\,dx}{\int_0^a (a^2 - x^2)\,dx} = \frac{3a}{8}.$$ ◄

PROBLEMS

In each of Problems 1 through 11, a region R in the plane is given. Describe R in set notation whenever it is not already described in such terms. Then find the center of mass of the solid generated by revolving R about the x axis.

1. R is bounded by $y = 0$, $x = h$, $y = (a/h)x$. (Note that the solid is a cone of altitude h and radius of base a.)
2. $R = \{(x, y) : 0 \leq x \leq b,\ 0 \leq y \leq \sqrt{2px}\}$; b, p are constants.
3. $R = \{(x, y) : 0 \leq x \leq a,\ 0 \leq y \leq (b/a)\sqrt{a^2 - x^2}\}$
4. R is bounded by $x = 0$, $x = \pi/3$, $y = 0$, $y = \sec x$.
5. R is bounded by $y = 0$, $x = 3$, $y = x/\sqrt{4 - x}$.
6. $R = \{(x, y) : 1 \leq x \leq 2,\ 0 \leq y \leq \ln x\}$
7. $R = \{(x, y) : 0 \leq x \leq \pi,\ 0 \leq y \leq \sin x\}$
8. $R = \{(x, y) : 0 \leq x \leq 1,\ 0 \leq y \leq \tfrac{1}{2}(e^x + e^{-x})\}$
9. R is bounded by the curves $y = x^2$ and $y^2 = x$.
10. R is bounded by $x + y = 5$ and $xy = 4$.
11. R is bounded by $y = x^2$ and $y = x + 2$.

In each of Problems 12 through 15, describe the region R in set notation when it is not already described in such terms. Then find the center of mass of the solid generated by revolving R about the y axis.

12. R is bounded by $y = 1$, $y = 3$, $x = 0$, $x = y^2$.
13. $R = \{(x, y) : 0 \leq x \leq \sqrt{y^2 + 1},\ 0 \leq y \leq 1\}$
14. R is bounded by $x = 0$ and $x + y^2 - 6y = 0$.
15. R is bounded by $x = 1/(y + 1)$, $y = 0$, $y = 3$, $x = 0$.
16. A cylindrical hole of radius 4 cm is bored through a solid hemisphere of radius 5 cm in such a way that the axis of the hole coincides with that of the hemisphere. Locate the center of mass of the remaining solid.
17. Using the principles stated and the result of Example 1, find the center of mass of a solid in the form of a right circular cylinder of altitude h and radius a, capped by a hemisphere of the same radius.

18. Using the principles stated and the results of Example 1 and Problem 1, find the center of mass of a solid "top" in the form of a cone of altitude h and base of radius a, with vertex downward, surmounted by a hemisphere of radius a.
19. The region $R = \{(x, y): 0 \leq x < \infty, 0 \leq y \leq e^{-x}\}$ is revolved about the x axis. Find the center of mass of the solid generated.*
20. Let S be a pyramid with a square base of side 2 and an altitude of 3. Suppose that the vertex of the pyramid is directly above the center of the square base. Use Principles 1' and 2' to find the center of mass of S.

9. CENTERS OF MASS OF WIRES AND SURFACES

Continuing the development of the problem of finding the centers of mass of various kinds of objects, we now show how to locate the center of mass of a thin wire. As usual, we assume that the material is uniform and the density constant and equal to 1; we idealize the problem further by supposing that the wire is one-dimensional. The total mass will then be numerically equal to the length of the curve.

Suppose that the wire is described by the equation $y = f(x)$ between the points $A(a, f(a))$ and $B(b, f(b))$. We make a decomposition of the arc

$$\{A = P_0, P_1, P_2, \ldots, P_{n-1}, P_n = B\}$$

and approximate each subarc by a straight-line segment (Fig. 15–48). *The center of mass of a straight-line segment is at its midpoint.* We use the analog of Principle 2 to approximate the center of mass $(\bar{x}, \bar{y})$ by

$$\frac{\sum (\Delta s) \cdot x}{\sum (\Delta s)}, \quad \frac{\sum (\Delta s) f(x)}{\sum (\Delta s)},$$

where Δs is the length of the arc $P_{i-1}P_i$, and abbreviated notation has been used (Fig. 15–49). Proceeding to the limit, we obtain

$$\bar{x} = \frac{\int_a^b x \, ds}{\int_a^b ds}, \quad \bar{y} = \frac{\int_a^b f(x) \, ds}{\int_a^b ds}.$$

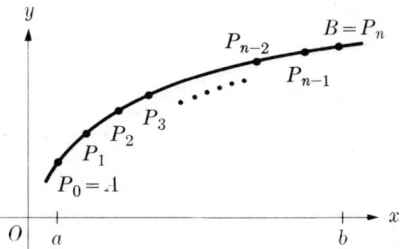

Figure 15–48

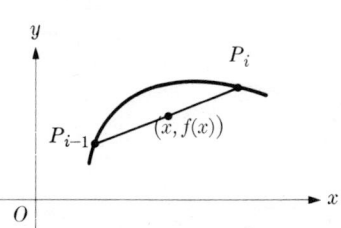

Figure 15–49

---

* $\lim_{h \to +\infty} h \cdot e^{-bh} = 0$ if $b > 0$.

Example 1. Find the center of mass of the arc of the circle $x^2 + y^2 = a^2$ which is in the first quadrant.

Solution. The line $y = x$ is a line of symmetry, and therefore $\bar{x} = \bar{y}$. We have

$$ds = \sqrt{1 + (dy/dx)^2}\, dx = \frac{a}{\sqrt{a^2 - x^2}}\, dx,$$

and therefore

$$\bar{x} = \frac{\int_0^a x(a/\sqrt{a^2 - x^2})\, dx}{\frac{1}{2}\pi a} = \frac{2a}{\pi} = \bar{y}.$$

In Fig. 15–50, note that the center of mass is not on the arc. ◀

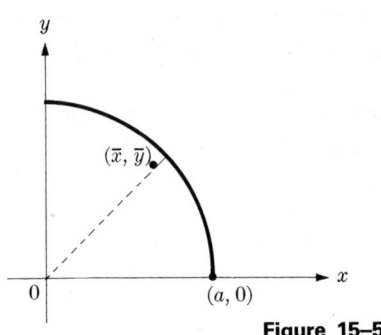

Figure 15–50

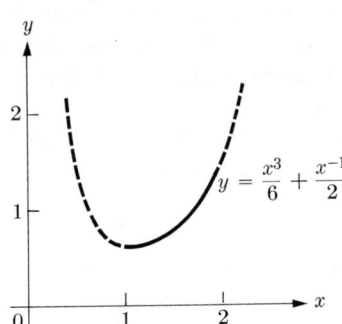

Figure 15–51

Example 2. Find the center of mass of the arc in Fig. 15–51:

$$y = \frac{1}{6}x^3 + \frac{1}{2x}, \qquad 1 \leq x \leq 2.$$

Solution. We subdivide the interval $1 \leq x \leq 2$. A typical section will have mass Δs, and its center of mass will be at $(x, \frac{1}{6}x^3 + 1/2x)$. As an approximation to $(\bar{x}, \bar{y})$, we obtain

$$\frac{\sum (\Delta s) \cdot x}{\sum \Delta s}, \qquad \frac{\sum (\Delta s)\left(\frac{x^3}{6} + \frac{1}{2x}\right)}{\sum (\Delta s)}.$$

Therefore

$$\bar{x} = \frac{\int_1^2 x\, ds}{\int_1^2 ds}, \qquad \bar{y} = \frac{\int_1^2 y\, ds}{\int_1^2 ds}.$$

We also have

$$ds = \sqrt{1 + (dy/dx)^2}\, dx = \frac{1}{2}\left(x^2 + \frac{1}{x^2}\right) dx,$$

and so

$$\bar{x} = \frac{\frac{1}{2}\int_1^2 [x^3 + (1/x)]\,dx}{\frac{1}{2}\int_1^2 [x^2 + (1/x^2)]\,dx}, \qquad \bar{y} = \frac{\frac{1}{2}\int_1^2 [\frac{1}{6}x^3 + (1/2x)][x^2 + (1/x^2)]\,dx}{\frac{1}{2}\int_1^2 [x^2 + (1/x^2)]\,dx}.$$

Computing the various integrals, we find

$$\bar{x} = \tfrac{45}{34} + \tfrac{6}{17}\ln 2, \qquad \bar{y} = \tfrac{141}{136}. \qquad \blacktriangleleft$$

When a thin wire is revolved about an axis, a shell of revolution is obtained. If we consider this as a two-dimensional object, the center of mass may sometimes be found by simple integrations. As usual, the density is assumed constant and equal to 1. Since the method follows the usual pattern, we shall limit the discussion to working examples.

Example 3. Find the center of mass of the surface of a hemisphere of radius a.

Solution. The hemisphere is obtained by revolving about the x axis the quarter circle $y = \sqrt{a^2 - x^2} \equiv f(x)$, $0 \le x \le a$, and the center of mass is on the x axis (Fig. 15–52). To locate it we divide the interval $[0, a]$ into small subintervals. The surface area of each slice generated by revolving the arc of the circle above a subinterval is

$$2\pi f(x)\Delta s$$

(using abbreviated notation). Using the analog of Principle 2', we now obtain, as an approximation for $\bar{x}$,

$$\frac{2\pi \sum [f(x)\Delta s] \cdot x}{2\pi \sum [f(x)\Delta s]}.$$

Proceeding to the limit, we find that

$$\bar{x} = \frac{2\pi \int_0^a x f(x)\,ds}{2\pi \int_0^a f(x)\,ds}.$$

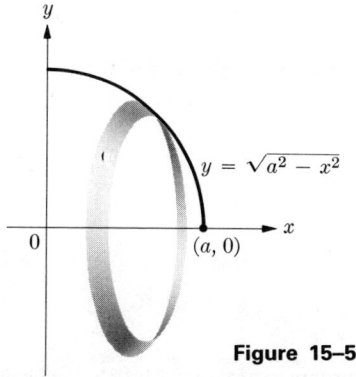

Figure 15–52

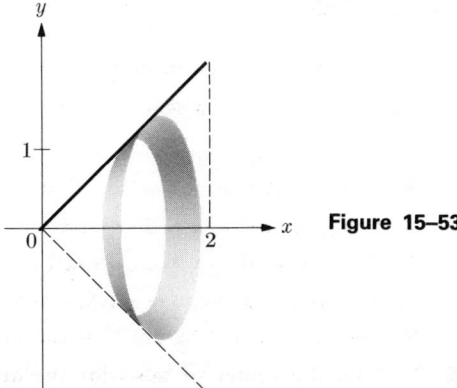

Figure 15–53

The denominator is simply the surface area of the hemisphere, which we know is $2\pi a^2$. Therefore

$$\bar{x} = \frac{1}{a^2} \int_0^a x\sqrt{a^2 - x^2} \frac{a}{\sqrt{a^2 - x^2}} dx = \frac{1}{2}a.$$

Example 4. The line segment $y = x$, $0 \le x \le 2$, is revolved about the x axis, generating the surface of a cone. Find the center of mass of this surface (Fig. 15–53).

Solution. With $f(x) = x$, we have the formula for $\bar{x}$:

$$\bar{x} = \frac{2\pi \int_0^2 xf(x)\, ds}{2\pi \int_0^2 f(x)\, ds}.$$

Therefore

$$\bar{x} = \frac{\int_0^2 x^2 \sqrt{2}\, dx}{2\sqrt{2}} = \frac{1}{2}\left[\frac{1}{3}x^3\right]_0^2 = \frac{4}{3}.$$

Suppose that f is a positive function for $a \le x \le b$. Examples 3 and 4 show that if this arc is revolved about the x axis, the center of mass is on the x axis and its coordinate is given by

$$\bar{x} = \frac{\int_a^b xf(x)\, ds}{\int_a^b f(x)\, ds}. \tag{1}$$

PROBLEMS

In Problems 1 through 6, in each case find the center of mass of the given arc C.

1. $C = \{(x, y): 1 \le x \le 2, y = \frac{1}{8}x^4 + \frac{1}{4}x^{-2}\}$
2. $C = \{(x, y): -a \le x \le a, y = \sqrt{a^2 - x^2}\}$
3. $C = \{(x, y): 0 \le x \le a, y^{2/3} = a^{2/3} - x^{2/3}, y \ge 0\}$
4. $C = \{(x, y): 0 \le x \le 1, y = \frac{1}{2}(e^x + e^{-x})\}$
5. $C = \{(x, y): 0 \le x \le 2, y = x^2\}$
6. $C = \{(x, y): x = \frac{1}{4}y^2 - \frac{1}{2}\ln y, 1 \le y \le 2\}$
7. Find $\bar{y}$ only for the arc $C = \{(x, y): 1 \le x \le 2, y = e^x\}$.
8. Find the center of mass for the arc C given in polar coordinates by the equation $r = a(1 - \cos\theta), 0 \le \theta \le \pi$.

In each of Problems 9 through 14, an arc C is revolved about the x axis. Find the center of mass of the surface generated.

9. $C = \{(x, y): 0 \leq x \leq 4, y^2 = 4x, y \geq 0\}$
10. $C = \{(x, y): 0 \leq x \leq h, y = ax/h\}, a > 0$
11. $C = \{(x, y): 0 \leq x \leq 3, y^2 = 3 - x, y \geq 0\}$
12. $C = \{(x, y): 0 \leq x \leq 1, y = \frac{1}{2}(e^x + e^{-x})\}$
13. $C = \{(x, y): 0 \leq x \leq 3, 4x^2 + 9y^2 = 36, y \geq 0\}$
14. $C = \{(x, y): 1 \leq x \leq 2, y = \frac{1}{4}x^2 - \frac{1}{2}\ln x\}$

In each of Problems 15 through 17, an arc C is revolved about the y axis. Find the center of mass of the surface generated.

15. $C = \{(x, y): 0 \leq x \leq 2, y = x^2\}$
16. $C = \{(x, y): 1 \leq x \leq 2, y = \frac{1}{6}x^3 + \frac{1}{2}x^{-1}\}$
17. $C = \{(x, y): 0 \leq x \leq a, y = \frac{a}{2}(e^{x/a} + e^{-x/a})\}$
18. Let C be an arc given by $y = f(x)$, $a \leq x \leq b$. If C is revolved about the y axis, find a formula similar to (1) on page 550 for the value of $\bar{y}$.

10. THEOREMS OF PAPPUS

The two theorems given (without proof) in this section are useful tools for finding volumes of solids of revolution and areas of surfaces of revolution. They provide interesting applications of the material on finding the center of mass.

Theorem 2 (First Theorem of Pappus). *If a region R lies on one side of a line l in its plane, the volume V of the solid generated by revolving R about l is equal to the product of the area A of R and the length of the path described by the center of mass of R. That is,*

$$V = 2\pi dA$$

where d is the distance of the center of mass of R to the line l.

Theorem 3 (Second Theorem of Pappus). *If an arc C in a plane lies on one side of a line l in the plane, the area S of the surface generated by revolving C about l is the product of the length of C and the length of the path described by the center of mass of C. That is,*

$$S = 2\pi dL$$

where L is the length of C and d is the distance from the center of mass of C to the line l.

Example 1. Find the volume of the solid generated by revolving about the y axis the region $R = \{(x, y): 0 \leq x \leq 4, 0 \leq y \leq 4x - x^2\}$. See Fig. 15–54.

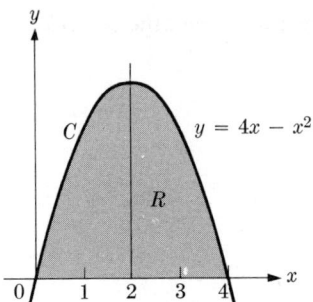

Figure 15–54

Solution. Since $y = 4x - x^2$ is a parabola with the line $x = 2$ as a line of symmetry, we have $\bar{x} = 2$. The area of the region R is

$$A = \int_0^4 (4x - x^2) \, dx = \tfrac{32}{3}.$$

Therefore the volume of the solid is

$$\frac{32}{3} \cdot 2\pi \cdot 2 = \frac{128\pi}{3}. \quad \blacktriangleleft$$

Example 2. The arc $C = \{(x, y) : 0 \le x \le 4, \, y = 4x - x^2\}$ is revolved about the y axis. (See Fig. 15–54.) Find the surface area generated.

Solution. By symmetry we have $\bar{x} = 2$. The length of the arc is

$$\int_0^4 \sqrt{1 + (4 - 2x)^2} \, dx = \frac{8\sqrt{17} + 2\ln(4 + \sqrt{17})}{4}.$$

The total surface area is

$$\frac{2\pi \cdot 2[8\sqrt{17} + 2\ln(4 + \sqrt{17})]}{4} = \pi[8\sqrt{17} + 2\ln(4 + \sqrt{17})].$$

PROBLEMS

In each of Problems 1 through 4, the region R is revolved about the axis stated. Use the First Theorem of Pappus to find the volume generated.

1. $R = \{(x, y) : 0 \le x \le 2, \, 0 \le y \le \sqrt{x}\}$; R is revolved about the x axis.
2. $R = \{(x, y) : 0 \le x \le \pi, \, 0 \le y \le \sin x\}$; R is revolved about the y axis.
3. R is the same as in Problem 2 but is revolved about the x axis.
4. $R = \{(x, y) : 0 \le x \le h, \, 0 \le y \le ax/h\}$; R is revolved about the x axis. (Use the result of Problem 1 of Section 8.)
5. The region of the Example in Section 8, is revolved about the x axis. Find the volume generated.
6. The region of the Example in Section 8, is revolved about the line $y = -3$. Find the volume generated.

7. $R = \{(x, y): 0 \leq x \leq 1, x^3 \leq y \leq x\}$. R is revolved about the x axis. Find the volume generated.
8. Find the volume and the surface area of a sphere of radius a by using the Theorems of Pappus.
9. Find the volume and surface area of the torus obtained by revolving about the y axis the circle $(x - b)^2 + y^2 = a^2$, where $a < b$.
10. Use the First Theorem of Pappus to find the volume of the solid generated by revolving about the x axis the figure bounded by $y = x^2$ and $y = 8 - x^2$.
11. Use the First Theorem of Pappus to find the volume of a right circular cylinder of height h and radius of base r.
12. Use the Second Theorem of Pappus to find the lateral surface area of a right circular cylinder of height h and radius of base r.

11. NEWTON'S LAWS OF MOTION. DIFFERENTIAL EQUATIONS

An equation which contains derivatives is called a **differential equation.** Since we defined velocity and acceleration in terms of derivatives (cf. Chapter 4, Section 5), we can say that any equation involving either of these quantities is a differential equation. In this section we shall present some applications of integration which are related to various physical concepts and the differential equations connecting them. We begin by stating Newton's first two laws (axioms) of motion.

First Law. *A body at rest remains at rest and a body in motion moves in a straight line with unchanging velocity, unless some external force acts on it.*

Second Law. *The rate of change of the momentum of a body is proportional to the resultant external force that acts on the body.*

In the discussion of motion in Chapter 4, Section 5, it was always assumed that the object moved in a straight line. We shall now study the behavior of objects which move along curved paths, with the restriction that *the motion lies in a plane.* The velocity will be a vector quantity (as described in Chapter 13, Section 10), which we denote by **v**. If m is the mass, the **momentum vector** is defined as the vector $m\mathbf{v}$. If the mass is constant during the motion—the only type of case we shall consider—Newton's second law becomes

$$m\frac{d\mathbf{v}}{dt} = m\mathbf{a} = k\mathbf{F},$$

where **a** is the acceleration vector, **F** is the force vector, and k is a proportionality constant that depends on the units used. A unit of force, called a **newton,** is the force exerted by gravity on a body with mass $1/g$ kilograms, where the number g is the acceleration due to gravity. The customary value used for g is 9.80 meters/sec^2. For motion near the surface of the earth the constant k in Newton's Second Law is the number g.

If a Cartesian coordinate system (x, y) is introduced, the vector equation $m\mathbf{a} = k\mathbf{F}$ becomes

$$m\frac{d^2x}{dt^2} = kF_X, \qquad m\frac{d^2y}{dt^2} = kF_Y,$$

where d^2x/dt^2, d^2y/dt^2 are the components of $\mathbf{a}$ along the x and y axes, and F_X and F_Y are the components of $\mathbf{F}$ along these axes.

Example 1. A ball is shot vertically upward with a speed of 100 m/sec from a point 40 meters above level ground. Express its height above the ground as a function of time; neglect air resistance.

Solution. According to Newton's law, the motion will take place in a vertical line. Let m be the mass of the ball. The only force acting on it is that of gravity, which acts directly downward, so that

$$F_Y = -m; \qquad F_X = 0.$$

Then we have

$$m\frac{d^2y}{dt^2} = -9.8m,$$

with the auxiliary conditions that

$$y = 40 \text{ when } t = 0; \qquad \frac{dy}{dt} = 100 \text{ when } t = 0.$$

Integrating the differential equation $d^2y/dt^2 = -9.8$, we obtain

$$\frac{dy}{dt} = -9.8t + C.$$

Inserting the auxiliary condition for the initial velocity, we find that

$$100 = -9.8(0) + C \quad \text{or} \quad C = 100.$$

The differential equation $dy/dt = -9.8t + 100$ may be integrated once more to give

$$y = -4.9t^2 + 100t + k.$$

The auxiliary condition for the initial position of the ball is now used to determine that $k = 40$. We finally obtain

$$y = -4.9t^2 + 100t + 40. \qquad \blacktriangleleft$$

Example 2. A projectile is fired from a gun with a velocity of v_0 meters/sec. The barrel of the gun is inclined at an angle α from the horizontal, as in Fig. 15–55. Assuming that there is no air resistance and that the motion of the projectile is in the vertical plane through the barrel of the gun, show that the equations of motion are

$$x = (v_0 \cos \alpha)t, \qquad y = (v_0 \sin \alpha)t - \tfrac{1}{2}gt^2,$$

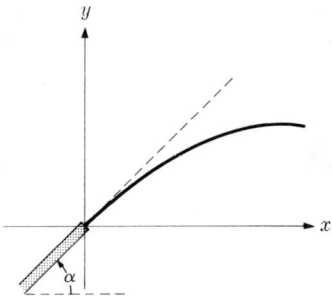

Figure 15-55

referred to Cartesian coordinates with origin at the muzzle of the gun, y axis vertical and x axis horizontal.

Solution. The components of the velocity vector in the x and y directions at time $t = 0$ (**initial velocity vector**) are $v_0 \cos \alpha$ and $v_0 \sin \alpha$, respectively. By hypothesis, the only force acting is that of gravity. Therefore $F_X = 0$, $F_Y = -mg$. The differential equations governing the motion are

$$\frac{d^2x}{dt^2} = 0, \qquad \frac{d^2y}{dt^2} = -g.$$

The auxiliary conditions are

$$x = 0 \quad \text{and} \quad y = 0 \quad \text{when } t = 0;$$

also,

$$\frac{dx}{dt} = v_0 \cos \alpha \quad \text{and} \quad \frac{dy}{dt} = v_0 \sin \alpha \quad \text{when } t = 0.$$

Integrating the differential equations, we find that

$$\frac{dx}{dt} = C_1, \qquad \frac{dy}{dt} = -gt + C_2.$$

The auxiliary conditions yield $C_1 = v_0 \cos \alpha$, $C_2 = v_0 \sin \alpha$. Integrating once again, we obtain

$$x = (v_0 \cos \alpha)t + C_3, \qquad y = -\tfrac{1}{2}gt^2 + (v_0 \sin \alpha)t + C_4.$$

Since $x = y = 0$ when $t = 0$, the constants C_3 and C_4 are both zero. ◀

A large class of problems in physics, chemistry, biology, economics, etc., involves a differential equation of the form

$$\frac{dy}{dt} = ky,$$

in which k is a constant and y is a quantity which is a *positive* function of the time t. The above differential equation expresses the fact that *the rate at which y changes is proportional to y itself.* To solve this equation, we write

$$\frac{dy}{y} = k\,dt$$

and integrate to obtain

$$\ln y = kt + C.$$

From the definition of natural logarithm we get

$$y = e^{kt+C}.$$

From the law of exponents, $e^{kt+C} = e^{kt} \cdot e^C$ and, writing A for the (positive) constant e^C, we obtain the more convenient form

$$y = Ae^{kt}.$$

The law is completely determined once the constants A and k are known. If k is positive, the law is one of **exponential growth.** If k is negative, the law is that of **exponential decay.** Applications of this law are illustrated by the following three examples.

Example 3. In a favorable environment the number of bacteria increases at a rate proportional to the number present. If 1,000,000 bacteria are present at a certain time and 2,000,000 are present an hour later, find the number present four hours later.

Solution. Let $y =$ number of bacteria present at time t. Then the growth law states that

$$\frac{dy}{dt} = ky;$$

integrating this equation, we find that

$$y = Ae^{kt}.$$

Letting $t = 0$ correspond to the time when 1,000,000 bacteria are present, we see that $1{,}000{,}000 = Ae^{k \cdot 0}$, and therefore $A = 1{,}000{,}000$. Further, at $t = 1$, $y = 2{,}000{,}000$, and so

$$2{,}000{,}000 = 1{,}000{,}000 e^{k \cdot 1} \quad \text{or} \quad 2 = e^k.$$

Taking logarithms, we find that $\ln 2 = k$. The equation for y is

$$y = 1{,}000{,}000 e^{t \ln 2}.$$

Since

$$t \ln 2 = \ln(2^t) \quad \text{and} \quad e^{\ln(2^t)} = 2^t,$$

the equation for y may be written

$$y = 1{,}000{,}000 \cdot 2^t.$$

Letting $t = 4$, we find that $y = 16{,}000{,}000$. ◂

Example 4. A radioactive substance decays at a rate proportional to the amount present. If one gram of a radioactive substance reduces to $\frac{1}{4}$ gram in four hours, find how long it will be until $\frac{1}{10}$ gram remains.

Solution. Letting y = amount of substance remaining at time t and using our knowledge of the differential equation $dy/dt = ky$, we write
$$y = Ae^{kt}.$$
The conditions $y = 1$ when $t = 0$ and $y = \frac{1}{4}$ when $t = 4$ yield
$$A = 1 \quad \text{and} \quad \tfrac{1}{4} = e^{4k}.$$
Therefore $k = \frac{1}{4}\ln\frac{1}{4} = -\frac{1}{4}\ln 4$, and we find $y = e^{-(1/4)t\ln 4}$. Since $-\frac{1}{4}t\ln 4 = \ln(4^{-t/4})$ and $e^{\ln 4^{-t/4}} = 4^{-t/4}$, we obtain y in the form
$$y = 4^{-t/4}.$$
We wish to find t when $y = 0.1$. Upon taking logarithms, we find that $\ln(0.1) = -(t\ln 4)/4$, and that
$$t = -\frac{4\ln 0.1}{\ln 4} = \frac{4\ln 10}{\ln 4} = 6.65 \text{ hours, approximately.} \quad \blacktriangleleft$$

Example 5. A tank initially contains 600 liters of brine in which 450 grams of salt are dissolved. Pure water is run into the tank at the rate of 15 liters/min, and the mixture, kept uniform by stirring, is withdrawn at the same rate. How many grams of salt remain after 20 min?

Solution. Let y be the number of grams of salt remaining after t minutes. At that instant there are $y/600$ grams of salt per liter in the mixture. Therefore at that instant y is decreasing at the rate of $15(y/600)$ gm/min. The corresponding differential equation is
$$\frac{dy}{dt} = -\frac{15y}{600} = -\frac{1}{40}y.$$
Integrating this equation, we have
$$y = Ae^{-t/40}.$$
Since $y = 450$ when $t = 0$, the resulting expression for y at time t is $y = 450e^{-t/40}$, and when $t = 20$, $y = 450(e^{-1/2}) = 273$ grams, approximately. $\quad \blacktriangleleft$

PROBLEMS

1. A ball is shot upward with a speed of 150 m/sec from a point 80 meters above level ground. Express its height above the ground as a function of time. What is the highest point it reaches? Neglect air resistance.
2. A ball is dropped from a balloon which is stationary at an altitude of 1000 meters. How long does it take for the ball to reach the ground? Neglect air resistance.

In Problems 3 through 7, assume that a particle moves along the x axis with the given value of its acceleration a_x (which may be positive or negative). Find x and $dx/dt = v$ in terms of t, given the stated auxiliary conditions.

3. $a_x = 2; v = 12$ and $x = 1$ when $t = 0$
4. $a_x = -t; v = 12$ and $x = -2$ when $t = 0$
5. $a_x = 2v; v = 5$ and $x = 1$ when $t = 0$
6. $a_x = -kv; v = 10$ and $x = 0$ when $t = 0$
7. $a_x = -kv^2; v = v_0$ and $x = x_0$ when $t = 0$
8. An automobile is traveling in a straight line at a speed of v_0 m/sec. Suddenly the driver applies the brakes and the car stops in T sec after traveling S meters. Assuming that the brakes produce a constant negative acceleration $-k$, find formulas for S and T in terms of v_0 and k.
9. A projectile is fired at an angle of 60° with the horizontal and with an initial velocity of 1500 m/sec. Determine how far the projectile travels and the length of time it takes to strike the ground. Neglect air resistance.
10. An airplane flying horizontally over level ground at an altitude of 3000 meters drops a projectile. If the plane is traveling 1000 km/hr find the equation of the trajectory in terms of the time t. What is the horizontal distance traveled by the projectile? Neglect air resistance.
11. An airplane is climbing at an angle of 30° with the horizontal at a speed of 750 km/hr. At the instant a projectile is dropped, the plane is 6000 meters above level ground. How long does it take for the projectile to strike the ground? Neglect air resistance.
12. A crystalline chemical present in a solution is such that crystals adhere to it at a rate proportional to the amount present. If there are 2 gm initially and 5 gm one hour later, find the amount of crystalline material present at any time t. What is the approximate amount after three hours?
13. The number of bacteria in a certain culture grows at a rate which is exactly equal to $\frac{1}{2}$ the number present. If there are 10,000 bacteria initially, find the number at any time t.
14. The *half-life* of a radioactive substance (see Example 4) is the time it takes for the original amount of material to reduce to one-half that amount. The half-life of radium is 1690 yrs. Find the approximate time it would take for 1 gm of radium to reduce to 0.1 gm.
15. Rework Problem 14, given that the radioactive substance has a half-life of 3.5 min.
16. If the half-life of a radioactive substance is 1 week, how long does it take to arrive at the point where only 1% of the original amount remains? (See Problem 14.)
17. Rework Example 5, given that the tank initially contains 800 liters of brine in which 400 grams of salt are dissolved, and pure water is run in at the rate of 20 liters/min, the solution being drawn off at the same rate.
18. Rework Example 5, given that the tank initially contains the mixture of the preceding problem, but a brine of 0.2 grams of salt per liter is run in at 20 liters/min, the solution being drawn off at the same rate.
19. Assume that the rate at which a body cools is proportional to the difference between its temperature and that of the surrounding air. A body originally at 40°C cools to 30°C in 10 min in air at 20°C. Find an expression for the temperature of the body at any time t.
20. A body cools at a rate (in minutes) which is exactly equal to $\frac{1}{3}$ the difference between its temperature and the temperature of the surrounding air. If the body is orginally at

50°C and the air is constantly at 20°C, find the temperature of the body after 20 min. How long will it take (approximately) for the body to reach a temperature of 21°C?

21. Carbon 14, a radioactive substance, has a halflife of approximately 5600 years. It is useful in geology since Carbon 14 in a living body begins to decay only after the death of its host. Therefore, in archaeological finds, an examination of the decayed Carbon 14 yields information about when death occurred. Find how long ago an animal died if its present Carbon 14 concentration is one third that of similar animal living now.

22. The growth of tissue in a body is a linear function of the amount of tissue present. That is, the amount y satisfies a differential equation of the form

$$\frac{dy}{dt} = k_1 y + k_2.$$

If t is measured in days, y in grams, then find the amount of tissue after ten days if there are three grams initially and $k_1 = 1.2$, $k_2 = 3$.

23. An example of exponential growth occurs when a bank offers to compound interest "continuously." (An offer to compound it "daily" is practically the same.) Find what interest rate is required if a sum of money compounded continuously is to double in twelve years.

12. APPROXIMATE INTEGRATION

In the preceding sections and in Chapter 7, we saw how many physical and geometrical quantities can be computed by evaluating definite integrals. The familiar antiderivative rule for the evaluation of definite integrals may be used if an antiderivative can be found. Unfortunately, there are still many functions for which an antiderivative cannot be found by any known method. But if a function is continuous on an interval $[a, b]$, we know that its definite integral has a specific value. In Chapter 6, Section 10 and Chapter 7, Section 5, we showed how to compute a definite integral approximately by means of sums of the type used in its definition; the midpoint rule (Chapter 7, Section 5) is often quite accurate. In most numerical work it is convenient to use values of the functions involved only at certain equally spaced points. In this section we shall introduce two new methods which involve the values of the integrand at such points. The first of these is known as **the Trapezoidal Rule**.

Method 1 (The Trapezoidal Rule). *Suppose that $f(x)$ is continuous for $a \leq x \leq b$, and let $\{a = x_0 < x_1 < x_2 < \cdots < x_n = b\}$ be a subdivision of $[a, b]$ into n equal intervals of length $h = (b - a)/n$. Then, approximately,*

$$\int_a^b f(x)\, dx = \frac{h}{2}[f(x_0) + 2f(x_1) + 2f(x_2) + \cdots + 2f(x_{n-1}) + f(x_n)].$$

In case $f(x) \geq 0$ on $[a, b]$, the right-hand side of the equation denotes the sum of the areas of the trapezoids indicated in Fig. 15–56. In general, it represents the integral of the "polygonal function" having its vertices on the locus of $y = f(x)$ "above" or "below" the points x_i.

560 APPLICATIONS OF THE INTEGRAL

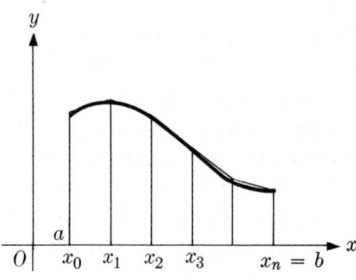

Figure 15–56

Example 1. Compute

$$\int_0^1 \frac{dx}{1+x^2},$$

using the Trapezoidal Rule with $n = 5$. Check the result by integrating and finding the exact value. Keep four decimals in computing each term, and round off the result to three.

Solution

$$\frac{b-a}{n} = \frac{1-0}{5} = 0.2 = h.$$

$$\begin{aligned} f(0) &= 1.0000 \\ 2f(0.2) &= 1.9231 \\ 2f(0.4) &= 1.7241 \\ 2f(0.6) &= 1.4706 \\ 2f(0.8) &= 1.2195 \\ f(1) &= \underline{0.5000} \\ \text{Sum} &= 7.8373 \end{aligned}$$

Therefore

$$\int_0^1 \frac{dx}{1+x^2} = \frac{0.2}{2}(7.8373) = 0.784 \text{ (approximately)}.$$

The exact value is obtained from

$$\int_0^1 \frac{dx}{1+x^2} = \arctan x \Big]_0^1 = \frac{\pi}{4} = 0.7854^-. \quad \blacktriangleleft$$

In order to derive the second method, known as Simpson's Rule, we establish the **Prismoidal Formula.**

Theorem 4 (Prismoidal Formula). *If $f(x)$ is a polynomial of degree three or less, then we have the exact formula*

$$\int_a^{a+2h} f(x)\,dx = \frac{h}{3}[f(a) + 4f(a+h) + f(a+2h)].$$

Proof. Let $f(x)$ be any polynomial of degree 3 or less, say

$$f(x) = a + bx + cx^2 + dx^3.$$

When we make the substitution $x = a + h + u$, we have

$$f(a + h + u) = g(u),$$

where $g(u)$ is still a polynomial of the same degree as f. We write

$$g(u) = A + Bu + Cu^2 + Du^3.$$

Then

$$\int_a^{a+2h} f(x)\,dx = \int_{-h}^{h} g(u)\,du = \left. Au + \frac{Bu^2}{2} + \frac{Cu^3}{3} + \frac{Du^4}{4} \right]_{-h}^{h}$$

$$= 2Ah + \frac{2Ch^3}{3}.$$

From the expression for $g(u)$, we have

$$g(0) = A, \qquad g(h) + g(-h) = 2A + 2Ch^2,$$

and therefore

$$C = \frac{g(h) + g(-h) - 2g(0)}{2h^2}.$$

We conclude that

$$\int_a^{a+2h} f(x)\,dx = 2g(0)h + \frac{h}{3}[g(h) + g(-h) - 2g(0)].$$

We now have $g(0) = f(a + h)$, $g(h) = f(a + 2h)$, $g(-h) = f(a)$ and, substituting these values in the last equation, we obtain the Prismoidal Formula.

Example 2. Evaluate $\int_1^3 x^3\,dx$, using the Prismoidal Formula. Check by integration.

Solution. $a = 1$, $a + 2h = 3$, $h = 1$, $a + h = 2$. Therefore

$$\int_1^3 x^3\,dx = \tfrac{1}{3}(1^3 + 4 \cdot 2^3 + 3^3) = 20.$$

On the other hand,

$$\int_1^3 x^3\,dx = \left[\tfrac{1}{4}x^4\right]_1^3 = \tfrac{81}{4} - \tfrac{1}{4} = 20. \qquad \blacktriangleleft$$

We now establish the second method, known as **Simpson's Rule**.

Method 2 (Simpson's Rule). *Suppose that $f(x)$ is continuous for $a \leq x \leq b$ and $\{a = x_0 < x_1 < x_2 < \cdots < x_{2n-1} < x_{2n} = b\}$ is a subdivision of $[a, b]$ into $2n$ intervals of length $h = (b - a)/2n$. Then, approximately,*

$$\int_a^b f(x)\,dx = \frac{h}{3}[f(x_0) + 4f(x_1) + 2f(x_2) + 4f(x_3) + 2f(x_4) + \cdots$$

$$+ 2f(x_{2n-2}) + 4f(x_{2n-1}) + f(x_{2n})].$$

Applying the Prismoidal Formula to the first two intervals, we obtain

$$\frac{h}{3}[f(x_0) + 4f(x_1) + f(x_2)].$$

Adding to this the result of applying the Prismoidal Formula to the third and fourth intervals, we get

$$\frac{h}{3}[f(x_0) + 4f(x_1) + 2f(x_2) + 4f(x_3) + f(x_4)].$$

Simpson's Rule results from a continuation of the process to include all n pairs of intervals.

Example 3. Compute

$$\int_0^1 \frac{dx}{1+x^2},$$

using Simpson's Rule with $n = 2$. Keep five decimal places in each term and round off the result to four decimals.

Solution. $(b-a)/2n = (1-0)/4 = 0.25 = h$. We have

$$\begin{aligned}
f(0) &= 1 = 1.00000 \\
4f(\tfrac{1}{4}) &= \tfrac{64}{17} = 3.76471 \\
2f(\tfrac{1}{2}) &= \tfrac{8}{5} = 1.60000 \\
4f(\tfrac{3}{4}) &= \tfrac{64}{25} = 2.56000 \\
f(1) &= \tfrac{1}{2} = 0.50000 \\
\text{Sum} &= 9.42471
\end{aligned}$$

Since $\tfrac{1}{12}(9.42471) = 0.78539^+$,

$$\int_0^1 \frac{dx}{1+x^2} = 0.7854^-. \qquad \blacktriangleleft$$

PROBLEMS

In Problems 1 through 6, use the Trapezoidal Rule with the given values of n to compute the approximate values of the given integrals. Keep the number of decimal places indicated in each term and round off to one less. Compute the exact values by integration.

1. $\int_1^4 x^2 \, dx$, $n = 6$, 2 dec
2. $\int_0^3 x\sqrt{16+x^2} \, dx$, $n = 6$, 2 dec
3. $\int_1^2 dx/x$, $n = 5$, 4 dec
4. $\int_0^{1/2} dx/\sqrt{1-x^2}$, $n = 5$, 4 dec
5. $\int_0^1 dx/\sqrt{1+x^2}$, $n = 5$, 4 dec
6. $\int_0^{\pi/2} \sin x \, dx$, $n = 10$, 4 dec

In Problems 7 through 10, compute the exact value of each integral by the Prismoidal Formula.

7. $\int_1^3 x^2 \, dx$
8. $\int_1^3 (x^3 - 2x^2 + 3x - 4) \, dx$
9. $\int_{-2}^0 (x^3 + x^2 - 3x + 4) \, dx$
10. $\int_{-1}^3 (2x^3 + 3x^2 - 4x - 5) \, dx$

15–12 APPROXIMATE INTEGRATION

In Problems 11 through 16, use Simpson's Rule with the given values of $2n$ to compute the approximate values of the given integrals. Keep the number of decimal places indicated in each term and round off to one less. Compute the exact values by integration.

11. $\int_1^2 dx/x$, $2n = 4$, 5 dec

12. $\int_0^{1/2} dx/\sqrt{1-x^2}$, $2n = 4$, 6 dec

13. $\int_0^1 dx/\sqrt{1+x^2}$, $2n = 4$, 5 dec

14. $\int_0^{\pi/2} \sin x \, dx$, $2n = 10$, 5 dec

15. $\int_0^1 dx/(x^2 + x + 1)$, $2n = 4$, 5 dec

16. $\int_0^2 x \, dx/(x^2 + 1)$, $2n = 10$, 5 dec

In Problems 17 through 22, compute approximate values for the integrals, using Simpson's Rule, as in Problems 11 through 16.

17. $\int_0^1 dx/(1 + x^3)$, $2n = 4$, 5 dec

18. $\int_0^1 dx/\sqrt{1 + x^3}$, $2n = 4$, 5 dec

19. $\int_0^1 \sqrt[3]{1 + x^2} \, dx$, $2n = 4$, 5 dec

20. $\int_0^1 e^{-x^2} \, dx$, $2n = 4$, 5 dec

21. $\int_0^1 x \, dx/\sqrt{1 + x^3}$, $2n = 4$, 5 dec

22. $\int_0^{\pi/2} dx/\sqrt{1 - \tfrac{1}{2}\sin^2 x}$, $2n = 4$, 5 dec

APPENDIXES

APPENDIX 1 TRIGONOMETRY REVIEW

APPENDIX 2 CONIC SECTIONS, QUADRIC SURFACES

APPENDIX 3 PROPERTIES OF HYPERBOLIC FUNCTIONS

APPENDIX 4 AXIOMS OF ALGEBRA. NUMBER SYSTEMS

APPENDIX 5 THEOREMS ON DETERMINANTS

APPENDIX 6 PROOFS OF THEOREMS 6, 10, 16, AND 17 OF CHAPTER 13

APPENDIX 7 INTRODUCTION TO A SHORT TABLE OF INTEGRALS

APPENDIX 1
TRIGONOMETRY REVIEW

1. BASIC FORMULAS

We assume that the reader has successfully completed a course in trigonometry. However, no matter how well he has learned the formulas of trigonometry, he can forget them unless he has used them frequently. We shall therefore devote this appendix to a review of the basic notions of trigonometry and a compilation of many of the elementary formulas, for we shall need to use them often. In any case, the formulas serve as handy tools for future use, and the student who has difficulty remembering them should relearn them systematically.

A circle of radius r is drawn and a Cartesian coordinate system constructed with origin at the center of this circle. An angle θ is drawn, starting from the positive direction of the x axis, and is measured counterclockwise. Angles in the first, second, third, and fourth quadrants are shown in Fig. T-1. The basic definitions of the trigonometric functions are

$$\sin \theta = \frac{y}{r}, \qquad \cos \theta = \frac{x}{r},$$

where (x, y) are the coordinates of the point where the terminating side of the angle θ intersects the circle. These definitions depend only on the angle θ and not on the size of the circle since, by similar triangles, the ratios are independent of the size of r. The quantity r is always positive, while x and y have the sign that goes with the quadrant: x and y are positive in the first quadrant; x is negative, y is positive in the second quadrant, and so on.

The remaining trigonometric functions are defined by the relations

$$\tan \theta = \frac{\sin \theta}{\cos \theta}, \qquad \cot \theta = \frac{\cos \theta}{\sin \theta},$$

$$\sec \theta = \frac{1}{\cos \theta}, \qquad \csc \theta = \frac{1}{\sin \theta}.$$

We also can define them by the formulas

$$\tan \theta = \frac{y}{x}, \qquad \cot \theta = \frac{x}{y}, \qquad \sec \theta = \frac{r}{x}, \qquad \csc \theta = \frac{r}{y}.$$

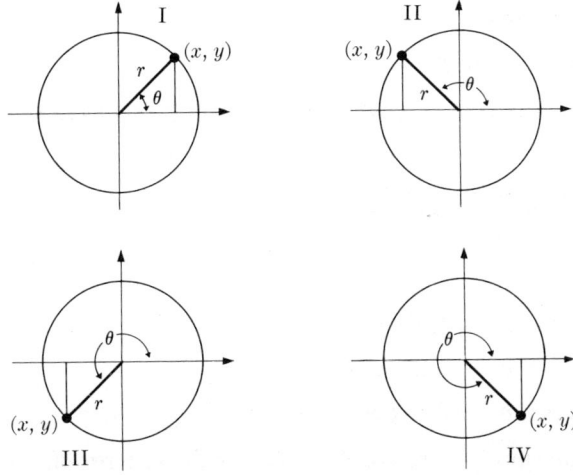

Figure T–1

For angles larger than 360° the trigonometric functions repeat, and for this reason they are called **periodic** functions. This means that for all θ

$$\sin(360° + \theta) = \sin\theta, \quad \cos(360° + \theta) = \cos\theta.$$

Negative angles are measured by starting from the positive x direction and going *clockwise*. Figure T–2 shows a negative angle $-\theta$ and its corresponding

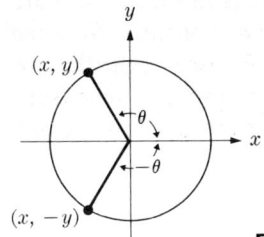

Figure T–2

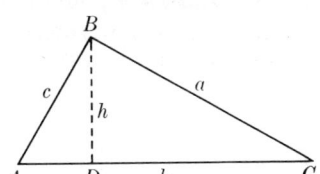

Figure T–3

positive angle θ. By congruent triangles it is easy to see that for *every* θ we have

$$\sin(-\theta) = -\sin\theta \quad \text{and} \quad \cos(-\theta) = \cos\theta,$$

which says that the sine function is an **odd** function of θ, while the cosine function is an **even** function of θ. It is easy to show that

$$\tan(-\theta) = -\tan\theta, \quad \cot(-\theta) = -\cot\theta,$$
$$\sec(-\theta) = \sec\theta, \quad \csc(-\theta) = -\csc\theta.$$

No matter what quadrant the angle θ is in, the Pythagorean theorem tells us that

$$x^2 + y^2 = r^2,$$

or, upon dividing by r^2,

$$\left(\frac{x}{r}\right)^2 + \left(\frac{y}{r}\right)^2 = 1.$$

This formula asserts that for every angle θ,

$$\cos^2 \theta + \sin^2 \theta = 1.$$

Let ABC be any triangle with sides a, b, c, as shown in Fig. T–3. We have the **Law of Sines:**

$$\frac{\sin A}{a} = \frac{\sin B}{b} = \frac{\sin C}{c},$$

and the **Law of Cosines:**

$$a^2 = b^2 + c^2 - 2bc \cos A,$$
$$b^2 = a^2 + c^2 - 2ac \cos B,$$
$$c^2 = a^2 + b^2 - 2ab \cos C.$$

The area of $\triangle ABC$ is: Area $= \frac{1}{2}ab \sin C = \frac{1}{2}bc \sin A = \frac{1}{2}ac \sin B$. The proofs of these statements should be reviewed. An ambitious student may want to rediscover these proofs for himself. As a help in establishing the law of sines, the student should drop the perpendicular from B to side b in Fig. T–3.

The most common unit for measurement of angles is the degree. The circumference of a circle is divided into 360 equal parts, and the angle from the center to two adjacent subdivision points is **one degree.** Each degree is divided into 60 **minutes** and each minute into 60 **seconds.**

A second type of unit for measuring angles is the *radian*. A circle is drawn and an arc is measured which is equal in length to the radius, as in Fig. T–4. Note that the arc $\widehat{AB}$ (*not* the chord) is r units in length. *The angle subtended from the center of the circle is* 1 **radian,** by definition. This angle is the same regardless of the size of the circle and so is a perfectly good unit for measurement. To measure any angle θ in radian units (Fig. T–5), we measure the arc s intercepted by a circle

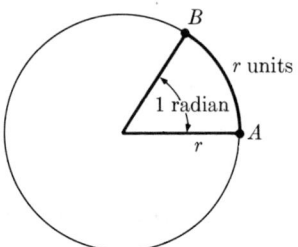
Figure T–4
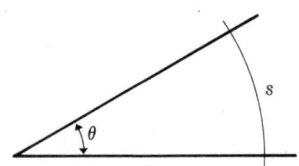
Figure T–5

of radius r. Then division by r gives the radian measure of the angle:

$$\theta = \frac{s}{r} \text{ radians.}$$

We are all familiar with the definition of the number π, which is

$$\pi = \frac{\text{length of circumference of circle}}{\text{length of diameter of circle}}.$$

If C is the length of the circumference and r is the radius, then
$$2\pi = C/r.$$
This is the same as saying that the radian measure of the angle we know as 360° is exactly 2π radians. We have the basic relation
$$\pi \text{ radians} = 180 \text{ degrees}.$$

The fundamental formulas upon which many of the trigonometric identities and relations are based are the **addition formulas for the sine and cosine.** These are
$$\sin(\phi + \theta) = \sin\phi\cos\theta + \cos\phi\sin\theta;$$
$$\sin(\phi - \theta) = \sin\phi\cos\theta - \cos\phi\sin\theta;$$
$$\cos(\phi + \theta) = \cos\phi\cos\theta - \sin\phi\sin\theta;$$
$$\cos(\phi - \theta) = \cos\phi\cos\theta + \sin\phi\sin\theta.$$

Proofs of these formulas should be reviewed. See Problems 4 and 5 at the end of this section. Many of the formulas listed below are special cases of the above formulas. Before listing them, we shall indicate some special values of the three principal trigonometric functions in a way that somewhat reduces the strain on memorization. The tangent function tends to infinity as $\theta \to 90°$ and we sometimes say, speaking loosely, that "tan 90° is infinite."

Function \ Angle	0°	30°	45°	60°	90°	120°	135°	150°	180°
sin	$\frac{\sqrt{0}}{2}$	$\frac{\sqrt{1}}{2}$	$\frac{\sqrt{2}}{2}$	$\frac{\sqrt{3}}{2}$	$\frac{\sqrt{4}}{2}$	$\frac{\sqrt{3}}{2}$	$\frac{\sqrt{2}}{2}$	$\frac{\sqrt{1}}{2}$	$\frac{\sqrt{0}}{2}$
cos	$\frac{\sqrt{4}}{2}$	$\frac{\sqrt{3}}{2}$	$\frac{\sqrt{2}}{2}$	$\frac{\sqrt{1}}{2}$	$\frac{\sqrt{0}}{2}$	$-\frac{\sqrt{1}}{2}$	$-\frac{\sqrt{2}}{2}$	$-\frac{\sqrt{3}}{2}$	$-\frac{\sqrt{4}}{2}$
tan	$\sqrt{\frac{0}{4}}$	$\sqrt{\frac{1}{3}}$	$\sqrt{\frac{2}{2}}$	$\sqrt{\frac{3}{1}}$	*	$-\sqrt{\frac{3}{1}}$	$-\sqrt{\frac{2}{2}}$	$-\sqrt{\frac{1}{3}}$	$-\sqrt{\frac{0}{4}}$

From the relation
$$\sin^2\theta + \cos^2\theta = 1,$$
we obtain, when we divide by $\cos^2\theta$, the relation
$$\tan^2\theta + 1 = \sec^2\theta.$$
If we divide by $\sin^2\theta$, we have the identity
$$1 + \cot^2\theta = \csc^2\theta.$$

The **double-angle formulas** are obtained by letting $\theta = \phi$ in the addition formulas. We find that

$$\sin 2\theta = 2 \sin \theta \cos \theta;$$
$$\cos 2\theta = \cos^2 \theta - \sin^2 \theta = 2 \cos^2 \theta - 1 = 1 - 2 \sin^2 \theta.$$

Since $\tan(\phi + \theta) = \sin(\phi + \theta)/\cos(\phi + \theta)$, we have

$$\tan(\phi + \theta) = \frac{\tan \phi + \tan \theta}{1 - \tan \phi \tan \theta};$$

$$\tan(\phi - \theta) = \frac{\tan \phi - \tan \theta}{1 + \tan \phi \tan \theta}.$$

The double-angle formula becomes

$$\tan 2\theta = \frac{2 \tan \theta}{1 - \tan^2 \theta}.$$

The **half-angle formulas** are derived from the double-angle formulas by writing $\theta/2$ for θ. We obtain

$$\sin^2 \frac{\theta}{2} = \frac{1 - \cos \theta}{2}, \qquad \cos^2 \frac{\theta}{2} = \frac{1 + \cos \theta}{2};$$

$$\tan \frac{\theta}{2} = \frac{1 - \cos \theta}{\sin \theta} = \frac{\sin \theta}{1 + \cos \theta}.$$

The following relations may be verified from the addition formulas:

$$\sin \phi + \sin \theta = 2 \sin\left(\frac{\phi + \theta}{2}\right) \cos\left(\frac{\phi - \theta}{2}\right);$$

$$\sin \phi - \sin \theta = 2 \cos\left(\frac{\phi + \theta}{2}\right) \sin\left(\frac{\phi - \theta}{2}\right);$$

$$\cos \phi + \cos \theta = 2 \cos\left(\frac{\phi + \theta}{2}\right) \cos\left(\frac{\phi - \theta}{2}\right);$$

$$\cos \phi - \cos \theta = -2 \sin\left(\frac{\phi + \theta}{2}\right) \sin\left(\frac{\phi - \theta}{2}\right).$$

The addition formulas, with one of the angles taken as a special value, yield the following useful relations. Starting with the formula

$$\sin\left(\frac{\pi}{2} - \theta\right) = \sin \frac{\pi}{2} \cos \theta - \cos \frac{\pi}{2} \sin \theta, \qquad \left(\frac{\pi}{2} = 90°\right),$$

and noting that $\sin 90° = 1$, $\cos 90° = 0$, we obtain

$$\sin\left(\frac{\pi}{2} - \theta\right) = \cos \theta.$$

In a similar way, we have

$$\cos\left(\frac{\pi}{2} - \theta\right) = \sin\theta;$$

$$\sin\left(\frac{\pi}{2} + \theta\right) = \cos\theta, \qquad \cos\left(\frac{\pi}{2} + \theta\right) = -\sin\theta;$$

$$\cos(\pi - \theta) = -\cos\theta, \qquad \sin(\pi - \theta) = \sin\theta;$$

$$\cos(\pi + \theta) = -\cos\theta, \qquad \sin(\pi + \theta) = -\sin\theta.$$

Problems

1. Prove the Law of Sines.
2. Prove the Law of Cosines.
3. Establish the formula: Area = $\frac{1}{2}ab \sin C$, for the area of a triangle.

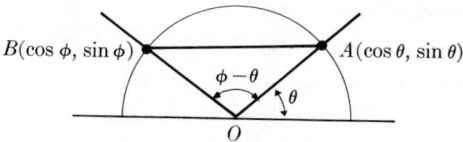

Figure T–6

4. Given the diagram in Fig. T–6 (circle has radius = 1). Apply the Law of Cosines to triangle ABO to obtain a formula for $\cos(\phi - \theta)$. Then use the distance formula for the length $|AB|$ to derive
$$\cos(\phi - \theta) = \cos\phi \cos\theta + \sin\phi \sin\theta.$$

5. Derive formulas for $\sin 3\theta$ and $\cos 3\theta$ in terms of $\sin\theta$ and $\cos\theta$.
6. Derive formulas for $\sin 4\theta$ and $\cos 4\theta$ in terms of $\sin\theta$ and $\cos\theta$.
7. Write out the derivation of the formulas for $\sin\theta/2$ and $\cos\theta/2$ in terms of $\cos\theta$.
8. Derive formulas for $\sin\frac{1}{4}\theta$ and $\cos\frac{1}{4}\theta$ in terms of $\cos\theta$.
9. Without using tables, find the values of the following:

$$\sin 15°, \quad \cos 15°, \quad \sin 22\tfrac{1}{2}°, \quad \cos 22\tfrac{1}{2}°, \quad \tan 75°, \quad \sec 75°.$$

10. Without using tables, find the values of the following:

$$\tan 22\tfrac{1}{2}°, \quad \sin 7\tfrac{1}{2}°, \quad \cos 7\tfrac{1}{2}°, \quad \sin 37\tfrac{1}{2}°, \quad \tan 37\tfrac{1}{2}°.$$

11. Write formulas in terms of $\sin\theta$ and $\cos\theta$ for

$$\sin\left(\frac{3\pi}{2} - \theta\right), \quad \cos\left(\frac{3\pi}{2} - \theta\right), \quad \sin\left(\frac{3\pi}{2} + \theta\right), \quad \cos\left(\frac{3\pi}{2} + \theta\right),$$

$$\sin(2\pi - \theta), \quad \cos(2\pi - \theta).$$

12. Since $\pi = 3.14^+$ approximately, let

$$x = 0, \ \frac{\pi}{6}, \ \frac{\pi}{4}, \ \frac{\pi}{3}, \ \frac{\pi}{2}, \ \frac{2\pi}{3}, \ \frac{3\pi}{4}, \ \frac{5\pi}{6}, \ \pi, \ \frac{7\pi}{6}, \ \text{etc.},$$

and plot the curve $y = \sin x$ on a Cartesian coordinate system. Extend the graph from -2π to $+4\pi$.

13. Work Problem 12 again for $y = \cos x$.
14. Work Problem 12 again for $y = \tan x$.
15. Derive (from the addition formulas) the identity
$$\sin \phi + \sin \theta = 2 \sin\left(\frac{\phi + \theta}{2}\right) \cos\left(\frac{\phi - \theta}{2}\right).$$
16. Work Problem 15 again for
$$\cos \phi + \cos \theta = 2 \cos\left(\frac{\phi + \theta}{2}\right) \cos\left(\frac{\phi - \theta}{2}\right).$$
17. In terms of $\sin \theta$ and $\cos \theta$, find
$$\sin\left(\frac{\pi}{6} - \theta\right), \quad \cos\left(\frac{\pi}{3} + \theta\right), \quad \cos\left(\frac{5\pi}{6} - \theta\right).$$
18. In terms of $\sin \theta$ and $\cos \theta$, find
$$\sin\left(\frac{\pi}{12} - \theta\right), \quad \cos\left(\frac{\pi}{3} + \frac{\pi}{4} - \theta\right), \quad \sin\left(\frac{\pi}{4} - \frac{\pi}{24} - \theta\right).$$
19. Sketch, in Cartesian coordinates, the graph of $y = \cot x$.
20. Sketch, in Cartesian coordinates, the graph of $y = \sec x$.
21. Sketch, in Cartesian coordinates, the graph of $y = \csc x$.

2. GRAPHS OF TRIGONOMETRIC FUNCTIONS

In the study of trigonometry we occasionally have use for a short table of the values of trigonometric functions in terms of radian measure. However, the most common measure used for angles is the degree, and most tables for the sine, cosine, etc., are in degrees, minutes, and perhaps seconds.

In calculus, however, the situation is reversed: the *natural unit* for the study of trigonometric and related functions is the *radian*. Measurement of angles in degrees is awkward and seldom used in calculus, as is apparent in Chapter 9. In this section we use radians to graph trigonometric functions, and in the process we shall become familiar with many of the properties of these functions.

The graph of the function $y = \sin x$ in a Cartesian coordinate system is made by drawing up a table of values and sketching the curve, as shown in Fig. T–7. It is necessary to make a table of values only between 0 and 2π. From the relation $\sin(2\pi + x) = \sin x$, we know that the curve then repeats indefinitely to the left and right.

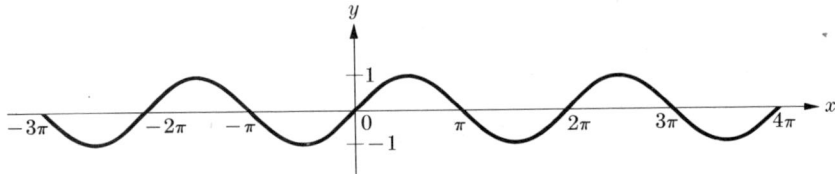

$y = \sin x$

Figure T–7

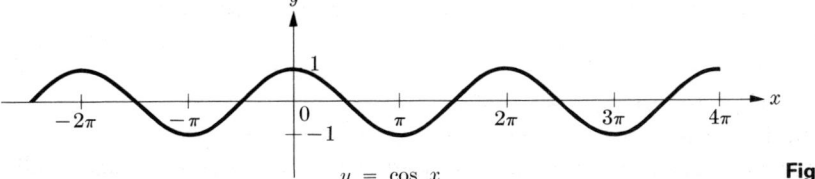

$y = \cos x$

Figure T–8

The graph of $y = \cos x$ is found in the same way. A table of values between 0 and 2π and a knowledge of the periodic nature of the function yield the graph shown in Fig. T–8. Note that the curves are identical except that the cosine function is shifted to the left by an amount $\pi/2$. This fact is not surprising if we recall the relation $\sin[(\pi/2) + x] = \cos x$.

x	0	$\pi/6$	$\pi/4$	$\pi/3$	$\pi/2$	$2\pi/3$	$3\pi/4$	$5\pi/6$	π
y	0	1/2	$\sqrt{2}/2$	$\sqrt{3}/2$	1	$\sqrt{3}/2$	$\sqrt{2}/2$	1/2	0

x	$7\pi/6$	$5\pi/4$	$4\pi/3$	$3\pi/2$	$5\pi/3$	$7\pi/4$	$11\pi/6$	2π
y	$-1/2$	$-\sqrt{2}/2$	$-\sqrt{3}/2$	-1	$-\sqrt{3}/2$	$-\sqrt{2}/2$	$-1/2$	0

A function f is called **periodic** if there is a number l such that

$$f(x + l) = f(x)$$

for all x (in the domain of f), in which case the number l is called a **period** of the function f. For the sine function, 2π is a period, but so are 4π, 6π, 8π, etc., since $\sin(x + 4\pi) = \sin x$, $\sin(x + 6\pi) = \sin x$, and so on. For periodic functions there is always a smallest number l for which $f(x + l) = f(x)$. This number is called the **least period** of f. The least period of $\sin x$ is 2π; the same is true for $\cos x$. The function $\tan x$ has period 2π, but this is not the least period. A graph of $y = \tan x$ is given in Fig. T–9, which shows clearly that π is a period of $\tan x$. This is the least period for the tangent function.

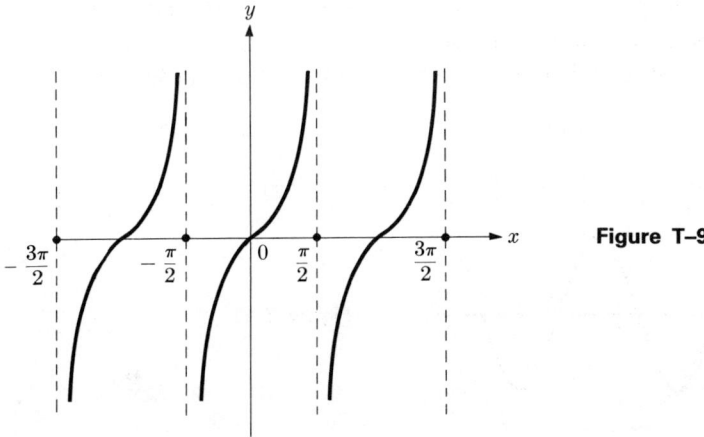

Figure T–9

The function $y = \sin kx$ repeats when kx changes by an amount 2π. If k is a constant, then $\sin kx$ repeats when x changes by an amount $2\pi/k$. The least period of the function $\sin kx$ is $2\pi/k$. The same is true for $y = \cos kx$. For example, $y = \cos 5x$ has as its least period $2\pi/5$.

A periodic function f, which never becomes larger than some number b in value and never becomes smaller than some number c, is said to have **amplitude** $\frac{1}{2}(b - c)$, provided that there are points at which f actually takes on the values b and c. For example, the function $y = 3 \sin 2x$ is never larger than 3 or smaller than -3. The amplitude is 3. Note that $y = 3$ when $x = \pi/4$. The tangent function has no amplitude.

Example 1. Find the amplitude and period of $y = 4 \cos(3x/2)$. Sketch the graph.

Solution. The amplitude is 4; the period is $2\pi/(\tfrac{3}{2}) = 4\pi/3$. The graph is sketched in Fig. T–10. ◀

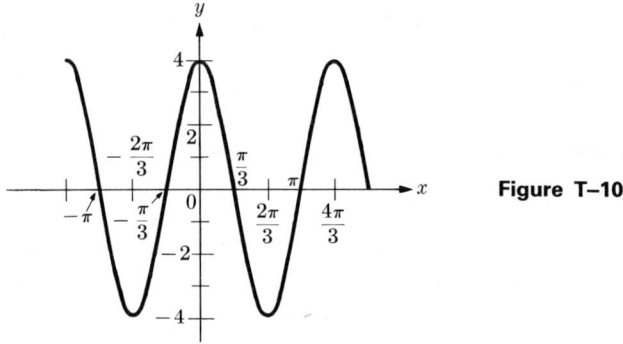

Figure T–10

Example 2. Find the amplitude and period of $y = 2.5 \sin[2x + (\pi/3)]$ and sketch its graph.

Solution. This function is similar to $2.5 \sin 2x$, except that it is shifted horizontally. We write $y = 2.5 \sin 2[x + (\pi/6)]$. When $x = -\pi/6$, $y = 0$, and we see that the "zero point" is moved to the left an amount $\pi/6$. The period is $2\pi/2 = \pi$, and the amplitude is 2.5. With this information it is now easy to sketch the curve, as shown in Fig. T–11. ◀

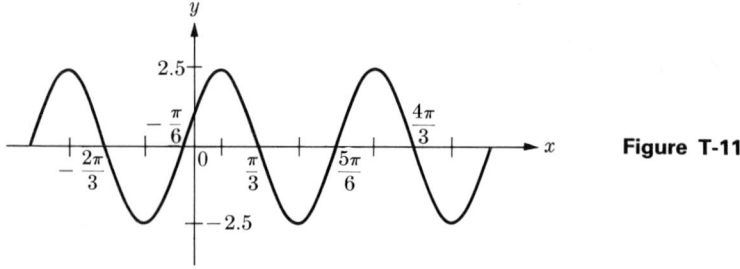

Figure T–11

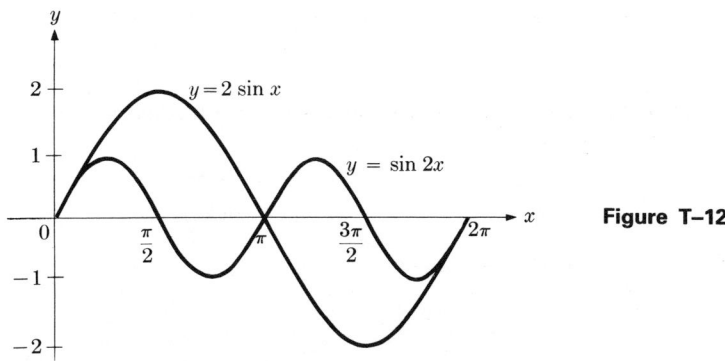

Figure T–12

Combinations of trigonometric functions may be sketched quickly by a method known as **addition of ordinates.** If the sum of two functions is to be sketched, we start by sketching each function separately. Then, at a particular value of x, the y values of each of the functions are added to give the result. Incidentally, this method is quite general and may be used for the sum or difference of any functions. An example illustrates the technique.

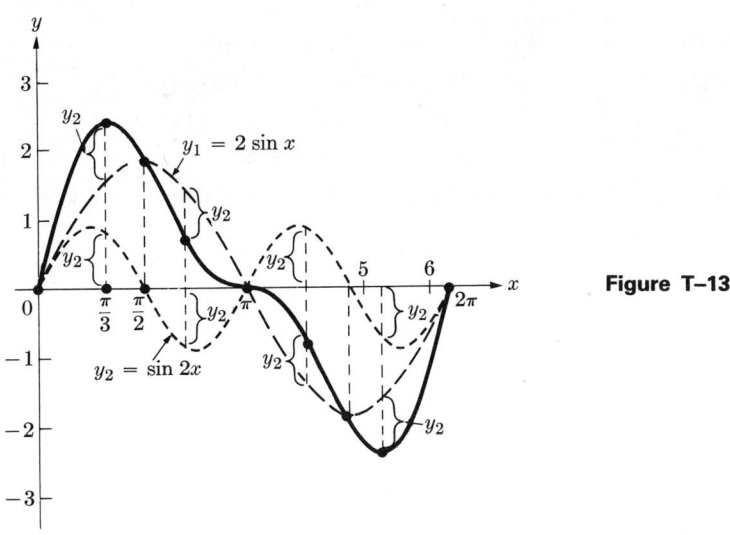

Figure T–13

Example 3. Sketch the graph of

$$y = 2 \sin x + \sin 2x.$$

Solution. The function $2 \sin x$ has period 2π and amplitude 2, while $\sin 2x$ has period π and amplitude 1. We sketch the graph of each of these functions on the same coordinate system from 0 to 2π, as shown in Fig. T–12. At convenient values of x, the y values of the two graphs are added and the results plotted either on the same graph or on a similar one, as shown in Fig. T–13. ◀

PROBLEMS

In Problems 1 through 16, find the amplitudes and the periods of the functions defined by the given expressions and sketch their graphs.

1. $2 \sin \frac{1}{2} x$
2. $3 \cos \frac{3}{2} x$
3. $4 \cos \frac{1}{2} \pi x$
4. $2.5 \sin \frac{1}{3} \pi x$
5. $3 \sin \frac{2}{3} x$
6. $5 \sin \frac{1}{2} \pi x$
7. $2 \cos \left(2x + \dfrac{\pi}{3}\right)$
8. $3 \sin \left(\frac{1}{2} \pi x + \frac{1}{4} \pi\right)$
9. $-2 \sin 2x$
10. $-3 \cos \frac{2}{3} \pi x$
11. $4 \sin \left(2x - \dfrac{\pi}{3}\right)$
12. $3 \cos \left(2x - \dfrac{\pi}{3}\right)$
13. $-3 \sin \left(\frac{1}{2} x + 7\right)$
14. $\frac{1}{3} \sin \left(2x + \frac{1}{2} \pi\right)$
15. $1 + \sin 2x$
16. $-3 + 2 \cos (x - 1)$

In Problems 17 through 24, find the periods and amplitudes (if any), and sketch the graphs.

17. $\tan 2x$
18. $\cot 3x$
19. $\sin x + \cos x$
20. $\sec 2x$
21. $\csc \frac{1}{2} x$
22. $2 \tan (3x - 1)$
23. $2 \cos 2x - 2\sqrt{3} \sin 2x$
24. $-4 \cot (2x - \pi)$

In Problems 25 through 32, sketch the graphs of the curves, using the method of addition of ordinates.

25. $y = 2 \sin x + \cos 2x$
26. $y = 3 \cos x - 2 \sin 2x$
27. $y = 3 \sin \frac{1}{2} x - \frac{3}{2} \sin x$
28. $y = 2 \sin \frac{1}{2} \pi x - 3 \cos \pi x$
29. $y = 3 \sin 2x - 2 \cos 3x$
30. $y = 2 \cos 3x - 3 \sin 4x$
31. $y = x + \sin x$
32. $y = 1 - x + \cos x$

33. Sketch the graph of $y = 2 \sin^2 x$. Show that it has period π. How does it compare with the graph of $\cos 2x$?

APPENDIX 2

CONIC SECTIONS, QUADRIC SURFACES

1. THE CIRCLE (continued)

We consider the general equation
$$x^2 + y^2 + Dx + Ey + F = 0$$
where D, E, F are any numbers. We perform the process of completing the square and obtain
$$x^2 + Dx + \frac{D^2}{4} + y^2 + Ey + \frac{E^2}{4} = -F + \frac{D^2}{4} + \frac{E^2}{4},$$
or
$$\left(x + \frac{D}{2}\right)^2 + \left(y + \frac{E}{2}\right)^2 = \tfrac{1}{4}(D^2 + E^2 - 4F).$$

We conclude that such an equation represents a circle with center at $(-D/2, -E/2)$ and radius $\tfrac{1}{2}\sqrt{D^2 + E^2 - 4F}$. But there is a problem: If the quantity under the radical is negative, there is no circle and consequently no graph. If it is zero, the circle reduces to a point. *The equation does represent a circle of radius $\tfrac{1}{2}\sqrt{D^2 + E^2 - 4F}$ if $D^2 + E^2 > 4F$, and only then.*

A circle is determined if the numbers D, E, and F are given. This means that in general three conditions determine a circle. We all know that there is exactly one circle through three points (not all on a straight line). The following example shows how the equation is found when three points are prescribed.

Example 1. Find the equation of the circle passing through the points $(-3, 4)$, $(4, 5)$, and $(1, -4)$. Locate the center and determine the radius.

Solution. If the circle has the equation
$$x^2 + y^2 + Dx + Ey + F = 0,$$
then each point on the circle must satisfy this equation. By substituting, we get

$$(-3, 4): \quad -3D + 4E + F = -25;$$
$$(4, 5): \quad 4D + 5E + F = -41;$$
$$(1, -4): \quad D - 4E + F = -17.$$

We subtract the second equation from the first and the third from the second and eliminate F, to obtain

$$7D + E = -16, \qquad 3D + 9E = -24.$$

We solve these simultaneously and get $D = -2$, $E = -2$. Substituting back into any one of the original three equations gives $F = -23$. The equation of the circle is

$$x^2 + y^2 - 2x - 2y - 23 = 0.$$

The center and radius are obtained by completing the square. We write

$$(x^2 - 2x \quad) + (y^2 - 2y \quad) = 23 \quad \text{and} \quad (x - 1)^2 + (y - 1)^2 = 25.$$

The center is at $(1, 1)$; the radius is 5. ◀

The problem of Example 1 could be solved in another way. (1) Find the perpendicular bisector of the line segment joining the first two points. (2) Do the same with the first and third (or second and third) points. (3) The point of intersection of the two perpendicular bisectors is the center of the circle. (4) The formula for the distance from the center of the circle to any of the original three points gives the radius. (5) Knowing the center and radius, we write the equation of the circle. (See Fig. A–1.) ◀

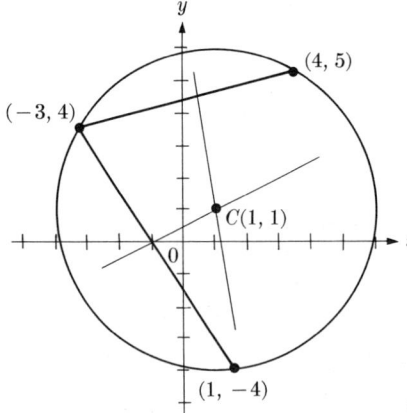

Figure A–1

The equation

$$x^2 + (y - k)^2 = 25$$

represents a **family of circles.** All circles of this family have radius 5 and, furthermore, the centers are always on the y axis. Since a line is determined by two conditions while a circle is fixed by three conditions, families of circles are more extensive than families of lines, from some points of view. For example, all circles of radius 3 comprise a "larger" class than all lines of slope 7. This has meaning in the following sense: consider any point in the plane; only one member of the above family of lines passes through this point; however, an infinite number of circles of radius 3 passes through this same point. The family of lines of slope 7 is called a *one-parameter family*. The equation $y = 7x + k$ has *one* constant k

which can take on any real value. The equation
$$(x - h)^2 + (y - k)^2 = 9$$
is a *two-parameter family* of circles. The numbers h and k may independently take on any numerical value.

Example 2. Find the equation of the family of circles with centers on the line $y = x$ and radius 7.

Solution. The center $C(h, k)$ must have $h = k$ if the center is on the line $y = x$. The required equation is
$$(x - h)^2 + (y - h)^2 = 49.$$
This is a one-parameter family of circles. ◂

Example 3. Find the equation of the family of circles with center at $(3, -2)$.

Solution. The equation is
$$(x - 3)^2 + (y + 2)^2 = r^2$$
This is a one-parameter family of concentric circles. ◂

The two circles
$$x^2 + y^2 + Dx + Ey + F = 0,$$
$$x^2 + y^2 + D'x + E'y + F' = 0,$$
may or may not intersect. If we subtract one of these equations from the other, we obtain
$$(D - D')x + (E - E')y + (F - F') = 0,$$
which is the equation of a straight line unless $D' = D$ and $E' = E$. This straight line has slope $-(D - D')/(E - E')$ (if $E \neq E'$). The circles have centers at $(-D/2, -E/2)$ and $(-D'/2, -E'/2)$, respectively. The line joining the centers has slope
$$\frac{(-E'/2) + (E/2)}{(-D'/2) + (D/2)} = \frac{E - E'}{D - D'},$$
which is the negative reciprocal of $-(D - D')/(E - E')$. *The equation of the line obtained by subtraction of the equations of two circles* is called the **radical axis,** and *it is perpendicular to the line joining the centers of the circles.* Furthermore, if the circles intersect, the points of intersection satisfy *both* equations of circles. These points must therefore be on the radical axis.

Example 4. Find the points of intersection of the circles:
$$x^2 + y^2 - 2x - 4y - 4 = 0, \qquad x^2 + y^2 - 6x - 2y - 8 = 0.$$

Solution. By subtraction we get
$$4x - 2y + 4 = 0,$$

and we solve this equation simultaneously with the equation of either circle. The substitution gives

$$5x^2 - 2x - 8 = 0 \quad \text{or} \quad x = \tfrac{1}{5} \pm \tfrac{1}{5}\sqrt{41}.$$

The points of intersection are

$$(\tfrac{1}{5} + \tfrac{1}{5}\sqrt{41}, \tfrac{12}{5} + \tfrac{2}{5}\sqrt{41}) \quad \text{and} \quad (\tfrac{1}{5} - \tfrac{1}{5}\sqrt{41}, \tfrac{12}{5} - \tfrac{2}{5}\sqrt{41}). \qquad \blacktriangleleft$$

PROBLEMS

1. Show that when $D = E$ the graph of $x^2 + y^2 + Dx + Ey + F = 0$ is always a circle if $F < D^2/2$. Note that if $F < 0$ the graph is always a circle, regardless of the sizes of D and E.

In Problems 2 through 5, in each case find the equation of the circle through the given points by the method used in Example 1. Find the center and radius.

2. $(-5, 2), (-3, 4), (1, 2)$
3. $(-3, 1), (-2, 2), (6, -2)$
4. $(7, -2), (3, -4), (0, -3)$
5. $(1, -1), (0, 1), (-3, -3)$

In Problems 6 and 7, find the equations of the circles by the method described after Example 1.

6. $(-2, -4), (2, -6), (4, 2)$
7. $(-2, -1), (0, 2), (4, 1)$

8. Determine whether or not the points $(2, 0), (0, 4), (2, 2)$, and $(1, 1)$ lie on a circle.
9. Find the equation of the circle or circles with center on the y axis and passing through $(-2, 3)$ and $(4, 1)$.
10. Find the equation of the circle with center on the line $x + y - 1 = 0$ and passing through the points $(0, 5)$ and $(2, 1)$.

In Problems 11 through 16, in each case write the equation of the family of circles with the given properties.

11. center on the y axis, radius 4
12. center on the line $x + y - 6 = 0$, radius 1
13. center on the x axis, tangent to the y axis
14. center on the line $y = x$, tangent to both coordinate axes
15. center on the line $y = 3x$, tangent to the line $x + 3y + 4 = 0$
16. center on the line $x + 2y - 7 = 0$ and tangent to the line $2x + 3y - 1 = 0$
17. Give a reasonable definition for the statement that two circles intersect at right angles. Do the circles $x^2 + y^2 - 6y = 0$ and $x^2 + y^2 - 6x = 0$ satisfy your definition at their intersection points?

In Problems 18 through 21, find the points of intersection of the two circles, if they intersect. If they do not intersect, draw a graph showing the location of the radical axis.

18. $x^2 + y^2 + 6x - 9y = 0,\ x^2 + y^2 - 2x + 3y - 7 = 0$
19. $x^2 + y^2 + x - 3y + 1 = 0,\ x^2 + y^2 - 4x + 2y - 3 = 0$
20. $x^2 + y^2 + 2x + 3y + 1 = 0,\ x^2 + y^2 - 8y - 7 = 0$
21. $2x^2 + 2y^2 + 4x - 2y - 1 = 0,\ 3x^2 + 3y^2 + 6x - 9y + 2 = 0$

22. Given the circles $x^2 + y^2 + 2x - 4 = 0$ and $x^2 + y^2 - x - y - 2 = 0$. What is the geometric property of the family of circles

$$(x^2 + y^2 - x - y - 2) + k(x^2 + y^2 + 2x - 4) = 0?$$

Sketch a few members of this family.

23. Find the graph of all points $P(x, y)$ such that the distance from $(3, 2)$ is always twice the distance from $(4, 1)$. Describe the graph.

24. Find the graph of all points $P(x, y)$ such that the distance from $(1, -4)$ is always $\frac{2}{3}$ the distance from $(2, 1)$. Describe the graph.

25. Given the three circles

$$x^2 + y^2 - x - y = 0,$$
$$x^2 + y^2 + 2x - y - 3 = 0,$$
$$x^2 + y^2 + x - 2y - 4 = 0,$$

show that the three radical axes obtained from each pair of circles are concurrent (i.e., the three lines meet in one point).

26. Work Problem 25 again for the circles

$$x^2 + y^2 + Dx + Ey + F = 0,$$
$$x^2 + y^2 + D'x + E'y + F' = 0,$$
$$x^2 + y^2 + D''x + E''y + F'' = 0.$$

What conditions are necessary for the result? [*Hint:* Can you think of a pair of circles which do not have a radical axis?]

---

2. THE PARABOLA (continued)

We consider a parabola and a point (x_1, y_1) on it. We show how to derive a formula which gives the equation of the line tangent to the parabola at (x_1, y_1). If, for example, the parabola has the equation

$$y^2 = 2px,$$

then the slope at any point is given (by implicit differentiation) as

$$2y \frac{dy}{dx} = 2p \quad \text{and} \quad \frac{dy}{dx} = \frac{p}{y}.$$

The equation of any line through (x_1, y_1) is

$$y - y_1 = m(x - x_1).$$

If the slope is p/y_1 the equation becomes

$$y - y_1 = \frac{p}{y_1}(x - x_1).$$

This formula can be simplified. Since (x_1, y_1) is on the parabola, we know that $y_1^2 = 2px_1$. Therefore the above equation may be written

$$y_1 y - y_1^2 = px - px_1$$

or
$$y_1 y - 2px_1 = px - px_1 \Leftrightarrow y_1 y = p(x + x_1).$$

Example 1. Find the equation of the line tangent to $y^2 = 20x$ at the point $(3, 2\sqrt{15})$.

Solution. In this case $x_1 = 3$, $y_1 = 2\sqrt{15}$, $p = 10$, and the equation is
$$2\sqrt{15}\, y = 10(x + 3). \blacktriangleleft$$

A somewhat more difficult problem is that of finding the line tangent to a parabola when the given point (x_1, y_1) is *not* on the parabola. There will be either two solutions or no solution to this problem, as Fig. A–2 shows. Let the parabola have equation $x^2 = 2py$, and suppose that $y - y_1 = m(x - x_1)$ is the equation of the desired line. The slope of the parabola at any point is
$$\frac{dy}{dx} = \frac{x}{p}.$$

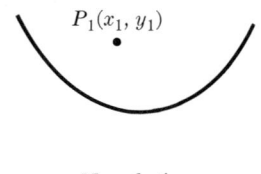

Two solutions
(a)

No solution
(b)

Figure A–2

Let (x_0, y_0) be a point of tangency. Since this point is on both the parabola and the line, we have $x_0^2 = 2py_0$ and $y_0 - y_1 = (x_0/p)(x_0 - x_1)$. We have two equations in the unknowns x_0 and y_0. Once these are solved simultaneously the problem is done, since the value x_0/p may be substituted in the equation of the line. The following example illustrates the method in a particular case.

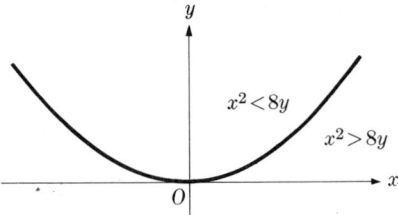

Figure A–3

Example 2. Find the equations of the lines which are tangent to the parabola $y^2 - y - 2x + 4 = 0$ and which pass through the point $(-5, 2)$.

Solution. Any line through $(-5, 2)$ has equation
$$y - 2 = m(x + 5)$$

and the slope of the curve at any point on the parabola is given by

$$2y\frac{dy}{dx} - \frac{dy}{dx} - 2 = 0 \quad \text{or} \quad \frac{dy}{dx} = \frac{2}{2y-1}.$$

If (x_0, y_0) is the point of tangency, then $m = 2/(2y_0 - 1)$. We have the equations

$$y_0^2 - y_0 - 2x_0 + 4 = 0 \quad \text{and} \quad y_0 - 2 = \frac{2}{2y_0 - 1}(x_0 + 5).$$

These are solved simultaneously by eliminating x_0, and the two solutions are $x_0 = 17$, $y_0 = 6$ and $x_0 = 5$, $y_0 = -2$; corresponding slopes are $m_0 = \frac{2}{11}, -\frac{2}{5}$. The desired lines have equations

$$y - 2 = \tfrac{2}{11}(x + 5) \quad \text{and} \quad y - 2 = -\tfrac{2}{5}(x + 5). \quad \blacktriangleleft$$

If there were no solution, the attempt to solve the quadratic equation for y_0 would give complex numbers for the roots. That there is no solution may also be determined in another way. As we know, a parabola such as $x^2 = 8y$ divides the plane into two regions; in one of them $x^2 > 8y$ and in the other $x^2 < 8y$. Any point (x_1, y_1) such that $x_1^2 > 8y_1$ is a point such that two tangents to the parabola may be constructed (Fig. A–3). If $x_1^2 < 8y_1$, no tangents can be drawn. More generally, if the equation of the parabola is

$$(x - h)^2 = 2p(y - k),$$

then two tangents may be drawn if $(x_1 - h)^2 > 2p(y_1 - k)$ and none can be drawn if $(x_1 - h)^2 < 2p(y - k)$.

Example 3. Given the parabola $-2x^2 + 3y - 2x + 1 = 0$ and the points $P(1, -2)$, $Q(2, 9)$, decide whether or not tangents to the parabola may be drawn from P and Q.

Solution. First complete the square, getting

$$(x + \tfrac{1}{2})^2 = \tfrac{3}{2}(y + \tfrac{1}{2}).$$

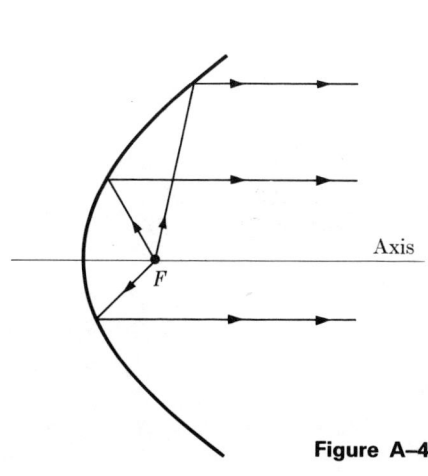

Figure A–4

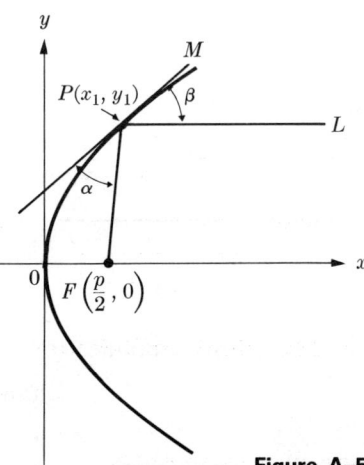

Figure A–5

Now substitute the coordinates of $P(1, -2)$ in the equation. This yields

$$\left(\tfrac{3}{2}\right)^2 > -\tfrac{9}{4},$$

so that two tangents may be drawn. For the point $Q(2, 9)$ the result is

$$\tfrac{25}{4} < \tfrac{57}{4},$$

so that no tangents can be constructed. ◂

The parabola has many interesting geometrical properties that make it suitable for practical applications. Probably the most familar application is the use of parabolic shapes in headlights of cars, flashlights, and so forth. If a parabola is revolved about its axis, the surface generated is called a **paraboloid.** Such surfaces are used in headlights, optical and radio telescopes, radar, etc., because of the following geometric property. If a source of light (or sound or other type of wave) is placed at the focus of a parabola, and if the parabola is a reflecting surface, then the wave will bounce back in a line parallel to the axis of the parabola (Fig. A–4). This creates a parallel beam without dispersion. (In actual practice there will be some dispersion, since the source of light must occupy more than one point.) Clearly the converse is also true. If a series of incoming waves is parallel to the axis of the reflecting paraboloid, the resulting signal will be concentrated at the focus (where the receiving equipment is therefore located).

We shall establish the reflecting property for the parabola $y^2 = 2px$. Let $P(x_1, y_1)$ be a point (not the origin) on the parabola, and draw the line segment from P to the focus (Fig. A–5). Now construct a line L through P parallel to the x axis and the line M tangent to the parabola at P. The law for the reflecting property of the parabola states that the angle formed by M and PF (α in the figure) must be equal to the angle formed by L and M (β in the figure). The slope of the line M is $\tan \beta$, which is also the slope of the parabola at (x_1, y_1). The result is

$$2y\frac{dy}{dx} = 2p \quad \text{and} \quad \tan \beta = \frac{p}{y_1}.$$

We now find $\tan \alpha$ from the formula for the angle between two lines. The slope of PF is

$$\frac{y_1 - 0}{x_1 - \frac{p}{2}}.$$

Therefore

$$\tan \alpha = \frac{\dfrac{y_1}{x_1 - \frac{p}{2}} - \dfrac{p}{y_1}}{1 + \dfrac{y_1}{x_1 - \frac{p}{2}} \cdot \dfrac{p}{y_1}} \quad \Leftrightarrow \quad \tan \alpha = \frac{y_1^2 - px_1 + \frac{1}{2}p^2}{x_1 y_1 - \frac{1}{2}py_1 + py_1}.$$

Since $y_1^2 = 2px_1$, this simplifies to

$$\tan \alpha = \frac{p}{y_1},$$

and the result is established.

The main cables of suspension bridges are parabolic in shape, since it can be shown that if the total weight of a bridge is distributed uniformly along its length, a cable in the shape of a parabola bears the load evenly. It is a remarkable fact that no other shape will perform this task.

PROBLEMS

In Problems 1 through 3, find the equation of the line tangent to the parabola at the given point P on the parabola.

1. $y^2 = -3x$, $P(-3, -3)$
2. $x^2 - 2y + 3x - 4 = 0$, $P(-2, -3)$
3. $-x^2 - 2y + 2x + 4 = 0$, $P(4, -2)$
4. Find the equation of the line tangent to the parabola $y^2 - 3y + 4x - 8 = 0$ at the point on the parabola at which $x = 1$.

In Problems 5 through 14, first decide whether lines tangent to the given parabola which pass through the given point P can be constructed. When they can, determine the equations of the lines.

5. $y^2 = 8x$, $P(-3, 1)$
6. $y^2 = 8x$, $P(3, 1)$
7. $y^2 = -4x$, $P(1, -5)$
8. $x^2 = -4y$, $P(4, 1)$
9. $x^2 = 12y$, $P(-1, -3)$
10. $x^2 - 4x + 2y - 1 = 0$, $P(-1, 3)$
11. $2y^2 - 3y + x - 4 = 0$, $P(1, -1)$
12. $x^2 - 4x + 2y - 1 = 0$, $P(1, -8)$
13. $x + 2y - 2x^2 + 1 = 0$, $P(1, 3)$
14. $x + 2y - 2x^2 + 1 = 0$, $P(1, -3)$

15. For the parabola $x^2 = 2py$, establish the "optical property." That is, if (x_1, y_1) is a point on the parabola and M is the line tangent at (x_1, y_1), show that the angle formed between M and a line from (x_1, y_1) through the focus is equal to the angle M forms with a line through (x_1, y_1) parallel to the axis.

16. A line through the focus of the parabola $x^2 = -12y$ intersects the parabola at the point $(5, -\frac{25}{12})$. Find the other point of intersection of this line with the parabola.

17. An arch has the shape of a parabola with vertex at the top and axis vertical. If it is 45 meters wide at the base and 18 meters high in the center, how wide is it 10 meters above the base?

18. The line through the focus of a parabola and parallel to the directrix is called the *line of the latus rectum*. The portion of this line cut off by the parabola is called the **latus rectum**. Show that for the parabola $y^2 = 2px$ the length of the latus rectum is $2p$. Is this true for any parabola where the distance from the focus to the directrix is p units?

19. Find the graph of all points whose distances from $(4, 0)$ are 1 unit less than their distances from the line $x = -6$. Describe the curve.

3. THE ELLIPSE (continued)

A mechanical construction of an ellipse can be made directly from the facts given in the definition, as shown in Fig. A–6. Select a piece of string of length $2a$. Fasten its ends with thumbtacks at the foci, insert a pencil so as to draw the string taut, and trace out an arc. The curve will be an ellipse.

The ellipse has many interesting geometrical properties. The paths of the planets about the sun and those of the man-made satellites about the earth are approximately elliptical; whispering galleries are ellipsoidal (an ellipse revolved about the major axis) in shape. If a signal (in the form of a light, sound, or other type of wave) has its source at one focus of an ellipse, *all* reflected waves will pass through the other focus.

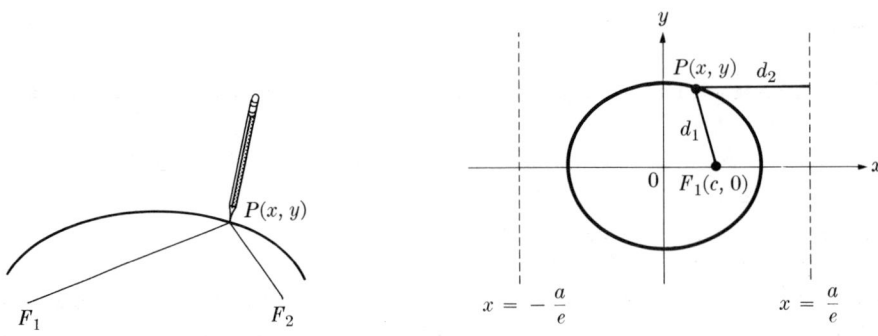

Figure A–6   **Figure A–7**

We have seen that a parabola has a focus and a directrix. The ellipse has two foci and also two directrices. If the ellipse has foci at $(\pm c, 0)$ and vertices at $(\pm a, 0)$, *the* **directrices** *are the lines*

$$x = \frac{a^2}{c} = \frac{a}{e} \quad \text{and} \quad x = -\frac{a^2}{c} = -\frac{a}{e}.$$

To draw an analogy with the parabola, we can show that such an ellipse is the graph of all points whose distances from a fixed point (F_1) are equal to e ($0 < e < 1$) times their distances from a fixed line ($x = a/e$). As Fig. A–7 shows, the conditions of the graph are

$$d_1 = ed_2 \quad \Leftrightarrow \quad \sqrt{(x-c)^2 + y^2} = e\left|x - \frac{a}{e}\right|.$$

If we square both sides and simplify, we obtain $(x^2/a^2) + (y^2/b^2) = 1$, which we know to be the equation of an ellipse. The same situation holds true if the focus $F_2(-c, 0)$ and the directrix $x = -a/e$ are used.

Example 1. Find the equation of the ellipse with foci at $(\pm 5, 0)$ and with the line $x = 36/5$ as one directrix.

Solution. We have $c = 5$, $a/e = 36/5$. Since $e = c/a$, we obtain $a^2/c = 36/5$, or $a = 6$. Then $b = \sqrt{36 - 25} = \sqrt{11}$, and the equation is

$$\frac{x^2}{36} + \frac{y^2}{11} = 1. \qquad \blacktriangleleft$$

The above example shows that if the foci and one directrix are specified, then the equation of the ellipse is determined. The next example shows that if the axes of the ellipse are specified, then two additional points on the ellipse may be used to find its equation.

Example 2. Find the equation of the ellipse with axes along the x axis and y axis which passes through the points $P(6, 2)$ and $Q(-4, 3)$.

Solution. Let the equation be
$$\frac{x^2}{A^2} + \frac{y^2}{B^2} = 1;$$
we don't know whether A is larger than B or B is larger than A. Since the points P and Q are on the ellipse, we have
$$36\left(\frac{1}{A^2}\right) + 4\left(\frac{1}{B^2}\right) = 1, \quad 16\left(\frac{1}{A^2}\right) + 9\left(\frac{1}{B^2}\right) = 1,$$
which gives us two equations in the unknowns $1/A^2$, $1/B^2$. Solving simultaneously, we obtain
$$\frac{1}{A^2} = \frac{1}{52}, \quad \frac{1}{B^2} = \frac{1}{13},$$
and the resulting ellipse is
$$\frac{x^2}{52} + \frac{y^2}{13} = 1.$$
The major axis is along the x axis. ◀

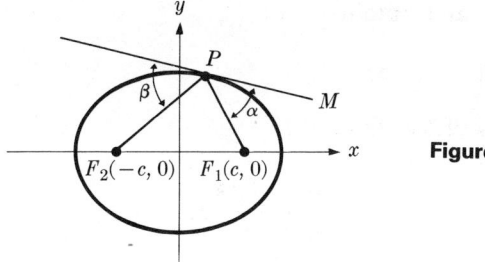

Figure A–8

The reflected-wave property of ellipses is exhibited by the following procedure. Let $P(x_1, y_1)$ be a point on the ellipse (not a vertex), as shown in Fig. A–8. Construct a tangent line M at this point. Then the angle made by M and the line PF_1 (α in the figure) is equal to the angle made by M and the line PF_2 (β in the figure). This result may be established by setting up the formulas for $\tan \alpha$ and $\tan \beta$ as the angles between two lines and showing them to be equal.

An ellipse divides the plane into two regions, one inside and one outside the curve itself. Every point inside the ellipse $(x^2/a^2) + (y^2/b^2) = 1$ satisfies the inequality
$$\frac{x^2}{a^2} + \frac{y^2}{b^2} < 1.$$

From any point outside the ellipse two tangents may be drawn, and clearly no tangents to the ellipse can be constructed which pass through a point inside the ellipse itself.

Example 3. Given the ellipse $(x^2/25) + (y^2/9) = 1$, and the points $P(1, 2)$, $Q(0, 6)$, and $R(3, 2)$. For each point, decide whether tangents to the ellipse passing through the point can be constructed. If they can, find the equations of the tangent lines.

Solution. For the point P we have

$$\frac{1^2}{25} + \frac{2^2}{9} < 1,$$

and no tangents exist. Similarly, for the point R we have $(9/25) + (4/9) < 1$. The point Q is outside the ellipse, since $(0^2/25) + (6^2/9) > 1$, and we proceed to find the tangent lines through Q. Any line through Q has the equation

$$y - 6 = m(x - 0).$$

The slope of the ellipse at any point is, by implicit differentiation,

$$\frac{2x}{25} + \frac{2y}{9}\frac{dy}{dx} = 0 \quad \text{or} \quad \frac{dy}{dx} = -\frac{9}{25}\frac{x}{y}.$$

Suppose that $P_1(x_1, y_1)$ is a point where the desired line is tangent to the ellipse. We have $m = -9x_1/25y_1$ and

$$\frac{x_1^2}{25} + \frac{y_1^2}{9} = 1, \quad y_1 - 6 = -\frac{9}{25}\frac{x_1}{y_1}x_1.$$

We solve these equations simultaneously and obtain

$$(\tfrac{5}{2}\sqrt{3}, \tfrac{3}{2}) \quad \text{and} \quad (-\tfrac{5}{2}\sqrt{3}, \tfrac{3}{2})$$

as the points of tangency. The equations of the lines are

$$y - 6 = -\frac{9}{25} \cdot \frac{\tfrac{5}{2}\sqrt{3}}{\tfrac{3}{2}} x, \quad y - 6 = -\frac{9}{25} \cdot \frac{-\tfrac{5}{2}\sqrt{3}}{\tfrac{3}{2}} x,$$

and, when we simplify, we find

$$3\sqrt{3}\,x + 5y - 30 = 0, \quad 3\sqrt{3}\,x - 5y + 30 = 0.$$

PROBLEMS

In Problems 1 through 6, find the equation of the ellipse satisfying the given conditions.
1. Foci at $(\pm 4, 0)$, directrices $x = \pm 6$
2. Eccentricity $\tfrac{1}{2}$, directrices $x = \pm 8$
3. Foci at $(0, \pm 4)$, directrices $y = \pm 7$
4. Vertices at $(\pm 6, 0)$, directrices $x = \pm 9$
5. Vertices at $(0, \pm 7)$, directrices $y = \pm 12$
6. Eccentricity $\tfrac{2}{3}$, directrices $y = \pm 9$
7. Prove the reflected-wave property of an ellipse, as described in this section.

8. Prove that the segment of a tangent to an ellipse between the point of contact and a directrix subtends a right angle at the corresponding focus.

9. A line through a focus parallel to a directrix is called a **latus rectum.** The length of the segment of this line cut off by the ellipse is called the *length of the latus rectum*. Show that the length of the latus rectum of
$$\frac{x^2}{a^2} + \frac{y^2}{b^2} = 1$$
is $2b^2/a$.

10. Prove that the line joining a point P_0 of an ellipse to the center and the line through a focus and perpendicular to the tangent at P_0 intersect on the corresponding directrix.

11. The tangent to an ellipse at a point P_0 meets the tangent at a vertex in a point Q. Prove that the line joining the other vertex to P_0 is parallel to the line joining the center to Q.

12. Given the ellipse $(x^2/4) + (y^2/2) = 1$. Decide whether or not the points $P(1, \frac{1}{2})$, $Q(1, 4)$, and $R(0, 1)$ are outside the ellipse. Construct the tangent lines passing through the points wherever possible.

13. Given the ellipse $(x^2/1) + (y^2/9) = 1$. Rework Problem 12 for the points $P(\frac{1}{2}, 1)$, $Q(2, 3)$, and $R(5, 0)$.

4. THE HYPERBOLA (continued)

The hyperbola, like the ellipse, has two directrices. For the hyperbola
$$\frac{x^2}{a^2} - \frac{y^2}{b^2} = 1, \quad \text{the directrices are the lines} \quad x = \pm \frac{a}{e},$$
and for the hyperbola
$$\frac{y^2}{a^2} - \frac{x^2}{b^2} = 1, \quad \text{the directrices are the lines} \quad y = \pm \frac{a}{e}.$$

A hyperbola has the property that it is the graph of all points $P(x, y)$ whose distances from a fixed point (F_1) are equal to e $(e > 1)$ times their distances from a fixed line (say $x = a/e$). As Fig. A–9 shows, the conditions of the graph are
$$d_1 = ed_2 \quad \Leftrightarrow \quad \sqrt{(x-c)^2 + y^2} = e\left|x - \frac{a}{e}\right|.$$

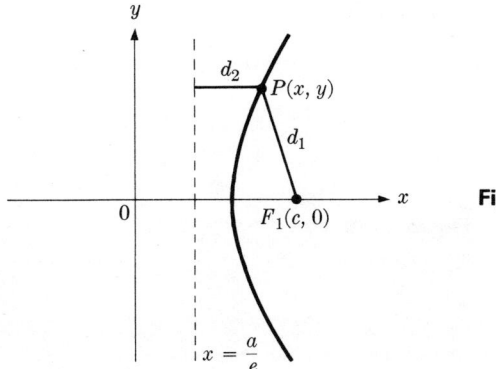

Figure A–9

If we square both sides and simplify, we obtain $(x^2/a^2) - (y^2/b^2) = 1$. The same is true if we start with $F_2(-c, 0)$ and the line $x = -a/e$.

Example 1. Find the equation of the hyperbola if the foci are at the points $(\pm 6, 0)$ and one directrix is the line $x = 3$.

Solution. We have $c = 6$ and $a/e = 3$. Since $e = c/a$, we may also write $a^2/c = 3$ and $a^2 = 18$, $a = \sqrt{18}$. In addition, $c^2 = a^2 + b^2$, from which we obtain $b^2 = 18$. The equation is

$$\frac{x^2}{18} - \frac{y^2}{18} = 1.$$ ◂

In studying the hyperbola

$$\frac{x^2}{a^2} - \frac{y^2}{b^2} = 1 \qquad (1)$$

we shall show that the lines

$$y = \frac{b}{a}x \quad \text{and} \quad y = -\frac{b}{a}x$$

are of special interest. For this purpose we consider a point $\bar{P}(\bar{x}, \bar{y})$ on the hyperbola (1) and compute the distance from $\bar{P}$ to the line $y = bx/a$. According to the formula for the distance from a point to a line (see Section 1 of Chapter 8), we have for the distance d:

$$d = \frac{|b\bar{x} - a\bar{y}|}{\sqrt{a^2 + b^2}}.$$

Suppose we select a point $\bar{P}$ which has both of its coordinates positive, as shown in Fig. A–10. Since $\bar{P}$ is a point on the hyperbola, its coordinates satisfy the equation $b^2\bar{x}^2 - a^2\bar{y}^2 = a^2b^2$, which we may write

$$(b\bar{x} - a\bar{y})(b\bar{x} + a\bar{y}) = a^2b^2 \quad \Leftrightarrow \quad b\bar{x} - a\bar{y} = a^2b^2/(b\bar{x} + a\bar{y}).$$

Substituting this last relation in the above expression for d, we find

$$d = \frac{a^2b^2}{\sqrt{a^2 + b^2}\,|b\bar{x} + a\bar{y}|}.$$

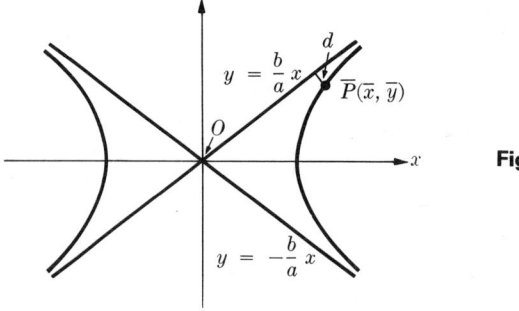

Figure A–10

Now a and b are fixed positive numbers and we selected $\bar{x}$, $\bar{y}$ positive. As P moves out along the hyperbola farther and farther away from the vertex, $\bar{x}$ and $\bar{y}$ increase without bound. Therefore, the number d tends to zero as $\bar{x}$ and $\bar{y}$ get larger and larger. That is, the distance between the hyperbola (1) and the line $y = bx/a$ shrinks to zero as $\bar{P}$ tends to infinity. We define the line $y = bx/a$ as an **asymptote of the hyperbola** $(x^2/a^2) - (y^2/b^2) = 1$. Analogously, the line $y = -bx/a$ (see Fig. A–10) is called an asymptote of the same hyperbola. We have only to recall the symmetry of the hyperbola to verify this statement.

Now we can see one of the uses of the central rectangle. *The asymptotes are the lines which contain the diagonals of the central rectangle.* In an analogous way the hyperbola $(y^2/a^2) - (x^2/b^2) = 1$ has the asymptotes

$$y = (a/b)x \quad \text{and} \quad y = -(a/b)x.$$

These lines contain the diagonals of the appropriately constructed central rectangle.

Example 2. Given the hyperbola $25x^2 - 9y^2 = 225$, find the foci, vertices, eccentricity, directrices, and asymptotes. Sketch the curve.

Solution. When we divide by 225, we obtain $(x^2/9) - (y^2/25) = 1$, and therefore $a = 3$, $b = 5$, $c^2 = 9 + 25$, and $c = \sqrt{34}$. The foci are at $(\pm\sqrt{34}, 0)$, vertices at $(\pm 3, 0)$; the essentricity is $e = c/a = \sqrt{34}/3$, and the directrices are the lines $x = \pm a/e = \pm 9/\sqrt{34}$. To find the asymptotes we merely write $y = \pm(b/a)x =$

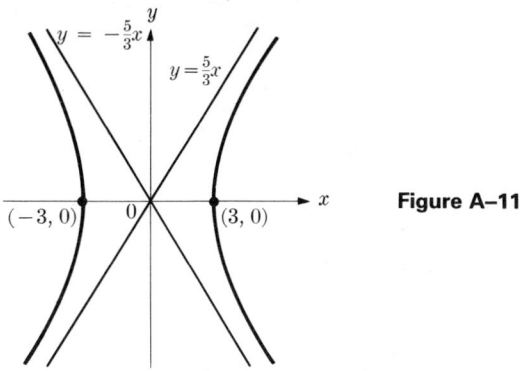

Figure A–11

$\pm \frac{5}{3}x$. Since the asymptotes are a great help in sketching the curve, we draw them first (Fig. A–11): knowing the vertices and one or two additional points helps us obtain a fairly accurate graph. ◂

An easy way to remember the equations of the asymptotes is to recognize that if the hyperbola is

$$\frac{x^2}{a^2} - \frac{y^2}{b^2} = 1, \quad \text{the asymptotes are} \quad \frac{x^2}{a^2} - \frac{y^2}{b^2} = 0,$$

and if the hyperbola is

$$\frac{y^2}{a^2} - \frac{x^2}{b^2} = 1, \quad \text{the asymptotes are} \quad \frac{y^2}{a^2} - \frac{x^2}{b^2} = 0.$$

A hyperbola divides the plane into *three* regions. For example, the hyperbola
$$S = \left\{(x, y): \frac{x^2}{4} - \frac{y^2}{16} = 1\right\}$$
divides the plane into the regions
$$S_1 = \left\{(x, y): \frac{x^2}{4} - \frac{y^2}{16} < 1\right\},$$
$$S_2 = \left\{(x, y): \frac{x^2}{4} - \frac{y^2}{16} > 1 \quad \text{and} \quad x > 2\right\},$$
$$S_3 = \left\{(x, y): \frac{x^2}{4} - \frac{y^2}{16} > 1 \quad \text{and} \quad x < -2\right\}.$$

For any given point P in the first of these regions, tangents to the hyperbola which pass through this point may be constructed, while no tangents can be constructed which pass through a point (such as Q) of the other two regions (Fig. A–12).

Example 3. Given the hyperbola $(x^2/4) - (y^2/3) = 1$. Decide whether or not tangents to the hyperbola can be constructed which pass through any of the points $P(3, 1)$, $Q(-4, 2)$, $R(1, 3)$. If they can, find the equations of the tangent lines.

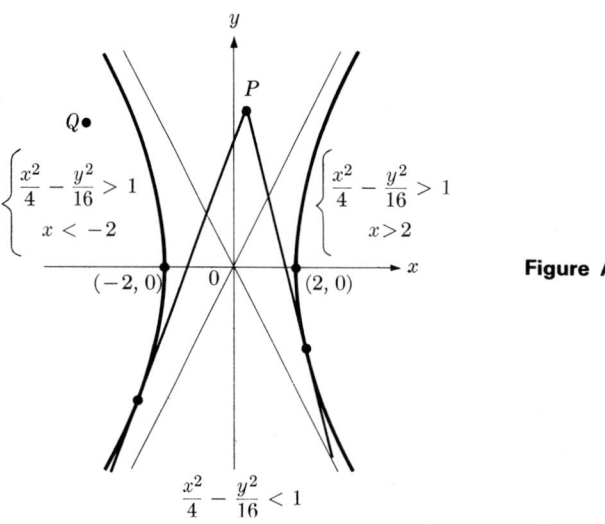

Figure A–12

Solution. For the point P we have $(3^2/4) - (1^2/3) > 1$, and no tangents can be constructed. Similarly, no tangents can be constructed which pass through Q, since $((-4)^2/4) - (2^2/3) > 1$. The point R satisfies the inequality $(1^2/4) - (3^2/3) < 1$, and we proceed to find the tangent lines. Any line through R has the equation
$$y - 3 = m(x - 1),$$

and the slope of the hyperbola at any point is

$$\frac{dy}{dx} = \frac{3}{4}\frac{x}{y}.$$

If (x_0, y_0) is a point of tangency, then $m = \frac{3}{4}(x_0/y_0)$, and the two equations

$$\frac{x_0^2}{4} = \frac{y_0^2}{3} + 1, \qquad y_0 - 3 = \frac{3}{4}\frac{x_0}{y_0}(x_0 - 1)$$

hold. We solve these simultaneously and find

$$\left(\frac{-4 + 12\sqrt{5}}{11}, \frac{-12 + 3\sqrt{5}}{11}\right), \qquad \left(\frac{-4 - 12\sqrt{5}}{11}, \frac{-12 - 3\sqrt{5}}{11}\right).$$

The equations of the lines are

$$(1 - 3\sqrt{5})x + (-4 + \sqrt{5})y + 11 = 0,$$
$$(1 + 3\sqrt{5})x - (4 + \sqrt{5})y + 11 = 0. \qquad \blacktriangleleft$$

PROBLEMS

In Problems 1 through 8, find the vertices, foci, eccentricity, equations of the asymptotes, and directrices. Construct the central rectangle and sketch the hyperbola.

1. $4x^2 - 9y^2 = 36$
2. $9x^2 - 4y^2 = 36$
3. $x^2 - 4y^2 = 4$
4. $9x^2 - y^2 = -9$
5. $x^2 - y^2 = 1$
6. $x^2 - y^2 = -1$
7. $25x^2 - 8y^2 = -200$
8. $2x^2 - 3y^2 = 8$

In Problems 9 through 17, find in each case the equation of the hyperbola (or hyperbolas) satisfying the given conditions.

9. Asymptotes $y = \pm 2x$, vertices $(\pm 6, 0)$
10. Asymptotes $y = \pm 2x$, vertices $(0, \pm 4)$
11. Eccentricity $\frac{4}{3}$, directrices $x = \pm 2$
12. Foci at $(\pm 6, 0)$, directrices $x = \pm 4$
13. Asymptotes $3y = \pm 4x$, foci at $(\pm 6, 0)$
14. Directrices $y = \pm 4$, asymptotes $y = \pm\frac{3}{2}x$
15. Asymptotes $2y = \pm 3x$, passing through $(5, 3)$
16. Directrices $x = \pm 1$, passing through $(-4, 6)$, center at origin
17. Directrices $y = \pm 1$, passing through $(1, \sqrt{3})$, center at origin
18. Prove that the lines $y = \pm(a/b)x$ are asymptotes of the hyperbola

$$(y^2/a^2) - (x^2/b^2) = 1.$$

19. A line through a focus of a hyperbola and parallel to the directrix is called a **latus rectum**. The length of the segment cut off by the hyperbola is called the *length of the latus rectum*. Show that for the hyperbola

$$(x^2/a^2) - (y^2/b^2) = 1$$

the length of the latus rectum is $2b^2/a$.

20. Find the graph of all points whose distances from the point $(0, \sqrt{41})$ are always $\sqrt{41}/5$ times their distances from the line $y = 25/\sqrt{41}$.
21. Find the graph of all points whose distances from the point $(8, 0)$ are always 3 times their distances from the line $x = 1$.
22. Show that an asymptote, a directrix, and the line through the corresponding focus perpendicular to the asymptote pass through a point.
23. Given the hyperbola $(y^2/4) - (x^2/6) = 1$. Decide whether or not tangents to this hyperbola can be drawn which pass through any of the points $P(0, 8)$, $Q(1, 5)$, $R(4, 0)$, and $S(-8, -9)$. Find the equations of the lines, wherever possible.

5. QUADRIC SURFACES

In the plane any equation of the form
$$Ax^2 + Bxy + Cy^2 + Dx + Ey + F = 0$$
is the equation of a curve. More specifically, we have found that circles, parabolas, ellipses, and hyperbolas, i.e., all conic sections, are represented by such second-degree equations.

In three-space the most general equation of the second degree in x, y, and z has the form
$$ax^2 + by^2 + cz^2 + dxy + exz + fyz + gx + hy + kz + l = 0, \qquad (1)$$
where the quantities $a, b, c, \ldots, l$ are positive or negative numbers or zero. The points in space satisfying such an equation all lie on a surface. Certain special cases, such as spheres and cylinders, were discussed in Chapter 12. Any second-degree equation which does not reduce to a cylinder, plane, line, or point corresponds to a surface which we call **quadric**. Quadric surfaces are classified into six types, and it can be shown that every second-degree equation which does not degenerate into a cylinder, a plane, etc., corresponds to one of these six types. The proof of this result involves the study of translation and rotation of coordinates in three-dimensional space, a topic beyond the scope of this book.

Definitions. *The x, y,* **and** *z* **intercepts** *of a surface are, respectively, the x, y, and z coordinates of the points of intersection of the surface with the respective axes. When we are given an equation of a surface, we get the x intercept by setting y and z equal to zero and solving for x. We proceed analogously for the y and z intercepts.*

The **traces** of a surface on the coordinate planes are the curves of intersections of the surface with the coordinate planes. When we are given a surface, we obtain the trace on the xz plane by first setting y equal to zero and then considering the resulting equation in x and z as the equation of a curve in the plane, as in plane analytic geometry. A **section of a surface** by a plane is the curve of intersection of the surface with the plane.

Example 1. Find the x, y, and z intercepts of the surface
$$3x^2 + 2y^2 + 4z^2 = 12.$$

Describe the traces of this surface. Find the section of this surface by the plane $z = 1$ and by the plane $x = 3$.

Solution. We set $y = z = 0$, getting $3x^2 = 12$; the x intercepts are at 2 and -2. Similarly, the y intercepts are at $\pm\sqrt{6}$, the z intercepts at $\pm\sqrt{3}$. To find the trace on the xy plane, we set $z = 0$, getting $3x^2 + 2y^2 = 12$. It is convenient to use set notation, and for this trace we write

$$\left\{(x, y, z): \frac{x^2}{4} + \frac{y^2}{6} = 1, \quad z = 0\right\}.$$

We recognize this curve as an ellipse, with major semi-axis $\sqrt{6}$, minor semi-axis 2, foci at $(0, \sqrt{2}, 0)$, $(0, -\sqrt{2}, 0)$. Similarly, the trace on the xz plane is the ellipse

$$\left\{(x, y, z): \frac{x^2}{4} + \frac{z^2}{3} = 1, \quad y = 0\right\},$$

and the trace on the yz plane is the ellipse

$$\left\{(x, y, z): \frac{y^2}{6} + \frac{z^2}{3} = 1, \quad x = 0\right\}.$$

The section of the surface by the plane $z = 1$ is the curve

$$\begin{Bmatrix} 3x^2 + 2y^2 + 4 = 12, \\ z = 1 \end{Bmatrix} \Leftrightarrow \begin{Bmatrix} \dfrac{x^2}{8/3} + \dfrac{y^2}{4} = 1, \\ z = 1 \end{Bmatrix}$$

which we recognize as an ellipse. The section by the plane $x = 3$ is the curve

$$\begin{Bmatrix} 27 + 2y^2 + 4z^2 = 12, \\ x = 3 \end{Bmatrix} \Leftrightarrow \begin{Bmatrix} 2y^2 + 4z^2 + 15 = 0, \\ x = 3 \end{Bmatrix}.$$

Since the sum of three positive quantities can never be zero, we conclude that the plane $x = 3$ does not intersect the surface. The section is empty. ◄

Definitions. *A surface is* **symmetric with respect to the** xy **plane** *if and only if the point* $(x, y, -z)$ *lies on the surface whenever* (x, y, z) *does; it is* **symmetric with respect to the** x **axis** *if and only if the point* $(x, -y, -z)$ *is on the graph whenever* (x, y, z) *is. Similar definitions are easily formulated for symmetry with respect to the remaining coordinate planes and axes.*

The notions of intercepts, traces, and symmetry are useful in the following description of the six types of quadric surfaces.

i) An **ellipsoid** is the locus of an equation of the form

$$\frac{x^2}{A^2} + \frac{y^2}{B^2} + \frac{z^2}{C^2} = 1.$$

The surface is sketched in Fig. A–13. The x, y, z intercepts are the numbers $\pm A$, $\pm B$, $\pm C$, respectively, and the traces on the xy, xz, and yz planes are, respectively,

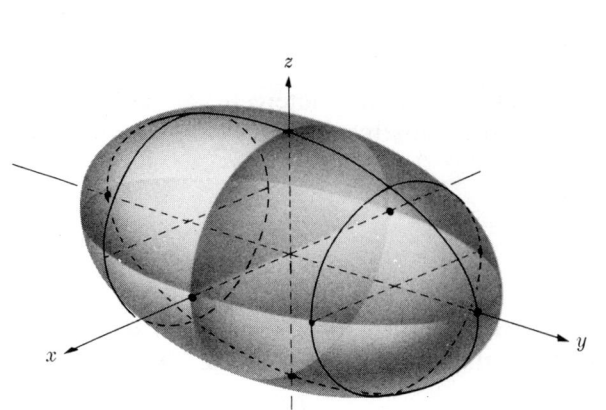

Figure A–13   Figure A–14

the ellipses

$$\frac{x^2}{A^2} + \frac{y^2}{B^2} = 1, \qquad \frac{x^2}{A^2} + \frac{z^2}{C^2} = 1, \qquad \frac{y^2}{B^2} + \frac{z^2}{C^2} = 1.$$

Sections made by the planes $y = k$ (k a constant) are the similar ellipses

$$\frac{x^2}{A^2(1 - k^2/B^2)} + \frac{z^2}{C^2(1 - k^2/B^2)} = 1, \qquad y = k, \qquad -B < k < B.$$

Several such ellipses are drawn in Fig. A–13.

If $A = B = C$, we obtain a sphere, while if two of the three numbers are equal, the surface is an **ellipsoid of revolution,** also called a **spheroid.** If, for example, $A = B$ and $C > A$, the surface is called a **prolate spheroid,** exemplified by a football. On the other hand, if $A = B$ and $C < A$, we have an **oblate spheroid.** The earth is approximately the shape of an oblate spheroid, with the section at the equator being circular and the distance between the North and South poles being smaller than the diameter of the equatorial circle.

ii) An **elliptic hyperboloid of one sheet** is the graph of an equation of the form

$$\frac{x^2}{A^2} + \frac{y^2}{B^2} - \frac{z^2}{C^2} = 1.$$

The graph of such a surface is sketched in Fig. A–14. The x intercepts are at $\pm A$ and the y intercepts at $\pm B$. As for the z intercepts, we must solve the equation $-z^2/C^2 = 1$, which has no real solutions. Therefore the surface does not intersect the z axis. The trace on the xy plane is an ellipse, while the traces on the yz and xz planes are hyperbolas. The sections made by any plane $z = k$ are the ellipses

$$\frac{x^2}{A^2(1 + k^2/C^2)} + \frac{y^2}{B^2(1 + k^2/C^2)} = 1, \qquad z = k,$$

and the sections made by the planes $y = k$ are the hyperbolas

$$\frac{x^2}{A^2(1 - k^2/B^2)} - \frac{z^2}{C^2(1 - k^2/B^2)} = 1, \qquad y = k.$$

iii) An **elliptic hyperboloid of two sheets** is the graph of an equation of the form

$$\frac{x^2}{A^2} - \frac{y^2}{B^2} - \frac{z^2}{C^2} = 1.$$

Such a graph is sketched in Fig. A–15. We observe that we must have $|x| \geq A$, for otherwise the quantity $(x^2/A^2) < 1$ and the left side of the above equation will

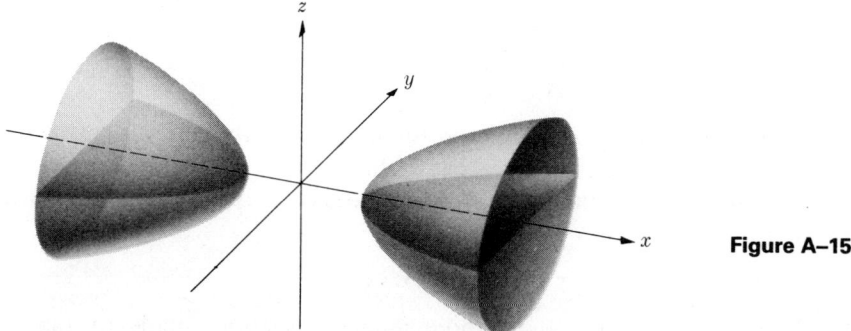

Figure A–15

always be less than the right side. The x intercepts are at $x = \pm A$. There are no y and z intercepts. The traces on the xz and xy planes are hyperbolas; there is no trace on the yz plane. The sections made by the planes $x = k$ are the ellipses

$$\frac{y^2}{B^2(k^2/A^2 - 1)} + \frac{z^2}{C^2(k^2/A^2 - 1)} = 1, \quad x = k, \quad \text{if} \quad |k| > A,$$

while the section is empty if $|k| < A$. The sections by the planes $y = k$ and $z = k$ are hyperbolas.

iv) An **elliptic paraboloid** is the graph of an equation of the form

$$\frac{x^2}{A^2} + \frac{y^2}{B^2} = z.$$

A typical elliptic paraboloid is sketched in Fig. A–16.

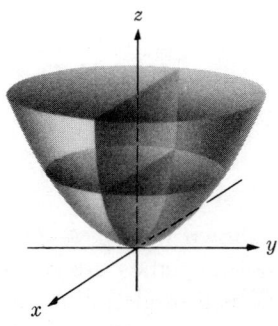

Figure A–16

All three intercepts are 0; the traces on the yz and xz planes are parabolas, while the trace on the xy plane consists of a single point, the origin. Sections made by planes $z = k$ are ellipses if $k > 0$, empty if $k < 0$. Sections made by planes $x = k$ and $y = k$ are parabolas.

If $A = B$, we have a **paraboloid of revolution,** and the sections made by the planes $z = k$, $k > 0$ are circles. The reflecting surfaces of telescopes, automobile headlights, etc., are always paraboloids of revolution. (See Section 2.)

v) A **hyperbolic paraboloid** is the graph of an equation of the form

$$\frac{x^2}{A^2} - \frac{y^2}{B^2} = z.$$

Such a graph is sketched in Fig. A–17. As in the elliptic paraboloid, all intercepts are zero. The trace on the xz plane is a parabola opening upward; the trace on the yz plane is a parabola opening downward; and the trace on the xy plane is the pair of intersecting straight lines

$$y = \pm(B/A)x.$$

As Fig. A–17 shows, the surface is "saddle-shaped"; sections made by planes $x = k$ are parabolas opening downward and those made by planes $y = k$ are parabolas opening upward. The sections made by planes $z = k$ are hyperbolas facing one way if $k < 0$ and the other way if $k > 0$. The trace on the xy plane corresponds to $k = 0$ and, as we saw, consists of two intersecting lines.

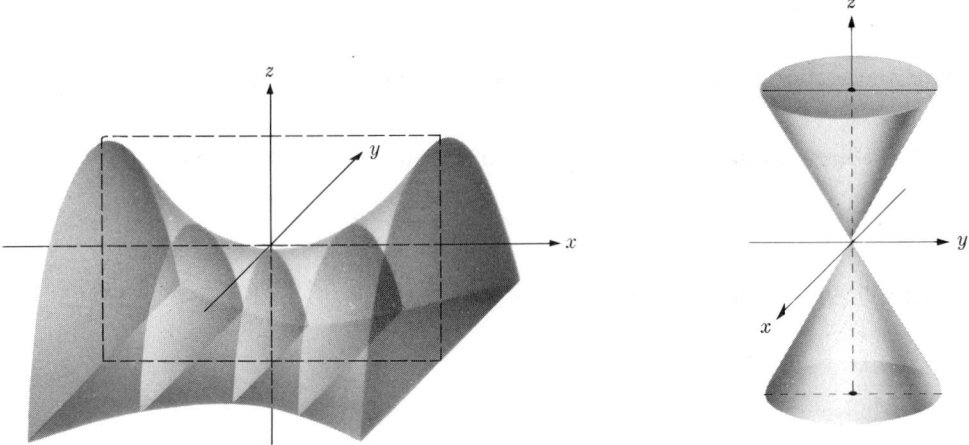

Figure A–17   **Figure A–18**

vi) An **elliptic cone** is the graph of an equation of the form

$$\frac{x^2}{A^2} + \frac{y^2}{B^2} = \frac{z^2}{C^2}.$$

A typical cone of this type is shown in Fig. A–18. Once again all intercepts are zero. The traces on the xz and yz planes are pairs of intersecting straight lines, while the trace on the xy plane is a single point, the origin. Planes parallel to the coordinate planes yield sections which are the familar conic sections of plane analytic geometry.

Example 2. Name and sketch the graph of $4x^2 - 9y^2 + 8z^2 = 72$. Indicate a few sections parallel to the coordinate planes.

Solution. We divide by 72, getting

$$\frac{x^2}{18} - \frac{y^2}{8} + \frac{z^2}{9} = 1,$$

which is an elliptic hyperboloid of one sheet. The x intercepts are $\pm 3\sqrt{2}$, the z intercepts are ± 3, and there are no y intercepts. The trace on the xy plane is the hyperbola

$$\frac{x^2}{18} - \frac{y^2}{8} = 1,$$

the trace on the xz plane is the ellipse

$$\frac{x^2}{18} + \frac{z^2}{9} = 1,$$

and the trace on the yz plane is the hyperbola

$$\frac{z^2}{9} - \frac{y^2}{8} = 1.$$

The surface is sketched in Fig. A–19.

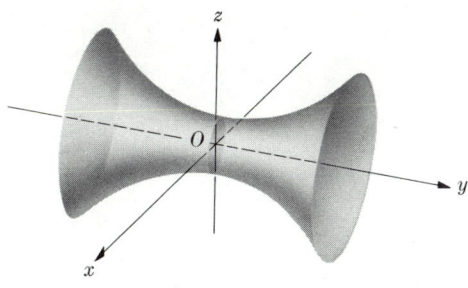

Figure A–19

PROBLEMS

Name and sketch the graph of each of the following equations.

1. $\dfrac{x^2}{16} + \dfrac{y^2}{9} + \dfrac{z^2}{4} = 1$

2. $\dfrac{x^2}{9} + \dfrac{y^2}{12} + \dfrac{z^2}{9} = 1$

3. $\dfrac{x^2}{16} + \dfrac{y^2}{20} + \dfrac{z^2}{20} = 1$

4. $\dfrac{x^2}{16} + \dfrac{y^2}{9} - \dfrac{z^2}{4} = 1$

5. $\dfrac{x^2}{16} - \dfrac{y^2}{9} - \dfrac{z^2}{4} = 1$

6. $\dfrac{x^2}{16} - \dfrac{y^2}{9} + \dfrac{z^2}{4} = 1$

7. $\dfrac{y^2}{9} + \dfrac{z^2}{4} = 1 + \dfrac{x^2}{16}$

8. $-\dfrac{x^2}{16} - \dfrac{y^2}{9} + \dfrac{z^2}{4} = 1$

9. $\dfrac{x}{4} = \dfrac{y^2}{4} + \dfrac{z^2}{9}$ 	 10. $z = \dfrac{y^2}{4} - \dfrac{x^2}{9}$

11. $y = \dfrac{x^2}{8} + \dfrac{z^2}{8}$ 	 12. $z^2 = 4x^2 + 4y^2$

13. $y^2 = x^2 + z^2$ 	 14. $z^2 = x^2 - y^2$

15. $2x^2 + 6y^2 - 3z^2 = 8$ 	 16. $4x^2 - 3y^2 + 2z^2 = 0$

17. $3x = 2y^2 - 5z^2$ 	 18. $\dfrac{y}{5} = 8z^2 - 2x^2$

19. Show that the intersection of the hyperbolic paraboloid $x^2 - y^2 = z$ with the plane $z = x + y$ consists of two intersecting straight lines. Establish the same result for the intersection of this quadric with the plane $z = ax + ay$, where a is any number.

*20. Using the result of Problem 19 as a guide, show that any hyperbolic paraboloid is "composed entirely of straight lines."

APPENDIX 3
PROPERTIES OF HYPERBOLIC FUNCTIONS

1. THE HYPERBOLIC FUNCTIONS

Exponential functions appear in many investigations in engineering, physics, chemistry, biology, and the social sciences. Certain combinations of them have been tabulated and given special names.

Definitions. *Two functions, denoted* sinh *and* cosh, *are defined by the formulas*

$$\sinh x = \tfrac{1}{2}(e^x - e^{-x}), \qquad \cosh x = \tfrac{1}{2}(e^x + e^{-x}).$$

We read these "hyperbolic sine of x" and "hyperbolic cosine of x." We shall see that these functions satisfy many relations which are reminiscent of the sine and cosine functions.

In analogy with trigonometric functions, there are four more hyperbolic functions defined in terms of sinh x and cosh x, as follows:

$$\tanh x = \frac{\sinh x}{\cosh x}, \qquad \coth x = \frac{\cosh x}{\sinh x},$$

$$\operatorname{sech} x = \frac{1}{\cosh x}, \qquad \operatorname{csch} x = \frac{1}{\sinh x}.$$

These are called "hyperbolic tangent of x," etc.

Sketches of the hyperbolic functions are shown in Fig. H–1. It is evident that, unlike the trigonometric functions, none of the hyperbolic functions is periodic. Sinh x, tanh x, coth x, and csch x are odd functions, while cosh x and sech x are even. To see this, we appeal directly to the definition, noting for example that

$$\sinh(-x) = \tfrac{1}{2}(e^{-x} - e^{-(-x)}) = -\tfrac{1}{2}(e^x - e^{-x}) = -\sinh x$$

and

$$\cosh(-x) = \tfrac{1}{2}(e^{-x} + e^{-(-x)}) = \tfrac{1}{2}(e^x + e^{-x}) = \cosh x.$$

The relation

$$\cosh^2 x - \sinh^2 x = 1$$

is established by observing that

$$[\tfrac{1}{2}(e^x + e^{-x})]^2 - [\tfrac{1}{2}(e^x - e^{-x})]^2 = \tfrac{1}{4}(e^{2x} + 2 + e^{-2x}) - \tfrac{1}{4}(e^{2x} - 2 + e^{-2x}) = 1.$$

The formulas
$$\tanh^2 x + \operatorname{sech}^2 x = 1, \qquad \coth^2 x - \operatorname{csch}^2 x = 1,$$
are then easily verified from the definitions of the functions and from the relation $\cosh^2 x - \sinh^2 x = 1$.

The addition formulas for the hyperbolic functions are:
$$\sinh(x \pm y) = \sinh x \cosh y \pm \cosh x \sinh y,$$
$$\cosh(x \pm y) = \cosh x \cosh y \pm \sinh x \sinh y,$$
$$\tanh(x \pm y) = \frac{\tanh x \pm \tanh y}{1 \pm \tanh x \tanh y}.$$

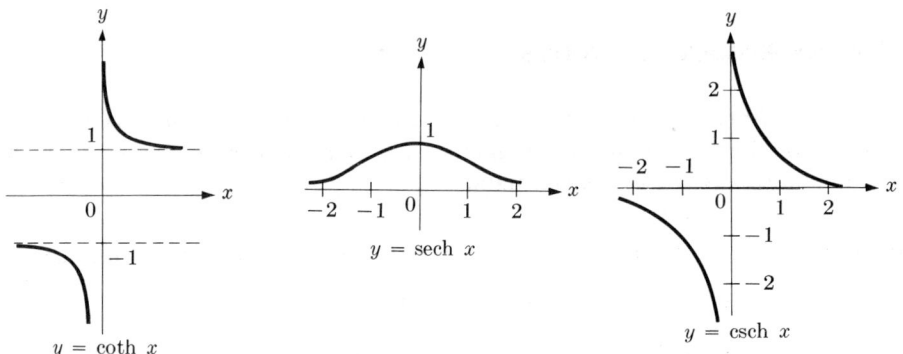

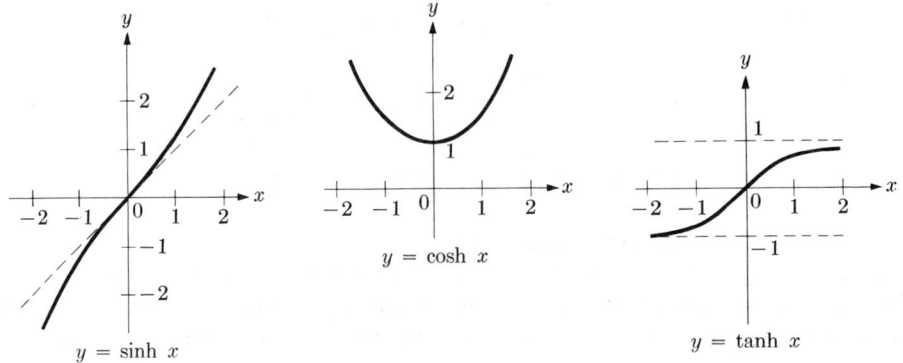

Figure H–1

These formulas are easier to establish than are the corresponding trigonometric addition formulas. For example, to prove the first one we merely substitute the appropriate exponentials in the left and right sides and see if they are equal. We have
$$\sinh(x + y) = \tfrac{1}{2}(e^{x+y} - e^{-x-y}),$$

sinh x cosh y + cosh x sinh y

$$= \tfrac{1}{2}(e^x - e^{-x})\tfrac{1}{2}(e^y + e^{-y}) + \tfrac{1}{2}(e^x + e^{-x})\tfrac{1}{2}(e^y - e^{-y})$$
$$= \tfrac{1}{4}(e^{x+y} + e^{x-y} - e^{-x+y} - e^{-x-y} + e^{x+y} - e^{x-y} + e^{-x+y} - e^{-x-y})$$
$$= \tfrac{1}{2}(e^{x+y} - e^{-x-y}) = \sinh(x+y).$$

The remaining formulas are obtained in a similar way.

Theorem 1. *The following differentiation formulas hold:*

i) $d \sinh u = \cosh u \, du$
ii) $d \cosh u = \sinh u \, du$
iii) $d \tanh u = \mathrm{sech}^2 u \, du$
iv) $d \coth u = -\mathrm{csch}^2 u \, du$
v) $d \,\mathrm{sech}\, u = -\mathrm{sech}\, u \tanh u \, du$
vi) $d \,\mathrm{csch}\, u = -\mathrm{csch}\, u \coth u \, du.$

Proof. To prove (i) we write

$$\sinh u = \tfrac{1}{2}(e^u - e^{-u}),$$
$$d \sinh u = \tfrac{1}{2} d(e^u - e^{-u}) = \tfrac{1}{2}(e^u + e^{-u}) \, du = \cosh u \, du.$$

The remaining formulas, (ii) through (vi), follow in the same way, by appeal to the definition and by differentiation of the exponential function.

Example 1. Given that $\sinh x = -\tfrac{3}{4}$, find the values of the other hyperbolic functions.

Solution. $\cosh^2 x = 1 + \sinh^2 x = \tfrac{25}{16}$. Since $\cosh x$ is always positive, we have $\cosh x = \tfrac{5}{4}$. Therefore

$$\tanh x = \frac{\sinh x}{\cosh x} = -\frac{3}{5}, \qquad \coth x = -\frac{5}{3},$$

$$\mathrm{sech}\, x = \frac{1}{\cosh x} = \frac{4}{5}, \qquad \mathrm{csch}\, x = \frac{1}{\sinh x} = -\frac{4}{3}. \quad\blacktriangleleft$$

Example 2. Given $f(x) = \sinh^3(2x^2 + 3)$, find $f'(x)$.

Solution. We have, by the Chain Rule,

$$df = 3 \sinh^2(2x^2 + 3) \, d[\sinh(2x^2 + 3)],$$

and from the formula for the derivative of the sinh function,

$$df = 3 \sinh^2(2x^2 + 3) \cosh(2x^2 + 3) \, d(2x^2 + 3),$$
$$f'(x) = 12x \sinh^2(2x^2 + 3) \cosh(2x^2 + 3). \quad\blacktriangleleft$$

PROBLEMS

In Problems 1 through 17, establish the formulas.

1. $\sinh(x - y) = \sinh x \cosh y - \cosh x \sinh y$
2. $\cosh(x + y) = \cosh x \cosh y + \sinh x \sinh y$

3. $\cosh(x - y) = \cosh x \cosh y - \sinh x \sinh y$
4. $\tanh(x + y) = \dfrac{\tanh x + \tanh y}{1 + \tanh x \tanh y}$
5. $\tanh(x - y) = \dfrac{\tanh x - \tanh y}{1 - \tanh x \tanh y}$.
6. $\sinh 2x = 2 \sinh x \cosh x$
7. $\cosh 2x = \cosh^2 x + \sinh^2 x = 2\cosh^2 x - 1 = 2\sinh^2 x + 1$
8. $\tanh 2x = \dfrac{2 \tanh x}{1 + \tanh^2 x}$
9. $\sinh A + \sinh B = 2 \sinh \dfrac{A+B}{2} \cosh \dfrac{A-B}{2}$
10. $\cosh A + \cosh B = 2 \cosh \dfrac{A+B}{2} \cosh \dfrac{A-B}{2}$
11. $\cosh(x/2) = \sqrt{(1 + \cosh x)/2}$
12. $\sinh(x/2) = \pm\sqrt{(\cosh x - 1)/2}$
13. $d \cosh u = \sinh u \, du$
14. $d \tanh u = \text{sech}^2 u \, du$
15. $d \coth u = -\text{csch}^2 u \, du$
16. $d \,\text{sech}\, u = -\text{sech}\, u \tanh u \, du$
17. $d \,\text{csch}\, u = -\text{csch}\, u \coth u \, du$

In Problems 18 through 23, find the values of the remaining hyperbolic functions.

18. $\sinh x = -\tfrac{12}{5}$
19. $\tanh x = -\tfrac{4}{5}$
20. $\coth x = 2$
21. $\text{csch}\, x = -2$
22. $\cosh x = 4, \ x > 0$
23. $\cosh x = 2, \ x < 0$

In Problems 24 through 29, examine the curves for maximum points, minimum points, and points of inflection. State where the curves are concave upward and where they are concave downward. Sketch the graphs.

24. $y = \sinh 2x$
25. $y = \cosh(x/2)$
26. $y = \tanh 3x$
27. $y = \coth \tfrac{1}{3}x$
28. $y = \text{sech}(x + 1)$
29. $y = \text{csch}(1 - x)$

2. THE INVERSE HYPERBOLIC FUNCTIONS

The function $f(x) = \sinh x$ is an increasing function for all values of x. This is so because the derivative of $\sinh x$ is $\cosh x$, a positive quantity for all values of x, and we know that a function with a positive derivative is an increasing function.

The inverse relation of a function which is increasing is a function (Inverse Function Theorem). We define

$$y = \text{argsinh}\, x$$

to be the inverse of the sinh function. Its domain consists of all real numbers. The graph is sketched in Fig. H–2.

Similarly, the functions $\tanh x$, $\coth x$, and $\text{csch}\, x$ are all increasing or decreasing functions for all values of x and so have inverse functions, which we denote by

$$\text{argtanh}\, x, \quad \text{argcoth}\, x, \quad \text{argcsch}\, x.$$

The domain of $\text{argtanh}\, x$ is the set of numbers $-1 < x < 1$, while the range consists of all real numbers. Its graph is sketched in Fig. H–3.

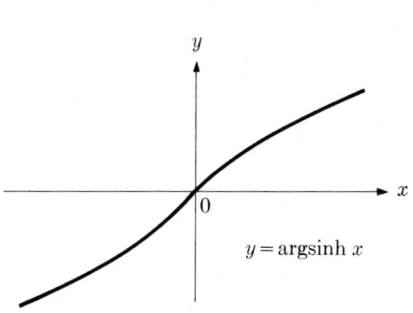

Figure H–2   **Figure H–3**

The function cosh x is an increasing function, and sech x is a decreasing function, for nonnegative values of x. We define

$$\text{argcosh } x, \quad \text{argsech } x$$

to be the inverse of the cosh and sech functions respectively, restricted to nonnegative values. The domain of $f(x) = \text{argcosh } x$ consists of all numbers $x \geq 1$, and the range is the set of nonnegative real numbers.

The hyperbolic functions are defined in terms of exponentials. The exponential and logarithmic functions are the inverse of each other. It is therefore natural to ask what relation, if any, exists between the logarithmic function and the inverse hyperbolic functions. The direct connection between the natural logarithm and the inverse hyperbolic functions is exhibited in the next theorem.

Theorem 2.

i) $\text{argsinh } x = \ln(x + \sqrt{x^2 + 1})$ *for all* x,
ii) $\text{argtanh } x = \frac{1}{2} \ln[(1 + x)/(1 - x)], \quad -1 < x < 1$,
iii) $\text{argcoth } x = \frac{1}{2} \ln[(x + 1)/(x - 1)], \quad |x| > 1$,
iv) $\text{argcosh } x = \ln(x + \sqrt{x^2 - 1}), \quad x \geq 1$.

Proof. To prove (i), we write $y = \text{argsinh } x$ and, equivalently, $x = \sinh y$. We wish to find an expression for y in terms of the logarithm. The definition of sinh y gives us

$$x = \tfrac{1}{2}(e^y - e^{-y}) \quad \Leftrightarrow \quad e^y - \frac{1}{e^y} - 2x = 0.$$

We solve this for e^y by first writing

$$(e^y)^2 - 2x(e^y) - 1 = 0$$

and then applying the quadratic formula to obtain

$$e^y = x \pm \sqrt{x^2 + 1}.$$

The plus sign must be chosen, since $e^y > 0$ always. Taking natural logs of both sides, we find that

$$y = \ln(x + \sqrt{x^2 + 1}),$$

which proves (i).

The proofs of (ii), (iii), and (iv) follow the same pattern: first, we express the hyperbolic function in terms of exponentials; second, we solve for the exponential by the quadratic formula; third, we make the correct choice of the plus or minus sign in the quadratic formula; and fourth, we take logarithms of both sides. We shall exhibit the steps again in the proof of (iv). Let $y = \operatorname{argcosh} x$, $x \geq 1$, and write $\cosh y = x$, $y \geq 0$: Then we have

$$e^y + e^{-y} = 2x, \quad y \geq 0,$$

and

$$(e^y)^2 - 2x(e^y) + 1 = 0, \quad y \geq 0.$$

Solving by the quadratic formula, we obtain

$$e^y = x \pm \sqrt{x^2 - 1}, \quad y \geq 0.$$

For $y \geq 0$ we must have $e^y \geq 1$. Since

$$(x + \sqrt{x^2 - 1})(x - \sqrt{x^2 - 1}) = 1,$$

it follows that for $x > 1$, $x + \sqrt{x^2 - 1} > 1$, and $x - \sqrt{x^2 - 1} < 1$. We therefore choose the plus sign. Taking logarithms, we get

$$y = \ln(x + \sqrt{x^2 - 1}), \quad x \geq 1.$$

Example 1. Express $\operatorname{argcosh} 2$, $\operatorname{argtanh}(-\tfrac{1}{2})$ in terms of the logarithm function.

Solution. From Theorem 2, we obtain

$$\operatorname{argcosh} 2 = \ln(2 + \sqrt{3}),$$

$$\operatorname{argtanh}\left(-\frac{1}{2}\right) = \frac{1}{2} \ln \frac{1/2}{3/2} = -\frac{1}{2} \ln 3 = \ln \frac{1}{\sqrt{3}}. \quad \blacktriangleleft$$

By expressing the inverse hyperbolic functions in terms of logarithms, we have simplified the problem of computing the derivatives of these functions.

Theorem 3. *The inverse hyperbolic functions have the following differentiation formulas:*

i) $d \operatorname{argsinh} u = \dfrac{du}{\sqrt{u^2 + 1}},$

ii) $d \operatorname{argcosh} u = \dfrac{du}{\sqrt{u^2 - 1}}, \quad u > 1,$

iii) $d \operatorname{argtanh} u = \dfrac{du}{1 - u^2}, \quad -1 < u < 1,$

iv) $d \operatorname{argcoth} u = \dfrac{du}{1 - u^2}, \quad |u| > 1,$

v) $d \operatorname{argsech} u = -\dfrac{du}{u\sqrt{1 - u^2}}, \quad 0 < u < 1,$

vi) $d \operatorname{argcsch} u = -\dfrac{du}{|u|\sqrt{u^2 + 1}}, \quad u \neq 0.$

Proof. To prove (i), we write
$$y = \text{argsinh}\, u = \ln(u + \sqrt{u^2 + 1}).$$
Therefore
$$dy = \frac{1}{u + \sqrt{u^2 + 1}} d(u + \sqrt{u^2 + 1})$$
$$= \frac{1}{u + \sqrt{u^2 + 1}}\left(1 + \frac{u}{\sqrt{u^2 + 1}}\right) du = \frac{du}{\sqrt{u^2 + 1}}.$$

The remaining formulas, (ii) through (vi), are proved in a similar manner.

Example 2. Given $f(x) = \text{argcosh}\,(x^2 + 2x + 2)$, find $f'(x)$.

Solution. Letting $u = x^2 + 2x + 2$, we have
$$df = \frac{du}{\sqrt{u^2 - 1}} = \frac{2(x + 1)\,dx}{\sqrt{(x^2 + 2x + 2)^2 - 1}},$$
and, since $x^2 + 2x + 2 = (x + 1)^2 + 1$,
$$f'(x) = \frac{2(x + 1)}{\sqrt{(x + 1)^4 + 2(x + 1)^2}} = \frac{2}{\sqrt{(x + 1)^2 + 2}}, \qquad x + 1 > 0. \quad \blacktriangleleft$$

Example 3. Evaluate
$$\int_{-1}^{2} \frac{dx}{\sqrt{1 + x^2}}$$
and express the result in terms of the logarithm function.

Solution. Since every differentiation formula carries with it an integration formula, Theorem 3(i) gives us
$$\int_{-1}^{2} \frac{dx}{\sqrt{1 + x^2}} = \text{argsinh}\, x \Big|_{-1}^{2} = \text{argsinh}\, 2 - \text{argsinh}\,(-1)$$
$$= \ln(2 + \sqrt{5}) - \ln(-1 + \sqrt{2}) = \ln\frac{2 + \sqrt{5}}{\sqrt{2} - 1}. \quad \blacktriangleleft$$

PROBLEMS

In Problems 1 through 6, express the given quantities in terms of the logarithm function.

1. $\text{argsinh}\,(\tfrac{1}{2})$
2. $\text{argtanh}\,(\tfrac{3}{5})$
3. $\text{argcosh}\, 2$
4. $\text{argcoth}\,(-2)$
5. $\text{argsech}\,(\tfrac{3}{5})$
6. $\text{argcosh}\,(2.6)$

In Problems 7 through 12, examine the curves for maximum points, minimum points, and points of inflection. State where the curves are concave upward and where they are concave downward. Sketch the graphs.

7. $y = \text{argsinh}\, 2x$
8. $y = \text{argcosh}\,(x + 1)$
9. $y = \text{argtanh}\,\tfrac{1}{2}x$
10. $y = \text{argcoth}\,(x - 1)$
11. $y = \text{argsech}\, 4x$
12. $y = \text{argcsch}\,(2x - 1)$

In Problems 13 through 20, perform the differentiations.

13. $f(x) = \text{argsinh } 2x$
14. $f(x) = x^2 \text{ argcosh } 3x$
15. $g(x) = x^{-1} \text{ argtanh } (x^2)$
16. $f(x) = e^{-2x} \text{ argsinh } (3x - 2)$
17. $G(x) = \text{argsinh } (\tan x)$
18. $F(x) = (\text{argsinh } \sqrt{x})/\sqrt{x}$
19. $F: x \to \text{argsinh}^2 (2x)$
20. $H: x \to xe^{-x} \text{ argcosh } (1 - x)$

In Problems 21 through 26, evaluate the integrals, giving the answers in terms of the logarithm function.

21. $\displaystyle\int_2^5 \frac{dx}{\sqrt{x^2 - 1}}$
22. $\displaystyle\int_{-1/2}^{3/5} \frac{dx}{1 - x^2}$
23. $\displaystyle\int_{-2}^{-1} \frac{dx}{\sqrt{x^2 + 1}}$
24. $\displaystyle\int_{-3}^{-2} \frac{dx}{1 - x^2}$
25. $\displaystyle\int_3^5 \frac{dx}{\sqrt{x^2 - 4}}$
26. $\displaystyle\int_1^3 \frac{dx}{16 - x^2}$

27. Prove that $\text{argtanh } x = \tfrac{1}{2} \ln [(1 + x)/(1 - x)]$, $\quad -1 < x < 1$.
28. Prove that $\text{argcoth } x = \tfrac{1}{2} \ln [(x + 1)/(x - 1)]$, $\quad |x| > 1$.
29. Sketch the graph of $y = \text{argcoth } x$; of $y = \text{argcsch } x$.
30. Prove that $d \text{ argcosh } u = \dfrac{du}{\sqrt{u^2 - 1}}$, $\quad u > 1$.
31. Prove that $d \text{ argtanh } u = \dfrac{du}{1 - u^2}$, $\quad -1 < u < 1$.
32. Prove that $d \text{ argcoth } u = \dfrac{du}{1 - u^2}$, $\quad |u| > 1$.
33. Prove that $d \text{ argsech } u = -\dfrac{du}{u\sqrt{1 - u^2}}$, $\quad 0 < u < 1$.
34. Prove that $d \text{ argcsch } u = -\dfrac{du}{|u|\sqrt{u^2 + 1}}$, $\quad u \neq 0$.

35. Let $x = \cosh t$, $y = \sinh t$. Make a table of values for x and y as t assumes values in the interval $-10 \leq t \leq 10$. Sketch the graph in the xy plane. Can you identify the curve?

APPENDIX 4

AXIOMS OF ALGEBRA. NUMBER SYSTEMS

1. THE AXIOMS OF ALGEBRA

In Chapter 1 we stated that the customary laws of algebra and the theorems on inequalities given there could be derived from a system of axioms. In this section we give the classical set of axioms of algebra and we derive a few simple theorems of algebra, ones which are direct consequences of the axioms.

We assume that we are given a set of objects which we call **real numbers.** The following axioms concern operations on such numbers.

Axioms of Addition and Subtraction

A–1. Closure Property. *If a and b are numbers, there is one and only one number, denoted by $a + b$, called their* **sum.**

A–2. Commutative Law. *For every two numbers a and b, we have*
$$b + a = a + b.$$

A–3. Associative Law. *For all numbers a, b, and c, we have*
$$(a + b) + c = a + (b + c).$$

A–4. Existence of a Zero. *There is one and only one number 0, called* **zero,** *such that $a + 0 = a$ for any number a.*

We remark that it is not necessary to assume in Axiom A–4 that there is *only one* number 0 with the given property. The uniqueness of this number may be established as follows: suppose that 0 and $0'$ are two numbers such that $a + 0 = a$ and $a + 0' = a$ for every number a. Then $0 + 0' = 0$ and $0' + 0 = 0'$. By Axiom A–2, $0 + 0' = 0' + 0$ and so $0 = 0'$; the two numbers are the same.

A–5. Existence of a Negative. *If a is any number, there is one and only one number x such that $a + x = 0$. This number is called the* **negative** *of a and is denoted by $-a$.*

As in Axiom A–4, it is not necessary to assume there is *only one* such number with the given property. The argument is similar to the one given after Axiom A–4.

Theorem 1. *If a and b are any numbers, then there is one and only one number x such that $a + x = b$. This number x is given by $x = b + (-a)$.*

Proof. We must establish two results: (i) that $b + (-a)$ satisfies $a + x = b$, and (ii) that no other number satisfies $a + x = b$. To prove (i), we suppose that $x = b + (-a)$. Then, using the commutative and associative laws, we have

$$a + x = a + [b + (-a)] = a + [(-a) + b]$$
$$= [a + (-a)] + b = 0 + b = b.$$

Therefore (i) holds. To prove (ii), we suppose that x is some number such that $a + x = b$. Adding $(-a)$ to both sides of this equation, we find that

$$(a + x) + (-a) = b + (-a).$$

Now,

$$(a + x) + (-a) = a + [x + (-a)] = a + [(-a) + x]$$
$$= [a + (-a)] + x = 0 + x = x.$$

We conclude that $x = b + (-a)$, and the uniqueness of the solution is established.

Notation. We denote the number $b + (-a)$ by $b - a$.

Thus far addition has been defined *only for two numbers.* By means of the associative law we can define addition for three, four, and in fact, any finite number of elements. Since $(a + b) + c$ and $a + (b + c)$ are the same, we define $a + b + c$ as this common value. The following lemma is an easy consequence of the associative and commutative laws of addition.

Lemma 1. *If a, b, and c are any numbers, then*

$$a + b + c = a + c + b = b + a + c = b + c + a$$
$$= c + a + b = c + b + a.$$

The formal details of writing out a proof are left to the student.
The next lemma is useful in the proof of Theorem 2 below.

Lemma 2. *If a, b, c, and d are numbers, then*

$$(a + c) + (b + d) = (a + b) + (c + d).$$

Proof. Using Lemma 1 and the axioms, we have

$$(a + c) + (b + d) = [(a + c) + b] + d$$
$$= (a + c + b) + d = (a + b + c) + d$$
$$= [(a + b) + c] + d = (a + b) + (c + d).$$

Theorem 2. i) *If a is a number, then $-(-a) = a$.* (ii) *If a and b are numbers, then*

$$-(a + b) = (-a) + (-b).$$

Proof. i) From the definition of negative, we have
$$(-a) + [-(-a)] = 0, \qquad (-a) + a = a + (-a) = 0.$$
Axiom A–5 states that the negative of $(-a)$ is *unique*. Therefore, $a = -(-a)$.

ii) From the definition of negative, we know that
$$(a + b) + [-(a + b)] = 0.$$
Furthermore, using Lemma 2, we have
$$(a + b) + [(-a) + (-b)] = [a + (-a)] + [b + (-b)]$$
$$= 0 + 0 = 0.$$
The result follows from the "only one" part of Axiom A–5.

Theorem 2 can be stated in words in the familiar form: (i) *the negative of* $(-a)$ *is a and* (ii) *the negative of a sum is the sum of the negatives.*

Axioms of Multiplication and Division

M–1. Closure Property. *If a and b are numbers, there is one and only one number, denoted by ab (or $a \times b$ or $a \cdot b$), called their* **product.**

M–2. Commutative Law. *For every two numbers a and b,*
$$ba = ab.$$

M–3. Associative Law. *For all numbers a, b, and c, we have*
$$(ab) \cdot c = a \cdot (bc).$$

M–4. Existence of a Unit. *There is one and only one number u, different from zero, such that $au = a$ for every number a. This number u is called the* **unit** *and (as is customary) will be denoted by* 1.

M–5. Existence of a Reciprocal. *For each number a different from zero there is one and only one number x such that $ax = 1$. This number x is called the* **reciprocal** *of a and is denoted by a^{-1} (or $1/a$).*

Remarks. We observe that Axioms M–1 through M–4 are the parallels of Axioms A–1 through A–4 with addition replaced by multiplication. However M–5 is not the exact analog of A–5, since there is the additional requirement $a \neq 0$. The reason for this is given below in Theorem 3 where it is shown that the result of multiplication of any number by zero is zero. In familiar terms, we say that *division by zero is excluded.*

Special Axiom

D. Distributive Law. *For all numbers a, b, and c, we have*
$$a(b + c) = ab + ac.$$

Remarks. The axioms of addition and multiplication and the distributive law are supposed to hold for all real numbers whether positive, negative, or zero. In fact, the axioms hold for many number systems of which the collection of real numbers is only one. For example, all the axioms stated so far hold for the system consisting of all *complex numbers*. Furthermore, there are many systems each consisting of only a finite number of elements which satisfy all the above axioms. Additional axioms are needed if we require the real numbers to be the *only* collection satisfying all the axioms.

Theorem 3. *If a is any number, then $a \cdot 0 = 0$.*

Proof. Let b be any number. Then $b + 0 = b$ and hence $a \cdot (b + 0) = a \cdot b$. From the distributive law, we conclude that

$$(a \cdot b) + (a \cdot 0) = (a \cdot b),$$

so that $a \cdot 0 = 0$ by Axiom A–4.

Theorem 4. *If a and b are numbers and $a \neq 0$, then there is one and only one number x such that $a \cdot x = b$. The number x is given by $x = ba^{-1}$.*

The proof is just like the proof of Theorem 1 with addition replaced by multiplication, 0 replaced by 1, and $-a$ replaced by a^{-1}. The details are left to the student.

We now establish the familiar principle which underlies the solution of quadratic equations by factoring.

Theorem 5. i) *We have $a \cdot b = 0$ if and only if $a = 0$ or $b = 0$ or both.* (ii) *We have $a \neq 0$ and $b \neq 0$ if and only if $a \cdot b \neq 0$.*

Proof. We must prove two statements in each of the parts (i) and (ii). To prove (i), if $a = 0$ or $b = 0$ or both it follows from Theorem 3 that $a \cdot b = 0$. Going the other way, suppose $a \cdot b = 0$. Then there are two cases: either $a = 0$ or $a \neq 0$. If $a = 0$, the result follows. If $a \neq 0$, then we see that

$$a^{-1}(ab) = (a^{-1}a)b = 1 \cdot b = b = a^{-1} \cdot 0 = 0.$$

Hence $b = 0$ and the result follows in both cases.

ii) Suppose $a \neq 0$ and $b \neq 0$. Then $a \cdot b \neq 0$ because $a \neq 0$ and $b \neq 0$ is the negation of the statement "$a = 0$ or $b = 0$ or both." Thus (i) applies. For the second part of (ii), suppose $a \cdot b \neq 0$. Then $a \neq 0$ and $b \neq 0$ for, if one of them were zero, Theorem 3 would apply to give $a \cdot b = 0$. ◀

We define abc as the common value of $(ab)c$ and $a(bc)$. The student can prove the following lemmas which are similar to Lemmas 1 and 2.

Lemma 3. *If a, b, and c are numbers, then*

$$abc = acb = bac = bca = cab = cba.$$

Lemma 4. *If a, b, c, and d are numbers, then*
$$(ac) \cdot (bd) = (ab) \cdot (cd).$$

Theorem 6. i) *If $a \neq 0$, then $a^{-1} \neq 0$ and $[(a^{-1})^{-1}] = a$.* (ii) *If $a \neq 0$ and $b \neq 0$, then $(a \cdot b)^{-1} = (a^{-1}) \cdot (b^{-1})$.*

The proof of this theorem is like the proof of Theorem 2 with addition replaced by multiplication, 0 replaced by 1, and $(-a)$, $(-b)$ replaced by a^{-1}, b^{-1}. The details are left to the student. We note that if $a \neq 0$, then $a^{-1} \neq 0$ because $aa^{-1} = 1$ and $1 \neq 0$. Then Theorem 5 (ii) may be used with $b = a^{-1}$.

Using Theorem 3 and the distributive law, we easily prove the **laws of signs** stated as Theorem 7 below. We emphasize that the numbers a and b may be positive, negative, or zero.

Theorem 7. *If a and b are any numbers, then* (i) $a \cdot (-b) = -(a \cdot b)$, (ii) $(-a) \cdot b = -(a \cdot b)$, (iii) $(-a) \cdot (-b) = a \cdot b$.

Proof. i) Since $b + (-b) = 0$, we find from the distributive law that
$$a[b + (-b)] = a \cdot b + a \cdot (-b) = 0.$$
On the other hand, the negative of $a \cdot b$ has the property:
$$a \cdot b + [-(a \cdot b)] = 0.$$
Hence it follows from Axiom A–5 that $a \cdot (-b) = -(a \cdot b)$. Part (ii) follows from (i) by interchanging a and b. The proof of (iii) is left to the student.

Corollary. *We have $(-1) \cdot a = -a$.*

We now show that the **laws of fractions,** as given in elementary algebra courses, follow from the axioms and theorems above.

Notation. We introduce the following symbols for $a \cdot b^{-1}$:
$$a \cdot b^{-1} = \frac{a}{b} = a/b = a \div b.$$

These symbols, representing an indicated division, are called **fractions.** The **numerator** and **denominator** of a fraction are defined as usual. A fraction with *denominator* zero has no meaning.

Theorem 8. i) *For every number a, we have $a/1 = a$.* (ii) *If $a \neq 0$, then $a/a = 1$.*

Proof. i) We have $a/1 = a \cdot (1)^{-1} = a \cdot 1 = a$.
ii) If $a \neq 0$, then $a/a = a \cdot a^{-1} = 1$, by definition.

Theorem 9. *If $b \neq 0$ and $d \neq 0$, then*
$$\left(\frac{a}{b}\right) \cdot \left(\frac{c}{d}\right) = \frac{a \cdot c}{b \cdot d}.$$

Proof. Using the notation for fractions, Lemma 4, and Theorem 6(ii), we find

$$\left(\frac{a}{b}\right) \cdot \left(\frac{c}{d}\right) = (a \cdot b^{-1}) \cdot (cd^{-1}) = (a \cdot c) \cdot (b^{-1} d^{-1})$$

$$= (a \cdot c)(bd)^{-1} = \frac{a \cdot c}{b \cdot d}.$$

The proofs of Theorems 10 through 14 are left to the student.

Theorem 10. *If $b \neq 0$ and $c \neq 0$ then*

$$\frac{a}{b} = \frac{a \cdot c}{b \cdot c}.$$

Theorem 11. *If $c \neq 0$, then*

$$\frac{a}{c} + \frac{b}{c} = \frac{a + b}{c}.$$

Theorem 12. *If $b \neq 0$, then $-b \neq 0$ and*

$$\frac{-a}{b} = \frac{a}{-b} = -\left(\frac{a}{b}\right).$$

Theorem 13. *If $b \neq 0$, $c \neq 0$, and $d \neq 0$, then $(c/d) \neq 0$ and*

$$\frac{a/b}{c/d} = \frac{a \cdot d}{b \cdot c} = \left(\frac{a}{b}\right) \cdot \left(\frac{d}{c}\right).$$

Theorem 14. *If $b \neq 0$ and $d \neq 0$, then*

$$\frac{a}{b} + \frac{c}{d} = \frac{ad + bc}{bd}.$$

PROBLEMS

1. Prove Lemma 1.
2. Prove Theorem 4.
3. Prove Lemma 3.
4. Prove Lemma 4.
5. Prove Theorem 6.
6. Complete the proof of Theorem 7.
7. Prove that $a(b + c + d) = ab + ac + ad$.
8. Assuming that $A + B + C + D$ means $(A + B + C) + D$, prove that $A + B + C + D = (A + B) + (C + D)$.
9. Assuming the result of Problem 8, prove that

$$(a + b) \cdot (c + d) = ac + bc + ad + bd.$$

10. Prove Theorem 10.
11. Prove Theorem 11.
12. Prove Theorem 12.
13. Prove Theorem 14.
14. Prove Theorem 13. [*Hint:* Use Theorem 10 appropriately.]
15. Prove that if $b \neq 0$, $d \neq 0$, $f \neq 0$, then

$$\left(\frac{a}{b}\right) \cdot \left(\frac{c}{d}\right) \cdot \left(\frac{e}{f}\right) = \frac{a \cdot c \cdot e}{b \cdot d \cdot f}.$$

2. NATURAL NUMBERS. SEQUENCES

Historically, the present-day concept of the real number system was built up by a sequence of successive enlargements. To begin with, the positive integers were extended to include positive rational numbers (quotients, or ratios, of integers); these were enlarged to include all the positive real numbers, and finally, all real numbers are obtained by the inclusion of negative numbers.

Since the system of axioms in Section 1 does not distinguish (or even mention) positive numbers, we shall discuss first the **natural numbers.** As we know, these turn out to be nothing but the positive integers.

Intuitively, the totality of natural numbers can be obtained by starting with 1 and then forming $1 + 1$, $(1 + 1) + 1$, $[(1 + 1) + 1] + 1$, and so on. We call $1 + 1$ the number 2, $(1 + 1) + 1$ is called 3, and in this way the collection of natural numbers is generated. Actually, it is possible to give a more logically satisfactory (but rather abstract) definition of a natural number which yields the same set. This process is frequently carried out in advanced mathematics courses. If we add to our system of axioms the **axiom of inequality,**[*] then all the usual properties of natural numbers can be proved. Since it would take us too far afield to establish all the necessary results, we shall just assume that the various applications to counting are known and continue from that point.

In Section 1 we defined the sum and product of three numbers. We wish now to indicate extensions of these notions to more terms. For our purposes, it is sufficient to think of a **sequence** as a set of numbers arranged in a specific order. For example, the numbers

$$1, \quad 3, \quad -2, \quad 1, \quad \pi, \quad 5, \quad \sqrt{2}$$

form a sequence. The numbers in the sequence, separated by commas, are called its **terms.** The reader is undoubtedly familiar with arithmetic and geometric progressions which are examples of sequences. A sequence may be **finite,** in which case there is a first and a last term, or a sequence may be **infinite,** in which case the terms continue without terminating. We shall indicate sequences by notations such as

$$a_1, a_2, \ldots, a_n \quad \text{or} \quad a_1, a_2, \ldots, a_n, \ldots,$$

---

[*] The axiom of inequality is simply the following: *Among the real numbers is a set called the* **positive numbers** *which satisfies the two conditions:* (i) *for any number a, exactly one of the following three alternatives holds: a is positive, $a = 0$, or $-a$ is positive;* (ii) *any sum or product of positive numbers is positive.*

the first sequence being finite and having n terms; the second sequence is infinite with the dots indicating the intervening and the succeeding terms. When $n = 1$ the sequence indicated has only one term a_1. In our notation a_i denotes the **ith term.**

We define the sum and product of a sequence $a_1, a_2, \ldots, a_n$ as follows: If $n = 2$, the sum is $a_1 + a_2$ and the product is $a_1 \cdot a_2$, as usual. For $n > 2$, we have

$$a_1 + a_2 + a_3 = (a_1 + a_2) + a_3,$$
$$a_1 + a_2 + a_3 + a_4 = (a_1 + a_2 + a_3) + a_4,$$

and so on. A similar definition may be given for products.

It is possible to prove the **Extended Commutative Law** for sums and products which states that:

The sum and product of a given finite sequence are independent of the order of its terms.

This has been proved for $n = 3$ in Lemma 1, Section 1. The associative law for sums and products may be extended as follows:

The sum of a finite sequence can be obtained by separating the given sequence into several shorter sequences, adding the terms in each of these and then adding the results; a similar statement holds for products.

For example,

$$1 + 5 + 3 + 7 + 4 + 1 + 6 = (1 + 5 + 3) + (7 + 4) + (1 + 6).$$

Finally, the distributive law may be extended so that

$$a \cdot (b_1 + \cdots + b_n) = (b_1 + \cdots + b_n) \cdot a = ab_1 + \cdots + ab_n.$$

This was proved for $n = 3$ in Problem 7, Section 1. This in turn leads to the general rule for multiplying two sums:

To multiply two sums, multiply each term of one sum by each term of the other and add all the results.

See Problem 9, Section 1 and the following example:

$$(a_1 + a_2)(b_1 + b_2 + b_3) = a_1(b_1 + b_2 + b_3) + a_2(b_1 + b_2 + b_3)$$
$$= a_1b_1 + a_1b_2 + a_1b_3 + a_2b_1 + a_2b_2 + a_2b_3.$$

We understand the signed sum

$$a - b - c + d - e,$$

for example, to mean

$$a + (-b) + (-c) + d + (-e).$$

Theorem 2(ii) (the negative of a sum is the sum of the negatives) can be generalized to sequences of any finite number of terms. Using Theorem 2(i) $[-(-a) = a]$, we obtain the usual rule for the negative of a signed sum. Then, since $A - B = A + (-B)$ by definition, we obtain the usual rule for subtraction of signed sums. With the aid of the sign laws for multiplication and the extended distributive law, we get the usual rules for the multiplication of signed sums. The symbol x^n, for n a natural number, is defined as usual, and the familiar laws of exponents follow. From these and the preceding rules follow the established rules for adding, subtracting, and multiplying polynomials. The validity of these rules, as well as those for the division and factoring of polynomials, will be assumed in the discussion below.

Finally, it can be shown that any natural number n has one and only one representation of the form

$$n = d_0 \cdot 10^k + d_1 \cdot 10^{k-1} + \cdots + d_k$$

in which each d_i is 0 or a natural number less than or equal to 9 (i.e., a **digit**) and $d_0 \neq 0$. By verifying the rules for addition and multiplication of digits, we can then derive the rules for the addition and multiplication of natural numbers from the corresponding rules for polynomials (with $x = 10$). We shall assume these rules.

We now define some frequently used terms.

Definitions. *A real number is an* **integer** *if and only if it is either zero, a natural number, or the negative of a natural number. A real number r is* **rational** *if and only if there are integers p and q, with $q \neq 0$, such that $r = p/q$.*

It is clear that the sum and product of a finite sequence of either integers or rational numbers is again an integer or a rational number, respectively, and that the quotient of two rational numbers is again rational.

We now sketch briefly how the general rules apply in specific cases with which the student is familiar. We use the customary notations and assume the laws of exponents for exponents which are positive integers.

Example 1. Add: $2x^3 - 3x^3 + 5x^3$.

Solution

$$2x^3 - 3x^3 + 5x^3 = 2x^3 + (-3)x^3 + 5x^3$$
$$= [2 + (-3) + 5] \cdot x^3 = 4x^3 \qquad \text{(distributive law)}. \quad \blacktriangleleft$$

Example 2. Multiply: $3x^3 \cdot (-5x^5)$.

Solution
$$3x^3 \cdot (-5x^5) = 3 \cdot x^3 \cdot (-5) \cdot x^5 \qquad \text{(extended commutative and associative laws)}$$
$$= 3 \cdot (-5) \cdot x^3 \cdot x^5 = [3 \cdot (-5)] \cdot (x^3 \cdot x^5) = (-15) \cdot x^8$$
$$= -15x^8. \qquad \blacktriangleleft$$

Example 3. Carry out the indicated operation:
$$(2x^3 - 3x^2 + x + 2) - (-x^3 - x^2 + 2x + 1).$$

Solution

$$2x^3 - 3x^2 + x + 2 - (-x^3 - x^2 + 2x + 1)$$
$$= 2x^3 - 3x^2 + x + 2 + [-(-x^3 - x^2 + 2x + 1)] \quad [A - B = A + (-B)]$$
$$= 2x^3 - 3x^2 + x + 2 + [x^3 + x^2 - 2x - 1] \quad \text{[negative of a sum is the sum}$$
$$\text{of the negatives; } -(-a) = a]$$
$$= [2x^3 + x^3] + [(-3)x^2 + x^2] + [x + (-2)x] + [2 + (-1)]$$
$$= 3x^3 - 2x^2 - x + 1. \quad \blacktriangleleft$$

Example 4. Multiply: $(2x^2 - x - 3) \cdot (x^3 - 2x^2 + 3x - 4)$.

Solution. Each term of one sum must be multiplied by each term of the other. In order to ensure this, we arrange the work (essentially) as in multiplication in arithmetic:

$$\begin{array}{r} x^3 + (-2)x^2 + 3x + (-4) \\ 2x^2 + (-1)x + (-3) \\ \hline 2x^5 + (-4)x^4 + 6x^3 + (-8)x^2 \\ (-1)x^4 + 2x^3 + (-3)x^2 + 4x \\ (-3)x^3 + 6x^2 + (-9)x + 12 \\ \hline 2x^5 - 5x^4 + 5x^3 - 5x^2 - 5x + 12 \quad \text{(Ans.)} \end{array}$$

$\blacktriangleleft$

Theorem 5 ($a \cdot b = 0$, etc.) can be extended to products of more than two terms. Then the rule for the multiplication of fractions (Theorem 9) can be extended as follows:

$$\frac{a_1}{b_1} \cdot \frac{a_2}{b_2} \cdots \frac{a_n}{b_n} = \frac{a_1 \cdot a_2 \cdots a_n}{b_1 \cdot b_2 \cdots b_n}.$$

Theorems 11 and 12, together with the definition of signed sums, etc., yield the following rule:

A sum of a finite sequence of fractions having a common denominator is equal to a single fraction of which the denominator is the common denominator and the numerator is the corresponding sum of the numerators.

When one adds only two fractions, Theorem 14 is useful. However, when more fractions are involved, it is best to find the least common denominator of all the fractions, express each fraction as a fraction with that denominator, and then add. We illustrate the procedure, giving a reason for each step and noting values which make a *denominator* zero: these must be excluded.

Any set of objects satisfying the axioms of Section 1 is called a **field.** The axioms given there imply only theorems concerned with the operations of addition, subtraction, multiplication, and division. They do not imply the *existence* of a number whose square is 2, for instance, since the set of all rational numbers is a field and there is no rational number whose square is 2.

PROBLEMS

In Problems 1 through 4, subtract the second polynomial from the first. Put in enough steps to indicate the fundamental principles being used.

1. $x^2 - 2xy + 3y^2$; $2x^2 - 3xy - y^2$
2. $x^3 + x - 2x^2 - 2$; $x^2 - x^3 + 4$
3. $4x^3 - 3x^2 - 2$; $x^2 - 3x - 3$
4. $4xy - 3x^2y - 2y^2 - 4x - 2$; $7 - 2x^2y - xy - 2y^2$

In Problems 5 through 8, multiply the two polynomials.

5. $2x - 3$; $x^3 - 2x^2 - 3x + 1$
6. $x^2 - x - 2$; $2x^3 - x + 1$
7. $2x^2 - x + 3$; $x^3 - 2x^2 - 3x + 3$
8. $x^3 - 3x^2 - 2x + 4$; $2x^4 - x^3 + 3x^2 - 1$

Reduce each of the following expressions to a single fraction in its lowest terms. Note excluded values and give a reason for each step.

9. $\dfrac{2}{a-b} + \dfrac{3}{b-a} + \dfrac{a}{a^2-b^2}$, $a \neq \pm b$

10. $\dfrac{x+y}{(y-z)(y-x)} + \dfrac{y+z}{(z-x)(z-y)} + \dfrac{z+x}{(x-y)(x-z)}$

11. $\dfrac{x+1}{2(x+4)} - \dfrac{4x-1}{3x-3} + \dfrac{7}{6}$

12. $\dfrac{x^2-x-2}{x^2+x-6} - \dfrac{2}{9-x^2} + 1 - \dfrac{5}{3-x}$

13. $\dfrac{x^2+6x+5}{x^2-1} \div \dfrac{x^2+3x+2}{x^3-1}$

14. $\dfrac{3x^2+8x+4}{2x^2+7x+3} \cdot \dfrac{x^2+2x-3}{9x^2+12x+4} \cdot \dfrac{2x^2-x-1}{3x^2-x-2}$

15. $\dfrac{\left(\dfrac{2x+y}{x+y} - 1\right)}{\left(1 - \dfrac{y}{x+y}\right)}$

16. $\dfrac{\left(1 - \dfrac{2}{a} + \dfrac{1}{a^2}\right)}{\left(1 - \dfrac{1}{a^2}\right)}$

17. $\dfrac{\left(\dfrac{a^2+b^2}{a^2-b^2} - \dfrac{a^2-b^2}{a^2+b^2}\right)}{\left(\dfrac{a+b}{a-b} - \dfrac{a-b}{a+b}\right)}$

APPENDIX 5

THEOREMS ON DETERMINANTS

DETERMINANTS OF SECOND AND THIRD ORDER

In this appendix we present an introduction to determinants of the second and third order. A determinant of the second order is denoted by

$$\begin{vmatrix} a & b \\ c & d \end{vmatrix},$$

in which a, b, c, and d are numbers. A determinant of the third order is denoted by

$$\begin{vmatrix} a & b & c \\ d & e & f \\ g & h & j \end{vmatrix},$$

in which $a, b, \ldots, j$ are numbers.

Determinants are actually functions of the elements which appear. That is, a second-order determinant is a function of the four quantities a, b, c, and d, while a third-order determinant is a function of the nine entries $a, b, c, \ldots, j$. Using the terminology of variables, we say that a second-order determinant is a function of four variables and a third-order determinant is a function of nine variables.

For a second-order determinant, we define

$$\begin{vmatrix} a & b \\ c & d \end{vmatrix} = ad - bc.$$

For example,

$$\begin{vmatrix} 2 & 1 \\ -5 & 6 \end{vmatrix} = 12 - (-5) = 17.$$

A third-order determinant is defined in terms of second-order determinants by the formula:

$$\begin{vmatrix} a & b & c \\ d & e & f \\ g & h & j \end{vmatrix} = c \begin{vmatrix} d & e \\ g & h \end{vmatrix} - f \begin{vmatrix} a & b \\ g & h \end{vmatrix} + j \begin{vmatrix} a & b \\ d & e \end{vmatrix}.$$

This formula is known as an **expansion in terms of the last column.** For example,

$$\begin{vmatrix} 2 & 1 & -1 \\ 0 & 3 & 2 \\ 1 & 4 & 3 \end{vmatrix} = -1 \begin{vmatrix} 0 & 3 \\ 1 & 4 \end{vmatrix} - 2 \begin{vmatrix} 2 & 1 \\ 1 & 4 \end{vmatrix} + 3 \begin{vmatrix} 2 & 1 \\ 0 & 3 \end{vmatrix}$$

$$= -(-3) - 2(8 - 1) + 3(6) = 7.$$

Theorem 1. **(Cramer's Rule for $n = 2$).** *Given the two equations*

$$\left. \begin{array}{c} ax + by = e, \\ cx + dy = f, \end{array} \right\} \tag{1}$$

if the determinant of the coefficients

$$\begin{vmatrix} a & b \\ c & d \end{vmatrix}$$

is not zero, then the two simultaneous equations (1) *have a unique solution given by*

$$x = \frac{\begin{vmatrix} e & b \\ f & d \end{vmatrix}}{\begin{vmatrix} a & b \\ c & d \end{vmatrix}}, \qquad y = \frac{\begin{vmatrix} a & e \\ c & f \end{vmatrix}}{\begin{vmatrix} a & b \\ c & d \end{vmatrix}}.$$

The proof is left to the reader.

Theorem 2. *Two sets of numbers $\{a_1, b_1, c_1\}$ and $\{a_2, b_2, c_2\}$, neither all zero, are proportional if and only if all three determinants*

$$\begin{vmatrix} a_1 & b_1 \\ a_2 & b_2 \end{vmatrix}, \qquad \begin{vmatrix} a_1 & c_1 \\ a_2 & c_2 \end{vmatrix}, \qquad \begin{vmatrix} b_1 & c_1 \\ b_2 & c_2 \end{vmatrix}$$

are zero.

Proof. If the sets are proportional, then there is a number k such that

$$a_2 = ka_1, \qquad b_2 = kb_1, \qquad c_2 = kc_1. \tag{2}$$

It is immediate that the three determinants vanish. Now suppose all determinants vanish. We wish to prove the sets are proportional. Suppose $c_1 \neq 0$, for instance. Then we have

$$\begin{vmatrix} a_1 & c_1 \\ a_2 & c_2 \end{vmatrix} = a_1 c_2 - a_2 c_1 = 0 \quad \Leftrightarrow \quad a_2 = \left(\frac{c_2}{c_1}\right) a_1,$$

$$\begin{vmatrix} b_1 & c_1 \\ b_2 & c_2 \end{vmatrix} = b_1 c_2 - b_2 c_1 = 0 \quad \Leftrightarrow \quad b_2 = \left(\frac{c_2}{c_1}\right) b_1.$$

We define $k = c_2/c_1$ and we see that (2) holds. If $c_1 = 0$, then either $a_1 \neq 0$ or $b_1 \neq 0$ and a similar argument leads to the same conclusion. ◀

It is clear that the value of a determinant depends on the arrangement of the elements which enter it. It is simpler to state theorems about determinants if we introduce a systematic notation for the entries. The standard method consists of a double subscript for each element, the first subscript identifying the row in which the element is situated, and the second subscript identifying the column. That is, we denote the element in the ith row and jth column by the symbol a_{ij}. The evaluation of a third-order determinant may now be written

$$\begin{vmatrix} a_{11} & a_{12} & a_{13} \\ a_{21} & a_{22} & a_{23} \\ a_{31} & a_{32} & a_{33} \end{vmatrix} = a_{13} \begin{vmatrix} a_{21} & a_{22} \\ a_{31} & a_{32} \end{vmatrix} - a_{23} \begin{vmatrix} a_{11} & a_{12} \\ a_{31} & a_{32} \end{vmatrix} + a_{33} \begin{vmatrix} a_{11} & a_{12} \\ a_{21} & a_{22} \end{vmatrix}.$$

Using the definition of second-order determinants, we obtain for the value of a third-order determinant:

$$\begin{vmatrix} a_{11} & a_{12} & a_{13} \\ a_{21} & a_{22} & a_{23} \\ a_{31} & a_{32} & a_{33} \end{vmatrix} = a_{11}a_{22}a_{33} + a_{12}a_{31}a_{23} + a_{13}a_{21}a_{32} \\ - a_{11}a_{32}a_{23} - a_{12}a_{21}a_{33} - a_{13}a_{22}a_{31}. \tag{3}$$

It is a simple matter to verify that the interchange of each a_{ij} with a_{ji} in the righthand side of (3) does not alter it in any way. Therefore we conclude the following result for third-order determinants.

Theorem 3. *If a new determinant is obtained from a given one by an interchange of its rows and columns, the value of the new determinant is equal to the value of the given one.*

Definitions. *Given a determinant*

$$D \equiv \begin{vmatrix} a_{11} & a_{12} & a_{13} \\ a_{21} & a_{22} & a_{23} \\ a_{31} & a_{32} & a_{33} \end{vmatrix}, \tag{4}$$

we define the quantity D_{ij} as the second-order determinant obtained from D by crossing out the ith row and jth column in D. For example,

$$D_{21} = \begin{vmatrix} a_{12} & a_{13} \\ a_{32} & a_{33} \end{vmatrix}.$$

The **cofactor** of the element a_{ij} in D is defined to be the number A_{ij} given by

$$A_{ij} = (-1)^{i+j} D_{ij}.$$

For example, the cofactor of a_{13} is

$$A_{13} = (-1)^{1+3} D_{13} = (+1) \begin{vmatrix} a_{21} & a_{22} \\ a_{31} & a_{32} \end{vmatrix} = a_{21}a_{32} - a_{31}a_{22}.$$

Theorem 4. *If D denotes the determinant in (4), we have*

$$D = a_{11}A_{11} + a_{12}A_{12} + a_{13}A_{13} = a_{11}A_{11} + a_{21}A_{21} + a_{31}A_{31}, \tag{5a}$$
$$D = a_{21}A_{21} + a_{22}A_{22} + a_{23}A_{23} = a_{12}A_{12} + a_{22}A_{22} + a_{32}A_{32}, \tag{5b}$$
$$D = a_{31}A_{31} + a_{32}A_{32} + a_{33}A_{33} = a_{13}A_{13} + a_{23}A_{23} + a_{33}A_{33}. \tag{5c}$$

Proof. We verify the first formula on the right:

$$a_{11}A_{11} + a_{21}A_{21} + a_{31}A_{31}$$
$$= a_{11}\begin{vmatrix} a_{22} & a_{23} \\ a_{32} & a_{33} \end{vmatrix} - a_{21}\begin{vmatrix} a_{12} & a_{13} \\ a_{32} & a_{33} \end{vmatrix} + a_{31}\begin{vmatrix} a_{12} & a_{13} \\ a_{22} & a_{23} \end{vmatrix}$$
$$= a_{11}(a_{22}a_{33} - a_{32}a_{23}) - a_{21}(a_{12}a_{33} - a_{32}a_{13}) + a_{31}(a_{12}a_{23} - a_{22}a_{13})$$
$$= D. \qquad \blacktriangleleft$$

Theorem 3 and a computation may be used to verify the remaining formulas in (5). The details are left to the student.

Definitions. *The first sums in (5a), (5b), and (5c) are called the* **expansions of** D **according to the first, second, and third rows,** *respectively. The second sums in (5a), (5b), and (5c) are called the* **expansions of** D **according to the first, second, and third columns,** *respectively.*

Example 1. Evaluate the following determinant by expanding it according to (i) the second column, and (ii) the third row:

$$\begin{vmatrix} 1 & -2 & 3 \\ 2 & 1 & -1 \\ -2 & -1 & 2 \end{vmatrix}.$$

Solution. i) Expanding according to the second column, we obtain

$$\begin{vmatrix} 1 & -2 & 3 \\ 2 & 1 & -1 \\ -2 & -1 & 2 \end{vmatrix} = -(-2)\begin{vmatrix} 2 & -1 \\ -2 & 2 \end{vmatrix} + 1 \cdot \begin{vmatrix} 1 & 3 \\ -2 & 2 \end{vmatrix} - (-1)\begin{vmatrix} 1 & 3 \\ 2 & -1 \end{vmatrix}$$
$$= 2(4 - 2) + (2 + 6) + (-1 - 6) = 5.$$

ii) Expanding according to the third row, we obtain

$$\begin{vmatrix} 1 & -2 & 3 \\ 2 & 1 & -1 \\ -2 & -1 & 2 \end{vmatrix} = (-2)\begin{vmatrix} -2 & 3 \\ 1 & -1 \end{vmatrix} - (-1)\begin{vmatrix} 1 & 3 \\ 2 & -1 \end{vmatrix} + 2\begin{vmatrix} 1 & -2 \\ 2 & 1 \end{vmatrix}$$
$$= (-2)(-1) + 1 \cdot (-7) + 2 \cdot 5 = 5. \qquad \blacktriangleleft$$

The definitions and theorems given for determinants of the second and third order may be extended to determinants of order n, where n is any positive integer. A determinant of order n has n rows and n columns. The evaluation of such determinants, a process beyond the scope of this appendix, is most easily accomplished by an induction technique. We now state a number of general properties of determinants, valid for any order, which are most useful in such evaluations. The proofs depend on the expansion given in Theorem 4 which we have proved only for $n = 3$. Actually the appropriate extension of Theorem 4 is valid for determinants of any order.

Theorem 5. *Let D be a determinant and suppose that each element of the kth row a_{kj} is the sum of two numbers: $a_{kj} = a'_{kj} + a''_{kj}$. Then $D = D' + D''$ where D' is the* determinant obtained from D by inserting a'_{kj} instead of a_{kj} for each element of the kth row. Similarly, D'' is obtained by inserting a''_{kj} for a_{kj}. The result also holds if the elements of a column are treated analogously.

Proof. We expand D according to the elements of the kth row:

$$\begin{aligned} D &= a_{k1}A_{k1} + a_{k2}A_{k2} + a_{k3}A_{k3} \\ &= (a'_{k1} + a''_{k1})A_{k1} + (a'_{k2} + a''_{k2})A_{k2} + (a'_{k3} + a''_{k3})A_{k3} \\ &= (a'_{k1}A_{k1} + a'_{k2}A_{k2} + a'_{k3}A_{k3}) + (a''_{k1}A_{k1} + a''_{k2}A_{k2} + a''_{k3}A_{k3}) \\ &= D' + D''. \end{aligned}$$
◄

Theorem 4 shows that the same analysis works for columns.

Theorem 5 shows that, for example,

$$\begin{vmatrix} a_{11} & a_{12} & a_{13} \\ a_{21} & a_{22} & a_{23} \\ a'_{31} + a''_{31} & a'_{32} + a''_{32} & a'_{33} + a''_{33} \end{vmatrix} = \begin{vmatrix} a_{11} & a_{12} & a_{13} \\ a_{21} & a_{22} & a_{23} \\ a'_{31} & a'_{32} & a'_{33} \end{vmatrix} + \begin{vmatrix} a_{11} & a_{12} & a_{13} \\ a_{21} & a_{22} & a_{23} \\ a''_{31} & a''_{32} & a''_{33} \end{vmatrix}.$$

Theorem 6. *If each element $a_{kj} = ca'_{kj}$ for k fixed and $j = 1, 2,$ and 3, then $D = cD'$ where D' is obtained from D as in Theorem 5.*

The proof is similar to that of Theorem 5. In words, Theorem 6 states that if all the elements of a row are multiplied by a constant, then the determinant is multiplied by that constant.

Theorem 7. *If D' is obtained from D by interchanging any two rows, then $D' = -D$. The same result holds for the interchange of two columns.*

Proof. The result is a direct consequence of formula (3). The details are left to the student.

Theorem 8. *If any two rows (or columns) of a determinant D are proportional, then $D = 0$.*

Proof. If two rows are identical, then interchanging them has no effect. On the other hand, Theorem 7 shows that $D' = -D$. Thus $D' = -D = D$ and $D = 0$. If two rows are proportional, one is c times the other, and by Theorem 6, $D = cD''$ where D'' has two identical rows. Hence $D = 0$. The proof for columns is similar.
◄

The next lemma and theorem are very useful in evaluating determinants, especially those of high order.

Lemma 1. *If D' is obtained from D by multiplying the kth row by the constant c and adding the result to the ith row, where $i \neq k$, then $D' = D$. The same result holds for columns.*

Proof. The elements of the ith row of D', denoted by a'_{ij}, have the form

$$a'_{ij} = a_{ij} + ca_{kj}.$$

Using Theorem 5, we find
$$D' = D + cD''$$
where D'' is obtained by replacing the ith row of D by the kth row. But then D'' has two identical rows (the ith and kth), so that $D'' = 0$. Hence
$$D' = D.$$

Theorem 9. *If D' is obtained from D by multiplying the kth row by c_i and adding the result to the ith row for all $i \neq k$ in turn, then $D' = D$. The same is true for columns.*

Proof. Each step in the process is one for which the lemma applies leaving the value unchanged. ◀

The next example shows how to use Theorem 9 to simplify the determinant before applying the expansion theorem.

Example 2. Simplify by using Theorem 9 and use the expansion theorem to evaluate the determinant
$$D = \begin{vmatrix} 2 & 1 & 3 \\ -1 & 2 & 2 \\ 2 & -3 & 1 \end{vmatrix}.$$

Solution. Multiplying the first row by -2 and adding to the second row, and then multiplying it by 3 and adding to the third row, we obtain
$$D = \begin{vmatrix} 2 & 1 & 3 \\ -5 & 0 & -4 \\ 8 & 0 & 10 \end{vmatrix}.$$
We can expand the new determinant according to the second column, thus getting
$$D = -1 \cdot \begin{vmatrix} -5 & -4 \\ 8 & 10 \end{vmatrix} = (-1) \cdot (-18) = 18. \quad ◀$$

The following theorem includes and supplements the expansion theorem (Theorem 4).

Theorem 10. *If D is any third-order determinant, then*

a) $a_{i1}A_{k1} + a_{i2}A_{k2} + a_{i3}A_{k3} = \begin{cases} D & \text{if } k = i, \\ 0 & \text{if } k \neq i, \end{cases}$

b) $a_{1j}A_{1k} + a_{2j}A_{2k} + a_{3j}A_{3k} = \begin{cases} D & \text{if } k = j, \\ 0 & \text{if } k \neq j. \end{cases}$

Proof. The cases $k = i$ in (a) and $k = j$ in (b) restate Theorem 4. If $k \neq i$ in (a), then we see from the expansion theorem that the left-hand side in (a) is the expansion of a determinant D' obtained from D by replacing the kth row by the ith row. But then D' has two identical rows and so is zero. The proof of (b) is the same.

Theorem 11. (Cramer's Rule for $n = 3$). *If the determinant*

$$D = \begin{vmatrix} a_{11} & a_{12} & a_{13} \\ a_{21} & a_{22} & a_{23} \\ a_{31} & a_{32} & a_{33} \end{vmatrix}$$

of the coefficients of the three equations

$$\left. \begin{aligned} a_{11}x + a_{12}y + a_{13}z &= b_1, \\ a_{21}x + a_{22}y + a_{23}z &= b_2, \\ a_{31}x + a_{32}y + a_{33}z &= b_3, \end{aligned} \right\} \quad (6)$$

is not zero, then the equations have one and only one solution given by

$$x = \frac{D_1}{D}, \quad y = \frac{D_2}{D}, \quad z = \frac{D_3}{D}, \quad (7)$$

where

$$D_1 = \begin{vmatrix} b_1 & a_{12} & a_{13} \\ b_2 & a_{22} & a_{23} \\ b_3 & a_{32} & a_{33} \end{vmatrix}, \quad D_2 = \begin{vmatrix} a_{11} & b_1 & a_{13} \\ a_{21} & b_2 & a_{23} \\ a_{31} & b_3 & a_{33} \end{vmatrix}, \quad D_3 = \begin{vmatrix} a_{11} & a_{12} & b_1 \\ a_{21} & a_{22} & b_2 \\ a_{31} & a_{32} & b_3 \end{vmatrix}.$$

Proof. a) Suppose the equations (6) hold. If we multiply the first, second, and third equations by A_{11}, A_{21}, and A_{31}, respectively, and then add, we get

$$Dx + (a_{12}A_{11} + a_{22}A_{21} + a_{32}A_{31})y + (a_{13}A_{11} + a_{23}A_{21} + a_{33}A_{31})z$$
$$= (b_1 A_{11} + b_2 A_{21} + b_3 A_{31}). \quad (8)$$

By Theorem 10, the coefficients of y and z are zero. The right-hand side of the above equation is the expansion of D_1 according to its first column. Hence $x = D_1/D$. Multiplying the equations (6) by A_{12}, A_{22}, A_{32} and proceeding similarly, we get the result $y = D_2/D$. The value for z is obtained in the same way.

b) Now assume that equations (7) hold and that $D \neq 0$. Then (8) holds; similarly for equations for y and z. Therefore, by addition,

$$D(a_{11}x + a_{12}y + a_{13}z) = b_1(a_{11}A_{11} + a_{12}A_{12} + a_{13}A_{13})$$
$$+ b_2(a_{11}A_{21} + a_{12}A_{22} + a_{13}A_{23})$$
$$+ b_3(a_{11}A_{31} + a_{12}A_{32} + a_{13}A_{33})$$
$$= Db_1.$$

The last equality is valid because of Theorem 10. In a similar way, we see that the remaining equations of (6) are satisfied.

Example 3. Use Cramer's rule to solve the following equations:

$$\begin{aligned} 3x - 2y + 4z &= 5, \\ x + y + 3z &= 2, \\ -x + 2y - z &= 1. \end{aligned}$$

Solution. To simplify and evaluate D, we multiply the second row by 2 and add it to the first row and then multiply the second row by -2 and add it to the third row. We obtain

$$D = \begin{vmatrix} 3 & -2 & 4 \\ 1 & 1 & 3 \\ -1 & 2 & -1 \end{vmatrix} = \begin{vmatrix} 5 & 0 & 10 \\ 1 & 1 & 3 \\ -3 & 0 & -7 \end{vmatrix}.$$

Now expanding according to the second column is easy. We get

$$D = 1 \cdot \begin{vmatrix} 5 & 10 \\ -3 & -7 \end{vmatrix} = -5.$$

Proceeding similarly for D_1, D_2, and D_3, we find

$$D_1 = \begin{vmatrix} 5 & -2 & 4 \\ 2 & 1 & 3 \\ 1 & 2 & -1 \end{vmatrix} = \begin{vmatrix} 9 & 0 & 10 \\ 2 & 1 & 3 \\ -3 & 0 & -7 \end{vmatrix} = 1 \cdot \begin{vmatrix} 9 & 10 \\ -3 & -7 \end{vmatrix} = -33,$$

$$D_2 = \begin{vmatrix} 3 & 5 & 4 \\ 1 & 2 & 3 \\ -1 & 1 & -1 \end{vmatrix} = \begin{vmatrix} 8 & 5 & 9 \\ 3 & 2 & 5 \\ 0 & 1 & 0 \end{vmatrix} = -1 \cdot \begin{vmatrix} 8 & 9 \\ 3 & 5 \end{vmatrix} = -13,$$

$$D_3 = \begin{vmatrix} 3 & -2 & 5 \\ 1 & 1 & 2 \\ -1 & 2 & 1 \end{vmatrix} = \begin{vmatrix} 5 & -2 & 9 \\ 0 & 1 & 0 \\ -3 & 2 & -3 \end{vmatrix} = 1 \cdot \begin{vmatrix} 5 & 9 \\ -3 & -3 \end{vmatrix} = 12.$$

Therefore $x = \frac{33}{5}$, $y = \frac{13}{5}$, $z = -\frac{12}{5}$. ◂

Corollary. *If there are numbers x, y, and z, not all zero, which satisfy Eqs. (6) with $b_1 = b_2 = b_3 = 0$, then $D = 0$.*

PROBLEMS

1. Prove Theorem 1.
2. Complete the proof of Theorem 4.

In Problems 3 through 5, evaluate the given determinant by expanding it according to (a) the second row, (b) the third column.

3. $\begin{vmatrix} 3 & 2 & -1 \\ -1 & 0 & 1 \\ 2 & 1 & -2 \end{vmatrix}$ 4. $\begin{vmatrix} 1 & 0 & 3 \\ 2 & -1 & -2 \\ 1 & 3 & 2 \end{vmatrix}$ 5. $\begin{vmatrix} 2 & 3 & 1 \\ 1 & 2 & -2 \\ -2 & 1 & 3 \end{vmatrix}$

In Problems 6 through 8, simplify the determinant, using Theorem 9, and then evaluate it by the expansion theorem.

6. $\begin{vmatrix} 2 & -1 & 3 \\ 3 & -1 & 2 \\ -1 & 2 & 3 \end{vmatrix}$ 7. $\begin{vmatrix} 1 & -1 & -2 \\ 2 & 3 & -2 \\ -1 & -3 & 2 \end{vmatrix}$ 8. $\begin{vmatrix} 2 & 1 & -2 \\ -1 & 3 & 2 \\ -3 & 1 & -2 \end{vmatrix}$

In Problems 9 through 11, solve by Cramer's Rule and check.

9. $2x - y + 3z = 1$
   $3x + y - z = 2$
   $x + 2y + 3z = -6$

10. $2x - y + z = -3$
    $x + 3y - 2z = 0$
    $x - y + z = -2$

11. $x + 3y - 2z = 4$
    $-2x + y + 3z = 2$
    $2x + 4y - z = -1$

APPENDIX 6

PROOFS OF THEOREMS 6, 10, 16, AND 17 OF CHAPTER 13

In this appendix we give statements and proofs of several of the more difficult theorems on vectors in three dimensions. The proofs make use of the material on determinants given in Appendix 5.

Theorem 6. *Suppose that A, B, C, and D are points in space and that a Cartesian coordinate system is introduced in space. Denote the coordinates of A, B, C, and D by (x_A, y_A, z_A), (x_B, y_B, z_B), and so forth. (i) If the coordinates satisfy the equations*

$$x_B - x_A = x_D - x_C, \quad y_B - y_A = y_D - y_C, \quad z_B - z_A = z_D - z_C, \quad (1)$$

then $\overrightarrow{AB} \approx \overrightarrow{CD}$. (ii) Conversely, if $\overrightarrow{AB} \approx \overrightarrow{CD}$, the coordinates satisfy the equations in (1).

Proof. i) We assume that the equations in (1) hold. Then also,

$$x_C - x_A = x_D - x_B, \quad y_C - y_A = y_D - y_B, \quad z_C - z_A = z_D - z_B. \quad (2)$$

From (1) and Corollary 2 on p. 397, it follows that either $AB \parallel CD$ or A, B, C, and D are on a line. From Eqs. (2), we conclude that either $AC \parallel BD$ or A, B, C, and D are on a line. Thus, either $ACDB$ is a parallelogram or A, B, C, and D are on a line $\overrightarrow{L}$, which we may assume is directed.

If $ACDB$ is a parallelogram then $\overrightarrow{AB} \approx \overrightarrow{CD}$ by definition. If A, B, C, and D are on a line L, let $\overrightarrow{L}$, have the parametric equations

$$x = x_0 + t \cos \alpha, \quad y = y_0 + t \cos \beta, \quad z = z_0 + t \cos \gamma \quad (3)$$

and let A, B, C, and D have t coordinates t_A, t_B, t_C, and t_D, respectively. Thus $x_A = x_0 + t_A \cos \alpha$, $x_B = x_0 + t_B \cos \alpha$, etc. Subtracting, we get

$$\left.\begin{array}{ll} x_B - x_A = (t_B - t_A) \cos \alpha, & x_D - x_C = (t_D - t_C) \cos \alpha, \\ y_B - y_A = (t_B - t_A) \cos \beta, & y_D - y_C = (t_D - t_C) \cos \beta, \\ z_B - z_A = (t_B - t_A) \cos \gamma, & z_D - z_C = (t_D - t_C) \cos \gamma. \end{array}\right\} \quad (4)$$

From the equations in (1) and (4), we conclude that

$$\left.\begin{array}{l} (t_B - t_A) \cos \alpha = (t_D - t_C) \cos \alpha, \\ (t_B - t_A) \cos \beta = (t_D - t_C) \cos \beta, \\ (t_B - t_A) \cos \gamma = (t_D - t_C) \cos \gamma. \end{array}\right\} \quad (5)$$

Since $\cos \alpha$, $\cos \beta$, and $\cos \gamma$ are never simultaneously zero (as $\cos^2 \alpha + \cos^2 \beta + \cos^2 \gamma = 1$), it follows from (5) that $t_B - t_A = t_D - t_C$, so that $\overline{AB} = \overline{CD}$ and hence $\overrightarrow{AB} \approx \overrightarrow{CD}$ in this case also.

ii) To prove the converse, we assume that $\overrightarrow{AB} \approx \overrightarrow{CD}$. Then either $ACDB$ is a parallelogram or A, B, C, and D are on a directed line $\overrightarrow{L}$. Let us first assume the former; we wish to show that the equations in (1) hold. Suppose they do not. It is clear that there are unique numbers x_E, y_E, z_E, coordinates of a point $E \neq D$ such that
$$x_E - x_C = x_B - x_A, \qquad y_E - y_C = y_B - y_A,$$
and
$$z_E - z_C = z_B - z_A.$$
Then, by part (i), we know that $ACEB$ is a parallelogram (since C is not on line AB because $ACDB$ is a parallelogram). But then D and E must coincide, thus contradicting the fact above that $D \neq E$. Accordingly, the equations in (1) must hold.

To consider the other case, let $\overrightarrow{L}$ have the parametric equations (3). If we use our previous notation, we conclude that the equations in (4) hold. But, since $\overrightarrow{AB} \approx \overrightarrow{CD}$, we know by definition that $\overrightarrow{AB} = \overrightarrow{CD}$, i.e., that $t_B - t_A = t_D - t_C$. But then the equations in (1) follow from those in (4), and the proof is complete.

◀

We saw in Chapter 13, Section 5, that the vectors **i**, **j**, and **k** corresponding to any given coordinate system in space are linearly independent. Consequently Theorem 10 is a special case of the following Theorem 10':

Theorem 10'. *Suppose that* $\mathbf{u}_1$, $\mathbf{v}_1$, *and* $\mathbf{w}_1$ *are linearly independent and suppose that*

$$\begin{aligned} \mathbf{u}_2 &= a_{11}\mathbf{u}_1 + a_{12}\mathbf{v}_1 + a_{13}\mathbf{w}_1, \\ \mathbf{v}_2 &= a_{21}\mathbf{u}_1 + a_{22}\mathbf{v}_1 + a_{23}\mathbf{w}_1, \\ \mathbf{w}_2 &= a_{31}\mathbf{u}_1 + a_{32}\mathbf{v}_1 + a_{33}\mathbf{w}_1, \end{aligned} \qquad D = \begin{vmatrix} a_{11} & a_{12} & a_{13} \\ a_{21} & a_{22} & a_{23} \\ a_{31} & a_{32} & a_{33} \end{vmatrix}. \qquad (6)$$

Then the set $\{\mathbf{u}_2, \mathbf{v}_2, \mathbf{w}_2\}$ *is linearly dependent* $\Leftrightarrow$ $D = 0$.

Proof. a) Suppose the set is linearly dependent. Then there are constants c_1, c_2, and c_3, not all zero, such that
$$c_1\mathbf{u}_2 + c_2\mathbf{v}_2 + c_3\mathbf{w}_2 = \mathbf{0}. \qquad (7)$$
If we substitute (6) into (7), we obtain
$$(c_1 a_{11} + c_2 a_{21} + c_3 a_{31})\mathbf{u}_1 + (c_1 a_{12} + c_2 a_{22} + c_3 a_{32})\mathbf{v}_1$$
$$+ (c_1 a_{13} + c_2 a_{23} + c_3 a_{33})\mathbf{w}_1 = \mathbf{0}. \qquad (8)$$
Since $\mathbf{u}_1$, $\mathbf{v}_1$, and $\mathbf{w}_1$ are linearly independent, their coefficients must all vanish. That is, we must have
$$\begin{aligned} a_{11}c_1 + a_{21}c_2 + a_{31}c_3 &= 0, \\ a_{12}c_1 + a_{22}c_2 + a_{32}c_3 &= 0, \\ a_{13}c_1 + a_{23}c_2 + a_{33}c_3 &= 0. \end{aligned} \qquad (9)$$
But if (9) holds with c_1, c_2, and c_3 not all zero, the determinant D' of the coefficients must vanish according to the Corollary to Cramer's Rule (Theorem 11, Appendix 5). But D' is obtained from D by interchanging rows and columns. Accordingly, $D = D' = 0$.

b) Now suppose $D = 0$. If all the cofactors A_{ij} are zero, any two rows of D and hence any two of the vectors $\mathbf{u}_2$, $\mathbf{v}_2$, and $\mathbf{w}_2$ are proportional and the set is linearly dependent. Otherwise, some $A_{pq} \neq 0$. By interchanging the order of the vectors, if necessary, we may assume that $p = 3$. From the expansion theorem (Theorem 10, Appendix 5), we conclude that

$$a_{11}A_{1q} + a_{21}A_{2q} + a_{31}A_{3q} = 0,$$
$$a_{12}A_{1q} + a_{22}A_{2q} + a_{32}A_{3q} = 0, \qquad (10)$$
$$a_{13}A_{1q} + a_{23}A_{2q} + a_{33}A_{3q} = 0.$$

Since $A_{3q} \neq 0$, we can solve equations (10) for the a_{3j}, obtaining

$$\left.\begin{array}{c} a_{31} = ka_{11} + la_{21}, \quad a_{32} = ka_{12} + la_{22}, \quad a_{33} = ka_{13} + la_{23}, \\ k = -A_{1q}/A_{3q}, \quad l = -A_{2q}/A_{3q}. \end{array}\right\} \qquad (11)$$

In this case it follows from (11) and (6) that $\mathbf{w}_2 = k\mathbf{u}_2 + l\mathbf{v}_2$, and the vectors $\mathbf{u}_2$, $\mathbf{v}_2$, and $\mathbf{w}_2$ are linearly dependent.

Theorem 16. *Suppose that $\mathbf{u}$ and $\mathbf{v}$ are any vectors, that $\{\mathbf{i}, \mathbf{j}, \mathbf{k}\}$ is a right-handed coordinate triple, and that t is any number. Then*

i) $\mathbf{v} \times \mathbf{u} = -\mathbf{u} \times \mathbf{v},$

ii) $(t\mathbf{u}) \times \mathbf{v} = t(\mathbf{u} \times \mathbf{v}) = \mathbf{u} \times (t\mathbf{v}),$

iii) $\mathbf{i} \times \mathbf{j} = -\mathbf{j} \times \mathbf{i} = \mathbf{k},$
    $\mathbf{j} \times \mathbf{k} = -\mathbf{k} \times \mathbf{j} = \mathbf{i},$
    $\mathbf{k} \times \mathbf{i} = -\mathbf{i} \times \mathbf{k} = \mathbf{j},$

iv) $\mathbf{i} \times \mathbf{i} = \mathbf{j} \times \mathbf{j} = \mathbf{k} \times \mathbf{k} = \mathbf{0}.$

Proofs. i) By definition, $|\mathbf{v} \times \mathbf{u}| = |\mathbf{u} \times \mathbf{v}|$ and $\mathbf{v} \times \mathbf{u}$ and $\mathbf{u} \times \mathbf{v}$ are both orthogonal to both $\mathbf{u}$ and $\mathbf{v}$ (or are both zero if $\mathbf{u}$ and $\mathbf{v}$ are proportional). Thus $\mathbf{v} \times \mathbf{u} = \pm \mathbf{u} \times \mathbf{v}$. If we let $\mathbf{w} = \mathbf{u} \times \mathbf{v}$, then $\{\mathbf{u}, \mathbf{v}, \mathbf{w}\}$ and $\{\mathbf{v}, \mathbf{u}, -\mathbf{w}\}$ are right-handed (see Theorem 15, Chapter 13), so $\mathbf{v} \times \mathbf{u}$ must equal $-\mathbf{w}$.

ii) If $t = 0$ or $\mathbf{u}$ and $\mathbf{v}$ are proportional, (ii) certainly holds. Otherwise, let us set $\mathbf{w} = \mathbf{u} \times \mathbf{v}$. Then (ii) follows since all the terms in (ii) have the same magnitude, all are orthogonal to both $\mathbf{u}$ and $\mathbf{v}$, and $\{t\mathbf{u}, \mathbf{v}, t\mathbf{w}\}$ and $\{\mathbf{u}, t\mathbf{v}, t\mathbf{w}\}$ are right-handed by Theorem 15, Chapter 13.

iv) This follows, since we must have $\mathbf{i} \times \mathbf{i} = -\mathbf{i} \times \mathbf{i} = \mathbf{0}$, etc.

iii) To prove (iii), we note that, since $\{\mathbf{i}, \mathbf{j}, \mathbf{k}\}$ is right-handed, $\theta = \pi/2$, $|\mathbf{i}| = |\mathbf{j}| = |\mathbf{k}| = 1$, and $\mathbf{k}$ is orthogonal to both $\mathbf{i}$ and $\mathbf{j}$, it follows that $\mathbf{i} \times \mathbf{j} = \mathbf{k}$. That $\mathbf{j} \times \mathbf{i} = -\mathbf{k}$ follows from this and from (i). Since $\{\mathbf{i}, \mathbf{j}, \mathbf{k}\}$ is a coordinate triple, it follows as above that $\mathbf{j} \times \mathbf{k} = \pm \mathbf{i}$. Setting

$$\mathbf{u}_1 = \mathbf{i}, \quad \mathbf{v}_1 = \mathbf{j}, \quad \mathbf{w}_1 = \mathbf{k}, \quad \mathbf{u}_2 = \mathbf{j}, \quad \mathbf{v}_2 = \mathbf{k}, \quad \mathbf{w}_2 = \mathbf{i},$$

we see that

$$\mathbf{u}_2 = 0 \cdot \mathbf{u}_1 + 1 \cdot \mathbf{v}_1 + 0 \cdot \mathbf{w}_1,$$
$$\mathbf{v}_2 = 0 \cdot \mathbf{u}_1 + 0 \cdot \mathbf{v}_1 + 1 \cdot \mathbf{w}_1,$$
$$\mathbf{w}_2 = 1 \cdot \mathbf{u}_1 + 0 \cdot \mathbf{v}_1 + 0 \cdot \mathbf{w}_1.$$

Since $\{\mathbf{i}, \mathbf{j}, \mathbf{k}\}$ was given as right-handed, it follows from the discussion in Chapter 13, Section 7, that $\{\mathbf{u}_2, \mathbf{v}_2, \mathbf{w}_2\}$ is right-handed since

$$D = \begin{vmatrix} 0 & 1 & 0 \\ 0 & 0 & 1 \\ 1 & 0 & 0 \end{vmatrix} = +1.$$

The proof that $\mathbf{k} \times \mathbf{i} = \mathbf{j}$ is similar.

Theorem 17 (Distributive law). *If $\mathbf{u}$, $\mathbf{v}$, and $\mathbf{w}$ are any vectors,*
  i) $\mathbf{u} \times (\mathbf{v} + \mathbf{w}) = (\mathbf{u} \times \mathbf{v}) + (\mathbf{u} \times \mathbf{w})$ *and*
  ii) $(\mathbf{v} + \mathbf{w}) \times \mathbf{u} = (\mathbf{v} \times \mathbf{u}) + (\mathbf{w} \times \mathbf{u})$.

Proof. Part (ii) follows from part (i) and part (i) of Theorem 16, for

$$(\mathbf{v} + \mathbf{w}) \times \mathbf{u} = -[\mathbf{u} \times (\mathbf{v} + \mathbf{w})] = -[(\mathbf{u} \times \mathbf{v}) + (\mathbf{u} \times \mathbf{w})]$$
$$= [-(\mathbf{u} \times \mathbf{v})] + [-(\mathbf{u} \times \mathbf{w})] = (\mathbf{v} \times \mathbf{u}) + (\mathbf{w} \times \mathbf{u}).$$

It is clear that (i) holds if $\mathbf{u} = \mathbf{0}$. Otherwise, let $\{\mathbf{i}', \mathbf{j}', \mathbf{k}'\}$ be a right-handed coordinate triple such that $\mathbf{u} = |\mathbf{u}|\mathbf{i}'$ (i.e., $\mathbf{i}'$ is the unit vector in the direction of $\mathbf{u}$). Suppose that

$$\mathbf{v} = a_1\mathbf{i}' + b_1\mathbf{j}' + c_1\mathbf{k}', \qquad \mathbf{u} \times \mathbf{v} = \mathbf{V} = A_1\mathbf{i}' + B_1\mathbf{j}' + C_1\mathbf{k}'.$$

We first find A_1, B_1, C_1 in terms of a_1, b_1, and c_1.

Since $\mathbf{V}$ is orthogonal to both $\mathbf{u}$ and $\mathbf{v}$, we must have

$$\mathbf{V} \cdot \mathbf{u} = A_1|\mathbf{u}| = 0, \qquad \mathbf{V} \cdot \mathbf{v} = A_1 a_1 + B_1 b_1 + C_1 c_1 = 0.$$

Thus

$$A_1 = 0 \quad \text{and} \quad b_1 B_1 + c_1 C_1 = 0 \quad \text{so that} \quad B_1 = -kc_1, \quad C_1 = kb_1$$

for some k. Moreover,

$$|\mathbf{V}| = |\mathbf{u}| \cdot \sqrt{a_1^2 + b_1^2 + c_1^2} \sin \theta$$

and

$$\mathbf{u} \cdot \mathbf{v} = |\mathbf{u}| \cdot \sqrt{a_1^2 + b_1^2 + c_1^2} \cos \theta = |\mathbf{u}| a_1.$$

Since $0 \leq \theta \leq \pi$ and $\cos \theta = a_1/|\mathbf{v}|$, it follows that

$$\sin \theta = \sqrt{b_1^2 + c_1^2}/|\mathbf{v}|.$$

Thus

$$|\mathbf{V}| = |k| \cdot \sqrt{b_1^2 + c_1^2} = |\mathbf{u}| \cdot \sqrt{b_1^2 + c_1^2} \quad \text{so} \quad k = \pm|\mathbf{u}|.$$

Finally $\{\mathbf{u}, \mathbf{v}, \mathbf{V}\}$ must be right-handed, so that

$$\begin{vmatrix} |\mathbf{u}| & 0 & 0 \\ a_1 & b_1 & c_1 \\ 0 & -kc_1 & kb_1 \end{vmatrix} = k|\mathbf{u}| \cdot (b_1^2 + c_1^2) > 0$$

(unless $\mathbf{v} = \mathbf{0}$ or $\mathbf{v}$ is proportional to $\mathbf{u}$). Hence $k = +1$ *and*
$$\mathbf{V} = |\mathbf{u}| \cdot (-c_1 \mathbf{j}' + b_1 \mathbf{k}'). \tag{12}$$
The result in (12) evidently holds also if $\mathbf{v} = \mathbf{0}$ or is proportional to $\mathbf{u}$.

In like manner, if we let
$$\mathbf{w} = a_2 \mathbf{i}' + b_2 \mathbf{j}' + c_2 \mathbf{k}', \qquad \mathbf{W} = \mathbf{u} \times \mathbf{w}, \qquad \mathbf{X} = \mathbf{u} \times (\mathbf{v} + \mathbf{w}),$$
we see that
$$\mathbf{W} = |\mathbf{u}| \cdot (-c_2 \mathbf{j}' + b_2 \mathbf{k}'), \qquad \mathbf{X} = |\mathbf{u}| \cdot [-(c_1 + c_2)\mathbf{j}' + (b_1 + b_2)\mathbf{k}']$$
from which (i) follows.

APPENDIX 7

INTRODUCTION TO A SHORT TABLE OF INTEGRALS

We recall some of the methods of integration with which the student should be familiar. These devices, together with the integrals listed below, enable the student to perform expeditiously any integration that is required in order to work the problems in this text.

1. Substitution in a Table of Integrals

Example. Letting $u = \tan x$, $du = \sec^2 x \, dx$, we find that $\int e^{\tan x} \sec^2 x \, dx = \int e^u \, du = e^u + C = e^{\tan x} + C$.

2. Certain Trigonometric and Hyperbolic Integrals

We illustrate with trigonometric integrals; the corresponding hyperbolic forms are treated similarly.

a) $\int \sin^m u \cos^n u \, du$.

i) **n an odd positive integer, m arbitrary.** Factor out $\cos u \, du$ and express the remaining cosines in terms of sines.

Example

$$\int \sin^4 2x \cos^3 2x \, dx = \tfrac{1}{2} \int \sin^4 2x (1 - \sin^2 2x) \cdot (2 \cos 2x \, dx)$$

$$= \tfrac{1}{2} \int (\sin^4 2x - \sin^6 2x) \, d(\sin 2x)$$

$$= \tfrac{1}{10} \sin^5 2x - \tfrac{1}{14} \sin^7 2x + C.$$

ii) **m an odd positive integer, n arbitrary.** Factor out $\sin u \, du$ and express the remaining sines in terms of cosines.

iii) **m and n both even integers ≥ 0.** Reduce the degree of the expression by the substitutions

$$\sin^2 u = \frac{1 - \cos 2u}{2}, \qquad \cos^2 u = \frac{1 + \cos 2u}{2}.$$

Example

$$\int \sin^4 u \, du = \tfrac{1}{4} \int (1 - \cos 2u)^2 \, du$$

$$= \tfrac{1}{4} \int (1 - 2 \cos 2u) \, du + \tfrac{1}{8} \int (1 + \cos 4u) \, du$$

$$= \frac{3u}{8} - \frac{\sin 2u}{4} + \frac{\sin 4u}{32} + C.$$

b) $\int \tan^m u \sec^n u \, du.$

i) **n an even positive integer, m arbitrary.** Factor out $\sec^2 u \, du$ and express the remaining secants in terms of $\tan u$.

Example

$$\int \frac{\sec^4 u \, du}{\sqrt{\tan u}} = \int (\tan u)^{-1/2}(1 + \tan^2 u) \cdot (\sec^2 u \, du)$$

$$= \int [(\tan u)^{-1/2} + (\tan u)^{3/2}] d(\tan u)$$

$$= 2(\tan u)^{1/2} + \tfrac{2}{5}(\tan u)^{5/2} + C.$$

ii) **m an odd positive integer, n arbitrary.** Factor out $\sec u \tan u \, du$ and express the remaining tangents in terms of the secants.

Example

$$\int \frac{\tan^3 u \, du}{\sqrt[3]{\sec u}} = \int (\sec u)^{-4/3} \tan^2 u \cdot (\sec u \tan u \, du)$$

$$= \int [(\sec u)^{2/3} - (\sec u)^{-4/3}] d(\sec u)$$

$$= \tfrac{3}{5} (\sec u)^{5/3} + 3(\sec u)^{-1/3} + C.$$

c) $\int \cot^m u \csc^n u \, du.$ These are treated like those in (b).

3. Trigonometric and Hyperbolic Substitutions

a) If $\sqrt{a^2 - u^2}$ occurs (or $a^2 - u^2$ occurs in the denominator), set $u = a \sin \theta$, $\theta = \arcsin(u/a)$, $du = a \cos \theta \, d\theta$, $\sqrt{a^2 - u^2} = a \cos \theta$.

Example

$$\int \sqrt{a^2 - u^2} \, du = a^2 \int \cos^2 \theta \, d\theta = \frac{a^2}{2} \int (1 + \cos 2\theta) \, d\theta$$

$$= \frac{a^2}{2} (\theta + \sin \theta \cos \theta) + C$$

$$= \frac{a^2}{2} \arcsin \frac{u}{a} + u\sqrt{a^2 - u^2} + C.$$

b) If $\sqrt{a^2 + u^2}$ occurs, set $u = a \tan \theta$, etc.

Example

$$\int u^3 \sqrt{a^2 + u^2} \, du = a^5 \int \tan^3 \theta \sec^3 \theta \, d\theta.$$

The last integral is of type (2)(b)(ii) above.

c) If $\sqrt{u^2 - a^2}$ occurs, set $u = a \sec \theta$, etc.

Example

$$\int \frac{\sqrt{u^2 - a^2}}{u} \, du = \int \frac{a \tan \theta}{a \sec \theta} \cdot a \sec \theta \tan \theta \, d\theta$$

$$= \int a \tan^2 \theta \, d\theta = a \int (\sec^2 \theta - 1) \, d\theta$$

$$= a (\tan \theta - \theta) + C$$

$$= \sqrt{u^2 - a^2} - a \, \text{arcsec} \, (u/a) + C.$$

A hyperbolic substitution is sometimes more effective:

Example. If we let $u = a \sinh v$, then

$$\int \sqrt{a^2 + u^2} \, du = \int a^2 \cosh^2 v \, dv = \frac{a^2}{2} \int (1 + \cosh 2v) \, dv$$

$$= \frac{a^2}{2} (v + \sinh v \cosh v) + C$$

$$= \frac{a^2}{2} \text{argsinh} \left(\frac{u}{a}\right) + u\sqrt{a^2 + u^2} + C.$$

4. Integrals Involving Quadratic Functions

Complete the square in the quadratic function and introduce a simple change of variable to reduce the quadratic to one of the forms $a^2 - u^2$, $a^2 + u^2$, or $u^2 - a^2$.

Example

$$\int \frac{(2x - 3) \, dx}{x^2 + 2x + 2} = \int \frac{(2x - 3) \, dx}{(x + 1)^2 + 1}.$$

Let $u = x + 1$. Then $x = u - 1$, $dx = du$, and

$$\int \frac{(2x - 3) \, dx}{(x + 1)^2 + 1} = \int \frac{(2u - 5) \, du}{u^2 + 1} = \ln (u^2 + 1) - 5 \arctan u + C$$

$$= \ln (x^2 + 2x + 2) - 5 \arctan (x + 1) + C.$$

5. Integration by Parts: $\int u \, dv = uv - \int v \, du$.

Example. If we let $u = x$ and $v = e^x$ in the integral $\int xe^x \, dx$, then the formula for integration by parts gives

$$\int xe^x \, dx = \int u \, dv = uv - \int v \, du = xe^x - \int e^x \, dx = xe^x - e^x + C.$$

6. Integration of Rational Functions (Quotients of Polynomials)

If the degree of the numerator $\geq$ that of the denominator, divide out, thus expressing the given function as a polynomial plus a "proper fraction." Each proper fraction can be expressed as a sum of simpler "proper partial fractions":

$$\frac{P(x)}{Q(x)} = \frac{P_1(x)}{Q_1(x)} + \cdots + \frac{P_n(x)}{Q_n(x)},$$

in which no two Q_i have common factors and each Q_i is of the form $(x - a)^k$ or $(ax^2 + bx + c)^k$; of course $Q = Q_1 \cdot Q_2 \cdots Q_n$. Each of these fractions can be expressed uniquely in terms of still simpler fractions as follows:

$$\frac{P_i(x)}{(x - a)^k} = \frac{A_1}{x - a} + \frac{A_2}{(x - a)^2} + \cdots + \frac{A_k}{(x - a)^k},$$

$$\frac{P_i(x)}{(ax^2 + bx + c)^k} = \frac{A_1 x + B_1}{ax^2 + bx + c} + \frac{A_2 x + B_2}{(ax^2 + bx + c)^2} + \cdots$$

$$+ \frac{A_k x + B_k}{(ax^2 + bx + c)^k}.$$

Each of these simplest fractions can be integrated by methods already described. The constants are obtained by multiplying up the denominators and either equating coefficients of like powers of x in the resulting polynomials or by substituting a sufficient number of values of x to determine the coefficients.

Example 1. Integrate

$$\int \frac{x^2 + 2x + 3}{x(x - 1)(x + 1)} \, dx.$$

According to the results above, there exist constants A, B, and C such that

$$\frac{x^2 + 2x + 3}{x(x - 1)(x + 1)} = \frac{A}{x} + \frac{B}{x - 1} + \frac{C}{x + 1}.$$

Multiplying up, we see that we must have

$$A(x - 1)(x + 1) + Bx(x + 1) + Cx(x - 1) \equiv x^2 + 2x + 3.$$

The constants are most easily found by substituting $x = 0$, 1, and -1 in turn in this identity, yielding $A = -3$, $B = 3$, $C = 1$.

Example 2. Integrate

$$\int \frac{3x^2 + x - 2}{(x - 1)(x^2 + 1)}.$$

There are constants A, B, and C such that

$$\frac{3x^2 + x - 2}{(x - 1)(x^2 + 1)} = \frac{A}{x - 1} + \frac{Bx + C}{x^2 + 1},$$

or
$$A(x^2 + 1) + Bx(x - 1) + C(x - 1) \equiv 3x^2 + x - 2.$$
Setting $x = 1, 0$, and -1 in turn, we find that $A = 1, C = 3, B = 2$.

Example 3. Show how to break up the fraction
$$\frac{2x^6 - 3x^5 + x^4 - 4x^3 + 2x^2 - x + 1}{(x - 2)^3(x^2 + 2x + 2)^2} = \frac{P(x)}{Q(x)}$$
into simplest partial fractions. Do not determine the constants.

Solution
$$\frac{P(x)}{Q(x)} = \frac{P_1(x)}{(x - 2)^3} + \frac{P_2(x)}{(x^2 + 2x + 2)^2}$$
$$= \frac{A_1}{x - 2} + \frac{A_2}{(x - 2)^2} + \frac{A_3}{(x - 2)^3}$$
$$+ \frac{A_4 x + A_5}{x^2 + 2x + 2} + \frac{A_6 x + A_7}{(x^2 + 2x + 2)^2}.$$

7. Three Rationalizing Substitutions

a) If the integrand contains a single irrational expression of the form $(ax + b)^{p/q}$, the substitution $z = (ax + b)^{1/q}$ will convert the integral into that of a rational function of z.

Example. If we let $z = (x + 1)^{1/3}$ so that $x = z^3 - 1$ and $dx = 3z^2\, dz$, then
$$\int \frac{\sqrt[3]{x + 1}}{x} dx = \int \frac{z}{z^3 - 1} \cdot 3z^2\, dz = \int 3\, dz + \int \frac{3\, dz}{(z - 1)(z^2 + z + 1)}.$$

b) If a single irrational expression of the form $\sqrt{a^2 - x^2}, \sqrt{a^2 + x^2}$, or $\sqrt{x^2 - a^2}$ occurs with an odd power of x outside, the substitution $z = \sqrt{a^2 - x^2}$ (or etc.) reduces the given integral to that of a rational function.

Example. If we let $z = \sqrt{a^2 - x^2}$, then
$$x^2 = a^2 - z^2, \qquad x\, dx = -z\, dz,$$
$$\int \frac{\sqrt{a^2 - x^2}}{x^3} dx = -\int \frac{z^2\, dz}{(z^2 - a^2)^2}.$$

c) In case a given integrand is a rational function of trigonometric functions, the substitution
$$t = \tan(\theta/2), \qquad \theta = 2 \arctan t, \qquad d\theta = \frac{2\, dt}{1 + t^2},$$
$$\cos \theta = \frac{1 - t^2}{1 + t^2}, \qquad \sin \theta = \frac{2t}{1 + t^2},$$
reduces the integral to one of a rational function of t.

Example. Making these substitutions, we obtain
$$\int \frac{d\theta}{5 - 4\cos \theta} = \int \frac{(2\, dt)/(1 + t^2)}{5 - 4[(1 - t^2)/(1 + t^2)]} = \int \frac{2\, dt}{1 + 9t^2}.$$

Table 4. A Short Table of Integrals

The constant of integration is omitted.

Elementary formulas

1. $\displaystyle\int u^n \, du = \frac{u^{n+1}}{n+1}, \quad n \neq -1$

2. $\displaystyle\int \frac{du}{u} = \ln |u|$

3. $\displaystyle\int e^u \, du = e^u$

4. $\displaystyle\int a^u \, du = \frac{a^u}{\ln a}, \quad a > 0, \quad a \neq 1$

5. $\displaystyle\int \sin u \, du = -\cos u$

6. $\displaystyle\int \cos u \, du = \sin u$

7. $\displaystyle\int \sec^2 u \, du = \tan u$

8. $\displaystyle\int \csc^2 u \, du = -\cot u$

9. $\displaystyle\int \sec u \tan u \, du = \sec u$

10. $\displaystyle\int \csc u \cot u \, du = -\csc u$

11. $\displaystyle\int \sinh u \, du = \cosh u$

12. $\displaystyle\int \cosh u \, du = \sinh u$

13. $\displaystyle\int \operatorname{sech}^2 u \, du = \tanh u$

14. $\displaystyle\int \operatorname{csch}^2 u \, du = -\coth u$

15. $\displaystyle\int \operatorname{sech} u \tanh u \, du = -\operatorname{sech} u$

16. $\displaystyle\int \operatorname{csch} u \coth u \, du = -\operatorname{csch} u$

17. $\displaystyle\int \frac{du}{\sqrt{a^2 - u^2}} = \arcsin \frac{u}{a}, \quad a > |u|$

18. $\displaystyle\int \frac{du}{a^2 + u^2} = \frac{1}{a} \arctan \frac{u}{a}, \quad a \neq 0$

19. $\displaystyle\int \frac{du}{u\sqrt{u^2 - a^2}} = \frac{1}{a} \operatorname{arcsec} \frac{u}{a}, \quad |u| > a$

20. $\displaystyle\int \frac{du}{\sqrt{a^2 + u^2}} = \begin{cases} \operatorname{argsinh} \dfrac{u}{a} \\ \ln(u + \sqrt{a^2 + u^2}) \end{cases}$

21. $\displaystyle\int \frac{du}{a^2 - u^2} = \begin{cases} \dfrac{1}{a} \operatorname{argtanh} \dfrac{u}{a}, & |u| < a \\ \dfrac{1}{a} \operatorname{argcoth} \dfrac{u}{a}, & |u| > a \end{cases} \text{ or } \begin{cases} \dfrac{1}{2a} \ln \dfrac{a+u}{a-u}, & |u| < a \\ \dfrac{1}{2a} \ln \dfrac{u+a}{u-a}, & |u| > a \end{cases}$

22. $\displaystyle\int \frac{du}{u\sqrt{a^2 - u^2}} = -\frac{1}{a} \operatorname{argsech} \frac{u}{a}, \text{ or } -\frac{1}{a} \ln \left(\frac{a + \sqrt{a^2 - u^2}}{u} \right), \quad 0 < u < a$

23. $\displaystyle\int \frac{du}{|u|\sqrt{u^2 + a^2}} = -\frac{1}{a} \operatorname{argcsch} \frac{u}{a}, \quad u \neq 0 \text{ or } -\frac{1}{a} \ln \left(\frac{a + \sqrt{u^2 + a^2}}{u} \right)$

24. $\displaystyle\int \frac{du}{\sqrt{u^2 - a^2}} = \begin{cases} \operatorname{argcosh} \dfrac{u}{a}, \\ \ln |u + \sqrt{u^2 - a^2}|, \end{cases} \quad |u| > a > 0$

Table 4 (cont.)

Algebraic forms

25. $\int \dfrac{u\,du}{a+bu} = \dfrac{u}{b} - \dfrac{a}{b^2}\ln(a+bu)$ 26. $\int \dfrac{du}{u(a+bu)} = \dfrac{1}{a}\ln\left|\dfrac{u}{a+bu}\right|$

27. $\int \dfrac{u\,du}{(a+bu)^2} = \dfrac{a}{b^2}\left(\dfrac{1}{a+bu} + \dfrac{1}{a}\ln|a+bu|\right)$

28. $\int \dfrac{du}{u(a+bu)^2} = \dfrac{1}{a(a+bu)} + \dfrac{1}{a^2}\ln\left|\dfrac{u}{a+bu}\right|$

29. $\int \dfrac{du}{u\sqrt{a+bu}} = \dfrac{1}{\sqrt{a}}\ln\left|\dfrac{\sqrt{a+bu}-\sqrt{a}}{\sqrt{a+bu}+\sqrt{a}}\right|$

30. $\int u\sqrt{a+bu}\,du = \dfrac{2(3bu-2a)\sqrt{(a+bu)^3}}{15b^2}$

31. $\int \dfrac{\sqrt{a+bu}}{u}\,du = 2\sqrt{a+bu} + a\int \dfrac{du}{u\sqrt{a+bu}}$

32. $\int \dfrac{u\,du}{\sqrt{a+bu}} = \dfrac{2(bu-2a)}{3b^2}\sqrt{a+bu}$

33. $\int \sqrt{a^2-u^2}\,du = \dfrac{u}{2}\sqrt{a^2-u^2} + \dfrac{a^2}{2}\arcsin\dfrac{u}{a},\quad |u|<a$

34. $\int \sqrt{u^2\pm a^2}\,du = \dfrac{u}{2}\sqrt{u^2\pm a^2} \pm \dfrac{a^2}{2}\ln|u+\sqrt{u^2\pm a^2}|$

35. $\int \dfrac{\sqrt{a^2\pm u^2}}{u}\,du = \sqrt{a^2\pm u^2} - a\ln\left|\dfrac{a+\sqrt{a^2\pm u^2}}{u}\right|$

36. $\int \dfrac{\sqrt{u^2-a^2}}{u}\,du = \sqrt{u^2-a^2} - a\arccos\dfrac{a}{u},\quad 0<a<|u|$

Trigonometric forms

37. $\int \tan u\,du = -\ln|\cos u|$

38. $\int \sec u\,du = \ln|\sec u + \tan u|$

39. $\int \sin^2 u\,du = \tfrac{1}{2}u - \tfrac{1}{4}\sin 2u$

40. $\int \sin^n u\,du = -\dfrac{\sin^{n-1} u \cos u}{n} + \dfrac{n-1}{n}\int \sin^{n-2} u\,du$

41. $\int \cos^n u\,du = \dfrac{\cos^{n-1} u \sin u}{n} + \dfrac{n-1}{n}\int \cos^{n-2} u\,du$

Table 4 (cont.)

42. $\displaystyle\int \frac{du}{\sin^n u} = -\frac{\cos u}{(n-1)\sin^{n-1} u} + \frac{n-2}{n-1} \int \frac{du}{\sin^{n-2} u}, \quad n \neq 1$

43. $\displaystyle\int \sin mu \sin nu \, du = \frac{\sin(m-n)u}{2(m-n)} - \frac{\sin(m+n)u}{2(m+n)}, \quad m \neq \pm n$

44. $\displaystyle\int \cos mu \cos nu \, du = \frac{\sin(m-n)u}{2(m-n)} + \frac{\sin(m+n)u}{2(m+n)}, \quad m \neq \pm n$

45. $\displaystyle\int \sin mu \cos nu \, du = -\frac{\cos(m-n)u}{2(m-n)} - \frac{\cos(m+n)u}{2(m+n)}, \quad m \neq \pm n$

46. $\displaystyle\int u^n \sin u \, du = -u^n \cos u + n \int u^{n-1} \cos u \, du$

47. $\displaystyle\int u^n \cos u \, du = u^n \sin u - n \int u^{n-1} \sin u \, du$

48. $\displaystyle\int \text{Arcsin } u \, du = u \text{ Arcsin } u + \sqrt{1-u^2}$

49. $\displaystyle\int \text{Arccos } u \, du = u \text{ Arccos } u - \sqrt{1-u^2}$

50. $\displaystyle\int u \text{ Arcsin } u \, du = \tfrac{1}{4}[(2u^2 - 1)\text{Arcsin } u + u\sqrt{1-u^2}]$

Logarithmic and exponential forms

51. $\displaystyle\int \ln|u| \, du = u(\ln|u| - 1)$

52. $\displaystyle\int (\ln|u|)^2 \, du = u(\ln|u|)^2 - 2u \ln|u| + 2u$

53. $\displaystyle\int u^n \ln|u| \, du = \frac{u^{n+1}}{n+1} \ln|u| - \frac{u^{n+1}}{(n+1)^2}, \quad n \neq -1$

54. $\displaystyle\int u^n e^u \, du = u^n e^u - n \int u^{n-1} e^u \, du$

55. $\displaystyle\int e^{au} \sin bu \, du = \frac{e^{au}(a \sin bu - b \cos bu)}{a^2 + b^2}$

56. $\displaystyle\int e^{au} \cos bu \, du = \frac{e^{au}(a \cos bu + b \sin bu)}{a^2 + b^2}$

Miscellaneous forms

57. $\displaystyle\int (a + bu)^n \, du = \frac{(a+bu)^{n+1}}{b(n+1)}, \quad n \neq -1$

58. $\displaystyle\int \frac{du}{u^2(a+bu)} = -(au)^{-1} + ba^{-2}[\log(a+bu) - \log u]$

Table 4 (cont.)

59. $\displaystyle\int\frac{du}{(a+bu)^{1/2}(c+du)^{3/2}} = \frac{2}{bc-ad}\left(\frac{a+bu}{c+du}\right)^{1/2},\quad bc-ad\neq 0$

60. $\displaystyle\int\frac{du}{u(a+bu^n)} = (an)^{-1}[\log(u^n) - \log(a+bu^n)],\quad n\neq 0$

61. $\displaystyle\int\frac{u\,du}{(a+bu)^{1/2}} = \frac{2(bu-2a)}{3b^2}(a+bu)^{1/2}$

62. $\displaystyle\int\frac{du}{u^2(u^2+a^2)^{1/2}} = -(a^2 u)^{-1}(u^2+a^2)^{1/2}$

63. $\displaystyle\int\frac{(a^2-u^2)^{1/2}}{u^2}\,du = -u^{-1}(a^2-u^2)^{1/2} - \arcsin\frac{u}{a}$

64. $\displaystyle\int\sin^4 u\,du = \frac{3u}{8} - \frac{\sin 2u}{4} + \frac{\sin 4u}{32}$

65. $\displaystyle\int\cot^3 u\,du = -\tfrac{1}{2}\cot^2 u - \log\sin u$

ANSWERS TO ODD-NUMBERED PROBLEMS

Chapter 1

Section 1–1 (pp. 6–7)

1. $x < 5$
3. $x > -3$
5. $x < \frac{14}{11}$
7. All x not in interval $[0, \frac{20}{3}]$
9. All x not in interval $[2, \frac{13}{4}]$
11. All x in interval $(\frac{5}{3}, 4)$
13. All x not in interval $[2, 3]$
15. All x in interval $(\frac{1}{3}, 4)$
17. All x in interval $(\frac{11}{5}, \frac{36}{5})$
19. All x in $(-\infty, 1)$, $(\frac{3}{2}, 2)$, $(\frac{17}{7}, \infty)$ satisfy both inequalities
21. All x in $(-\infty, 0)$ and $(\frac{3}{5}, \frac{5}{4})$ satisfy inequalities
23. $-4 < x < 0$
27. *Hint*: multiply by x and transfer all terms to left side, getting $x^2 - 2x + 1 \geq 0$.
29. *Hint*: Set $x/y = u$ and use the binomial formula for $(a + b)^4$.

Section 1–2 (pp. 10–11)

1. $-\frac{5}{2}, \frac{3}{2}$
3. $\frac{1}{2}, \frac{7}{2}$
5. $\frac{14}{3}, -\frac{16}{3}$
7. $-\frac{3}{4}, \frac{7}{10}$
9. $3, -3$
11. $-\frac{13}{2}, -\frac{17}{4}$
13. $-5 < x < 3$
15. $-4 < x < -1$
17. $-\frac{13}{5} < x - \frac{11}{5}$
19. All x not in $[-\frac{11}{2}, -\frac{5}{6}]$
21. $-\frac{1}{3} \leq x \leq 7$
23. $|x| < 2$
25. $(x, y) = (\frac{11}{5}, -\frac{6}{5}), (-1, 2), (1, -2), (-\frac{11}{5}, \frac{6}{5})$

Section 1–3 (pp. 14–15)

1. $M = 14$
3. $M = 196$
5. $M = 10$
7. $M = 98.5$
9. $M = 10$
11. None

Chapter 2

Section 2–3 (pp. 27–28)

1. $f(-3) = -2$, $f(-2) = -4$, $f(-1) = -4$, $f(0) = -2$, $f(1) = 2$, $f(2) = 8$, $f(3) = 16$, $f(a + 2) = a^2 + 7a + 8$
3. $f(-4) = 2$, $f(-3) = \frac{7}{3}$, $f(-2) = 4$, $f(-1) = -1$, $f(0) = \frac{2}{3}$, $f(1) = 1$, $f(-1000) = \frac{2998}{1997}$, $f(1000) = \frac{3002}{2003}$, $f[f(x)] = \frac{13x + 12}{12x + 13}$, No.
5. $f(-1000) = \frac{-2000}{1,000,001}$, $f(-3) = \frac{-3}{5}$, $f(-2) = \frac{-4}{5}$, $f(-1) = -1$, $f(0) = 0$, $f(1) = 1$, $f(2) = \frac{4}{5}$, $f(3) = \frac{3}{5}$, $f(1000) = \frac{2000}{1,000,001}$

7.
x	-1	0	1	2	3
$f(x)$	0	1	$\sqrt{2}$	$\sqrt{3}$	2

, domain: $x \geq -1$

9.
x	-4	-3	-2	-1	0	1	2
$f(x)$	0	1	2	3	2	1	0

11. Domain: $-1 - \sqrt{3} \leq x \leq -1 + \sqrt{3}$

13. Domain: All $x \neq 2, 6, -2$; $f(x) = G(x) = x + 8$ if $x \neq -2, 2, 6$

15. $4x + 2h + 1$

17. $\dfrac{-1}{x(x+h)}$

19. $\dfrac{1}{\sqrt{x+h} + \sqrt{x}}$

23. $f[g[h(x)]] = 18x^2 + 12x + 5$

Section 2-4 (pp. 34-35)

	Intercepts	Symmetry		Intercepts	Symmetry
1.	$(0, -4), (\pm 2, 0)$	y axis	3.	$(0, 0)$	x axis
				$y_1 = +\sqrt{3x}$	
				$y_2 = -\sqrt{3x}$	
5.	$(0, \pm\sqrt{2}), (-1, 0)$	x axis	7.	$(0, \pm 1), (-1, 0)$	x axis
	$y_1 = +\sqrt{2x+2}$			$y_1 = +\sqrt{x+1}$	
	$y_2 = -\sqrt{2x+2}$			$y_2 = -\sqrt{x+1}$	
9.	$(0, \pm\sqrt{6}), (\pm 3, 0)$	origin,	11.	$(0, \pm 2)$	origin
	$y_1 = +\sqrt{\frac{1}{3}(18 - 2x^2)}$	x axis,		$y_1 = x + \sqrt{x^2 + 4}$	
	$y_2 = -\sqrt{\frac{1}{3}(18 - 2x^2)}$	y axis		$y_2 = x - \sqrt{x^2 + 4}$	
13.	$(0, \pm\sqrt{3})$	origin,	15.	$(0, \pm\sqrt{3}), (\pm\sqrt{3}, 0)$	origin
	$y_1 = +\sqrt{x^2 + 3}$	x axis,		$y_1 = \dfrac{-x + \sqrt{12 - 3x^2}}{2}$	
	$y_2 = -\sqrt{x^2 + 3}$	y axis		$y_2 = \dfrac{-x - \sqrt{12 - 3x^2}}{2}$	
17.	$(0, 0)$	origin	19.	$(0, 1), (1, 0)$	none w.r.t.
				$y_1 = 1 + \sqrt{x}$,	axis or
				$y_2 = 1 - \sqrt{x}$	origin
21.	$(0, 0)$	origin,	23.	$(0, 0), (0, \pm 2), (-4, 0)$	x axis
	$y_1 = \sqrt{x}$,	x axis,		$y_1 = \sqrt{x+4}$,	
	$y_2 = \sqrt{-x}$,	y axis		$y_2 = -\sqrt{x+4}$	
	$y_3 = -\sqrt{x}$			$y_3 = \sqrt{-x}$	
	$y_4 = -\sqrt{-x}$			$y_4 = -\sqrt{-x}$	

25. $(0, 0)$, $(0, \pm 2)$ origin,
$$y_1 = +\sqrt{2 + \sqrt{4 - x^2}} \quad x \text{ axis,}$$
$$y_2 = \sqrt{2 - \sqrt{4 - x^2}} \quad y \text{ axis}$$
$$y_3 = -\sqrt{2 + \sqrt{4 - x^2}}$$
$$y_4 = -\sqrt{2 - \sqrt{4 - x^2}}$$

Section 2–5 (p. 40)

	Intercepts	Symmetry	Domain	Range	Asymptote
1.	$(0, 0)$	y axis	$x \in R$	$y \geq 0$	none
3.	$(0, 0)$	x axis	$x \leq 0$	$y \in R_1$	none
5.	$(0, 3)$ $(\pm\sqrt{3}, 0)$	y axis	$x \in R_1$	$y \leq 3$	none
7.	none	y axis	$x \neq 0$	$y > 0$	$x = 0, y = 0$
9.	$(\pm\sqrt{3}, 0)$	origin, x axis, y axis	$|x| \geq \sqrt{3}$	$y \in R_1$	no vert. or horiz.
11.	$(0, \pm\sqrt{2})$ $(\pm\sqrt{3}, 0)$	origin, x axis, y axis	$|x| \leq \sqrt{3}$	$|y| \leq \sqrt{2}$	none
13.	none	x axis	$x > 0$	$y \neq 0$	$x = 0, y = 0$
15.	$(0, \pm 2\sqrt{3})$ $(\pm 2\sqrt{3}, 0)$	origin	$|x| \leq 4$	$|y| \leq 4$	none
17.	$(0, \pm 2)$	x axis	$x > -1$	$y \neq 0$	$x = -1, y = 0$
19.	none	origin, x axis, y axis	$x \neq 0$	$|y| > 2$	$x = 0$, $y = 2, -2$
21.	$(0, \pm 2)$	origin, x axis, y axis	$x \in R_1$	$|y| \leq 2$, $y \neq 0$	$y = 0$
23.	$(0, \frac{4}{3})$	none	$x \neq 1, 3$	$y \leq -4$ or $y > 0$	$x = 1, 3$, $y = 0$
25.	$(0, 0)$	origin	$x \in R_1$	$y \neq \pm 2$	$x = 0$, $y = 2, -2$
27.	$(0, 0)$	x axis	$-1 < x \leq 0$, $x > 1$	$y \in R_1$	$y = 0$, $x = 1, -1$
29.	$(0, 0)$	origin, x axis, y axis	$x \in R_1$	$y \in R_1$	none
31.	$(-2, 0)$	x axis	$-2 \leq x < -1$, $x > 1$	$y \in R_1$	$x = +1, -1$ $y = 0$
33.	$(-\frac{3}{5}, 0)$ $(0, \frac{2}{5})$	none	$x \neq 1, 3$	$y \geq -\frac{1}{5}$ or $y \leq -5$	$x = 1, 3$, $y = 0$

35. Range: all of R_1; intercepts: $(0, 3), (0, -1), (3, 0), (1, 0)$; no symmetry
37. Domain: all of R_1; range all of R_1; intercepts: $(0, \pm\sqrt{3}), (-3, 0), (-1, 0)$; symmetry: x axis

Chapter 3

Section 3–1 (pp. 43–44)

1. $|AB| = 2\sqrt{2}, |BC| = 3\sqrt{5}, |AC| = \sqrt{41}$
3. $|AB| = \sqrt{26}, |BC| = 5, |AC| = \sqrt{13}$

ANS–4   ANSWERS TO ODD-NUMBERED PROBLEMS

5. $\bar{x} = \frac{9}{2}, \bar{y} = \frac{1}{2}$ 7. $(\frac{19}{4}, 2), (\frac{5}{2}, 3), (\frac{1}{4}, 4)$ 9. $(-16, -1)$
11. $\frac{1}{2}\sqrt{149}, \frac{1}{2}\sqrt{389}, \frac{1}{2}\sqrt{338}$ 13. $|AB| = |BC| = \sqrt{41}$
15. $|AB| = \sqrt{13}, |BC| = \sqrt{65}, |AC| = \sqrt{52}; |AC|^2 + |AB|^2 = |BC|^2; \therefore$ right triangle
19. $(\frac{3}{2}, 2)$
23. $M_{AB} = (\frac{1}{2}(x_1 + x_2), \frac{1}{2}(y_1 + y_2)), M_{BC} = (\frac{1}{2}(x_2 + x_3), \frac{1}{2}(y_2 + y_3))$,
  $|M_{AB}M_{BC}| = \frac{1}{2}\sqrt{(x_3 - x_1)^2 + (y_3 - y_1)^2} = \frac{1}{2}|AC|$. Other sides similar.
25. $x^2 + y^2 - 2x - 2y - 2 = 0$. Circle, center $(1, 1)$, radius 2
27. $(y + 1)^2 = 8(x - 2)$
29. $(x - 5)^2 + (y + 1)^2 = 4$, circle, center $(5, -1)$, radius 2

Section 3–2 (pp. 51–52)

1. Parallel   3. Perpendicular   5. Parallel
7. Yes   9. No   11. $3x - 4y - 23 = 0$
13. Parallelogram   15. Trapezoid   17. None of these
19. None of these   21. Rhombus   23. None of these
25. Completely general since axes may always be chosen so that origin is at A and the base is along the x axis.
27. Completely general for reasons similar to above answer. Point of bisection is
$$\left(\frac{a + b + e}{4}, \frac{c + f}{4}\right).$$
29. Coordinates are: $(a, \frac{1}{2}a\sqrt{3}), (\frac{1}{2}a, a\sqrt{3}), (-\frac{1}{2}a, a\sqrt{3}), (-a, \frac{1}{2}a\sqrt{3})$.

Section 3–3 (pp. 56–57)

1. $2x - y + 6 = 0$ 3. $2x + 3y - 11 = 0$ 5. $x - 2y - 8 = 0$
7. $y = -7$ 9. $y = -2$ 11. $y + 5 = 0$
13. $4x - 2y - 9 = 0$ 15. $y + 3x = 0$ 17. $y + 5 = 0$
19. $m = \frac{2}{3}, b = -\frac{7}{3}$ 21. $3x - 7y - 15 = 0$ 23. $4x + 5y - 29 = 0$
25. $8x - 3y - 38 = 0$ 27. $15x - y + 78 = 0$ 29. $7x - y - 15 = 0$
35. $\left(\frac{a + b}{3}, \frac{c}{3}\right)$ 37. $\left(b, \frac{ab - b^2}{c}\right)$

Chapter 4

Section 4–1 (pp. 64–65)

1. -4 3. $\frac{32}{5}$ 5. 3   7. $\frac{1}{4}$ 9. $-\frac{2}{3}$
11. 6   13. 15   15. 8   17. $\sqrt{5}/20$ 19. $\frac{4}{3}$
21. $1/\sqrt{b}$ 23. -6144 25. $1/(2\sqrt[4]{8})$
27. Limit is 1.   29. Graph indicates limit must be 1.

Section 4–2 (pp. 68–69)

1. 24   3. 6   5. $4a$ 7. -4
9. $-\frac{1}{9}$ 11. $-2x$ 13. 29   15. 0
17. $\frac{3}{2}\sqrt{x}$ 19. $2c + 3$ 21. $-3x^2/(x^3 + 1)^2$
23. $1 + \dfrac{1}{2\sqrt{2}}$ 25. $2/(x + 1)^2$

Section 4-3 (pp. 70-71)

1. $6x$
3. $4x - 3$
5. $-4x^3$
7. $\dfrac{-1}{(x-1)^2}$
9. $\dfrac{-3}{x^4}$
11. $\dfrac{1}{(1-x)^2}$
13. $\dfrac{4x}{(x^2+1)^2}$
15. $\dfrac{-1}{2x^{3/2}}$
17. $\tfrac{3}{2}\sqrt{x}$
19. $\dfrac{x-2}{2(x-1)^{3/2}}$
21. $\dfrac{-x}{\sqrt{4-x^2}}$
23. $\dfrac{3x^2-6}{(x^2+x+2)^2}$
25. $\dfrac{2x+1}{2\sqrt{x^2+x+1}}$

Section 4-4 (pp. 76-77)

1. Tangent: $3x - y - 2 = 0$; normal: $x + 3y - 4 = 0$
3. Tangent: $4x - y - 5 = 0$; normal: $x + 4y - 14 = 0$
5. Tangent: $9x + y - 17 = 0$; normal: $x - 9y - 11 = 0$
7. Tangent: $x - 2y + 2 = 0$; normal: $2x + y - 6 = 0$
9. Tangent: $x + 16y - 14 = 0$; normal: $32x - 2y - 191 = 0$
11. Tangent: $y = 4$; normal: $x = 1$
13. Decreasing if $x \leq 2$; increasing if $x \geq 2$; rel. min. at $x = 2$
15. Increasing if $x \leq -1$, $x \geq 2$; decreasing if $-1 \leq x \leq 2$; rel. max. at $x = -1$, rel. min. at $x = 2$
17. Decreasing if $x \leq 0$, $x \geq 2$; increasing if $0 \leq x \leq 2$; rel. min., $x = 0$; rel. max., $x = 2$
19. Decreasing if $x \leq -1$, $1 \leq x \leq 2$; increasing if $-1 \leq x \leq 1$, $x \geq 2$; rel. min., $x = -1$, 2; rel. max. $x = 1$
21. $\Delta y = -0.09$, $\dfrac{\Delta y}{\Delta x} = -0.9$
23. $\Delta y = \dfrac{.000901}{1.0001}$, $\dfrac{\Delta y}{\Delta y} = \dfrac{.0901}{1.0001}$
25. $\dfrac{4x}{(x^2+1)^2}$, at $x = 1$, value is 1.

Section 4-5 (pp. 82-83)

1. $v = 4$, moving right
3. $v = 14$, right
5. $v = 26$, moving right
7. $v = 2t - 4$, $a = 2$; $v = 0$ at $t = 2$
9. $v = t^2 - 4t + 3$, $a = 2t - 4$, $v = 0$ at $t = 1, 3$
11. $v = 2bt + c$, $a = 2b$, $v = 0$ if $t = -c/2b$
13. $v = 2t + 3$, $a = 2$; left if $t < -\tfrac{3}{2}$, right if $t > -\tfrac{3}{2}$; $a = 2$
15. $v = 2t - 4$, $a = 2$; moving left if $t < 2$, right if $t > 2$
17. $v = 6(t^2 - t - 2)$, $a = 6(2t - 1)$; moving left if $-1 < t < 2$, right if $t < -1$ or $t > 2$
19. $v = \dfrac{(1-t)(1+t)}{(1+t^2)^2}$, $a = -2t(3-t^2)(1+t^2)^{-3}$; moving left if $t < -1$ or $t > 1$, right if $-1 < t < 1$
21. $v = 2bt + c$, $a = 2b$
23. Max. height is $\tfrac{245}{4}$ meters, reached when $t = \tfrac{3}{2}$ sec. The stone hits the ground when $t = 5$ sec.
25. $g = 1.6$ m/sec^2

Section 4-6 (p. 87)

1. $\delta = 0.0005$
3. $\delta = 0.005$
5. $\delta = 0.029701$
7. $\delta = 0.0398$
9. $\delta = \tfrac{84}{121}$
11. $\delta = \tfrac{1}{101}$

13. $\delta = \frac{200}{2601}$
15. $\frac{\sqrt{101} - 10}{10}$
17. $\delta = 0.05$, although it is not the largest possible δ.
19. $\delta = \epsilon_1(2\sqrt{2} - \epsilon_1)$, where ϵ_1 is the smaller of ϵ and $\sqrt{2}$.
21. $\delta = \frac{9\epsilon_1}{1 + 3\epsilon_1}$, where ϵ_1 is the smaller of ϵ and $\frac{1}{3}$.

Section 4-7 (pp. 94-95)

1. 9
3. $-\frac{1}{2}$
5. 12
7. $\frac{3}{4}$
9. $\frac{3}{2}$
11. 2
13. $\sqrt{3}$
15. $1/(2\sqrt{x})$
17. $-\frac{1}{2}$
19. 80

Section 4-8 (pp. 100-102)

1. 0
3. 0
5. 1
7. 4
9. Both limits 0. Yes
11. Continuous, $-4 < x < 7$
13. Discontinuous at $x = 2$
15. Continuous, $0 < x < 2$
17. Continuous, $-\infty < x < \infty$
19. Continuous
21. Discontinuous at $x = 2$
23. Continuous
25. Continuous

Section 4-9 (pp. 106-108)

1. $\frac{2}{3}$
3. 0
5. -1
7. ∞
9. ∞
11. $+\infty$
13. $+\infty$
15. $+\infty$
17. Discontinuous at $x = -3, 3$
19. Discontinuous at $x = 2$
21. 0
23. No limit

27. $\lim_{x \to \infty} \frac{P(x)}{Q(x)} = \begin{cases} 0 & \text{if } m > n \\ a_n/b_n & \text{if } m = n \\ \infty & \text{if } m < n \end{cases}$

29. If for some i and j it is time that $d_i = c_j$, then limit is as in Problem 27. If $d_i \neq c_j$ for all i, j, then limit is ∞.

Section 4-10 (pp. 113-114)

1. 0
3. $-\frac{3}{2}$
5. $+\infty$
7. 0
9. 2
11. 0
13. 0
15. $1/(3\sqrt{2})$
19. Hint: $n = 1 \cdot 2 \cdot \frac{3}{2} \cdot \frac{4}{3} \cdot \frac{5}{4} \cdots \frac{(n-1)}{(n-2)} \cdot \frac{n}{(n-1)}$ and $\frac{n}{3^n} = \frac{1}{3} \cdot \frac{2}{3} \cdot \frac{\frac{3}{2}}{3} \cdot \frac{\frac{4}{3}}{3} \cdots \frac{n/(n-1)}{3}$

21. Yes, limit is 1.

Chapter 5

Section 5-1 (pp. 120-121)

1. $2x + 3$
3. $14x^{13} + 20x + 7$
5. $-2x^{-3} + 20x^{-6} - 24x^{-9}$
7. $3x^2 + 4x - 3 + 2x^{-2} - 8x^{-3}$
9. $2x + 2 + 2x^{-3}$
11. $-\frac{21}{4}x^{-8}$
13. $9x^2 + 14x + 2$
15. $(3x^2 + 12x - 2)(x^2 + 3x - 5) + (x^3 + 6x^2 - 2x + 1)(2x + 3)$
$= 5x^4 + 36x^3 + 33x^2 - 70x + 13$
17. $2 - 4x^{-2}$
19. $3x^{-2} - 8x^{-3} + 6x^{-4}$
21. $\frac{2}{(x + 1)^2}$
23. $\frac{13}{(2x + 3)^2}$

25. $(1 - x^2)/(1 + x^2)^2$

27. $-\dfrac{x^2 + 2x}{(x^2 + 2x + 2)^2}$

29. $\dfrac{4x^2 + 12x + 1}{(2x + 3)^2}$

31. a) $4(x + 1)(x^2 + 2x - 1)$
    b) $2(x^3 + 7x^2 - 8x - 6)(3x^2 + 14x - 8)$
    c) $2(x^7 - 2x + 3x^{-2})(7x^6 - 2 - 6x^{-3})$

33. $f'(x) = \dfrac{u \cdot v \cdot w' + u' \cdot v \cdot w - u \cdot v' \cdot w}{v^2}$

   a) $\dfrac{1}{(3x + 1)^2}(3x^2 + 2x + 32)$

   b) $\dfrac{2}{(2x + 6)^2}(3x^4 + 12x^3 + 4x^2 + 24x + 5)$

Section 5-2 (pp. 124–125)

1. $18(2x + 3)^8$
3. $-20(2 - 5x)^3$
5. $-8(2x + 1)^{-5}$
7. $4(x^3 + 2x - 3 + x^{-2})^3(3x^2 + 2 - 2x^{-3})$
9. $-1(x^2 + 2 - x^{-2})^{-2}(2x + 2x^{-3})$
11. $2x(x^2 + 1)(x^2 - 2)(5x^4 - 3x^2 - 2)$
13. $\dfrac{(x^3 + 2x - 6)^6}{x^2}(23x^6 + 39x^4 - 32x^3 + 14x^2 - 12x - 6)$
15. $\dfrac{2x(x^2 + 1)^2(x^2 + 4)}{(x^2 + 2)^3}$
17. $-\dfrac{(x^2 + 2)(2x^5 + 12x^3 - 7x^2 - 6)}{x^2(x^2 + x^{-1})^4}$
19. $\dfrac{x^3 - 2x^{-2}}{(x^2 + x^{-1})^2}(4x^4 + 7x + 12x^{-1} + 6x^{-4})$
21. $(x + 5)(3x - 6)^2(7x^2 + 1)^3(273x^3 + 735x^2 - 1665x + 33)$
25. $f'(x) = g'(u + v) \cdot (u' + v');\quad F'(x) = G'(u \cdot v)(uv' + vu');$
    $H'(x) = h'\left(\dfrac{u}{v}\right) \cdot \left(\dfrac{vu' - uv'}{v^2}\right)$

Section 5-3 (pp. 128–130)

1. $\tfrac{5}{3}x^{2/3} + \tfrac{8}{3}x^{1/3} + x^{-4/3}$
3. $-\tfrac{2}{3}x^{-5/3} - \tfrac{4}{3}x^{-7/3} - \tfrac{8}{7}x^{-3/7}$
5. $x^{1/3} + x^{-1/2} + 2x^{-2}$
7. $\tfrac{1}{10}(3x^{1/2} - 2x^{-1/2} - 4x^{-3/2})$
9. $3x^{1/2} + 2x^{-1/3} - 7x^{2/5}$
11. $\tfrac{3}{4}x^{1/2} - \tfrac{3}{4}x^{-1/2} - \tfrac{1}{2}x^{-3/2}$
13. $\tfrac{20}{3}(2x + 3)^{7/3}$
15. $3(x + 1)(x^2 + 2x + 3)^{1/2}$
17. $-(2x^2 - 2x + 1)(2x^3 - 3x^2 + 3x - 1)^{-4/3}$
19. $2(x^2 + 1)(x^3 + 3x + 2)^{-1/3}$
21. $\tfrac{3}{10}(2x + 1)(x^2 + x - 3)^{1/2}$
23. $(2x + 3)^3(3x - 2)^{4/3}(38x + 5)$
25. $(2x - 1)^{3/2}(7x - 3)^{-4/7}(41x - 18)$
27. $\tfrac{1}{6}(x + 1)^{-1/3}(x - 1)^{-1/2}(7x - 1)$
29. $-2t^{-2}(2 + 6t)^{-2/3}(1 + 2t)$
31. $\tfrac{7}{2}(y^2 + 3y + 4)^{-3/2}$
33. $\dfrac{1}{\sqrt{(\tau - 1)(\tau + 1)^3}} = \dfrac{1}{(\tau + 1)\sqrt{\tau^2 - 1}}$
35. $-6(3t + 4)^{-2/3}(3t - 2)^{-4/3}$
37. 0
39. $\tfrac{3725}{6}$
41. Tangent: $5x - y - 9 = 0$; normal: $x + 5y - 7 = 0$
43. Tangent: $7x + 3y - 64 = 0$; normal: $3x - 7y + 72 = 0$
45. Tangent: $7x + 12y - 32 = 0$; normal: $24x - 14y - 27 = 0$
47. $u^{m-1} \cdot v^{n-1}[mu'v + nuv']$; (a) $(2x + 1)^5(x^3 - 6)^{1/3}(20x^3 + 4x^2 - 72)$;
    (b) $(x^2 + 2x - 1)^{1/2}(x^4 - 3)^2(15x^5 + 27x^4 - 12x^3 - 9x - 9)$

49. $f'(x) = \frac{1}{2}x^{-1/2}$

51. $f'(x) = \dfrac{1}{q} x^{(1-q)/q}$

Section 5-4 (pp. 133–134)

1. $y' = x/2y$
3. $y' = -x^3/2y^3$
5. $y' = (2x - 3y)/(3x - 2y)$
7. $y' = -(x^2 + 2y)/(2x + 5y^2)$
9. $y' = (6x^2 - 6xy + 2y^2)/(3x^2 - 4xy + 3y^2)$
11. $y' = (6x^5 + 6x^2y - y^7)/(-2x^3 + 7xy^6)$
13. $y' = (x - y)/x;\ y = \frac{1}{2}x - \frac{3}{2}x^{-1}$
15. $y' = -4x^{-2} = -y/x;\ y = 4x^{-1}$
17. $y' = -x^{-1/3}y^{1/3};\ y = \pm(a^{2/3} - x^{2/3})^{3/2}$
19. $y' = -1/y;\ y = \pm(8 - 2x)^{1/2}$
21. $y' = 2x/y;\ y = \pm(2x^2 - 1)^{1/2}$
23. $y' = (4x - 3y)/(3x + 8y);\ y = (-3x \pm \sqrt{41x^2 - 80})/8$
25. $y' = -\frac{1}{2};\ y = -(2x)^{1/2}$
27. $y' = 3;\ y = -(10 - x^2)^{1/2}$
29. $y' = \frac{1}{6};\ y = \frac{3}{4}x - \frac{1}{4}(16 - 7x^2)^{1/2}$
31. $y' = 3y/(2y - 3x)$

Chapter 6

Section 6-1 (pp. 140–141)

1. Max. at (1, 1), (3, 1); min. at (0, 0), (2, 0)
9. $(-1, -2)$
11. Max., (1, 2); min., (3, −6)
13. Max., (2, 2); $f'(2) = 0$
15. No

Section 6-2 (pp. 145–146)

1. $x_0 = 1$
3. $x_0 = \frac{11}{3}$
5. $x_0 = -2 + \sqrt{6}$
7. $x_0 = \pm\frac{1}{2}\sqrt{2}$
9. $x_0 = 0, \frac{1}{2}, 1$
11. $x_0 = 1$
13. The equation $\frac{5}{2} = -5/(3x + 2)^2$ has no solution.
15. There is no number x_0. f becomes infinite at $x = 1$; Theorem of the Mean not applicable.
19. Theorem of Mean applicable as f and f' are continuous for $0 < x < 2$.

Section 6-3 (pp. 152–153)

1. Decreasing for $x \leq -\frac{3}{2}$, increasing for $x \geq -\frac{3}{2}$, rel. min. at $x = -\frac{3}{2}$
3. Increasing for $x \leq \frac{1}{4}$; decreasing for $x \geq \frac{1}{4}$, rel. max. at $x = \frac{1}{4}$
5. Increasing for $x \leq -2,\ x \geq 1$; decreasing for $-2 \leq x \leq 1$; rel. max. at $x = -2$, rel. min. at $x = 1$
7. Increasing for $x \leq \dfrac{-2 - \sqrt{13}}{3}$;

   decreasing for $\dfrac{-2 - \sqrt{13}}{3} \leq x \leq \dfrac{-2 + \sqrt{13}}{3}$;

   increasing for $x \geq \dfrac{-2 + \sqrt{13}}{3}$;

   rel. max. at $x = \dfrac{-2 - \sqrt{13}}{3}$; rel. min. at $x = \dfrac{-2 + \sqrt{13}}{3}$

9. Increasing for all x; $f'(x) = 0$ if $x = -2$
11. Decreasing for $x \leq \frac{1}{3},\ x \geq 1$; increasing for $\frac{1}{3} \leq x \leq 1$; $x = \frac{1}{3}$, rel. min.; $x = 1$, rel. max.
13. Decreasing, $x \leq -1$; increasing, $-1 \leq x \leq 0$; decreasing, $0 \leq x \leq 2$; increasing $x \geq 2$; rel. min. at $x = -1, 2$; rel. max. at $x = 0$

15. Decreasing, $x \leq -\frac{3}{2}$; increasing, $x \geq -\frac{3}{2}$; rel. min. $x = -\frac{3}{2}$
17. Increasing for $x \leq -1$, $x \geq 1$; decreasing for $-1 \leq x < 0$, $0 < x \leq 1$; rel. max. at $x = -1$; rel. min. at $x = 1$; $x = 0$ is a vertical asymptote
19. Increasing for $x < -1$; increasing for $x > -1$; $x = -1$ a vertical asymptote; $y = 1$, a horizontal asymptote
21. Decreasing for $-1 < x \leq 0$; increasing for $x \geq 0$; rel. min. at $x = 0$; $x = -1$ is vertical asymptote
23. Increasing for $x \leq -2$; decreasing for $-2 \leq x \leq 0$; increasing for $x \geq 0$; rel. max. at $x = -2$; rel. min. at $x = 0$; vertical tangent at $x = -3$
25. Increasing if $x \leq 2$; decreasing if $2 \leq x \leq 3$; rel. max. at $x = 2$
27. Decreasing for $-\sqrt{2} \leq x \leq -1$, $1 \leq x \leq \sqrt{2}$; increasing for $-1 \leq x \leq 1$; rel. min., $x = -1$; rel. max., $x = 1$
29. Increasing for all x; $x = 2$, a vertical asymptote
31. Decreasing for $x < -1$, increasing for $x > -1$, rel. min. at $x = -1$

Section 6–4 (pp. 158–159)

1. Minimum at $x = \frac{3}{2}$, concave upward for all x
3. Rel. max. at $x = -3$, rel. min. at $x = 3$; inflection point at $x = 0$; concave downward, $x \leq 0$; concave upward, $x \geq 0$
5. $x = -3$, rel. min.; $x = -2$, inflection point; $x = 0$, inflection point; $x \leq -2$, concave upward; $-2 \leq x \leq 0$, concave downward; $x \geq 0$, concave upward
7. $x = -1, 1$, rel. minima; concave upward for all x; $x = 0$, vertical asymptote
9. $x = -1$, rel. min.; $x = 1$, rel. max.; points of inflection $x = 0, +\sqrt{3}, -\sqrt{3}$; concave downward, $x < -\sqrt{3}$, $0 < x < \sqrt{3}$; concave upward, $-\sqrt{3} < x < 0$ and $x > \sqrt{3}$
11. Rel. max., $x = -1$; rel. min., $x = 2$; inflection point, $x = \frac{1}{2}$; concave downward, $x \leq \frac{1}{2}$; concave upward, $x \geq \frac{1}{2}$
13. Rel. max., $x = \frac{2}{3}$; rel. min., $x = 2$; inflection point, $x = \frac{4}{3}$; concave downward, $x \leq \frac{4}{3}$; concave upward, $x \geq \frac{4}{3}$
15. Inflection point $x = \frac{1}{3}$; concave downward, $x \leq \frac{1}{3}$; concave upward, $x \geq \frac{1}{3}$
17. Rel. min., $x = -1$; inflection points $x = 0, 2$; concave upward $x \leq 0$, $x \geq 2$; concave downward, $0 \leq x \leq 2$; horizontal tangent at $x = 2$
19. Rel. minima $x = 0$, $(-15 - \sqrt{33})/8$; rel. max., $x = (-15 + \sqrt{33})/8$; concave upward $x \leq -2$, $x \geq -\frac{1}{2}$; concave downward, $-2 \leq x \leq -\frac{1}{2}$; points of inflection $x = -2, -\frac{1}{2}$
21. $x = -2$, rel. min.; $x = 2$, rel. max.; concave downward for $x \leq -2\sqrt{3}$ and $0 \leq x \leq 2\sqrt{3}$; concave upward for $-2\sqrt{3} \leq x \leq 0$ and $x \geq 2\sqrt{3}$; $x = -2\sqrt{3}, 0, 2\sqrt{3}$, inflection points
23. $x = -2$, rel. min.; $x = 2$, rel. max.; concave upward, $-2\sqrt{2} \leq x \leq 0$; concave downward, $0 \leq x \leq 2\sqrt{2}$; $x = 0$, inflection point
25. $x = -4$, rel. max.; $x = 0$, rel. min.; concave downward $-5 \leq x \leq -4 + \frac{2}{3}\sqrt{6}$; concave upward, $x \geq -4 + \frac{2}{3}\sqrt{6}$; inflection point at $x = -4 + \frac{2}{3}\sqrt{6}$
27. $x = 2$, rel. max.; concave downward for $x < -2$, $-2 < x \leq 2 - \sqrt{6}$; concave upward for $2 - \sqrt{6} \leq x \leq 0$; concave downward for $0 \leq x \leq 2 + \sqrt{6}$; concave upward for $x \geq 2 + \sqrt{6}$; inflection points at $x = 2 - \sqrt{6}, 2 + \sqrt{6}$. Note that the curve is concave upward on one side of $x = 0$ and concave downward on the other. There is a vertical asymptote at $x = -2$.
29. Rel. max. at $x = (-2 - \sqrt{10})/3$; rel. min. at $x = 0$

Section 6–5 (pp. 161–162)

1. Min. at $(-1, -5)$; max. at $(3, 11)$
3. Max. at $(1, 3)$; min. at $(-2, -18)$
5. Minima $(-5, -1)$ or $(1, -1)$; max., $(6, 274)$
7. Min. at $(1, 0)$; max. at $(5, 576)$
9. Min. at $(-\frac{1}{2}, -1)$; max. at $(1, \frac{1}{2})$
13. No maximum or minimum. [For interval $-3 \leq x \leq 3$, max. is $(3, 11)$, min. is $(-3, -49)$.]

15. No maximum or minimum 17. Max. at $(2, \frac{1}{3})$; no minimum
19. $f'(x) = 0 \Leftrightarrow x = (-a \pm \sqrt{a^2 - 3b})/3$; if $a^2 < 3b$, then $f'(x) \neq 0$ for all x and max and min. must be at endpoints.

Section 6–6 (pp. 167–170)

1. Length = width = 30 meters; length = width = $\frac{1}{4}L$
3. $h = \frac{2}{3}\sqrt{3}R$; $r = \frac{1}{3}\sqrt{6}R$
5. 10, −10
7. Radius of semicircle = height of rectangle = $12/(4 + \pi)$ meters
9. Width = $(2 + 2\sqrt{5})$ cm; height = $(3 + 3\sqrt{5})$ cm
11. $r = \sqrt[6]{(360)^2/2\pi^2}$ cm; $h = \sqrt[3]{720/\pi}$ cm; $r = \sqrt[6]{9V^2/2\pi^2}$ cm; $h = \sqrt[3]{6V/\pi}$ cm
13. Land at $\frac{5}{3}\sqrt{3}$ km down beach; walk $6 - \frac{5}{3}\sqrt{3}$ km; time ≈ 3.66 hr.
15. (1, 2)
17. (a) base = 13/2; height = 30/13; (b) base = $H/2$; height = $h/2$
19. Each number is 10.
21. (a) $r = h = \sqrt[3]{16/\pi}$ cm; (b) $r = h = \sqrt[3]{V/\pi}$ cm
23. From (5, 0) to (0, 10)
25. Radius = $\frac{1}{2}R\sqrt{2}$; height = $R\sqrt{2}$
27. (a) radius of circle = $\dfrac{L\sqrt{3}}{18 + 2\pi\sqrt{3}}$, length of side of triangle = $\dfrac{3L}{9 + \pi\sqrt{3}}$; (b) circle only, $r = L/2\pi$
29. Height = $4R/3$; radius of base = $2\sqrt{2}R/3$
31. $r = (9V^2/2\pi^2)^{1/6}$; $h = (6V/\pi)^{1/3}$
33. $r = (3V/20\pi)^{1/3}$; $h = 6(3V/20\pi)^{1/3}$
35. $4(1 + \sqrt[3]{4})^{3/2}$ meters
37. 3500

Section 6–7 (pp. 174–175)

1. $df(x, h) = (4x^3 + 6x - 2)h$
3. $df(x, h) = \dfrac{(4x^5 + 2x)h}{\sqrt{x^4 + 1}}$
5. $df(x, h) = \dfrac{(8x^2 + 3x + 4)h}{3(x^2 + 2)^{1/2}(2x + 1)^{2/3}}$
7. $df = 0.03$; $\Delta f = 0.0301$
9. $df = 0.18$; $\Delta f = 0.180901$
11. $df = -0.01250$; $\Delta f = -0.01220$
13. $df = -0.05$; $\Delta f = -0.04654$
15. 8.0625
17. 1.02000
19. 1.98750
21. 0.049
23. $1\frac{2}{3}\%$ (approximately)
25. $6a^2 t$ cm^3 of paint

Section 6–8 (pp. 178–179)

1. $15x(x^3 - 3x^2 + 2)^4(x - 2)dx$
3. $-5(x^3 + 4)^{-6}(3x^2)dx$
5. $(5x^2 + 6x)(2x + 3)^{-1/2}dx$
7. $2(x + 1)(2x - 1)^2(5x + 2)dx$
9. $2(1 - x^2)(x^2 + 1)^{-2}dx$
11. $\frac{2}{3}x^{-1/3}(x + 1)^{-5/3}dx$
13. $-(x + 1)^{-1/2}(x - 1)^{-3/2}dx$
15. $(4x + y + 2)/(-x + 2y + 3)$
17. $-y^{1/2}/x^{1/2}$
19. $(6x^2 - y^2 + 2)/(2xy + 3y^2 + 1)$
21. $-8(2u^2 + 2u + 1)r(r^2 + 5)^3/(u^2(u + 1)^2)$
23. $-(5x^3 + 6x^2 - 8x - 17)(3s^2 - 8)/((y - 1)^2(x + 1)^{3/2})$
25. $-4x(x^2 + y)/((2x^2 - 4y^3 - 3)\sqrt{2t + 3})$
27. $-9x^2 t^{5/2}/(2(2y^3 + 1)(2t^{3/2} - 1))$

Section 6–9 (pp. 182–184)

1. $(2/\pi)$ m/min
3. $5\sqrt{193}$ km/hr

5. 11 cm²/min   7. 4/3 m/sec
9. (a) $-(150/\sqrt{117})$ ft/sec;   (b) $(25/\sqrt{10})$ ft/sec
11. $\frac{6}{5}\sqrt{10}$ 13. $\frac{dz}{dt} = \pm\frac{7}{6}$
15. $-2/5$ m/sec   17. 0.12 m/min
19. $(1/12\pi)$ m/min   21. $\frac{5}{4}\sqrt{5}$ m²/min
23. 2.4 m/sec   25. $39/\sqrt{217}$ m/sec
27. -8.75 gr/cm²/sec   29. -0.2 amp/sec

Section 6–10 (pp. 191–192)

1. $f(x) = \frac{1}{5}x^5 - \frac{1}{2}x^4 + \frac{7}{2}x^2 - 6x + \frac{24}{5}$
3. $f(x) = \frac{1}{8}(x+2)^8 + 1$ 5. $f(x) = (2x+1)^3(x+2)^2$
7. $s = \frac{1}{3}t^3 + \frac{1}{2}t^2 - 2t - \frac{1}{6}$ 9. $s = 6t - t^2 - t^3$
11. Rises for 8 sec; stays in air for 16 sec
13. Sum = 2.000; exact area = 2   15. Sum = 2.700; exact area = $\frac{8}{3}$
17. $\frac{1}{3}$ 19. $\frac{4}{3}$ 21. $\frac{13}{6}$ 23. $\frac{13}{6}$ 25. 2.330   27. 0.786

Chapter 7

Section 7–2 (p. 197)

1. 78   3. 12   5. $\frac{137}{60}$ 7. 2046   11. 3/8

Section 7–3 (pp. 203–204)

1. $\underline{S}(\Delta) = 9.006, \bar{S}(\Delta) = 12.444$ 3. $\underline{S}(\Delta) = 0.803; \bar{S}(\Delta) = 0.873$
5. $\underline{S}(\Delta) = 0.1588; \bar{S}(\Delta) = 0.2702$ 7. $\underline{S}(\Delta) = 0.0322; \bar{S}(\Delta) = 2.1710$
9. $\underline{S}(\Delta) = 1.000; \bar{S}(\Delta) = 7.000$

Section 7–4 (pp. 207–208)

1. Smallest value = 6; largest value = 24
3. Smallest value = -3; largest value = 24
5. Smallest value = $\frac{3}{4}$; largest value = $\frac{4}{5}$
7. Smallest value = 0; largest value = $\frac{64}{17}$
9. Smallest value = $3\sqrt{3}$; largest value = $3\sqrt{6}$
11. Smallest value = 0, largest value = 4
13. Smallest value = 0, largest value = $5\sqrt{5}$
15. Smallest value = 0, largest value = $\pi/4$

Section 7–5 (pp. 212–214)

1. $-1\frac{1}{6}$ 3. $-3\frac{11}{15}$ 5. $\frac{38}{5}\sqrt{3} - \frac{26}{15}$ 7. $118\frac{2}{15}$
9. $\frac{100}{3}$ 11. $\frac{446}{135}$ 13. $x^3/3$
15. Approx. = 2.330; exact = $2\frac{1}{3}$
17. Approx. = 0.497; exact = 0.5   19. Approx. = 10; exact = 10
21. Approx. = exact = $(A/2)(b^2 - a^2) + B(b - a)$
23. Approx. = 10.156; exact = $10\frac{1}{6}$ 25. Approx. = 0.692⁻
27. Approx. = 0.7930⁻   29. Approx. = 2.221

Section 7–6 (pp. 219–221)

1. $f(2) = 9$ 3. Theorem does not apply.
5. Theorem does not apply.   7. $f(\frac{1}{4}) = \frac{1}{2}$, function is continuous
9. $\xi = \frac{7}{2}$ 11. $\xi = (3 + \sqrt{129})/3$ 13. $\xi = (3 \pm \sqrt{57})/12$
15. $\xi = 0, \pm 1, \pm\sqrt{3}$ 17. $\sqrt{14} - \sqrt{6}$ 19. $\frac{2}{3}$

Section 7–7 (pp. 224–225)

1. $\frac{1}{18}(3x - 2)^6 + C$
3. $\frac{1}{3}(2 - x)^{-3} + C$
5. $\frac{1}{5}(2y + 1)^{5/2} + C$
7. $\frac{3}{40}(2x^2 + 3)^{10/3} + C$
9. $-\frac{3}{4}(3 - 2x^2)^{1/3} + C$
11. $-\frac{1}{6}(3x^2 + 2)^{-1} + C$
13. $\frac{2}{9}(x^3 + 1)^{3/2} + C$
15. $\frac{1}{2}(x^3 + 3x^2 + 1)^{2/3} + C$
17. $\frac{1}{40}(x^2 + 1)^4(4x^2 - 1) + C$
19. $\frac{5}{612}(x^3 + 1)^{12/5}(12x^3 - 5) + C$
21. $\frac{33}{5}$
23. $\frac{1}{4}$
25. $\frac{2}{3}(\sqrt{15} - 1)$
27. $\frac{3}{5}[(\sqrt[3]{2} + 2)^5 - 243]$
29. $u = x - (3/x)$ is not continuous in $[-1, 3]$

Section 7–8 (pp. 231–232)

1. $\frac{74}{3}$ 3. 45 5. 6 7. $\frac{3}{2}$ 9. $\frac{4}{15}$
11. $\frac{4}{3}(2\sqrt{2} - 1)$ 13. $\frac{10}{3}$ 15. 72 17. $\frac{5}{12}$ 19. $10\frac{2}{3}$
21. $7\frac{7}{48}$ 23. $\frac{3}{2}$ 25. a) $\frac{5}{4} - \frac{3}{4}\sqrt[3]{4}$ b) $\frac{3}{2} - \frac{3}{4}\sqrt[3]{4}$
27. $9\frac{1}{2}$ 29. $\frac{1}{3}$
31. $1 - (1/a)$; as $a \to +\infty$, area $\to 1$
33. If $p > 1$, area $= \dfrac{1}{1 - p}\left(\dfrac{1}{a^{p-1}} - 1\right)$, as $a \to +\infty$, $A \to \dfrac{1}{p - 1}$.

If $0 < p < 1$, area $= \dfrac{1}{1 - p}(a^{1-p} - 1)$, as $a \to \infty$, $A \to \infty$.

35. If $p_1, p_2, \ldots, p_m$ are the lengths of the subintervals where f is positive and $q_1, q_2, \ldots, q_r$ are the lengths of the subintervals where f is negative, then integral $= \sum_{i=1}^m p_i - \sum_{i=1}^r q_i$.

Section 7–9 (pp. 236–237)

1. $\frac{165}{4}$ 3. $\frac{1}{3}[10^{3/2} - 2^{3/2}]$ 5. $\frac{7}{192}$ 7. 180 ergs
9. a) 40 ergs b) 120 ergs 11. a) 2700 ergs b) 8100 ergs
13. 400,000 kg-m
15. $6562.5w$ kg-m, where $w =$ weight of one m^3 of water
17. $\frac{40}{3}w$ kg-m, where $w =$ weight of one m^3 of water
19. $4w$ kg-m, where $w =$ weight of one m^3 of water
21. $\frac{7}{5}$ 23. $E/24$

Section 7–10 (pp. 241–242)

Throughout $w = 1000$ kg, the weight of a cubic meter of water.
1. $10,000w$ 3. $2000w$ 5. $90,000w$ 7. $14,000w$ 9. $62.5\sqrt{2}w$
11. 4800 kg 13. $96w$ 15. $226.5w$ 17. $11,250w\sqrt{3}$
19. $\frac{1}{3}(169,600\sqrt{3})w$ 23. a) $v = 10 \times 4^{5/7}$ b) $1500(1 - 4^{-2/7})$ joules

Chapter 8

Section 8–1 (pp. 245–246)

1. $\sqrt{5}$ 3. $\frac{15}{4}\sqrt{2}$ 5. $\frac{16}{13}\sqrt{13}$ 7. 2 9. 2
11. $\sqrt{61}/61$ 13. $\dfrac{9}{\sqrt{5}}, \dfrac{9}{2}$ 15. $\dfrac{11}{\sqrt{5}}, 11$ 17. $(\frac{16}{3}, 0), (-\frac{4}{3}, 0)$
19. $(64, -44), (4, -4)$ 21. $x - 2y + 1 \pm 3\sqrt{5} = 0$ 23. $(1, 1)$
25. $x - 3y + 6 = 0$; $3x + y = 0$

Section 8–2 (p. 248)

1. $2x + 3y + k = 0$ 3. $y = mx + 5$ 5. $y + 3 = m(x - 2)$
7. $(x/a) + y/(10 - a) = 1$ 9. $3x + 5y - 10 = 0$ 11. $4x - y + 5 = 0$

13. $8x + y + 3 = 0$
15. $(17 \pm \sqrt{65})x + (25 \mp \sqrt{65})y - 6 \pm 12\sqrt{65} = 0$
17. $x + 17y + k = 0$
19. $2x + 3y = 4$

Section 8-3 (pp. 252-253)

1. $\tan \phi = \frac{11}{10}$
3. $\tan \phi = \frac{1}{12}$
5. $\tan \phi = -\frac{5}{12}$
7. $\tan \phi = -\frac{3}{2}$
9. $\tan \angle CAB = 8$; $\tan \angle ABC = \frac{2}{3}$; $\tan \angle BCA = 2$
11. $\tan \angle CAB = 1$; $\tan \angle ABC = 2$; $\tan \angle BCA = 3$
15. $3x + 4y - 5 = 0$; $x = -1$
17. $(\sqrt{3} - 1)x + (\sqrt{3} + 1)y + 3\sqrt{3} + 1 = 0$;
    $(\sqrt{3} + 1)x + (\sqrt{3} - 1)y + 3\sqrt{3} - 1 = 0$
19. -6; no
21. $\frac{8}{15}$
23. $4x - 7y = 0$; $7x + 4y = 0$
25. $2x + y = 0$; $x - 2y = 0$
27. $x - 7y + 5 = 0$
29. $x - (3 + \sqrt{10})y + 10 + 3\sqrt{10} = 0$
31. $y = 3x + b$; $y = -\frac{1}{3}x + b$

Section 8-4 (pp. 256-257)

1. $x^2 + y^2 + 6x - 8y = 0$
3. $x^2 + y^2 - 4x - 6y + 9 = 0$
5. $x^2 + y^2 - 12x - 8y + 44 = 0$
7. $x^2 + y^2 - 6x - 8y + 9 = 0$
9. $x^2 + y^2 - 4x - 2y - 5 = 0$
11. $x^2 + y^2 - 4x - 6y - 3 = 0$
13. $x^2 + y^2 + 12x - 12y + 36 = 0$
15. Circle; center $(-3, 4)$, radius 5
17. Point $(2, -1)$
19. Circle; center $(-\frac{3}{2}, \frac{5}{2})$, radius 3
21. Circle; center $(-\frac{3}{4}, -\frac{5}{4})$, radius $\frac{3}{4}\sqrt{2}$
23. $2x + y - 5 = 0$
25. $9x - 8y - 26 = 0$
27. $x + 2y + 4 = 0$
29. $5x + 8y - 38 = 0$
31. $3x + 5y - 13 = 0$, $5x - 3y + 35 = 0$
33. No tangent lines as $P(3, 0)$ is inside circle

Section 8-5 (pp. 264-265)

1. Focus: $(2, 0)$; directrix, $x = -2$
3. Focus: $(-3, 0)$; directrix, $x = 3$
5. $y^2 = -16x$
7. $x^2 = -8y$
9. $2y^2 = 9x$
11. Vertex: $(1, 2)$; focus: $(1, 2\frac{1}{4})$; directrix: $y = 1\frac{3}{4}$; axis: $x = 1$
13. Vertex: $(\frac{3}{8}, \frac{9}{16})$; focus: $(\frac{3}{8}, \frac{1}{2})$; directrix: $y = \frac{5}{8}$; axis: $x = \frac{3}{8}$
15. Vertex: $(\frac{3}{2}, -\frac{11}{8})$; focus: $(\frac{3}{2}, -\frac{15}{8})$; directrix: $y = -\frac{7}{8}$; axis: $x = \frac{3}{2}$
17. $y^2 = 12x - 36$
19. $x^2 = 32 - 8y$
21. $4x^2 + 4xy + y^2 + 4x + 32y + 16 = 0$
25. $(2 \pm 6\sqrt{3})x - y - 5 = 0$
27. $3x - 2y + 3 = 0$
29. $3x - 2y - 4 = 0$
31. $8x - y + 9 = 0$

Section 8-6 (pp. 270-271)

	a	b	c	e	foci	vertices
1.	5	4	3	$\frac{3}{5}$	$(\pm 3, 0)$	$(\pm 5, 0)$
3.	4	3	$\sqrt{7}$	$\sqrt{7}/4$	$(\pm\sqrt{7}, 0)$	$(\pm 4, 0)$
5.	$\sqrt{3}$	$\sqrt{2}$	1	$1/\sqrt{3}$	$(0, \pm 1)$	$(0, \pm\sqrt{3})$
7.	$\sqrt{5}$	$\sqrt{2}$	$\sqrt{3}$	$\sqrt{3}/5$	$(0, \pm\sqrt{3})$	$(0, \pm\sqrt{5})$
9.	$\sqrt{11}/2$	$\sqrt{11}/3$	$\sqrt{11}/6$	$1/\sqrt{3}$	$(\pm\sqrt{11}/6, 0)$	$\sqrt{11}/2, 0)$

11. $9x^2 + 25y^2 = 225$
13. $100x^2 + 36y^2 = 3600$
15. $25x^2 + 9y^2 = 225$
17. $7x^2 + 15y^2 = 247$
19. $x^2 + 2y^2 = 18$
21. $16x^2 + 7y^2 = 688$
23. $3x^2 + 4y^2 - 36x = 0$
25. $9x^2 + 5y^2 = 180$
27. $16x^2 + 7y^2 - 16y = 176$, or major axis: $x = 0$; minor axis: $y = \frac{8}{7}$
29. $x + 3y = 1$
31. $25x + 15y + 27 = 0$
35. max $\approx 152, 550$; min $\approx 147, 450$

Section 8-7 (pp. 276-277)

	a	b	c	e	foci	vertices
1.	3	4	5	5/3	$(\pm 5, 0)$	$(\pm 3, 0)$
3.	5	12	13	13/5	$(0, \pm 13)$	$(0, \pm 5)$
5.	$\sqrt{3}$	$\sqrt{2}$	$\sqrt{5}$	$\sqrt{5/3}$	$(\pm \sqrt{5}, 0)$	$(\pm \sqrt{3}, 0)$
7.	$\sqrt{2}$	$\sqrt{5}$	$\sqrt{7}$	$\sqrt{7/2}$	$(\pm \sqrt{7}, 0)$	$(\pm \sqrt{2}, 0)$
9.	$1/\sqrt{3}$	$1/\sqrt{2}$	$\sqrt{5/6}$	$\sqrt{5/2}$	$(\pm \sqrt{5/6}, 0)$	$(\pm(1/\sqrt{3}), 0)$

11. $24x^2 - 25y^2 = 600$ 13. $9x^2 - 7y^2 + 343 = 0$ 15. $4x^2 - y^2 = 32$
17. $y^2 - x^2 = 5$ 19. $4y^2 - 5x^2 = 19$ 21. $15x^2 - y^2 = 15$
23. $252(x - 6)^2 - 4y^2 = 63$
25. Tangent: $5x + 6y + 16 = 0$; normal line: $12x - 10y + 75 = 0$
27. Tangents are $x - y - 2 = 0$ or $3x + 5y + 2 = 0$
29. $7x^2 - 9y^2 = 63$

Section 8-8 (pp. 281-282)

1. $(7, -1)$, $(1, 3)$, $(-3, -\frac{5}{2})$, $(-2, -6)$ 3. $3x' - 7y' = 0$, $2x' - 5y' = 0$
5. Ellipse; center $(-1, 2)$; vertices $(-1, 2 \pm 5)$; foci $(-1, 2 \pm 3)$; $e = \frac{3}{5}$; directrices $y = 2 \pm \frac{25}{3}$
7. Hyperbola: center $(-1, 2)$; vertices $(-1, 2 \pm 3)$; foci $(-1, 2 \pm \sqrt{13})$; $e = \sqrt{13}/3$; directrices $y = 2 \pm 9/\sqrt{13}$; asymptotes $(y - 2) = \pm \frac{3}{2}(x + 1)$
9. Ellipse; center $(-1, 2)$; vertices $(-1 \pm 3, 2)$; foci $(-1 \pm \sqrt{5}, 2)$; $e = \sqrt{5}/3$; directrices $x = -1 \pm (9/\sqrt{5})$
11. Ellipse; center $(1, -1)$, vertices $(1, -1 \pm \frac{3}{2})$; foci $(1, -1 \pm (\sqrt{5}/2))$; $e = \sqrt{5}/3$, directrices $y = -1 \pm (9/(2\sqrt{5}))$
13. Ellipse; center $(2, -1)$; vertices $(2 \pm 2, -1)$; $e = \frac{1}{2}$; foci $(2 \pm 1, -1)$; directrices $x = 2 \pm 4$
15. Hyperbola; center $(-1, -2)$; vertices $(-1 \pm 2, -2)$; $e = \sqrt{10}/2$; foci $(-1 \pm (\sqrt{10}/2), -2)$; directrices, $x = -1 \pm (2/5)\sqrt{10}$; asymptotes: $(y + 2) = \pm \frac{\sqrt{6}}{2}(x + 1)$
17. Two intersecting lines: $y - 2 = \pm(2/\sqrt{3})(x + 1)$
19. $16x^2 + 7y^2 - 64x - 14y - 41 = 0$
21. $x^2 - 5y^2 - 2x + 10y - 20 = 0$ 23. $y^2 - 6x + 2y + 22 = 0$
25. Ellipse: $4x^2 + 3y^2 - 16x - 12y + 16 = 0$
27. Circle: $x^2 + y^2 + 12x - 10y + 29 = 0$

Section 8-9 (pp. 289-290)

1. $5(x')^2 = 9$; two lines $x + 2y = \pm 3$
3. $(x')^2 - (y')^2 = 10$; hyperbola; vertices $(3, 1)$, $(-3, -1)$
5. $x'^2 = -4\sqrt{2}y'$; parabola; focus $(1, -1)$; directrix $x - y + 2 = 0$
7. $17x'^2 + 4y'^2 = 884$; ellipse; vertices $(\mp 3\sqrt{17}, \pm 2\sqrt{17})$
9. $y'^2(1 + \sqrt{5}) - x'^2(-1 + \sqrt{5}) = 20$; hyperbola; vertices $(\pm(2\sqrt{5})^{1/2}, \mp(3\sqrt{5} - 5)^{1/2})$
11. $(\sqrt{13} - 3)x'^2 - (\sqrt{13} + 3)y'^2 = 10$; hyperbola;

vertices $\left(\pm \dfrac{(\sqrt{13} + 3)}{2} \left(\dfrac{5\sqrt{13}}{13}\right)^{1/2}, \left(\dfrac{5\sqrt{13}}{13}\right)^{1/2} \right)$

13. $4x''^2 - y''^2 = 0$; two lines: $3x + y - 5 = 0$, $x + 3y + 1 = 0$
15. $x''^2 = -4\sqrt{5}y''$; parabola; focus $(0, 1)$; vertex $(-1, 3)$; and directrix: $x - 2y + 12 = 0$
17. $28x''^2 + 3y''^2 = 0$; locus a single point $(-\frac{3}{2}, 2)$

19. $9x''^2 + 4y''^2 = 36$; ellipse; center $(2, -4)$
21. $3y''^2 - x''^2 = 12$; hyperbola; center $(-\sqrt{3}, 0)$; vertices $(-2\sqrt{3}, 1), (0, -1)$
27. (a) $x^2 - y^2 = 0$; (b) $x^2 - 3x + 2 = 0$; (c) $x^2 - 2x + 1 = 0$; (d) $x^2 + y^2 = 0$

Chapter 9

Section 9–2 (pp. 296–297)

1. 5   3. $\frac{2}{5}$ 5. $\frac{1}{8}$ 7. 1   9. 0   11. $\frac{1}{8}$
13. $3 \sec^2 3x$ 15. $2s \sec(s^2) \tan(s^2)$
17. $-\csc^3 \frac{1}{3}y \cot \frac{1}{3}y$ 19. $-2 \cot^2 2x + 2$
21. $\sec^2 2s/\sqrt{\tan 2s}$ 23. $x \tan \frac{1}{2}x(x \sec^2 \frac{1}{2}x + 2 \tan \frac{1}{2}x)$
25. $2 \sec 2x \tan^3 2x$ 27. $2(x \cos 2x - \sin 2x)/x^3$
29. $-2 \cos 3x(1 + x^2)^{-2}[x \cos 3x + 3(1 + x^2)(\sin 3x)]$
31. $-6 \cos 3x \csc^3 2x (\sin 3x + \cos 3x \cot 2x)$
33. $-2 \cot ax[(1 + x^2)a \csc^2 ax + x \cot ax]/(1 + x^2)^2$
35. $\sin 4x(1 + \sin^2 2x)^{-1/2}$ 37. $-\sin x \cos(\cos x)$
39. $-[x(1 + \cos^2 3x) + (1 + x^2) \sin 6x]/((1 + x^2)^{3/2}(1 + \cos^2 3x)^{2/3})$
45. Rel. max. at $x = \frac{1}{4}\pi + 2n\pi$; rel. min. at $x = \frac{5}{4}\pi + 2n\pi$, $n = \pm 1, \pm 2, \ldots$
47. Rel. max. at $x = \frac{1}{3}\pi + 2n\pi$; rel. min. at $x = -\frac{1}{3}\pi + 2n\pi$, $n = \pm 1, \pm 2, \ldots$
49. $f'' > 0$ for $\frac{1}{6}\pi < x < \frac{1}{2}\pi$

51. Let x_0 be the solution in the interval $\dfrac{\pi}{2} < x < \pi$ of the equation $\tan x_0 = x_0$. Rel. max. and min. alternate at the points $x = 1 \Big/ \left(\dfrac{1}{\pi}x_0 + n\right)$, $n = \pm 1, \pm 2, \ldots$.

Section 9–3 (pp. 299–300)

1. $-\frac{1}{6} \cos 6x + C$ 3. $-\frac{3}{2} \cot((2x + 1)/3) + C$ 5. $\frac{1}{6} \sec^2 3x + C$
7. $-\frac{1}{2} \cot(2x - 6) - x + C$ 9. $-\cos x(1 - \frac{2}{3}\cos^2 x + \frac{1}{5}\cos^4 x) + C$
11. $-\frac{1}{2} \cos(x^2) + C$ 13. $\frac{1}{4} \sec^4 x + C$ 15. $\frac{1}{6} \tan^6 x + C$
17. $-\frac{1}{6} \tan^3(3 - 2x) + C$ 19. $\frac{1}{2}x - \frac{1}{12} \sin 6x + C$
21. $\frac{1}{4} \cos 2x - \frac{1}{16} \cos 8x + C$ 23. 1   25. $1 - \dfrac{1}{\sqrt{3}}$
27. $(24 - 9\sqrt{3})/24$

Section 9–4 (pp. 305–306)

1. $g(x) = \dfrac{x - 2}{3}$; domain $-\infty < x < \infty$

3. Inverse exists; domain $-\infty < x < \infty$

5. No inverse function

7. Inverse: $y = -1 + \sqrt{4 - x}$; domain, $-\infty < x < 3$

9. Inverse: $y = \sqrt{\dfrac{1}{x}}$; domain: $1 \leq x < \infty$

11. Domain of f: $-\infty < x \leq -2$; inverse: $g(x) = -2 - \sqrt{x + 5}$, $x \geq -5$
     Domain of f: $-2 \leq x < \infty$; $g(x) = -2 + \sqrt{5 + x}$, $x \geq -5$

13. Domain of f: $-\infty < x \leq -\frac{3}{4}$;
     Inverse $g(x) = \dfrac{-3 - \sqrt{8x - 23}}{4}$, $x \geq 23/8$

     Domain of f: $-\frac{3}{4} \leq x < \infty$; inverse $g(x) = \dfrac{-3 + \sqrt{8x - 23}}{4}$, $x \geq 23/8$

15. Domain of f: all $x \neq -1$; inverse $g(x) = \dfrac{x}{1-x}, x \neq 1$

17. Domain of f: all $x \neq -2$; inverse $g(x) = \dfrac{1+2x}{2-x}, x \neq 2$

19. If domain is $-\infty < x \leq -3$, inverse function has domain: $-\infty < x \leq 34$.
    If domain is $-3 \leq x \leq 1$, inverse function has domain: $2 \leq x \leq 34$.
    If domain is $1 \leq x < \infty$, inverse function has domain: $2 \leq x < \infty$.

21. Domain: $-\infty < x < \infty$; inverse function has domain $-\infty < x < \infty$

27. Domain of g is $0 \leq x < \infty$, range is $[0, \infty)$; g is differentiable except at $n = 0, 1, 2, \ldots$.

Section 9–5 (pp. 313–314)

5. a) $\pi/3$,   b) $3\pi/4$,   c) $-\pi/4$
7. a) $\pi/6$,   b) $\pi/3$,   c) $-\pi/6$
9. $f'(x) = 2/(1 + 4x^2)$
11. $\phi'(y) = -1/2\sqrt{y - y^2}$
13. $y'(x) = \arccos x - (x/\sqrt{1 - x^2})$
15. $f'(x) = -[3(1 + x^2) + 2x(1 + 9x^2) \operatorname{arccot} 3x]/((1 + 9x^2)(1 + x^2)^2)$
17. $f'(x) = 1/(1 + x^2)$
19. $f'(x) = 1/(1 + x^2)$
21. $f'(x) = \arcsin 2x$
23. $f'(x) = \sqrt{x^2 - 4}/x$ $(|x| > 2)$
25. $f'(x) = 3/(5 + 4 \cos x)$
27. $f'(x) = 0, f(x) = $ constant
31. $a = -1, b = \pi/2; f(x) = -x + \pi/2$

Section 9–6 (pp. 315–316)

1. $\frac{1}{4} \arcsin 4x + C$
3. $\operatorname{Arcsec} 3x + C$
5. $(1/\sqrt{7}) \arctan \sqrt{7} x + C$
7. $\frac{1}{2} \operatorname{arcsec} \frac{1}{2} x + C$
9. $x - \frac{3}{2} \arctan \frac{1}{2} x + C$
11. $(1/\sqrt{10}) \operatorname{arcsec} (x/\sqrt{5}) + C$
13. $\pi/2$
15. $\pi/6$
17. $\pi/2$
19. $(3\pi - 2)/12\sqrt{2}$

Section 9–7 (pp. 322–324)

1. $\ln 4 = 1.38630$; $\ln 5 = 1.60944$; $\ln 6 = 1.79176$; $\ln 8 = 2.07945$; $\ln 9 = 2.19722$
3. $\ln 24 = 3.17806$; $\ln 25 = 3.21888$; $\ln 27 = 3.29583$; $\ln 30 = 3.40120$; $\ln 32 = 3.46575$
5. $\ln \sqrt{30} = 1.70060$; $\ln \sqrt[3]{270} = 1.86614$; $\ln (\sqrt{2/3}) = -0.20273$; $\ln \sqrt[4]{7.2} = 0.49352$
7. $\frac{3}{2} < \ln 7 < \frac{147}{60} < \frac{11}{4}$
9. $\frac{43}{48} < \ln 3.2 < \frac{47}{30}$
11. $-1 < -0.75 < \ln 0.5 < -0.5$
13. $-500 < \ln 0.001 < -0.999$
15. $-0.005/0.995 < \ln 0.995 < -0.005$
17. $f'(x) = 4/x$
19. $f'(x) = 3(x^2 - 1)/(x^3 - 3x + 1)$
21. $h'(x) = 2 \tan 2x$
23. $F'(x) = 4 (\ln x)^3/x$
25. $G'(x) = 1 + \ln x$
27. $f'(x) = (x^2 + a^2)^{-1/2}$
29. $H'(x) = -4x/(x^4 - 1)$
31. $f'(x) = 2 \ln x(1 + x^2 - x^2 \ln x)/(x(1 + x^2)^2)$
33. 0.91629   35. 0.47001   37. 1.09861   39. 0.47001
43. $y = 2 \ln 3 + \frac{8}{3}(x - 2)$
45. $\frac{5}{6} \ln 15 - \frac{1}{2}$
47. Domain of f: $1 < x < \infty$; range of f: $-\infty < y < \infty$

Section 9–8 (pp. 328–329)

15. a) 4;   b) 2;   c) -5
17. a) 0;   b) -2;   c) 1
23. $\ln 8/\ln 3$; $\ln 12/\ln 7$; $\ln 15/\ln 2$
25. 2.522
27. -1.113
29. 1.613   33. 0
35. Domain of f: all x not in interval $-1 \leq x \leq 1$; range is $-\infty < y < \infty$, $y \neq 0$. Inverse $g(x) = (1 + e^x)/(1 - e^x)$; domain is $-\infty < x < \infty$, $x \neq 0$; range is all y not in $-1 \leq y \leq 1$.

Section 9–9 (pp. 333–334)

1. $f'(x) = -3e^{-3x}$
3. $f'(x) = x^2 e^{4x}(4x + 3)$
5. $H'(x) = e^{3x}\left(3 \ln |x| + \dfrac{1}{x}\right)$
7. $g'(x) = (\sec^2 x)e^{\tan x}$
9. $f'(x) = 3^{5x} 5 \ln 3$
11. $f'(x) = 2^{-7x}[3x^2 - 7(x^3 + 3) \ln 2]$
13. $F'(x) = x^{-1+\sin x}(\sin x + x \cos x \ln x)$
15. $G'(x) = (G(x)/2\sqrt{x})(2 + \ln x)$
17. $H'(x) = H(x)[\ln (\ln x) + (1/\ln x)]$
19. $f'(x) = f(x)[(2/x) + (1/(2x + 3)) - (8x/(x^2 + 1))]$
21. $g'(x) = \tfrac{1}{2}g(x)[(1/(x + 2)) + (1/(x + 3)) - (1/(x + 1))]$
23. $g'(x) = g(x)[(2/x) + (1/[(1 + x^2) \arctan x]) - (2x/(1 + x^2))]$
27. $-\ln |\sec (1/x) + \tan (1/x)| + C$
29. $-\tfrac{1}{3}e^{-3x} + C$
31. $-e^{-\sin x} + C$

Section 9–10 (p. 335)

1. $G(\tfrac{1}{5}) = 2.488$; $G(\tfrac{1}{6}) = 2.522$; $G(\tfrac{1}{7}) = 2.547$;
   $G(\tfrac{1}{8}) = 2.566$; $G(\tfrac{1}{20}) = 2.653$; $G(\tfrac{1}{50}) = 2.691$

Section 9–11 (pp. 337–340)

1. Tangent: $x - 2y - 1 + \pi/2 = 0$; normal: $2x + y - 2 - \pi/4 = 0$
3. Tangent: $x + 2\sqrt{3}y + \tfrac{4}{3}\pi\sqrt{3} + 2 = 0$;
   Normal: $2\sqrt{3}x - y - \tfrac{2}{3}\pi + 4\sqrt{3} = 0$
5. Tangent: $x - ey = 0$; normal: $ex + y - e^2 - 1 = 0$
7. Tangent: $2x - y - e = 0$; normal: $x + 2y - 3e = 0$
9. $(\tfrac{1}{2}\sqrt{2}, \pi/4)$; $\tan \theta = 2\sqrt{2}$
11. $(\ln 2, 4)$; $\tan \theta = \tfrac{4}{33}$
13. $(1, \pi/4)$; $\tan \theta = 2$
15. $(0, 0)$, $\tan \theta = 1$; $(1, e^{-1})$, $\tan \theta = e^{-1}$
17. Rel. max. at $(1, e^{-1})$; point of inflection, $(2, 2e^{-2})$; f increasing for $-\infty < x \le 1$; decreasing for $1 \le x < \infty$; concave downward for $-\infty < x \le 2$; concave upward for $2 \le x < \infty$.
19. Rel. max. at $(0, 1)$; points of inflection $(\pm\tfrac{1}{2}\sqrt{2}, e^{-1/2})$; increasing for $-\infty < x \le 0$; decreasing for $0 \le x < \infty$; concave upward for $-\infty < x \le -\tfrac{1}{2}\sqrt{2}$, $\tfrac{1}{2}\sqrt{2} \le x < \infty$; concave downward for $-\tfrac{1}{2}\sqrt{2} \le x \le \tfrac{1}{2}\sqrt{2}$
21. Rel. max. at $(\pm 1, e^{-1})$; rel. min. at $(0, 0)$; increasing for $-\infty < x \le -1$, $0 \le x \le 1$; decreasing for $-1 \le x \le 0$, $1 \le x < \infty$; points of inflection at $x_1 = -\tfrac{1}{2}(5 + \sqrt{17})^{1/2}$, $x_2 = -\tfrac{1}{2}(5 - \sqrt{17})^{1/2}$, $x_3 = \tfrac{1}{2}(5 - \sqrt{17})^{1/2}$, $x_4 = \tfrac{1}{2}(5 + \sqrt{17})^{1/2}$; concave upward for $-\infty < x \le x_1$, $x_2 \le x \le x_3$, $x_4 \le x < \infty$; concave downward for $x_1 \le x \le x_2$, $x_3 \le x \le x_4$; domain is $x > 0$.
23. Rel. min. $(e^{-1}, -e^{-1})$; decreasing for $0 < x \le 1/e$; increasing for $1/e \le x < \infty$; concave upward for all $x > 0$.
25. Rel. max. $(e^{-2}, 8e^{-2})$; rel. min. $(1, 0)$; increasing for $0 < x \le e^{-2}$, $1 \le x < \infty$; decreasing for $e^{-2} \le x \le 1$; inflection point $(e^{-1}, 2e^{-1})$; concave downward for $0 < x \le e^{-1}$; concave upward $e^{-1} \le x < \infty$.
27. Rel. max. at $(-\pi/4, \tfrac{1}{2}\sqrt{2}e^{\pi/4})$; rel. min. at $(3\pi/4, -\tfrac{1}{2}\sqrt{2}e^{-3\pi/4})$; decreasing for $-\pi/4 \le x \le 3\pi/4$; increasing for $3\pi/4 \le x \le 7\pi/4$. Inflection points at $(0, 1)$ and $(\pi, -e^{-\pi})$; concave upward for $0 \le x \le \pi$; concave downward for $\pi \le x \le 2\pi$. Pattern repeats in each interval of length 2π, but function f is not periodic.
29. $(52\pi/3)$ km/min
31. 440 m/sec
33. $(3/26)$ rad/sec
35. Width $\ge 5\sqrt{5}$ m
37. $\tfrac{1}{3}\pi(6 - 2\sqrt{6})$ rad for any radius

39. $\dfrac{ds}{dt} = \dfrac{10t}{\sqrt{1 - \left(\dfrac{25 - 5t^2}{25}\right)^2}}$

At $t = 1$, $ds/dt = (50/3)$ m/sec
As $t \to 0^+$, $ds/dt \to (50/\sqrt{10})$ m/sec

41. $(-11\pi/10)$ m/hr

43. $t = \dfrac{1}{c} \ln 2$

45. $I = 2e^{-Rt/L}$

Chapter 10

Section 10–1 (pp. 347–349)

1. $x = 3 - (7t/\sqrt{218})$, $y = -7 + (13t/\sqrt{218})$
3. $x = 4 - (5t/\sqrt{29})$, $y = -2 + (2t/\sqrt{29})$
5. $x = 3 + (3t/\sqrt{13})$, $y = 1 + (2t/\sqrt{13})$
7. $5x + 2y = 0$ 	9. $xy = 1$
11. $4(x + 1)^2 + (y - 2)^2 = 4$
13. $(x/2)^{2/3} + (y/2)^{2/3} = 1$
15. $x^2 - 2xy + y^2 - 7x + 6y + 11 = 0$
17. $x^2 + y^2 = 25$ 	19. $xy = 1$ 	21. $x^2 - y^2 = 4$
23. (i) $y = x^2, -1 \le x \le 1$; 	(ii) $y = x^2, 0 \le x \le 1$
    (iii) $y = x^2, x > 0$; 	(iv) $x^2 = y, x \le 0$
25. $T = (v_0 \sin \alpha)/5$, $R = (v_0^2 \cos \alpha \sin \alpha)/5$
    $y = x \tan \alpha - 5x^2(v_0)^{-2} \sec^2 \alpha$
27. $x = a(3 \cos \theta + \cos 3\theta)$, $y = a(3 \sin \theta - \sin 3\theta)$
29. $y^2 = x^2(a + x)/(a - x)$

Section 10–2 (pp. 352–353)

1. $\dfrac{dy}{dx} = -\dfrac{5}{4}$; $\dfrac{d^2y}{dx^2} = 0$

3. $\dfrac{dy}{dx} = -\dfrac{3}{2}t - 4 - t^{-1}$; $\dfrac{d^2y}{dx^2} = \dfrac{3}{4t} - \dfrac{1}{2t^3}$

5. $\dfrac{dy}{dx} = -3te^{2t}$; $\dfrac{d^2y}{dx^2} = \dfrac{3}{2}e^{4t}(1 + 2t)$

7. $\dfrac{dy}{dx} = -\cos t$; $\dfrac{d^2y}{dx^2} = -\dfrac{1}{4}$

9. $\dfrac{dy}{dx} = -\tan \theta$; $\dfrac{d^2y}{dx^2} = \dfrac{1}{3a \sin \theta \cos^4 \theta}$

11. $\dfrac{dy}{dx} = 2t$; $\dfrac{d^2y}{dx^2} = \dfrac{2}{3t^2 + 1}$; $\dfrac{d^3y}{dx^3} = \dfrac{-12t}{(3t^2 + 1)^3}$; $\dfrac{d^4y}{dx^4} = 12\dfrac{(15t^2 - 1)}{(3t^2 + 1)^5}$

13. Tangent: $7x + 2y - 11 = 0$; normal: $2x - 7y - 94 = 0$
15. Tangent: $x + 2y - 5 = 0$; normal: $4x - 2y - 5 = 0$
17. Tangent: $x + ey - 2 = 0$; normal: $e^2x - ey + 2 = 0$
19. Tangent: $x + y - \sqrt{2} = 0$; normal: $x - y = 0$
21. x increasing on $(-\pi/2, \pi/2)$, $(\pi/2, 3\pi/2)$; y decreasing on $(-\pi/2, 0]$ and $[\pi, 3\pi/2)$; y increasing on $[0, \pi/2)$ and $(\pi/2, \pi]$; $4y^2 - x^2 = 4$
23. x decreasing for $t \le -1$; x increasing for $t \ge -1$; y decreasing for $t \le -\frac{1}{2}$, y increasing for $t \ge -\frac{1}{2}$; $x^2 - 2xy + y^2 + x - 2y = 0$

25. x increasing for all t; y increasing for all t; $(1-y)^2 = 1/(x-1)$

27. $\dfrac{dy}{dx} = \dfrac{-2}{[1+f(t)]^2}$

29. (a) $y = (\tan \alpha)x - 5x^2/v_0^2 \cos^2 \alpha$;   (b) $s'(1) = (v_0^2 - 20v_0 \sin \alpha + 100)^{1/2}$, $s'(2) = (v_0^2 - 40v_0 \sin \alpha + 400)^{1/2}$;   (c) distance travelled $= \frac{1}{10}v_0^2 \sin 2\alpha$

Section 10–3 (pp. 358–359)

1. $\frac{3}{4} + (\ln 2)/2$ 3. $\frac{14}{3}$ 5. $6 \arcsin \frac{2}{3}$ 7. $\frac{3}{2}(4^{1/3} - 1)$

9. $\sqrt{2}(1 - e^{-\pi/2})$ 11. $s = \displaystyle\int_1^2 \sqrt{1+9x^4}\, dx$

13. $s = \dfrac{1}{2}\displaystyle\int_1^2 \dfrac{x+1}{\sqrt{x}}\, dx$ 15. $s = \displaystyle\int_1^5 \sqrt{9t^4 + 16t^2 - 4t + 5}\, dt$

17. $s = \dfrac{1}{5}\displaystyle\int_0^4 \sqrt{\dfrac{625 - 9x^2}{25 - x^2}}\, dx$

19. 3.61   21. 2.04   23. 1.54   25. 1.06

27. $8/\sqrt{3}$

29. Let b be any positive number and choose $a = 3b$.

31. If c is the point where the jump occurs, find the length from a to c, then from c to b and add.

Section 10–4 (pp. 364–365)

1. $\kappa(x) = \dfrac{4}{(1 + 16x^2)^{3/2}}$, $R(x) = \dfrac{(1 + 16x^2)^{3/2}}{4}$

3. $\kappa(x) = \dfrac{6}{(4x + 9)^{3/2}}$, $R(x) = |\frac{1}{6}(4x + 9)^{3/2}|$

5. $\kappa(x) = \dfrac{-\sin x}{(1 + \cos^2 x)^{3/2}}$, $R(x) = |\csc x|\,(1 + \cos^2 x)^{3/2}$

7. $\kappa(x) = \dfrac{ab^2 e^{bx}}{\{1 + a^2 b^2 e^{2bx}\}^{3/2}}$, $R(x) = \dfrac{(1 + a^2 b^2 e^{2bx})^{3/2}}{|ab^2 e^{bx}|}$

9. $\kappa(x) = x(2 - x^2)^{-3/2}$, $R(x) = \dfrac{1}{|\kappa(x)|}$

11. $\kappa(x) = -a^4 b(a^4 - c^2 x^2)^{-3/2}$ with $c^2 = a^2 - b^2$

13. $\kappa(y) = -2\csc^2 y \cot y (1 + \csc^4 y)^{-3/2}$, $R(y) = |\tfrac{1}{2}\sin^2 y \tan y|\,(1 + \csc^4 y)^{3/2}$

15. $\kappa(y) = -\dfrac{1}{a}$, $R(y) = a$ $(a > 0)$

17. $\kappa(t) = |t|^{-1}(1 + t^2)^{-3/2}$, $R(t) = \dfrac{1}{|\kappa(t)|}$

19. $\kappa(t) = \dfrac{1}{\sqrt{2}e^t}$, $R(t) = |\sqrt{2}e^t|$

21. $\kappa(t) = -1/3\,|a \sin \theta \cos \theta|$, $R(t) = 3\,|a \sin \theta \cos \theta|$

23. $\kappa(x) = \dfrac{1}{(1 + x^2)^{3/2}}$, max. at $x = 0$

25. $\kappa(x) = 2x^3(x^4 + 1)^{-3/2}$, max. at $x = 1$; min. at $x = -1$

29. $\kappa = -\dfrac{4}{7\sqrt{7}}$, $x_c = \dfrac{\pi}{3} - \dfrac{7\sqrt{3}}{4}$, $y_c = -3$

31. $\kappa = \dfrac{1}{2\sqrt{2}}$, $x_c = -2$, $y_c = 3$

33. $x_c = \dfrac{1}{a}(\sec^3 t)(a^2 + b^2)$, $y_c = -\dfrac{1}{b}(\tan^3 t) \cdot (a^2 + b^2)$

37. Minimum at $-\tfrac{1}{2} + \tfrac{3}{2}\sqrt{2}$ 39. $\kappa(t) = -1/9 |\sin t \cos t|$

Chapter 11

Section 11-1 (p. 368)

1. $(2\sqrt{3}, 2)$, $\left(-\dfrac{3\sqrt{2}}{2}, \dfrac{3\sqrt{2}}{2}\right)$, $(-2, 0)$, $(1, 0)$, $(2, 0)$

3. $(-1, 0)$, $(\sqrt{3}, -1)$, $(2, -2\sqrt{3})$, $\left(\dfrac{3\sqrt{2}}{2}, -\dfrac{3\sqrt{2}}{2}\right)$, $(0, 0)$

5. $\left(3\sqrt{2}, \dfrac{\pi}{4}\right)$, $\left(4, \dfrac{\pi}{2}\right)$, $\left(2, \dfrac{2\pi}{3}\right)$, $\left(1, \dfrac{-\pi}{2}\right)$, $(2, 0)$

7. $\left(2\sqrt{2}, \dfrac{3\pi}{4}\right)$, $\left(3\sqrt{2}, \dfrac{-\pi}{4}\right)$, $\left(2, \dfrac{5\pi}{6}\right)$, $\left(4, \dfrac{\pi}{6}\right)$, $\left(4, \dfrac{\pi}{3}\right)$

9. $r = 5$ is circle of radius 5, center at pole; $\theta = \pi/3$ is straight line through pole, slope $= \sqrt{3}$; curves $r =$ constant and $\theta =$ constant intersect at right angles.

11. $\sqrt{13}$ 13. $\bar{r} = \tfrac{3}{2}\sqrt{5 + \sqrt{6} + \sqrt{2}}$, $\tan \bar{\theta} = \dfrac{\sqrt{2} + 1}{\sqrt{6} + 1}$

Section 11-2 (pp. 372–373)

1. Symmetric with respect to the x axis
3. Symmetric with respect to the y axis
5. None
7. Symmetric with respect to the x axis
9. Symmetric with respect to the y axis
11. Symmetric with respect to the x axis
13. Symmetric with respect to the y axis
15. Symmetric with respect to the x axis, y axis, and pole
17. Symmetric with respect to the x axis, y axis, and pole
19. Symmetric with respect to the x axis
21. Symmetric with respect to the x axis
23. Symmetric with respect to the x axis, y axis, pole
25. Symmetric with respect to the x axis, y axis, and pole
27. Symmetric with respect to the y axis
29. Symmetric with respect to the x axis
31. Symmetric with respect to the y axis
33. None   35. None

39. $(0, 0)$, $\left(\tfrac{1}{2}\sqrt{3}, \dfrac{\pi}{3}\right)$, $(\tfrac{1}{2}\sqrt{3}, \tfrac{2}{3}\pi)$ 41. $(0, 0)$, $(1, 0)$

43. $(\tfrac{12}{5}, \pm \arccos(2/5))$ 45. $\left(1 + \tfrac{1}{2}\sqrt{2}, \dfrac{\pi}{4}\right)$, $\left(1 - \tfrac{1}{2}\sqrt{2}, \dfrac{5\pi}{4}\right)$

Section 11-3 (p. 375)

1. $r \cos \theta = -2$ 3. $\theta = \pi/4$
5. $r(2 \cos \theta + \sqrt{2} \sin \theta) = 4$ 7. $r^2 \sin 2\theta = 8$
9. $r + 2 \cos \theta - 4 \sin \theta = 0$ 11. $r + 2 \cos \theta + 6 \sin \theta = 0$
13. $x^2 + y^2 = 49$ 15. $x^2 + y^2 - 3x = 0$

17. $x = 5$
21. $x^2 - y^2 = 4$
25. $x^2 = 2y$
29. $(x^2 + y^2)^2 = a(3x^2y - y^3)$
33. $(x^2 + y^2)^3 - (x^2 + y^2)^2 + 8x^2y^2 = 0$; degree 6

19. $x + y = 2\sqrt{2}$
23. $(x^2 + y^2)^2 = 2xy$
27. $3x^2 - y^2 + 8x + 4 = 0$
31. $(x^2 + y^2 + 2y)^2 = x^2 + y^2$

Section 11-4 (pp. 379–380)

1. $r \cos(\theta - \pi/3) = 2$
5. $r \cos \theta = 3$
9. $r = 8 \cos \theta$
11. $r^2 - 8r \cos(\theta - \pi/3) - 49 + 14(\sqrt{2} + \sqrt{6}) = 0$
13. $r = \frac{10}{3}\sqrt{3} \cos\left(\theta - \frac{\pi}{3}\right)$
17. $r = \dfrac{1}{2 + \sin \theta}$
*21. $d = |p - r_1 \cos(\theta_1 - \alpha)|$
25. $e = 1$; $r \sin \theta = -2$
29. $e = \frac{1}{6}$; $r \cos(\theta + \pi/2) = -4$
31. $r = \dfrac{ep}{1 + e \sin \theta}$ or $r = \dfrac{-ep}{1 - e \sin \theta}$
    Equation in Cartesian coordinates: If $e = 1$, $x^2 = -2py$.
    If $e \ne 1$: $\dfrac{(1 - e^2)^2}{e^2p^2}(y + h)^2 + \dfrac{1 - e^2}{e^2p^2}x^2 = 1$, $h = \dfrac{e^2p}{1 - e^2}$

3. $r \sin \theta = -1$
7. $r^2 - 6r \cos(\theta - \pi/6) + 5 = 0$

15. $r = \dfrac{10}{1 - 2 \cos \theta}$
19. $r = \dfrac{6}{1 - 2 \cos(\theta + \frac{4}{3}\pi)}$
23. $e = \frac{1}{2}$; $r \cos \theta = -9$
27. $e = 2$; $r \cos(\theta - \pi/3) = -\frac{5}{2}$

Section 11-5 (pp. 384–385)

1. $\dfrac{ds}{d\theta} = |2a|$, $\cot \psi = \cot \theta$
3. $\dfrac{ds}{d\theta} = \left|6 \sin \dfrac{\theta}{2}\right|$, $\cot \psi = \cot \dfrac{\theta}{2}$
5. $\dfrac{ds}{d\theta} = 5\sqrt{1 + \theta^2}$, $\cot \psi = \dfrac{1}{\theta}$
7. $\dfrac{ds}{d\theta} = \sqrt{13 + 12 \cos \theta}$, $\cot \psi = -2 \sin \theta/(3 + 2 \cos \theta)$
9. $\dfrac{ds}{d\theta} = e^{5\theta}\sqrt{26}$, $\cot \psi = 5$
11. $\dfrac{ds}{d\theta} = \dfrac{2}{(2 - \sin \theta)^2}\sqrt{5 - 4 \sin \theta}$, $\cot \psi = \dfrac{\cos \theta}{2 - \sin \theta}$
13. $\dfrac{ds}{d\theta} = \dfrac{2\sqrt{5 + 4 \cos \theta}}{(1 + 2 \cos \theta)^2}$, $\cot \psi = \dfrac{2 \sin \theta}{1 + 2 \cos \theta}$
*15. $16\sqrt{2} - 5\sqrt{5}$
17. $\dfrac{3\pi}{4}$
21. $\kappa = \dfrac{1}{|a|}$
23. $x = \frac{1}{4}$
*25. Line tangent to $r = f(\theta)$ at (r_1, θ_1) is $x \cos \alpha + y \sin \alpha = p$ where $\alpha = \theta_1 + \psi_1 - \pi/2$ and $p = f(\theta_1) \sin \psi_1$

Section 11-6 (pp. 389)

1. $\dfrac{2\pi^3}{3}$
3. $\frac{1}{4}(e^{3\pi} - 1)$
5. $\dfrac{\pi^5}{10}$
7. $\dfrac{1}{\pi}$
9. π
11. 2
13. 6π
15. $\dfrac{\pi}{4}$ if n is odd, $\dfrac{\pi}{2}$ if n is even.
17. $\frac{2}{3}(2\pi + 3\sqrt{3})$
19. $4\left(\sqrt{3} - \dfrac{\pi}{3}\right)$
21. $\frac{1}{4}(9\pi - 12\sqrt{3})$
23. $\frac{1}{2}(2\pi - 3\sqrt{3})$

Chapter 12

Section 12-1 (pp. 394-395)

1. $|AB| = \sqrt{30}$; $|BC| = \sqrt{19}$; $|AC| = \sqrt{21}$; scalene
3. $|AB| = \sqrt{17}$; $|AC| = \sqrt{18}$; $|BC| = \sqrt{35}$; right triangle
5. $|AB| = \sqrt{21}$; $|AC| = \sqrt{51}$; $|BC| = \sqrt{126}$
7. $(-4, \frac{3}{2}, \frac{9}{2})$
9. $(\frac{9}{4}, -\frac{3}{2}, \frac{7}{4}), (\frac{7}{2}, 3, \frac{7}{2}), (\frac{19}{4}, \frac{15}{2}, \frac{21}{4})$
11. $\frac{1}{2}\sqrt{61}, \frac{1}{2}\sqrt{61}, \frac{1}{2}\sqrt{10}$
13. $\frac{1}{2}\sqrt{74}, \frac{1}{2}\sqrt{74}, \frac{1}{2}\sqrt{26}$
15. $P_2(4, 5, -6), Q(1, 3, 0)$
17. Not on a line
19. Not on a line
21. Block extending indefinitely in x and z directions, bounded by planes $y = -2$ and $y = 5$.
23. Interior of sphere of radius 1
27. $6x + 4y - 8z - 5 = 0$; plane

Section 12-2 (pp. 399-400)

1. Dir. nos.; 1, 4, −6; dir. cosines: $\frac{1}{\sqrt{53}}, \frac{4}{\sqrt{53}}, \frac{-6}{\sqrt{53}}$
3. Dir. nos.: 2, −6, −3; dir. cosines: $\frac{2}{7}, \frac{-6}{7}, \frac{-3}{7}$
5. $(3, 5, 2)$
7. $(0, 4, 2)$
9. On line
11. Not on line
13. Parallel
15. Not parallel
17. Perpendicular
19. $\sqrt{2}/3$
21. $41/3\sqrt{190}$
23. $4\sqrt{10}$

Section 12-3 (pp. 402-404)

1. $x = 1 + t, y = 3 - 4t, z = 2 + 2t$
3. $x = 4 - t, y = -2 + 4t, z = -t$
5. $\frac{x-1}{2} = \frac{y}{1} = \frac{z+1}{-3}$
7. $\frac{x-4}{2} = \frac{y}{-1} = \frac{z}{-3}$
9. $\frac{x-3}{2} = \frac{y+1}{0} = \frac{z+2}{0}$
11. Perpendicular
13. Not perpendicular
15. $\frac{x-4}{1} = \frac{y}{-2} = \frac{z-2}{0}$; $\frac{x-3}{0} = \frac{y-1}{1} = \frac{z-4}{-2}$; $\frac{x-2}{1} = \frac{y-5}{-3} = \frac{z}{2}$
17. $\frac{x-2}{a} = \frac{y+1}{b} = \frac{z-5}{-2a+3b}$
19. $(0, \frac{-5}{2}, \frac{-5}{2}), (\frac{-5}{3}, 0, \frac{10}{3}), (\frac{-5}{7}, \frac{-10}{7}, 0)$
21. $x = 3 - t, y = 1 + 3t, z = 5 + t$
23. $A'B': \frac{x-2}{5} = \frac{y}{0} = \frac{z-6}{2}$; $A'C': \frac{x-2}{3} = \frac{y}{0} = \frac{z-6}{1}$; $B'C': \frac{x+3}{8} = \frac{y}{0} = \frac{z-4}{3}$

Section 12-4 (pp. 406-408)

1. $3x + y - 4z + 1 = 0$
3. $3y - 2z - 4 = 0$
5. $2x - z = 1$
7. $9x + y - 5z = 16$
9. $2x + 3y - 4z + 11 = 0$
11. $3x - 2z = 10$
13. $\frac{x+2}{2} = \frac{y-3}{3} = \frac{z-1}{1}$
15. $\frac{x+1}{1} = \frac{y}{0} = \frac{z+2}{2}$
17. $3x + 2y - z = 0$
19. $x - 2y - 3z + 5 = 0$
21. $\frac{x-2}{3} = \frac{y+1}{-2} = \frac{z-3}{4}$
23. $\frac{x-1}{2} = \frac{y+2}{-1} = \frac{z}{4}$
25. $2x - 2y - z - 4 = 0$
27. $2x - 3y - 5z = 7$

29. $2x - 2y - z - 6 = 0$

35. $\begin{vmatrix} A_1 & B_1 & C_1 \\ A_2 & B_2 & C_2 \\ A_3 & B_3 & C_3 \end{vmatrix} = 0$

33. $aA + bB + cC = 0$

Section 12-5 (pp. 411-413)

1. 8/21
3. $\sqrt{14/17}$
5. $x = 11 - 5t, y = -19 + 8t, z = t$
7. $x = -1 - 2t, y = t, z = 3$
9. $(3, 2, -1)$
11. $(-\frac{3}{5}, 0, -\frac{6}{5})$
13. 1
15. $8/\sqrt{29}$
17. $10x - 17y + z + 25 = 0$
19. $14x + 8y - 13z + 15 = 0$
21. $y + x = 1$
23. $(\frac{91}{57}, \frac{31}{57}, \frac{25}{57})$
25. No intersection
27. $x = 3 + 4t, y = -1 + 5t, z = 2 - t$
29. $x = 29t, y = 2 + t, z = 4 - 22t$
31. $5x + 2y + 3z - 3 = 0$
33. $\sqrt{14}; 2\sqrt{14}/\sqrt{15}$

Section 12-6 (pp. 416-417)

1. $x^2 + y^2 + z^2 - 2x - 8y + 4z + 12 = 0$
3. $x^2 + y^2 + z^2 - 2y - 8z - 19 = 0$
5. Sphere: $C(-1, 0, 2), r = 2$
7. No locus
9. Sphere: $C(3, -2, -1), r = 2$
11. $x^2 + y^2 + z^2 + 5x + 7y - 12z + 41 = 0$
13. Plane
15. Plane
17. Parabolic cylinder
19. Elliptic cylinder
21. Circular cylinder
23. Circular cylinder
25. Circle, $C(0, 0, 3), r = 4$
27. No intersection

Section 12-7 (p. 420)

1. (a) $\left(3\sqrt{2}, \frac{\pi}{4}, 7\right)$ b) $(4\sqrt{5}, \arctan 2, 2)$ c) $\left(\sqrt{13}, \arctan\left(\frac{-3}{2}\right), 1\right)$
3. (a) $\left(2\sqrt{3}, \frac{\pi}{4}, \arccos \frac{1}{\sqrt{3}}\right)$ b) $\left(2\sqrt{3}, \frac{-\pi}{4}, \arccos \frac{1}{\sqrt{3}}\right)$ c) $\left(2\sqrt{2}, \frac{2\pi}{3}, \frac{\pi}{4}\right)$
5. (a) $\left(4, \frac{\pi}{3}, 0\right)$ b) $\left(1, \frac{2\pi}{3}, -\sqrt{3}\right)$ c) $\left(\frac{7}{2}, \frac{\pi}{2}, \frac{7}{2}\sqrt{3}\right)$
7. $r^2 + z^2 = 9$
9. $r^2 = 4z$
11. $r^2 = z^2$
13. $r^2 \cos 2\theta = 4$
15. $r = 4 \sin \theta$
17. $\rho = 4 \cos \phi$
19. $\phi = \frac{\pi}{4}$
21. $\rho = \dfrac{\mp 2}{(1 \pm \cos \phi)}$

Chapter 13

Section 13-2 (p. 428)

1. $\mathbf{v} = -2\mathbf{i} + 7\mathbf{j}$
3. $\mathbf{v} = -5\mathbf{i} + 4\mathbf{j}$
5. $\mathbf{v} = -12\mathbf{i} - 4\mathbf{j}$
7. $\mathbf{u} = -\frac{5}{13}\mathbf{i} - \frac{12}{13}\mathbf{j}$
9. $\mathbf{u} = \dfrac{-2}{\sqrt{29}}\mathbf{i} + \dfrac{5}{\sqrt{29}}\mathbf{j}$
11. $B(4, 6)$
13. $A(4, 7)$
15. $A(-3, \frac{1}{2}), B(-5, \frac{7}{2})$
17. $\pm 2\mathbf{i} + 2\mathbf{j}$
19. $-\mathbf{i} + \mathbf{j}$

Section 13-3 (pp. 433-434)

1. $|\mathbf{v}| = 5, |\mathbf{w}| = 5, \cos\theta = -\frac{7}{25}$, proj. $\mathbf{v}$ on $\mathbf{w} = -\frac{7}{5}$
3. $|\mathbf{v}| = 5, |\mathbf{w}| = 13, \cos\theta = -\frac{33}{65}$, proj. $\mathbf{v}$ on $\mathbf{w} = -\frac{33}{13}$
5. $|\mathbf{v}| = \sqrt{13}, |\mathbf{w}| = \sqrt{13}, \cos\theta = 0$, proj. $\mathbf{v}$ on $\mathbf{w} = 0$
7. -1
9. $\dfrac{6}{\sqrt{17}}$
11. 0
13. $\cos\theta = \dfrac{+1}{\sqrt{5}}, \cos\alpha = \dfrac{1}{\sqrt{5}}$
15. $\cos\theta = \dfrac{2}{\sqrt{13}}, \cos\alpha = \dfrac{11}{\sqrt{130}}$
17. $\cos\theta = -1, \cos\alpha = 1$
19. $a = \frac{3}{2}$
21. Impossible
23. $a = \dfrac{(-240 + \sqrt{(240^2 + 69 \cdot 407)})}{407}$
27. $|\overrightarrow{BC}| = \sqrt{93}$; proj. of $\overrightarrow{AB}$ on $\overrightarrow{BC} = -30/\sqrt{93}$; proj. of $\overrightarrow{AC}$ on $\overrightarrow{BC} = 63/\sqrt{93}$
29. Proj. of $\overrightarrow{AC}$ on $\overrightarrow{AB} = \dfrac{33}{5}$; proj. of $\overrightarrow{BC}$ on $\overrightarrow{AB} = \dfrac{-17}{5}$
31. $\mathbf{v}[\overrightarrow{DE}] = \frac{1}{2}\mathbf{v}[\overrightarrow{AC}] - \frac{2}{3}\mathbf{v}[\overrightarrow{AB}]$
33. $\mathbf{v}[\overrightarrow{EF}] = \frac{1}{3}\mathbf{v}[\overrightarrow{AC}] - \frac{1}{2}\mathbf{v}[\overrightarrow{AB}]$

Section 13-4 (pp. 439-440)

1. $-\mathbf{i} + 4\mathbf{j} - 6\mathbf{k}$
3. $5\mathbf{i} - \mathbf{j} + 3\mathbf{k}$
5. $-2\mathbf{i} - 3\mathbf{k}$
7. $\dfrac{1}{\sqrt{29}}(3\mathbf{i} + 2\mathbf{j} - 4\mathbf{k})$
9. $\dfrac{1}{\sqrt{21}}(2\mathbf{i} - 4\mathbf{j} - \mathbf{k})$
11. $B: (3, 3, -4)$
13. $A: (-1, -2, 0)$
15. $A: (\frac{3}{2}, 0, 3), B: (\frac{5}{2}, -2, 5)$
17. $A: (\frac{7}{4}, -\frac{3}{4}, \frac{7}{2}), B: (\frac{3}{4}, \frac{1}{4}, \frac{3}{2})$
19. $4\mathbf{i} - 5\mathbf{j} + \mathbf{k}$

Section 13-5 (p. 443)

1. Linearly independent
3. Linearly dependent
5. Linearly dependent
7. $13\mathbf{u} + 7\mathbf{v} - 2\mathbf{w}$
9. $\frac{1}{4}\mathbf{u} + \frac{3}{8}\mathbf{v} - \frac{9}{8}\mathbf{w}$
11. $-\frac{2}{5}\mathbf{u} + 3\mathbf{v} + \frac{8}{5}\mathbf{w}$
17. $\dfrac{x-1}{-1} = \dfrac{y+4}{4} = \dfrac{z}{7}$

Section 13-6 (pp. 446-448)

1. $\dfrac{16}{7\sqrt{6}}$
3. $\dfrac{1}{\sqrt{3}}$
5. $\dfrac{-11}{29}$
7. $-\frac{2}{7}$
9. $-23/\sqrt{50}$
11. 96
13. 35
15. $\frac{1}{7}(2\mathbf{i} - 6\mathbf{j} + 3\mathbf{k})$
17. $\dfrac{1}{\sqrt{62}}(3\mathbf{i} - 2\mathbf{j} + 7\mathbf{k})$
19. $k = 2, h = \frac{2}{5}$
21. $k = 3, h = \frac{42}{145}$
25. $3g + 5h = 0$
27. $4g - 9h = 0$
29. $h = 1, g = -1$
31. Only if $\mathbf{u} = k\mathbf{v}$ for some $k > 0$
33. $\dfrac{2}{\sqrt{6}}$

Section 13-7 (pp. 452-454)

1. $2\mathbf{i} - 3\mathbf{j} - 7\mathbf{k}$
3. $-4\mathbf{i} - 2\mathbf{j} - 5\mathbf{k}$
5. $20\mathbf{i} + 30\mathbf{j} - 16\mathbf{k}$
7. $\dfrac{7\sqrt{3}}{2}, x - y - z = 0$
9. $\dfrac{3\sqrt{35}}{2}, 11x + 5y + 13z - 30 = 0$

11. $\dfrac{\sqrt{421}}{2}$, $9x + 12y + 14z = 32$ 13. $\dfrac{107}{\sqrt{1038}}$

15. $\dfrac{x+1}{2} = \dfrac{y-3}{-11} = \dfrac{z-2}{-7}$ 17. $\dfrac{x-1}{1} = \dfrac{y+2}{-1} = \dfrac{z-3}{1}$

19. $\dfrac{x-3}{-2} = \dfrac{y}{1} = \dfrac{z-1}{3}$ 21. $\dfrac{x+2}{2} = \dfrac{y-1}{11} = \dfrac{z+1}{8}$

23. $8x + 14y + 13z + 37 = 0$ 25. $2x - z + 1 = 0$

27. $5x - 3y - z = 6$ 29. $4x - 3y - z + 9 = 0$

31. $x - y - z + 6 = 0$ 33. $\dfrac{x-3}{2} = \dfrac{y+2}{1} = \dfrac{z}{2}$

Section 13-8 (pp. 456-457)

1. $V = 20$ 3. Plane: $x + 2y - 5 = 0$
7. $\mathbf{i} + 5\mathbf{j} - 2\mathbf{k}$ 9. $8\mathbf{i} + 10\mathbf{j} - 14\mathbf{k}$

13. $\dfrac{x-2}{160} = \dfrac{y+1}{-45} = \dfrac{z-3}{37}$ 15. $-16\mathbf{v} + 12\mathbf{w}$

17. $[(\mathbf{t} \times \mathbf{u}) \cdot \mathbf{w}]\mathbf{v} - [(\mathbf{t} \times \mathbf{u}) \cdot \mathbf{v}]\mathbf{w}$ 19. $\mathbf{v} = |\mathbf{a}|^2\, p\mathbf{a} - |\mathbf{a}|^{-2}\,(\mathbf{a} \times \mathbf{b})$

Section 13-9 (pp. 460-461)

1. $\mathbf{f}'(t) = 2t\mathbf{i} + (3t^2 - 3)\mathbf{j};\ \mathbf{f}''(t) = 2\mathbf{i} + 6t\mathbf{j}$
3. $\mathbf{f}'(t) = 3\sec^2 3t\,\mathbf{i} - \pi \sin \pi t\,\mathbf{j};\ \mathbf{f}''(t) = 18\sec^2 3t \tan 3t\,\mathbf{i} - \pi^2 \cos \pi t\,\mathbf{j}$
5. $\mathbf{f}'(t) = 2e^{2t}\mathbf{i} - 2e^{-2t}\mathbf{j};\ \mathbf{f}''(t) = 4e^{2t}\mathbf{i} + 4e^{-2t}\mathbf{j}$

7. $\dfrac{2t(t^2+1)}{(t^2+2)^3} + 4t$ 9. $\dfrac{2\sin 4t - 3\sin 6t}{2(\sin^2 2t + \cos^2 3t)^{1/2}}$

11. 0 13. $\dfrac{3t^2 + 40}{t^6}$

15. $\dfrac{d\theta}{dt} = -\dfrac{3}{13t^2 + 4t + 1}$ if $t > 0$; $\dfrac{d\theta}{dt} = \dfrac{3(2t+1)}{|2t+1| \cdot (13t^2 + 4t + 1)}$ if $t < 0,\ t \ne -\tfrac{1}{2}$

19. Straight line, $x - y - 1 = 0$
21. $\mathbf{f}$ and $\mathbf{f}'$ are perpendicular (or $\mathbf{f} \perp \mathbf{f}'$), or the tangent is $\perp$ radius vector.

Section 13-10 (pp. 463-464)

1. $\mathbf{v}(t) = 2t\mathbf{i} - 3\mathbf{j};\ |\mathbf{v}(t)| = s'(t) = \sqrt{4t^2 + 9};$
   $\mathbf{a}(t) = 2\mathbf{i};\ |\mathbf{a}(t)| = 2;\ s''(t) = \dfrac{4t}{\sqrt{4t^2+9}}$

3. $\mathbf{v}(t) = 3\mathbf{i} + t^{-2}\mathbf{j};\ |\mathbf{v}(t)| = s'(t) = \sqrt{9 + t^{-4}};$
   $\mathbf{a}(t) = -2t^{-3}\mathbf{j};\ |\mathbf{a}(t)| = |2t^{-3}|;\ s''(t) = \dfrac{-2}{t^3\sqrt{9t^4+1}}$

5. $\mathbf{v}(t) = \tan t\,\mathbf{i} + \mathbf{j};\ |\mathbf{v}(t)| = s'(t) = |\sec t|;$
   $\mathbf{a}(t) = \sec^2 t\,\mathbf{i};\ |\mathbf{a}(t)| = \sec^2 t;\ s''(t) = \sec t \tan t$

7. $\mathbf{v}(t) = 2e^t\mathbf{i} - 3e^{-t}\mathbf{j};\ |\mathbf{v}(t)| = s'(t) = \sqrt{4e^{2t} + 9e^{-2t}};$
   $\mathbf{a}(t) = 2e^t\mathbf{i} + 3e^{-t}\mathbf{j};\ |\mathbf{a}(t)| = \sqrt{4e^{2t} + 9e^{-2t}};\ s''(t) = \dfrac{4e^{2t} - 9e^{-2t}}{\sqrt{4e^{2t} + 9e^{-2t}}}$

9. $\mathbf{v}(1) = \mathbf{i};\ \mathbf{a}(1) = 2\mathbf{i} + 2\mathbf{j};\ s'(1) = 1;\ s''(1) = 2$

11. $\mathbf{v}\left(\dfrac{\pi}{6}\right) = 2\mathbf{i} + \tfrac{8}{3}\mathbf{j};\ \mathbf{a} = \dfrac{10}{\sqrt{3}}\mathbf{i} + \dfrac{16}{3\sqrt{3}}\mathbf{j};\ s' = \dfrac{10}{3},\ s'' = \dfrac{154}{15\sqrt{3}}$

13. $\mathbf{v} = \tfrac{1}{3}\mathbf{i} - \tfrac{3}{4}\mathbf{j};\ \mathbf{a} = -\tfrac{1}{9}\mathbf{i} + \tfrac{3}{4}\mathbf{j};\ s' = \dfrac{\sqrt{97}}{12};\ s'' = \dfrac{-259}{36\sqrt{97}}$ 17. $\mathbf{T} = \dfrac{4e^{2t}\mathbf{i} - 3e^{-2t}\mathbf{j}}{(16e^{4t} + 9e^{-4t})^{1/2}}$

Section 13-11 (pp. 467-468)

1. $\mathbf{f}' = 2t\mathbf{i} + 2t\mathbf{j} - 3\mathbf{k}$; $\mathbf{f}'' = 2\mathbf{i} + 2\mathbf{j}$
3. $\mathbf{f}' = -2(\sin 2t)\mathbf{i} + 2(\cos 2t)\mathbf{j} + 2\mathbf{k}$
   $\mathbf{f}'' = -4(\cos 2t)\mathbf{i} - 4(\sin 2t)\mathbf{j}$
5. $\mathbf{f}' = \dfrac{2t}{(t^2+1)^2}\mathbf{i} - \dfrac{2t}{(t^2+1)^2}\mathbf{j} + 2t\mathbf{k}$
   $\mathbf{f}'' = \dfrac{-6t^2+2}{(t^2+1)^3}\mathbf{i} - \dfrac{-6t^2+2}{(t^2+1)^3}\mathbf{j} + 2\mathbf{k}$
7. $\mathbf{f}' = -(\sin t)\mathbf{i} + (\sec^2 t)\mathbf{j} + (\cos t)\mathbf{k}$
   $\mathbf{f}'' = -(\cos t)\mathbf{i} + 2(\sec^2 t \tan t)\mathbf{j} - (\sin t)\mathbf{k}$
9. $f' = 9t^2 + 2t + 2$ 11. $f'(t) = 1$ 13. 14
15. $\dfrac{\pi}{8}\sqrt{\pi^2+8} + \ln(\pi + \sqrt{\pi^2+8}) - \tfrac{3}{2}\ln 2$
17. $\mathbf{v}(t) = (\sin t + t\cos t)\mathbf{i} + (\cos t - t\sin t)\mathbf{j} + \mathbf{k}$;
    $s'(t) = \sqrt{t^2+2}$; $\mathbf{a}(t) = (2\cos t - t\sin t)\mathbf{i} - (2\sin t + t\cos t)\mathbf{j}$

Section 13-12 (p. 474)

1. $(3t^2\mathbf{i} - \mathbf{j} + 2\mathbf{k})/\sqrt{5+9t^4}$
3. $(-e^{-2t}\mathbf{i} + e^{2t}\mathbf{j} + t\mathbf{k})/\sqrt{e^{-4t}+e^{4t}+t^2}$
5. $\dfrac{(\cos t + \sin t)\mathbf{i} + e^t(2\cos t - \sin t)\mathbf{j} - e^{-2t}\mathbf{k}}{[(1+2\cos t \sin t) + e^{2t}(2\cos t - \sin t)^2 + e^{-4t}]^{1/2}}$
7. $\mathbf{T} = (1/\sqrt{6})(\mathbf{i} - \mathbf{j} + 2\mathbf{k})$; $\mathbf{B} = \mathbf{N} =$ not defined; $\kappa = 0$;
   $(x-4)/1 = y/(-1) = (z-10)/2$; osculating plane not defined.
9. $\mathbf{T} = (1/\sqrt{3})(\mathbf{i} + \mathbf{j} + \mathbf{k})$; $\mathbf{B} = (1/\sqrt{6})(-\mathbf{i} - \mathbf{j} + 2\mathbf{k})$;
   $\mathbf{N} = (1/\sqrt{2})(-\mathbf{i} + \mathbf{j})$; $\kappa = \sqrt{2}/3$; $(x-1)/1 = y/1 = (z-1)/1$;
   $x + y - 2z + 1 = 0$
11. $\mathbf{T} = (1/\sqrt{5})(\mathbf{j} + 2\mathbf{k})$; $\mathbf{N} = \mathbf{i}$; $\mathbf{B} = (1/\sqrt{5})(2\mathbf{j} - \mathbf{k})$; $\kappa = \tfrac{1}{10}$;
    $(x-2)/0 = y/1 = z/2$; $2y - z = 0$
13. $\mathbf{v} = \mathbf{i} + 6\mathbf{j} + 18\mathbf{k}$; $\mathbf{a} = 3\mathbf{j} + 18\mathbf{k}$; $s' = 19$; $\mathbf{T} = \tfrac{1}{19}\mathbf{v}$; $R = \tfrac{361}{3}$;
    $\mathbf{N} = -\tfrac{1}{19}(6\mathbf{i} + 17\mathbf{j} - 6\mathbf{k})$; $a_T = 18$; $a_N = 3$
15. $\mathbf{v} = \mathbf{i} + 2\mathbf{j} - 2\mathbf{k}$; $\mathbf{a} = 2\mathbf{i} + 4\mathbf{k}$; $s' = 3$; $\mathbf{T} = \tfrac{1}{3}\mathbf{v}$; $R = \tfrac{9}{4}$;
    $\mathbf{N} = \tfrac{1}{3}(2\mathbf{i} + \mathbf{j} + 2\mathbf{k})$; $a_T = -2$; $a_N = 4$

Chapter 14

Section 14-1 (pp. 479-480)

1. $-\sqrt{3-2x} + C$
3. $-\tfrac{1}{5}\cos 5x + C$
5. $\tfrac{1}{2}\tan 2y + C$
7. $-\tfrac{1}{2}\ln|\cos 2x| + C$
9. $2e^{x/2} + C$
11. $\tfrac{1}{12}\arctan\dfrac{3x}{4} + C$
13. $-2\cos\sqrt{x} + C$
15. $\tfrac{1}{2}e^{x^2+1} + C$
21. $\tfrac{1}{3}\sin^3 x + C$
23. $\tfrac{1}{4}\tan^2 2x + C$
25. $\tfrac{1}{2}\ln|1 + (\ln x)^2| + C$
27. $\tfrac{1}{2}\ln|\sin 2x| + C$
29. $\tfrac{1}{4}(\ln x)^4 + C$
31. $-(e^x + 2)^{-1} + C$
33. $\tfrac{1}{14}\sec^7 2x + C$
35. $\ln|e^x + 1| + C$
37. $y - \ln|1 + e^y| + C$
39. $2(\sqrt{7} - 1)$
41. $\tfrac{7}{3}$
43. $\pi/12$
45. $\tfrac{1}{4}$
47. $\tfrac{3}{2}$
49. $\tfrac{1}{3}(e^2 + 1)^{3/2} - \dfrac{2\sqrt{2}}{3}$

Section 14–2 (pp. 482–483)

1. $\frac{1}{2}\ln|x^2 + 2x + 2| + C$
3. $\ln|1 + \tan x| + C$
5. $\frac{1}{2}\ln|\sin 2x| + C$
7. $\frac{x^2}{2} + 2x - 4\ln|x + 1| + C$
9. $\frac{x}{2} + \frac{1}{4}\ln|2x - 1| + C$
11. $\frac{1}{2\sqrt{3}}\arctan\frac{\sqrt{3}}{2}x + C$
13. $-\frac{1}{2}(x^2 + 1)^{-1} + C$
15. $\frac{1}{2}\frac{2^{x^2}}{\ln 2} + C$
17. $\frac{-x^2}{1} - 5x - 14\ln|3 - x| + C$
19. $x + \ln|x^2 + 1| + C$
21. $2x + \ln|x^2 + 2x - 2| + C$
23. $e^{\tan x} + C$
25. $\arcsin\left(\frac{e^x}{\sqrt{3}}\right) + C$
27. $2\sec\sqrt{x} + C$
29. $-\frac{1}{2}\ln|2 + \cos 2x| + C$
31. $-\sqrt{5 - x^2} - 2\arcsin\left(\frac{x}{\sqrt{5}}\right) + C$
33. $-\frac{1}{2}(2x + \cos x)^{-2} + C$
35. $-\frac{\sqrt{3}}{6}\arctan[(\cos 2x)/\sqrt{3}] + C$
37. $\ln 4$
39. 0
41. $\frac{1}{3}\tan^2\frac{3\pi}{8} - \frac{1}{3}$
43. $\ln 8$
45. $1 + \ln(2 + e^2) - \ln(1 + 2e)$

Section 14–3 (pp. 486–487)

1. $\frac{1}{8}(\frac{1}{3} - \sqrt{3})$
3. $-2^{-1/4}(\frac{9}{5}) + \frac{8}{5}$
5. $\frac{\sqrt{3}}{27}$
7. $\frac{\pi}{4}$
9. $\frac{3\pi}{16}$
11. $-2\cos\frac{x}{2} + \frac{4}{3}\cos^3\frac{x}{2} - \frac{2}{5}\cos^5\frac{x}{2} + C$
13. $\ln|\sin x| - \frac{1}{2}\sin^2 x + C$
15. $\frac{1}{4}\tan^2 2x + C$
17. $\frac{1}{2}\tan 2x + \frac{1}{6}\tan^3 2x + C$
19. $-\frac{1}{7}\csc^7 x + \frac{2}{5}\csc^5 x - \frac{1}{3}\csc^3 x + C$
21. $-x + \tan x + C$
23. $x - \frac{1}{3}\cot^3 x + \cot x + C$
25. $-\frac{1}{2}\cos(x^2) + \frac{1}{6}\cos^3(x^2) + C$
27. $\frac{1}{3}\sin^3 x + C$
29. $\frac{1}{3}\tan^3 x + C$
31. $-\frac{1}{4}\cos^4 x + C$
35. $(15/154)2^{-3/4}$

Section 14–4 (pp. 489–490)

1. $\frac{\pi}{6}$
3. $-\frac{\pi}{18}$
5. $\frac{32}{3}$
7. $(2/\sqrt{5})\operatorname{arcsec}(x/\sqrt{5}) + C$
9. $\frac{1}{3}(9 - x^2)^{3/2} - 9(9 - x^2)^{1/2} + C$
11. $\frac{1}{12}\sqrt{2x^2 + 7}(2x^2 - 14) + C$
13. $\frac{1}{5a^4}(a^2u^2 + b^2)^{5/2} - \frac{b^2}{3a^4}(a^2u^2 + b^2)^{3/2} + C$
15. $\sqrt{x^2 - a^2} - a\cdot\operatorname{arcsec}\frac{x}{a} + C$
17. $-(1/a^2x)\sqrt{x^2 + a^2} + C$
19. $-a^{-4}\left(\frac{\sqrt{a^2 - x^2}}{x} + \frac{(a^2 - x^2)^{3/2}}{3x^3}\right) + C$

21. $\sqrt{2}/32$

23. $3^{-12} - \dfrac{\pi}{6}$

25. $\tfrac{2}{9}(1 - 3^{-1/2})$

27. $\dfrac{25\pi}{16}$

29. For type (b), $-\dfrac{\pi}{2} < \theta < \dfrac{\pi}{2}$;

for type (c), $-\pi \le \theta < -\dfrac{\pi}{2}$ or $0 \le \theta < \dfrac{\pi}{2}$

31. $-\dfrac{1}{3} \dfrac{(1 - x^2)^{3/2}}{x^3} + C$

33. n is a nonnegative even integer

Section 14-5 (pp. 492–493)

1. $\arctan(x + 2) + C$

3. $x - \ln|x^2 + 2x + 5| - \tfrac{3}{2}\arctan\dfrac{(x + 1)}{2} + C$

5. $\operatorname{arcsec}(x + 2) + C$

7. $\ln|x - 1 + \sqrt{5 - 2x + x^2}| + C$

9. $\tfrac{1}{2}\ln|x^2 + 2x + 5| + \arctan\dfrac{x + 1}{2} + C$

11. $\tfrac{1}{3}x^3 - \tfrac{3}{2}x^2 + 8x - (21/2)\ln|x^2 + 3x + 1| - \dfrac{47}{2\sqrt{5}}\ln\left|\dfrac{\sqrt{5} + 2x + 3}{\sqrt{5} - 2x - 3}\right| + C$

13. $3\sqrt{x^2 + 2x} + \ln|x + 1 + \sqrt{x^2 + 2x}| + C$

15. $\dfrac{11x - 19}{4(x^2 - 2x + 3)} + \dfrac{11\sqrt{2}}{8}\arctan\dfrac{x - 1}{\sqrt{2}} + C$

17. $\dfrac{8x - 17}{23(2x^2 + 3x + 4)} + \dfrac{16}{(23)^{3/2}}\arctan\dfrac{4x + 3}{\sqrt{23}} + C$

19. $\dfrac{9x - 13}{\sqrt{x^2 - 2x + 2}} + C$

21. $\dfrac{5x - 3}{4\sqrt{x^2 + 2x - 3}}$

23. $\tfrac{2}{7}\sqrt{7}(\arctan\sqrt{7} - \arctan\tfrac{3}{7}\sqrt{7})$

25. $\arctan(3) - (\pi/4)$

27. $\tfrac{3}{40} + \tfrac{5}{16}\arctan(\tfrac{1}{2})$

29. Integrand is infinite at $x = \pm 1$. Integrable over any interval not containing $x = 1$, $x = -1$.

Section 14-6 (pp. 497–499)

1. $\left(\dfrac{x^2}{2}\right)\ln x - \dfrac{x^2}{4} + C$

3. $-x^2 \cos x + 2x \sin x + 2\cos x + C$

5. $x(\ln x)^2 - 2x \ln x + 2x + C$

7. $x \arctan x - \tfrac{1}{2}\ln|1 + x^2| + C$

9. $x \operatorname{arcsec} x - \ln|x + \sqrt{x^2 - 1}| + C$

11. $\tfrac{1}{4}(2x^2 - 1)\arcsin x + \tfrac{1}{4}x\sqrt{1 - x^2} + C$

13. $-2x \cot(x/2) + 4\ln|\sin(x/2)| + C$

15. $3x \tan 3x - \dfrac{9x^2}{2} + \ln|\cos 3x| + C$

17. $2x^3 \arcsin 2x + \tfrac{1}{4}\sqrt{1 - 4x^2} - \tfrac{1}{12}(1 - 4x^2)^{3/2} + C$

19. $-\sqrt{2}x \cos\sqrt{2}x + \sin\sqrt{2}x + C$

21. $e^{x^2}(x^2 - 1) + C$

23. $\frac{1}{5}e^x(\sin 2x - 2\cos 2x) + C$ 25. $\frac{1}{10}e^{-x}(3\sin 3x - \cos 3x) + C$
27. $e^{ax}(a^2 + b^2)^{-1}(a\sin bx - b\cos bx) + C$
29. $\frac{1}{3}\sin^3 2x + C$
31. $\frac{5x}{16} - \frac{1}{4}\sin 2x + \frac{3}{64}\sin 4x + \frac{1}{48}\sin^3 2x + C$
33. -2 35. $-\frac{\pi}{12} + \frac{1}{2}\left(\frac{\pi^2}{9} + \frac{1}{4}\right)\arctan\frac{2\pi}{3}$
37. $\frac{4}{25}(1 + e^{3\pi/4})$
39. $-\cos x\left(\frac{\sin^{n-1}x}{n} + \frac{(n-1)\sin^{n-3}x}{n(n-2)} + \frac{(n-1)(n-3)}{n(n-2)(n-4)}\sin^{n-5}x\right)$
$\qquad\qquad\qquad\qquad\qquad\qquad + \frac{(n-1)(n-3)(n-5)}{n(n-2)(n-4)}\int \sin^{n-6}x\,dx$
41. $\frac{b}{a+b}(g'(x)f(x) - g(x)f'(x)) + C$
43. $\int x^5(\log x)^2\,dx = \frac{1}{6}x^6[(\log x)^2 - \frac{1}{3}(\log x) + \frac{1}{18}] + C$
47. $a_0 = -\frac{2}{\ln 2}\left[1 + \frac{1!}{\ln 2} + \frac{2!}{(\ln 2)^2} + \frac{3!}{(\ln 2)^3} + \cdots + \frac{(n-1)!}{(\ln 2)^{n-1}}\right]$;
$\quad a_1 = 1, a_2 = 1, a_3 = 2!, \ldots, a_n = (n-1)!$

Section 14–7 (pp. 505–506)

1. $\frac{x^2}{2} + 5x + 14\ln|x - 2| + C$
3. $\frac{x^2}{2} - 4x + 13\ln|x + 2| - \ln|x + 1| + C$
5. $x + 11\ln|x - 2| - 6\ln|x - 1| + C$
7. $x + (x - 2)^{-1} + 2\ln|x - 2| + C$
9. $-\frac{8}{3}\ln|x + 2| + \frac{5}{2}\ln|x + 1| + \frac{1}{6}\ln|x - 1| + C$
11. $2\ln|x + 3| - \frac{1}{3}\ln|x + 2| + \frac{1}{3}\ln|x - 1| + C$
13. $\frac{1}{2}\ln|x + 1| + \frac{1}{4}\ln|x^2 + 1| + \frac{5}{2}\arctan x + C$
15. $\frac{1}{9}\ln|x + 1| - \frac{4}{3}(x + 1)^{-1} - \frac{1}{9}\ln|x - 2| + C$
17. $-\frac{1}{9}\ln|x + 2| + \frac{1}{9}\ln|x - 1| - \frac{5}{3}(x - 1)^{-1} + C$
19. $-\frac{3}{4}\ln|x + 1| + \frac{7}{4}\ln|x - 1| + \frac{1}{2}(x - 1)^{-1} - \frac{1}{2}(x - 1)^{-2} + C$
21. $\ln|x^2 + 4| + \frac{11}{6}\arctan\frac{x}{2} - \frac{1}{2}\ln|x^2 + 1| - \frac{2}{3}\arctan x + C$
23. $\frac{1}{4}\ln|x - 1| - \frac{1}{8}\ln|x + 1| + \frac{7}{8}\ln|x - 3| - \ln|x - 2| + C$
25. $-\frac{4}{25}\ln|x - 1| - \frac{2}{5}(x - 1)^{-1} + \frac{2}{25}\ln|x^2 + 4| + \frac{19}{50}\arctan\frac{x}{2} + C$
27. $\frac{1}{9}(x + 1)^{-1} - \frac{5}{27}\ln|x + 1| + \frac{32}{27}\ln|x - 2| - \frac{5}{9}(x - 2)^{-1} + C$
29. $2\ln|x^2 + 4| + \frac{5}{4}\arctan\frac{x}{2} + \frac{1}{2}(16 - 11x)(x^2 + 4)^{-1} + C$
31. $\frac{14}{27}\ln|x - 3| + \frac{13}{54}\ln|x^2 + 3x + 9| + \frac{7}{3\sqrt{27}}\arctan\frac{2x + 3}{\sqrt{27}}$
33. $\frac{51}{256}\arctan\frac{x}{2} - \frac{77x}{128(x^2 + 4)} + \frac{17x}{16(x^2 + 4)^2}$

Section 14–8 (pp. 508–509)

1. $\frac{4}{3}(x + 2)^{3/2} - 2(x + 2)^{1/2} + C$
3. $\frac{2}{5}(x + 1)^{5/2} - \frac{2}{3}(x + 1)^{3/2} + C$

5. $\frac{1}{12}(3x - 1)^{4/3} - \frac{5}{3}(3x - 1)^{1/3} + C$
7. $\frac{3}{64}(2x + 1)^{8/3} - \frac{3}{20}(2x + 1)^{5/3} + \frac{3}{16}(2x + 1)^{2/3} + C$
9. $2(x + 4)^{1/2} + 4 \ln |{-2} + \sqrt{x + 4}| - 2 \ln |x| + C$
11. $x + 6\sqrt{x} + 6 \ln |\sqrt{x} - 1| + C$
13. $\frac{1}{3}(x^2 - 4)^{3/2} + 4(x^2 - 4)^{1/2} + C$
15. $\frac{1}{5}(x^2 + 4)^{5/2} - 2(x^2 + 4)^{3/2} + 8(x^2 + 4)^{1/2} + C$
17. $-5(16 - x^2)^{3/2} + \frac{1}{5}(16 - x^2)^{5/2} + C$
19. $\frac{1}{a} \ln |a - \sqrt{a^2 - x^2}| - \frac{1}{a} \ln |x| + C$
21. $\sqrt{a^2 - x^2} + a \ln |a - \sqrt{a^2 - x^2}| - a \ln |x| + C$
23. $\sqrt{x^2 - a^2} - a \arctan (\sqrt{x^2 - a^2}/a) + C$
25. $3x^{1/3} - 6x^{1/6} + 6 \ln |1 + x^{1/6}| + C$
27. $\frac{1}{32} \ln \left| \frac{\sqrt{x^2 + 4} + 2}{\sqrt{x^2 + 4} - 2} \right| - \frac{\sqrt{x^2 + 4}}{8x^2} + C$
29. $\frac{1}{6} \ln |x| - \frac{1}{6} \ln |\sqrt{x^2 + 9} + 3| - \frac{1}{2}x^{-2}\sqrt{x^2 + 9} + C$
33. $-\frac{1}{4} \cot^2 \left(\frac{x}{2}\right) - \frac{1}{2} \ln \left| \tan \frac{x}{2} \right| + C$
35. $\frac{1}{2} \ln \left| \frac{1 + \tan \frac{\theta}{2}}{1 - \tan \frac{\theta}{2}} \right| - \frac{\tan \frac{\theta}{2}}{\left(1 + \tan \frac{\theta}{2}\right)^2} + C$ 37. $\frac{1}{3} \ln |4 \tan (\frac{1}{2}\theta) + 2| - \frac{1}{3} \ln |4 \tan (\frac{1}{2}\theta) + 8|$

Chapter 15

Section 15-1 (pp. 516–517)

1. 160π
3. $\frac{\pi}{54}$; $R = \{(x, y): 0 \leq x \leq \frac{1}{2}, 0 \leq y \leq \frac{1}{3}(1 - 2x)\}$
5. $\frac{4\pi}{3} a^3$
7. $\frac{512}{15} \pi$; $R = \{(x, y): -2 \leq x \leq 2, 0 \leq y \leq 4 - x^2\}$
9. $\frac{\pi}{3}(3\sqrt{3} - \pi)$; $R = \{(x, y): 0 \leq x \leq \frac{\pi}{3}, 0 \leq y \leq \tan x\}$
11. $2\pi[(\ln 2)^2 - 2 \ln 2 + 1]$
13. $R = \{(x, y): 0 \leq x \leq 4, \sqrt{x} \leq y \leq 2\}; 8\pi$
15. $\frac{16\pi}{3}(2\sqrt{2} - 1)$
17. 16π; $R = \{(x, y): y^2 \leq x \leq 8 - y^2, 0 \leq y \leq 2\}$
19. $\frac{5\pi}{14}$; $R = \{(x, y): 0 \leq x \leq 1, x^3 \leq y \leq \sqrt{x}\}$
21. $\frac{\pi}{2}\left(4 + \frac{e^4 - e^{-4}}{2}\right)$; $R = \left\{(x, y): -2 \leq x \leq 2, 0 \leq y \leq \frac{e^x + e^{-x}}{2}\right\}$
23. $\frac{45}{4} \pi$; $R = \{(x, y): \frac{1}{2}(y^2 - 3y) \leq x \leq y - 2, 1 \leq y \leq 4\}$
25. $5\pi^2 a^3$
27. $2\pi^2 a^2 b$
29. 4
31. $\frac{3}{2}\sqrt{3}$

Section 15-2 (pp. 520–521)

1. $\frac{32\pi}{5}$
3. $\frac{\pi}{7}$; $R = \{(x, y): 0 \leq x \leq 1, 0 \leq y \leq x^3\}$
5. $2\pi ((\ln 2)^2 - 2 \ln 2 + 1)$
7. $2\pi^2$
9. $\frac{256\pi}{3}$

11. $\dfrac{2\pi}{5}$; $R = \{(x, y): 0 \leq x \leq 1, x^3 \leq y \leq \sqrt{x}\}$ 13. $\dfrac{4\pi}{3}$

15. $\dfrac{128\pi}{3}$ 17. $2\pi^2 + 4\pi$ 19. $V = \pi \displaystyle\int_a^b \{[f(x) + c]^2 - c^2\}\, dx$

21. $2\pi \displaystyle\int_\alpha^\beta (y + \gamma)\, g(y)\, dy$

Section 15-3 (pp. 526–527)

1. 1
3. 2
5. Divergent
7. $1/(1 - p)$, $p < 1$
9. Divergent
11. 2
13. 1
15. 9
17. 0
19. $\tfrac{1}{2}$
21. Divergent
23. 4
25. $2(2\sqrt{2} - 1)$
27. π
29. Divergent
31. $2\sqrt{2}$
33. $\dfrac{4\pi\sqrt{3}}{9}$
35. π
37. $\tfrac{1}{2}$

Section 15-4 (pp. 529–530)

1. $\tfrac{1}{4}[2\sqrt{5} + \ln(2 + \sqrt{5})]$ 3. $3a$
5. $2\ln(1 + \sqrt{2})$ 7. $\tfrac{1}{27}[(13)^{3/2} + (40)^{3/2} - 16]$
9. $\tfrac{1}{4}(4\sqrt{17} - \sqrt{2}) + \tfrac{1}{4}\ln\left(\dfrac{4 + \sqrt{17}}{1 + \sqrt{2}}\right)$
11. Not finite   13. $\sqrt{3}$
15. $\tfrac{1}{4}[2\sqrt{5} + \ln(2 + \sqrt{5})]$ 17. $\ln(\sqrt{3} + \sqrt{2})$
19. $l_1(T) \leq l(T)$ for all T on $[a, b] \Leftrightarrow \{[F'(t)]^2 + [G'(t)]^2\}$
$\leq \{[f'(t)]^2 + [g'(t)]^2\}$ for all t on $[a, b]$.

Section 15-5 (pp. 534–535)

1. $\dfrac{\pi}{9}(17\sqrt{17} - 1)$ 3. $\dfrac{8\pi}{3}(5\sqrt{5} - 2\sqrt{2})$

5. $\dfrac{13\pi}{3}$ 7. $\dfrac{1179\pi}{256}$

9. $\dfrac{\pi}{4}(4 + e^2 - e^{-2})$ 11. $\dfrac{\pi}{4}(15 + 4\ln 2)$

13. $\dfrac{12\pi a^2}{5}$ 15. $\dfrac{6\pi}{5}(\sqrt{2} + 1)$

17. $\dfrac{64\pi a^2}{3}$

19. $\dfrac{2\pi ab}{e}(\arcsin e + e\sqrt{1 - e^2})$ where $e = \dfrac{c}{a}$, $a^2 - b^2 = c^2$

21. $\pi\left[-\dfrac{\sqrt{17}}{4} + \ln|\sqrt{17} + 4| + \sqrt{2} - \ln|\sqrt{2} + 1|\right]$

23. $4\pi^2 ab$ 25. $\dfrac{\pi}{4}(15 + 4\ln 2)$

27. $\dfrac{\pi}{2}[4\sqrt{17} + \ln|4 + \sqrt{17}|]$

29. $2\pi \displaystyle\int_a^b [f(x) + k]\sqrt{1 + [f'(x)]^2}\, dx \equiv 2\pi \displaystyle\int_a^b (y + k)\, ds$

31. 16π 33. $2\pi \displaystyle\int_{\theta_1}^{\theta_2} f(\theta) \cos\theta \sqrt{[f(\theta)]^2 + [f'(\theta)]^2}\, d\theta$

Section 15-6 (pp. 538-539)

1. $(\frac{13}{4}, 0)$
3. $(\frac{2}{5}, 0)$
5. $(\frac{11}{9}, 0)$
7. $(\frac{7}{8}, \frac{11}{8})$
9. $(\frac{23}{17}, \frac{35}{17})$
11. With axes on outer sides of figure, $\bar{x} = 2, \bar{y} = 1.5$
13. Origin at point of tangency of circle and rectangle;
$$\bar{x} = \frac{4\pi - 5}{2\pi + 5}, \bar{y} = \frac{-5}{4\pi + 10}$$

Section 15-7 (p. 544)

1. $(\bar{x}, \bar{y}) = \left(\frac{18}{5}, \frac{3\sqrt{6}}{4}\right); R = \{(x, y): 0 \le x \le 6, 0 \le y \le \sqrt{4x}\}$
3. $(\bar{x}, \bar{y}) = \left(\frac{\pi}{2}, \frac{\pi}{8}\right); R = \{(x, y): 0 \le x \le \pi, 0 \le y \le \sin x\}$
5. $(\bar{x}, \bar{y}) = \left(0, \frac{4a}{3\pi}\right); R = \{(x, y): -a \le x \le a, 0 \le y \le \sqrt{a^2 - x^2}\}$
7. $\bar{x} = \frac{67}{18(8\sqrt{2} - 7)}, \bar{y} = \frac{2(72\sqrt{2} - 53)}{15(8\sqrt{2} - 7)}$
9. $\bar{x} = \frac{14 \ln 2 - 3}{6 \ln 2 - 2}; \bar{y} = [7(\ln 2)^2 - 6 \ln 2 + 2]/6(6 \ln 2 - 2)$
11. $\bar{x} = \frac{5}{32}; \bar{y} = \frac{5}{64}$
15. $\bar{x} = \frac{3}{5}, \bar{y} = \frac{12}{35}$

Section 15-8 (pp. 546-547)

1. $\bar{x} = \frac{3h}{4}; R = \left\{(x, y): 0 \le x \le h, 0 \le h, 0 \le y \le \frac{ax}{h}\right\}$
3. $\bar{x} = \frac{3a}{8}$
5. $\bar{x} = \frac{256 \ln 2 - 150}{64 \ln 2 - 33}; R = \left\{(x, y): 0 \le x \le 3, 0 \le y \le \frac{x}{\sqrt{4 - x}}\right\}$
7. $\bar{x} = \frac{\pi}{2}$
9. $\bar{x} = \frac{5}{9}; R = \{(x, y): 0 \le x \le 1, x^2 \le y \le \sqrt{x}\}$
11. $\bar{x} = \frac{25}{32}; R = \{(x, y): -1 \le x \le 2, x^2 \le y \le x + 2\}$
17. On axis: $\frac{6h^2 + 8ah + 3a^2}{12h + 8a}$
19. $\frac{1}{2}$

Section 15-9 (pp. 550-551)

1. $\bar{x} = \frac{263}{165}, \bar{y} = \frac{393}{352}$
3. $\bar{x} = \bar{y} = \frac{2a}{5}$
5. $\bar{x} = \frac{17\sqrt{17} - 1}{3[4\sqrt{17} + \ln(4 + \sqrt{17})]}, \bar{y} = \frac{132\sqrt{17} - \ln(4 + \sqrt{17})}{16[4\sqrt{17} + \ln(4 + \sqrt{17})]}$
7. $\begin{cases} \bar{y} = \frac{L\bar{y}}{L} \text{ where} \\ L = \sqrt{1 + e^4} - \sqrt{1 + e^2} - \ln\left(\frac{1 + \sqrt{1 + e^4}}{e^2}\right) + \ln\frac{1 + \sqrt{1 + e^2}}{e} \\ \text{and} \quad L\bar{y} = \frac{1}{2}[e^2\sqrt{1 + e^4} + \ln(e^2 + \sqrt{1 + e^4}) - e\sqrt{1 + e^2} - \ln(e + \sqrt{1 + e^2})] \end{cases}$
9. $\bar{x} = \frac{2(25\sqrt{5} + 1)}{5(5\sqrt{5} - 1)}$
11. $\bar{x} = \frac{169\sqrt{13} - 31}{130\sqrt{13} - 10}$
13. $\bar{x} = \frac{38}{10 + 9\sqrt{5} \arcsin \sqrt{5}/3}$

15. $\bar{y} = \dfrac{391\sqrt{17} + 1}{10(17\sqrt{17} - 1)}$

17. $\bar{y} = \dfrac{a(6 + e^2 - 3e^{-2})}{16(1 - e^{-1})}$

Section 15-10 (pp. 552-553)

1. 2π 3. $\dfrac{\pi^2}{2}$ 5. $\dfrac{242\pi}{5}$

7. $\dfrac{8\pi}{105}$ 9. $V = 2a^2b\pi^2$, $S = 4ab\pi^2$ 11. $\pi r^2 h$

Section 15-11 (pp. 557-559)

1. $y = -4.9t^2 + 150t + 80$; highest point $= 1228$ m
3. $x = t^2 + 12t + 1$; $v = 2t + 12$ 5. $x = \tfrac{1}{2}(5e^{2t} - 3)$; $v = 5e^{2t}$
7. $x = x_0 + (1/k) \log(1 + v_0 kt)$; $v = v_0/(1 + v_0 kt)$
9. Distance $= 112.5 \times 10^4 \sqrt{3}/g$, $t = 1500\sqrt{3}/g$; $g = 9.78$
11. $T = (375 + \sqrt{(375)^2 + 12{,}000\,g})/g$; $g = 9.78$
13. $10^4 e^{t/2}$ 15. $[(3.5)\log 10]/\log 2$
17. $400 e^{-1/2}$ grams   19. $20 + 20 \cdot 2^{-t/10}$
21. $(5600 \log 3)/\log 2$ years   23. 5.8%

Section 15-12 (pp. 562-563)

1. Approx $= 21.1$; exact $= 21$
3. Approx $= 0.696$; exact $= \ln 2 \approx 0.693$
5. Approx $= 0.880$; exact $= \ln(1 + \sqrt{2}) \approx 0.8813$
7. $\tfrac{26}{3}$ 9. $\tfrac{38}{3}$
11. Approx $= 0.6933$; exact $= \ln 2 = 0.69315$
13. Approx $= 0.8814$; exact $= \ln(1 + \sqrt{2}) \approx 0.8814$
15. Approx $= 0.6045$; exact $= \pi\sqrt{3}/9 \approx 0.6046$
17. Approx $= 0.8358^-$
19. Approx $= 1.0948$
21. Approx $= 0.4298^+$

Appendix 1

Section 1 (p. App-7)

5. $\sin 3\theta = 3\sin\theta - 4\sin^3\theta$; $\cos 3\theta = 4\cos^3\theta - 3\cos\theta$
7. $\sin(\theta/2) = \pm\sqrt{(1 - \cos\theta)/2}$; $\cos(\theta/2) = \pm\sqrt{(1 + \cos\theta)/2}$
9. $\sin 15° = \tfrac{1}{2}\sqrt{2 - \sqrt{3}}$; $\cos 15° = \tfrac{1}{2}\sqrt{2 + \sqrt{3}}$;

   $\sin 22\tfrac{1}{2}° = \tfrac{1}{2}\sqrt{\sqrt{2} - 1}$; $\cos 22\tfrac{1}{2}° = \tfrac{1}{2}\sqrt{\sqrt{2} + 1}$;

   $\tan 75° = 2 + \sqrt{3}$; $\sec 75° = 2/\sqrt{2 - \sqrt{3}}$

11. $\sin\left(\dfrac{3\pi}{2} - \theta\right) = -\cos\theta$; $\cos\left(\dfrac{3\pi}{2} - \theta\right) = -\sin\theta$;

    $\sin\left(\dfrac{3\pi}{2} + \theta\right) = -\cos\theta$; $\cos\left(\dfrac{3\pi}{2} + \theta\right) = \sin\theta$;

    $\sin(2\pi - \theta) = -\sin\theta$; $\cos(2\pi - \theta) = \cos\theta$

17. $\sin\left(\dfrac{\pi}{6} - \theta\right) = \tfrac{1}{2}\cos\theta - \tfrac{1}{2}\sqrt{3}\sin\theta = \cos\left(\dfrac{\pi}{3} + \theta\right)$;

    $\cos\left(\dfrac{5\pi}{6} - \theta\right) = -\tfrac{1}{2}\sqrt{3}\cos\theta + \tfrac{1}{2}\sin\theta$

Section 2 (p. App–12)

1. Amp = 2; per = 4π
3. Amp = 4; per = 4
5. Amp = 3; per 3π
7. Amp = 2; per = π
9. Amp = 2; per = π
11. Amp = 4; per = π
13. Amp = 3; per = 4π
15. Amp = 1; per = π
17. Per = $\pi/2$
19. Amp = $\sqrt{2}$; per = 2π
21. Per = 4π
23. Amp = 4; per = π

Appendix 2

Section 1 (p. App–16)

3. $x^2 + y^2 - 2x + 4y - 20 = 0$; $C(1, -2)$; $r = 5$
5. $x^2 + y^2 + 3x + 2y - 3 = 0$; $C(-\frac{3}{2}, -1)$; $r = 5/2$
7. $7x^2 + 7y^2 - 19x + 15y - 58 = 0$
9. $x^2 + y^2 + 2y - 19 = 0$
11. $x^2 + (y - k)^2 = 16$
13. $(x - h)^2 + y^2 = h^2$
15. $(x - h)^2 + (y - 3h)^2 = \frac{1}{10}(10h + 4)^2$
19. $x = \frac{1}{10}(1 \pm \sqrt{39})$; $y = \frac{1}{10}(9 \pm \sqrt{39})$
21. Do not intersect
23. Circle; $3x^2 + 3y^2 - 26x - 4y + 55 = 0$; $C(\frac{13}{3}, \frac{2}{3})$; $r = 2\sqrt{2}/3$
25. Point of intersection: $(1, -2)$

Section 2 (App–21)

1. $x - 2y - 3 = 0$
3. $3x + y - 10 = 0$
5. $x + y + 2 = 0$; $2x - 3y + 9 = 0$
7. $2x + (-5 \pm \sqrt{29})y - 27 \pm 5\sqrt{29} = 0$
9. $(1 \pm \sqrt{37})x + 6y + 19 \pm \sqrt{37} = 0$
11. $x - 3y - 4 = 0$, $x - 11y - 12 = 0$
13. No tangents
17. 30 meters
19. $y^2 = 18x + 9$, parabola; $V(-\frac{1}{2}, 0)$, $F(4, 0)$, dir. $x = -5$

Section 3 (App–24)

1. $x^2 + 3y^2 = 24$
3. $7x^2 + 3y^2 = 84$
5. $144x^2 + 95y^2 = 4655$
13. Tangents through Q: $y = 3$, $4x - y - 5 = 0$
    Tangents through R: $3x \pm 2\sqrt{6}y - 15 = 0$

Section 4 (App–29)

	a	b	c	e	foci	vertices	directrices	asymptotes
1.	3	2	$\sqrt{13}$	$\dfrac{\sqrt{13}}{3}$	$(\pm\sqrt{13}, 0)$	$(\pm 3, 0)$	$x = \pm\dfrac{9}{\sqrt{13}}$	$y = \pm\dfrac{2x}{3}$
3.	2	1	$\sqrt{5}$	$\dfrac{\sqrt{5}}{2}$	$(\pm\sqrt{5}, 0)$	$(\pm 2, 0)$	$x = \pm\dfrac{4}{\sqrt{5}}$	$y = \pm\dfrac{x}{2}$
5.	1	1	$\sqrt{2}$	$\sqrt{2}$	$(\pm\sqrt{2}, 0)$	$(\pm 1, 0)$	$x = \pm\dfrac{1}{\sqrt{2}}$	$y = \pm x$
7.	5	$\sqrt{8}$	$\sqrt{33}$	$\dfrac{\sqrt{33}}{5}$	$(0, \pm\sqrt{33})$	$(0, \pm 5)$	$y = \pm\dfrac{25}{\sqrt{33}}$	$y = \pm\dfrac{5x}{\sqrt{8}}$

9. $4x^2 - y^2 = 144$
13. $400x^2 - 225y^2 = 5184$
17. $y^2 - x^2 = 2$
23. $R: 11y \pm \sqrt{22}x \mp 4\sqrt{22} = 0$

11. $63x^2 - 81y^2 = 448$
15. $9x^2 - 4y^2 = 189$
21. $8x^2 - y^2 - 2x - 55 = 0$

Section 5 (App–35)

1. Ellipsoid
5. Elliptic hyperboloid of two sheets
9. Elliptic paraboloid; axis; x axis
13. Circular cone about y axis
17. Hyperbolic paraboloid
19. a) $x + y = 0, z = 0; x - y = 1, z = x + y$
    b) $x + y = 0, z = 0; x - y = a, z = a(x + y)$

3. Ellipsoid (oblate)
7. Elliptic hyperboloid of one sheet
11. Circular paraboloid; axis; y axis
15. Elliptic hyperboloid of one sheet

Appendix 3

Section 1 (App–39)

19. $\cosh x = \frac{5}{3}$, $\sinh x = -\frac{4}{3}$, $\coth x = -\frac{5}{4}$, $\text{sech } x = \frac{3}{5}$, $\text{csch } x = -\frac{3}{4}$

21. $\sinh x = -\frac{1}{2}$, $\cosh x = \frac{\sqrt{5}}{2}$, $\tanh x = \frac{-1}{\sqrt{5}}$, $\coth x = -\sqrt{5}$, $\text{sech } x = \frac{2}{\sqrt{5}}$

23. $\sinh x = -\sqrt{3}$, $\tanh x = \frac{-\sqrt{3}}{2}$, $\coth x = \frac{-2}{\sqrt{3}}$, $\text{sech } x = \frac{1}{2}$, $\text{csch } x = \frac{-1}{\sqrt{3}}$

25. Minimum at $(0, 1)$, concave upward for all x.
27. No. max or min; concave downward for $x < 0$; concave upward for $x > 0$; $x = 0$ and $y = \pm 1$ are asymptotes.
29. No max. or min; concave downward for $x > 1$; concave upward for $x < 1$.

Section 2 (App–43)

1. $\ln\left(\frac{1}{2} + \frac{\sqrt{5}}{2}\right)$
3. $\ln(2 + \sqrt{3})$
5. $\ln 3$

7. No max or min; point of inflection at $(0, 0)$; concave upward for $x \leq 0$; concave downward for $x \geq 0$.
9. No max. or min; $(0, 0)$ is point of inflection; concave upward for $0 \leq x < 2$; concave downward for $-2 < x \leq 0$.
11. No rel. max. or min; point of inflection at $x = \sqrt{2}/8$; concave downward for $\sqrt{2}/8 \leq x < \frac{1}{4}$, concave upward for $0 < x \leq \sqrt{2}/8$.

13. $f'(x) = 2/\sqrt{4x^2 + 1}$

15. $g'(x) = -\frac{1}{x^2}\text{argtanh } x^2 + \frac{2}{1 - x^4}$

17. $G'(x) = \sec x$

19. $f'(x) = \frac{4}{\sqrt{4x^2 + 1}} \text{argsinh } 2x$

21. $\ln(5 + 2\sqrt{6}) - \ln(2 + \sqrt{3})$

23. $\ln(\sqrt{2} - 1) - \ln(\sqrt{5} - 2)$

25. $\ln(5 + \sqrt{21}) - \ln(3 + \sqrt{5})$

Appendix 4

Section 2 (App–55)

1. $-x^2 + xy + 4y^2$
5. $2x^4 - 7x^3 + 11x - 3$

3. $4x^3 - 4x^2 + 3x + 1$
7. $2x^5 - 5x^4 - x^3 + 3x^2 - 12x + 9$

9. $\dfrac{-b}{a^2 - b^2}$, $a \neq \pm b$

11. $\dfrac{2x^2 - 9x - 23}{6(x - 1)(x + 4)}$; $x \neq 1, -4$

13. $\dfrac{(x + 5)(x^2 + x + 1)}{(x + 1)(x + 2)}$; $x \neq -1, -2$

15. 1; $x \neq 0, -y$

17. $\dfrac{ab}{a^2 + b^2}$; $a \neq \pm b$, $a \neq 0$, $b \neq 0$

Appendix 5

(App–63)

3. -2
5. 24
7. 8
9. $(1, -2, -1)$
11. $\left(-\dfrac{73}{19}, \dfrac{21}{19}, -\dfrac{43}{19}\right)$

INDEX

a^x, definition of, 326
Absolute value, 7–10
Acceleration, 80
   average, 80–81
   instantaneous, velocity, speed, and, 77–82
   vector, 467, 472, 553
   in plane, 461–463
Addition of ordinates, APP 11
Addition of vectors, 424, 427, 436–438
Algebra, axioms of, APP 45–50
   addition and subtraction, APP 45–47
   fractions, laws of, APP 49–50
   multiplication and division, APP 47–50
   number systems and, APP 51–54
Algebraic functions, differentiation of, 115–133
Angle, between bisecting lines, in space, 397–398
   between lines, in plane, and bisectors of, 249–252
   negative, APP 3
   between planes, 408
   sum of, sine and cosine of, 282–283
   units for measurement, APP 4
   between vectors, 429, 443
Antiderivatives, 188
   notation for, 221
Approximate integration, 559–562
Approximate percentage error, 174
Approximation, to area under curve, 187
   differential and, 170–174
   Midpoint Rule, 212
Arc length, 353–358, 466
   formula, 356–357
   in polar coordinates, 383–384
   theorem, 360
Arc notation, 307n
Arccos function, 308, 310–311, 313, 315
Arccot function, 309–310, 312–313
Arccsc function, 310, 312–313
Archimedes, spiral of, 372, 381

Arcsec function, 309–310, 312–314
Arcsin function, 307, 310–311, 313–314
Arctan function, 309–310, 312–314
Area, between curves, 225–230
   definite integral and, 208
   differentiation and, 185–191
   formulas, 162
   inner and outer, 195
   of parallelogram and triangle, 451
   in polar coordinates, 385–388
   procedures for defining, 193–196
   of the region, 196
   of sector of circle, 385
   of surface of revolution, 530–534
Argcosh, APP 41–42
Argcoth, APP 40–42
Argcsch, APP 40, 42
Argsech, APP 42
Argsinh, APP 40–42
Argtanh, APP 40–42
Arithmetic progression, 112
Associative law, in addition and subtraction, APP 45
   extended, APP 52
   in multiplication and division, APP 47
   vectors and, 427, 438
Asymptotes
   domain, range, and, 35–39
   horizontal, 37–38
   of hyperbola, APP 27
   locating, 38
   negative exponents and, 125
   vertical, 37
Attitude numbers, 405
Axes, coordinate, 390
   major and minor, of ellipse, 267
   of parabola, 259
   polar, 366
   radical, APP 15
   of revolution, 544
   rotation of, general equation of second degree and, 282–289

of symmetry, center of mass and, 539
translation of, 277–281
transverse and conjugate, of hyperbola, 273.
See also x and y axis; z axis
Axiom of Continuity (Axiom C), 110–111
Axiom of Inequality, APP 51n
Axioms, algebraic, APP 45–50

Base of directed segment, in plane, 421
in three space, 435
Binomial Theorem, 614
Binormal vector, 469
Bisectors of angles, 250
angles between two lines and, 249–252
Branches of inverse relation, 302, 304–305

Cardioids, 371
Cartesian coordinate system, 41, 391
cylindrical coordinates and, 417
polar coordinates and, 366–368
spherical coordinates and, 418
Center, of curvature, 469
of ellipse, 267
of hyperbola, 273
of sphere, 413
Center of mass, 535–538
deriving formulas, 541
of plane region, 539–544
of solid of revolution, 544–546
of straight-line segment, 547
of wires and surfaces, 547–550.
Central rectangle of hyperbola, 275
Chain Rule, 122
applications of, 121–124
in terms of differentials, 176
Change of variable, 221–224
Circle, 253–256, APP 13–16
area formula, 162
of curvature, 362
equation in polar coordinates, 377
equations, 253, 281, 288
involute of, 348
lines, conics, and, 243–289
solid of revolution generated by, 511
Closed interval, 3
Closure property, in addition and subtraction, APP 45
in multiplication and division, APP 47
Cofactor, APP 58
Colatitude of point in spherical coordinates, 420
Common logarithms, 326
Commutative law, in addition and subtraction, APP 45
extended, APP 52
in multiplication and division, APP 47
vectors and, 427, 438

Completing the square, 254, 279, 490
Component of **v** along **w**, 445
Concavity, orientation and, 361
upward and downward, 153–154
Cone, 418
elliptic, APP 34
right circular, 162
lateral surface area of frustum of, 530–531
Conic sections, 288, APP 13–29
circle, 253–256, APP 13–16
ellipse, 265–270, APP 21–24
equations in polar coordinates, 378–379
hyperbola, 271–276, APP 25–29
lines, circles, and, 243–289
parabola, 257–263, APP 17–21
second degree equations and, APP 30.
See also sections by name
Conjugate axis of hyperbola, 273
Constant, density, 537, 539, 544, 547, 549
derivative of, 115
derivative of, times function, 117
of integration, 210–211, 221
Continuity, 95
Axiom of (Axiom C), 110–111
on closed interval, 99
Extreme Value Theorem and, 135–136
one-sided limits and, 95–100
on right or left at endpoint, 99
uniform, 202–203
for vector function, 458, 464
Convergent integral, 522, 524–525
Coordinate axes, 390
Coordinate plane, 390
Coordinates and coordinate systems, 18
Cartesian, 41, 391
cylindrical, 417–418
left- and right-handed, 390
polar, 366–388
rectangular, 391
spherical, 418–420
in three-dimensional geometry, 417–420
in three-dimensional number space, 390–393
transformation of, 277–281, 417–418.
See also coordinate systems by name
Cosecant, APP 2
Cosh (hyperbolic cosine), APP 37
Cosines, APP 2–5
law of, 398, APP 4
of sum of two angles, 282–283
Cotangent, APP 2
Coth (hyperbolic cotangent), APP 37
Cramer's Rule, 442
for $n = 2$, APP 57
for $n = 3$, APP 62
Critical point, of one-variable function, 148

Critical value of function, 148
Cross product, 448-452
Csch (hyperbolic cosecant), APP 37
Curvature, center of, 469
  circle of, 362
  formulas, 360-361
  in plane, 359-363
  radius of, 362, 469
  and rate of change, 360
  in three space, 469
  vector, 469
Curves, area between, 225-230
  concavity of, 153-154
  degenerate, 279
  length, 353-358
  of second degree, 253-289
  in three-dimensional space, 464-467.
  See also curves by name
Cycloid, 346-347
Cylinders, 414
  concentric, 517
  parabolic, 415
  right circular, 162, 511-512
  sphere and, 413-415
Cylindrical coordinate system, 417-418
Cylindrical shell, 517-518
Cylindrical surface, 414, 415

Decay, exponential, 556
Decreasing on the interval, 146-147
Definite integral, 187, 193-241
  antiderivatives and, 185-191
  area, 193-196
  area between curves, 225-230
  denotation of, 199
  evaluation, 208-212
  evaluation by integration by parts, 494
  fluid pressure, 237-241
  integrability and, 198-203
  properties of, 204-207
  sums, notation for, 196-197
  work, 232-236
Degenerate curves, 279, 288-289
Degree, of angle, APP 4
  of polynomial, 115
Delta (Δ), 198
  f, 116, 171-172
  sub i of x, 186
  x, 67
Dependence and independence, linear, 440-442
Derivatives, 69-71
  applications, tools for, 135-140
  of constant, 115
  of constant, time function, 117
  differential equation and, 553
  differentials and, 176
  of exponential function, 330
  geometric interpretation of, 72-76
  integral and, 209, 219
  of inverse trigonometric functions, 311-313
  left-hand and right-hand, 141
  notation, 76, 81
  parametric equations and, 349-351
  partial (*see* Partial derivatives)
  in polar coordinates, 381-384
  of product of two functions, 118
  of quotient of two functions, 119
  second, acceleration as, 81
  second, differentiation applications using, 153-158
  of sine and cosine functions, 293-295
  of sum, 117
  three-step rule for, 70
  usual and partial, 621
  vector function in plane and, 457-460.
  See also Differential; Differentiation; Differentiation Applications
Determinants of second and third order, APP 56-63
  Cramer's Rule, APP 57, APP 62
Differential, 171
  approximation and, 170-174
  arc length formula using, 357
  derivatives and, 176
  formulas for derivatives of inverse trigonometric functions expressed in terms of, 313
  formulas for differentiating trigonometric functions expressed in terms of, 295
  notation, 175-178
  of product, 493.
  See also Derivatives; Differentiation
Differential equations, 553
Differentiation, 115-133
  of exponential functions, 329-333
  formulas, 127, 331
  formulas for trigonometric functions, 295
  of functions involving scalar and vector products, 465
  Fundamental Lemma of, 122
  of hyperbolic functions, formulas, APP 39
  implicit, 130-133
  integration and, 219
  of inverse hyperbolic functions, formulas, APP 42
  logarithmic, 332-333
  notation, 130-131
  power function, 125-128
  of trigonometric functions, 293-296.
  See also Derivatives; Differential; Differentiation applications

Differentiation applications, 135–191, 335–337
  definite integral, antiderivatives, and, 185–191
  differential, approximation and, 170–174
  differential notation, 175–178
  Extreme Value Theorem, 135
  to graphs of functions, 146–152
  of maxima and minima, 162–167
  maximum and minimum values of a function on an interval, 159–161
  related rates, 179–182
  Rolle's Theorem, 141–142
  Theorem of the Mean, 143–145
  tools for, 135–140
  using second derivative, 153–158
Direct substitution, 89
Directed distance, 42–43
Directed line segments, 394–395, 421
  vectors in plane and, 421–423
  vectors in three dimensions and, 435
Directed tangent line, 466
Direction, of arc, 360
Direction angles, 395
Direction cosines, 395
  direction numbers and, 394–398
Direction numbers, 396
  direction cosines and, 394–398
Directrix (directrices), of conic, 379
  of ellipse, APP 22
  of hyperbola, APP 25
  of parabola, 257
Discontinuous function, 95
Disk method, for volume of solid of revolution, 511–515
Distance, formula, in R_3 space, 392–393
  formula, in two dimensions, 42, 353
  midpoint formula and, 41–44
  between parallel lines, 245
  perpendicular, from point to plane, 410
  from point to line, 243–245
Distributive law, APP 47
  vectors and, 427, 431, 438, APP 68
Divergent integral, 522, 524
Division, simplification by, 481, 500
Domain, of exponential function, 324
  of inverse relation, 301
  of ln function, 319
  of one-variable function, 23–24
  range, asymptotes, and, 35–39
  of relation, 29, 35, 300–301
  of sequence, 108
  of two-variable function, 170
  of vector function in plane, 458
Dot product, 431, 443–446
Double-angle formulas, 285, APP 6
Double inequalities, 3
Dynes, 232

e, number, 326, 327, 334–335
Eccentricity, of ellipse, 267
  of hyperbola, 274
  in polar coordinates, 378
Elements of set, 16
Eliminating the parameter, 343–344
Ellipse, 265–270, APP 21–24
  axes of, 267
  as conic section, 288
  construction of, APP 21
  eccentricity of, 267
  equations, 266–267, 281, 288
  geometric properties, APP 21–22
  in polar coordinates, 378
  reflected-wave property, APP 23
  slope of, 269
  tangent line to, 269
Ellipsoid, APP 31–32
Elliptic cone, APP 34
Elliptic hyperboloid, of one sheet, APP 32
  of two sheets, APP 33
Elliptic paraboloid, APP 33–34
Empty set, 16
Equations, in Cartesian and polar coordinates, 373–375
  of circle, 253
  of cylindrical surface, 415
  differential, Newton's laws of motion and, 553–557
  of ellipse, 266–267
  first-degree, 4, 55
  general, of second degree, 282–289
  graph of, 19
  of hyperbola, 273
  of line (*see* Line(s))
  linear, 55
  of parabola, 259
  parametric, of line, two-point form, 400
  of plane, 405
  quadratic and trigonometric, 4
  for relationship between Cartesian and polar coordinates, 367
  second-degree, 253–289
  of sphere, 413–414
Equivalence class, 423
  addition of vectors and, 424
  in three space, 435
Equivalent directed line segments, in plane, 422
  in three space, 435
Ergs, 232
Errors, approximation and, 174
  percentage, 174
  proportional, 174
  in using Taylor's Theorem, 602
Euclidean plane, 18–19
Expansion, of D according to first, second, and third columns, APP 59

of D according to first, second, and
    third rows, APP 59
  in terms of last column, APP 56-57
Exponential decay, 556
Exponential function, 324-328
  differentiation of, 329-333
  formulas, 330
  properties of, 324
  trigonometric functions and, 291-340
Exponential growth, 556
Exponents, irrational, 317
  law of, 316
  negative, 120, 125
  power function, 125-128
  properties of, 326
  rational number as, 127
Extended Associative and Commutative
    Laws, APP 52
Extreme Value Theorem, 135

Factors, of denominator, 501-503
  linear and quadratic, 499-503
Family, of circles, APP 14
  of lines, 246-248
  one-parameter, APP 14
  two-parameter, APP 15
Field, APP 54
"Figure eights," 371-372
Finite interval of integration, 523
Finite sequence, 108, 570, APP 51-52
Fluid pressure, 237-241
Focus (foci), of conic, 379
  of ellipse, 265, APP 22
  of hyperbola, 271
  of parabola, 257
Force, 232, 237
  total, 238
  vector, 553
Four-leaved rose, 371
Fractions, laws of, APP 49-50, APP 54
  partial, method of integration, 500-504
Free vector, 423
Frustum of cone, lateral surface area of,
    530-531
Functions, 23, 170
  applications, 335-337
  composite, limit of, 93
  continuous and discontinuous, 95
  critical value and critical point of, 148
  differentiation of exponential, 329-333
  equal, limit of, 89
  exponential, 324-332
  functional notation and, 22-27
  functional notation in evaluating limits, 65
  graphs of, differentiation applications to, 146-152
  implicitly defined, 29, 132
  inverse, 300-305
  maximum and minimum values of, 159-161
  negatively infinite, 105
  number e, 334-335
  of one variable, 23, 170
  positively infinite, 105
  quadratic, integrands involving, 490-491
  on R_1 and R_2, 170
  rational, 116
  rational, integration of, 499-504
  relations, graphs, and, 16-39
  trigonometric, 291-299, APP 2-11
  of two variables, 170
  vector, in plane, and their derivatives, 457-460
  vector, in space, 464-467
Fundamental Lemma on Differentiation,
    for one-variable functions, 122, 627
Fundamental Theorem of Calculus, 219
  first form, 209-210
  second form, 218

General equation of second degree, 284
  rotation of axes and, 282-289
Generator of cylinder, 414-415
Geometric plane, 18-19
Geometric progression, 112
  limit theorem, 112-113
Geometric shapes, area and volume
    formulas, 162
Gradient vector, 640-641
Graphs, 18
  arc, 354
  circle, 253
  ellipse, 265
  of equation, defined, 19
  hyperbola, 271
  parabola, 257
  of parametric equations, 342
  in polar coordinates, 368-372
  relations, functions, and, 16-39
  of second-degree equations, 281
  of trigonometric functions, APP 8-11
Gravity, 234
Growth, exponential, 556

Half-angle formulas, 28, APP 6
Half-open interval, 3, 135
Head of directed line segment, in plane, 421
  in three space, 425
Helix, 466
Hemisphere, 519
Hooke's Law, 233
Hyperbola, 271-276, APP 25-29
  asymptote of, APP 27
  axes, 273

central rectangle, 275
  as conic section, 288
  eccentricity of, 274
  equations, 273, 281, 288
  in polar coordinates, 378
  tangent line to, 275–276
Hyperbolic functions, 340, APP 37–43
Hyperbolic integrals, APP 70–71
Hyperbolic paraboloid, APP 34
Hyperboloid, elliptic, of one sheet, APP 32
  elliptic, of two sheets, APP 33

"If and only if," 17, 244n
Image of x, 24
Imaginary numbers, 326n
Implicit definition of function, 29, 132
Implicit differentiation, 130–133
  method of differentials in, 177
Improper integrals, 521–526
Inclination of line, 45–46, 249
Increasing on the interval, 146–147
Indefinite integrals, 221
  change of variable and, 221–224
  finding, 483–486
  formulas for trigonometric functions, 298
Independence and dependence, linear, 440–442
Independence of the path, 704
Index of summation, 196
Inequality, axiom of, APP 51n
Inequalities, 1–14
Infinite limits, 102–106
Infinite sequence, 108–109, 570, APP 51–52
Infinite slope, 47–48, 52
Infinity, 3
  limits at, 102–106
Inflection point, 156
Initial line, 366
Inner area of region, 195
Inner product, 431, 443–446
Inner volume, 513n, 713
Integers, APP 53
  even, odd, positive, and trigonometric integrals, 483–485
  positive and negative, 120
Integrable functions, 199
Integrability of functions, 201–202, 712, 714.
  See also Definite integral
Integral, algebraic forms, APP 76
  convergent and divergent, 522, 524–525
  definite, 193–241
  derivatives and, 209, 219
  elementary formulas, APP 75

formula for arc length, 357
improper, 521–526
indefinite, 221–224
logarithmic and exponential forms, APP 77
miscellaneous forms, APP 77–78
negative value of, 208
for one-variable functions, 199
short table of, APP 75–78
for total work, 234
trigonometric, 483–486, APP 76–77.
See also Integral applications; Integration
Integral applications, 511–562
Integrands, 199
  infinite, 526
  involving quadratic functions, 490–491, APP 72
Integration, approximate, 559–562
  constant of, 210–211, 221
  formulas and methods of, 475–510, APP 70–74
  formulas for trigonometric functions, 298, 331
  interval of, finite, 523
  interval of, unbounded, 523, 525
  limits of, 199
  multiple (see Multiple integration)
  partial fractions method, 500–504
  by parts, 493–497, APP 72
  of rational functions, 499–504, APP 73–74
  rationalizing substitutions, 506–508, APP 74
  by substitution, 475–482, APP 71
  of trigonometric functions, 297–299, 314–315
  trigonometric integrals, 483–486, APP 70
Intercepts, 32
  relations, symmetry, and, 28–34
  x, y, and z, 32, 54, APP 30
Interior point, of interval, 137–138
Intermediate Value Theorem, 214–216
Intersecting lines, angles between, 249–252
  as conic section, 288
  equations and, 281, 288
Intersection of sets, 16–17
Interval, 3
  closed, continuity on, 99
  closed and open, 3, 135
  of convergence, 594
  decreasing on, 146–147
  half-open, 135
  improper integrals and, 523–526
  increasing on, 146
  maximum and minimum values of a function on, 159–161

symbols for, 3
uniform continuity on, 202–203
Invariant, under rotation of axes, 288
Inverse Function Theorem, 302–303
Inverse functions, derivatives of
    trigonometric, 311–313
  exponential and logarithmic, 324–325
  hyperbolic, APP 40–43
  relations and, 300–305
  trigonometric, 306–313, 314–315
Inverse mapping, 303
Inverse of the inverse, 301
Inverse operator, 303
Inverse relation, 301
Inverse trigonometric formulas for
    integration, 476
Involute of circle, 348

Joules, 232

Least period, APP 9
Left-hand derivative, 141
Left-handed system, 390
Left-handed triple, 448
Lemniscates, 371–372
Length, of arc, 353–358, 466
  of major and minor axes of ellipse, 267
  of transverse and conjugate axes of
    hyperbola, 273
  of vector, in plane, 423, 425–426
  of vector, in three space, 436, 438
Limaçon, 370–371
Limit of Composite Function, 93
Limit of Constant, 89
Limit of Equal Functions, 89
Limit of Product, 90
Limit of Quotient, 91
Limit of Sum, 90
Limits, 58–113
  area of region and, 185–191
  axiom of continuity, 110
  definition of, 83–87
  derivative, 69–71, 72–76
  finding, 65–68
  infinite and at infinity, 102–106
  of infinite sequence, 571
  instantaneous velocity, speed, and
    acceleration, 77–82
  intuitive, 58–64
  lower, 197, 199
  notation for, 60–61
  one-sided, continuity and, 95–100
  of sequences, 108–113
  special, of trigonometric and exponential
    functions, 291–293
  of sums, 90
  theorems, 88–94, 97, 100, 103, 112, 138n
  upper, 197, 199

Line(s), 41–56
  angle between two, bisectors of angles
    and, 249–252
  circle, 253–256
  circles, conics, and, 243–289
  directed and undirected, 394–395
  distance formula and midpoint formula,
    41–44
  distance from point to, 243–245
  ellipse, 265–270
  equation of, point-slope form, 53
  equation of, slope-intercept form, 54
  equation of, symmetric form, 402
  equation of, two-point form, 54
  equations of, in three-dimensional
    geometry, 400–402
  equations of, in two dimensions, 53–54
  families of, 246–248
  hyperbola, 271–276
  intersecting (*see* Intersecting lines)
  parabola, 257–263
  parametric equations of, standard form,
    343
  parametric equations of, two-point form,
    400
  rotation of axes, general equation of
    second degree, and, 282–289
  slope (*see* Slope of line)
  straight, 52–56 (*see also* Straight lines).
  See also Parallel lines; Perpendicular
    lines
Linear combination of vectors, 440
Linear dependence and independence,
    440–442
Linear equation, 55
Linear function, 55
ln x, 317, 326 (*see also* Logarithm
    function)
Logarithm function, 316–322
  formulas, 330
  integration formulas in terms of, 479
Logarithmic differentiation, 332–333
Logarithmic spiral, 372, 383
Logarithms, common, 326
  laws of, 317
  Napierian, 326
  natural, 326, 330
  notation, 327
  properties of, 326
Longitude of point in spherical
    coordinates, 419

Magnitude, of directed line segment, 421,
    435
Major axis of ellipse, 267
Mapping, 303
Mass. *See* Center of mass

Maximum and minimum, applications of, 162–167
   Extreme Value Theorem, 135
   of functions on an interval, 159–161
   relative (*see* Relative maximum and minimum)
   rule to obtain, of continuous function on interval, 160
Mean, Theorem of the, 143–145, 564
   for Integrals, 216
Midpoint formula, in plane geometry, 41–44
   in three-dimensional geometry, 393
Midpoint rule, 212
Minimum. *See* Maximum and minimum
Minor axis of ellipse, 267
Minutes of angle, APP 4
Moment, 742
Moment of mass, 535–536
   of rectangle, 537
Momentum vector, 553
Motion, along curve, 461–463
   instantaneous velocity, speed, and acceleration, 77–82
   Newton's laws of, 553–557
   along straight line in one direction, 232–236
Moving trihedral, 470
   tangential and normal components and, 468–473

Napierian logarithms, 326
Nappes, 288
Natural logarithms, 326
Natural numbers, APP 51
Nearly equal, 173
"Necessary and sufficient," 17
Newton-meters, 232
Newtons, 232, 553
Newton's laws of motion, 232. *See also* Differential equations
Norm of the subdivision, definite integral and, 198
   double integral and, 711
   of interval, 186, 355
Normal, to the curve, 74
Normal components, 471
   tangential components, moving trihedral, and, 468–473
Number e, 326, 327, 334–335
Number plane (R_2), 18, 390
   graphs and, 18–21
Number space (R_3), 390–393
Number systems, APP 51–54
Numbers, direction, direction cosines and, 394–398
   imaginary, 326n
   natural, sequences and, APP 51–54

   negative, 126, APP 45
   positive, APP 51n
   rational, 127, APP 53
   real, 67, APP 45–50

Oblate spheroid, APP 32
Obvious Limit, 89
Odd function, 228
One-sided limits, 96–97
   continuity and, 95–100
   from right and from left, 97, 105
Open interval, 3
Operator, 303
   inverse, 303
"Or," in sets, 17
Ordered pair, 18
Ordered triple of vectors, 448
   oppositely oriented sets of, 448
   similarly oriented sets of, 448–449
Ordinates, addition of, APP 11
Orientation, of arc, 360
   similar and opposite, 448–449
Origin, 18–19
   polar coordinates and, 366
   symmetry with respect to, 32–33
Orthogonal vectors, in plane, 423
   in three space, 436, 444, 446
Osculating plane, 469
Outer area of region, 195
Outer volume, 513n, 713

Pairs, of numbers, 18
   ordered, 170
Pappus, Theorems of, 551–552
Parabola, 257–263, APP 17–21
   axis of, 259
   as conic section, 288
   equations, 259, 281, 288
   formulas for, 262
   in polar coordinates, 378
   properties and applications, APP 20–21
   reflecting property, APP 20
   standard positions, 258–260
   tangent to, 263
Parabolic cylinder, 415
Paraboloid, 514, APP 20
   elliptic, APP 33–34
   hyperbolic, APP 34
   of revolution, APP 34
Parallel lines, 45–46
   direction cosines and, 395
   distance between, 245
   equations, 281, 288
   proportional direction numbers and, 397
   slope of line, perpendicular lines, and, 45–51
   in three-dimensional space, 405

Parallel planes, 391-392
  theorem, 405
Parallel vectors, 429, 443
Parallelepiped, partial derivatives and, 692
Parameter, 341
  eliminating, 343-344
  one-parameter family, APP 14
  two-parameter family, APP 15.
  See also Parametric equations
Parametric equations, 341-347
  arc length and, 341-363
  derivatives and, 349-351
  of line in standard form, 343
  theorem, 342
Partial fractions, method of, 500-504
Percentage error, 174
Periodic functions, APP 3, APP 9
Periodicity, trigonometric and hyperbolic functions and, 340
Perpendicular lines, 48-49
  slope of line, parallel lines, and, 45-51
  in three-dimensional space, 398, 405
Perpendicular planes, 404, 409
Perpendicular vectors, 423, 436
Plane(s), coordinate, 390
  distance from point to, 410-411
  equation of, 405
  Euclidean, 18-19
  geometric, 18-19
  intersection of, 402
  number (R_2), 18-21
  osculating, 469
  parallel, 405
  perpendicular, 404, 409
  in three-dimensional geometry, 404-406
  through three points, finding, 451
  vector functions in, their derivatives and, 457-460
  vectors in, and in three space, 421-473
Point, 18-19
  as conic section, 289
  distance from, to line, 243-245
  distance from, to plane, 410-411
  equations and, 281, 288
  of intersection, 409
Point of division formula, 404
Point of inflection, 156
Point-slope form for equation of line, 53
Polar axis, 366
Polar coordinates, 366-388
  area in, 385-388
  Cartesian coordinates and, 366-368, 373-375
  derivatives in, 381-384
  graphs in, 368-372
  rules of symmetry, 369-370
  straight lines, circles, and cones, 376-379

Pole, 366
  symmetry with respect to, 370
Polynomials, 115
  decomposition of, 499
Power, 125-128
Pressure, fluid, 237-241
Principal normal vector, 469
Prismoidal Formula, 560
Product, absolute value of, 11
  closure property and, APP 47
  extended commutative and associative laws, APP 52
  limit of, 90
  scalar (inner or dot), 443-446
  of three vectors, 454-456
  of two functions, derivative of, 118
  vector or cross, 448-452
Progression. *See* Sequences
Projection, of vector, 430, 445
  on x and y axes, 23, 301
Prolate spheroid, APP 32
Proportional error, 174
Proportional sets of number triples, 395, 397
Proportional vectors, 429, 440, 443
Pyramid, regular, 399
Pythagorean theorem, 41-42, APP 3

Quadrants, positive and negative signs in, APP 2
  tangent function in first and second, 46
Quadratic expression, square root of, in integrand, 489
Quadratic factors, 499, 503
Quadratic functions, integrals involving, APP 72
  integrands involving, 490-491
Quadric surfaces, APP 30-35
Quantity as function of time, 179, 181
Quotient, absolute value of, 11
  limit of, 91
  of polynomials, APP 73-74
  remainder and, 500
  of two functions, derivative of, 119

Radians, 291-294, 385, APP 4, APP 8
Radical axis, APP 15
Radical plane, 416
Radius, of sphere, 413
Radius of curvature, 362, 469
Range. *See* Domain
Rational functions, 116
  integration of, 499-504, APP 73-74
Rational number, 127, APP 53
Rationalizing substitutions, 506-508
"Rationalizing the numerator or denominator," 63

Real number, algebraic axioms for, APP 45–50
  derivative of, 126–127
  designating, 67
Reciprocals, 292
  existence of, APP 47
  negative, normal line, tangent line, and, 74
  negative, perpendicularity and, 49
Rectangular coordinate system, 391
Rectangular grids, 196
Rectifiable arc, 355
Reflected-wave property of ellipse, APP 23
Reflecting property of parabola, APP 20
Regular pyramid, 399
Regular tetrahedron, 399
Related rates, 179–182
  problem-solving procedures, 181
Relations, 29, 300
  defined by decomposition, 131
  distinguished from function, 301
  domain, range, and asymptotes, 35–39
  functions, graphs, and, 16–39
  functions and functional notation, 22–27
  intercepts, symmetry, and, 28–34
  inverse, 301
  inverse functions and, 300–305
  number plane, and graphs, 18–21
  in R_2, 341
  sets and set notation, 16–17
Relative maximum and minimum, 76, 147
  finding, 149–150
Representative of vector, 423, 436
Revolution about x or y axis, 532. *See also* Solid of revolution
Right circular cone, 162
Right circular cylinder, 162, 511–512
Right-hand derivative, 141
Right-handed system, 390
Right-handed triple, 448–449
Rolle's Theorem, 141–142
Rose curves, 371, 387
Rotation of axes, general equation of second degree and, 282–289

"Sandwiching theorem," 292n
Scalar product, 431
  multiplication, 443–446
  plane vector operations and, 429–433
Secant, 72, APP 2
Sech (hyperbolic secant), APP 37
Second derivative, 81
  differentiation applications using, 153–158
  notation, 177–178
  parametric equations and, 349–350
Second Derivative Test, 155

Seconds of angle, APP 4
Section of surface, APP 30
Semicircle, solid of rotation generated by, 511
Sequences, APP 51
  arithmetic progression, 112
  axiom of continuity, 110–111
  finite and infinite, 108–109, APP 51–52
  geometric progression, 112
  limit theorem, 112–113
  limits of, 108–113
  natural numbers and, APP 51–54
Sets, of all real numbers, 18
  intersection of, 16–17
  in plane and solid geometry, 391
  set notation and, 16–17
  union of, 16–17
Shell method for volume of solid of revolution, 517–520
Sigma (Σ), 196
Simpson's Rule, 561
Sines, APP 2–5
  of sum of two angles, 282–283
Sinh (hyperbolic sine), APP 37
Skew, 397
Slope, of ellipse, 269
  of line, 45–51, 74, 249
Slope-intercept form for equation of line, 54
Slope of line, 46, 249
  derivative as, 74
  formula, 48
  parallel and perpendicular lines and, 45–51
  positive and negative, 46–47
Solid of revolution, 511
  center of mass, 544–546
  volume of, disk method, 511–515
  volume of, shell method, 517–520
Solution set, of equation in two unknowns, 19, 29
  for inequality, 4
Space curves, vector functions, tangents, arc length, and, 464–467
Speed, 79
  instantaneous velocity, acceleration, and, 77–82
  vector velocity and, 462, 467
Sphere, 413
  area and volume formulas, 162
  cylinders and, 413–415
  equation of, 413–414
  as solid of revolution, 511
Spherical coordinate system, 418–420
Spheroid, APP 32
Spiral of Archimedes, 372, 381

Spirals, 372, 381, 383
Square grid, 193-196
Straight line, 52-56
   center of mass of, 547
   as conic section, 289
   equation in Cartesian coordinates, 342, 376
   equation in normal form, 376
   equation in polar coordinates, 376-377
   equations and, 288
   length, 353
   parametric equations of, in standard form, 343.
   See also Line(s)
Strophoid, 348
Subdivision, norm of
   of interval, 198
Subscripts for functions, 23
Substitution, hyperbolic and trigonometric, APP 71-72
   integration method, 475-479
   rationalizing, 506-508
   in table of integrals, APP 75
   trigonometric, 487-488
Subtraction, algebraic axioms and theorems, APP 45-47
   of convergent series, 574
   from inequalities, 2
Summation notation, 196-197
Summation sign, 196
Surface of revolution, 530
   area of, 530-534
Symmetric form for equations of line, 402
Symmetry
   axis of, center of mass and, 539
   parabola line of, 259
   relations, intercepts, and, 28-34
   rules of, in Cartesian coordinates, 32-33
   rules of, in polar coordinates, 369-370
   of surface, APP 31

Tangent formulas, of angle from L_1 to L_2, 249-250
   of difference between two angles, 249
Tangent function, APP 2
   properties of, 46
   in right triangle, 47
Tangent line, to arc, 466
   to circle, 72, 254
   to curve, 73-74, 154, 359
   directed, 466
   to ellipse, 269
   to hyperbola, 275-276
   to parabola, 263, APP 17-18
   in polar coordinates, 382
   vector functions in space, space curves, arc length, and, 464-467

Tangential components, 471
   normal components, moving trihedral, and, 468-473
Tanh (hyperbolic tangent), APP 37
"Tending to infinity," 102
   notation, 109n
Terms, in sequence, APP 51
Tetrahedron, regular, 399
Theorem of the Mean, 143-145
   for Integrals, 216
Third derivative, parametric equations and, 350
Three-dimensional space (three space), 390
   coordinates, distance formula, and, 390-393
   vectors in, 434-439
   vectors in, and in plane, 421-473
Three-step rule for differentiation, 70
Time, functions of, 179, 181
Torus, 511
Total force, 238
Total work, 233, 707
Traces of surface on coordinate planes APP 30
Transformation of coordinates, 277-281
   between Cartesian and cylindrical, 417
   between Cartesian and spherical, 418
Translation of axes, 277-281
Transverse axis of hyperbola, 273
Trapezoid, area formula, 162
"Triangle inequality," 202
Trigonometric functions, differentiation of, 293-296
   exponential functions and, 291-340
   formulas, 476, APP 2-11
   graphs of, APP 8-11
   integration of, 297-299
   integrations yielding inverse, 314-315
   inverse, 306-313
   limits, special, 291-293
   relations and inverse functions, 300-305
Trigonometric identities, 298-299
Trigonometric integrals, 483-486, APP 70
Trigonometric substitution, 487-489
Trihedral, moving, 470
   tangential and normal components and, 468-473
Two-point form, for equation of line, 54
   of parametric equations of line, 400
Two-sided limit, 97

Unbounded intervals, 523, 525
Undirected line, 394-395
Uniform continuity, 202-203
Union of sets, 16-17
Uniqueness of Limits, 88

Unit, existence of, APP 47
Unit tangent vector, 466, 469
Unit vector, 423, 436
   in plane, 424–425
   in three space, 437–438

Value, improper integrals and, 526
Variable, change of, 221–224
Variable of integration, 199
Vector acceleration, 462
   vector velocity and, in plane, 461–463
Vector functions, 457, 464
   in plane, their derivatives and, 457–460
   in space, space curves, tangents, arc length, and, 463–467
Vector product, 448–452
Vector velocity, 462
   vector acceleration and, in plane, 461–463
Vectors, 423, 436
   linear dependence and independence, 440–442
   notation, 432
   in plane, directed line segments and, 421–423
   plane operations with 424–427
   plane, operations with, scalar product and, 429–433
   in plane and three space, 421–473
   products of three, 454–456
   proofs of theorems, APP 65–69
   scalar (inner or dot) product, 443–446
   tangential and normal components, moving trihedral and, 468–473
   in three dimensions, 434–439
   vector product or cross product, 448–452
Velocity, average, 78
   instantaneous, speed, acceleration, and 77–82
   related rates and, 179
   vector, in plane, 461–463
Velocity vector, 467
   initial, 555
Vertex (vertices), of ellipse, 267
   of hyperbola, 273
   of parabola, 258, 260
Volume, 513n
   of cylindrical shell, 517–518
   formulas for, 162
   inner and outer, 513n
   of parallelepiped, 454, 511
   of right circular cylinder, 512
   solid of revolution (*see* Solid of revolution)

Weight, 234
Wires, center of mass of, 547–550
Witch of Agnesi, 348
Work, 232
   definite integral and, 232–236
   formula, 233
   principles, 232, 234
   problem-solving procedures, 235
   scalar product applied to, 445

x and y axes, 18–19
   hyperbola with transverse axis on, 274
   line parallel to, equation of, 52
   moment of mass with respect to, 536
   moment of rectangle with respect to, 537
   negative exponents and, 125
   projection on, 23, 29, 301
   revolution about, area and, 532
   symmetry with respect to, in Cartesian coordinates, 32–33
   symmetry with respect to, in polar coordinates, 369
   symmetry with respect to, quadric surfaces, APP 31
x intercept, 32, APP 30

y intercept, 32, 54, APP 30

z axis, 390
   in cylindrical coordinate system, 417
   generators parallel to, 415
   in spherical coordinate system, 418
z intercept, APP 30
Zero, existence of, APP 45
Zero derivative, 115, 132, 148
   in graphing, 149
Zero factorial, 577
Zero slope, 46, 48, 52
Zero vector, multiplication of vectors and, 424
   in plane, 423
   in three space, 436
Zero velocity, direction reversal and, 79–80

Table 3 Natural Logarithms of Numbers

n	$\log_e n$	n	$\log_e n$	n	$\log_e n$
0.0	*	4.5	1.5041	9.0	2.1972
0.1	7.6974	4.6	1.5261	9.1	2.2083
0.2	8.3906	4.7	1.5476	9.2	2.2192
0.3	8.7960	4.8	1.5686	9.3	2.2300
0.4	9.0837	4.9	1.5892	9.4	2.2407
0.5	9.3069	5.0	1.6094	9.5	2.2513
0.6	9.4892	5.1	1.6292	9.6	2.2618
0.7	9.6433	5.2	1.6487	9.7	2.2721
0.8	9.7769	5.3	1.6677	9.8	2.2824
0.9	9.8946	5.4	1.6864	9.9	2.2925
1.0	0.0000	5.5	1.7047	10	2.3026
1.1	0.0953	5.6	1.7228	11	2.3979
1.2	0.1823	5.7	1.7405	12	2.4849
1.3	0.2624	5.8	1.7579	13	2.5649
1.4	0.3365	5.9	1.7750	14	2.6391
1.5	0.4055	6.0	1.7918	15	2.7081
1.6	0.4700	6.1	1.8083	16	2.7726
1.7	0.5306	6.2	1.8245	17	2.8332
1.8	0.5878	6.3	1.8405	18	2.8904
1.9	0.6419	6.4	1.8563	19	2.9444
2.0	0.6931	6.5	1.8718	20	2.9957
2.1	0.7419	6.6	1.8871	25	3.2189
2.2	0.7885	6.7	1.9021	30	3.4012
2.3	0.8329	6.8	1.9169	35	3.5553
2.4	0.8755	6.9	1.9315	40	3.6889
2.5	0.9163	7.0	1.9459	45	3.8067
2.6	0.9555	7.1	1.9601	50	3.9120
2.7	0.9933	7.2	1.9741	55	4.0073
2.8	1.0296	7.3	1.9879	60	4.0943
2.9	1.0647	7.4	2.0015	65	4.1744
3.0	1.0986	7.5	2.0149	70	4.2485
3.1	1.1314	7.6	2.0281	75	4.3175
3.2	1.1632	7.7	2.0412	80	4.3820
3.3	1.1939	7.8	2.0541	85	4.4427
3.4	1.2238	7.9	2.0669	90	4.4998
3.5	1.2528	8.0	2.0794	95	4.5539
3.6	1.2809	8.1	2.0919	100	4.6052
3.7	1.3083	8.2	2.1041		
3.8	1.3350	8.3	2.1163		
3.9	1.3610	8.4	2.1282		
4.0	1.3863	8.5	2.1401		
4.1	1.4110	8.6	2.1518		
4.2	1.4351	8.7	2.1633		
4.3	1.4586	8.8	2.1748		
4.4	1.4816	8.9	2.1861		